FRACTURE MECHANICS OF CONCRETE

Developments in Civil Engineering

DEVELOPMENTS IN CIVIL ENGINEERING, 7

FRACTURE MECHANICS OF CONCRETE

Edited by

FOLKER H. WITTMANN

Ecole Polytechnique Fédérale de Lausanne, Lausanne, Switzerland

ELSEVIER
Amsterdam — Oxford — New York — Tokyo   **1983**

ELSEVIER SCIENCE PUBLISHERS B.V.
1 Molenwerf
P.O. Box 211, 1000 AE Amsterdam, The Netherlands

Distributors for the United States and Canada:

ELSEVIER SCIENCE PUBLISHING COMPANY INC.
52, Vanderbilt Avenue
New York, NY 10017

Library of Congress Cataloging in Publication Data
Main entry under title:

Fracture mechanics of concrete.

    (Developments in civil engineering ; v. 7)
    Bibliography: p.
    Includes indexes.
    1. Concrete--Fracture. 2. Fracture mechanics.
I. Wittmann, F. H. (Folker H.) II. Series.
TA440.F73 1983   620.1'366   83-11634
ISBN 0-444-42199-8 (U.S.)

ISBN 0-444-42199-8 (Vol. 7)
ISBN 0-444-41715-X (Series)

Printed in The Netherlands

PREFACE

The classical concept of fracture mechanics has proved to be a useful mathematical formulation of crack propagation. Since the pioneering paper of Griffith in 1920 fracture mechanics has been applied to a vast range of different materials and components and by now has become an important branch of engineering science. On the other hand its application to concrete has been cautious and its success is still rather modest.

One basic element in fracture mechanics is the assumption of one slitlike crack in a linear elastic isotropic continuum. Under simplifying conditions such as remotely applied tensile load fracture energy can be related to the surface energy of the material. In a porous material such as hardened cement paste this idea has to be revised because all cracks will propagate through a three-dimensional network of colloidal particles. Thus an effective surface energy of hardened cement paste depends, among other influences, on its characteristic porosity. Concrete is a composite material and as a consequence cracks are arrested when they encounter aggregates. This is one of the origins for the development of a rather large fracture process zone and it explains why fracture mechanics parameters such as fracture toughness depend on crack length.

Here we have chosen just a few examples to illustrate why linear elastic fracture mechanics cannot be applied to concrete. At the moment we are still far from a comprehensive fracture mechanics approach for concrete. Nevertheless recently advanced experimental techniques have been applied and sophisticated theoretical concepts have been developed. This warrants a summarizing description. Relevant and important results have been compiled in this volume. Computerized structural design has served as a strong driving force for the development of realistic fracture mechanics concepts.

The aim of this publication is twofold : first of all an engineer should find the potential as well as the limits of application of fracture mechanics to concrete and second those active in research will find for the first time different trends, partially diverging views, and the broad field of application in numerical analysis combined in one volume. It is hoped that this state-of-the-art report together with the annotated bibliography will stimulate and enhance further work in this field.

The preparation of this publication has arisen from work done in a RILEM Technical Committee (50-FMC). It is my pleasure to thank all the authors for their enormous effort to enable the publication of this volume. In addition I gratefully aknowledge help of my collaborators in the editorial work.

Lausanne, March 1983   F.H. WITTMANN

TABLE OF CONTENTS

Fracture Mechanics of Concrete,
edited by F.H. Wittmann, 1983
Elsevier Science Publishers B.V., Amsterdam — Printed in The Netherlands

1

Chapter 1.1

THE APPLICATION OF FRACTURE MECHANICS TO CEMENT AND CONCRETE:
A HISTORICAL REVIEW

by Sidney MINDESS

1   INTRODUCTION

The cracking of hardened cement paste (hcp) and concrete has been studied
seriously at least since the work of Richart et al. (ref. 1) in 1928, who
studied the development of cracks in concrete under compressive loading. Since
then, a great deal of work has been carried out in detailed investigations of
the crack patterns that develop in hcp and concrete under load, or due to thermal and shrinkage stresses. In addition, with the development of the scanning
electron microscope (SEM), other studies have considered the morphology of
cracks and fracture surfaces. However, while the energy-balance concept of
fracture was proposed by Griffith (ref. 2) in 1920, it took almost forty years
before the concepts of fracture mechanics were applied to cementitious materials, and there is still a great deal of controversy as to the applicability
of fracture mechanics to these materials.

The purpose of this paper is to review the historical development of the
application of fracture mechanics to hcp, mortar and concrete, through a review
of the principal literature in this area. The focus is on the difficulties
that have arisen in trying to establish suitable fracture criteria. A companion
paper (ref. 3) deals specifically with the applications of fracture mechanics to
fibre reinforced and polymer impregnated concrete. A more comprehensive description of the literature can be found in a separate annotated bibliography
(ref. 4), containing about 400 entries. In order to put the fracture mechanics
studies in context, two other groups of papers are also referred to. The first
deals with the crack patterns that develop in hcp and concrete under load, and
such phenomena as rate-of-loading and sustained loading effects. The second
deals with the morphology of cracks and fracture surfaces, and the microstructural phenomena involved in crack propagation.

2   CRACK GROWTH AND FRACTURE

The cracking processes in hcp and concrete under load are now fairly well
understood. Both hcp and concrete may be modelled as two-phase materials, consisting of strong particles dispersed in a weaker matrix. In the case of hcp,
unhydrated cement grains are dispersed in cement hydrates (C-S-H); for concrete, aggregate particles are dispersed in hcp. As shown in Fig. 1, the stress-
strain (σ-ε) curve for aggregate in compression is essentially linear to the

2

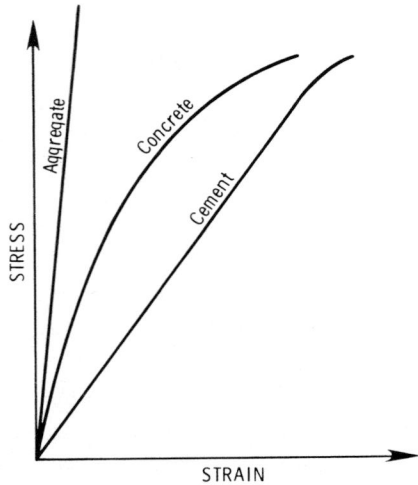

Fig. 1. Stress-strain curves for aggregate, cement and concrete in compression

point of failure. The σ-ε curve for hcp is also essentially linear until about
90-95% of its ultimate strength. However, the σ-ε curve for concrete is highly
non-linear. This non-linearity is due, in part, to the composite action of the
concrete, since there is a highly imperfect bond between the aggregate and the
hcp. In addition, a considerable amount of progressive cracking takes place as
the load increases (ref. 5). The much more linear nature of the σ-ε curve for
hcp is largely due to the significantly better bond between the unhydrated
cement grains and the C-S-H.

The σ-ε curve for concrete in compression may be divided into the four regions shown in Fig. 2 (ref. 6). Even prior to loading, cracks already exist
at the cement-aggregate interface, due to bleeding, volume changes in the
cement during hydration, and as a result of drying shrinkage after the concrete
has hardened. Below about 30% of the ultimate stress (σ_{ult}) there is very
little extension of these cracks for normal rates of loading, and the σ-ε curve
is essentially linear. Beyond this point, the σ-ε curve becomes increasingly
non-linear as the interfacial cracks begin to grow under load, partly due to
differences in the elastic constants for hcp and aggregate, and partly due to
the high stress concentrations which occur in these regions (refs. 7,8). At
about $0.5\sigma_{ult}$, in addition to further growth of the interfacial cracks, cracks
begin to extend through the cement matrix, bridging between the coarse aggregate
particles, approximately parallel to the axis of loading, but still in a stable
fashion. (Although it is generally assumed that these matrix cracks are simply

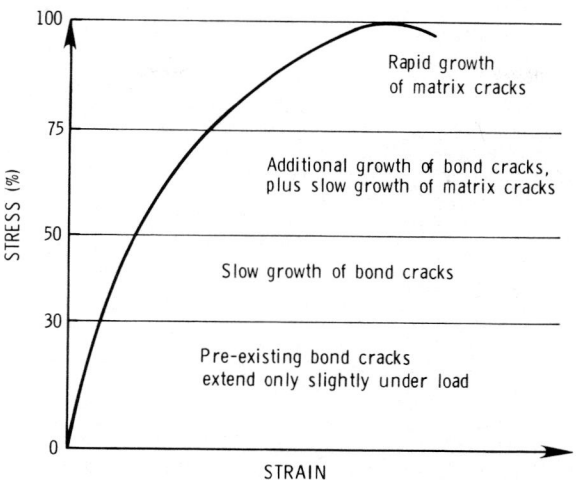

Fig. 2. Stress-strain curve for concrete showing different regions of crack growth (ref. 6).

due to extension of the bond cracks, Yoshimoto et al. (ref. 9) have suggested, from studies involving high speed photography, that the longitudinal matrix cracks are due to spontaneous tensile splitting.) Finally, beyond about 0.75 σ_{ult}, the matrix cracks begin to form a much more extensive network, though there is still enough redundancy in the system for it to remain relatively stable under short-term loads. Eventually, this network becomes so extensive that failure occurs. Static fatigue (delayed failure, or failure under sustained stress) in concrete is generally associated with loading within this region. It should be noted that, even at σ_{ult}, the crack pattern still leaves a structure that has some load-carrying capacity; with very stiff testing machines, the σ-ε curves in both tension (ref. 10) and compression (ref. 11) show a considerable descending branch.

Some of these features of cracking have been observed directly by Mindess and Diamond (refs. 12-14) on mortars. They developed a wedge-loaded compact tension device (refs. 12,13) and a compression device (refs. 14,14a) which could be used within the sample chamber of an SEM, so that cracking could be monitored under load. Their studies confirmed that cracks develop preferentially along the sand-cement interfaces, and that a considerable amount of branch or multiple cracking occurs; the crack surfaces also appear to be very tortuous. Typical crack patterns that they obtained are shown in Fig. 3. Similar results were obtained by Tait and Bohm (ref. 15), who carried out double-torsion tests on mortar in an SEM.

4

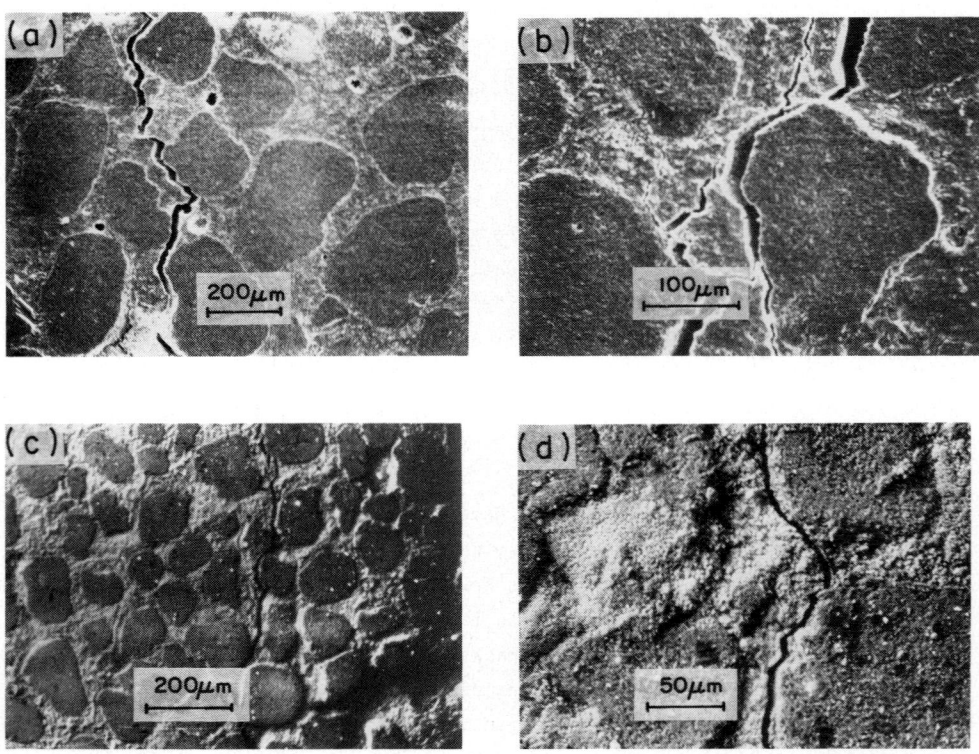

Fig. 3. Crack patterns obtained from mortars loaded within the sample chamber of an SEM: (a),(b) compact tension specimens; (c),(d) compression specimens.

The process of crack growth has been modelled by Zaitsev and Wittmann (ref. 16) using computer simulations. Fig. 4 shows the crack patterns that develop in hcp, normal concrete, high strength concrete, and lightweight aggregate concrete. In the latter two materials, cracks may run through the aggregate particles as well, leading to more linear σ-ε curves.

3 MECHANISMS OF FRACTURE

On a microstructural scale, the mechanism of crack growth through the hcp itself is less well understood. At early ages, there is some evidence that fracture propagates preferentially through the high porosity C-S-H phase (refs. 17,18) since the calcium hydroxide corresponds to a low porosity region and acts as a rigid inclusion, as shown in Fig. 5 (ref. 19). The $Ca(OH)_2$ sometimes even seems to act as a crack arrestor. However, in more mature hcp, this discrimination is lost as the matrix becomes more homogeneous and the fracture path becomes more direct. On the other hand, it has been suggested (refs. 20,21) that, even in mature hcp, fracture occurs preferentially along the weakly bonded basal planes of the $Ca(OH)_2$.

There is considerable evidence that the presence of water plays an important role in determining fracture behaviour. A number of investigators (refs. 22-28) have noted that, under sustained load, the time to failure decreases as the relative humidity at which the specimens were equilibrated increases. In other words, the presence of water appears to enhance subcritical crack growth, and this has indeed been confirmed experimentally (ref. 29). Similar effects were noted by Cook and Haque (refs. 30-32) on sustained loading of specimens that were first completely dried and then subjected to resaturation. In addition, it was noted that the fracture surface energy (estimated from the area under the σ-ε curve) was less for wet than for dry specimens (ref. 33), and that the critical strain energy release rate, G_c, also decreased considerably as specimens were dried, particularly below 20% relative humidity (ref. 34).

The reasons for this behaviour are not entirely clear. Shah and Chandra (ref. 22) suggested that a stress corrosion mechanism, with water as the corrosive agent, was responsible. On the other hand, in a series of studies, Husak, Barrick, and Krokosky (refs. 23-25, 27,28) argued that, while there was indeed a stress corrosion mechanism acting, the corrosive agent was not simply water. On the basis of extensive test results (ref. 25), which showed that the time to failure not only decreased as the relative humidity increased, but also increased as the temperature increased (Table 1), they postulated a stress corrosion mechanism dependent on the presence of $Ca(OH)_2$. In their view, static fatigue is due to hydroxyl ion attack on highly stressed Si-O bonds. Since the solubility of $Ca(OH)_2$ decreases with increasing temperature, the longer times to

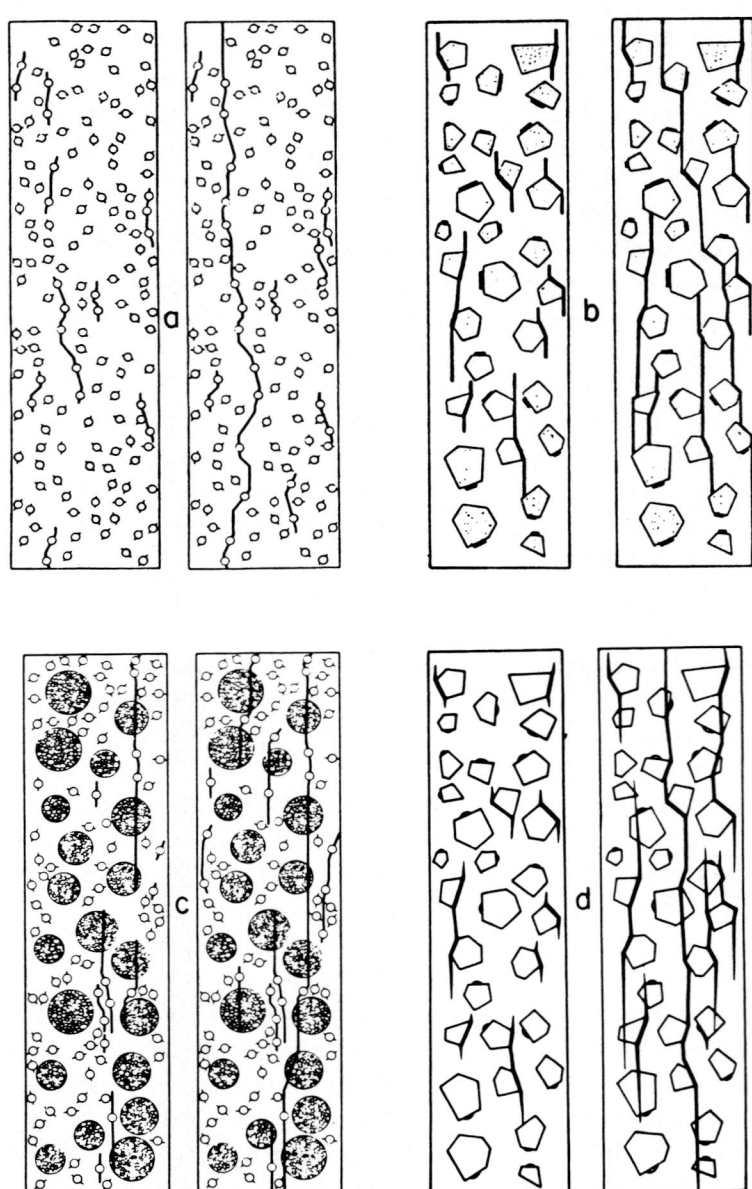

Fig. 4. Computer simulations of the crack patterns that develop in hcp and concrete under compressive loading. (a) hardened cement paste: cracks bridge between pores; (b) normal concrete: cracks go around aggregates; (c) light-weight concrete: cracks go through aggregate; (d) high strength concrete: some cracks go through aggregate. After Zaitsev and Wittmann (ref. 16).

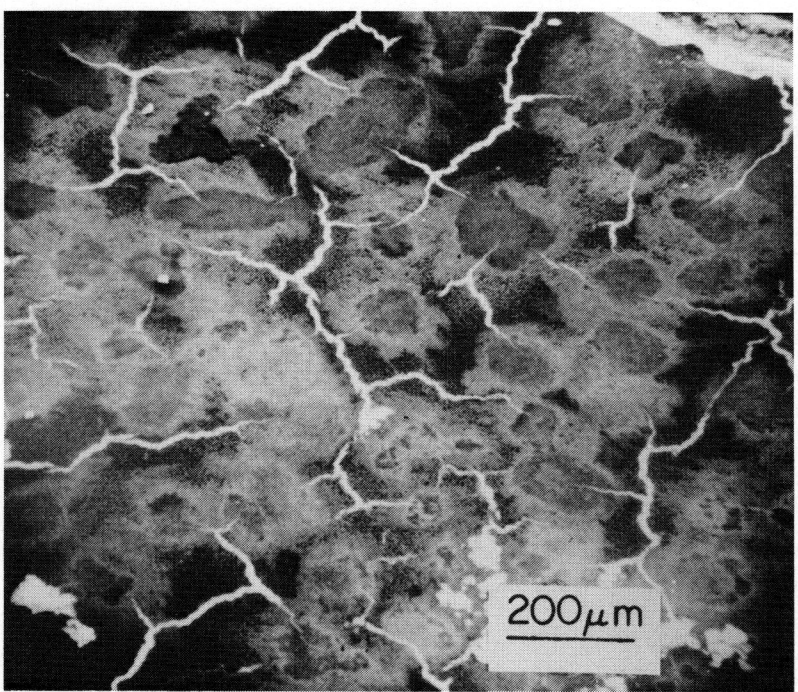

Fig. 5.   Cracking in a C_3S paste hydrated for 14 days: The gray areas are $Ca(OH)_2$. The cracks ("lightning bolts") appear to be largely in the C-S-H phase (ref. 19).

TABLE 1

Median failure times (hours) under sustained load at different temperatures. After Barrick (ref. 25).

Relative humidity	Temperature		
	25°C	35°C	60°C
	Time to failure (hours)		
10%	2.505	114.700	--
55%	0.106	2.879	--
96%	0.016	0.664	327.400

failure at higher temperatures indicate that the rate-determining step is the production of OH⁻ions from the dissociation of $Ca(OH)_2$ rather than a direct water attack.

(Related to the subject of static fatigue is the observation that sustained loads less than those required to cause failure due to static fatigue have a strengthening effect on concrete (refs. 22,35). This has been attributed to an increase under pressure in the van der Waals attraction between C-S-H particles in the hcp.)

It has also been found that, under short term loading, the strengths of hcp and concrete decrease with increasing relative humidity. Feldman and Sereda (ref. 36) showed that the greatest decrease in strength occurred on increasing the relative humidity from 0 to 15%, with little change in strength on further wetting, similar to the results of Glucklich and Korin (ref. 34). The explanation of Feldman and Sereda (ref. 36) was that further cracking will occur at crack tips through straining of Si-O-Si bonds; this strain energy will contribute to the greater ease of formation of hydroxyl groups $(-\overset{|}{\underset{|}{Si}}-OH \quad OH-\overset{|}{\underset{|}{Si}}-)$ in the presence of water vapor. No further decreases in strength will occur beyond the relative humidity at which a sufficient quantity of water can diffuse to the crack tip. In their view, the decrease in strength is due to a change in the crack tip environment, rather than to changes in the bonding between the solid units which make up hcp.

Attempts have also been made to explain the influence of water on strength by using thermodynamic considerations. Cook and Haque (refs. 31,32) suggested that the adsorption of water dilates the structure of the C-S-H, and hence reduces the surface energy, while Glucklich and Korin (ref. 34) felt that the reduction in surface energy due to adsorption of water was due to bond saturation within the C-S-H. In a much more detailed explanation, Wittmann (ref. 37) modelled the structure of hcp as a xerogel (a 3-dimensional network of colloidal particles), in which both primary and secondary bonds contribute to strength. Above about 50% relative humidity, the disjoining pressure of the water separates surfaces held together only by van der Waals bonding. Using the Griffith equation, the change in strength can be related to the change in interfacial energy. The strength should decrease as the moisture content (and hence the disjoining pressure) increases, at least in the range where changes in relative humidity affect surface energy. Quantitatively, by interpreting sorption data in the light of the Griffith equation, Setzer (ref. 38) proposed that strength changes could be related to changes in the surface free energy by

$$\left(\frac{\sigma}{\sigma_o}\right)^2 = 1 - \frac{\Delta\gamma}{\gamma_o} \tag{1}$$

where σ_o and γ_o are the strength and surface free energy, respectively, at saturation, σ is the strength at a given relative humidity, and $\Delta\gamma$ is the corresponding change in surface free energy. However, there is still no concensus as to the relative contributions of changes in surface energy and stress corrosion in determining the effects of water on the strength of hcp.

4 APPLICATIONS OF LINEAR ELASTIC FRACTURE MECHANICS

There are a number of phenomena regarding the behaviour of hcp and concrete under load which suggest that a fracture mechanics approach can be used with these materials:

(i) Fracture generally occurs in a brittle fashion, though more so for hcp than for concrete.

(ii) The strengths of both hcp and concrete increase with increased rates of loading (refs. 39-41) and, as mentioned above, these materials are also subject to static fatigue (refs. 23-25,42). Both of these phenomena can be related to subcritical (slow) crack growth.

(iii) The tensile strengths of hcp and concrete are approximately 1/10 of their compressive strengths (the exact values depend on the method of measurement). This is quite close to the value of 1/8 predicted by Griffith (ref. 43) for a brittle material.

(iv) The stress concentrations which have been noted near the cement-aggregate interfaces (refs. 7,8) could easily be of such a magnitude as to exceed the local theoretical strength in these regions.

(v) Almost all investigations of notch sensitivity show that hcp is a highly notch sensitive material (refs. 44-48); only in one investigation (ref. 49) was it concluded that hcp is not notch sensitive. Mortars and concretes appear to be considerably less notch sensitive than hcp (refs. 45,48,50,51). However, here the evidence is mixed, as several investigators (refs. 44,49) have found concrete to be notch insensitive. Some of the data on notch sensitivity of hcp, mortar and concrete are shown in Fig. 6. Ziegeldorf et al. (ref. 48) expressed notch sensitivity as the ratio of net failure stress to the modulus of rupture of an unnotched specimen. For a notched beam in four-point bending, they showed that notch sensitivity was a function of relative crack length, specimen depth, and the ratio of fracture toughness to tensile strength:

$$\frac{\sigma_n}{\sigma_f} = \frac{K_{IC}}{\sigma_f} \cdot \frac{1}{\sqrt{a}(1 - \frac{a}{b})^2 F(a/b)} \leq 1 \tag{2}$$

where σ_n = net stress at the crack tip of a flexure specimen, σ_f = tensile strength, a = crack length, K_{IC} = fracture toughness, b = specimen depth, and $F(a/b)$ = correction factor for the particular specimen geometry. They showed

10

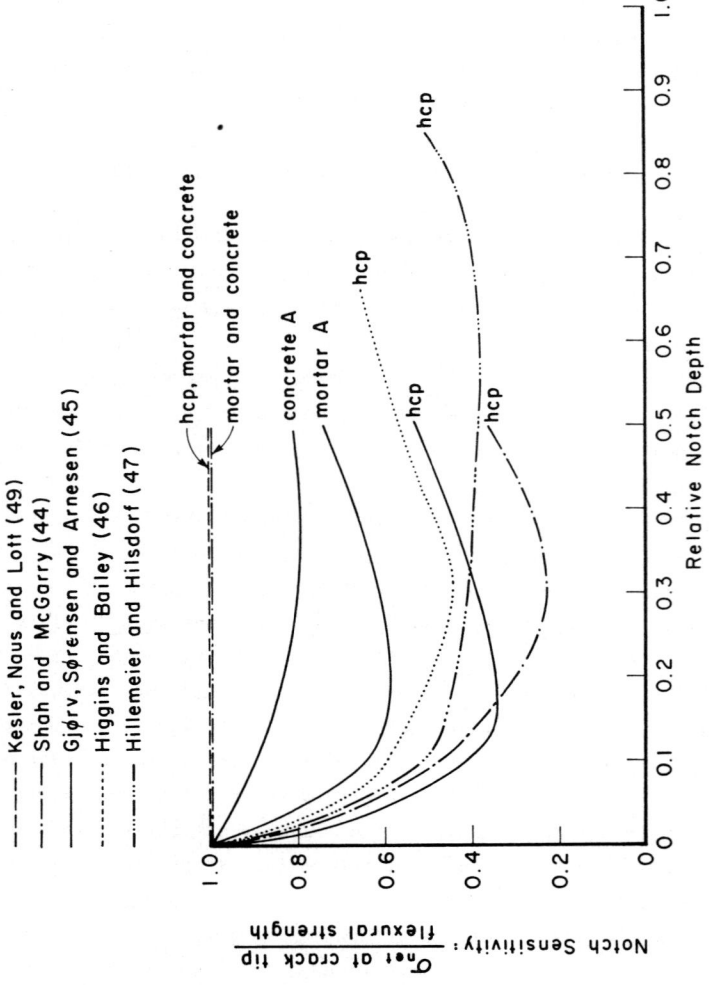

Fig. 6. Effect of relative notch depth on notch sensitivity.

that notch sensitivity is a necessary, though not sufficient, condition for the applicability of linear elastic fracture mechanics to cementitious materials.

4.1 Measurements of G_c and K_c

The first application of the Griffith theory to concrete appears to have been made by Neville (ref. 52), who suggested that the effects of specimen size on the strength of concrete could be related to the random distribution of Griffith flaws. He argued that the Griffith theory would only be approximate for conc- rete, since the strength might be governed by a limiting strain rather than a limiting stress criterion. However, the first experimental study of the appli- cability of fracture mechanics to concrete was conducted by Kaplan (ref. 53), who carried out bending tests on single-edge notched (SEN) beams of mortar and concrete, and calculated G_c, the critical strain energy release rate. Some of his results are summarized in Table 2. It is apparent from these results that larger beams gave higher values of G_c, as did center-point loading as compared to third-point loading. In addition, he found that G_c of the mortar was appro- ximately equal to twelve times the value of 2γ where 2γ was the estimated sur- face energy of the mortar. Nonetheless, Kaplan concluded "that the Griffith concept of a critical strain-energy release rate being a condition for rapid crack propagation and consequent fracture, is applicable to concrete".

It is of historical interest to note that there were three published dis- cussions of Kaplan's (ref. 53) paper. Blakey and Beresford (ref. 54) rejected the notion that fracture mechanics could be applied to concrete. They suggested that Kaplan's data could be better explained by "simpler" ideas, such as the size effect, variations in the type of loading, and the energy dissipated as heat. Glucklich (ref. 55) supported the fracture mechanics approach, but pointed out three limitations: (i) the necessity to account for slow crack growth; (ii) uncertainty with regard to the most severe crack in a concrete member; and (iii) the effect of the degree of indeterminacy of the system. Finally, Irwin (ref. 56) also supported the fracture mechanics approach, and reiterated the need to estimate the true crack depth at the critical load. He also added (with amusing condescension) that "a smaller degree of sophistica- tion in measurement procedures than seems necessary for metals would be appro- priate for crack toughness measurements in concrete"! These three discussions have foreshadowed almost all of the subsequent arguments as to the applicability of fracture mechanics to concrete; each of the views noted above has its strong proponents today.

Following the pioneering work of Kaplan (ref. 53), Glucklich (refs. 6,57-59) developed the Griffith theory for concrete in much greater detail. He stated that $G_c \gg 2\gamma$ because the fracture of concrete was not limited to the propaga-

TABLE 2

Critical strain energy release rates, G_c, from Kaplan (ref. 53).

Size of beam, mm	Method of loading	Ratio of notch depth to beam depth	G_c, J/m^2		
			Mortar E = 27,945 MPa	Concrete A E = 37,743 MPa	Concrete B E = 28,911 MPa
76.2 x 101.6 x 406.4	Third-point	0.17	13.9	14.0	11.6
		0.33	15.4	15.3	10.3
		0.50	13.7	14.4	10.2
	Center-point	0.17	16.1	16.5	11.4
		0.33	17.2	19.6	10.9
		0.50	15.3	17.4	11.9
152.4 x 152.4 x 1248.0	Third-point	0.17	19.6	21.4	15.4
		0.33	21.2	28.2	18.6
		0.50	18.9	26.3	16.0
	Center-point	0.17	25.4	31.7	16.1
		0.33	27.2	33.9	22.3
		0.50	24.9	27.2	17.7

tion of a single crack; instead, a multitude of microcracks formed in the highly stressed zone, and therefore the true fracture surface area was much greater
than the apparent one. He also showed that "high strength" areas in concrete,
such as aggregate particles, act as crack arrestors because they increase the
energy demand. Due to microcracking and the influence of aggregate particles
the energy demand must then increase as the crack grows. For an ideal
"Griffith" material, the energy belance is shown in Figure 7a. The energy
requirement curve is linear with a slope of 2γ, while the energy release curve
is a second power parabola; instability occurs when the slopes of these curves
are equal, the crack length at this point being the critical crack length. For
concrete, however, the energy balance is shown in Figure 7b. A crack of initial length C_0 will start to grow, but will be stopped at one of the high
strength areas mentioned above, which increase the energy demand. A higher
stress is now required to reactivate the crack growth, leading to a step-wise
growth until a new critical crack length is reached, at which point the slope
of the energy requirement curve has stopped increasing. This has been confirmed
experimentally by Brown (ref. 60), who showed that in both hcp and mortars crack
growth is initiated at a lower K than that needed to keep the crack running.

In addition to this theoretical development, a large number of experimental
studies were carried out, to try to determine the effects of various paramenters
on measured values of K_c or G_c. In line with the above theoretical considerations, a number of studies (refs. 61-66) showed that, for normal weight concrete, where the aggregate was stronger than the hcp, K_c and G_c increased with
increasing aggregate volume. On the other hand, for lightweight aggregates
which are weaker than hcp, an increase in aggregate volume led to a decrease in
fracture toughness (refs. 63-65). It was also generally found that K_c or G_c increased as the maximum size of coarse aggregate increased (refs. 45,50,62,65,67),
though Petersson (ref. 66) saw no effect. Alford and Poole (ref. 68) expressed
the aggregate shape and texture of seventeen different rocks in terms of an
"angularity factor". Although there was considerable scatter, they found that
G_c generally increased as the angularity factor increased; they suggested that
more angular aggregate particles are better crack arrestors, since they have a
better bond with the hcp, and also cause more microcracking.

As is the case with strength, K_c and G_c increased as the w/c ratio decreased
(refs. 46,65,68-70). However, other investigators (refs. 61,62) found that this
was true only for cements and mortars, while the fracture toughness of concrete
was independent of the w/c ratio. Similarly, Wittmann (ref. 71) showed that the
surface energy of hcp increased as the w/c ratio decreased, from about 0.66 J/m^2
at w/c = 0.6 to about 1.75 J/m^2 at w/c = 0.3. On the other hand, Morita (ref.
72) found that the work of fracture was essentially independent of the w/c
ratio.

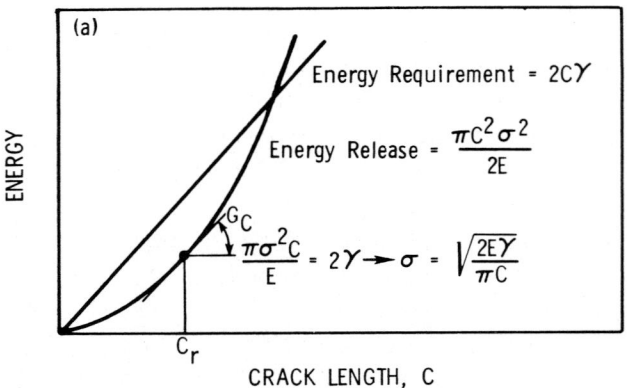

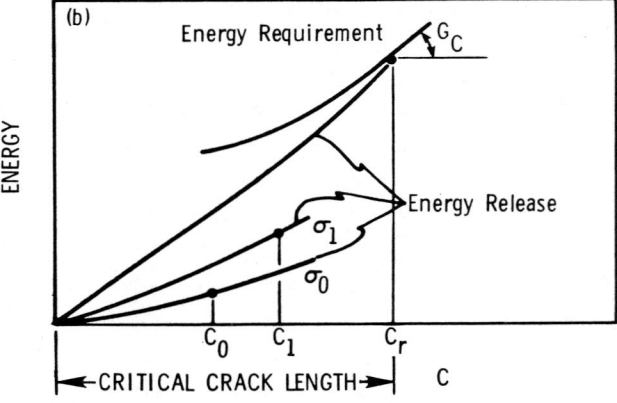

Fig. 7. Crack growth (a) according to Griffith theory, and (b) modified Griffith theory for concrete, after Glucklich (ref. 57).

For normal cements and concretes, K_c was found to decrease with increasing air content (ref. 62), presumably because of the reduced strength of the matrix. Kayyali et al. (ref. 73) also noted that freezing and thawing cycles applied to cement pastes less than one day old reduced G_c considerably, and caused an explosive mode of fracture, while air entrainment made the fracture much less explosive. However, these effects disappeared if the cyclic freezing and thawing began after the hcp was more than one day old.

In a most interesting modern development, Birchall et al. (ref. 74) argue that the Griffith theory applies adequately to hcp, and that the strength of hcp is controlled not by the total porosity, but by the presence of large voids, which act as Griffith flaws. For ordinary hcp, they measured a fracture energy of 19 J/m^2; when large voids were removed by a special sample preparation, the fracture energy increased to 30 J/m^2. By reducing the maximum pore size, they were able to obtain flexural strengths of over 70 MPa. Alford (ref. 75) also suggested that the strength of hcp is controlled by the critical flaw, i.e. the largest pore.

As would be expected, K_c increased with age (refs. 46,62,66,67,76) although, unlike strength, it seemed to reach its maximum value within about one month. K_c also increased slightly as the strain rate increased (refs. 46,46a).

For reasons that will be discussed below, there is a great deal of variability in measurements of K_c of G_c, which is due in large part to the fact that different specimen geometries and measuring techniques appear not to be directly comparable. However, in general, the measurements of K_c cited above gave values in the range of 0.10 to 0.50 $MNm^{-1.5}$ (100 to 450 psi $\sqrt{in}$) for hcp; for a wide variety of concretes, values ranged from 0.45 to 1.40 $MNm^{-1.5}$ (400 to 1300 psi $\sqrt{in}$).

4.2 Subcritical crack growth

A somewhat different approach has been the application of fracture mechanics to obtain subcritical crack growth data, in particular to determine the relationship between the stress intensity factor, K_I, at a crack tip and the crack velocity, V. This data is generally presented in the form of a log V vs. log K_I plot, the first part of which is linear, and so can be defined by the slope (n) and the intercept (A). This data can then be used, at least in principle, to characterize the time-dependent fracture properties. The first study of this type on hcp and mortar was carried out by Nadeau et al. (ref. 77), using double-torsion specimens. They found that, for hcp, the $V-K_I$ relationship could be expressed as

$$\log V = 36 \log K_I - 200 \tag{3}$$

Also, using the equation

$$K_{IC} = Y\sigma_a C^{1/2} \tag{4}$$

where C is the critical flaw size, σ_a is the apparent fracture stress, and Y is a parameter depending on the specimen geometry, they found the critical flaw size to be of the order of 0.10 mm. This was similar to the value (0.02-0.2 mm) obtained by Hansen (ref. 78) by inserting the estimated surface energy of hcp into the Griffith equation. Subsequent work (ref. 29) indicated that resistance to subcritical crack growth could be improved by reducing the w/c ratio. However, high pressure steam curing made the specimens more susceptible to subcritical crack growth. Finally, as mentioned earlier, subcritical crack growth was found to be enhanced by the presence of water. Using mortar double torsion specimens, Evans et al. (ref. 79) obtained similar V-K_I plots. They found that neither the age nor the w/c ratio had much effect on the fracture parameters.

Using much larger hcp double torsion specimens, Yam and Mindess (ref. 80) obtained values of the fracture parameters similar to those obtained on small specimens (ref. 77). Also, using very large mortar double torsion specimens, Wecharatana and Shah (refs. 81,82) were able to obtain V-K_I plots which showed both low and high velocity crack growth, unlike the earlier studies (refs. 29,77, 79) in which data were obtained only for the low velocity range of crack growth.

The studies cited above attempted to measure subcritical crack growth either directly or indirectly. In addition, Zaitsev and his co-workers (refs. 83-86) argued that the fracture mechanics approach could also be used to predict the static fatigue behaviour of hcp and concrete, as long as the effects of creep, stress relaxation near the crack tip, and the time dependence of the modulus of elasticity were taken into consideration.

These subcritical crack growth studies (refs. 29,77,79-82) are, unfortunately, not all consistent with other data. In principle, it should be possible to obtain the crack growth parameter n in three different ways: (i) by direct measurement of crack growth, (ii) by measuring the time to failure under constant load, and (iii) by determining the effect of strain rate on strength. These three methods are shown schematically in Fig. 8. However, Mindess and Nadeau (ref. 39) found that the n-values obtained from method (i) did not agree with those from method (iii) for cement paste, though there was fairly good agreement for mortar. The reasons for the discrepancy are not understood.

4.3 Does linear elastic fracture mechanics apply to hcp and concrete?

As the amount of experimental fracture mechanics data accumulated, a number of inconsistencies began to appear, particularly with regard to the effects of specimen type, specimen size, and the ratio of notch depth to specimen depth

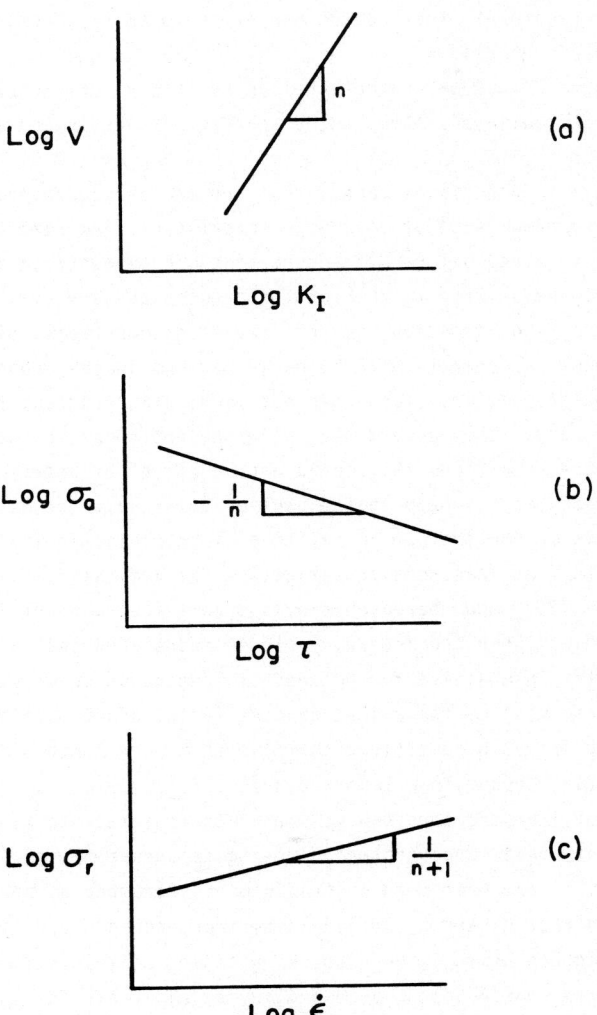

Fig. 8. Three ways of determining n, the slope of the log V vs. log K_I plot: (a) direct observation of the velocity of crack growth vs. the stress intensity factor; (b) applied stress vs. time to failure; (c) failure stress vs. rate of loading.

(a/d ratio). Even though, on reviewing the then available literature, Welch and Haisman (ref. 87) could conclude that "although methods of test and assumptions used in calculations of G_c and K_c appear to influence their magnitude significantly, G_c and K_c nevertheless may be considered to be material constants for a particular concrete", the problem cannot be so easily dismissed. There are really two basic questions:

1. Can the strength of hcp or concrete be controlled by artificially introduced notches, or is it always controlled by inherent cracks in the material (ref. 88)?

2. If artificially induced flaws can control the fracture of hcp and concrete, is there some minimum size of specimen, or ratio of flaw size to specimen size, that is required before a fracture test can be considered to be valid?

Unfortunately, the answers to both of these questions are still uncertain. With regard to the first question, several investigators (refs. 89,90), on the basis of tests done on concrete with holes of various shapes, concluded that due to stress redistributions, holes did not particularly affect the structural behaviour of concrete. They argued that the inherent flaws in the concrete introduced stress concentrations that could not be offset by artificially induced flaws. In addition, as has been stated earlier (ref. 48), notch sensitivity is a necessary condition for the applicability of linear elastic fracture mechanics to hcp or concrete. At least one investigation has indicated that hcp is notch insensitive (ref. 49). And, based on previous work (refs. 53,62,91) as well as on their own studies, Shah and McGarry (ref. 44) concluded that both mortar and concrete were notch insensitive for notches of lengths up to at least a few inches. Nonetheless, most of the recent studies (refs. 45,48,50,51,92) indicate not only that hcp is notch sensitive, but also that even mortars and concretes are notch sensitive, though to a lesser degree.

Therefore, most investigators now assume that the answer to the first question is "yes"; they have concentrated on trying to answer the second question. Yet here the results are even more contradictory. A number of investigators have indeed found that G_c and K_c appear to be independent of crack length. For 14 mm deep notched hcp beams in bending, Higgins and Bailey (ref. 46) found K_c to be independent of notch depth, as did Gjørv et al. (ref. 45) for 50 mm deep beams. Based on tests of 38x38x250 mm notched beams and 50x100x350 mm double cantilever beams, Brown (ref. 60) and Brown and Pomeroy (ref. 70) found that K_c was independent of crack growth for hcp, though the two types of tests gave somewhat different K_c values. For mortars, Evans et al. (ref. 79), using 105 mm long double-torsion specimens, concluded that the fracture mechanics parameters were independent of crack length for crack lengths greater than 20 mm. Similarly, from flexure tests on 76x76x292 mm beams, Kim et al. (ref. 92) found K_c

to be independent of the (a/d) ratio. For concrete, Yokomichi et al. (ref.69) tested 150 mm deep notched beams on a 450 mm span; G_c was not influenced by the notch depth. Using smaller notched beams in bending, Kitagawa and Suyama (ref. 93) found that K_c was approximately constant for different crack lengths (decreasing slightly with increasing crack length) and specimen sizes, though K_c increased slightly with crack length for mortar. Mazars (ref. 94) tested 600x340x80 mm concrete plates in tension, both with central cracks and with single-edge cracks, as well as 150x100x100 mm notched beams in bending. He found that G_c was essentially independent both of crack length and of the type of specimen.

Other investigations have produced quite different results. Ohigashi (ref. 95) found that for small mortar beams, the fracture energy decreased markedly with increasing notch depth. Kesler et al. (ref. 49) tested hcp, mortar and concrete plates with a center crack, 50 mm deep x 300 mm high, and varying in length from 460 to 915 mm. They also reported that K_c decreased considerably with increasing crack length, though the effect was less pronounced for hcp. Indeed, their results led them to conclude that "the concepts of linear elastic fracture mechanics are not directly applicable to cement pastes, mortars and concretes". (However, it should be noted here that Saouma et al. (ref. 96) have recently re-analyzed these results, using modern computational techniques and a better stress intensity analysis. They argued that the above conclusion of Kesler et al. (ref. 49) is incorrect, and that proper interpretation of their data suggests that linear elastic fracture mechanics can indeed be applied to hcp, mortar and concrete.) Hillemeier and Hilsdorf (ref. 47), from tests on wedge-loaded compact tension specimens with dimensions 105x100x40 mm, found that K_c decreased with increasing crack depth only up to an (a/d) ratio of about 0.6, beyond which the values stabilized. Similar results were obtained by Watson (ref. 97) from tests on single-edge notched hcp beams of dimensions 20x20x250 mm though he obtained different G_c values for different beam spans. Gjørv et al. (ref. 45) found that K_c decreased with increasing notch depth for mortar and concrete, but not, as mentioned above, for hcp.

Still other investigations have produced completely opposite results. For mortar, Brown (ref. 60) showed that K_c increased with crack growth over the range 10-100 mm. Romualdi and Batson (ref. 98) carried out tension tests on concrete plates (810x610x64 mm) with centrally located slots ranging in length from 50 to 300 mm, and found that G_c increased with increasing crack length. Desayi (ref. 99), from compression tests on 150x150x300 mm concrete prisms with central notches, also concluded that G_c appeared to increase with increasing crack length.

Finally, some investigations have indicated that the fracture parameters depend not on the relative notch depth, but on overall specimen size or specimen type. Kitagawa et al. (ref. 100), from tests on radially cracked mortar discs in diametral compression found that K_c increased as the disc diameter increased from 100 to 150 mm; K_c obtained from bending tests was different again. Higgins and Bailey (ref. 46) showed that for hcp K_c increased with increasing specimen depth. Strange and Bryant (refs. 50,101) carried out bending tests on notched beams of hcp, mortar and concrete, ranging in size from 12x12x60 mm to 200x100x1000 mm, and also found that K_c increased as the beam size increased. Examples of these different types of behaviour are shown in Fig. 9 (refs. 46, 47,60).

Somayaji (ref. 102) compared the results of many of the experimental studies which had been carried out to measure K_c and G_c. He concluded that the results obtained were incompatible with each other, as the values obtained appeared to depend on (i) beam dimensions, (ii) notch dimensions, (iii) concrete properties, and (iv) the method of analysis used. These contradictory results have led a number of investigators to question the applicability of linear elastic fracture mechanics to hcp and concrete. Apart from the statement by Kesler et al. (ref. 49) quoted earlier, Strange and Bryant (ref. 50) argued that linear stress analysis was not applicable in the crack tip region. Kim et al. (ref. 92) concluded that the existence of a true K_c for concrete was uncertain. Similarly, Swartz et al. (ref. 103) concluded that for concrete "The failure mechanism appears to be a combination of aggregate fracture and bond failure, and does not depend on particle size. Because of this combined effect, it is felt that the fracture toughness as normally used is not a pertinent material property". However, since an even larger group of investigators continue to believe in the applicability of fracture mechanics, an attempt should be made to explain at least some of these discrepancies.

Clearly, some of the problems are inherent in the experimental techniques themselves. Most of the studies described above did not account for any slow crack growth that may have taken place during the test (largely because it would have been extremely difficult to do so experimentally); G_c or K_c were simply calculated from the maximum load and the length of the pre-formed notch. Since some slow crack growth undoubtedly does occur during a test (as was discussed in sections 1.2 and 1.3), such studies would have underestimated the effective crack length at maximum load. On the other hand, those studies that did try to account for slow crack growth did so largely on the basis of compliance measurements. However, the compliance of a specimen with a cast or sawn notch cannot really be compared to that of a "naturally" cracked specimen. In the first case, there is no interaction between the sides of the crack; in the second

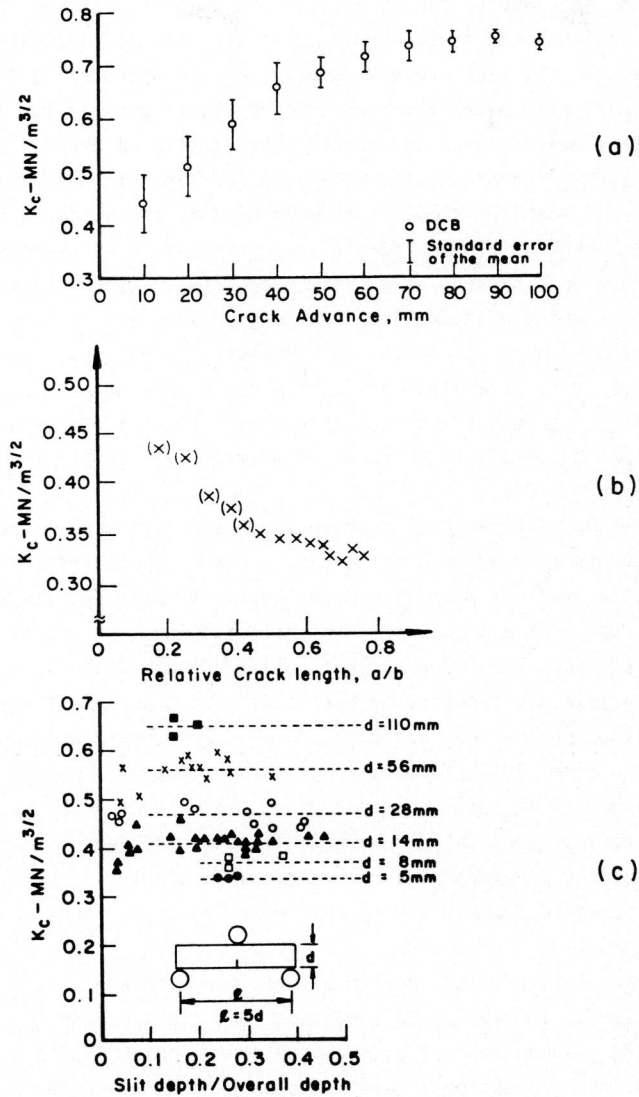

Fig. 9. Effects of notch depth and specimen size on K_C, according to
(a) Brown (ref. 60); (b) Hillemeier and Hilsdorf (ref. 47); (c) Higgins
and Bailey (ref. 46).

case, due to tortuosity and surface roughness, there may be considerable inter-
actions between the two sides of the crack, particularly near the crack tip.
Therefore, as was concluded by Kim et al.(ref. 92), the compliance method for
estimating slow crack growth is inaccurate. At the present time there is no
real way to determine the true crack length in hcp or concrete. Indeed, considering the very complex microstructure of these materials, it would be very
difficult to define exactly what constitutes the "end" of a crack.

In addition, the different studies have been carried out on a wide variety of
testing machines and loading frames, with wide differences in their stiffness or
compliance. The amount of elastic strain energy stored in a system may play a
part in determining what happens as a crack propagates. Glucklich and Cohen
(ref. 104) have presented a theoretical analysis of size effects insofar as they
govern the amount of stored strain energy. However, recent data obtained by
Mindess and Bentur (ref. 104a) indicate that machine stiffness has no effect on
K_c. However, where fracture energy is defined as the total energy under the
σ-ε curve, then machine stiffness becomes very important, as pointed out by
Brown and Hudson (ref. 104b).

Perhaps an even more fundamental problem is that of defining a specimen size
which is large enough to provide a valid test. Gjørv et al. (ref. 45) concluded
that linear elastic fracture mechanics might apply to small specimens of hcp or
lightweight concrete, but did not apply to small specimens of ordinary mortar or
concrete. However, Cook and Crookham (ref. 51), in their discussion of ref. 45,
argued that linear elastic fracture mechanics did apply to relatively small
concrete beams (100x100x500 mm), as long as slow crack growth was accounted for.
Mindess and Nadeau (ref. 105), from bending tests on notched hcp and mortar
beams that were only 51 mm deep, concluded that the assumptions of a plane
strain stress condition were valid. Nevertheless, it has become increasingly
clear that much of the fracture data reported, particularly in the earlier
literature, was obtained from specimens that were too small to be valid fracture
specimens.

The first person to study the problem of specimen size explicitly appears to
have been Walsh (refs. 106-108). He suggested that the specimen must be large
enough so that the assumed zone of disturbance at the crack tip is small compared to the specimen dimensions. Based on beam bending tests, he concluded
that a minimum specimen size for concrete should be about 230 mm (ref. 106), or
more generally that the requirement can be expressed (ref. 107) as

$$(\frac{K_c}{\sigma_f})^2 > 50 \text{ mm}$$

(5)

where σ_f is the nominal failure stress in flexure. He subsequently proposed
(ref. 108) that for proper fracture toughness tests on notched beams, the beam

must be of sufficient depth so that the crack length is much greater than 78 (f_c'/σ_f^2), where f_c' is the cylinder compressive strength, and σ_f is the modulus of rupture. For the concrete tested, this yielded a suggested minimum beam depth of 230 mm.

Even these estimates may be low. Bazant (ref. 109) has recently stated that for true Griffith behaviour, the cross-sectional dimensions of the member should be at least 100 times the maximum aggregate size, which even for a maximum aggregate size of only 10 mm, would lead to a beam depth of at least 1000 mm. However, for high strength concrete, which behaves in a more brittle manner, the minimum specimen size could be decreased.

Quite different types of studies have also indicated that large specimens are required. Entov and Yagust (ref. 109a) tested specimens up to 100x2000x2500 mm in size, and found a microfracture process gave about 80-100 mm in length. Sok et al. (ref. 110), from tests on large double canti-lever beams of concrete with some longitudinal prestressing, concluded that for valid fracture tests, the specimens must be large enough so that more than 0.5 m of crack propagation can take place. Subsequent work (ref. 111) showed that a microcracked region about 150-200 mm in size existed ahead of the apparent crack tip; this led to the observation that K_c reached a constant value only after about 1 m of crack propagation. For mortar, Visalvanich and Naaman (ref. 112) found from tests on large double cantilever beam specimens, that K_c reached a constant value only after about 200 mm of crack growth. Similarly, for double-torsion specimens of mortar, Wecharatana and Shah (ref. 82) found that unstable crack propagation only occurred at a crack length of about 305-380 mm.

From consideration of the "fictitious crack model" (see below), Modeer (ref. 113) defined a new parameter called the characteristic length (ℓ_{ch}):

$$\ell_{ch} = G_c \, E/\sigma_t^2 \tag{6}$$

where σ_t is the tensile strength. He argued that for linear elastic fracture mechanics tests to be valid, notched beams should have a depth of at least $10\ell_{ch}$. This leads to enormous beam depths: 50-100 mm for hcp, 1-2 m for mortar, and 2-3 m for concrete! To the writer's knowledge, no tests on mortar or concrete beams even approaching these sizes have ever been carried out.

Finally, Carpinteri (refs. 114-116) defined a non-dimensional parameter, the test brittleness number, s, obtained by applying Buckingham's theorem for physical similitude and scale modelling to cementitious materials:

$$s = K_c/\sigma_u \ b^{1/2} \qquad\qquad\qquad\qquad\qquad\qquad (7)$$

where σ_u is the ultimate tensile stress and b is the specimen depth. Carpinteri argued that for s > 0.50, concrete is no longer notch-sensitive, and tests for K_c and G_c lose their meaning. This would indicate minimum specimen depths of about 100 mm for hcp, and 650 mm for concrete.

Considering all of these contradictions, the weight of the evidence would suggest that linear elastic fracture mechanics is probably applicable to hcp. For mortar and concrete, the evidence is still ambiguous, partly because it may be that in view of the problem of specimen size discussed above, almost no valid fracture tests have ever been carried out. Thus, there is still no concensus as to whether linear elastic fracture mechanics is even a reasonable approximation when applied to concrete, or, if it is, what constitutes a valid test.

1.5 NON-LINEAR FRACTURE CRITERIA

Partly in response to these difficulties in applying linear elastic fracture mechanics, a number of non-linear fracture criteria have been tried. Mindess et al. (ref. 117) and Halvorsen (refs. 118,119) suggested that the J-integral might be an appropriate fracture criterion for concrete. Carrato (ref. 120) showed that the J-integral was independent of the notch geometry, and that in notched beam tests the notch depth became important only when the maximum aggregate size approached the size of the uncracked ligament. However, as was pointed out by all of the investigators, the variability in determinations of the J-integral is very large, and this posed a difficulty in adopting the J-integral as a suitable fracture criterion.

Bazant (refs. 109,121) and Bazant and Cedolin (refs. 122-125) used a "smeared" crack to model cracking in concrete. In this model, the crack front is assumed to consist of a diffuse zone of microcracks. The size of this zone is related to the maximum aggregate size. An energy criterion is used for crack propagation, which can be generalized for non-linear material behaviour. This type of model is particularly suitable for finite element analysis.

A number of investigations (refs. 81,82,110,126,127) have used R-curve analysis to characterize the fracture of both hcp and concrete. Wecharatana and Shah (ref. 81) showed that the calculated R-curves appear to be independent of the test configuration, and so this appears to be a fruitful line of research.

Finally, in another extremely interesting development, Hillerborg and his colleagues (refs. 66, 113,128-131) have proposed a tied-crack model (which they have named the "fictitious crack model"), in which stresses are assumed to act across a crack as long as it is only narrowly opened. Thus the region just behind the crack tip corresponds to a microcracked zone with some remaining ligaments for stress transfer. The proposed model, and its finite element representation, are shown in Fig. 10. A relationship between the applied stress, σ, and the crack width, w, is developed; the area under the σ-w curve is assumed to be equal to G_c. This model is also particularly suitable for finite element analysis. Subsequently, Petersson (ref. 132) and Hillerborg and Petersson (ref. 133) extended this model by introducing as a fracture criterion the complete tensile stress-strain curve.

6 CONCLUSIONS

From the discussion above, it seems clear that a great deal of further research, both experimental and theoretical, into the fracture of hcp and concrete is warranted. First, the physical process of cracking needs further study; the simplistic models of crack propagation which have so far been used are not a sufficiently good representation of the fracture process. Second, researchers are still far from agreement as to what constitutes a valid fracture test, or how to analyze the fracture data that they obtain. It can only be hoped that these problems will soon be solved, so that fracture analysis can become a useful tool for the analysis of concrete structures.

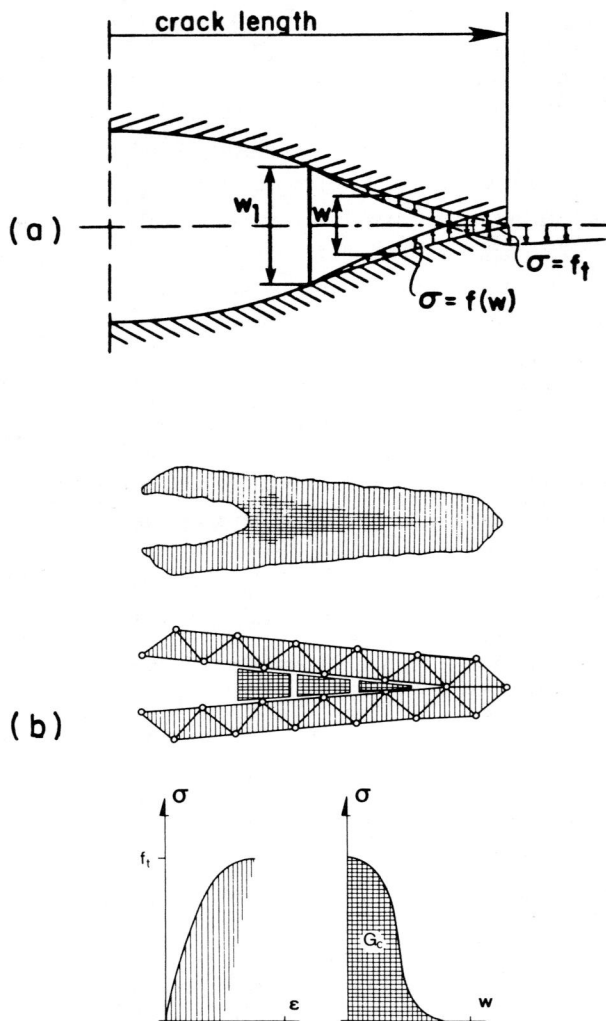

Fig. 10. (a) proposed crack tip model for hcp (ref. 128); (b) actual tensile crack, its finite element representation, and material properties (ref. 113).

REFERENCES

1   F.E. Richart, A. Brandtzaeg and R.L. Brown, Bulletin No. 185, Engineering
     Experiment Station, University of Illinois, 1928, 104pp.
2   A.A. Griffith, Phil. Trans. R. Soc., A221 (1920) 163-198.
3   S. Mindess, in F.H. Wittmann (Ed.), Fracture Mechanics of Concrete,
     Elsevier, Amsterdam, 1982, pp. 481-502.
4   S. Mindess, in F.H. Wittmann (Ed.), Fracture Mechanics of Concrete,
     Elsevier, Amsterdam, 1982, pp. 539-661.
5   T.T.C. Hsu, F.O. Slate, G.M. Sturman and G. Winter, J. Amer. Concr.
     Inst., 60 (1963) 209-224.
6   J. Glucklich, in A.E. Brooks and K. Newman (Eds.), Proc. Int. Conf.
     The Structure of Concrete, London, Sept., 1965, Cement and Concrete
     Association,London, 1968, pp. 176-189.
7   P. Dantu, Annales de L'Inst. Tech. Bat. et des Trav. Publ., 11 (1958)
     55-77.
8   R.N. Swamy, in M. Te'eni (Ed.), Proc. Southampton 1969 Civil Eng. Mat.
     Conf. Structure, Solid Mechanics and Engineering Design, Southampton,
     April 1969, Wiley-Interscience, 1971, pp. 301-315.
9   A. Yoshimoto, K. Kawasaki and M. Kawakami, in Rev. 29th Gen. Meet., The
     Cement Association of Japan, Tokyo, 1975, pp. 243-244.
10   R.H. Evans and M.S. Marathe, Mater. Constr. (Paris), 1 (1968) 61-64.
11   P.T. Wang, S.P. Shah and A.E. Naaman, J. Amer. Concr. Inst., 75 (1978).
12   S. Mindess and S. Diamond, Cem. Concr. Res., 10 (1980) 509-519.
13   S. Mindess and S. Diamond, Mater. Constr. (Paris), 15 (1982) 107-113.
14   S. Mindess and S. Diamond, Cem. Concr. Res., in press.
14a S. Diamond, S. Mindess and J. Lovell, Cem. Concr. Res., in press.
15   R.B. Tait and H. Bohm, Proc. Electronmicroscopy Soc. of South Africa,
     10 (1980) 17-18.
16   Y.B. Zaitsev and F.H. Wittmann, Mater. Constr. (Paris), 14 (1981) 357-365.
17   R.L. Berger, Science, 175 (1972) 626-629.
18   R.L. Berger, F.V. Lawrence and J.F. Young, Cem. Concr. Res., 3 (1973)
     497-508.
19   R.L. Berger, personal communication.
20   D. Walsh, M.A. Otooni, M.E. Taylor and M.J. Marcinkowski, J. Mat. Sci.,
     9 (1974) 423-429.
21   B. Marchese, Cem. Concr. Res., 7 (1977) 9-18.
22   S.P. Shah and S. Chandra, J. Amer. Concr. Inst., 67 (1970) 816-825.
23   A.D. Husak, Static Fatigue of Portland Cement Concrete, Ph.D. Thesis,
     Carnegie-Mellon University, Pittsburgh, 1969.
24   J.E. Barrick, The Effects of Temperature and Relative Humidity on Static
     Fatigue of Hydrated Portland Cement, Ph.D. Thesis, Carnegie-Mellon
     University, Pittsburgh, 1972.
25   J.E. Barrick and E.M. Krokosky, J. Test. Evaln., 4 (1976) 61-73.
26   P.L. Domone, Mag. Concr. Res., 26 (1974) 144-152.
27   A.D. Husak and E.M. Krokosky, J. Amer. Concr. Inst., 68 (1971) 263-271.
28   E.M. Krokosky, Mater. Constr. (Paris), 6 (1973) 447-452.
29   S. Mindess, J.S. Nadeau and J.M. Hay, Cem. Concr. Res. 4 (1974) 953-965.
30   D.J. Cook and M.N. Haque, Cem. Concr. Res., 4 (1974) 367-379.
31   D.J. Cook and M.N. Haque, Mater. Constr. (Paris), 7 (1974) 191-196.
32   D.J. Cook and M.N. Haque, Cem. Concr. Res., 4 (1974) 735-744.
33   G.A. Cooper and J. Figg, J. Brit. Ceram. Soc., 71 (1972) 1-4.
34   J. Glucklich and U. Korin, J. Amer. Ceram. Soc., 58 (1971) 517-521.
35   A. de Sousa Coutinho, Mater. Constr. (Paris) 2 (1969) 49-57.
36   R.F. Feldman and P.J. Sereda, Eng. J. (Canada),53 (1970) 53-59.
37   F.H. Wittmann, in Hydraulic Cement Pastes: Their Structure and Properties,
     Proceedings of a Conference at University of Sheffield, April 8-9, 1976,
     Cement and Concrete Association, Wexham Springs, 1976, pp. 96-117.
38   M.J. Setzer, Cem. Concr. Res., 6 (1976) 37-48.

39 S. Mindess and J.S. Nadeau, Amer. Ceram. Soc. Bull., 56 (1977) 429-430.
40 B. Zech and F.H. Wittmann, J. Amer. Concr. Inst., 77 (1980) 358-362.
41 H. Mihashi and F.H. Wittmann, Heron, 25, No. 3 (1980) 54 pp.
42 M.A. Al-Kubaisy and A.G. Young, Mag. Concr. Res., 27 (1975) 171-178.
43 A.A. Griffith, in C.B. Biezeno and J.M. Burgers (Eds.), Proc. 1st. Int.
    Congr. Appl. Mech., Delft, 1924, J. Waltman Jr., Delft, 1924, pp. 55-63.
44 S.P. Shah and F.J. McGarry, ASCE Proc., J. Eng. Mech. Div., 97 (1971)
    1663-1676.
45 O.E. Gjørv, S.I. Sorensen and A. Arnesen, Cem. Concr. Res., 7 (1977)
    333-344.
46 D.D. Higgins and J.E. Bailey, J. Mat. Sci., 11 (1976) 1995-2003.
46a B. Barr and T. Bear, Concrete, 10 (1976) 25-27.
47 B. Hillemeier and H.K. Hilsdorf, Cem. Concr. Res., 7 (1977) 523-536.
48 S. Ziegeldorf, H.S. Muller and H.K. Hilsdorf, Cem. Concr. Res., 10 (1980)
    589-599.
49 C.E. Kesler, D.J. Naus and J.L. Lott, in Proc. Int. Conf. Mechanical
    Behavior of Materials, Kyoto, Aug. 15-20, 1971, The Society of Materials
    Science, Japan, 1972, Vol. IV, pp. 113-124.
50 P.C. Strange and A.H. Bryant, ASCE Proc., J. Eng. Mech. Div., 105 (1979)
    337-343.
51 D.J. Cook and G. Crookham, Cem. Concr. Res., 8 (1978) 387-388.
52 A.M. Neville, Civ. Eng. (London), 54 (1959) 1153-1156; 54 (1959)
    1308-1310; 54 (1959) 1435-1439.
53 F.M. Kaplan, J. Amer. Concr. Inst., 58 (1961) 591-610.
54 F.A. Blakey and F.D. Beresford, J. Amer. Concr. Inst., 59 (1962) 919-923.
55 J. Glucklich, J. Amer. Concr. Inst., 59 (1962) 923-929.
56 G.R. Irwin, J. Amer. Concr. Inst., 59 (1962) 929.
57 J. Glucklich, ASCE Proc., J. Eng. Mech. Div., 89 (1963) 127-138.
58 J. Glucklich, in Proc. 1st Int. Conf. on Fracture, Japan, 1965, The
    Japanese Society for Strength and Fracture of Materials, 1966, Vol. 3,
    pp. 1343-1382.
59 J. Glucklich, in Proc. Int. Conf. Mechanical Behavior of Materials, Kyoto,
    Aug. 15-20, 1971, The Society of Materials Science, Japan, 1972, pp. 104-
    112.
60 J.H. Brown, Mag. Concr. Res., 24 (1972) 185-196.
61 J. Lott and C.E. Kesler, in Symposium on Structure of Portland Cement
    Paste and Concrete, Special Report 90, Highway Research Board,
    Washington, 1966, pp. 204-218.
62 D.J. Naus and J.L. Lott, J. Amer. Concr. Inst., 66 (1969) 481-489.
63 W. Koyanagi and K. Sakai, in Rev. 25th General Meeting, The Cement
    Association of Japan, Tokyo, 1971, 153-157.
64 K. Okada and W. Koyanagi, in Proc. Int. Conf. Mechanical Behavior of
    Materials, Kyoto, Aug. 15-20, 1971, The Society of Materials Science,
    Japan, 1972, pp. 72-83.
65 K. Togawa, T. Satoh and K. Araki, in Rev. 27th General Meeting, The Cement
    Association of Japan, Tokyo, 1973, pp. 117-120.
66 P.E. Petersson, Cem. Concr. Res., 10 (1980) 91-101.
67 A.A. Khrapkov, L.P. Trapesnikov, G.S. Geinats, V.I. Paschenko and
    A.P. Pak, in Fracture 1977, ICF4, Waterloo, Ontario, 1977, Vol. 3,
    pp. 1211-1217.
68 N. McN. Alford and A.B. Poole, Cem. Concr. Res., 9 (1979) 583-589.
69 H. Yokomichi, Y. Fujita and N. Saeki, in Rev. 27th General Meeting, The
    Cement Association of Japan, Tokyo, 1973, pp. 144-147.
70 J.H. Brown and C.D. Pomeroy, Cem. Concr. Res., 3 (1973) 475-480.
71 F.H. Wittmann, Mater. Constr. (Paris), 1 (1968) 547-552.
72 K. Morita, in 35th Ann. Meeting of the Civil Engineering Institute of
    Japan, 5 (1980) 269-270.
73 O.A. Kayyali, C.L. Page and A.G.B. Ritchie, J. Amer. Concr. Inst., 76
    (1979) 1217-1225.

74   J.D. Birchall, A.J. Howard and K. Kendall, Nature, 289 (1981) 388-390.
75   N. McN. Alford, Cem. Concr. Res., 11 (1981) 605-610.
76   F. Moavenzadeh and R. Kuguel, J. Mater., 4 (1969) 497-519.
77   J.S. Nadeau, S. Mindess and J.M. Hay, J. Amer. Ceram. Soc., 57 (1974) 51-54.
78   T.C. Hansen, in Causes, Mechanism and Control of Cracking in Concrete, SP-20, American Concrete Institute, Detroit, 1968, pp. 43-66.
79   A.G. Evans, J.R. Clifton and E. Anderson, Cem. Concr. Res., 6 (1976) 535-548.
80   A.S.-T. Yam and S. Mindess, Int. J. Cem. Comp. and Light. Concr., in press.
81   M. Wecharatana and S.P. Shah, in W.F. Chen and E.C. Ting (Eds.), Fracture in Concrete, Proc. ASCE Session, Hollywood, Florida, Oct. 27-31, 1980, American Society of Civil Engineers, New York, 1980, pp. 82-105.
82   M. Wecharatana and S.P. Shah, Cem. Concr. Res., 10 (1980) 833-844.
83   Ju.B. Zaitsev, Cem. Concr. Res., 1 (1971) 329-344.
84   Ju.B. Zaitsev, Cem. Concr. Res., 1 (1971) 437-447.
85   Ju.V. Zaitsev and F.H. Wittmann, Cem. Concr. Res., 3 (1973) 389-395.
86   E.N. Scerbakov and Ju.V. Zaitsev, Cem. Concr. Res., 6 (1976) 515-528.
87   G.B. Welch and B. Haisman, Mater. Constr. (Paris) 2 (1969) 171-177.
88   G.E. Blight, in 1st Australian Conf. on Engineering Materials, University of New South Wales, Sydney, 1974, pp. 757-789.
89   R.H. Evans and M.S. Marathe, Mater. Constr. (Paris), 1 (1968) 57-60.
90   W. Wright and J.G. Byrne, Nature, 203 (1964) 1374-1375.
91   I.D.C. Imbert, in Highway Res. Rec., No. 324, Highway Research Board, Washington, 1970, pp. 54-65.
92   M.M. Kim, H.-Y. Ko and K.H. Gerstle, in W.F. Chen and E.C. Ting (Eds.), Fracture in Concrete, Proc. ASCE Session, Hollywood, Florida, Oct. 27-31, 1980, American Society of Civil Engineers, New York, 1980, pp. 1-14.
93   H. Kitagawa and M. Suyama, in Proc. 19th Japan Congress on Materials Research, The Society of Materials Science, Tokyo, 1976, pp. 156-159.
94   J. Mazars, in Fracture 1977, ICF4, Waterloo, Ontario, 1977, Vol. 3, pp. 1205-1209.
95   T. Ohigashi, in R.N. Swamy (Ed.), Testing and Test Methods of Fibre Cement Composites, RILEM Symposium, Sheffield, 1978, Construction Press, Lancaster, 1978, pp. 67-78.
96   V.E. Saouma, A.R. Ingraffea and D.M. Catalano, ASCE Proc., J. Eng. Mech. Div., in press.
97   K.K. Watson, Cem. Concr. Res., 8 (1978) 651-656.
98   J.P. Romualdi and G.B. Batson, ASCE Proc., J. Eng. Mech. Div., 89 (1963) 147-168.
99   P. Desayi, Mater. Constr. (Paris), 10 (1977) 139-144.
100 H. Kitagawa, S. Kim and M. Suyama, in Proc. 19th Japan Congress on Materials Research, The Society of Materials Science, Tokyo, 1976, pp. 160-163.
101 P.C. Strange and A.H. Bryant, J. Mat. Sci., 14 (1979) 1863-1868.
102 S. Somayaji, in W.F. Chen and E.C. Ting (Eds.), Fracture in Concrete, Proc. ASCE session, Hollywood, Florida, Oct. 27-31, 1980, American Society of Civil Engineers, New York, 1980, pp. 36-49.
103 S.E. Swartz, K.-K. Hu and G.L. Jones, ASCE Proc., J. Eng. Mech. Div., 104 (1978) 789-800.
104 J. Glucklich and L.J. Cohen, Int. J. Fract. Mech., 3 (1967) 278-289.
104a S. Mindess and A. Bentur, in preparation.
104b E.T. Brown and J.A. Hudson, ASCE Proc., J. Eng. Mech. Div., 98 (1972) 1310-1312.
105 S. Mindess and J.S. Nadeau, Cem. Concr. Res., 6 (1976) 529-534.
106 P.F. Walsh, Indian Concr. J., 46 (1972) 469-470,476.
107 P.F. Walsh, Mag. Concr. Res., 25 (1973) 220-221.
108 P.F. Walsh, Mag. Concr. Res., 28 (1976) 37-41.

109   Z.P. Bazant, in S.P. Shah (Ed.), High Strength Concrete, Proc. Workshop,
      Univ. of Illinois at Chicago Circle, 1979, Published 1980,
      pp. 79-92.
109a V.M. Entov and V.I. Yagust, Izv. AKad. Nauk SSR, Mekh. Tverd. Tela.,
      4 (1975) 93-103.
110   C. Sok, J. Baron and D. Francois, Cem. Concr. Res., 9 (1979) 641-648.
111   S. Chhuy, M.E. Benkirane, J. Baron and D. Francois, in Advances in
      Fracture Research, ICF5, Cannes, 1981, Pergamon Press, 1981, Vol.4,
      pp. 1507-1514.
112   K. Visalvanich and A.E. Naaman, in W.F. Chen and E.C. Ting (Eds.), Frac-
      ture in Concrete, Proc. ASCE session, Hollywood, Florida, Oct. 27-31,
      1980, American Society of Civil Engineers, New York, 1980, pp. 65-81.
113   M. Modeer, A Fracture Mechanics Approach to Failure Analysis of Concrete
      Materials, Report TVBM-1001, Division of Building Materials, University
      of Lund, Sweden, 1979.
114   A. Carpinteri, Eng. Fract. Mech., in press.
115   A. Carpinteri, in Advances in Fracture Research, ICF5, Cannes, 1981,
      Pergamon Press, 1981, Vol. 4, pp. 1491-1498.
116   A. Carpinteri, Mater. Constr. (Paris), 14 (1981) 151-162.
117   S. Mindess, F.V. Lawrence and C.E. Kesler, Cem. Concr. Res., 7 (1977)
      731-742.
118   G.T. Halvorsen, Toughness of Portland Cement Concrete, Ph.D. Thesis,
      University of Illinois at Urbana-Champaign, 1979.
119   G.T. Halvorsen, Int. J. Cem. Comp., 2 (1980) 143-148.
120   J.L. Carrato, Experimental Evaluation of the J-Integral, M.S. Thesis,
      University of Illinois at Urbana-Champaign, 1980.
121   Z.P. Bazant, in A.P.S. Selvadurai (Ed.), Mechanics of Structured Media,
      Proc. Int. Symp. Mechanical Behaviour of Structured Media, Carleton
      University, Ottawa, 1981, Elsevier, 1981, Part B, pp. 3-35.
122   Z.P. Bazant and L. Cedolin, ASCE Proc., J. Eng. Mech. Div., 105 (1979)
      297-315
123   Z.P. Bazant and L. Cedolin, in W.F. Chen and E.C. Ting (Eds.), Fracture in
      Concrete, Proc. ASCE session, Hollywood, Florida, Oct. 27-31, 1980,
      American Society of Civil Engineers, New York, 1980, pp. 28-35.
124   Z.P. Bazant and L. Cedolin, ASCE Proc., J. Eng. Mech. Div., 106 (1980)
      1287-1306.
125   Z.P. Bazant and L. Cedolin, in Advances in Fracture Research, ICF5,
      Cannes, 1981, Pergamon Press, 1981, Vol. 4, pp. 1523-1529.
126   C. Sok, Bull. Liais. Lab. Ponts Chaussees, 98 (1978) 73-84.
127   S.P. Shah, in Proc. Engineering Foundation Conference on Cement Production
      and Use, Rindge, N.H., 1979, pp. 187-199.
128   A. Hillerborg, M. Modeer and P.E. Petersson, Cem. Concr. Res., 6 (1976)
      773-782.
129   A. Hillerborg, A Model for Fracture Analysis, Report TVBM-3005, Division
      of Building Materials, Lund Institute of Technology, Sweden, 1978.
130   A. Hillerborg, presented at Int. Conf. on Fracture Mechanics in Engineering
      Applications, Bangalore, India, 1979.
131   P.E. Petersson, Cem. Concr. Res., 10 (1980) 78-89.
132   P.E. Petersson, Crack Growth and Development of Fracture Zones in Plain
      Concrete and Similar Materials, Report TVBM-1006, Division of Building
      Materials, Lund Institute of Technology, Sweden, 1981.

Fracture Mechanics of Concrete,
edited by F.H. Wittmann, 1983
Elsevier Science Publishers B.V., Amsterdam — Printed in The Netherlands

Chapter 1.2
PHENOMENOLOGICAL ASPECTS OF THE FRACTURE OF CONCRETE
by S. Ziegeldorf

1.   INTRODUCTION

In the simplest case concrete may be modeled as a composite material consisting of two phases: hardened cement paste and aggregates. This model yields valuable information concerning some
physical and mechanical properties of concrete. The model fails,
however, to describe properly the fracture behavior of concrete,
as seen in Fig. 1, where stress-strain-diagrams of a normal aggregate, hardened cement paste and a normal concrete consisting of
both components are given. It is evident that the aggregate as
well as the hardened cement paste exhibit a brittle behavior. The
σ-ε-curve of concrete, on the other hand (ref. 1,2), deviates from
linearity even at low loads and has a descending slope after maximum load.

It was recognized rather early that the marked difference bet-

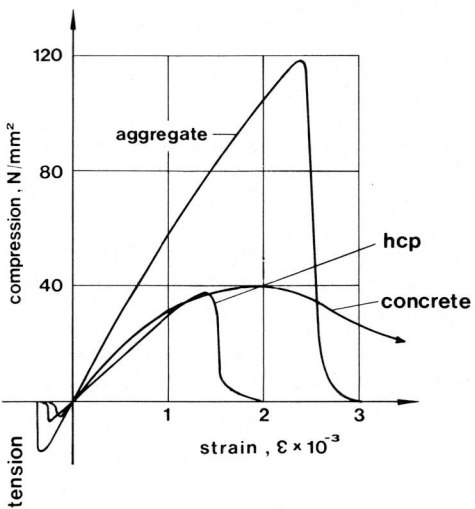

Fig. 1:   Stress-strain-diagrams of hardened cement paste, aggregate and concrete

ween the deformation of concrete and the deformation of its constituents is mostly due to crack formation, even though part of
the observed plasticity may be attributed to a viscous behavior of
the hardened cement paste as was recently proposed by Spooner et al.
(ref. 3).

This paper gives a short review of some phenomenological observations concerning the fracture process in hardened cement paste,
mortar and concrete. Such knowledge may serve as a starting point
as well as a guide for developing theories of concrete fracture
and strength. Before looking at fracture under load, crack formation in concrete prior to loading is treated since such cracking
is an important characteristic of concrete influencing the subsequent behavior under load.

2.   CRACK FORMATION IN CONCRETE PRIOR TO LOADING

There have been many observations of bond cracks at the interface between mortar and coarse aggregates in unloaded concrete
specimens (ref. 4-7). Usually such crack formation is explained
by shrinkage of hydrating cement paste. Since shrinkage of the
hardened cement paste predominantly would lead to radial (mortar)
cracks, this explanation seems to be incorrect. It must be re-
called, however, that under laboratory conditions concrete specimens are often cured under water. Such water curing has the effect of swelling of the hardened cement paste and indeed gives
rise to tangential interface cracks (ref. 8).

It has been shown furthermore that even autogenous shrinkage
alone can be sufficient to cause cracking, in which case the
cracks emanate radially from the aggregates (ref. 9).

Another possibility of explaining spontaneous crack formation
in unloaded concrete is related to the observation that water
lenses develop under coarse aggregates during setting of the
fresh concrete. This explanation is supported by experimental work
by Dhir and Sangha (ref. 7), who observed that the crack density
is greatest in the horizontal direction at all stress levels. As
a conclusion it seems that both effects, shrinkage and bleeding,
seem to be responsible for the pre-existing cracks. Thus the pattern of cracks illustrated in Fig. 2 must be expected in concrete.

It is noteworthy that the pre-existing cracks constitute a major proportion of the cracks present through most of the loading

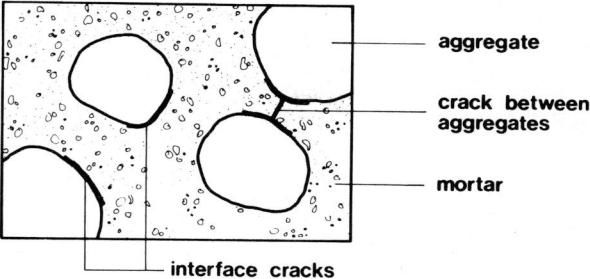

aggregate

crack between
aggregates

mortar

interface cracks

Fig. 2:   Cracks in unloaded concrete (ref. 18)

range (ref. 10, 11). According to Stroeven (ref. 11) the increase
in the specific crack surface area over the complete loading range
amounts to 26 per cent.

3.   FRACTURE BEHAVIOR OF CONCRETE UNDER TENSION

   Though concrete normally is used in compression, its fracture
behavior in tension is of as great an importance as fracture under compression which is actually initiated by tensile forces.
Unfortunately only few experiments on concrete loaded in tension
with complete σ-ε-diagrams are presented in the literature (ref.
12-16). From these experiments the following conclusions may be
drawn (Fig. 3):

-   the σ-ε-diagram is linear up to stresses of 80 per cent of β_t;
-   after the maximum of the σ-ε-diagram the curve decreases gra-
    dually with increasing strain;
-   lower aggregate contents and smaller aggregates produce more
    abrupt decreases in the σ-ε-curve after maximum load.

   On a microscopical scale it has been observed that the deviation of the stress-strain-diagram from linearity is linked to the
enlargement of pre-existing bond cracks. At higher loads continuous cracks - i.e. cracks combining at least two bond cracks by
one mortar crack - are formed. Under constant load such continuous
cracks normally would lead to failure of the specimen. When the
load is decreased after maximum load additional cracks develop.
Krishnaswamy (ref. 6) states that the crack density of specimens
which have failed in tensile rupture is even greater (0.2 mm/mm²)
than that of specimens which have failed under compressive forces
(0.05 mm/mm²). Similarly the bond cracks in specimens loaded in

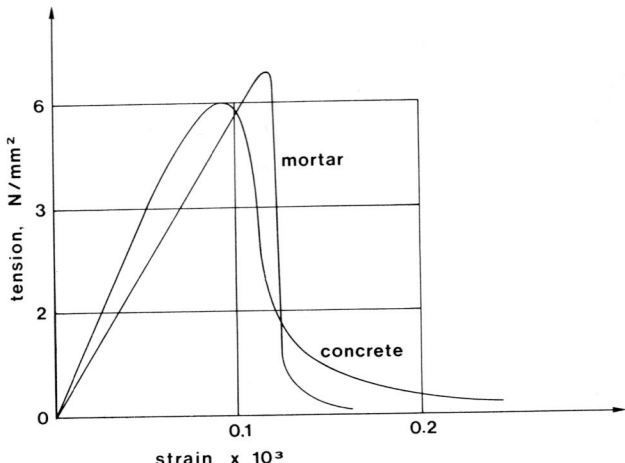

Fig. 3: Stress-strain diagrams of mortar and concrete loaded in
          tension

tension have been observed to be more open than those of specimens loaded in compression (ref. 6).

On a microscopical scale the tensile rupture of concrete is
characterized by the formation of a surface which is perpendicular
to the direction of tensile loading. For normal concrete with a
compressive strength up to approximately 30 N/mm² the crack only
propagates in the matrix or along paste/aggregate interfaces. For
light weight aggregate concretes or for concretes with a higher
compressive strength crack propagation through aggregates may
occur.

Crack propagation in concrete can be linked with the heterogeneity of the material on different levels. It has been postulated
(ref. 17) that even in specimens made of hardened cement paste the
cracks follow on the microscopic level a tortuous path (Fig. 4).
Furthermore numerous cracks are formed in the vicinity of an
initial crack (Fig. 5) rather than one single crack. Thus the
crack density in a small zone adjacent to the fracture surface is
markedly greater than that of the deeper parts of a specimen.

In mortar and concrete specimens multiple crack formation occurs. In addition, propagating cracks are arrested by aggregate
particles resulting in the meandering and branching of cracks

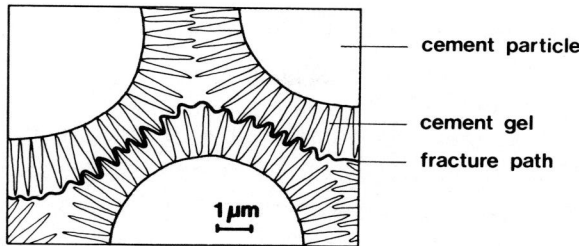

Fig. 4: Tortuosity of a fracture path in hardened cement paste (ref.18)

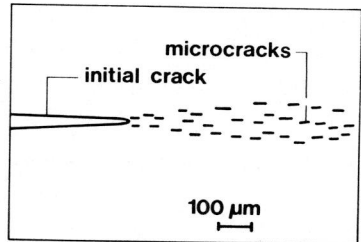

Fig. 5: Microcrack formation in hardened cement paste (ref. 18)

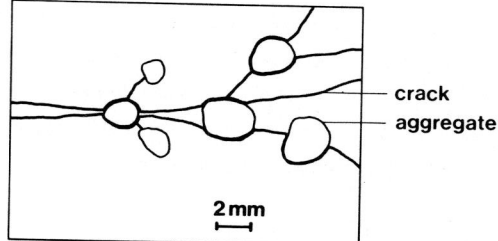

Fig. 6: Crack arrest and crack branching at aggregates (ref. 18)

(Fig. 6). Thus, the true crack surface in concrete is consider-ably larger than the apparent crack surface. Studies by Mcaven-zadeh et al. (ref. 19) and Higgins et al. (ref. 20) lead to the assumption of the following approximate ratios of the true and ap-parent fracture surface area:

- hcp: (1 - 2) : 1
- mortar: (5 - 10) : 1
- concrete: (15 - 20) : 1.

4. FRACTURE BEHAVIOR OF CONCRETE UNDER COMPRESSION

Crack propagation in a normal concrete loaded in compression is described on the basis of the shape of a relevant stress-strain-diagram (Fig. 7). The pre-existing bond cracks initially are stable up to external stresses of 30 - 40 per cent of β_c. At this stage, the bond cracks begin to propagate in a stable manner along the interface. At the same time, their number and length increase. At 70 - 80 per cent of β_c there is a significant increase in the number of mortar cracks, and by joining with nearby bond cracks, these begin to form continuous cracks. Their orientation is mainly parallel to the direction of the external load.

Under constant load such continuous cracks will join with smaller microcracks to form large cracks. Thus, sooner or later a crack surface of critical size will develop causing unstable fracture.

As long as macroscopically unstable crack propagation is prevented, the stress-strain-diagram exhibits a pronounced descending portion. It has been shown that at all stress levels bond cracks predominate and that failure of the contact zone of sand particles rarely occurs (ref. 7).

Moavenzadeh et al. (ref. 21) have observed that the structural features which were most prone to crack initiation were, in order:

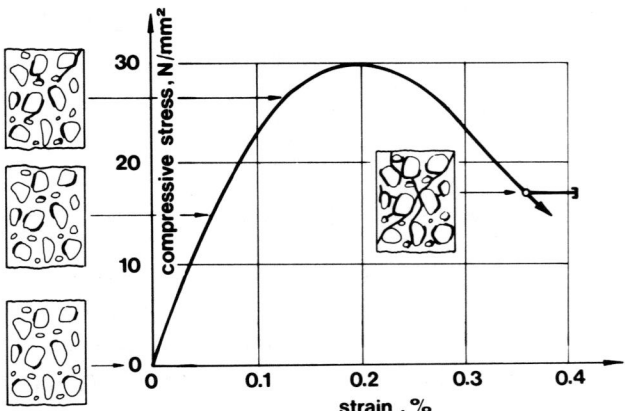

Fig. 7: Stress-strain relationship and crack propagation of concrete loaded in compression

1) aggregate cracks which were inclined at 45° to the applied
stress, 2) large aggregates with bond cracks, 3) air voids,
4) local segregation of large aggregates.

The type of the failure surface is strongly dependent on the
type of loading, i.e. on the use of rigid or frictionless load
bearing platens. Rigid platens generally lead to the well-known
double pyramid with inclined fracture surfaces of the specimen.
Various authors (ref. 11, 22, 23) have observed that the larger
aggregates often are covered with a double cone of mortar. When
using frictionless load bearing platens thus eliminating contraction of area, vertical crack surfaces are predominant (ref. 24).

For higher strength concretes the stress-strain-curve is
steeper and linear up to a higher stress-strength ratio than in
normal concretes because of a decrease in the amount and extent
of bond cracking. Further, the mortar cones on the coarse aggregates do not occur in such concretes leading to the formation of
a smoother and more vertical failure plane (ref. 23).

The crack density of concrete loaded in compression increases
markedly at the onset of pronounced mortar cracking (Fig. 8).
Awad et al. (ref. 25) and Diaz et al. (ref. 26) have shown that
the crack density at fracture is higher for repeated or sustained
load than for monotonic loading and that crack density increases
with decreasing maximum stress at failure (Fig. 9).

The stress-strain diagram of concrete normally does not exhibit
distinct features which could be related in a simple manner with
certain stages of crack formation. Several authors have proposed
therefore to characterize the increasing destruction of the concrete structure in the course of a compression experiment by defining certain parameters which are linked to measured values
which relate to crack formation and crack propagation.

Thus, several authors have pointed out (ref. 27, 28) that at
about 40 - 55 per cent of β_c, Poisson's ratio begins to increase
rather markedly for concrete, however, this does not occur for
hardened cement paste, mortar or high strength concrete. The critical point was called "initiation stress" and was related to internal microcrack propagation (Fig. 10). In Fig. 11 the stresses
are plotted against the volumetric strain. Newman (ref. 29) observed an inflection point at 50 - 60 per cent of β_c which he
called "discontinuity point" and identified it as the onset of

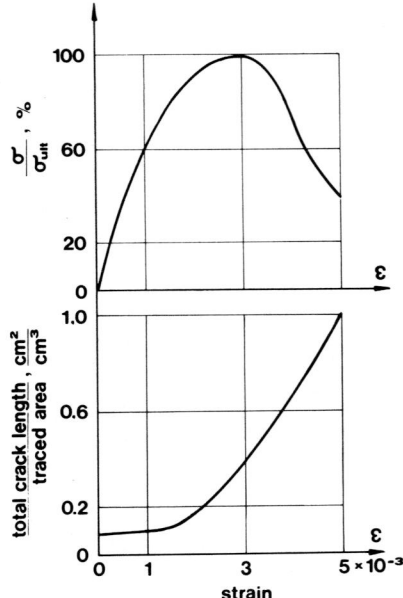

Fig. 8: Crack density of concrete loaded in compression (ref. 26)

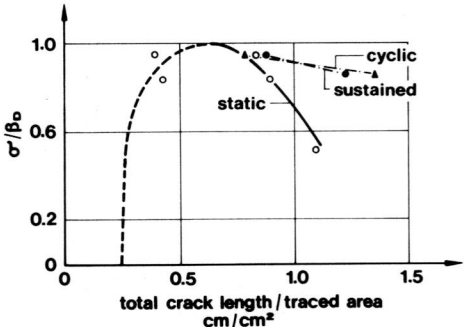

Fig. 9: Effect of load history on crack density of concrete at
         failure (ref. 25)

considerable microcracking. Other authors defined the point of onset of considerable microcracking to be between 40 – 60 per cent
(ref. 30) and 50 – 75 per cent of β_c (ref.31), respectively. When
the load is increased beyond the discontinuity point, a load is
reached after which the volume no longer decreases but increases.
Brandtzaeg (ref.32) called this point the "critical stress".

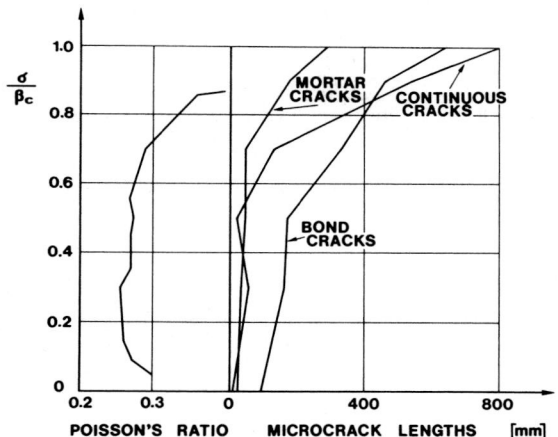

Fig. 10: Relation between Poisson's ratio and microcrack propa-
          gation in concrete (ref. 27)

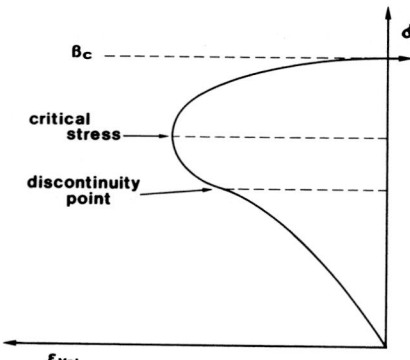

Fig. 11: Characteristic points in σ-ε_{vol}-diagram

Since it is generally accepted that this increase in volume is
due to pronounced crack formation, it is not surprising that a
similar minimal volume has not been observed in experiments on
hardened cement paste or high strength concrete (ref. 23, 33).

   A detailed classification of characteristic stress values for
concretes with compression strengths between 15 N/mm² and 35 N/mm²
has been proposed by Schickert in a recent paper (ref. 24):
- infσ (at about 64 per cent of β_c): lower limit of structural
   loosening; (determined by increase of Poisson's ratio).

- mic σ (at about 77 per cent of β_c): border line for stable and unstable crack formation (point of inflection of a $\sigma - \ln|\varepsilon|$ - diagram.
- sup σ (at about 93 per cent of β_c): upper limit of structural loosening; (minimum specimen volume).
- mac σ (mac σ > sup σ): border line for macrocrack formation; (determined by acoustic emission).

The validity of the utilization of characteristic stress values is supported by experimental evidence indicating a close correlation between these values and fatigue strength or sustained strength (ref. 38, 39).

5.   CONCLUSION

The phenomenological description of the fracture process of concrete has delivered valuable information. But many fundamental problems are still unsolved;

1) The experimental results concerning the effect of aggregate concentration on concrete strength are rather contradictory (ref. 34 - 36).

2) The effect of aggregate - paste bond strength on concrete strength is not well understood (ref.37).

What generally is missing is a theoretical basis capable of explaining such effects. Fracture mechanics will certainly be an important tool in developing such a theory.

6.   REFERENCES

1 Sangha, C.M.; Dhir, R.K.; Mat. Constr. 5 (1972) pp.361-370.
2 Wang, P.T.; Shah, S.P.; Naaman, A.E.; ACI Journal Nov.(1978) pp.603-611.
3 Spooner, D.C.; Pomeroy, C.D.; Dougill, J.W.; Mag. Concr. Res. 28 (1976) pp.21-29.
4 Slate, F.O.; Olsefsky, S.; ACI Journal, 60 (1963) pp.575-588.
5 Hsu, T.T.C.; Slate, F.O.; Sturman, G.M.; Winter, G.; ACI Journal, Feb. (1963) pp.209-223.
6 Krishnaswamy, K.T.; Ind. Concr. J., May (1971) pp. 204-222.
7 Dhir, R.K.; Sangha, C.M.; Mat. Constr. 7 (1974) pp. 17-23.
8 L'Hermite, R.; Cahiers de la Recherche. No. 12, edr.: Eyrolles, Paris, 1961.
9 Ziegeldorf, S.; Müller, H.S.; Hilsdorf, H.K.; 5th Int. Conf. on Fracture, Cannes (1981), pp.2243-2252.
10 Shah, S.P.; Winter, G.; ACI Journal, (1966) pp.6-28.
11 Stroeven, P.; Conference on "The influence of volume change on the design and technology of modern building structures", Karlovy Vary, CSSR (1975).

12 Rüsch, H.; J. ACI 57 (1960) pp.1-28.
13 Hughes, B.P., Chapman, G.P., Bulletin RILEM 30 (1966) pp.95-97.
14 Evans, R.H.; Marathe, M.S.; Mat. Constr. 1 (1968) pp.61-64.
15 Heilmann, H.G.; Hilsdorf, H.K.; Finsterwalder, K.; Deutscher
    Ausschuß für Stahlbeton, Heft 203 (1969).
16 Hillerborg, A.; Petterson, P.E.; 5th Int. Conf. on Fracture,
    Cannes (1981) pp.1515-1522.
17 Grudemo, Å.; Cem. Concr. Res. 9 (1979) pp.19-34.
18 Hilsdorf, H.K.; Ziegeldorf, S.; NATO ARI Symposion on Adhesion
    Problems in the Recycling of Concrete, Paris (1980).
19 Moavenzadeh, F.; Kuguel, R.; J. of Mat., JMLSA, 4 (1969) pp.
    497-519.
20 Higgins, D.D.; Bailey, J.E.; Hydraulic Cement Pastes-Structure
    Properties, Univ. Sheffield, 1976.
21 Moavenzadeh, F.; Bremner, T.W.; Int. Conf. on Structure, Solid
    Mechanics and Engineering Design in Civil Engineering, South-
    hampton Univ. (1969).
22 Perry, C.; Gillott, J.E.; Cem. Concr. Res. 7 (1977) pp.553-564.
23 Carrasquillo, R.L.; Slate, F.O.; Nilson, A.H.; ACI Journal 78
    (1981) pp.179-186.
24 Schickert, G.; Deutscher Ausschuß für Stahlbeton, Heft 312,
    (1980).
25 Awad, M.E.; Hilsdorf, H.K.; Univ. of Illinois, Urbana, (1971).
26 Diaz, S.J.; Hilsdorf, H.K.; Univ. of Illinois, Urbana, (1971).
27 Shah, S.P.; Chandra, S.; ACI Journal, Sep. (1968) pp.770-781.
28 Popov, L.N.; Ippolitov, E.N.; 5th Int. Conf. on Fracture,
    Cannes (1981) pp.2287-2292.
29 Newman, K.; Newman, J.; Conference "Structure, Solid Mechanics
    and Engineering Design", Southhampton, (1969) Wiley-Inter-
    science, London, pp.963-995.
30 Uppal, J.; Kemp, K.; Mag. Concr. Res. 25 (1973) pp.21-30.
31 L'Hermite, R.; RILEM Bulletin 18 (1954) pp.27-39.
32 Richart, F.; Brandtzaeg, A.; Brown, R.; Univ. of Illinois,
    Bulletin Nr. 190, (1929).
33 Inse, D.; Stegbauer, A.; Deutscher Ausschuß für Stahlbeton,
    Heft 254, (1976).
34 Kawakami, H.; 14th Jap. Congr. Mat. Res. (1970) pp.102-104.
35 Swamy, R.N.; Kameswara Rao, C.V.S.; Cem. Concr. Res. 3 (1973)
    pp.413-427.
36 Stock, A.F.; Hannant, D.J.; Williams, R.I.T.; Mag. Concr. Res.
    10 (1980) pp.225-234.
37 Struble, L.; Skalny, J.; Mindess, S.; Cem. Concr. Res. 10
    (1980) pp.277-286.
38 Kotsovos, M.D.; Materials and Structures 12 (1979) pp.431-437.
39 Schickert, G.; Personal note.

Fracture Mechanics of Concrete,
edited by F.H. Wittmann, 1983
Elsevier Science Publishers B.V., Amsterdam — Printed in The Netherlands

Chapter 2

STRUCTURE OF CONCRETE WITH RESPECT TO CRACK FORMATION

by F.H. WITTMANN

1.   INTRODUCTION

There are many papers on specific details of the structure of hardened cement paste and concrete. The wide range of literature begins with microscopical observation and purely phenomenological description and follows through to real mathematical models. In this contribution, we will try to summarize and discuss only those structural aspects which are directly linked with crack formation and failure processes. A more general treatment of modelling the structure and performance of concrete has been published recently (ref. 1).

Until now many structural details of the porous and composite materials were not known well enough. Therefore the existing knowledge has been condensed into several models. This is one possibility to describe the real behaviour in a simplified and approximative way. As new information is available these models can be adjusted continuously.

It has been proven to be advantageous to subdivide the structure of concrete into different levels. In Table I a hierarchic system of three different levels is shown. This means that the models on the different levels are interrelated in a systematic way, or more precisely models on a given level are based on the results of the previous level.

At the micro-level the structure of hardened cement paste is treated. So far, the only model which gives us quantitative results on the micro-level is the Munich Model (ref. 2). Therefore we will choose this model among the different existing materials science models as the basis for further discussion.

The observed behaviour of concrete cannot be linked directly with microstructural mechanisms because there are additional factors within the hierarchic structural system which interfere. In concrete the most important factors are pores, cracks, and inclusions. These structural details of a composite material will be introduced on the meso-level.

The final aim of the hierarchic structural system is, of course, to characterize the macroscopically observed behaviour of a given material in a realistic and rather general way. On the macro-level the information which results from the two previous levels will be used to describe the materials behaviour in

TABLE I

Characteristic structural features of hardened cement paste
and concrete subdivided into three different levels and
corresponding types of models.

Structural Level	Characteristic Features	Type of Models
Micro-Level	Structure of Hardened Cement Paste, Xerogel	Materials Science Models
Meso-Level	Pores, Cracks, Inclusions, Interfaces	Materials Engineering Models, Mechanical and Numerical Models
Macro-Level	Geometry of Structural Elements	Structural Engineering Models, Material Laws

such a way that it can be used directly in structural engineering and design.
In this context macroscopic fracture mechanics parameters have to be considered.

In the following sections the three different structural levels will be dealt
with consecutively. The interrelationship of results obtained at different
structural levels will be outlined in particular.

2.   MICRO-LEVEL : HARDENED CEMENT PASTE

2.1   Introductory rermarks

In concrete, a heterogeneous multi-phase material, different kinds of aggregates may be cemented by hydraulic cement paste to form an artificial stone.
The properties of the resulting material depend both on the type of aggregate
and the hardening matrix. Most typical aggregates may be assumed to react in
reasonable approximation as linearly elastic. Hardened cement paste, however,
is known to be a viscoelastic material. Depending on the relative humidity of
the surrounding atmosphere the microporous structure of hardened cement paste
contains a considerable amount of adsorbed and capillary condensed water. As
the water content is changed shrinkage or swelling occurs. The hygral length
change of concrete, however, is restrained by the inert aggregates. As a consequence a complex state of internal stresses is built up. The coefficient of
thermal dilatation of aggregate and hardened cement paste may vary by a factor
of up to 5. It is well known that the elastic moduli also differ to a great
extent in most concretes. Therefore a complex state of internal stresses is
also created when the temperature is changed or an external load is applied.

Most aggregates used in concrete technology, being approximately linearly elastic within the range of service loads, show no drastic influence of relative humidity of the surrounding air or of temperature on the elastic modulus, on strength and on the coefficient of thermal expansion. The properties of those materials can be realistically described with the help of simple and idealized expressions.

The highly dispersed hydration products of portland cement react to changes of the environment in a far more complicated way. Nearly all characteristic properties of hardened cement paste can be directly linked with the mutual interaction of colloidal particles and/or with the interaction of the total gel structure with water. The presence of aggregates in concrete moderates the actual behaviour of pure hardened cement paste. Therefore on the micro-level we will discuss the structure of hardened cement paste and the interaction of the xerogel with water.

Among the essential reasons why concrete has become the most extensively used building material of our day is the fact that it is comparatively cheap and also that even unskilled personnel under certain conditions may produce concrete with satisfactory serviceability. This situation is not really a challenge for research in this field. As a consequence it is not surprising that so far, very little on the physical basis of the materials properties is really known.

Concrete structures, however, have become increasingly more sophisticated. Cementitious materials are now used under extreme conditions : high temperatures, low temperatures, impact loading conditions, etc. It is essential for a simultaneously economic and safe application of any material that its behaviour while under service can be accurately predicted. This aim, however, can only be achieved via a better understanding of the structure of hardened cement paste and its relation to the macroscopically observed behaviour. As mentioned above this problem will be approached by introducing three different hierarchic levels of structure.

Some investigators tried to develop models of the microstructure of hardened cement paste on the basis of comparatively crude mechanical tests on concrete. These early models had a purely phenomenological character and they can only be used to describe the observed behaviour under given conditions. As fundamental research on the microstructure proceeded, new and more realistic models were created. With the help of these models the actual complex situation within a gel were to be described in a simplified manner. Even these models should not be taken too literally but they should rather be judged by the contribution they provide for a more comprehensive understanding of the total system.

Models will have to be abandoned completely as soon as we have sufficiently detailed and reliable data. So far, however, we definitely need appropriate models. This is the only possibility to accumulate most of the information available. In this way, finally the gap between the research on the microstructure and the engineering properties will be bridged. New results necessarily have to be incorporated in suitable models.

Research on the structure of hardened cement paste can be subdivided into two groups. In one group, mainly a chemical and mineralogical approach is used. In this way the hydration process, the composition, and the crystal structure of hydration products have been studied. An excellent summary of the results obtained in this way has been published by J.F. Young (ref. 3). On the other hand there is a more physics-orientated approach. In the latter mentioned connection the mutual coupling of gel particles and the interaction of gel particles with absorbed vapours are of prime interest. Of course, there is no strict borderline separating these two areas of research.

In this contribution emphasis is placed upon the physical approach because it seems easier to link physical models of the xerogel with crack formation and failure, but some chemical and mineralogical aspects of the structure will be treated first.

2.2   Chemical and mineralogical aspects of the structure

The main compounds in Portland cement are calcium silicates, i.e. $3CaO \cdot SiO_2$ and $2CaO \cdot SiO_2$. The aluminate and ferrite phases are of minor importance with respect to the structure and the mechanical properties of the resulting cementitious material. The hydration of the calcium silicates may be described in a simplified way by the following equations :

$$2(3CaO \cdot SiO_2) + 6H_2O \rightarrow 3CaO \cdot 2SiO_2 \cdot 3H_2O + 3Ca(OH)_2$$
$$2(2CaO \cdot SiO_2) + 4H_2O \rightarrow 3CaO \cdot 2SiO_2 \cdot 3H_2O + Ca(OH)_2$$

These simple expressions, however, are somewhat misleading because the hydration process and the exact determination of the hydration products becomes extremely complex if one goes into detail. Probably the first comprehensive treatise of the chemistry of hydration is provided by the proceedings of the Washington Symposium (ref. 4). In the meantime, a number of additional results on the hydration products have been gathered. Most essential findings have been presented in subsequent meetings in Tokyo (ref. 5), Moscow (ref. 6) and Paris (ref. 7).

It is quite obvious that progress in this field is comparatively slow. This
may be explained by the fact that there are few techniques which may be successfully applied to study the structure of hydration products of cements. Only limited information can be deduced from X-ray diffraction. The same applies to the
application of electron microscopy. Much effort has been spent to develop appropriate chemical methods. An early break-through has been achieved by sorption
methods (ref. 8). As there are no prevailing methods so far, the actual state of
knowledge has to be discussed by combining the relevant information of different
approaches.

From the early micrographs (ref. 9 and 10) taken from hydrated cement and
hydrated clinker components it became clear that a variety of different particles are to be found. Needles, crumbled or rolled foils and tubes have been
observed in hardened cement pastes. The variability of layered silicate hydrates
is known from natural minerals. If the silicate tetraeders match the $\{Me(O,OH)_6\}$
-octaeders completely plane sheets are formed. This principle of geometrical
fitting has been discussed among others by Liebau (ref. 11). If the just mentioned subunits do not fit completely structures like corrugated iron (crumbled)
sheets or rolled foils are formed. In the case of calcium silicate hydrates,
aluminium and ferric ions among other ions may substitute for both silicon and
calcium. In addition sulfur may substitute for silicon. The morphology of the
gel particles changes as calcium or silicon are substituted (ref. 12).
Richartz and Locher (ref. 13) prepared on this basis a vivid graphic model of
different morphologies in hardened pastes.

Kantro, Brunauer and Weise (ref. 14) distinguish three stages of the hydration of cement. According to these authors, immediately after mixing with water
a skin adhering to the particles of the anhydrous compounds is formed. This
first product has a high C/S ratio. In the second stage splitting-off of particles of the skin is observed. The second product formed in this way has a C/S
ratio of 1.0 to 1.5. In the electron microscope one detects foils or platelets
at this stage. Finally these particles grow to a thickness of about three layers
with a C/S ratio of 1.5 to 2.0. Although modified by several authors this concept basically proved to be correct (ref. 15, 16 and 17).

Depending on the C/S mole ratio calcium silicate hydrates may be subdivided
into two groups. In aqueous suspensions at room temperature C-S-H (I) with a
C/S ratio of 0.8 to 1.5 is formed. The semi-crystalline calcium silicate hydrate
with a C/S ratio of 1.5 or above generally is termed C-S-H (II). In hydrated
cement paste semi-crystalline and near amorphous products are found. X-ray

powder patterns usually show only three broadened lines. Sometimes this phase was called tobermorite gel. This name was chosen because it was concluded that the hydration products of cement were degenerate varieties of tobermorite. This statement has often been questioned and in fact there is some evidence that some products are degenerate structures of jennite. Therefore it seems reasonable to follow a suggestion made by Taylor (ref. 18) and to use a more general expression i.e. calcium silicate hydrate or in abbreviated form C-S-H. This general term also takes into consideration the fact that there is not a finite number of phases (each of a definite composition and structure) but instead a large continuous range.

Essentially based on sorption experiments, Powers and Brownyard (ref. 8) determined an average particle radius of 140 $\overset{o}{A}$. In this calculation it was assumed that the microstructure is composed of equal spheres. Later Powers (ref. 19) calculated a mean value of $6.6. \cdot 10^6$ $\overset{o}{A}^3$. This value leads to a revised radius of 117 $\overset{o}{A}$. There is no doubt that these values should not be taken too literally but they are a strong indication that hydration products are of colloidal dimensions.

More recently Grudemo (ref. 20) suggested a different structure of the hydration products. The fact that basal reflexions are not observed in the region of 9 - 15 $\overset{o}{A}$ may indicate that layered or lamellar structures are rare. Therefore Grudemo concludes that cement gel is a submicrocrystalline mixture of structural elements. Some of them are related to tobermorite and some are related to jennite whereas others are related to CH-portlandite. In this concept, gel pores are formed as silica chains and are left out during the growth of the structure. If this assumption holds true pores with a diameter of 9 $\overset{o}{A}$ or multiples thereof must be expected.

If the pore size distribution is determined from sorption data a maximum at a radius of about 18 $\overset{o}{A}$ is found (ref. 21 and 22). At much lower radii this method to determine pore size distributions looses its significance. It may be mentioned, however, that often a second maximum is recorded at about 8 to 10 $\overset{o}{A}$. This fact would confirm the Grudemo's concept. The observed pore size distribution, however, depends on the preparation of the hydrate sample. Therefore it might be tempting to compare experimentally determined pore size distribution with results of morphology studies of C-S-H.

Another promising approach to the microstructure of hydration products has been put forward by Tamas (ref. 23). Tamas studied polymerization of $Si\,O_4^{4-}$ monomers. In the meantime, this interesting approach to study the microstructure has been considerably extended (see f.e. ref. 24). Rio and his co-workers (ref. 25) also studied the hydration process on a macromolecular level. They tried to

correlate the mechanical behaviour and the morphology with the degree of condensation of the silicate hydrates.

Based on sorption data the internal surface of fully hydrated cement paste is
found to be 100 - 200 cm^2/g. Considerably higher values have been calculated by
Winslow (ref. 26) who used small angle X-ray scattering. His findings are discussed in detail with respect to the structure and properties of hardened cement
paste by Copeland and Verbeck (ref. 27).

The actual state of knowledge on the structure and composition of hydrates
has been reviewed by Taylor and Roy (ref. 28). Structure formation and development in hardened cement pastes have been discussed by Sereda and co-workers
(ref. 29). The major components of the microstructure of a xerogel, i.e. solid
phase, pores, and water, have been treated separately and discussed in connection with properties of hardened cement paste by Wittmann (ref. 30).

This compilation of information on the microstructure of hardened cement
paste is, by no means, complete and it may even seem to be arbitrarily selected in
some respect. For the present purpose we may conclude, however, that although
there is a wide field of active research, there are not enough well established
data available to understand the microstructure in full detail. Lack of knowledge is the main reason why we have to introduce simplifying models on the
micro-level. In the following sections therefore we have to deal with the development of appropriate models.

2.3   Earlier models

Models described in the literature can be subdivided into two groups : Inductive models and deductive models. By taking into consideration all relevant
information on the structure such as pore size distribution, mutual interaction
of colloidal particles in a xerogel and the characteristic properties of adsorbed water films an inductive model can be developed. The validity of such a
model must be checked by a critical comparison of the predictions of the model
with the actually observed behaviour of the system. On the other hand one may
deduce a model of the microstructure from the experimentally determined macroscopic data. Models of this type have to be tested by comparing them with results of more fundamental work on the structure. Some deductive models are
suggested with the aim of covering just one specific point such as influence of
moisture content on strength.

A typical example for a deductive model has been given by Ishai (ref. 31).
The mechanical behaviour under load has been analyzed and on the basis of the
comprehensive results obtained in this way a structural model of hardened ce-

ment paste has been established. There is no doubt that the most detailed information on the structure of hardened cement paste has been gained by sorption methods. The mean particle size and the pore size distribution have been estimated by Powers (ref. 19) as indicated above. Powers summarized his findings in a geometrical model and a physical model (ref. 32 and 33). With the help of the geometrical model the pores of the structure are subdivided into gel pores and capillary pores. The physical model of Powers serves as a basis for a thermodynamic treatment of partially water filled micro pores. Within the framework of this inductive model, Powers deals with three basic mechanisms :

&) Change of surface tension of the colloidal particles.

b) Change of disjoining pressure in narrow gaps.

c) Change of hydrostatic tension.

The most important feature of this model probably is the load bearing capacity of water adsorbed in zones of hindered adsorption.

Within the thermodynamic models it is presumed that colloidal particles do not change their structure and/or composition significantly as the moisture content of the system changes. Bernal (ref. 34) has demonstrated, however, that the structure of C-S-H (I) changes during drying. The c-spacing of a unit cell goes down from 14 to 9 Å. It may be concluded that during severe drying, this material loses interlayer water. On this basis Feldman and Sereda proposed another model for hydrated portland cement (ref. 35-37). With the help of this model it is tried to link shrinkage, creep and the influence of moisture content on the elastic modulus with the exchange of interlayer water.

All models described above have been modified by various authors. In some cases major components of a model have been adapted and used for the interpretation of experimental findings. Hope and Brown (ref. 38), for instance, used Feldman and Sereda's model to postulate a possible mechanism of creep.

For many years there has been a lively controversy on the validity of different models. So far, no generally accepted agreement could be reached. But it seems that by now most people concerned have realized that different models may contribute to progress in various ways and that one single model is not able to characterize the complex situation.

Kondo and Daimon (ref. 39) have developed another inductive pore model for C-S-H gel. In this essentially geometrical model clusters of crystallites are separated by inter gel particle pores. In each gel particle there are inter-crystallite pores and finally intra-crystallite pores are found in individual crystallites. This model may partially bridge the gap between the two opposing models of Powers-Brunauer and Feldman-Sereda if it is realistic to compare the

inter-crystallite pores with traditional micro pores and the intra-crystallite
pores with interlayer space. But the more important question about the extent
to which water in micro pores and in interlayer space influences the mechanical
behaviour and in particular crack formation and strength still remains
untouched.

2.4   The Munich Model

2.4.1   New data forming the basis of the model. With respect to the xerogel of

C-S-H the methods used so far may be subdivided into two groups :
  a) Direct observations of characteristic properties of the gel.
  b) Investigations into the properties of water adsorbed in the
     colloidal system.
Within the first group the determination of van der Waals forces at low distances, of the surface energy, and of the coupling of individual particles in
the gel are of primary interest. The study of the mobility and of the disjoining
pressure of adsorbed films in micro-pores plays a dominant role in the second
group mentioned above. In the following paragraphs some recent results are
briefly summarized.

The van der Waals experiments have been restricted to the observation
of the interaction at comparatively large distances ($d > 1000$ Å). The properties
of a xerogel, however, are only affected by attractive forces acting between
solid surfaces which are separated by micro-pores. That means, they are separated
by distances of a few Ångströms. If these micro-pores will be separated by a
crack, van der Waals forces contribute to the energy consumed.

The theoretical background as well as experimental techniques had to be extended so that the range of short distances could be investigated (ref. 40). By
evaluating the bending line of a thin quarz plate which was mounted on a solid
quarz support at a certain distance it was possible to determine the van der
Waals attractive force down to about 80 Å (ref. 41). With the help of theoretical considerations it is possible to extrapolate beyond the range of experimental data with a reasonable degree of reliability to colloidal distances.

The van der Waals attraction is strongly dependent on the dielectric properties of the medium which is between the interacting surfaces. As a consequence the van der Waals attraction is diminished as water is adsorbed on two
opposing surfaces. The influence of adsorbed films on van der Waals forces has
been studied carefully as a function of film thickness (ref. 41). Adsorbed water
reduces the attractive force approximately by one order of magnitude.

There is a close correlation between van der Waals attraction and surface free energy (see e.g. ref. 42). Therefore the dependence of surface free energy on the thickness of adsorbed water films may be directly deduced from the observed decrease of van der Waals attractive force. By comparison of results obtained by independent methods this is shown in references (41) and (43). In a more direct way the change of surface free energy can be calculated with the help of a thermodynamic approach from sorption data (ref. 22). Starting from the dry state the surface energy decreases sharply as the water vapour pressure is increased. Above $p/p_0 = 0.5$ the resulting change, however, is comparatively small. We will compare this result directly with the influence of moisture content on strength.

Primary bonds and secondary bonds both contribute to the mutual coupling of gel particles within a xerogel. The mechanical properties of the system of course depend primarily on these coupling forces. It could be shown that Mössbauer's spectroscopy is a powerful tool to investigate the coupling of gel particles (ref. 44). If the water vapour pressure is raised above a certain level, the disjoining pressure of water separates surfaces which are exclusively held together by van der Waals bonds. The disjoining pressure may be subdivided into different components each having a different physical origin (ref. 45). By extending the van der Waals experiments mentioned above to the range of high vapour pressures the action of disjoining pressure could be observed directly (ref. 46). Stockhausen has later discussed the disjoining pressure and its meaning for C-S-H (ref. 47).

The mobility of adsorbed water is a decisive factor in a number of models. If the first adsorbed layers had an ice-like structure the creep mechanism could be traced back to displacements within this modification. Using the Debye theory the viscosity of liquids can be deduced from dielectric measurements. Schlude and Wittmann determined the complex permittivity in the range of microwave frequencies (ref. 48), and Zech and Wittmann in the range of Hertzian spectroscopy (ref. 49). By comparing these results with NMR measurements one can conclude that the viscosity increases with decreasing thickness of the adsorbed layer. In the region of a monolayer, however, the definition of viscosity looses its significance. In this case a high mobility of water molecules along the surface is observed but the molecules are hindered from leaving the surface. Under these conditions, the adsorbed films are called a two-dimensional van der Waals gas.

In the next section some general relations will be introduced. With the help of these relations it is possible to describe the xerogel of hydrated portland cement realistically. In this context, of course, drastic simplifications are still inevitable.

2.4.2 _Elements of the model._ It is well known that a liquid droplet having a radius r is under a hydrostatic pressure P :

$$P = \frac{2\gamma}{r} \tag{1}$$

In equation (1) γ represents the surface tension of the liquid. In a liquid, surface tension and surface energy are numerically equal. In solids these two values are at least in the same order of magnitude. In a colloidal system non-spherical particles can exist. Flood (ref. 50) has shown that the mean pressure in solid particles created by surface tension in such a system can be estimated with the help of the following equation :

$$P = 2\gamma \frac{S}{3} \tag{2}$$

S stands for the specific surface area and has to be expressed as cm^2/cm^3 in this connection. If the specific surface area is introduced instead of the radius the actual particle size distribution is neglected or rather replaced by a mean value. In C-S-H there are particles which are large enough to ensure that no appreciable internal pressure will be created by surface tension. Other particles in the same system will experience comparatively high pressures. The overall response of a system with active and inactive particles has been calculated by Krasilnikov and co-workers (ref. 51). In their paper it is pointed out that expansion of gel particles in a heterogeneous system is not linearely related to the expansion of the total system but a geometrical magnification factor has to be taken into consideration.

Well aware of the implied simplifications we may go back to equation (1). Now r has to be looked at as a characteristic value of a given xerogel and P as a mean internal pressure. The resulting internal pressure changes as the surface energy is changed. The surface energy or rather the interfacial energy of a colloidal system may be changed by adsorption of gases or vapours. If a film of thickness Γ is adsorbed at a given vapour pressure P the interfacial energy measured in vacuum decreases by $\Delta\gamma$ (ref. 52) :

$$\Delta\gamma = \gamma_0 - \gamma = RT \int_0^p \Gamma \, d(\ln p) \tag{3}$$

If γ of equation (3) is inserted into equation (1) the change of internal pressure caused by a changing surface energy can be calculated. Each individual gel particle expands as the internal pressure is reduced. Bangham and co-workers showed that within certain limits, a linear relation exists between the change of interfacial energy and the resulting length change (ref. 53) :

$$\frac{\Delta l}{l} = \lambda \; \Delta\gamma \tag{4}$$

Later Hiller expressed λ in terms of properties of the colloidal system (ref. 54). It is assumed that in the range of low RH the hygral length change can be described semi-phenomenologically by utilizing equation (4). A more quantitative application of equation (4) is not possible as decisive factors such as the particle size distribution are not sufficiently well known.

As the relative humidity is raised above 50% some surfaces will be separated by disjoining pressure (ref. 46). This leads to additional expansion of the colloidal system. This length change is not caused by a corresponding change of surface energy. Therefore equation (4) cannot be applied in this range. Simultaneously the total structure is weakened by the action of the disjoining pressure.

The Griffith-criterion has proved to be very successful in describing fracture phenomena in hardened cement paste and concrete (ref. 55). According to this concept the square of the related strength is equal to the related interfacial energy :

$$\left(\frac{\sigma}{\sigma_0}\right)^2 = \frac{\gamma}{\gamma_0} = 1 - \frac{\Delta\gamma}{\gamma_0} \tag{5}$$

By inserting equation (3) into equation (5) the relative strength decrease as a function of moisture content can be estimated (ref. 56). By using Bangham's equation (4), equation (5) may be rewritten :

$$\left(\frac{\sigma}{\sigma_0}\right)^2 = 1 - \frac{1}{\lambda\gamma_0} \cdot \frac{\Delta l}{l} \tag{6}$$

Equations (5) and (6) indicate a linear relationship between the square of the related strength and change of interfacial energy and length change respectively. This statement, of course, is only valid in the range of RH in which drying or rewetting changes the interfacial energy only. As mentioned above at high moisture content the action of disjoining pressure cannot be neglected and therefore additional weakening of the structure has to be anticipated.

In conclusion we can say that the Munich Model introduces two terms which can
be related to strength and failure of concrete :
   a) Interfacial energy of the xerogel.
   b) Disjoining pressure of adsorbed water films.

2.4.3   Comparison with experimental results. Theoretical predictions of the
Munich Model are compared with experimental results in references (2) and (57).
There the influence of moisture content on hygral length change (swelling and
shrinkage), on modulus of elasticity, on damping, and on creep, is discussed.
Here we will concentrate on strength exclusively.

   Wittmann has measured strength of hardened cement paste as function of moisture content (ref. 56). In this paper (ref. 56) it has been shown that the
square of the related strength decreases linearely with increasing swelling. This
relation is predicted by equation (6). Above 50% RH, however, there is additional length change observed due to the action of disjoining pressure.

   If we replot experimental data of reference (56) as function of change of interfacial energy we find the relation plotted in Figure 1. For two different
values of water/cement ratio, a straight line is obtained in the low humidity
region. Above 50% RH the disjoining pressure further weakens the microstructure and hence causes a further decrease of the related strength.

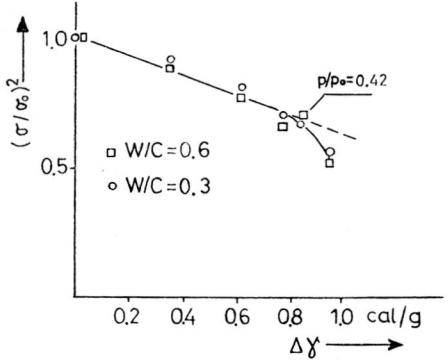

Fig. 1. Square of the related compressive strength
of hardened cement paste cylinders as function of
change of interfacial energy. Results of two test
series (ref. 56) with differing water/cement ratio
are shown.

   Comparable test series have been carried out under carefully controlled conditions by Pihlajavaara (ref. 58). Experimentally determined values of flexural

strength of mortar prisms have been replotted in Figure 2. Within the range of
accuracy again a straight line is obtained in the low humidity range and additional weakening is caused by disjoining pressure at humidities above 50%.

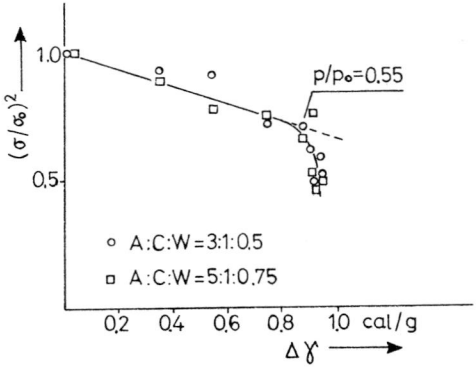

Fig. 2. Square of the related flexural strength
of mortar prisms as measured by Pihlajavaara
(ref. 58) replotted as function of change of
interfacial energy.

The linear relation between the square of the related strength and the change
in interfacial energy as documented in Figures 1 and 2 has been predicted by
equation (5), a basic relation of the Munich Model.

2.5   Porosity
     It has often been tried to relate strength of a porous material directly to
the total porosity. In the literature therefore a number of different equations
relating strength and porosity can be found. All of them predict an increase of
strength as the total porosity decreases. We will not deal with these empirical
equations here. By applying these formulae tacitly one assumes that the pore
size distribution either is independent of porosity or changes in an analogous
way.
     This assumption is not fulfilled in many porous building materials. In Figure
3, the pore size distributions of two natural sandstones are shown. In addition
the corresponding strengths are indicated in Figure 3. If we based the prediction
on the total porosity of the samples only, we had concluded that sample A must
have the higher strength. In fact the contrary is observed. The reason for this
is that the pore size distribution functions are different. Sample A has more
coarse pores than sample B. It is obvious that we have to take into considera-

tion both total porosity and pore size distribution in deriving strength of a
porous material.

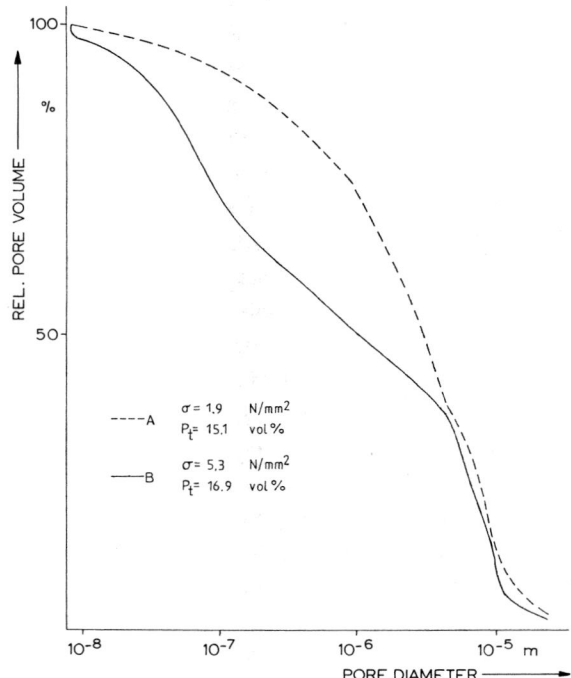

Fig. 3. Pore size distribution of two different
types A and B of porous natural stone (sand
stone). In addition the total porosity P_t and
the biaxial flexural strength R_b of the two
materials are indicated.

In concrete technology the water/cement ratio plays a dominant role. In
Figure 4 strength of hardened cement paste is shown as function of the water/
cement ratio. In addition in Figure 4 the elastic modulus is shown as a function
of water/cement ratio.

In Figures 5 and 6 the pore size distribution of hardened cement paste are
given as a function of water/cement ratio and as a function of duration of
hydration respectively.

Even though the Griffith criterion is a simplified description of the actual
failure process we can still use it to discuss the influence of porosity and
pore size distribution on strength of hardened cement paste. If we use as usual,
σ for strength, E for elastic modulus, γ for the fracture surface energy and 2c

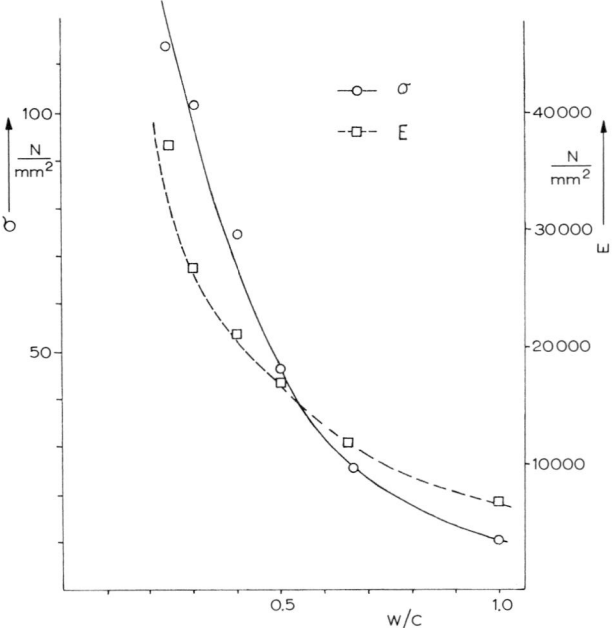

Fig. 4. Influence of water/cement ratio on strength
and modulus of elasticity of hardened cement paste.

for the crack length, the Griffith equation reads as follows :

$$\sigma = \sqrt{\frac{2E\gamma}{\pi C}} \qquad\qquad (7)$$

From Figure 4 we know the influence of water/cement ratio on the elastic
modulus E. In fact it can be shown that the elastic modulus depends essentially
on the total porosity P :

$$E = E'(P) \qquad\qquad (8)$$

In fact, the influence of porosity can be adequately predicted by models of
composite materials. The fracture surface energy γ is well defined for a non-
porous material. If a crack spreads across a porous structure the energy which
is consumed depends on the fracture surface energy of the non-porous material
and the number N of particles to be cracked per unit area. In other words, this
means that only the part of the fracture surface which goes across solid par-

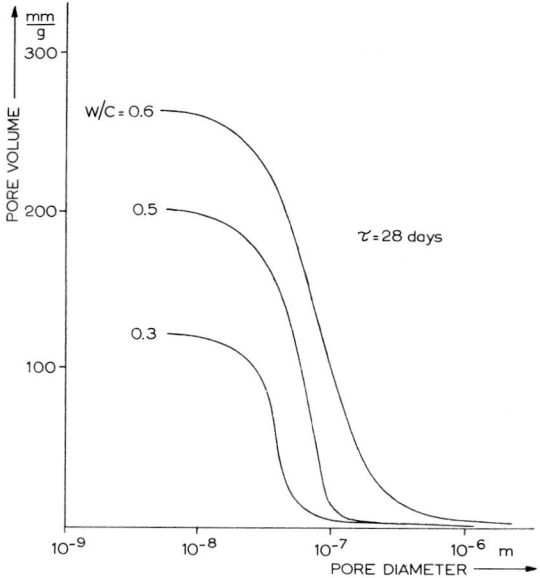

Fig. 5. Pore size distribution of hardened cement paste after 28 days of hydration for three different water/cement ratios.

ticles contributes to the energy balance. This is shown schematically in Figure 7. It is obvious that the effective surface energy depends essentially on the total porosity. In material (A) in Figure 7 the crack passes approximately through 55% of solid matter in its way whereas in material (B) only about 25% of the crack length has passed solid particles. In fact this relation holds true even if one considers a turtuous crack path as has been suggested by Higgins and Bailey (ref. 59).

Then we can redefine the fracture surface energy of a porous material in the following way :

$$\gamma_p = \gamma \, N = \gamma \, (1-P) \qquad (9)$$

The fracture mechanism in a xerogel such as hardened cement paste as suggested by Higgins and Bailey (ref. 59), is schematically shown in Figure 8. The meaning of crack length 2c in equation (7) will be further discussed in the following section on structural features of the meso-level.

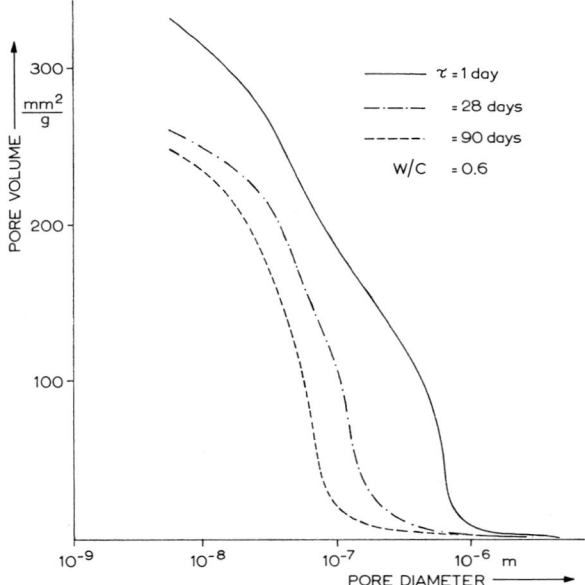

Fig. 6. Pore size distribution of hardened cement paste having a water/cement ratio of 0.6 after 1, 28, and 90 days of hydration.

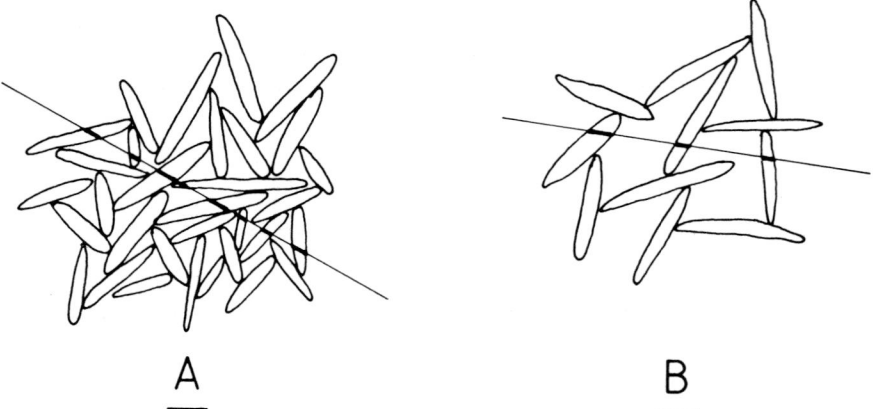

Fig. 7. Schematic representation of fracture energy consumed by a crack running through a material (A) with low porosity and a material (B) with high porosity.

Fig. 8. Fracture mechanism in a xerogel
such as hardened cement paste according
to Higgins and Bailey (ref. 59).

3. MESO-LEVEL

3.1 General remarks

In the preceeding chapter we have described the microstructure of hardened
cement paste and in particular the interaction of the xerogel with adsorbed
water. It has been shown that it is possible to predict qualitatively the influence of moisture content on strength of hardened cement paste on this basis.

So far we have neglected, however, the heterogeneous structure of the material. Therefore we will introduce big pores, pre-existing cracks, and inclusions
as the main characteristic features of the meso-level.

3.2 Big pores

In Figure 5 the pore size distributions of hardened cement paste with different water/cement ratios are shown. Hardened cement paste prepared under normal
conditions always has some big pores which are not covered by the distribution
functions shown in Figure 5. These big pores, however, considerably reduce the
strength of the material. Recently it has been tried to limit big pores to a
minimum, thus increasing the strength.

The crack length 2c in the idealized Griffith criterion (see equation 7) is
defined to be the length of one single crack in an otherwise homogeneous infinite plate. In a real porous material it stands for an effective maximum crack
length $\bar{C}$ or an equivalent big pore. This value $\bar{C}$ must not necessarily be identical with the existing maximum crack length because there is interaction
between large neighboring cracks. In any case it means that C depends sensiti-
vely on the pore size distribution (PSD) and in particular on the probability of
finding large pores.

$$C = \bar{C}\ (PSD) \tag{10}$$

Then we can rewrite equation (7) by using equations (8) to (10) in the following way :

$$\sigma = \sqrt{\frac{2 \cdot E'(P) \cdot \gamma(1-P)}{\pi \, \bar{C} \, (PSD)}} \tag{11}$$

In doing so we can link structural details of hardened cement paste directly with strength. In many cases $\bar{C}$ has to be determined from fracture tests. Possibly in the future we will be able to estimate the effective maximum crack length $\bar{C}$ by fitting the pore size distribution to an extreme value distribution function.

3.3   Cracks

It is well known that under usual climatic conditions in hardened cement paste, mortar, and concrete, numerous cracks exist before a load is applied. There are different causes for the development of these cracks, which must be looked upon to be an important feature of the structure of concrete with respect to behaviour under load and with respect to failure.

Considering the total lifetime of a concrete member, the earliest cracks are formed by incomplete compaction. Insufficient compaction can lead to local zones of high porosity which act under load like precracked areas. In Table II different stages of strength development and the corresponding crack formation are compiled.

TABLE II
Characteristic periods in the lifetime of concrete
and corresponding crack formation.

Stage within Strength Development	Typical Discontinuity
Pouring and Compaction	Compaction Pores
Fresh Concrete	Bleeding Cavities
Hardening Concrete	Thermal Cracks, Chemical and Capillary Shrinkage Cracks
Drying Concrete	Hygral Shrinkage Cracks
Loaded Concrete	Interfacial Cracks, Crack Growth

Immediately after pouring and before hardening, sedimentation (bleeding) takes place. This process causes water filled pockets under coarse aggregates. As a consequence horizontal cracks are formed. This effect obviously causes a certain degree of anisotropy.

As the hydration continues heat of hydration is liberated. Under normal conditions this causes a time-dependent temperature gradient. It can be shown that in many concrete elements thermal cracking takes place in the outer cooler zones. The orientation of these cracks depends on the geometry. There are measures, however, to prevent excessive thermal cracking.

After demoulding, the surface of concrete begins to dry and reaches equilibrium with the environmental humidity very quickly, whereas the centre of a given specimen may remain saturated for many years. This hygral gradient induces shrinkage cracks. Again the orientation of these cracks depends on geometry.

In hardened concrete the interfaces between hardened cement paste and coarse aggregates remain weak for a long time. Therefore comparatively moderate loads, far below the design load, may cause interfacial cracking.

In summary we can conclude that in a concrete specimen some of the causes mentioned in this section and compiled in Table II or combinations of these different causes will introduce cracks into the structure under normal conditions. Some of these cracks are oriented at random and others initiate a certain degree of anisotropy. The observed strength therefore does not only depend on concrete composition but to a large degree on curing conditions.

3.4.   Inclusions

According to the Griffith criterion a crack spreads catastrophically once the load becomes critical. There is no stable crack growth and no crack arresting. In a composite material therefore the unmodified Griffith equation is not a realistic approximation for a composite material.

Consider for a moment the situation in normal concrete, where the aggregates are stronger than the matrix. If a crack starts to spread under a given load from a big pore in the weaker matrix or from an interface there is a chance that it will meet an inclusion and then will be stopped. This simplified approach is schematically shown in Figure 9. $2C_M$ is supposed to be the length of the initial crack in the matrix. As soon as σ_M is reached the crack will spead in an unstable way according to :

$$\sigma_M = \frac{2 E \gamma_M}{\pi C_M} \tag{12}$$

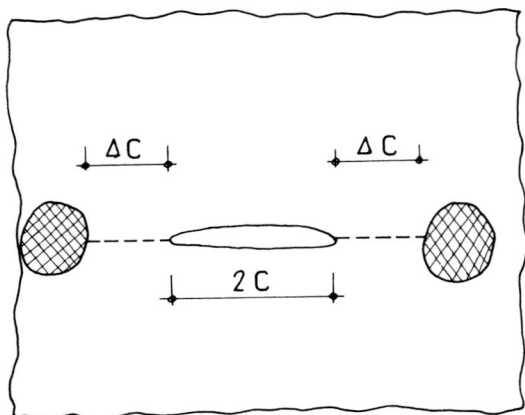

Fig. 9. Schematic representation of a Griffith
crack. Once the crack spreads it meets two
inclusions after a crack length increase of
ΔC.

γ_M denotes the fracture surface energy of the matrix. In this simplified
approach we consider in fact a few isolated particles in an otherwise homogeneous matrix, i.e. a diluted system.

Now we suppose that the crack meets an inclusion when it has grown by ΔC. At
the crack tip the condition for further crack growth has now changed :

$$\sigma_A = \frac{2\ E\ \gamma_A}{\pi(C_M + \Delta C)} \tag{13}$$

In this equation γ_A stands for the fracture surface energy of the aggregate.
In normal concrete this value is higher than the one of the matrix ($\gamma_A > \gamma_M$) and
therefore the resulting curve is shifted towards the right as shown in Figure 10.
In the example chosen, the crack runs from point P_1 and is arrested at point P_2.
Further crack growth is only possible if the load is increased to σ_A. In point
P_3 the condition of crack propagation through the inclusion is fulfilled.

If, however, the crack has increased at least by $\Delta C'$ before it meets an inclusion, the crack will not be arrested by the aggregate. This means that crack
arresting is not only dependent on the mechanical properties of matrix and aggregates but also on the geometrical distribution. If a crack has already reached
a critical length before it meets the aggregate crack, arresting has become impossible. This also explains that beyond a certain critical crack length this

crack arresting mechanism does not work any more and the composite material
fails finally in an unstable way.

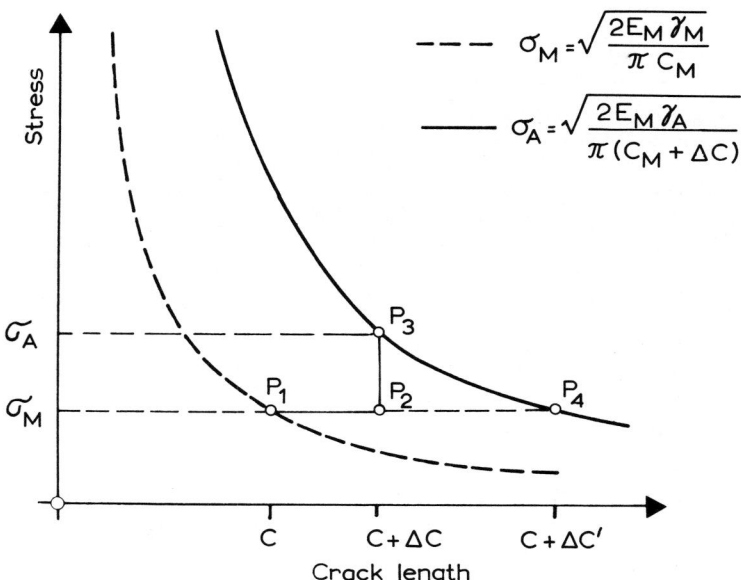

$$\sigma_M = \sqrt{\frac{2E_M \gamma_M}{\pi \, C_M}}$$

$$\sigma_A = \sqrt{\frac{2E_M \gamma_A}{\pi \, (C_M + \Delta C)}}$$

Fig. 10. Schematic representation of equations (12) and (13).
If the crack in the matrix becomes critical it will spread
from P_1 to P_2. At a higher load σ_A the crack can penetrate
the inclusion. If ΔC is large enough so that point P_4 is
reached there will be no crack arresting.

3.5   Structural models of Grudemo

In section 3.4 we have discussed the role of hard and strong aggregates as
crack arresting inclusions. A sharp crack can also be arrested if it runs into
a wide rounded pore. Grudemo has summarized his observations with the electron
microscope in the form of simplifying structural sketches (ref. 60 and 61). In
this way he designed a possible element of the composite structure of hardened
cement paste in which a crack runs into the weak zone around an anhydrous cement
particle embedded in hydration products (ref. 61). The sharp crack tip has become rounded and thus the crack is arrested. This situation is schematically
shown in Figure 11a.

With the help of another structural model Grudemo points out that a crack
will pass different zones such as the lamellar CH phase, inner and outer gel and
remaining anhydrous nuclei while it spreads. This simplifying model is shown in

Figure 11b and underlines again the fact that hardened cement paste has to be
looked upon to be a heterogeneous composite material. This means that in principle there is no difference between crack formation in hardened cement paste,
mortar, and concrete. It is a difference in scale only.

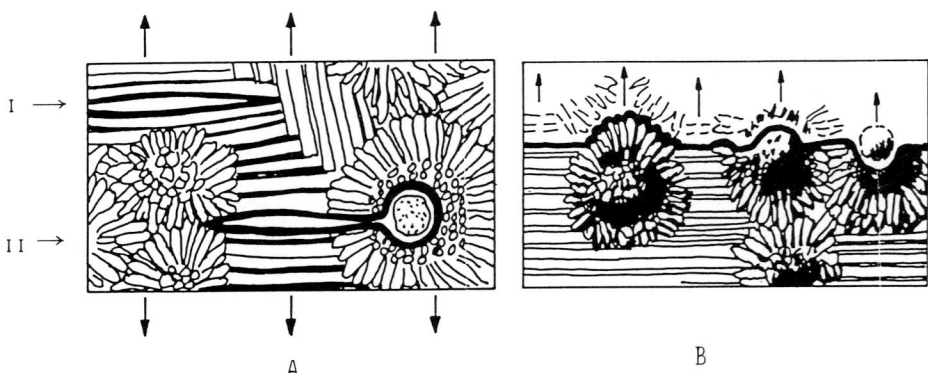

Fig. 11. Idealized details of the gel structure according
to Grudemo (ref. 61) :
a) Cracks are arrested either as they meet stronger zones (I)
   or as they run into rounded-off spaces.
b) The crack runs along cleavage surfaces in lamelar CH phase,
   zones of contact between gel aggregates, outer and inner
   gel, and remaining anhydrous nuclei.

3.6 Micromechanics and simulation of composite structures

So far we have given a qualitative description of crack propagation and
crack arresting in a composite material. The micromechanics of concrete deal
with this subject in more detail and in a quantitative way. In fact a crack, if
it is arrested, can propagate through the inclusion or along the interface
around it (ref. 62 and 63). The micromechanics of concrete is treated in the
theoretical chapter by Zaitsev. Therefore it is not necessary to repeat it here
in detail.

We shall, however, mention at this point the possibility of generating composite random structures to simulate the real concrete structure. These computer
generated random structures form a basis for the application of micromechanics.
In this way many aspects of failure of concrete can be studied systematically.
In Figure 12 a computer simulated structure of a matrix with polygonial inclusions is shown. More details are given in chapter 4.2. For this type of a composite material, crack growth can be calculated analytically by means of micromechanics. In addition, typical results are shown as different stages of stable
crack growth in Figure 12.

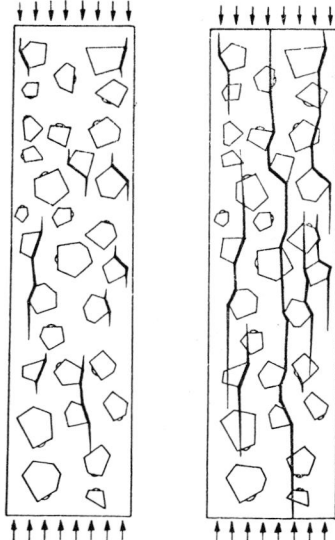

Fig. 12. Simulated composite structure
and calculated crack pattern for two
different load levels, according to
reference (60).

Similar structures can be generated by taking into consideration the random
geometry and the size distribution of aggregates (ref. 62). A typical example is
shown in Figure 13. In this case, however, crack propagation and crack arresting
cannot be treated analytically any more. By means of numerical methods such as
finite element analysis it is possible, however, to obtain close approximations
of the real fracture process in composite structures.

As a last example for computer generated composite structures, Figure 14
shows a model of concrete with spherical aggregates of different size where the
size distribution of the aggregates follows the Fuller function. It is possible
to generate in this way three-dimensional structures with spherical aggregates.
These three-dimensional composite structures allow us to study crack formation
under uniaxial and under multiaxial states of stress.

4.   MACRO-LEVEL
On the macro-level the actual macroscopically observed behaviour of a given
material has to be described. For numerous materials this can be done by introducing appropriate fracture mechanics parameters. If possible these fracture
mechanics parameters have to be linked with structural aspects such as porosity,

68

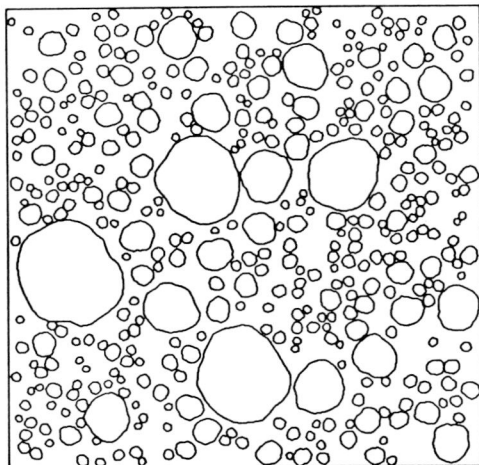

Fig. 13. Simulated structure of concrete
with aggregates having random geometry
(ref. 62).

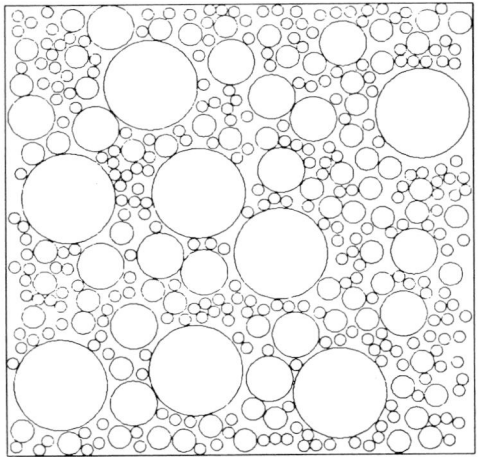

Fig. 14. Simulated structure of concrete
with aggregate size distribution following
the Fuller curve (ref. 62).

cement content, aggregate size, and aggregate geometry.

In Table III some values for K_{IC} as determined on hardened cement paste, different aggregates and the range observed with concrete specimens are given. Further details on the experimental techniques as well as the results obtained are described in the corresponding chapters of this volume.

TABLE III
Critical fracture toughness as determined for different
hardened cement pastes, aggregates and concrete.

Material		K_{IC} (MN/m$^{3/2}$)	Ref.
Hardened Cement Paste	W/C = 0.30	0.40	
	W/C = 0.36	0.35	66
	W/C = 0.50	0.29	
Aggregates	Limestone	0.70 - 1.00	67
	Marble	1.9	
	Quartz	3.4	68
Concrete		0.70 - 1.50	69

There is an obvious influence of water/cement ratio on K_{IC} (see also ref. 63).
This is related to porosity and pore size distribution. K_{IC} for different types
of aggregates varies within a wide range. The influence of the age of concrete
and of mix proportions on K_{IC} has been studied by various authors (ref. 68). Pak
and Trapeznikov have determined experimentally the influence of maximum aggregate size on K_{IC} (ref. 69).

At first glance it seems as if strength and failure process can be adequately
described by simple fracture mechanics parameters such as K_{IC} and G_{IC}. Unfortunately this is not the case and these values must be considered to be rough
approximations in the case of concrete in particular.

In an attempt to overcome these difficulties, more complex fracture mechanics
parameters such as crack opening displacement (COD) and J-integral (see f.e. ref.
70) have been introduced. So far, however, it is not yet possible to discuss the
values in relation with the structure of the material.

In addition, fracture mechanics parameters of concrete are not independent of
the crack history. As a crack extends a damaged zone the so-called fracture process zone is created ahead of the crack. As a consequence, fracture mechanics parameters increase with crack length until the damaged zone has fully developed.
This situation is described by means of the R-curve approach (ref. 71 and 72).
It is obvious that the development of the fracture process zone will be stress
and time dependent. At present it is unknown in which way the fracture zone depends on the structure of concrete. Further details are described in the contri-

bution of Shah to this volume.

Hillerborg and co-workers developed another approach to describe the failure process of concrete, i.e. the fictitious crack model (ref. 73-75). There is a special section in this volume on this approach. We mentioned already in context with the J-integral and the R-curve that at this moment we cannot discuss these fracture mechanics parameters in terms of structural details mainly because of a lack of suitable experimental data. The same is true for the fictitious crack model. Extensive studies will be necessary to provide a solid basis for a realistic link between the structure of the composite material and fracture mechanics parameters.

On the macro-level appropriate material laws should be formulated. At present it is not yet possible to predict crack resistance if the composition of concrete is known.

5. CONCLUSIONS

The heterogeneous structure of the composite material concrete can be described on three different levels. The physical basis of the xerogel, the binding agent in concrete, is described on the micro-level (Munich Model). This model has proven to be applicable to other cement based xerogels such as aerated concrete (ref. 76). Additional structural details such as pores, cracks and inclusions are introduced on the meso-level. Finally, the macroscopically observed behaviour of concrete is represented on the macro-level.

The influence of porosity, cracks, and inclusions on strength and failure of a composite material can be described in a quasi-quantitative way. K_{IC} and G_{IC} must be looked upon to be rough approximations in the case of concrete. There are different attempts to come to a more realistic description of the failure process such as COD, J-integral, R-curve, and fictitious crack model (FCM). So far, however, existing experimental data do not allow us a rigorous comparison between these more complex fracture mechanics parameters and structural details.

REFERENCES

1 F.H. Wittmann, Modelling of concrete behaviour, Proc. European Concrete Research, Swedish Cement and Concrete Research Institute, Stockholm, 1981, pp. 171-189.
2 F.H. Wittmann, Grundlagen eines Modells zur Beschreibung charakteristischer Eigenschaften des Betons, Deutscher Ausschuss für Stahlbeton, Report Nr. 292, Wilhelm Ernst & Sohn, Berlin (1976).
3 J.F. Young, The microstructure of hardened Portland cement paste, to be published in Z.P. Bazant and F.H. Wittmamm (Ed.) : Creep and Shrinkage in Concrete Structures, Wiley 1982

4   Chemistry of Cement, Proceedings of the 4th International Symposium Washington (1960) edited in 1962.
5   Chemistry of Cement, Proceedings of the 5th International Symposium Tokyo (1968) edited in 1969.
6   Chemistry of Cement, Proceedings of the 6th International Symposium Moscow (Sept. 1974).
7   Chemistry of Cement, Proceedings of the 7th International Symposium Paris (August 1980).
8   T.C. Powers and T.L. Brownyard, Studies of the physical properties of hardened Portland cement paste, Research Laboratories of the Portland Cement Association, Bulletin 22 (1948).
9   A. Grudemo, The microstructures of cement gel phases, Trans. Royal Inst. Techn., Stockholm, Nr.242 (1965).
10 A. Grudemo, An electronmicroscopic study of the morphology and crystallization properties of calcium silicate hydrates, Proceedings Swedish Cement and Concrete Institute Stockholm, Nr. 26 (1955).
11. F. Liebau, Ein Beitrag zur Kristallchemie der Schichtsilikate, Acta Cryst. 824 (1968) pp. 690-699.
12 L.E. Copeland, E. Bodor, T.N. Chang and C.H. Weise, Reactions of tobermorite gel with aluminates, ferrites and sulfates, Res. and Development Laboratories of the Portland Cement Association, Bulletin 211 (1967).
13 W. Richartz and F.M. Locher, Ein Beitrag zur Morphologie und Wasserbindung von Calciumsilikathydraten und zum Gefüge des Zementsteins, Zement-Kalk-Gips, 18 (1965) pp. 449-459.
14 D.L. Kantro, S. Brunauer and C.H. Weise, Development of surface in the hydration of calcium silicates - II. Extension of investigations to earlier and later stages of hydration, J. Phys. Chem., 65 (1962) p. 1804.
15 R.Sh. Mikhail and S.H. Abo-El-Enein, Studies of water and nitrogen adsorption on hardened cement pastes - I. Development of surface in low porosity pastes, Cem. Concr. Res., 2 (1972) pp. 401-414.
16 S. Diamond, Identification of hydrated cement constituents using a scanning electron microscope - Energy dispersive X-ray spectrometer combination, Cem. Concr. Res., 2 (1972) pp. 617-632.
17 U. Ludwig, Investigations on the hydration mechanism of clinker minerals (see ref. 6).
18 H.F.W. Taylor, Crystal chemistry of Portland cement hydration products (see ref. 6).
19 T.C. Powers, Physical properties of cement paste, Research and Development Laboratories of the Portland Cement Association, Res. Dept., Bull. 154 (1960).
20 A. Grudemo, On the development of hydrate crystal morphology in silicate cement binders, Liaisons de Contact dans les Matériaux Composites Utilisés en Génie Civil, RILEM-INSA-Coll., Toulouse, France (November 22-24, 1972).
21 F. Wittmann and G. Englert, Bestimmung der Mikroporenverteilung im Zement-stein, Mat. Sci. Eng., 2 (1967) p. 14.
22 M.J. Setzer and F.H. Wittmann, Modified method to calculate pore size distribution using sorption data, Proceedings RILEM-IUPAC International Symposium, Pore Structure and Properties of Materials, Prague (September 18-21, 1973).
23 F.D. Tamas and T.G. Varadi, Role of poly-reactions in the hydration of cement (see ref. 6).
24 L.S. Dent-Glasser, E.E. Lachowski, K. Mohan and H.F.W. Taylor, A multi-method study of C_3S hydration, Cem. Concr. Res., 8 (1978) p. 733.
25 A. Rio and A. Saini, L'industria Italiana del cemento, 39 (1969) p. 867.
26 D.N. Winslow, The specific surface of hardened Portland cement paste as measured by low angle X-ray scattering, Thesis, Purdue University, La Fayette, Indiana, USA (1973).

27 L.E. Copeland and G.J. Verbeck, Structure and properties of hardened cement pastes (see ref. 6).

28 H.F.W. Taylor and D.M. Roy, Structure and composition of hydrates (see ref. 7).

29 P.J. Sereda, R.F. Feldman and V.S. Ramachandran, Structure formation and development in hardened cement pastes (see ref. 7).

30 F.H. Wittmann, Properties of hardened cement paste (see ref. 7).

31 O. Ishai, The time-dependent deformational behaviour of cement paste, mortar and concrete, Proceedings International Conference, The Structure of Concrete and its Behaviour under Load, London (September, 1965) p. 345.

32 T.C. Powers, Mechanism of shrinkage and reversible creep of hardened cement paste, Proceedings International Conference, The Structure of Concrete and its Behaviour under Load, London (September 1965) p. 319.

33 T.C. Powers, The thermodynamics of volume change and creep, Materials and Structure, 1 (1968) pp. 487-507.

34 J.D. Bernal, Proceedings 3th International Symposium on the Chemistry of Cement, London (1952) p. 216.

35 R.F. Feldman and P.J. Sereda, A model for hydrated Portland cement paste as deduced from sorption - Length change and mechanical properties, Materials and Structures, 1 (1968) pp. 509-520.

36 R.F. Feldman and P.J. Sereda, A new model for hydrated Portland cement and its pratical implications, Engineering Journal, 53 (1970) p. 53.

37 R.F. Feldman, Sorption and length change scanning isotherms of methanol and water on hydrated Portland cement, Proceedings 5th International Symposium on the Chemistry of Cement, Tokyo, Vol. III (1968) p. 53.

38 B.B. Hope and N.H. Brown, A model for the creep of concrete, Cem. Concr. Res., 5 (1975) pp. 577-586.

39 R. Kondo and M. Daimon, Phase composition of hardened cement paste (see ref. 6).

40 F. Wittmann, H. Splittgerber and K. Ebert, Z. Physik, 245 (1971) p. 354.

41 H. Splittgerber and F. Wittmann, Einfluss adsorbierter Wasserfilme auf die van der Waals Kraft zwischen Quarzglasoberflächen, Surface Science, 41 (1974) p. 504.

42 H. Krupp, Particles adhesion, theory and experiments, Advan. Colloid Interfaces Sci., 1 (1967) p. 116.

43 M.J. Setzer and F.H. Wittmann, Surface energy and mechanical behaviour of hardened cement paste. Appl. Physics, 3 (1974) pp. 403-409.

44 H. Ubelhack and F.H. Wittmann, Debye-Waller-facktor of colloidal particles in hydro- and xerogels, Proceedings International Conference on Mössbauer Spectroscopy, Cracow, Vol. 1 (1975) pp. 349-350.

45 B.V. Derjaguin, J. Colloid Interfaces Sci., 49 (1974) p. 249.

46 H. Splittgerber, Spaltdruck zwischen Festkörpern und Auswirkungen an Probleme in der Technik, Cem. Concr. Res., 6 (1976) pp. 29-36.

47 N. Stockhausen, Van der Waals interaction and disjoining pressure between solid surfaces, Proceedings Conference, Hydraulic Cement Pastes; their Structure and Properties, Sheffield (April, 1976) pp. 219-226.

48 F. Schlude and F.H. Wittmann, Uber ein Verfahren zur raschen Bestimmung der komplexen DK im Mikrowellenbereich, Nachrichtentechn. Zeitung, 27 (1974) pp. 365-368.

49 B. Zech and F.H. Wittmann, Studium des dielektrischen Verlhaltens von dünnen adsorbierten Wasserfilmen, Z. Phys. Chemie, NF 92 (1974) pp. 45-62.

50 E.A. Flood, Adsorption potentials, adsorbent self-potentials and thermodynamic equilibria, solid surfaces and the gas-solid interface, Advances in Chemistry Series, Nr. 33 (1961) p. 249.

51 K.G. Krasilnikov, A.M. Podvalny and A.E. Segalov, Self-induced deformations in porous bodies, Kolloidnyi Zhurnal, 36 (1974) pp. 266-271.

52 J.W. Gibbs, Collected works, Yale University Press, New Haven (1957).

53 D.H. Bangham and N. Fakhoury, The swelling of charcoal, Part I : Preliminary experiments with water vapour, carbon dioxide, amonia and sulphur dioxide, Proceedings Royal Society, A130 (1931) pp. 81-89.

54 K.H. Hiller, Strength reduction and length changes in porous glass caused by water vapour adsorption, J. Appl. Phys., 35 (1964) pp. 1622-1628.

55 F.H. Wittmann and J. Zaitsev, Verformung und Bruchvorgang poröser Baustoffe bei kurzzeitiger Belastung und Dauerlast, DAfStb Report, Nr. 232, Wilhelm Ernst & Sohn, Berlin (1972).

56 F. Wittmann, Surface tension, shrinkage and strength of hardened cement paste, Materials and Structures, 1 (1968) pp. 547-552.

57 F.H. Wittmann, The structure of hardened cement paste - A basis for a better understanding of the materials properties, Proceedings Conference, Hydraulic Cement Pastes; their Structure and Properties, Sheffield (April, 1976) pp. 96-117.

58 S.E. Pihlajavaara, A review of some of the main results of a research on the ageing phenomena of concrete : effect of moisture conditions on strength, shrinkage and creep of mature concrete, Cem. Concr. Res., 4 (1974) pp. 761-771.

59 D.D. Higgins and J.E. Bailey, A microstructural investigation of the failure behaviour of cement paste, Proceedings Conference, Hydraulic Cement Pastes; their Structure and Properties, Sheffield (April, 1976) pp. 283-296.

60 A. Grudemo, Strength-structure relationships of cement paste materials, Part 1 and Part 2 CBI Research Reports 6:77 and 8:79, Stockholm (1977 and 1979).

61 A. Grudemo, Microcracks, fracture mechanism, and strength of the cement paste matrix, Cem. Concr. Res., 9 (1979) pp. 19-34.

62 J.B. Zaitsev and F.H. Wittmann, Simulation of crack propagation and failure of concrete, Mat. and Struct., 14 (1981) pp. 357-365.

63 F.H. Wittmann, Mechanisms and mechanics of fracture of concrete, Adv. in Fracture Research, ICF-5, Vol. 4 (1981) pp. 1467-1487.

64 P.E. Roelfstra and H. Sadouki, Simulation des structures composites, Internal Report, Laboratory for Building Materials Science, Swiss Federal Institute of Technology, Lausanne (1981).

65 D.D. Higgins and J.E. Bailey, Fracture measurements on cement paste, J. Mat. Sci., 11 (1976) pp. 1955-2003.

66 R.A. Schmidt, Exp. Mech., 15 (1976) pp. 161-167.

67 B. Hillemeier and H.K. Hilsdorf, Fracture mechanics studies on concrete compounds, Cem. Concr. Res., 7 (1977) pp. 523-536.

68 R.N. Swamy, Fracture mechanics applied to concrete, Chapter 6, in Developments in Concrete Technology, edited by F.D. Lydon, Applied Science Publishers (1979).

69 A.P. Pak and L.P. Trapeznikov, Experimental investigations based on the Griffith-Irwin theory process of the crack development in concrete, Adv. in Fracture Research, ICF-5, Vol. 4 (1981) pp. 1531-1539.

70 S. Mindess, F.V. Lawrence and C.E. Kesler, The J-integral as a fracture criterion for fiber reinforced concrete, Cem. Concr. Res., 7 (1977) pp. 731-742.

71 C. Sok, J. Baron and D. François, Mécanique de la rupture appliquée au béton hydraulique, Cem. Concr. Res., 9 (1979) pp. 641-648.

72 S. Chhuy, M.E. Benkirane, J. Baron and D. François, Crack propagation in prestressed concrete, Interaction with reinforcement, Adv. in Fracture Research, ICF-5, Vol. 4 (1981) pp. 1507-1514.

73 A. Hillerborg, Analysis of fracture by means of the fictitious crack model, particularly for fibre reinforced concrete, Int. J. of Cement Composites, 2 (1980) pp. 177-184.

74 A. Hillerborg and P.E. Petersson, Fracture mechanical calculations, test
    methods and results for concrete and similar materials, Adv. in Fracture
    Research, ICF-5, Vol. 4 (1981) pp. 1515-1522.
75 P.E. Petersson, Crack growth and development of fracture zones in plain
    concrete and similar materials, Lund Institute of Technology, Report
    TVBM-1006 (1981).
76 Y. Houst, F. Alou and F.H. Wittmann, Influence of moisture content on
    mechanical properties of autoclaved aerated concrete, in F.H. Wittmann (Ed.)
    Autoclaved Aerated Concrete, Moisture and Properties, Elsevier, Amsterdam
    (1983) pp. 219-234.

Fracture Mechanics of Concrete,
edited by F.H. Wittmann, 1983
Elsevier Science Publishers B.V., Amsterdam — Printed in The Netherlands

Chapter 3.1

MICROSCOPIC OBSERVATION OF CRACKS IN CONCRETE, WITH EMPHASIS ON TECHNIQUES
DEVELOPED AND USED AT CORNELL UNIVERSITY

by Floyd O. SLATE

1.   SOME HIGH-LIGHTS OF PAST USE OF MICROSCOPY

The microscope has been a powerful tool in the study of cement and concrete
since the days of development of these materials. Perhaps the earliest intensive and effective use was by Le Chatelier [1] in his remarkable work; his
efforts undoubtedly greatly influenced later workers in use of the microscope.
His use was apparently entirely for study of the chemical and physical aspects
of hydration and setting, primarily by powder mounts and transmitted light,
and not for study of cracks. Following some less successful earlier work,
Tavasci [2] reintroduced with great success the use of polished sections, again
for study of composition and structure but not of cracks. However, his
work set the stage for studies of cracks in the 1960's, on interior surfaces of
cut specimens.

Many workers over the years, such as Richard et al. [3], have used magnification, usually at low levels and often on specimens during load tests, to
observe presence and growth of exterior cracks. The word exterior must be
emphasized; cracks on an exterior surface may be different from cracks in the
interior of concrete and many conclusions about a mass of concrete but based
on surface observations only, may be misleading or wrong. Much or most of the
current studies of concrete by fracture mechanics people depend on observation
of such surface cracks. However, much has been learned from such observations,
which have the advantage of being non-destructive in nature and permitting continuing observation on a given specimen as its loading is changed. Among early
uses of the microscope to observe internal cracks on cut specimens, using both
thin and polished sections, was the work on autogenous healing of cracks by
Lauer and Slate [4].

Important publications on the use of the microscope for studying concrete
and for studying concrete aggregates are those by Mather [5] and Mielenz [6].
Both place emphasis on use of petrography and use of the petrographic microscope; although their objectives were generally the study of things other than
cracks, the basic techniques they use are valuable for all aspects of structure, including cracks. As an example, Lauer and Slate [4] used the petrographic microscope, with crossed Nicol prisms, to study formation of calcium
hydroxide and calcium carbonate in cracks in cement paste. The reference lists

of Mather [5] and Mielenz [6] are of great help in finding other publications on the subject.

The electron microscope was apparently first used by Eitel [7, 8], and by Eitel with his co-workers [9] on the concrete system, to study hydration processes. Another important pioneer in the use of high magnifications, including the electron microscope, was Grudemo [10]. Again, although these studies were not directed specifically toward cracks, they were partially directed toward structure, and led the way to later studies of cracks.

More recently, use of the scanning electron microscope has introduced a powerful tool for studying cracks, along with other features of structures. Recent work by Diamond and Mindess [11] has used the scanning EM to observe the growth of surface cracks during loading, using magnifications generally from 35X to 450X. It must be emphasized that removal of air to form a high vacuum in the chamber holding the specimen will cause removal of water from the system, resulting in shrinkage and potentially in shrinkage cracking. Such shrinkage effects are imposed on strains from external loading, and complicate the picture, since the method of observation influences the results. Earlier, initiation and propagation of cracks during loading was studied by Slate and co-workers [12, 13] by use of X-rays, as is described in this report in the section on X-rays, Chapter 3, Section 2.

A recent development is use of stereology in connection with observations of concrete by the microscope. An example is the thesis by Stroeven [14].

2. SIZE LIMITS OF MICROCRACKS

Microcracks have not been generally defined as to size. The author favors an upper limit of 0.1 mm for microcracks, to distinguish them from the larger macrocracks; this size represents an average limit of the unaided human eye, below which magnification is generally needed. A lower limit for microcracks would seem to be the smallest crack-like discontinuity that can be detected. This may be the tensile strain limit beyond which elementary building units of the brittle material (usually ions or dipoles for brittle materials) will not return to their original positions when load is released, thus resulting in a permanent separation between particles. This size is in the order of magnitude of ionic diameters. Thus the question of when and where a crack actually starts remains as an important gap in our knowledge. It is likely that we first detect cracks in concrete only after they are far larger than when they were first formed.

3. EXTERIOR VS. INTERIOR CRACKS

The matter of surface or exterior cracks vs. internal cracks must be put forward. The great majority of observations of cracks by magnification have

been made on external surfaces only. Cracks and crack patterns on surfaces
must be different, and sometimes much different, from internal cracking. After
all, a free surface has an entirely different stress field than an interior
region unless the entire body is stress free--which is never true in concrete.
Therefore, loading will cause different stresses on a surface than in an
interior, and thus different cracking. Further, any drying or shrinkage at a
surface will cause different cracking at that surface as compared to the
interior. Most experiments on concrete allow some surface drying. To make
matters worse, often a thin layer of laitance, or low water/cement ratio paste,
will be present from contact with forms or from finishing, and often is not
removed by grinding or cutting. Such a layer of laitance is especially sensitive to cracking from drying. Much of the work done on fracture mechanics of
concrete involves use of a microscope to observe lengths of cracks on exterior
surfaces of notched beams. Qualitatively, it is expected that the lengths of
these exterior cracks will be longer than internal cracks above the notch,
particularly if significant drying shrinkage has occurred. This matter should
be given serious consideration.

4.   TECHNIQUES DEVELOPED AND USED AT CORNELL UNIVERSITY
4.1.   Background and motivation

Starting about 1960, intensive study of cracks in concrete was begun at
Cornell University. The first motivation was to attempt to explain the shape
of the stress-strain curve in terms of the internal structure and in terms of
changes in the internal structure with loading. It was felt necessary to
observe the interior of the concrete, instead of the surface of specimens.

A decision was made to use direct observation of internal structures and
cracks, and not to use the interesting but indirect methods involving sound
energy and sound emission upon cracking, as developed and used by Jones [15]
and by Rüsch [16]. This resulted in the necessity of using destructive tests,
to expose the internal structure. Thus specimens were loaded or treated in the
desired manner, then cut open for observation.

4.2.   Methods studied and discarded

Several methods were studied, either briefly or extensively--all methods
studied involved radiant energy. Among methods studied briefly and then
discarded were the following:

(1) Cut and polish a surface, allow a hydrophilic tracer liquid to
penetrate cracks and voids in the slightly dried surface, lightly grind
the surface to remove the surface film of the tracer liquid, then examine
the surface for cracks and voids as shown by the tracer liquid which had
penetrated them. Tracers tried were fluorescent aqueous systems, and mildly

radioactive aqueous solutions; the former could be observed or photographed directly in a darkroom under ultraviolet radiation, and the latter could be recorded by placing photographic film directly on the surface with cracks containing radioactive materials that emitted short wave radiation. The fluorescence method was somewhat useful, but cumbersome, and not conducive to detailed prolonged study such as can be done readily with the optical microscrope. The radioactive method did not work well.

(2) Cut a slice of perhaps one-half to one centimeter thickness, allow it to dry a little, place the bottom face in a container of an aqueous dye that did not reach to the top of the slice, allow capillary rise of the dye to bring it to the top surface through cracks, then observe or photograph the colored cracks. This did not work well because the internal surfaces of the cracks absorbed most of the dyes during capillary rise, and because (as was later learned) many or most of the cracks may not be continuous for an appreciable distance, even through the thickness of the thin specimen.

(3) Cut a thin slice, partially dry it, allow capillary rise of a saturated water solution of a lead salt, then X-ray to see cracks and voids that were then relatively opaque. This did not work well.

After other minor trials, two major methods were developed, one involving use of transmitted X-rays and the other involving optical (microscopic) study of a surface in which a dye had penetrated cracks and voids open to the surface. Apparently the X-ray technique had not been used previously for study of concrete; it is described in detail by Slate and Olsefski [17] in their 1963 paper on "X-rays for Study of Internal Structure and Microcracking of Concrete." This paper also describes the optical method involving dye absorption. Another paper in 1963 by Hsu, Slate, et al. [18] describes the optical dye method in somewhat different terms and refers briefly to the X-ray method. Details of the microscope technique with dye are given below.

4.3. Microscope technique with dye

The following description is adapted from Slate and Olsefski [17]:

After processing to cause cracking, or after simply curing, a diamond rotary saw was used to cut horizontal or vertical slices 0.150 to 3 in. thick. The thinner slices were used when both microscope and X-ray techniques were to be used on the same specimen, as checks.

The coolant for the saw was a mixture of equal parts by volume of de-odorized kerosene and light flushing oil. If the specimens are even slightly damp, the internal porous structure is hydrophilic and will not absorb the hydrophobic organic coolant; the coolant on the surface can easily be slushed away by flowing water. It is vital to remove the powder from the cutting process that has been packed into the cracks; this can be done by flushing out the powder (or paste) with a jet of water. Failure to do this can result in failure to detect cracks; in the author's opinion, this has frequently happened.

Direct microscopic observation of the cut surfaces of the slices was used. A saw-cut face of a specimen 0.150 to 3 in. thick was ground wet with abrasives (aluminum oxide or silicon carbide), starting with coarser abrasives and ending with No. 180 grit, on a plate glass surface, washed clean, then surface dried and painted with carmine drawing ink, which penetrated into voids and cracks. After the ink became surface dry on all the paste portions, the inked surface was ground wet with No. 180 or finer grit aluminum oxide or silicon carbide on a flat glass surface, until only a faint pink color could be seen by the naked eye. The drying time for the ink was about 15 min. Shorter drying time resulted in loss of most of the color into the water of the grinding medium; longer drying times resulted in formation of a tough surface film by the binder in the ink, and flaking and lifting of the colored material from larger cracks as the surface film was removed. Cracks and voids thus were dyed carmine, with great contrast to the rest of the surface of the specimen. This was the most successful dyeing technique tried.

A stereomicroscope was used, at 4X to 40X. Higher magnifications, and a monocular microscope, were employed only as checks on the technique; the stereomicroscope was much easier to use, and gave results substantially as good, or better for some purposes (e.g., depth).

Cracks were drawn on a sketch or a photograph of the surface being observed, as shown in Figs. 1, 2 and 3, and often compared with the X-ray plate of the same slice as a control.

4.4. Check for introduction of cracks by sawing, processing, and handling

The following description is from Slate and Olsefski [17]:

Slices 2 or 3 in. thick were cut from undisturbed cylinders fresh from the moist room or the curing tank. The sawed surface was stained, studied under the microscope, and a map made of all cracks present. A parallel cut was then made 0.150 in. from the first cut. The surface originally studied was restained, reground, and restudied with the microscope, and a new map of cracks made. No appreciable additional cracking was found on the second examination, after the thin slice had been sawed out of the larger mass. [This procedure required partial wetting and air-drying cycles--partial surface drying for application of ink, wetting by wet grinding, and partial surface drying (or inundation) for observation--which by themselves could cause cracking; the amount of drying was carefully kept to the absolute minimum needed by the techniques.] Further, any cracking caused by sawing

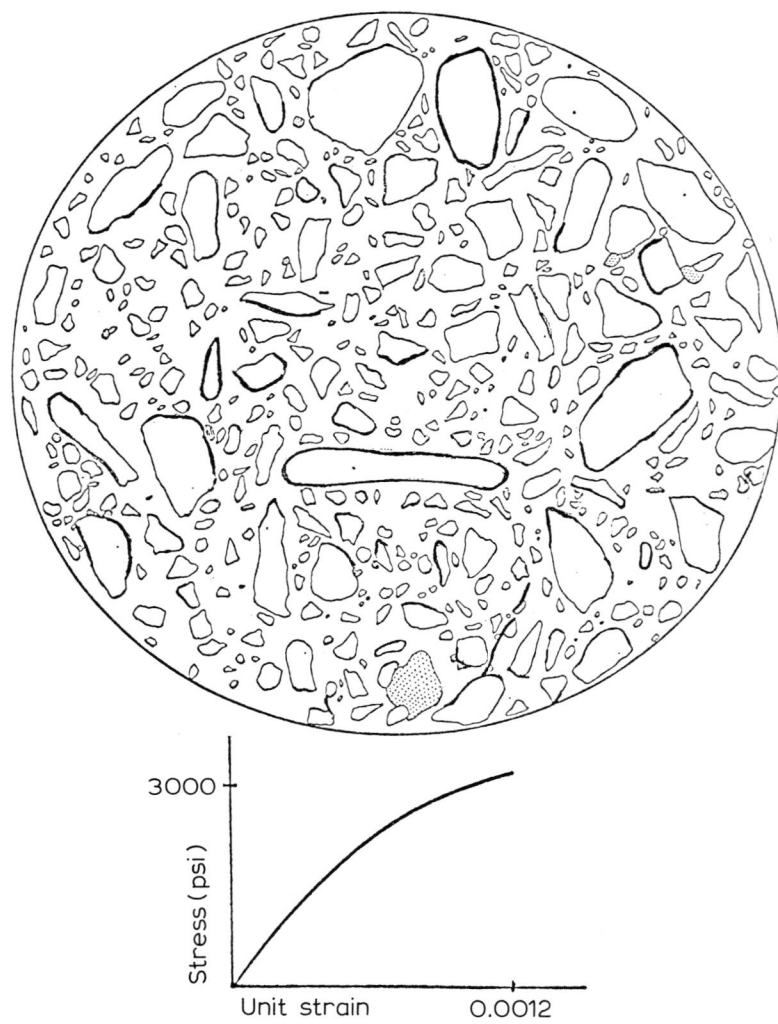

Fig. 1. Sketch of a polished surface of concrete with cracks enhanced in ink by thick black lines, as observed through the microscope using the microscope-dye method; the section is horizontal and perpendicular to compressive load on a 4-inch diameter cylinder. The unit strain was 0.0012, as shown by the accompanying stress-strain curve.

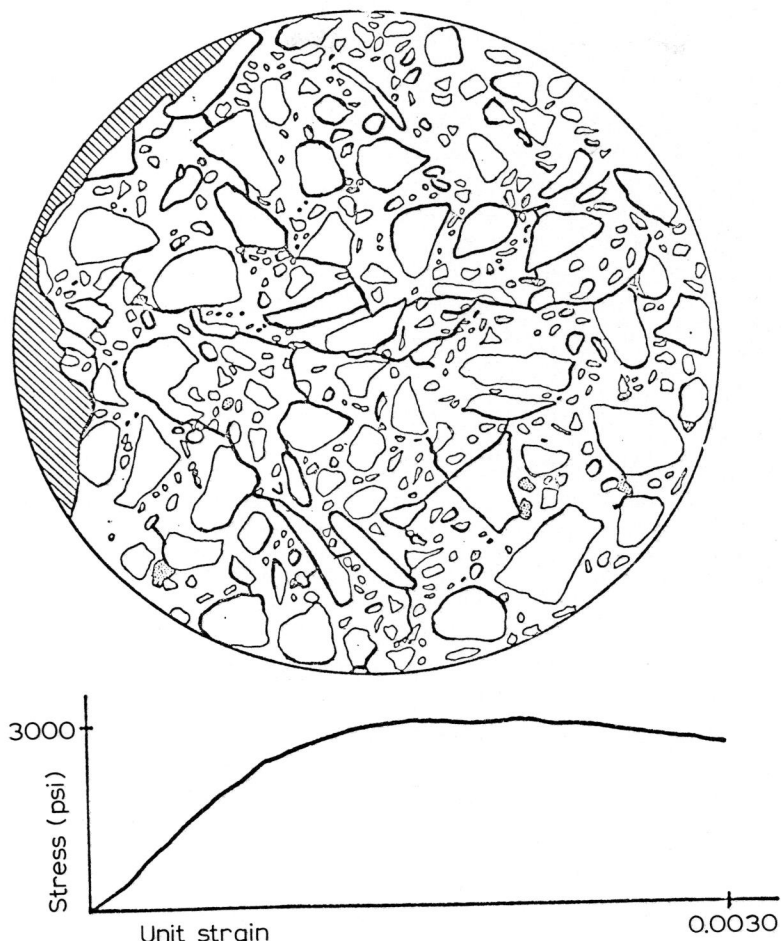

Fig. 2. Sketch of a polished surface of concrete with cracks enhanced in ink by thick black lines, as observed through the microscope using the microscope-dye method; the section is horizontal and perpendicular to compressive load on a 4-inch diameter cylinder. The unit strain was 0.0030, as shown by the accompanying stress-strain curve.

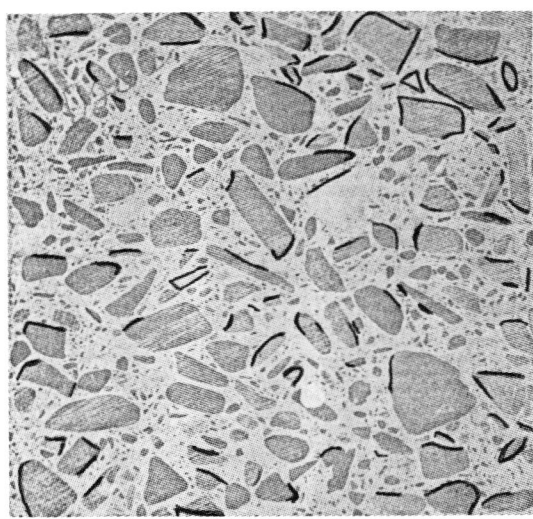

Fig. 3. Faint (low exposure) photograph of a polished surface of concrete with cracks enhanced in ink by thick black lines, as observed through the microscope using the microscope-dye method; the section is from a 4-inch square prism, is perpendicular to compressive load, and shows cracking associated with early creep at stress about 60 percent of ultimate.

must necessarily show directional effects from the pressure feed and direction of cut; careful examination showed the absence of any such directional pattern of cracking. Therefore, no appreciable additional cracking was introduced by sawing, processing, and handling. It should be emphasized that reasonable care must be taken in these steps, or cracking may be introduced.

4.5. General comments

Slices of concrete are routinely stored in plastic bags, until further processing, to prevent drying and shrinkage cracking. Extended storage should be avoided to prevent autogenous healing of cracks. However, for the dyeing technique, surface drying must occur, but appreciable drying at depth should be avoided (unless drying shrinkage is being studied).

For recording cracking, it was found very helpful to photograph and make a faint print of the surface, then to draw the cracks on the print from study with the microscope, as shown in Fig. 3.

Little change has been made in subsequent years on the procedures described above, because they continue to work well.

REFERENCES

1   H. Le Chatelier, Experimental Researches on the Constitution of Hydraulic
    Mortars, translated by J.L. Mack, McGraw, New York, 1905. Also: Compt.
    rend., 94 (1882) 13; and J. Soc. Chem. Ind., 1 (1882) 151.
2   B. Tavasei, Structures of Hydrated Portland Cement, Zement 30 (1941) 43-8,
    55-6.
3   F.E. Richart, A. Brandtzaeg and R.L. Brown, A Study of the Failure of
    Concrete under Combined Compressive Stresses, Bull. No. 185, Univ. of
    Illinois Engr. Expt. Sta., Urbana, Ill., 1928.
4   K.R. Lauer and F.O. Slate, Autogenous Healing of Cement Paste, J. Am.
    Concrete Inst., Proc. 52 (1956) 1083-97.
5   K. Mather, Petrographic Examination [of hardened concrete]. Significance
    of Tests and Properties of Concrete and Concrete-Making Materials, Am.
    Soc. for Testing and Materials, Special Technical Publn. 169-A, ASTM,
    Philadelphia, 1966, pp. 125-43.
6   R.C. Mielenz, Petrographic Examination [of concrete aggregates] Signifi-
    cance of Tests and Properties of Concrete and Concrete-Making Materials,
    Am. Soc. for Testing and Materials, Special Technical Publn. 169-A, ASTM,
    Philadelphia, 1966, pp. 381-403.
7   W. Eitel, Recent Results of Cement Research, Z. angew. Chem. 54 (1941)
    185-92.
8   W. Eitel, Electron Microscopic Cement Research, Zement 31 (1942) 489-95.
9   O.E. Radczewski, H.O. Müller and W. Eitel, Ultra-Microscopic Investigation
    of the Hydration of Free Lime, Zement 28 (1939) 693-7.
10 A. Grudemo, The Microstructure of Hardened Cement Paste, Proc. Fourth
    International Symp. on the Chemistry of Cement, Vol. 2 (1960) 615-658,
    Washington, D.C.; and U.S. National Bureau of Standards, Monograph 43,
    Washington, D.C.
11 S. Diamond and S. Mindess, Scanning Electron Microscope Observations of
    Cracking in Portland Cement Paste, Proc. Seventh International Symp. on
    the Chemistry of Cement, Vol. 3, VI-114-119 (1980), Paris; see also, a
    Preliminary SEM Study of Crack Propagation in Mortar, Cement and Concrete
    Research 10 (1980) 509-19.
12 O. Buyukozturk, A.H. Nilson and F.O. Slate, Stress-Strain Response and
    Fracture of a Concrete Model in Biaxial Loading, J. of American Concrete
    Inst., Proc. 68 (1971) 590-9.
13 T.C.Y. Liu, A.H. Nilson and F.O. Slate, Stress-Strain Response and
    Fracture of Concrete as Uniaxial and Biaxial Compression, J. of American
    Concrete Inst., Proc. 69 (1972) 291-5.
14 P. Stroeven, Some Aspects of the Micromechanics of Concrete, PhD Thesis,
    Stevin Laboratory, Technological University of Delft, 1973, 325 pp.
15 R. Jones, "A Method of Studying the Formation of Cracks in a Material
    Subjected to Stresses," British Journal of Applied Physics, (London),
    V. 3, No. 7 (1952) 229-32.
16 H. Rüsch, Physikalische Fragen der Betonprufung, Zement-Kalk-Gips 12
    No. 1 (1959) 1-9.
17 F. Slate and S. Olsefski, X-Rays for Study of Internal Structure and
    Microcracking of Concrete, J. of American Concrete Inst., Proc. 60
    (1963) 575-88.
18 T.T.C. Hsu, F. Slate, G. Sturman and G. Winter, Microcracking of Plain
    Concrete and the Shape of the Stress-Strain Curve, J. of American Concrete
    Inst., Proc. 60 (1963) 209-24.

Fracture Mechanics of Concrete,
edited by F.H. Wittmann, 1983
Elsevier Science Publishers B.V., Amsterdam — Printed in The Netherlands

Chapter 3.2

X-RAY TECHNIQUE FOR STUDYING CRACKS IN CONCRETE, WITH EMPHASIS ON METHODS DEVELOPED AND USED AT CORNELL UNIVERSITY

by Floyd O. SLATE

1. INTRODUCTION

As far as the author can determine, the first use of X-radiography to study cracks and other internal structural features of concrete (in contrast to X-ray diffraction and similar techniques for analytical study of composition) was that developed and used by Slate and Olsefski [1] and later used by their co-workers and associates Hsu et al. [2], Sturman et al. [3], Sturman et al. [4], Slate and Matheus [5], Shah and Slate [6], Meyers et al. [7], Buyukozturk et al. [8], Buyukozturk et al. [9], Liu et al. [10], Carino and Slate [11], Tasuji et al. [12], Carrasquillo et al. [13], and Ngab et al. [14]. Robinson [15] used X-rays at about the same time as Slate and Olsefski [1], but unfortunately used specimens that were too thick, (7 cm) resulting in radiographs that were of limited value, with impaired clarity and a confusing multitude of details. More recently, Isenberg [16], Bhargava [17], and Stroeven [18], among others, have done important work involving X-rays to study concrete. The technique is now well established.

The reader is referred to Chapter 3, Section 1 on use of the microscope for discussion of size limit of microcracks and exterior vs. interior cracks.

2. TECHNIQUES DEVELOPED AND USED AT CORNELL UNIVERSITY
2.1. Background and motivation

Chapter 3, Section 1, on Microscopic Observation of Cracks in Concrete, reports on background and motivation of the Cornell studies, for X-rays as well as for the microscope.

Since the Cornell researchers felt strongly that meaningful observations and studies of cracks must be on interior cracks within the concrete rather than on cracks on an exterior surface, the idea of using X-rays was developed for observations at depth within concrete. It was correctly anticipated that the technique could be used either for thin specimens (up to about 2 cm thick, but best for specimens about 0.4 to 0.5 cm thick), or for slices of concrete carefully cut from the interior of specimens. Thus this technique met the specific requirements for observation that had been established, both for a destructive test involving cutting open concrete, and for a non-destructive test on thin intact specimens.

2.2. Details of the X-ray technique

The following is adapted from Slate and Olsefski [1]:

After processing to cause cracking, or after simply curing, thin (plate) specimens were X-rayed directly. For other specimens, when it was desired to observe the interior portion of a mass of concrete (such as a concrete test cylinder or prism), a diamond rotary saw was used to cut horizontal or vertical slices 0.150 in. thick, plus or minus a maximum of 0.005 in. With care, the saw was capable of an accuracy of plus or minus 0.002 in. Specimens that were too thick showed so much detail as to cause confusion in interpretation of the X-ray plate, or caused loss of definition of very small cracks extending only a small part of the way through the specimen; specimens that were too thin sometimes had cracks introduced by the sawing process. Nonparallel faces lead to variation in darkness of different parts of the plate.

The coolant for the saw was a mixture of equal parts by volume of de-odorized kerosene and light flushing oil. If the specimens are even slightly damp, the internal porous structure is hydrophilic and will not absorb the hydrophobic organic coolant; the coolant on the surface can easily be flushed away by flowing water. The specimen slices can be X-rayed wet or dry, as desired. Normally, they are kept wet, stored immediately in plastic bags with some water, and X-rayed while remaining in the bags (to prevent drying and shrinkage cracking).

An industrial X-ray unit was used, with a rating of 150 kilovolts (kv). The tube had a tungsten target, a focal length of 3 ft, an inherent filtration equal to approximately 0.5 mm Al, and a limit of 6 milliamperes at 150 kv. An exposure time of 2 min at 40 kv and 5 milliamperes was used. A fine-grained X-ray film was used, to make possible enlargements of the images. The film was developed as instructed by the manufacturer. However, developing time was increased or decreased to increase or decrease contrast, as desired for special observations. Enlargements of 3 to 10X were sometimes made from the X-ray plates, to aid in location of small cracks and other fine structure. The plates or enlargements were examined on a light box. A 4 to 10X hand lens was sometimes used to magnify a portion of the image for close study. An example of an X-ray image of concrete near failure is shown in Fig. 1.

2.3. Check for introduction of cracks by sawing, processing, and handling

The following description is from Slate and Olsefski [1]:

Slices 2 or 3 in. thick were cut from undisturbed cylinders fresh from the moist room or the curing tank. The sawed surface was stained, studied under the microscope, and a map made of all cracks present. A parallel cut

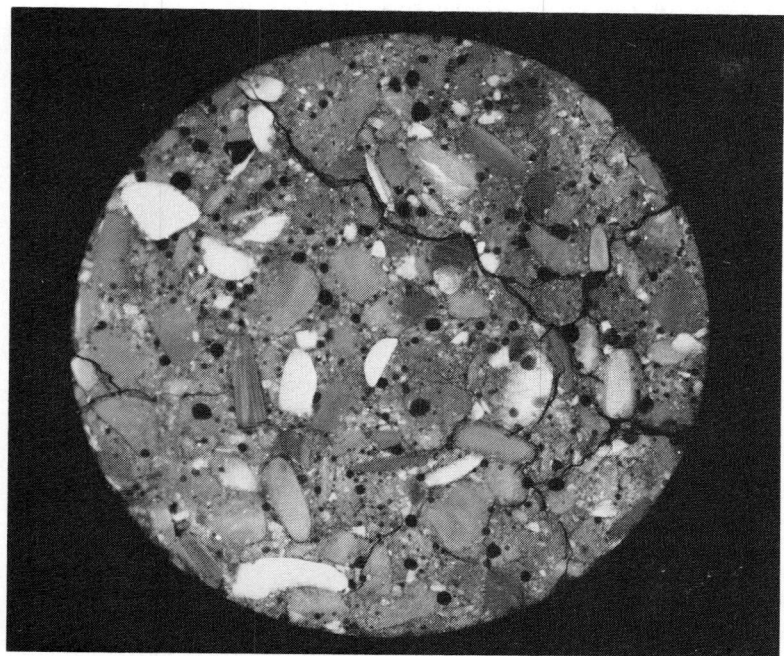

Fig. 1. X-ray image of a 0.15-inch thick slice of concrete cut from a 4-inch diameter cylinder and perpendicular to compressive loading. This image is a double reversal of the original X-ray film, so the cracks appear in black, as in the original. No enhancement of cracks or other structures was used. The concrete is near failure, and shows extensive cracking through the hydrated paste, as well as at paste-aggregate interfaces.

was then made 0.150 in. from the first cut. The surface originally studied was restained, reground, and restudied with the microscope, and a new map of cracks made. No appreciable additional cracking was found on the second examination, after the thin slice had been sawed out of the larger mass. [This procedure required partial wetting and air-drying cycles—partial surface drying for application of ink, wetting by wet grinding, and partial surface drying (or inundation) for observation—which by themselves could cause cracking; the amount of drying was carefully kept to the absolute minimum needed by the techniques.] Further, any cracking caused by sawing must necessarily show directional effects from the pressure feed and direction of cut; careful examination showed the absence of any such directional pattern of cracking. Therefore, no appreciable additional cracking was introduced by sawing, processing, and handling. It should be emphasized that reasonable care must be taken in these steps, or cracking may be introduced.

2.4. Problems and solutions

Even relatively small variations in thickness of the specimens (especially for the thin slices) cause significant differences in darkness of the X-ray film. With such variations of darkness, it may be impossible to achieve sufficient contrast to show cracks clearly in the dark portion of a film and still have sufficient exposure to see cracks in the light portion of the film. Therefore, care must be used to make the cut faces parallel to each other.

It is vital, for the X-ray techniques as well as for the microscope technique, that after saw cutting all surfaces be carefully cleaned to remove powder caused by the cutting process from the cracks and other void spaces. The saw cutting process will always create a powder (or a paste, with a liquid saw coolant) and this powder or paste will be packed into the cracks and voids. If this filler is not removed, the cracks may not be shown by X-rays because the filler has about the same opacity to X-rays as does the uncracked paste. A jet of water directed carefully over all surfaces will remove the powder or paste. If doubt remains as to completeness of removal, the specimen surface can be checked for cracks by microscope at intervals during the washing-out process.

Routinely, as a check for either technique (when slices were used) an X-ray image was checked against observation of the slice by microscope, and conversely a crack map made from observation by microscope was checked against an X-ray image of the same slice. This is particularly valuable in cases where there is doubt whether or not a crack is present. For very careful work, both techniques were used to prepare a final crack map.

It must be noted that an X-ray image will show a crack pattern that is not identical with that shown by the microscope-dye method. There are occasionally cracks at the paste-aggregate interface that are in the interior of a slice of concrete and do not extend to the surface, even in thin slices of concrete. Such cracks will be shown by the X-ray method but not by the microscope-dye method. Also, the microscope can show very small cracks that may not be seen in an X-ray film, particularly within the mortar or paste regions. Cracks that are inclined steeply from the perpendicular with respect to the cut face of a slice may not show up via X-ray. Thus the microscope may show slightly more mortar or paste cracking, and X-rays may show slightly more bond or interface cracking, but such differences are small.

Both techniques tend to show bands or regions of high water-cement ratio, as well as clear-cut cracks. These are appreciably transparent to X-rays, and will absorb dye for the microscopic method. Such bands are frequently found at the interfaces between aggregate and paste; they are somewhat diffuse in appearance and less dark than clear-cut cracks in X-ray images, and dye absorption is less than in a clear-cut crack, resulting in a less clearly defined line when observed with the microscope. For example, such bands of higher water-cement ratio are often found under coarse aggregate particles, from segregation, and although weak, may not be cracks until a physical separation occurs. In the opinion of the author, such "crack-like" voids and bands should not be recorded as cracks unless physical separation has actually occurred. In a few cases, it is very difficult to distinguish between the two cases. Since the interpretation becomes somewhat subjective and personal, it is best that one particular observer make the decisions (a crack or not a crack) in a given study. Thus different observers may report somewhat different results, but the differences are usually small, for experienced and careful observers.

For X-ray study, a faint photo print (double reversal so cracks will be in black, not white) from the X-ray film is convenient for recording cracking, as illustrated by Fig. 2. This print will, of course, look somewhat different from the surface or a photograph of the surface, since it shows details at depth, and may be a bit confusing at first.

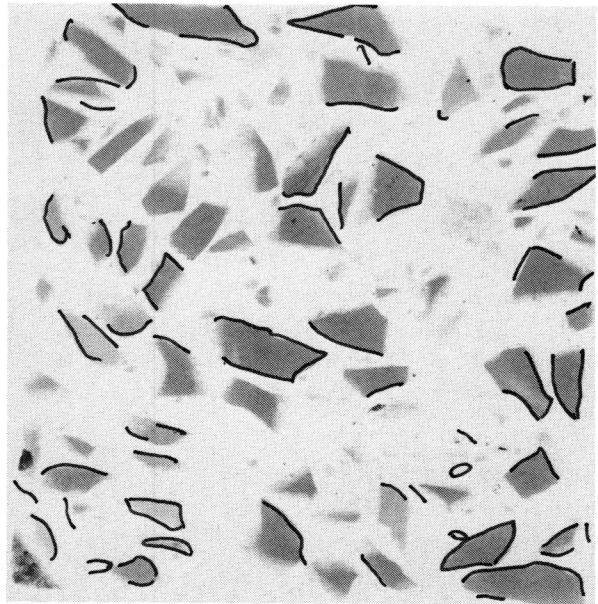

Fig. 2. Faint photographic print from an X-ray image of a slice of concrete cut from a 4-inch square prism tested under sustained load; the cracks are enhanced by thick black lines in ink. This image is a single reversal of the original X-ray film, so the cracks (before inking) appear in white and the more opaque aggregate particles appear darker.

After cracks are marked on a photocopy (from either the X-ray or the micro-scope-dye method) their individual and total length can be conveniently and accurately measured by using a magnetic digitizing table (readily available with an accuracy of 0.1 mm) to compute automatically the x-y coordinates of any point along a crack. Carrisquillo et al. [13] used this method and the digitized coordinates were automatically punched on computer cards. Each crack length was approximated as the length of up to four straight segments. A simple computer program was used to compute the length of each crack. The consistency and reliability of the test results were checked and found to be well within acceptable limits.

3.   RECENT DEVELOPMENTS

Stroeven [19] has worked on stereological techniques for calculating three-dimensional structures of concrete from two-dimensional images, and has worked extensively on X-ray, microscopic, and other techniques of observation. He has applied some of the techniques to study of steel-fiber concrete [20].

At Cornell, thin (1/2-inch) plates of models of concrete were loaded in uniaxial and biaxial compression and tension, X-ray film was placed under the specimens, and radiation was passed through the specimens to expose film after film during loading, thus resulting in a non-destructive test showing pro-gressive cracking and failure, as described by Buyukozturk et al. [8, 9], by Liu et al. [10], and by Carino and Slate [11]. Fig. 3 shows cracking maps drawn from X-ray images of a model of concrete during progressive loading. Work was done earlier by Robinson [15] on such a technique, with thicker specimens, about 2-3/4 inches thick; while having the advantage of being non-destructive, the work with thick specimens resulted in less clarity and a somewhat confusing multitude of details.

More recently, stereo X-ray images have been used at Cornell to study various aspects of the internal structure of concrete, such as cracking by Slate and Olsefski [21] and structures of steel-fiber concrete by Alvi [22]. Fig. 4 shows one X-ray image (of a stereo pair) of steel fiber concrete, and illustrates fiber spacing and orientation. The use of the true stereo technique of X-radiography seems to have considerable promise for direct observation in three dimensions.

4.   SUMMARY

The X-ray technique has become a major tool for study of internal structure of concrete, particularly cracks. The film makes an excellent original and permanent record. The technique is less time consuming than use of the microscope, and easier for most people. It can readily be used as a non-destructive test on thin plates of concrete, to trace initiation and progress

progress of cracks. True stereo radiography to study internal structure of
concrete, including cracks, has been used successfully [21, 22], and shows
promise for the future.

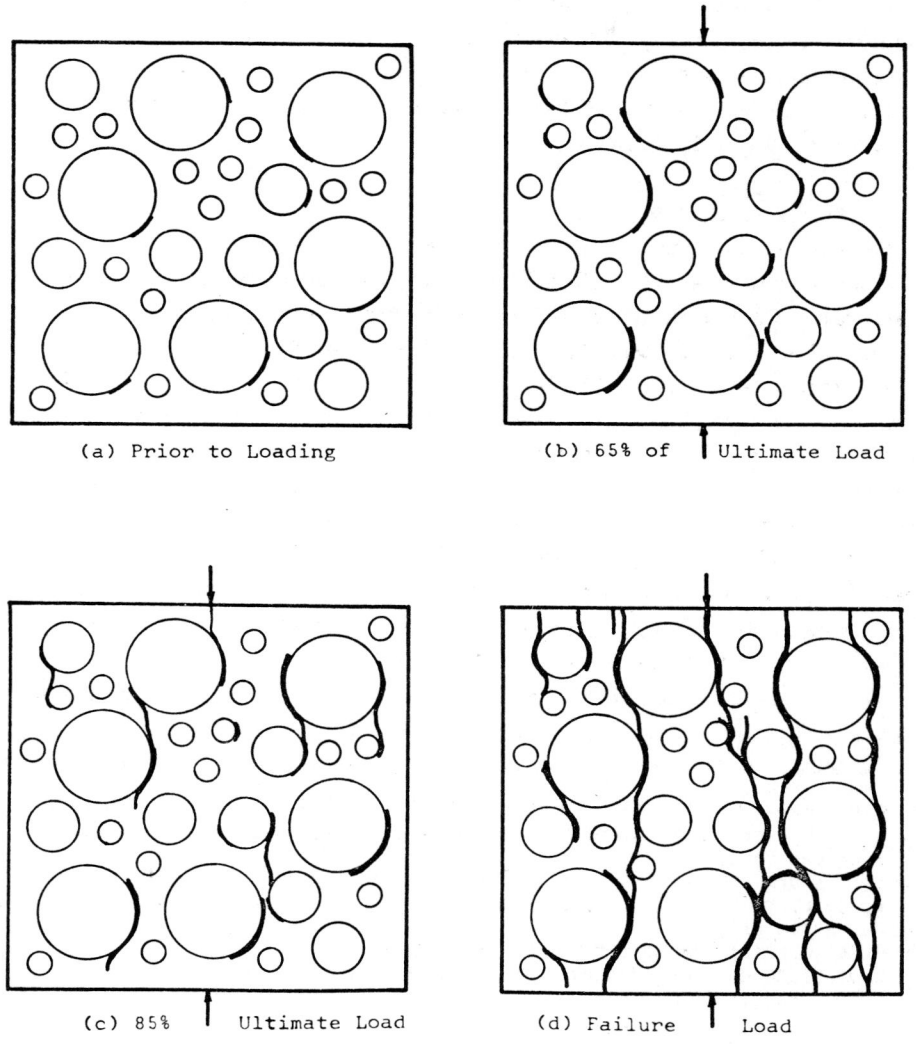

(a) Prior to Loading   (b) 65% of Ultimate Load

(c) 85% Ultimate Load   (d) Failure Load

Fig. 3. Sketches of progressive cracking of a 1/2-inch thick plate of a model
of concrete, made from X-ray images of the cracking, at various stages of
uniaxial compressive loading.

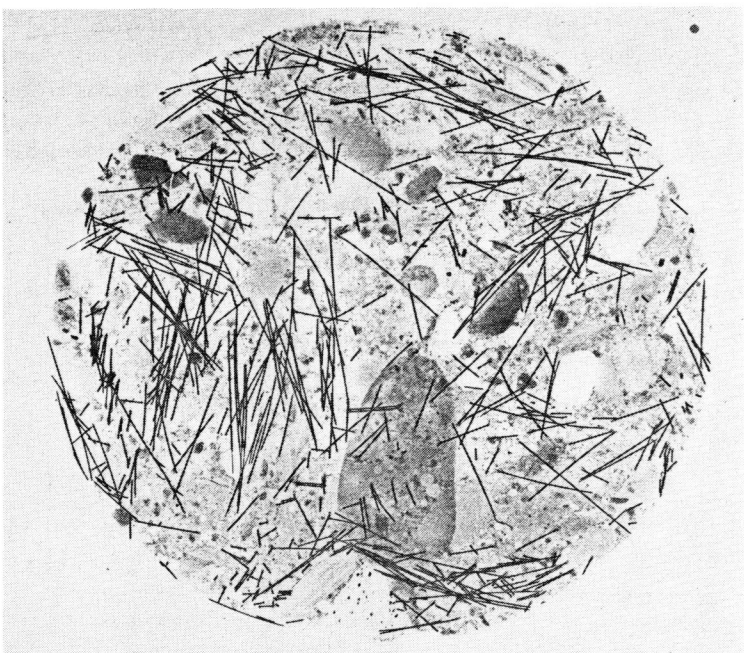

Fig. 4. Photographic print from an X-ray image of steel fiber concrete, showing distribution and orientation of the fibers. This is a single reversal of the X-ray image, so the opaque steel fibers appear in black. This is a single image of a stereo pair, which showed the third dimension dramatically.

REFERENCES

1   F.O. Slate and S. Olsefski, X-rays for Study of Internal Structure of Concrete, J. of American Concrete Inst., Proc. 60 (1963) 575-88.
2   T.T.C. Hsu, F.O. Slate, G.M. Sturman and G. Winter, Microcracking of Plain Concrete and the Shape of the Stress-Strain Curve, J. of American Concrete Inst., Proc. 60 (1963) 209-24.
3   G.M. Sturman, S.P. Shah and G. Winter, Effects of Flexural Strain Gradients on Microcracking and Stress-Strain Behavior of Concrete, J. of American Concrete Inst., Proc. 62 (1965) 805-22.
4   G.M. Sturman, S.P. Shah and G. Winter, Microcracking and Inelastic Behavior of Concrete, Proc. of the International Symposium on Flexural Mechanics of Reinforced Concrete, Miami, Florida, 1964, American Concrete Inst. Special Publication 12 (1965) 473-99.
5   F.O. Slate and R.E. Matheus, Volume Changes on Setting and Curing of Cement Paste and Concrete from Zero to Seven Days, J. of American Concrete Inst., Proc. 64 (1967) 34-9.
6   S.P. Shah and F.O. Slate, Internal Microcracking, Mortar-Aggregate Bond and the Stress-Strain Curve of Concrete, Proc. of the International Conf. on the Structure of Concrete, Imperial College, London, 1965. Cement and Concrete Assoc., London, 1968, 82-92.
7   B.L. Meyers, F.O. Slate and G. Winter, Relationship between Time-Dependent Deformation and Microcracking of Plain Concrete, J. of American Concrete Inst., Proc. 66 (1969) 60-8.

8   O. Buyukozturk, A.H. Nilson and F.O. Slate, Stress-Strain Response and
     Fracture of a Concrete Model in Biaxial Loading, J. of American Concrete
     Inst., Proc. 68 (1971) 590-9.
9   O. Buyukozturk, A.H. Nilson and F.O. Slate, Deformation and Fracture of a
     Particulate Composite, J. of Engineering Mech. Div., Proc. of American
     Soc. of Civil Engineers 98 (1972) 581-93.
10   T.C.Y. Liu, A.H. Nilson and F.O. Slate, Stress-Strain Response and
     Fracture of Concrete in Uniaxial and Biaxial Compression, J. of American
     Concrete Inst., Proc. 69 (1972) 291-5.
11   N.J. Carino and F.O. Slate, Limiting Tensile Strain Criterion for Failure
     of Concrete, J. of American Concrete Inst., Proc. 73 (1976) 160-5.
12   M.E. Tasuji, F.O. Slate and A.H. Nilson, Stress-Strain Response and
     Fracture of Concrete in Biaxial Loading, J. of American Concrete Inst.,
     Proc. 75 (1978) 306-12.
13   R.L. Carrasquillo, F.O. Slate and A.H. Nilson, Microcracking and Behavior
     of High Strength Concrete Subject to Short-Term Loading, J. of American
     Concrete Inst., Proc. 78 (1981) 179-86.
14   A.S. Ngab, F.O. Slate and A.H. Nilson, Microcracking and Time-Dependent
     Strains in High Strength Concrete, J. of American Concrete Inst., Proc. 78
     (1981) 262-8.
15   G.S. Robinson, Methods of Detecting the Formation and Propagation of
     Microcracks in Concrete, Proc. of the International Conf. on the Structure
     of Concrete, Imperial College, London, 1965. Cement and Concrete Assoc.,
     London, 1968, 131-45.
16   J. Isenberg, A Study of Cracks in Concrete by X-Radiography, RILEM Bull.
     (New Series) 30 (1966) Mar., 107-14.
17   J. Bhargava, Nuclear and Radiographic Methods for the Study of Concrete,
     Acta Polytechnica Scandinavica, Civil Engineering and Bldg. Construction
     Series 60, The Royal Swedish Acad. of Engr. Sciences, Stockholm, 1969,
     103 pp.
18   P. Stroeven, Morphometry of Fiber Reinforced Cementitious Materials.
     Part I: Efficiency and Spacing in Idealized Structures, Materiaux et Con-
     structions 11 No. 61 (1978) 31-38; Part II: Inhomogeneity, Segregation
     and Anisometry of Partially Oriented Fiber Structures, Ibid, 12 No. 67
     (1979) 9-20.
19   P. Stroeven, Some Aspects of the Micromechanics of Concrete, PhD Thesis,
     Stevin Laboratory, Technological University of Delft, 1973, 325 pp.
20   P. Stroeven, Micro- and Macromechanical Behavior of Steel Fiber Reinforced
     Mortar in Tension, Heron 24 No. 4 (1979) 7-40.
21   F.O. Slate and S. Olsefski, Unpublished work.
22   A.Q. Alvi, Steel Fiber Reinforced Concrete in Axial Compression, PhD
     Thesis, Cornell University, 1977, 258 pp.

Fracture Mechanics of Concrete,
edited by F.H. Wittmann, 1983
Elsevier Science Publishers B.V., Amsterdam — Printed in The Netherlands

Chapter 3.3

INVESTIGATIONS INTO CRACKED REINFORCED CONCRETE STRUCTURAL ELEMENTS WITH THE
AID OF PHOTOELASTIC METHODS

by J. LIERSE and M. RINGKAMP

1. INTRODUCTION

In order to investigate the real bearing capacity of reinforced concrete
structural elements, a knowledge of the crack behaviour is of fundamental importance. The following problems of detail are particulary significant :
- the origin of the cracks,
- the propagation of cracks due to increasing stress, under constant load or
  under dynamic influences,
- the crack pattern,
- the crack spacing,
- the crack width under stress as well as
- the residual crack width, if necessary also during just a partial removal of
  the load.

As the development of cracks is not dependent alone on the exceeding of critical stresses but is determined in the main by materially induced contingencies
in the case of heterogeneous reinforced concrete bonded material, it will always
be irregular and never predeterminable.

For experimental investigation into sheet-like structural elements, photoelasticity offers special advantages. This is because an almost unlimited number of individual data (measurement data) are stored in just a few photographic
images, the interpretations of which nevertheless require considerable time and
effort.

Attention is drawn to the pertinent references, i.e. (ref. 1, 2), for the fundamental of photoelasticity.

2. TRANSMITTED LIGHT METHOD

Initially, the transmitted light method was used for the static model investigation of irregular reinforced concrete supporting units.

The models consisted of photoelastically active, homogeneous, isotropic and
almost ideally elastic material (e.g. Araldite), so that with this method,
stresses in an 'uncracked' state could be simply determined (see Fig. 2a). The
necessary cross-sections of the reinforcing steel were then calculated from the
resulting tensile forces of the photoelastically determined tensile stresses,
(ref. 3) to (ref. 6).

Boiten (ref. 7), Hiltscher-Müller (ref. 8), Müller (ref. 9) and Steffens
(ref. 10) also investigated reinforced concrete structural elements in the
'cracked' state and simulated the effect of reinforcement, using this method.

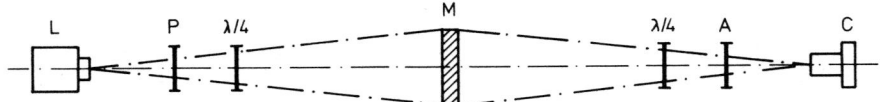

Fig. 1. Test set-up for the transmitted light method (L = Light source, P =
Polarizer, λ/4 = λ/4-plate, M = Model, A = Analyser, C = Camera).

They cut their models along previously observed compressive strength trajecto-
ries in a more or less arbitrarily selected crack spacing and joined the individual parts together again, sometimes inserting reinforcements. Steffens (ref. 10)
used a special resin mortar, whereby the tensile and compressive strength values
differed, similar to concrete. Aluminium, brass and glass fibre reinforced
plastics can be used as a reinforcement.

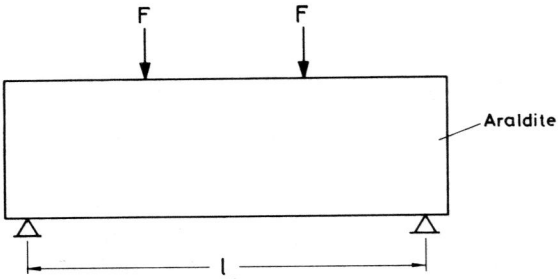

Fig. 2a. Investigation of a beam model in an uncracked state,(e.g. ref. 3 to 6).

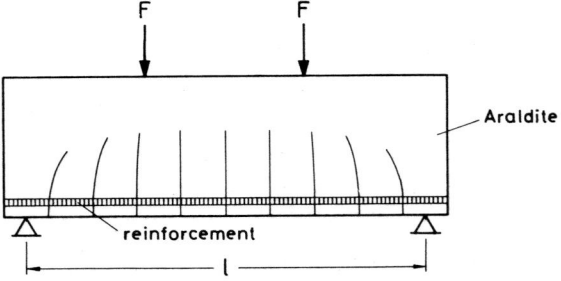

Fig. 2b. Investigation of a beam model in a cracked state, (e.g. ref.7 to 10).

   All the methods hitherto mentioned were disadvantageous in that for most of
the in fact characteristic phenomena of reinforced concrete building material,
it was possible only to simulate them inadequately or not at all.

---

3. THE REFLECTION METHOD

The principle of the reflection method, the application of which was first reported by Mesnager (ref. 11), is shown in Fig. 3.

A photoelastically active plastic foil is glued directly onto the object to be investigated (structural element or model) and the polarised light rays are reflected onto the structural element surface. The element itself is therefore loaded and errors,as a consequence of varying properties of model and construction materials are avoided.

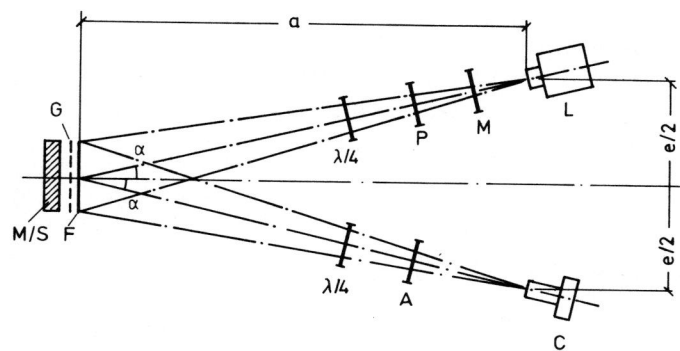

Fig. 3. Test set-up for the reflection method (L = Light source, M = Monocromator, P = Polarizer, $\lambda/4$ = $\lambda/4$-plate, F = Photoelastic foil, G = Glue with reflecting surface, M/S = Model or Structural element, A = Analyser, C = Camera). Type of light: white light, respectively yellow light with $\lambda = 0,586 \cdot 10^{-3}$ mm .

Prefabricated epoxide resin or polycarbonate plates of high photoelastic sensitivity in the range of 0.09 to 0.15/order were used as foils. Foil thickness varied between 0,25 to 3,0 mm. These were glued onto the surface of the test specimen, carefully smoothed beforehand, with the aid of cold hardening epoxide resin. Aluminium powder is mixed with the glue for light reflection purposes. A review of normal material combinations is given in Table 1 and 2.

Using this method on concrete test specimens, Dantu (ref. 12) was successful in making visible the very irregular strain and stress state on the concrete surface, which had been caused by the interaction and the most varied stiffness of cement mortar and concrete aggregate. Oppel (ref. 13) and Swamy (ref. 14) drew attention for the first time to the possibility of making crack development photoelastically visible. Reitblat and Baikow (ref. 15) as well as Cubaud, Jullien and Lemaire (ref. 16) in actual size experiments apply the reflected light method to reinforced concrete beams under bending but do not, however, confine themselves mainly to bending in the pressure zone. On the basis of lo-

FOILS	$\sigma_{Z,u}$	$\varepsilon_{Z,u}$	E	k	$S_{589,3}$	$Q=\dfrac{k}{E}$	Resin/ Hard-ener	Melt-ing-point	max. Hard.-temp.	State at Room-temp.
	$\dfrac{N}{mm^2}$	%	$\dfrac{N}{mm^2}$	Order	$\dfrac{N}{mm\ Order}$	10^{-5}Order N/mm^2	weight-parts	Resin °C	°C	
Araldite B Hardener 901 CIBA	40 50	3 5	3200 3800	0.10 0.13	10 13	~4	100/25 100/40	100 120	160 200	solid powder
Araldite F Hardener HY 951 CIBA	40 80	3 5	3100 3200	0.11 0.13	13 14	~4	100/10 100/12		20 60	liquid liquid
Araldite D Hardener 951 CIBA	30 40	3 5	2500 4200	0,09 0.14	12 17	~5	100/9 100/10		20 60	liquid liquid
Lekutherm x 30 Hardener PSA	40 50	3 5	3000 3500	0.10 0.13	10 12	~4	100/30	otherwise like Araldite B		
PS 1 VISHAY	40 50	10 10	2100 2200	0.15 0.15	6 7	~7	Polycarbonate, prefabricated plates			
PS 2 VISHAY	40 50	3 3	3100 3200	0.13 0.13	11 12	~4	prefabricated plates			
PS 3 VISHAY	44 6	30 30	210 300	0.02 0.02	4 5	~8	Epoxide resin system, prefabricated plates			
Araldite B SPR Hardener 901	0 3	0 0,01	~4000	0.08 0.13	~10	~4	100/4 100/7	100 120	160 200	solid powder

Table 1. Photoelastic coatings for the reflection method.

GLUE	Max. Elongation	E	Resin/ Hardener	Hard.-temp.	Hardening-time	Viscosity at Room-temp.
	%	$\dfrac{N}{mm^2}$	weightparts	°C	h	
PC 1 VISHAY PC 8 VISHAY PC 6 VISHAY	3 – 5 10 50	3200 3500 210	100/10 50/50	24 24 24	12 48 24	slight medium

Table 2. Glues for photoelastic coatings.

cally developed isochromatic concentrations in the tension zone, Khesin and Sakharov (ref. 17) infer appertaining crack widths and through subsequent not-ching of the glued-on resin foil in the vicinity of suspected cracks, try to carry through the crack propagation. However, as they use an elastic foil which tolerates considerably greater tensile strain than the concrete does, statements will by necessity be uncertain especially concerning the vicinity where the crack ends. In (ref. 17), Khesin and Sakharov also show that photoelastic investigations in the cracked tension zone only permit realistic statements when the coating foil is satisfactorily adjusted to the crack behaviour of the concrete.

Finally, Jullien and Cubaud (ref. 18) apply underpolymerised Araldite B, a brittle coating material with high photoelastic sensitivity, with which they could study the crack behaviour of a reinforced concrete beam, right up to failure. Bosc, Pera and others (ref. 19) investigate a series of single-span reinforced concrete beams for crack development, shear effects, crack serration and the dowelling effect with the aid of the brittle foil. The relatively great dispersal of internal forces for evaluated sections, were traced back by Ringkamp (ref. 20) to inacurrate photoelastic measurements and to errors which could occur later in the evaluation. It is known (ref. 20, 21), that with coating material from underpolymerised Araldite B the characteristic values of the material disperse particularly widely and as a result the photoelastic data can contain corresponding errors of measurement. In addition, the measured values are falsified to a greater or lesser degree due to stiffening and thickening effects, whereby all influences are included in the thickening effects, which result from the inconsistent expansion of the photoelastic foil beyond its own thickness.

4.   APPLICATION OF THE REFLECTION METHOD TO MODEL TESTS

Ringkamp (ref. 20) applies the reflection method to model tests as these are considerably more economical than actual size experiments as:
- a model as an object of measurement requires considerably lower costs than larger structural elements in terms of manufacture and setting up,
- due to smaller loads each test needs only loading devices which are very much cheaper to buy,
- during the investigation of larger structures, a model test mainly allows coverage with one photograph of the entire isochromatic pattern (stress state).

It is nevertheless beyond all doubt that the stiffening and thickening effects which falsify measurements in model tests are of particular significance. For this reason, they have been systematically analysed by Ringkamp (ref. 20). Further, he indicates calibration contingencies.

Through appropriate selection of model materials, e.g. resin mortar and wrought copper alloy as reinforcement (ref. 22), a two to three times exaggeration is achieved in contrast to the actual size experiment, i.e. the principal version, so that with equal photoelastic effects thinner foil with correspondingly lesser stiffening effects can be used, which only have a minor influence on the global bearing capacity. These effects on the evaluation process are eliminated simply by Ringkamp in that he relates the measured values to the loaded elements, with due consideration for the varying stiffness and lateral strain of foil and model material. On the other hand, correction of the thickness effect requires considerably more time and effort.

As a complete conformity of the critical cracking strain of construction materials, respectively model material and coating foil can never be attained, due in particular to the greater dispersal in the case of concrete tensile strength, three cases are possible (Fig. 4):

1. uncracked (elastic) model and cracked (brittle) foil
2. cracked model and cracked foil (normal case)
3. cracked model and uncracked (elastic) foil.

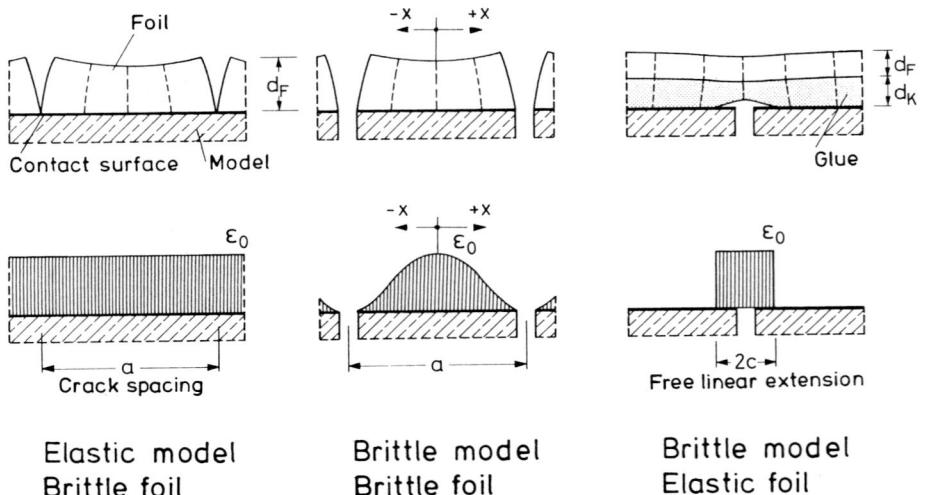

Fig. 4. Methods for determining the thickening effects.

The following symbols are used in the Figs. 5 to 7:

a = crack spacing
$2c$ = free linear extension
d_F = foil thickness
n_o = unfalsified isochromatic order
n_x = observed isochromatic order
n_{max} = maximum isochromatic order between tow cracks

In accordance with Duffy (ref. 23), Ringkamp (ref. 20) determined extension integrals over the foil thickness at the point X for the cases shown above. With these integrals one then obtains the calibration factors against the observed

isochromatic patterns. The displacements occuring between the foil and the structural element surface are corrected with the aid of further factors. These
correction values are obtained from the ratio of average, i.e. the observed foil
extension, to the actual value on the structural element surface.

In (ref. 20), these values for the three cases according to Fig. 4 are calculated. Some diagrams are given in Figs. 5 to 7.

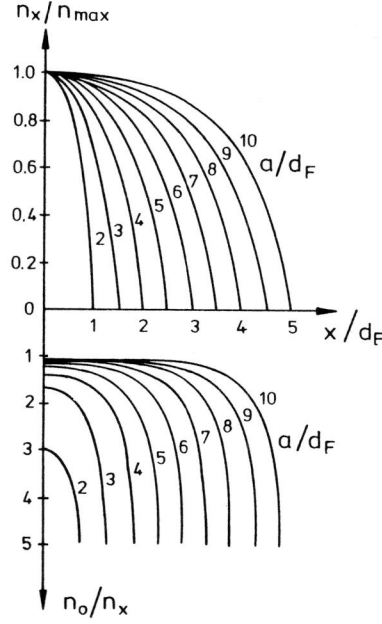

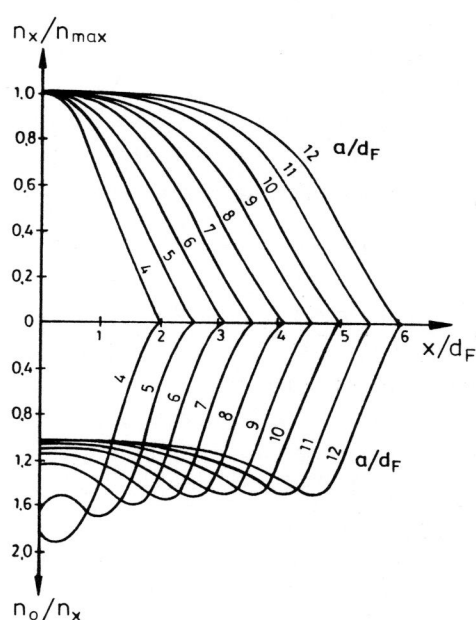

Fig. 5. Correction factors for the
case of an uncracked model and
cracked foil.

Fig. 6. Correction factors for normal
case - model and foil are cracked.

With the parameters a/d_F and x/d_F as well as the read off isochromatic
arrangements n_x, the arrangements n_o are determined which are no longer falsified by thickening effects.

In the case of elastic foil over a very small crack, the crack width can be
determined in relation to the free linear extension (disturbed bonded area in
the vicinity of both crack surfaces) of the isochromatic hill present over the
crack (Fig. 4). The crack width is given by

$$w_c = \varepsilon_o \cdot 2c,$$

whereby the free linear extension 2c present in each case can be found from a
comparison of the read off isochromatic courses with the curves n_x/n_o, in
Fig. 7.

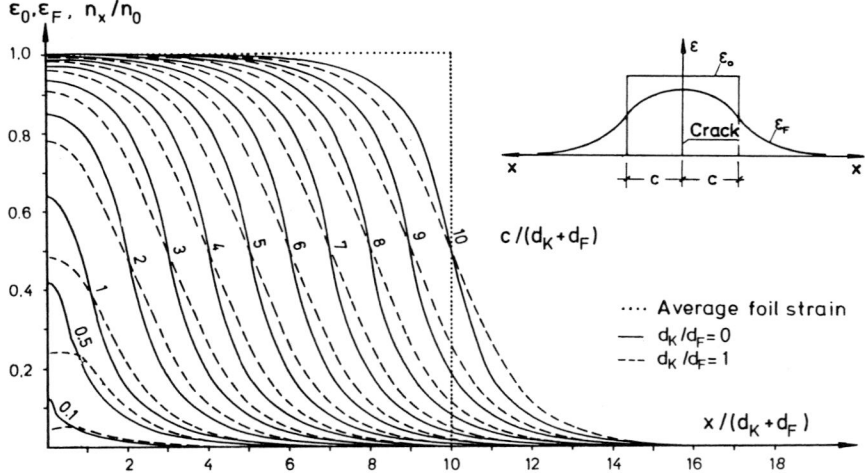

Fig. 7. Isochromatic curves for the case - cracked model and uncracked foil.

5.   DEVELOPMENT OF A NUMERICAL EVALUATION PROCESS

A serious disadvantage that photoelastic measuring methods have is the considerable input of personal time and effort for interpreting the photography.
To circumvent this shortcoming, interpretation procedures have been developed
in many places with the aid of digital computers, e.g. (ref. 24), which are intended for the assumption of time consuming manual evaluation tasks.

General requirements are:
- simple operation, minimum of input data,
- consideration of the differing material properties of model material and foil
  material,
- no time-consuming complementary measures, so that with consideration for
  creep, shorter measuring times can be achieved,
- interpretation of individual, particulary distinctive areas,
- fixing a matrix, with consideration for minor crack development and
- consideration of the non-linear, respectively actual material behaviour.

Taking these requirements as a basis, Ringkamp (ref. 20) develops a numerical interpretation program based on a generalised integration method with due
consideration for non-linear material laws for concrete, respectively the model
material. Along orthogonal intersecting lines the balance of all effective forces is formed on the concrete-foil bonding element (see Fig. 8).

With due consideration to all side and transitional conditions, the required
extension and stress states are iteratively calculated from the measured isochromatic patterns and isoclines. Here the non-linear material behaviour is
also fed in, whereby it is possible to choose between two applications, the
anisotropic of Link (ref. 25) or the isotropic of Grünberg (ref. 26), adapted

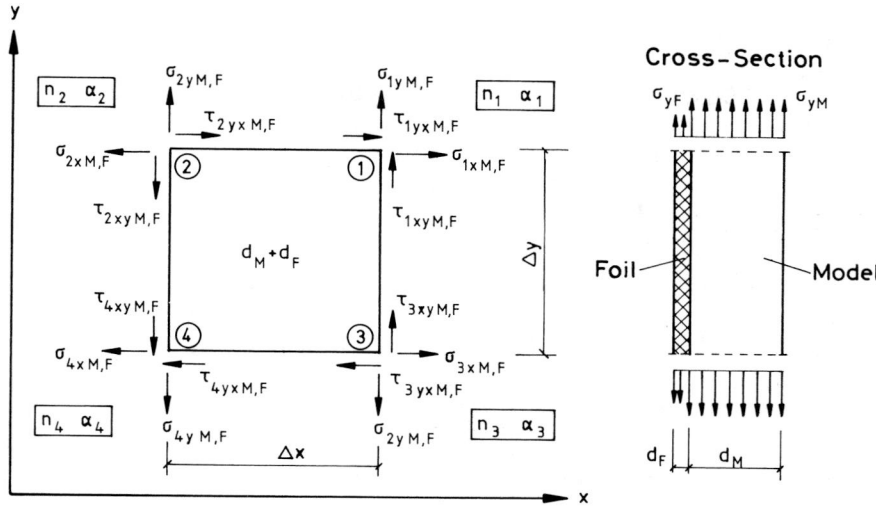

Fig. 8. Balance on the element for the integration process.

to the real stress-extension line in the compression and tension range.

After establishing a matrix and input of the geometry including the crack pattern for the area to be interpreted, isochromatic and isoclinal values are fed into a large computer. Basing on known boundary values, the strains, displacements, normal stresses, shear and principal stresses are determined according to dimension and direction, by staged iteration of all matrix dots and can also be called up for graphic display. As the matrix dot spacing and the area for interpretation can be freely selected interpretation is possible over almost the whole area, at least for some particulary interesting zones. By changing the direction of iteration the areas between or "under" the cracks can also be interpreted (see Fig. 9).

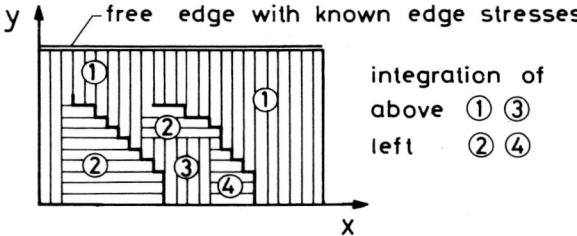

Fig. 9. Iterative Calculation of the internal forces from the photoelastic measurements.

6. PHOTOELASTIC INVESTIGATIONS ON REINFORCED CONCRETE MODELS

In conclusion, the capabilities of the improved reflection methods are demonstrated with two model tests. The measurements in Fig. 10 were selected in
order to be able to compare individual measurement results with those from an
actual size experiment, here from (ref. 27). The model beams are covered on the

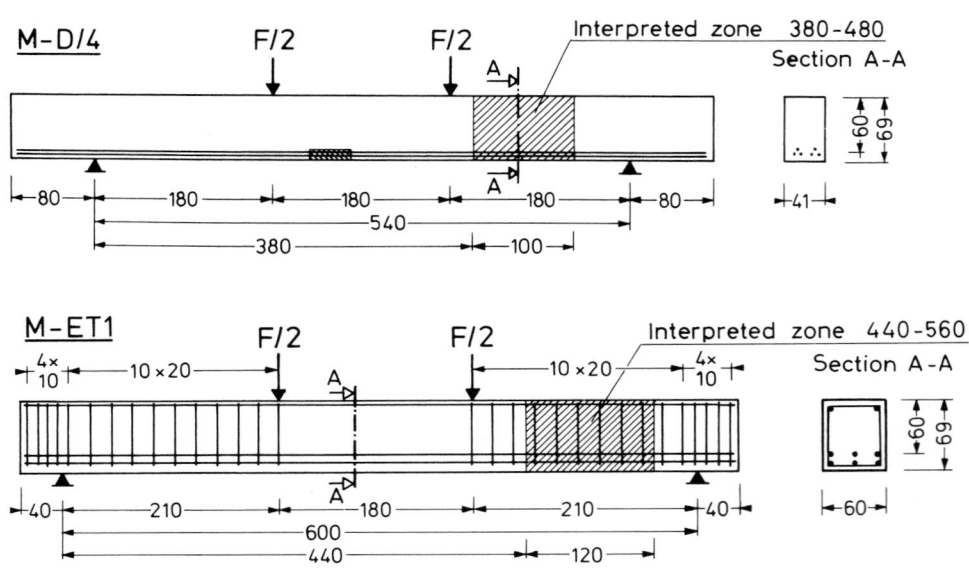

Fig. 10. Dimensions of two models (mm) M-D/4 and M-ET/1.

Test	Foil	Glue	Foil-Thickness	Foil-Properties	
				E-Modulus	Optical-Stress-Constants
			mm	N/mm^2	mm/Order
M-D/4	Araldite B-SPR 7	PC 1	3.08	4000	$5.86 \cdot 10^{-3}$
	PS 1	PC 1	0.29	2000	$3.91 \cdot 10^{-3}$
M-ET1	Araldite B-SPR 7	PC 1	3.40	4000	$5.33 \cdot 10^{-3}$

Table 3. Foils and glues used in the beam tests (PS1, PC1: firm Vishay GmbH,
München; B-SPR 7: manufactured of Institut for Massivbau, University Hannover)

facing side with a brittle foil (Araldite B - SPR 7) and beam M-D/4 on the rear
side with an elastic foil (PS1). The data for foils used are contained in
Table 3.

The model loads also result analogous to the actual size experiments in stages, with specific interim reduction of load.

The crack pattern follows as a direct result from the photoelastic photographs with elastic foil (see Fig. 11).

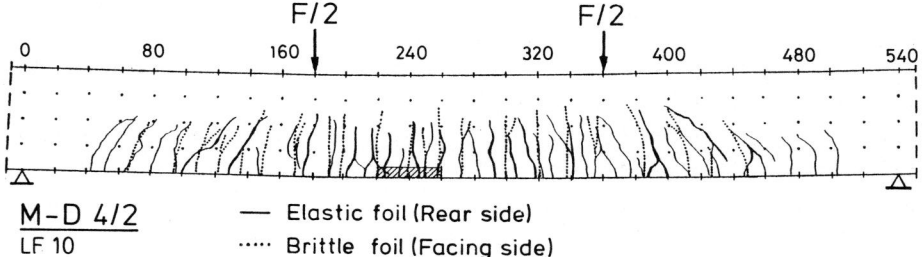

M-D 4/2   —— Elastic foil (Rear side)
LF 10   ····· Brittle foil (Facing side)

Fig. 11. Crack development of model beam without stirrup reinforcement.

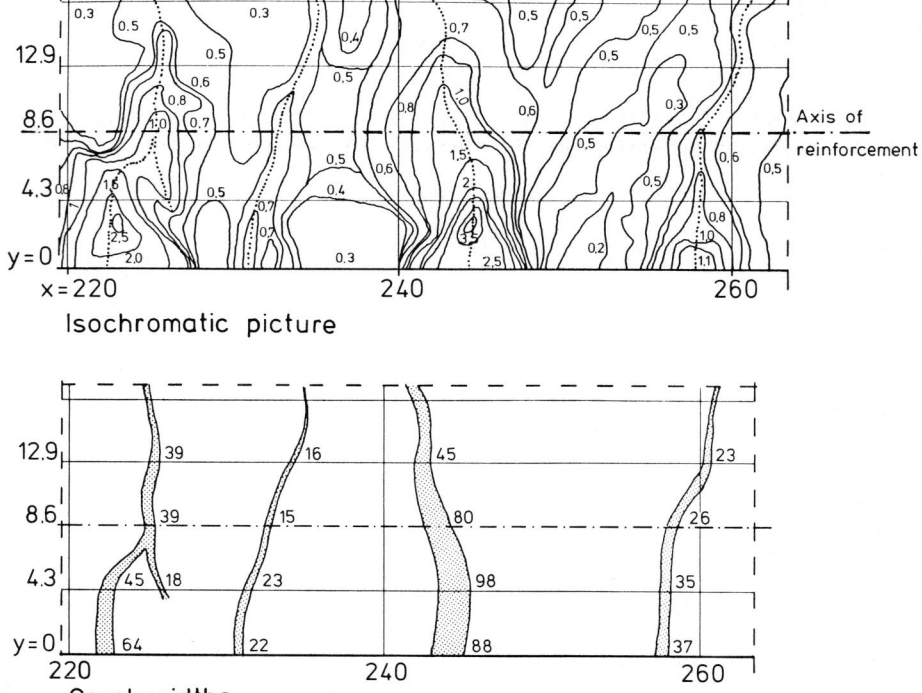

Isochromatic picture

Crack widths

Fig. 12. Determination of crack widths (1/1000 mm) on the basis of the isochromatic pattern.

Bases on the isochromatic photograph the course of the cracks and crack widths are given as contours for a section in the beam area, with constant external bending moment (Fig. 12).

The sum of the crack widths related to the abscissa difference of 260-220 = 40 mm in the range of the reinforcement insertions agreed to within 5 % with the strain measurements on the reinforcing inserts. Further, it arises that in the case of correspondingly careful interpretation, crack widths of less than 1/100 mm can, with satisfactory accuracy of measurement, still be recognized.

Fig. 11 shows the complete crack pattern. This includes not only the main cracks, those which can be established with a normal crack magnifier, even intermediate cracks are also clearly visible.

The following illustrations (Figs.13 to 16) show the interpretation for zone 380 to 480. The isochromatic patterns (Fig.13) are fed dot-wise into the computer for the numerical interpretation. The strains resulting in the longitudinal direction of the beam (Fig. 14) are checked through strain gauges on the beam upper side or on the reinforcements. The values coincide very well. Further, not only from the photoelasticity but also on the basis of the strain measurements the same position is given for the neutral line (stress-free line), whereby the cracks, the tips of which lie in the area of the stress-free line can be followed dependably.

On the under side of the beam there are as expected only smaller strains because of the cracks, apart from the value at abscissa 480 which however also agreed very well with the checking measurements. The stresses σ_x (Fig. 15) show the anticipated course. With increasing strain values towards the beam centre

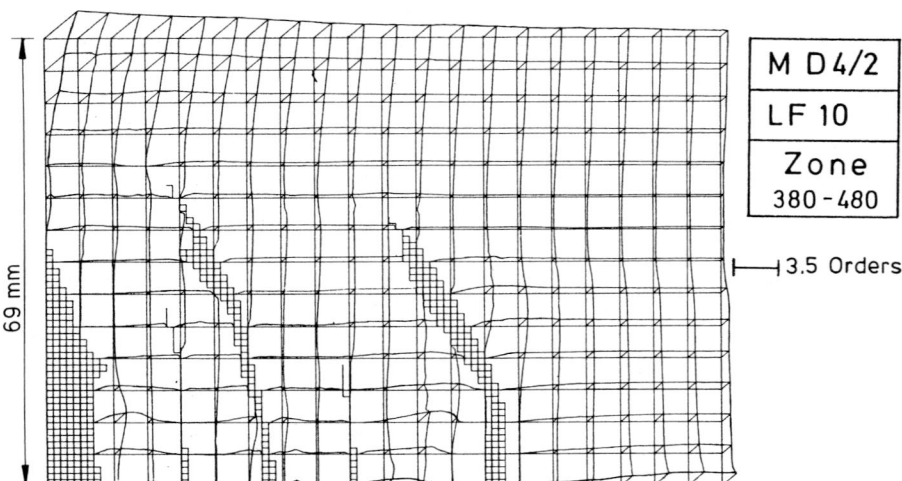

Fig. 13. Isochromatic pattern for load case 10 in sector 380 to 480 of model test M-D/4 (F = 11.1 kN).

(from section 480 to 380), the stresses increase and furthermore, are distinctly non-linear in the individual sections. Further, as expected, the stresses in the X-direction below the neutral line almost equate to zero, i.e. the effect of tensile strength remains a subordinated order of magnitude.

In the case of shear stresses (Fig. 16) the unexpected distribution across the cross-section height is obvious. In the compression zone however, the maximum values in the case of reinforced beams without stirrup reinforcement occur

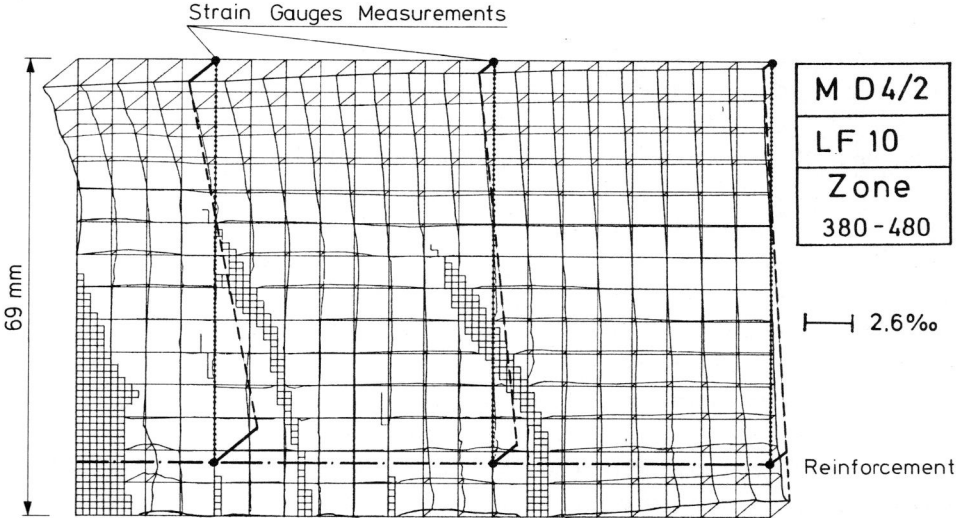

Fig. 14. Strains ε_x in the beam longitudinal direction (F = 11,1 kN).

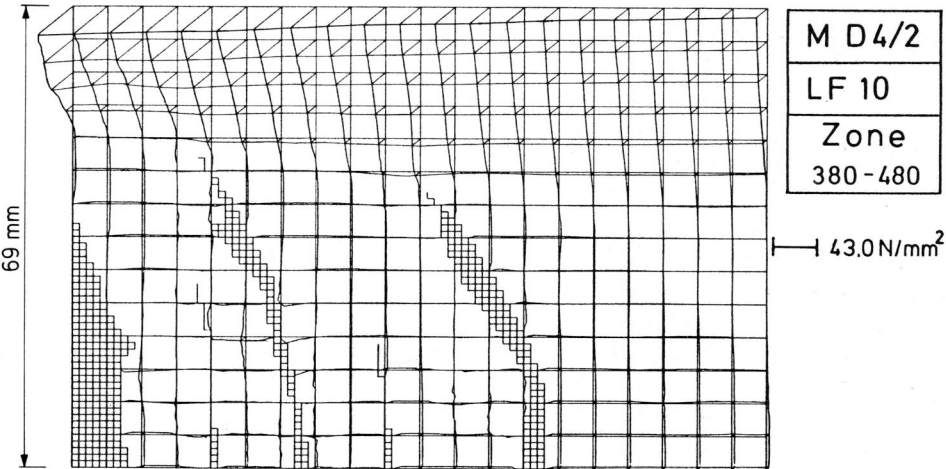

Fig. 15. Stresses σ_x in the beam longitudinal direction (F = 11,1 kN).

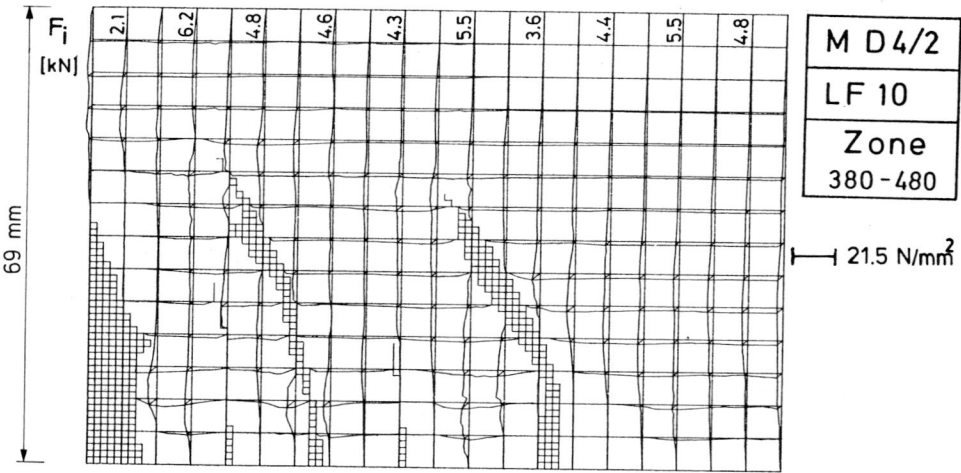

$F_e = 5.6$ [kN]

Fig. 16. Shear stresses τ_{xy} (F = 11,1 kN).

in the lower area and particularly in the vicinity of cracks. From this, considerable crack serration forces and dowelling forces can be derived due to an
absence of shear reinforcement, which were also established in other institutes,
(ref. 19).

Fig. 17 shows that the distribution of shear stresses is greatly influenced
by stirrup reinforcement. Here the model for beam ET 1 (ref. 27) is dealt with.

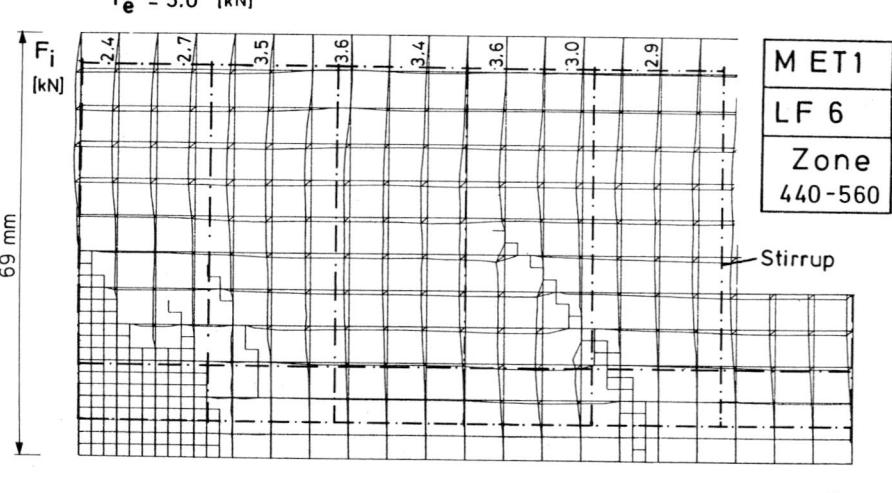

$F_e = 3.0$ [kN]

Fig. 17. Shear stresses for load case 6 in sector 440 to 560 of the model test
M-ET/1 (F = 6,0 kN).

In the cracked area of the beam the maxima lie above the stress-free line. To-wards the support, i.e. in the as yet uncracked area, the maximum values shift towards the beam axis. Accordingly, in the case of stirrup-reinforced concrete beams the crack serration is of negligible magnitude. As a test of stability, the integral was determined over the shear stresses (F_i) in several sections. It agreed completely with the external shear force (F_e) to within 10 %.

The principal stresses show a very even course (Fig. 18) not only in terms of dimension but also in terms of their direction. The stress redistributions in the shear reinforced cross-section are insignificant. Below the cracks, a well developed brace is positioned which clearly supports itself on the longi-tudinal reinforcement and is suspended from above.

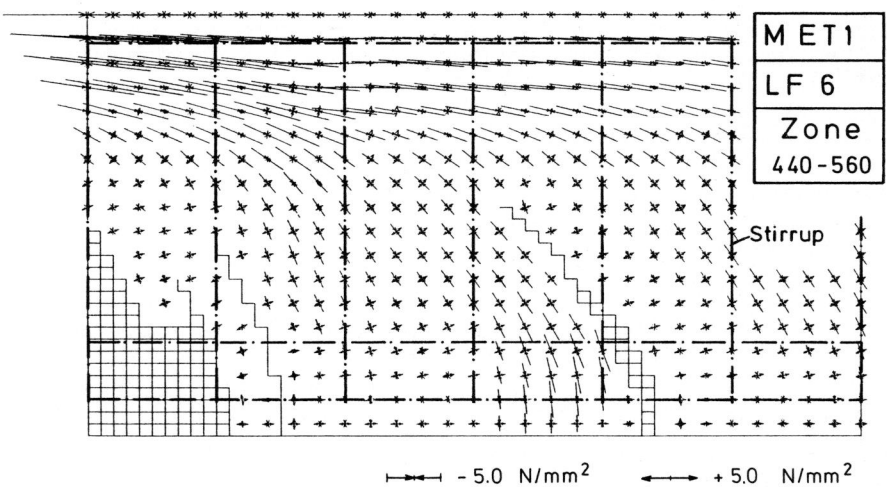

M ET1
LF 6
Zone
440-560

Stirrup

$\longmapsto$ - 5.0 N/mm^2 $\longleftrightarrow$ + 5.0 N/mm^2

Fig. 18. Principal stresses according to dimension and direction (F = 6,0 kN).

7.   CONCLUSION

With the reflection method a photoelastically active foil is glued to the structural unit or model and the light rays are reflected. In this way, in-vestigations can be carried out as required on materials which in themselves are nontransparent ans photoelastically inactive. With this method, therefore, the actual bearing capacity of reinforced concrete structural elements can be followed and analysed in all phases of load up to failure. Through the use of foils with higher photoelastic sensitivity, cracks can be detected early, i.e. even at the time of origin and crack widths of less than 1/100 mm can still be clearly detected. This method therefore allows crack investigations even into the sphere of microcracks.

Through the use of foil, reliable strain measurements are also possible in the cracked area. Further, the use of the reflection method in model simulations provides a satisfactory harmony with reality (actual size experiments), when the measured values resulting from the stiffening and thickening effects can be corrected.

Finally, it is emphasized that the method of investigating reinforced concrete structural elements in models has particular economical advantages.

8.   REFERENCES

1   Frocht, M.M.: Photoelasticity. 1. Bd., 3. Druck, 1949/2. Bd., 1948, New York, J. Wiley.

2   Wolf, H.: Spannungsoptik, Band 1. Berlin/Heidelberg/New York 1976.

3   Kuhn, R.: Bemessung einer Stahlbetonkonstruktion mit Hilfe der Spannungsoptik. Der Bauingenieur 26 (1951) 177 - 181, 205 - 207.

4   Zellerer, E., Thiel, H.: Über das Kraftfeld einer Stahlbetonwand mit einer Türöffnung unter unsymmetrisch wirkender Einzellast. Beton- und Stahlbetonbau 51 (1956) 267 - 274.

5   Frühauf, H.: Beitrag zur Ermittlung der Bewehrung von Stahlbetonscheiben mit einem rechteckigen Loch. Dissertation, Stuttgart 1968.

6   Niedenhoff, H.: Untersuchungen über das Tragverhalten von Konsolen und kurzen Kragarmen. Dissertation, Karlsruhe 1961.

7   Boiten, R.G.: Photoelastic investigation of armoured models with an application to bending bars with cracks on the tensile side. Appl. sci. Res. Section A, Vol. 5 (1955) 359 - 373.

8   Hiltscher, R., Müller, R.K.: Bemessung der Bewehrung von Stahlbetonkonstruktionen mit Hilfe des spannungsoptischen Modellversuchs. Beton- und Stahlbetonbau 54 (1959) 263 - 271.

9   Müller, R.K.: Ein Beitrag zur spannungsoptischen Untersuchung von bewehrten Balkenmodellen. Dissertation, Darmstadt 1960.

10 Steffens, K.: Spannungsoptische Modelluntersuchung von Stahlbetonbauteilen, Düsseldorf 1974.

11 Mesnager, M.: Sur la dêtermination optique des tensions intêrieurs dans des solides à trois dimensions. Comptes Rendus, Paris 190 (1930), 1249 - 1250.

12 Dantu, P.: Etude des contraintes dans les milieux hêtêrogênes application au bêton. Ann. de l'Institut Technique du Batiment et des Travaux Publics 11 (1958) No. 121, 55 - 77.

13 Oppel, G.: Application of photoelasticity to the investigations of the properties of concrete. Problemy Prochnosti w Mashinostrojenij 8 (1962) 44 - 51.

14 Swamy, R.N.: Application of photoelastic coating techniques to the determination of internal strain distribution in cementitious materials. Proceedings of the 4. Intern. Conf. on Experimental Stress Analysis 1970, Paper 36, 58 - 67.

15 Reitblat, Z.V., Baikov, V.V.: The photoelastic coating method and its application in investigations of steel-reinforced concrete structures. Trudy. Nauchno-issledovatel'skij institut poligraficheskogo mashinostroeniga, No. 4, Moskau 1965.

16 Cubaud, J.C., Jullien, J.F., Lemaire, M.: Application de la photoélastici-
    métrie à l'étude du béton armé soumis à la flexion. Revue Française de
    Mécanique No. 37 (1971) 77 - 86.

17 Khesin, G.L., Sakharov, V.N.: Methods of Strain Measurement on the Surface
    of Concrete and Reinforced Concrete Constructions by Means of Photoelastic
    Coatings. Proceedings of the 4. Intern. Conf. on Experimental Stress Ana-
    lysis, 1970, Paper 20, 47 - 57.

18 Jullien, J.F., Cubaud, J.C.: Sur la détermination expérimentale de la dis-
    stribution des contraintes dans la zone fissurée d'une structure en béton
    armé au moyen de vernis photoélastiques fragiles. Comptes Rendus de l'Acad.
    Sciences, Paris 274 (1972), A, 1181 - 1184.

19 Bosc, J.L., Pera, J., e.a.: Utilisation de Revetements photoélastiques pour
    l'Etude du cisaillement dans les poutres en béton armé. RILEM-Symposium
    'Testing in situ of concrete structures' Budapest 1977, Part II, 216 - 228.

20 Ringkamp, M.: Modellstatisches Verfahren zur Untersuchung von Stahlbeton-
    bauteilen mit der spannungsoptischen Auflichtmethode. Düsseldorf 1982.

21 Bignell, V.F., Smalley, V., Roberts, N.P.: A new photoelastic material for
    use in problems concerning reinforced concrete. Magazine of Concrete Re-
    search 15 (1963) Nr. 45, 171 - 176.

22 Bieger, K.W., Reich, E.: Systematische Untersuchung der mechanischen Eigen-
    schaften eines Kunstharzmörtels für die Nachbildung des Werkstoffes Beton
    bei Modellversuchen für Stahlbetontragwerke.Forschungsbericht Nr. 7839,
    Institut für Massivbau, Universität Hannover, 1978.

23 Duffy, J.: Effects of the Thickness of Birefringent Coatings. Experimental
    Mechanics 1 (1961) 74 - 82.

24 Müller, R.K., Saackel, L.R.: Complete Automatic Analysis of photoelastic
    Fringes. Experimental Mechanics 19 (1979) 245 - 251.

25 Link, J.: Eine Formulierung des zweiachsialen Verformungs- und Bruchver-
    haltens von Beton und deren Anwendung auf die wirklichkeitsnahe Berechnung
    von Stahlbetonplatten. DAfStb, Heft 270, Berlin 1976.

26 Grünberg, J.: Berechnung von ebenen Stahlbetonflächentragwerken im geris-
    senen Zustand mit der Methode der finiten Elemente. Düsseldorf 1973.

27 Leonhardt, F., Walther, R.: Schubversuche an einfeldrigen Stahlbetonbalken
    mit und ohne Schubbewehrung. DAfStb, Heft 151, Berlin 1962.

Fracture Mechanics of Concrete,
edited by F.H. Wittmann, 1983
Elsevier Science Publishers B.V., Amsterdam — Printed in The Netherlands

Chapter 3.4

SPECKLE METROLOGY AND HOLOGRAPHIC INTERFEROMETRY APPLIED TO THE STUDY
OF CRACKS IN CONCRETE

P. JACQUOT and P.K. RASTOGI

1 INTRODUCTION

Concrete is by far the material most solicited in construction work. It
has been called upon to withstand loads in the most adverse of circumstances.
Yet, somehow, its fracture mechanics behaviour is the least understood. A better
understanding of concrete, its response to loads and phenomenon leading to its
fracture, is thus needed to save the spectre of a catastrophic failure and to
avoid the demolition and reconstruction of an important work.

Microcracking has been recognized to be one of the important phenomenon in
the failure of the concrete. This aspect has been treated in the other chapters
and is shown to adversely affect the integrity of the structures. This high-
lights the necessity to visualize the process of crack initiation and development at both macro and micro levels. To fit this need, a method has to be
capable to identify cracks on the concrete and monitor the displacement fields
across the cracks and the regions surrounding it; and further to bring the
identified cracks under minute inspection permitting to postulate on their
growth and other geometrical characteristics.

Holographic and speckle interferometry are well suited to the study of the
mechanisms of crack formation and propagation. The techniques are non invasive
and permit a cartographic visualization of the displacement fields over the object surface. This visualization is achieved in the form of fringes translatable
in terms of displacements/deformations in accordance to the optical system employed for the study. The specific characteristics of these fringes in crack
analysis will be considered in Section 2.

The holographic and speckle interferometry provide one with global information : maps on displacement fields and cracks. For a local and detailed
examination of the detected cracks, it is desirable to combine an interferometric
system with microscopic observation - thereby permitting one to exploit simultaneously the data on the displacement fields provided by the interferometer as
well the high resolution capability furnished by the optical imaging system.
Holographic microscope is particularly suitable as it enables high resolution
and significant object depth recordings. Alternatively, the displacement fields
in the vicinity of the cracks can be investigated using holographic interfer-

114

ometry microscopy.

All these techniques when used conjointly make holography and speckle methods a highly versatile tool in fracture mechanics. However, the powerful potential that is inherent in these techniques in the study of the crack process and its growth in concrete has gone largely unexploited. This is explained in part by the relative newness of these techniques, the time stretch needed for their theoretical evaluation followed by a period devoted to the development of adapted equipments. Hence, the aim here is to transmit a simple yet rigorous picture of the mechanisms underlying the holography and speckle techniques. Different optical set-ups capable of being employed in crack analysis in concrete are examined and the significance of fringes in each case is explained. Few important applications are briefly reviewed, mostly in way of referencing to the concerned works. Finally holographic microscopy is dealt in detail and its possibilities are outlined.

2 PRINCIPAL CHARACTERISTICS OF THE INTERFEROMETRIC METHODS IN CRACK ANALYSIS

A crack does not manifest itself only by the appearance of tiny openings on the considered surface. It distinguishes itself also by discontinuities in a displacement field. A crack results due to the combination of three types of discontinuities :
- Mode I   (opening mode) : the internal surfaces of the crack move perpendicularly to each other.
- Mode II   (edge sliding mode) : the surfaces move in the same plane and in a direction parallel to the crack.
- Mode III (tearing mode) : the surfaces move in the same plane and in a direction perpendicular to the crack (see Fig.1 (a) and (b)).

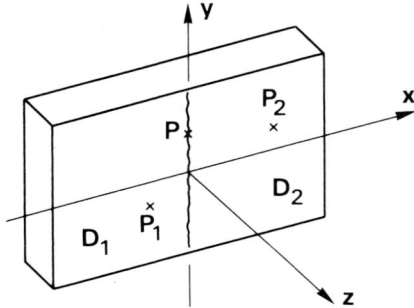

Fig.1(a) - A rectilinear portion of the crack coinciding with the y-axis is envisaged. The displacement field on the two sides of the crack (region D_1 and D_2) is defined by $\vec{d}_1(P_1) = (u_1,v_1,w_1)$ and $\vec{d}_2(P_2) = (u_2,v_2,w_2)$.

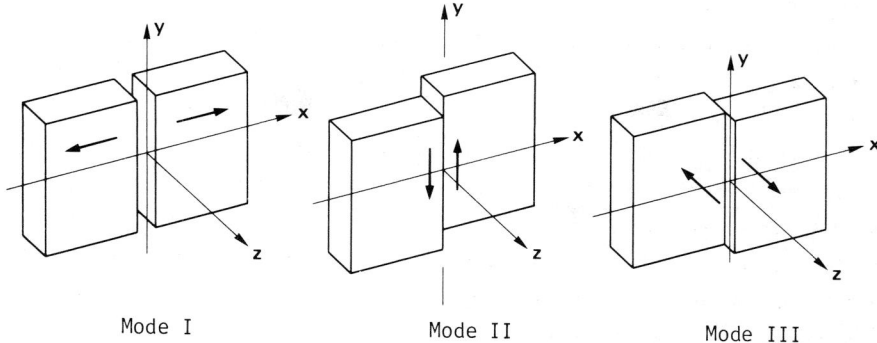

Mode I   Mode II   Mode III

Fig.1(b) - Illustration of the three modes of crack formation.

The interferometric methods furnish, in the form of interference fringes, the lines of iso-displacement (u, v or w) - with a sensitivity lying in the range of a fraction of wavelength - or their derivatives. When applied to crack detection (ref. 1 - 5), these methods permit to visualize the crack paths by the discontinuities present in the fringe pattern : the crack passing through those points where a fringe breaks off bluntly (discontinuity in the fringe order) or presents an angular bent (discontinuity in the fringe tangent).

The principal characteristics of the interferometric methods can be summarized as follows :

(i) The trace of the crack is visualized in a discrete way. To interpolate its path between two consecutive discontinuities reduces to the problem of fringe interpolation. To determine the fractional fringe order, three types of solutions are at our disposal :
  - a simple or heterodyning photoelectric detection;
  - the recourse to the multi-pass interferometry or the fringe multiplication technique;
  - the addition of a known and appropriate displacement field to densify the obtained interferograms.

(ii) The aspect of discontinuity in the fringe pattern depends on :
  - the mode of crack formation;
  - the behaviour of the displacement component on the two sides of the crack front;
  - the choice of the interferometric method in function of its sensitivity to an unique component of the displacement field.

As indicates the Fig.2, the discontinuity in the fringe pattern is not directly apparent in the case of symmetric displacements at the crack edges,

116

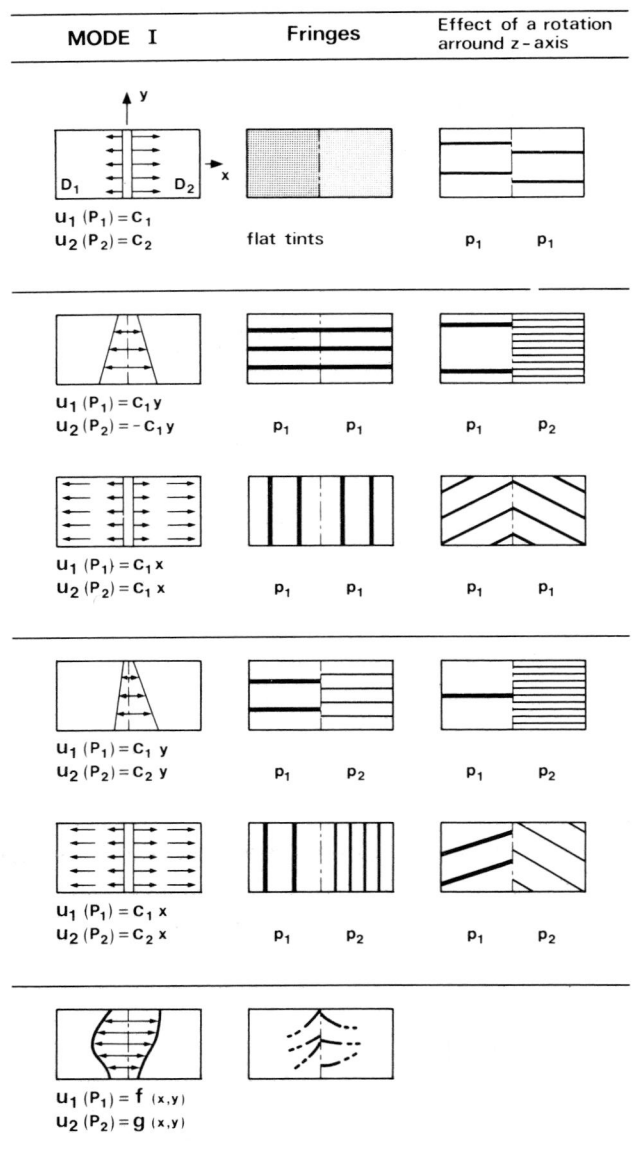

Fig.2 - Discontinuity aspects in the fringe pattern for the mode I and different behaviour of the component u undergoing the preponderant discontinuity.

as certain devices (for example, compact tension specimen) tend to pre-
scribe. In this hypothesis, the superposition of an appropriate rigid body
rotation to the model deformation is sufficient in the majority of the

cases to make visible the discontinuity. The envisaged rotation is carried
out by slightly modifying the beam inclination in the reference arm of the
interferometer.

In Fig.2, the fringes are supposed to be solely sensitive to the u displacement component, or in other words the component undergoing the preponderant jump in the Mode I of crack formation. It is not excluded that
the jump on the other components - smaller by definition - be more
interesting to visualize. Sometimes, in effect, the variations and the
discontinuities of the secondary displacement components are better
adapted than the predominant component in the range and the sensitivity of
the interferometric methods.

As we will see later on, certain interferometric methods are simultaneously sensitive to several components of displacement with a selective
coefficient of sensitivity. At the detection stage of the crack, this
property, well exploited, constitutes an advantage to the extent where it
permits to put in glaring evidence the discontinuity in the fringes, in
order and in inclination. As to the interpretation, on the contrary, it
complicates the analysis.

(iii) The output of an interferometric system consists of an image modulated by
the interference fringes. Whereas the image can be exploited to extract
the resolution information permitting a detailed examination of cracks and
its surrounding microstructure, the modulating interference fringes provide the visualization of crack paths by way of fringe discontinuities.
Though the resolution conditions are quite relaxed in the exploitation of
fringes, they are considerably demanding in the exploitation of the image.
This is due to the fact that in the former case one is concerned with resolving the interference fringes of low spatial frequencies as compared
to crack itself in the latter case. Consequently, the fringe patterns do
not provide an ultimate precision in determining the position and the
geometry of the cracks. On the contrary, they present the unique feature
of permitting the detection and the instantaneous localization of cracks,
few microns wide, over the fields ranging from 1 dm^2 to 1 m^2. Thus a complete information on the crack geometry and its location is obtained by
using a high resolution interferometric system combined to microscopic
observation.

(iv) The last basic characteristic hinges on the possibility of measuring the
deformations in the neighborhood of the crack. This point is capital and
clearly responds to the longfelt need formulated in fracture mechanics.

From fringe parameters, one can deduce - at least partially - the state of strain and the redistribution of stresses in the region around the crack or even further away. Recent studies have put in evidence the interest of possessing this information much ahead of the zone where the crack incarnates on the surface (ref. 6).

The classical interferometric methods will not be evoked elsewhere than this Section. They necessitate a special surface preparation (depositing a reflection coating of excellent planity), operation which is avoided in speckle metrology and holographic interferometry.

Finally, the concrete depolarises nearly totally an incident linearly polarised vibration. For the interference conditions to be satisfied, necessitates the examination of a single direction of polarisation. Though not figuring explicitly in the proposed optical set-ups, a linear polariser is always interposed between the object and the receptor. The outcoming beam from the laser is supposed to be linearly polarised.

3   CHARACTERISTICS PARTICULAR TO SPECKLE METROLOGY AND HOLOGRAPHIC INTERFEROMETRY

In laser light, the natural roughness of the object surface is at the origin of the speckle phenomenon - random granulation accompanying the image of the surface or its diffraction pattern. This phenomenon is due to the multiple interference of the wavelets issuing from a random division, by the object surface, of the incident beam of illumination. Its statistical properties, well known (ref. 7-8), are of interest in measurement applications.

The speckle amplitude fluctuates in magnitude and in phase respectively around a zero mean and equiprobably in the interval $[-\pi,\pi]$. The amplitude is correlated over a small volume in space of lateral $\langle \rho_\ell \rangle$ and longitudinal dimensions $\langle \rho_L \rangle$:

$$\langle \rho_\ell \rangle = \frac{\lambda D}{\emptyset} \quad ; \quad \langle \rho_L \rangle = \frac{8\lambda D^2}{\emptyset^2} \tag{1}$$

where λ is the wavelength, D the distance object-observation plane or the distance lens-image plane depending on whether we observe the diffraction pattern or the image, $\emptyset$ the size of the object illuminated or alternatively the lens diameter. These represent the mean dimensions of the grains dotting the intensity of the speckle pattern.

Initially considered as a noise, this phenomenon has rapidly been recognized as being susceptible to carry the information on displacement-deformation of the object (ref. 9-11).

In effect, on the local scale and when the object is deformed (see Fig.3),

119

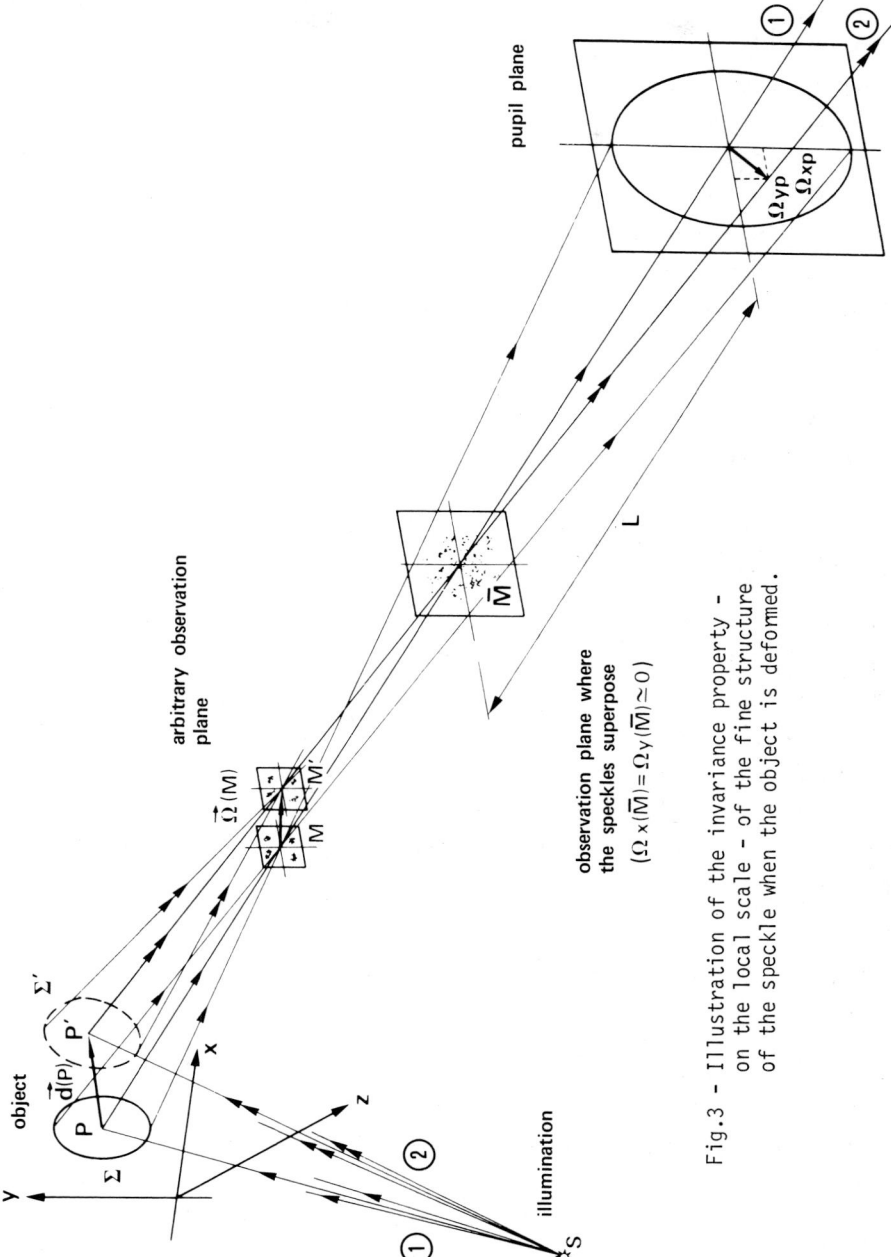

Fig.3 - Illustration of the invariance property -
on the local scale - of the fine structure
of the speckle when the object is deformed.

a remarquable property of the speckle amplitude is to conserve unchanged its fine distribution in the space, leaving aside a shift factor $\Omega(M)$ (ref. 12-16). The speckle moves in the space without modification to its structure, in function of the displacement-deformation parameters of the object surface and the illumination and the observation geometry. Schematically, these relations can be written in the form (ref. 17-18) :

$$
\begin{bmatrix} \Omega_x(M) \\ \Omega_y(M) \\ \Omega_z(M) \end{bmatrix} = [C_1] \begin{bmatrix} u(P) \\ v(P) \\ w(P) \end{bmatrix} + [C_2] \begin{bmatrix} \bar{R}_1(\bar{P}) \\ \bar{R}_2(\bar{P}) \\ \bar{R}_3(\bar{P}) \end{bmatrix} + [C_3] \begin{bmatrix} \varepsilon_{xx}(P) \\ \gamma_{xy}(P) \\ \varepsilon_{yy}(P) \end{bmatrix} \tag{2}
$$

where $(\Omega_x, \Omega_y, \Omega_z)$ are the components of speckle translation at the point M of observation; (u,v,w) the displacement components of the centre P of the object region which contributes to the illumination in M ; (R_1, R_2, R_3), the three local rotations of the surface element surrounding P ; $(\varepsilon_{xx}, \gamma_{xy}, \varepsilon_{yy})$, the in-plane strain coefficients at the point P. The matrices C_i [3×3] have for elements the terms which are function of the geometrical parameters of the recording system.

The displacement field attached to the object surface is coded by the speckle displacement field and the knowledge of one is sufficient to decode the other.

The access to the speckle displacement field is gained through specialized systems, which, if desired, permit to stock the information permanently.

Three types of procedures exist, which respectively, put in memory through a photosensitive support :
- the speckle intensity in a given plane (speckle photography);
- the intensity and phase of the speckle in a given plane (speckle interferometry);
- the amplitude and phase of the speckle in volume (holographic interferometry).

The observation and the direct measurement of the movements of isolated grains using an optical sighting system represents a further means to gather data on the speckle displacement field. Not belonging to the category of whole field methods, this technique is not considered here.

3.1   Speckle photography (ref. 9-11, 19-21)

The method requires exposing twice a photographic plate to the speckle intensities corresponding to the two states of deformation of the object surface. For visualizing the speckle displacement field, frozen in the photographic emulsion, two techniques are used.

(i)   The point-by-point filtering technique consists in forming the diffraction

pattern of a small region on the recorded plate (see Fig.4). This pattern
is composed of :

- a_diffraction_halo of diameter $\emptyset_H$

$$\emptyset_H = 2 \frac{\lambda_r}{\lambda_e} \frac{d_r}{F(1+g)} \qquad (3)$$

with λ_r and λ_e : recording and the reconstruction wavelength;
      d_r : distance between the plate and the observation screen;
      F and g   : aperture number and the magnification employed.
The photometric profile of the mean intensity in the halo assumes the
form of the autocorrelation of the square of the pupil function limiting
the lens aperture.

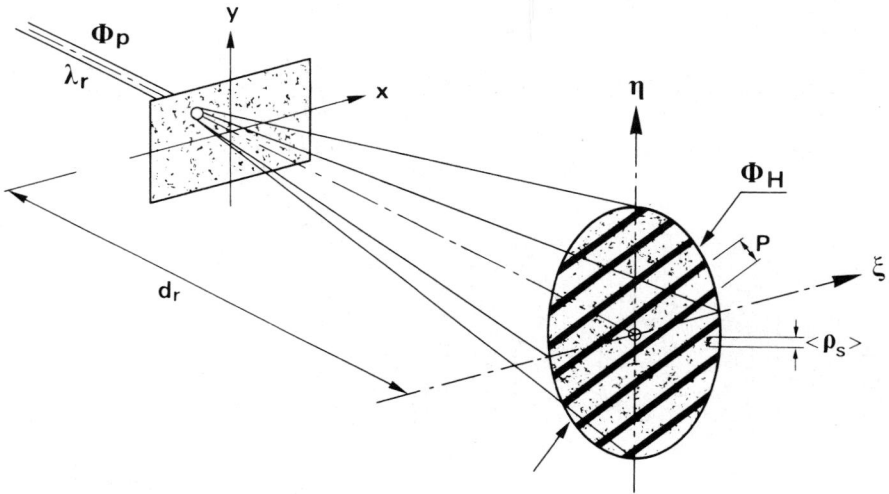

Fig.4 - Point-by-point filtering set-up.

- Young's_fringes whose direction is perpendicular to the speckle displacement vector in the illuminated region and whose period p is related
to the magnitude of this displacement by the relation

$$p = \frac{\lambda_r d_r}{(\Omega_x^2 + \Omega_y^2)^{\frac{1}{2}}} \qquad (4)$$

If the imaging lens is supposed to be free of aberrations and the displacement of speckles in the pupil plane small compared to its size, the
fringes are strictly parallel, equidistant and fill completely the halo.

In presence of strong aberrations, the Young's fringes can be distorted (ref. 22-23).

- Secondary speckles which serve as noise, their mean dimension being given by

$$<\rho_s> = \lambda_r d_r / \emptyset_P \tag{5}$$

where $\emptyset_P$ represents the diameter of the illuminated region.

The diameter of the halo and the mean size of the secondary speckle impose limits to the speckle displacement fields finally possible to measure :

$$<\rho_s> \; < p < \; \emptyset_H \tag{6}$$

Lately, numerous techniques for automatic analysis of the halo fringes have been proposed (ref. 24-26).

(ii) The whole field filtering technique consists in selecting a narrow band of spatial frequencies recorded on the plate in placing a small aperture in the transform plane in a Fourier filtering set-up (see Fig.5). The fil-

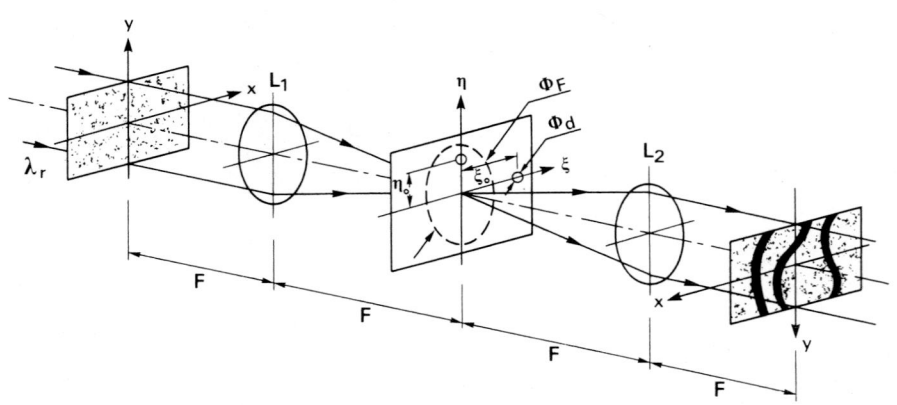

Fig.5 - Whole-field filtering set-up.

tered image is covered by fringes of equal displacement in the plane of the photographic plate, in a direction and with a sensitivity related to the position of the aperture :

$$- \underline{fringes} : \quad \Omega_x(x,y) = \frac{n \lambda_r f}{\xi_0} \tag{7}$$

if the aperture is placed at a distance ξ_0 from the optical axis in a direction parallel to and along Ox ;

- fringes : $\Omega_y(x,y) = \dfrac{n \lambda_r f}{n_0}$ (8)

if the aperture is in a direction parallel to and along Oy ; λ_r and f are the reconstruction wavelength and the focal length of the lens; n is an integer representing the fringe order.

The extent of the diffraction pattern, $\emptyset_F$, and the presence of a secondary speckle, $<\rho_s>$, limit, here also, the range of measurements :

$$n_0 \text{ and } \xi_0 \leqslant \frac{\emptyset_F}{2} = \frac{\lambda_r f}{\lambda_e F(1+g)}$$

$$p \geqslant <\rho_s> = \frac{\lambda_r r}{\emptyset_d}$$

(9)

where $\emptyset_d$ is the aperture diameter and p the distance between two consecutive fringes. Here, equally, important aberrations of the lens and the shift between speckles in the pupil plane during the recording affect the fringes in the filtered image.

Whatever the mode of visualization chosen, the relation between speckle displacement and object displacement and deformation depends on the optical set-up used at the recording. Certain specific set-ups which simplify considerably the relation (2) will be described in the following Sections.

Lastly, one final limitation owes its allegiance to the invariance property of the speckle. Since the speckles extend in volume, an imperative condition to fulfil is that the longitudinal displacement of the speckle remains smaller than its longitudinal size

$$\Omega_z \ll <\rho_L>$$ (10)

3.2   Speckle interferometry (ref. 9-11, 27-28)

A photographic plate records the interference between a speckle originating from the object (referred to object speckle hereafter) and a "witness" speckle. This second speckle plays the role of a reference and can be both affected or not by the object deformation. The speckle resulting from the interference codes and memorizes in a plane the intensity and the phase of the object speckle relative to the reference speckle.

If one takes the precaution to choose the size of each of these speckles much smaller than their displacement components, when the object is deformed, then, contrary to the previous case, the lateral movement will not be perceived, the speckles remaining coincided :

$$(\Omega_{x0}^2 + \Omega_{y0}^2)^{\frac{1}{2}} << <\rho_{\ell 0}> \quad ; \quad \Omega_{z0} << <\rho_{L0}>$$

$$(\Omega_{xr}^2 + \Omega_{yr}^2)^{\frac{1}{2}} << <\rho_{\ell r}> \quad ; \quad \Omega_{zr} << <\rho_{Lr}> \tag{11}$$

the suffixes 0 and r refer to the object speckle and the reference speckle respectively. These conditions are met by reducing the imaging lens aperture (see
equation (1)).

On the contrary, the intensity of a given speckle evolves in function of
the variation of the phase difference between the two speckles, provoked by the
object deformation. In those regions of the photographic plate where the variation of the path difference between the object and the reference speckle is an
integral multiple of the wavelength, the resulting final speckle matches exactly
with the resulting initial speckle. The resulting initial and final speckles add
coherently and thus have features characteristic of a high contrast speckle. For
a variation of half the integral wavelength, the two interfering speckles are
out of phase. In this case, the resulting initial and final speckles combine in-
coherently. In consequence the corresponding regions are covered by low contrast
speckles.

In other words, supposing that Eq.(11) is satisfied, it becomes possible
to visualize the fringes

$$\Omega_{zr}(x,y) - \Omega_{z0}(x,y) = n\lambda$$
$$\text{or}$$
$$= \frac{(2n+1)\lambda}{2} \tag{12}$$

where Ω_{zr} and Ω_{z0} are the longitudinal displacement components, induced by the
object deformation, of the object speckle and the reference speckle respectively. These two components can be calculated using Eq.(2).

It is evident that the methods described in Section 3.1, can no longer be
applied to visualize these fringes; they leave place to two other techniques.

(i)   The double exposure technique superposes on the same plate, before devel-
      opment, the two resulting speckles, initial and final. The regions under-
      going an integral variation in the path difference have the features
      characteristic to a granular pattern of theoretical contrast unity,
      whereas the regions undergoing the variation of half the integral wave-
      length possess a more uniform intensity distribution having a theoretical
      contrast of $1/\sqrt{2}$ (ref. 8).

(ii) In the mask technique one observes the resulting final speckle field
      through the repositioned negative of the resulting initial speckle field.

The regions undergoing integral variations appear in the form of dark fringes dotted with few bright grains whereas the regions undergoing half integral variations conserve their typical speckle features.

In the two cases, different photographic procedures permit to accentuate the visual difference between these two regions. The mask technique is intrinsically superior to the double exposure. As in the case of speckle photography, the speckles necessarily present in the final interferogram limit by their size the spatial resolution of these methods. Conjointly to Eq.(11), it is necessary that the fringe periodicity in the whole field pattern does not go below the lateral dimensions of the speckle in the image. In speckle interferometry, it is difficult to predict the photometric profile of the fringes with any good precision as the non linear processes frequently employed for obtaining the fringes are generally not well controlled. As a result, all attempts at fringe interpolation while using classical microdensitometric techniques are prone to large errors.

Often, the reference speckle has the same origin as the object speckle. They distinguish uniquely by the distinct optical paths (frequently symmetric) they traverse in their propagation in space : illumination source - object surface - receptor. The principal set-ups which aim at rendering the path difference $\Omega_{zr}-\Omega_{z0}$ exclusively sensitive to certain specific components of the object deformations will be examined in Section 5.

3.3   Holographic Interferometry (ref. 16, 29-30)

The method compares in amplitude, in phase and in volume, the two speckle fields associated to the two states of object deformation, by means of an holographic recording.

The process of holographic recording and reconstruction being well known (ref. 31-32), it suffices to recall here that
- the recording consists of superposing on a high resolution photographic plate a reference wave (plane or spherical to be easily reproductible) and the wave diffracted by the object in the direction of the plate. Due to the interference, the plate records a complicated and dense system of micro-fringes (modulation);
- the system of micro-fringes, when illuminated by a wave identical to the reference wave, diffracts the incident light so to recreate the original object wave - the same as present during the recording (demodulation).

The only working notion we will retain from the holography is that it provides a means to dispose simultaneously two non commitant waves, as shown in

Fig. 3.

On the basis of the invariance and the state of correlation between the two speckles, it can be argued that the interference between the two object speckles will give rise to macroscopic fringes only in that region of the space where they are shifted by a quantity smaller than the speckle size in that region :

$$(\Omega_x^2(M) + \Omega_y^2(M))^{\frac{1}{2}} \ll \ <\rho_\ell(M)> \quad ; \quad \Omega_z(M) \ll \ <\rho_L(M)> \tag{13}$$

which is analogue to Eq.(11).

This condition, the same as in speckle interferometry, however, gives rise to a different interpretation and introduces the notion of *localization* of the interference fringes produced holographically (ref. 12, 14, 16, 33-35). Since it is now possible to explore in volume the potential fringe fields, it is no longer indispensable to impose a priori large granulation to the speckle field. The apertures used in holographic interferometry are optimal since it is permissible to adjust them during reconstruction in function of the necessity expressed by Eq.(13), and in consequence much larger than in speckle interferometry. A net improvement in the quality of the interferograms thus results.

In a localization region, where Eq.(13) is satisfied, the fringes can be relatively simply described as the fringes of equal difference of altitude between the two speckles (see Fig.3) :

$$\Omega_{z0}(\overline{M}) + \left[\frac{\Omega_{xp}}{L}\right] x + \left[\frac{\Omega_{yp}}{L}\right] y = n\lambda \tag{14}$$

where Ω_{z0} is the longitudinal translation between the two speckles at the point $\overline{M}$ and in its neighborhood; the coefficients of x and y represent the relative inclinations of the two waves, expressed by the ratio of the translation component in the pupil plane (Ω_{xp} and Ω_{yp}) to the distance L between the pupil and the observation plane.

In effect the relations (14) and (2) used in parallel underscore a complex liaison between the fringe characteristics and the displacement-deformation parameters of the object surface : in the general case, these factors combined influence the period, the direction, the fringe order - entire or fractional - and the localization of the considered fringe.

Confronted to the state-of-the-things, three reactions are possible :
- accept the inconvenience related to the simultaneous recording of several holograms, but in return, harvest a complete information on the displacement-deformation of the object. This approach is described in the references (15-

16, 36-38);

- proceed to a precise determination of the lines or the surface of localization by a photometric analysis of the fringe contrast permitting one to directly measure the strain components (ref. 16, 39-41);

- accept a partial information, simple to access, acquired in particular optical systems which strive to simplify the relation (14), in leaving only a re-strained number of displacement-deformation parameters of the object (ref. 42-50). This second point of view will be adopted in Section 6, not for excluding the first two methodologies from the field of applications in fracture mechanics, but for the sake of simplicity.

In comparison with the methods in speckle metrology, the visualization of the interferograms is immediate :

- in double exposure, the hologram is successively exposed to two speckle wave-fields corresponding to the two deformation states of the object. At reconstruction, the two fields are reproduced simultaneously and interfere in the conditions described in the foregoing paragraphs. The same procedure is used in double-pulse holography with a Q-switched laser, the reconstruction being carried out by a CW laser;

- in the real-time mode, the hologram is exposed only once and memorizes a wave-field corresponding to a state of object deformation which subsequently serves as reference. Developed and repositioned exactly in its original position, it permits the recorded field to interfere at each instant with the field issuing from the object. The interferogram, observable directly, evolves in function of time.

In this *mode opératoire* it is particularly judicious to incorporate the translation and the rotation micrometer stages to the hologram support, the reference and the illumination source, in order to apply the compensation techniques (ref. 16-17, 51). By the intermediate of appropriate compensatory movements, these techniques aim at assuring a superposition as perfect as possible of the two interfering speckle fields in a given observation plane, all in leaving untouched their relative inclination. Given the smallness of the volume of correlation when one works at large apertures, they appear very useful in the cancellation of the effect of parasite object displacements and contribute to the improvement of the contrast of the interferograms.

The limitations on holographic interferometry, compared to that evoked for speckle metrology, are due more to the high stability requirements imposed on the optical set-ups and the environment than to the presence of the secondary speckles in the interferograms.

To conclude this Section, the principal characteristics of the speckle metrology and the holographic interferometry methods can be summarized as follows :
- these methods do not require any special preparation of the object surface. In addition, they do not call for a physical model or an identical mechanical twin of the object;
- contrary to the classical interferometry where certain displacement components are not discerned, these methods are, by nature, sensitive to all the displacement and deformation components of the object surface;
- inspiring from the same basic principles, these methods offer a vast range of variety and extreme flexibility as to the measurement possibilities that can be explored; using the same standard equipments, they give way to multi-purpose and complementary systems;
- the interferograms are always modulated by a secondary speckle whose size is proper to each method and which imposes the ultimate limitations on the sensitivity and the resolution that can be achieved;
- these techniques possess the advantage of being independent of linear elastic or non linear elastic behavior as well as of the homogeneity and isotropic characteristics of the material studied.

The visibility of the interferograms plays equally an important role in the appreciation of these methods. This point has already been taken up partially : the interfering speckles are considered to be either perfectly correlated or totally decorrelated resulting in a visibility plunge towards either unity or zero. The theoretical developments relative to the intermediate case, met in practice, of the state of partial correlation and hence of visibility lying between 0 and 1 are not reviewed here but appear in the cited references and in few recent publications (ref. 52-53).

In the following Sections, the aim will be to harness these methods to the study of cracks in the concrete. Only those particular variants of these methods will be presented which have either been effectively used as such or which appear as potentially applicable.

4   APPLICATION OF SPECKLE PHOTOGRAPHY
4.1   Focused Speckle Photography (ref. 19-21)
Of all the methods proposed here, it presents the simplest of optical configuration (see Fig.6.a). The photographic plate records the speckle intensity distribution onto the object surface. In this case, Eq.(2) simplifies considerably : the nullity of the distance between the object and the observation

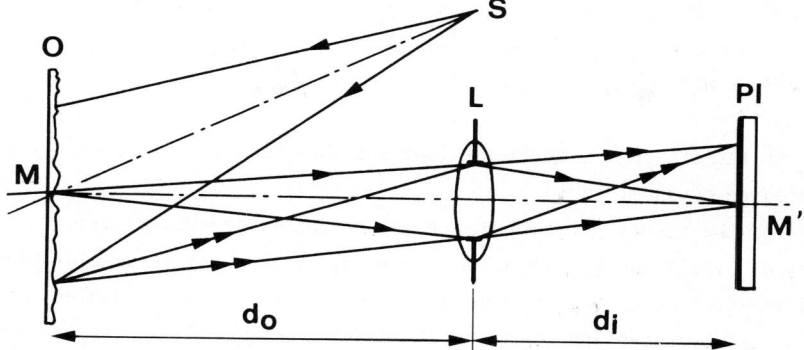

Fig.6.a - Schematic of Focused Speckle Photography.

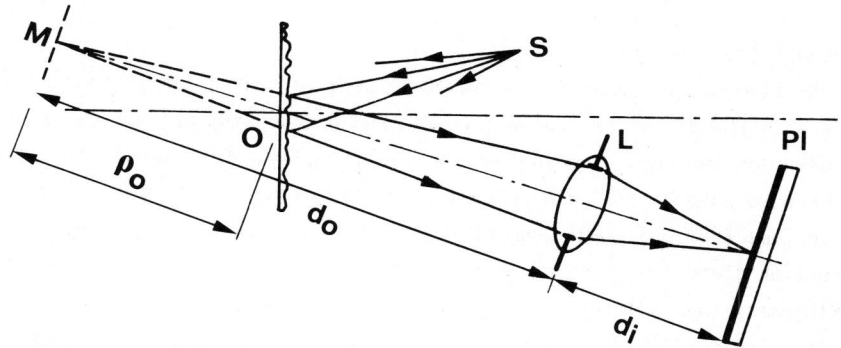

Fig.6.b - Schematic of Gregory system.

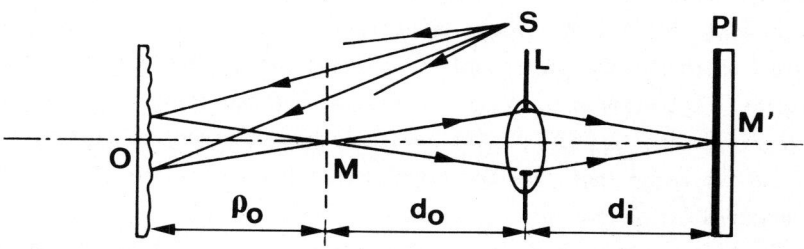

Fig.6.c - Schematic of Chiang system.

plane ($\rho_0=0$) results in the annulation of several elements of matrices $[C_i]$. The speckle displacements (Ω_x, Ω_y) at any point $M(x,y)$ on the object surface are then given by

$$\Omega_x(M) = u(M) + w(M) \frac{x}{d_0}$$

$$\Omega_y(M) = v(M) + w(M) \frac{y}{d_0}$$

(15)

where d_0 is the distance between the lens and the object, and the magnification is supposed to be equal to 1.

If for the points lying in the neighborhood of the optical axis the field angles $(x/d_0, y/d_0)$ can be considered to be small, then the speckle displacement is equal to the in-plane displacements at these points. One can neglect the influence of the out-of-plane components whenever the objective characteristics (field angle and focal length) render the approximation :

$$\frac{x}{d_0} \simeq \frac{y}{d_0} \simeq 0$$

(16)

acceptable (ref. 54-55).

An alternative consists in simultaneously recording two photographic plates with the aid of two objectives centred on two distinct optical axes. One thus disposes two sets of relations of the type (15), which permit to solve for the three components of the displacement vector at any given point on the object surface (ref. 56). The mode of recording supposes that one freezes discrete deformation states by brief exposures of the plate. The analysis of the recorded specklegram either point-by-point or whole field is possible only in differed time (see Section 3.1). These two characteristics render the method time consuming. However, new recording supports, other than silver emulsions, are presently undergoing rapid development (photothermoplastic, electro-optic crystals, etc.) which aim at diminishing or even eliminating the dead-time between the recording and the analysis stage (ref. 57). Alternatively, sandwich speckle photography can be used which consists in recording a photograph of each deformation state on a separate photographic plate (ref. 58). The superposition of any two plates a posteriori permits to obtain relative displacements undergone by the object between the two recordings. The technique thus opens way to the study of transient phenomena.

Pflug has reported on the successful use of focused speckle photography for crack analysis of an unnotched concrete beam ($1000 \times 100 \times 50$ mm^3) subjected to a three point bending load (ref. 59). The point-by-point analysis (see Fig.7) has permitted him to position the cracks on the concrete surface at different loads, by detection of the regions which manifest by a jump in the orientation and the period of the Young's fringes. These observations are in conformity with

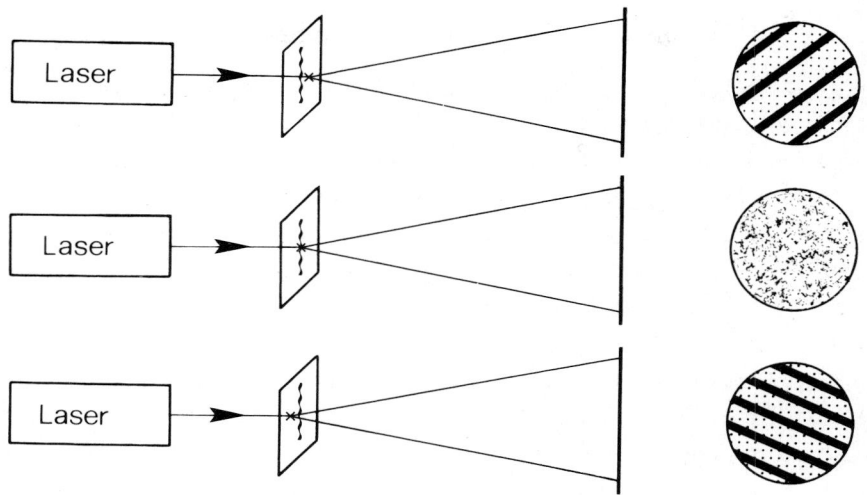

Fig.7 - Visualization of the crack path and subsequent
measurement using speckle photography.

those of the direct microscopic observation.

De Backer has studied more quantitatively the cracking of a beam (120 ×
25 × 15 mm^3) cantilevered from a concrete wall (ref. 60). The Young's fringes
have allowed him to measure crack widening with stepwise increasing load. In introducing a fictitious displacement of the photographic plate between the two
exposures, the widening as low as 0.35 μm has been detected. The references
(61-64) describe equally the success obtained in the application of this method
to the study of the cracks in material other than concrete.

4.2 Defocused Speckle Photography (ref. 65-66)

The recording configuration turns around several variants which consist
in recording the speckle intensity in a plane different from the object surface
- either it be behind or in front of the object. In this way, the resulting
speckle displacements are rendered predominantly sensitive to the local rotations or the slopes of the deformed surface. In the two optical systems shown
(see Fig.6.b & c) the relations between the speckle displacements (Ω_x, Ω_y), observed at the point M, and the slopes of the surface element $\{\overline{0}\}$, centred at 0
on the optical axis, can be expressed by :

$$\Omega_x(M) = 2 \, \rho_0 \, \frac{\partial w}{\partial x} \, (0)$$

$$\Omega_y(M) = 2 \, \rho_0 \, \frac{\partial w}{\partial y} \, (0)$$

(17)

where ρ_0 is the distance between the observation and the object plane and for the above equation to hold the illumination and the observation directions are supposed to lie very close to the object normal.

As soon as one deviates from these conditions, the examination of Eq.(2) shows that corrective terms should also be taken into consideration in the equation (17) - terms which indicate the contribution due to the in-plane and the out-of-plane object displacements. A detailed analysis of these equations is given in a recent publication (ref. 67).

Gregory has made an important use of defocused speckle photography in crack analysis of pressure vessels and for the detection of flaws in aeronautical structures (ref. 68). He has shown that the flaws are often better visualized by an anomaly in the slopes than in the displacement patterns. However, to our knowledge, this group of methods has as yet not been used in the study of cracks in concrete.

4.3   Tandem Speckle Photography (ref. 69)

Stetson has proposed to record a focused plane and several tandem speckle-grams simultaneously from different directions. Subsequent solving of the linear system of equations (see Eq.(2)) allows one to determine the displacements, slopes, shear and surface strains undergone by the object. This method, though difficult to implement, is the only one capable to determine completly the surface deformation tensor.

4.4   White Light Speckle Photography (ref. 56, 70-71)

In principle and in the optics required, this method is the exact twin of the focused speckle photography except that the source of illumination in the later is a laser whereas here a conventional lamp with a large chromatic content is sufficient. However, the granular structure serving to memorize the position and thereafter the evolution of each region of the object, must be created artificially : either by depositing a special covering on the surface, or in resolving with sufficiently high contrast the surface irregularities, if the surface is rugged like concrete. The principal advantage associated with this technique resides in the possibility to illuminate large surfaces at minimum of cost. The limitations on the measurement range and resolution are governed by the same law as in Eqs.(6,9,10). Evidently, the possibilities offered by the defocused speckle photography are inexistant in the present case.

Chiang has applied the white light speckle method to the determination of displacement field around a crack tip in a plexiglas specimen and to the subsequent calculation of the stress intensity factor (ref. 72). According to the author, the technique is applicable to almost all engineering structural material.

5 APPLICATION OF SPECKLE INTERFEROMETRY (ref. 9-11, 27-28)

Three types of optical systems are equally envisaged here. From the point of view of application, a third means of visualizing the interference fringes is the Electronic Speckle Pattern Interferometer - ESPI - system (ref. 73-74). The object and the reference speckle (Section 3.2) are picked up by the tube of a video system. The instrument then performs a chain of operations : recording on the magnetoscope, play-back, addition, filtering and improvement of the image contrast. The acquisition of an initial interference state is very rapid (of the order of few seconds). The camera tube having a low spatial resolution, it is no longer the relation (11) which governs the mean size of the speckles in the image plane, but rather the resolution of the tube. It thereby results that the speckles are much larger than strictly necessary and the quality of the final interferogram is in consequence affected. On the contrary, the observation is quasi-immediate and effected in real-time.

5.1 Fringes of equal out-of-plane displacement (ref. 28, 74)

The corresponding optical system is schematized in Fig.8.a. The reference speckle, independent of the object deformation, is in reality a spherical wavefront propagating in the same direction as the speckle issuing from the object. In these conditions, it can be shown that the variation of the path difference between the reference and the object speckles is essentially due to the out-of-plane displacement components of the object. In the neighborhood of the optical axis, the relation (12) becomes :

$$w(M) = \frac{n\lambda}{2} \tag{18}$$

Once again, the observation on the surface plane ($\rho_0 = 0$) nullifies several elements of matrices $[C_1]$ in Eq.(2). As one goes away from the optical axis, the in-plane displacement components equally participate in the variation of the optical path length, although their effect is very small : the coefficient of

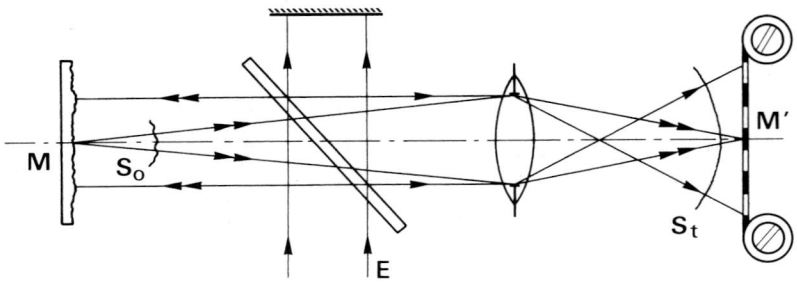

Fig.8.a - System for obtaining the fringes of equal out-of-plane displacement.

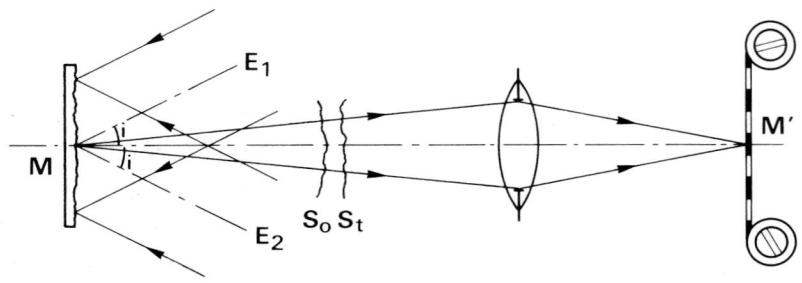

Fig.8.b - System for obtaining the fringes of equal in-plane displacement.

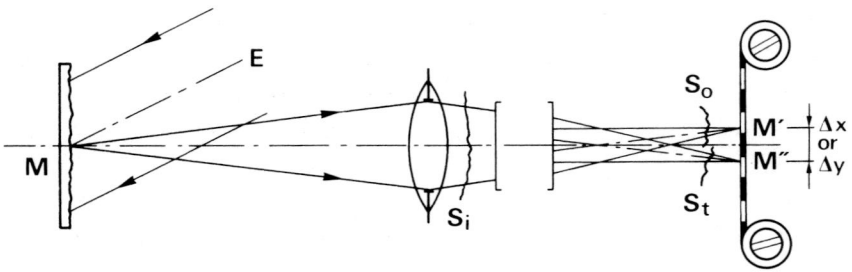

Fig.8.c - Speckle shearing interferometric set-up.

ponderation being equal to the sine of the semi-field angle (ref. 67).

This system is extremely sensitive, permitting to visualize the fringes of equal out-of-plane displacements with an intrinsic sensitivity of the order of a quarter of a micron per fringe.

This high sensitivity is very useful at the detection stage of the first crack appearance. However, the noisy aspect of the image renders difficult the precise determination of the crack location and its shape. This drawback is compensated by the easiness to obtain interferograms in real-time.

5.2 Fringes of equal in-plane displacement (ref. 27, 74)

A symmetric dual-beam illumination of the object (Fig.8.b) gives rise to the object and the reference speckles and permits to render the variation of the optical path length between these two interfering speckles uniquely sensitive to the in-plane displacement component of the object, parallel to the projection of the illumination directions on the object. The fringes corresponding to

$$
\begin{matrix} u(M) \\ \\ v(M) \end{matrix} = \frac{n\lambda}{2 \sin i} \tag{19}
$$

are generated, where i is the angle of incidence of the illuminated beams (see Eqs.(2) & (12)). For collimated illumination, the method does not suffer from any alteration in sensitivity over the object field : the out-of-plane deformation components having no influence, whatsoever, even at points far away from the optical axis. The quantitative study is therefore easy and simple to carry out. The other particular characteristics of this method are identical to those described in Section 5.1.

A variant of this method consists in replacing the symmetric dual illumination by a symmetric dual observation by means of an opaque screen containing two small holes and placed in the pupil of the imaging lens (ref. 75).

Another interesting extension of this method consists in illuminating a transparent model by a thin sheet of laser light at the angle of incidence $i = \pi/2$. It thus becomes possible to investigate the stress intensity factor along a crack front in any plane in the interior of the transparent body (ref. 76).

The two interferometers described in Sections 5.1 and 5.2 have found extensive applications in nondestructive testing (ref. 11, 73-74). But, to our knowledge, this group of methods has as yet not been used in the specific case of crack analysis in concrete.

5.3 Speckle Shearing Interferometer (ref. 77-78)

The photographic plate receives two laterally shifted interfering speckle fields both originating from the object. In practice, these two operations are performed simultaneously by means of a classical interferometer (Fig.8.c), interposed between the object and its image, before or after the imaging lens. A large variety of optical means exists, viz., use of Michelson interferometer, Young's double apertures, bi-prisms, etc.

When the interferometer is adjusted so as to introduce a symmetrical lateral shift (Δx or Δy), with respect to the optical axis, between the two speckles, it can be shown that the variation of the optical path length is a linear combination of the partial derivatives of the object displacements in the two directions. In the same manner as described in Section 3.2, one visualizes the fringes :

$$\sin i \frac{\partial u}{\partial x} (M) + (1 + \cos i) \frac{\partial w}{\partial x} (M) = \frac{n\lambda}{\Delta x}$$

or, (20)

$$\sin i \frac{\partial v}{\partial y} (M) + (1 + \cos i) \frac{\partial w}{\partial y} (M) = \frac{n\lambda}{\Delta y}$$

where i represents the angle of incidence of the illumination beam placed in the (x0z) or (y0z) plane.

This technique, in general, suffers from the drawback of a limited object field; away from the optical axis the corrective terms introduce in the relation (20), (ref. 67, 79-80).

In properly choosing the angle of incidence of the illumination beam, it is possible to eliminate the influence of in-plane strain $\frac{\partial u}{\partial x}$ or $\frac{\partial v}{\partial y}$. While employing a more elaborate system, requiring the simultaneous recording of two plates in using two different illumination directions, it is possible to separate the two families of fringes (in-plane and out-of-plane derivatives) on a point-by-point basis (ref. 81).

The applications of speckle shearing interferometry to the problem of flaw detection and crack analysis are to be found in literature (see for example ref. (11, 82)), although concerned with other materials than concrete. Moreover in the reference (82) and as in defocused speckle photography, Hung considers that it is more practical to correlate flaws with strain anomalies rather than displacement anomalies.

6   APPLICATION OF HOLOGRAPHIC INTERFEROMETRY (ref. 29-30, 36-38, 43-50)

In holographic interferometry as in speckle metrology, a large variety of set-ups exist offering multiple possibilities of measurements. This diversity has already been pointed out in Section 3.3. In that follows, the two systems most often employed and which permit to visualize the in-plane and the out-of-plane displacement fields are described.

6.1 Holographic set-up for the measurement of the out-of-plane displacement

The system shown in Fig.9 is the most classic of those used in holographic interferometry. The illumination and the observation directions are symmetric with respect to the normal to the object surface, and at the limit, coincide with the surface normal. If, in addition, the following conditions are met :
- collimated illumination of the object;
- pupil of the observation system at infinity;
- observation of the interferogram on the surface of the object ($\rho_0 = 0$, i.e. fringes localized on the object),

then in this case, the fringes of equal difference of altitude (Eqs.(2) & (14)) interpret equally as the fringes of equal out-of-plane displacement. The interferogram visualizes the fringes :

$$w(M) = \frac{n\lambda}{2 \cos i} \tag{21}$$

The maximum sensitivity is attained for $i = 0$, i.e. around one fourth of micron, as in the case of the speckle interferometric system shown in Fig.8.a. As indicated in Section 3.3, the interferograms possess an excellent defintion.

This technique seems to have been the most often employed in fracture mechanics of the concrete. In the same loading configuration as in Section 4.1, Pflug (ref. 59) was able to detect the appearance of a crack as early as the first instants of its formation, at low loads ; follow its evolution with increasing load; observe the redistribution of strains around the crack tips and finally retrace the evolution with unloading, (see Fig.10, 11). The interferograms were visualized in real-time by means of a video-system.

A second example is presented in Fig.12. Here it concerns the first results obtained in the study of crack formation in small cylindrical concrete models under the influence of shrinkage[*]. The first cracks appear a few minutes after the start of the drying process, developing and progressing by abrupt and sudden jumps.

Light and Luxmoore have reported the application of the holographic out-of-plane displacement fringes to the study of crack formation in concrete cubes (of 100 mm side) and cylinders (of 150 mm diameter) in compression (ref. 83-84). The first crack was detected by discontinuities in the fringes at around one-

[*] Collaboration between the Laboratory of Stress Analysis (Professor L. Pflug) and the Laboratory of Construction Materials (Professor F.H. Wittmann) of the Swiss Federal Institute of Technology, 1015-LAUSANNE; January 1981. Swiss National Foundation grant.

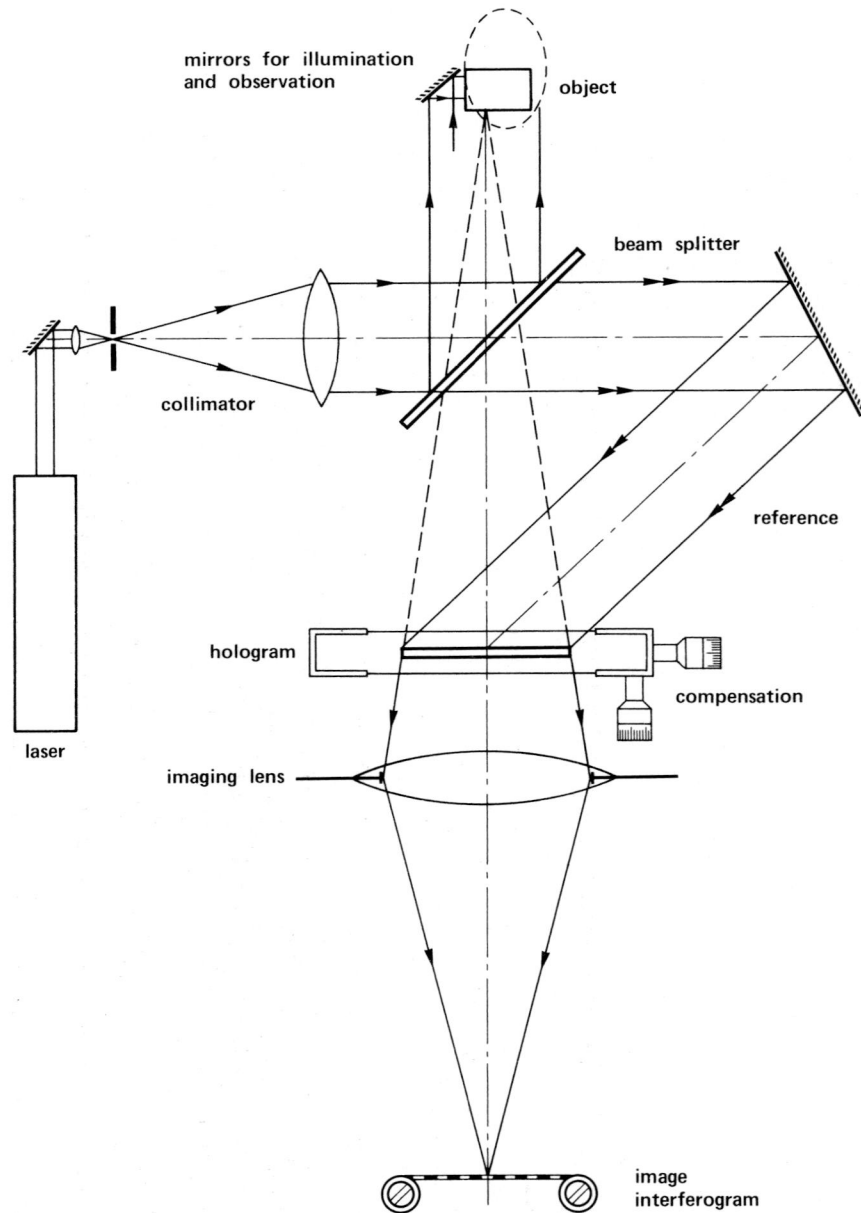

Fig.9. Holographic system for obtaining the out-of-plane
displacement fringes, in its real-time version.

quarter to one-third of the ultimate load. At no time before failure could any
of these cracks be detected visually. Their aim was to estimate the density of

139

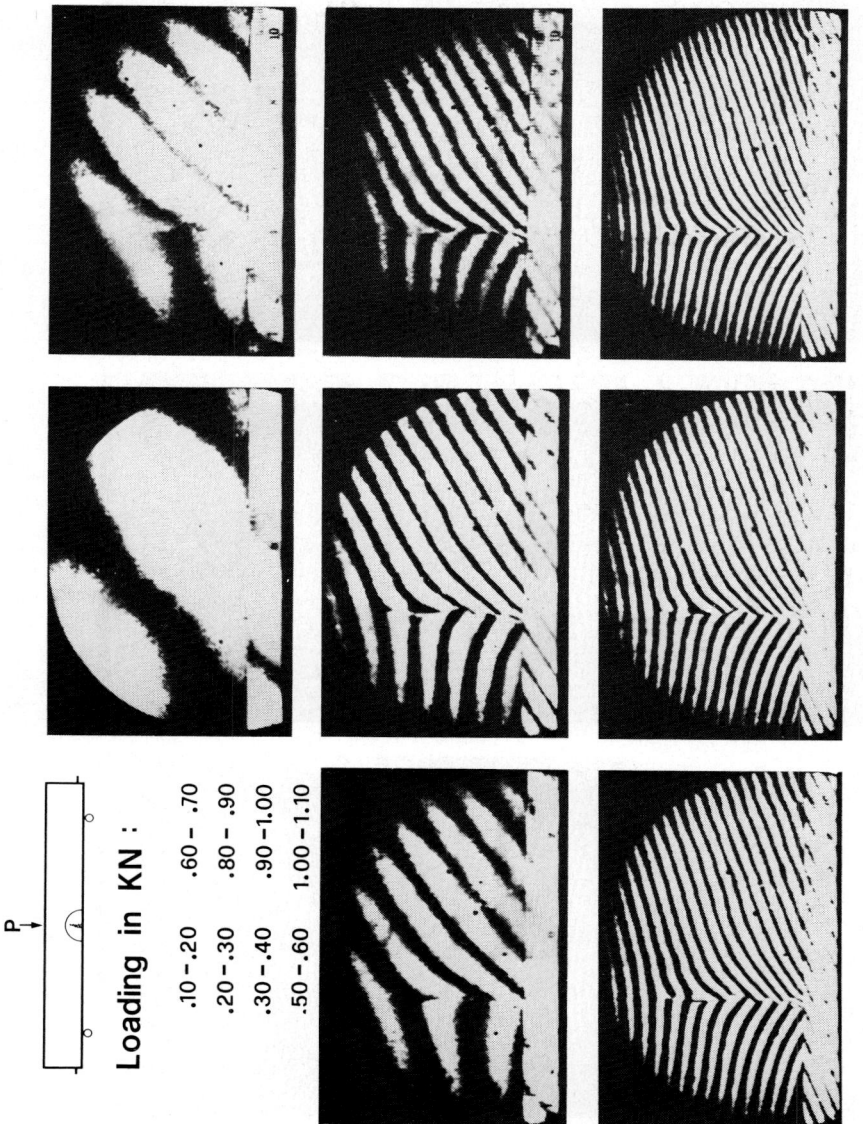

Loading in KN :

.10 – .20   .60 – .70
.20 – .30   .80 – .90
.30 – .40   .90 –1.00
.50 – .60   1.00 –1.10

Fig.10 – Crack propagation in a concrete beam using holographic out-of-plane displacement fringes.

140

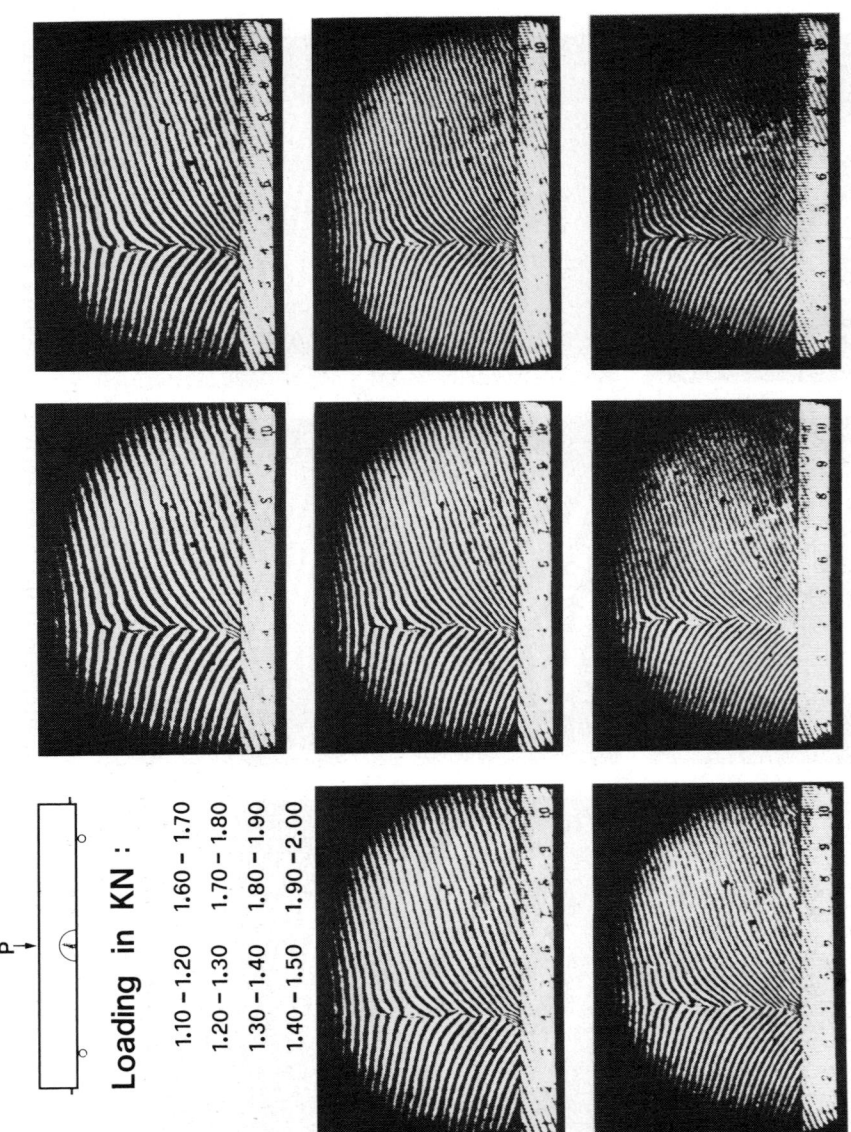

Loading in KN :

1.10 – 1.20   1.60 – 1.70
1.20 – 1.30   1.70 – 1.80
1.30 – 1.40   1.80 – 1.90
1.40 – 1.50   1.90 –2.00

Fig.10 – continued

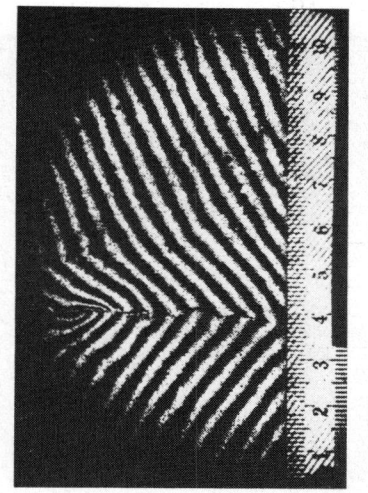
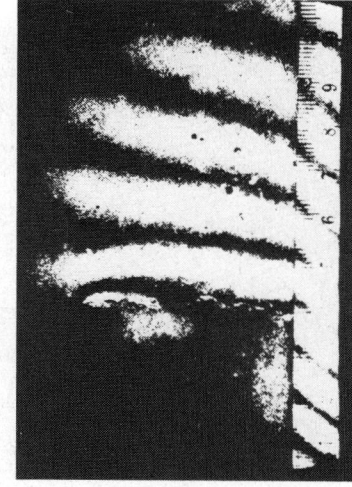
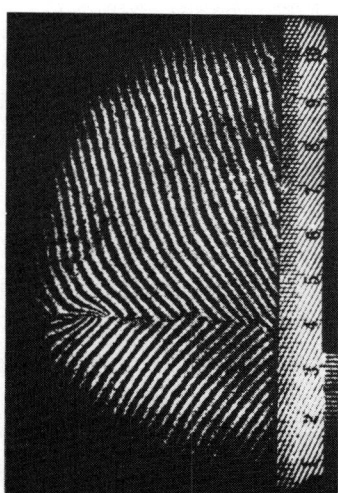
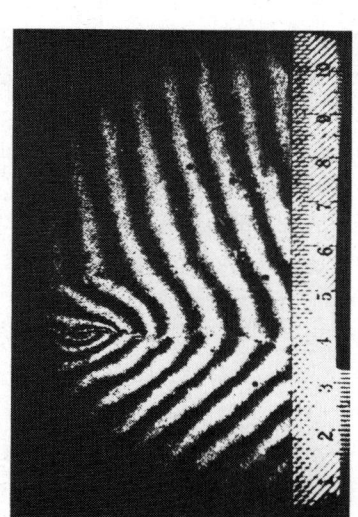

Fig. 11 - Evolution during unloading.

142

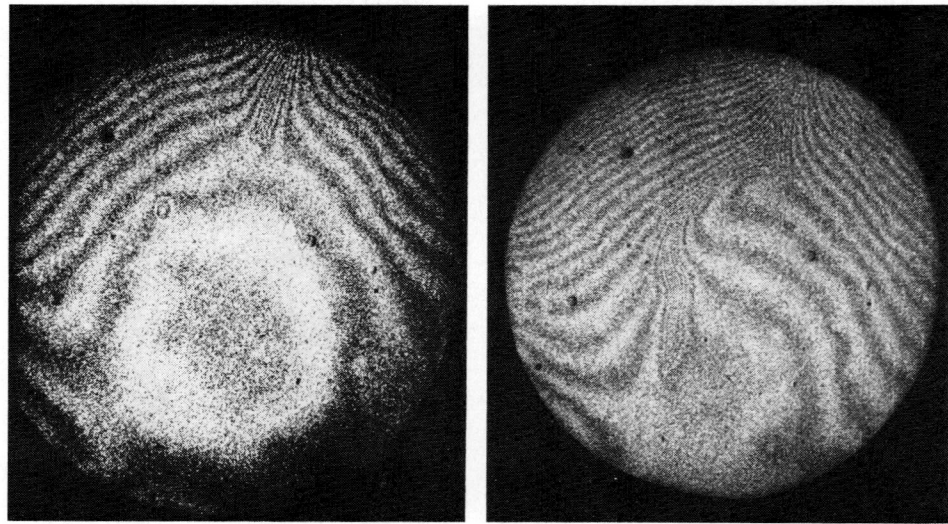

Fig.12 - Holographic observation of the out-of-plane displacement field and the crack formation accompanying the shrinkage during the drying of concrete.

cracking at predetermined values of the percentage of the failure load.

De Haas and Stroeven have conducted tests on small blocks of steatite concrete in uniaxial compression and on a notched beam of mortar under flexion (ref. 85-86). The experiments were carried out in real-time using out-of-plane displacement fringes. The development of crack was observed by discontinuities in the interference patterns. Their other aim was to control, by means of fringes, load increments just large enough to open slightly the crack while avoiding any sudden crack extension. Thus the measurement of the crack length as a function of the crack opening and the external load could be performed quite accurately.

For the materials other than concrete, the works of Vest (ref. 87) and Dudderar (ref. 88) are to be considered as pioneering in the field. Numerous quantitative studies (ref. 89-94) have been carried out on materials as steel, aluminium, plastic, rocks, etc., aiming to measure the fracture mechanics parameters such as the stress intensity factor K, the strain energy release rate G, or the J integral, including dynamic effects using pulsed laser techniques. It goes without saying that the analysis of crack in the concrete can benefit directly from these studies.

6.2 <u>Holographic set-up for the measurement of the in-plane displacements</u>
(ref. 49, 52, 95-99)

As shown in Fig.13, the object is illuminated by two collimated beams
placed symmetrically with respect to the surface normal. This system differs
from that of Section 5.2, Fig.8.b, uniquely by the presence of a reference beam
superposed to the object image.

The in-plane displacements, in the direction containing the two beams,
are visualized as a beat or a moiré between the two interferograms due to each
illumination beam. The obtaining of the moiré is subject to two conditions :
- each interferogram should be well localized in the observation (image) plane;
- to obtain a good quality moiré, it is necessary to densify the interferograms,
  by the addition of a fictitious state of deformation, so that a sufficient
  number of carriers may participate in the formation of a beat (moiré).

The local spatial frequency of the moiré is classically the difference of
the spatial frequencies of the two interferograms in the region under consideration. The precedent conditions being satisfied, each interferogram visualizes
the fringes

(i) $\sin i \, u(M) + [1 + \cos i]w(M) + f_r(M) = n\lambda$

$$(22)$$

(ii) $-\sin i \, u(M) + [1 + \cos i]w(M) + f_r(M) = n'\lambda$

where $f_r(M)$ represents at each point the effect of the residual contribution due
to the other displacement-deformation parameters of the object, and as well the
effect of the addition of the fringes.

The moiré appears as a family of fringes

$$u(M) = \frac{N\lambda}{2 \sin i}$$

or, (23)

$$v(M) = \frac{N\lambda}{2 \sin i}$$

The contrast of the moiré being a fraction of the contrast of the initial
interferograms, it is imperative that these last are of high contrast. In addition, and in particular for the application of crack detection, it is natural
to preserve the advantage of a good fringe definition, proper to holographic interferometry, by working at large lens apertures. These two contradictory requirements are discussed in detail in the reference (52). The micrometric compensation stages, indispensable here, are solicited in any efficient manner only
in real-time operation. The visualization of the moiré on the TV monitor greatly

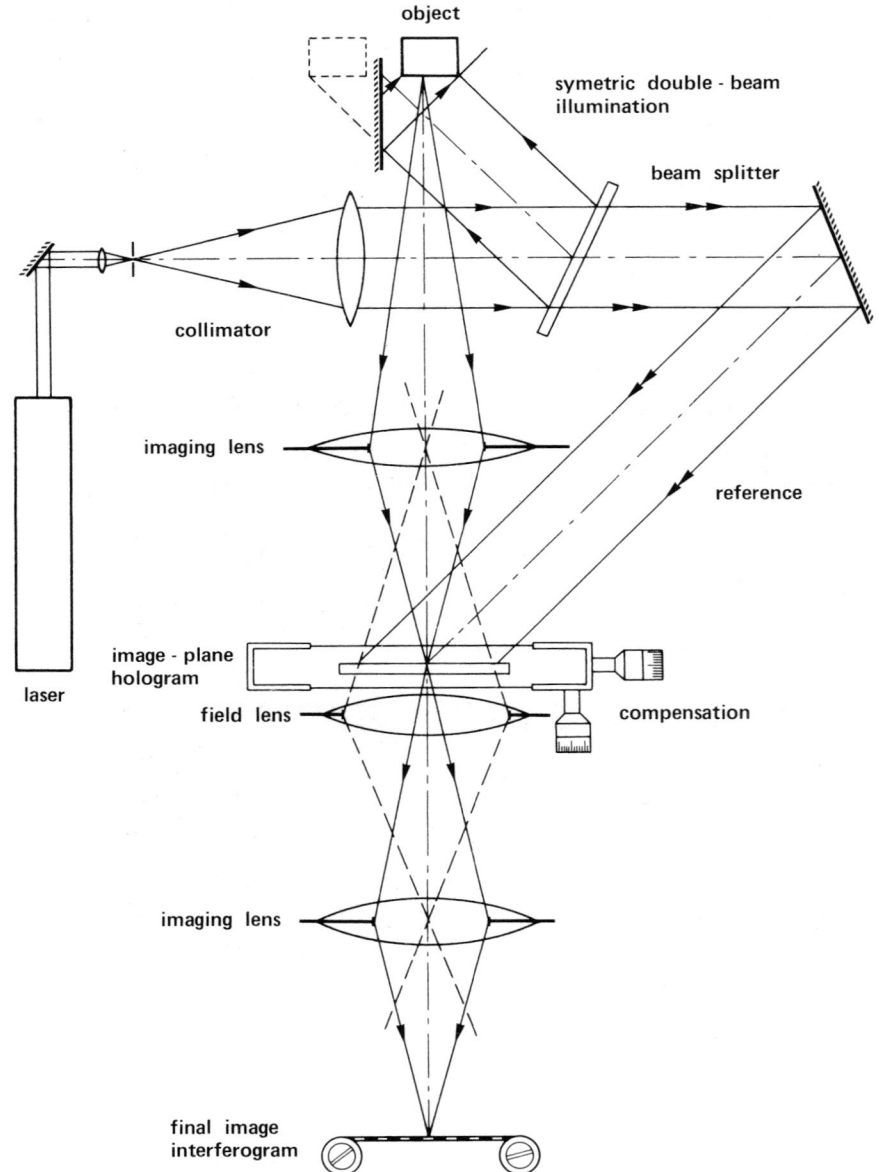

object

symetric double - beam
illumination

beam splitter

collimator

imaging lens

reference

laser

image - plane
hologram

field lens

compensation

imaging lens

final image
interferogram

Fig.13 - Real-time holographic system for the measurement
of the in-plane displacements.

facilitates these interventions in permitting an immediate control on the effect of each manipulation. This ensures a rapid optimization of the fringe visibility by the trial and error procedure (ref. 98).

In the same loading configurations as in Sections 4.1 and 6.1, Pflug,(ref. 59), has obtained interferograms permitting to follow the evolution of the crack by the intermediate of the discontinuities introduced in the displacement field u, (see Fig.14). The discontinuities appear principally in the slopes and less in the fringe order. This leads to infer that the in-plane displacement field presents a symmetry on both sides of the crack. For this particular study, Pflug finally concludes that of the three techniques used, it is the out-of-plane holographic technique which gives the fringes of best quality, whereas the in-plane holographic technique provides the most interesting information on the displacement fields.

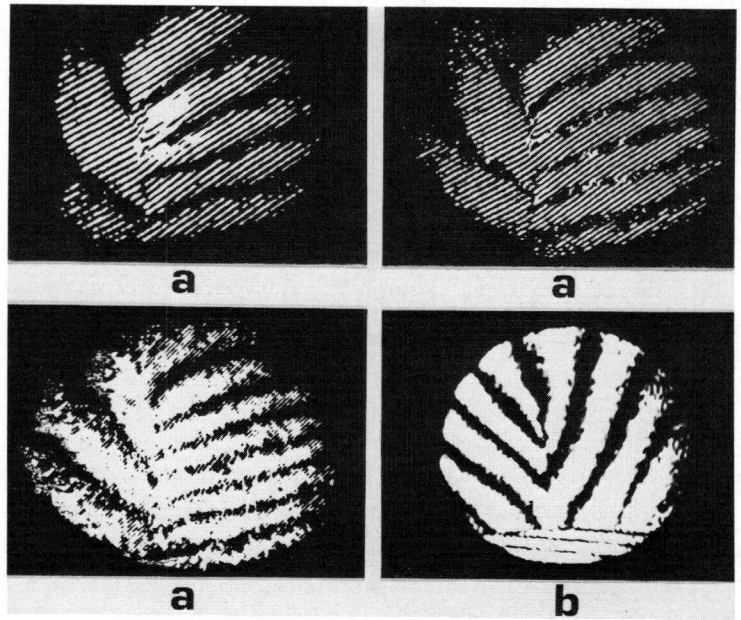

Fig.14 - Analysis of crack in the concrete beam using holographic
moiré technique. In-plane displacement fringes obtained
(a) on the TV monitor (b) after filtering.

Other applications of this technique are described in the literature, concerning the study of materials other than concrete (ref. 91-92).

6.3 Comments on the flexibility of holographic interferometry

The two preceding examples possess their exact counterparts in speckle interferometry (Sections 5.1 and 5.2) - from the point of view of the nature of the components visualized and the sensitivity of the measurement. Under these conditions, why resort to an holographic system, reputed to be more delicate ?

It is advisable to regroup under the title of flexibility the advantages decisive and proper to the holographic interferometry - flexibility which is not apparent to a comparable degree in speckle photography and interferometry. This flexibility is essentially due to the existence of the reference beam which facilitates several operations, sometimes possible but difficult to implement in speckle photography and interferometry. The versatility of the holographic process is most conspicuous when it comes to, for example, fringe compensation, fringe multiplication, fringe interpolation by heterodyning (see ref. 100), analogic and optical derivation of the interferograms (see ref. 50), etc..

A second important element in favour of holographic interferometry in the crack analysis is that it permits to best combine the function of high resolution image formation and fringe formation in relaxing the severity of the depth of field problem. This aspect is treated in the following Section.

7   HOLOGRAPHIC MICROSCOPY

7.1   Principle

Being the principal optical element employed in holographic microscopy, some of the useful characteristics possessed by a conventional microscope are first given. The separation between two object points which are just resolved by a microscope is given by :

$$R = \frac{0.61 \ \lambda}{N} \tag{24}$$

where λ is the wavelength of the object illuminating wave and N is the numerical aperture of the microscope objective. The ultimate limit on resolution is set by the numerical aperture of the objective used. Another parameter of interest is the depth of field δ associated with the microscope. The expression for the depth of field depends on the criterion chosen for ensuring a given quality of the image. In first approximation,

$$\delta = \frac{\lambda}{N^2} \tag{25}$$

or, in terms of R and λ the depth of field can be written as

$$\delta = \frac{2.69 \ R^2}{\lambda} \tag{26}$$

This equation is of great importance in designing an adequate holo-microscopy system suited for a particular job.

Holographic microscopy set-ups can be divided into two main categories, (ref. 101-105). The first consists in recording a hologram of the magnified image of the object. This magnification is achieved by means of a high power microscope objective. The reconstructed image can then be examined at leisure at different focal depths, depending on the recorded depth of field. In the other variant, hologram of the object is recorded and which is viewed later on with the aid of a microscope. A combination of the pre- and post-magnification techniques has also been proposed. The relative advantages of these variants and the possibilities to eliminate the aberration of the lenses has been adequately discussed in the literature.

In the set-ups which employ a pre-imaging system (relay systems and virtual image reconstructions come under this category) before subjecting the reconstructed/relayed image to microscopic examination, the constraints on resolution are more severe, being determined in major part by the pre-imaging optics. In order to resolve the image details, the numerical aperture of the microscope objective N should be such that

$$N \geqslant \frac{1}{2F(1+g)} \tag{27}$$

where F and g are the aperture number and magnification of the imaging system respectively, These constraints become less restrictive for small F and small magnification lens and in using the right microscopic objective satisfying Eq. (27). Moreover, the imaging optics provides to the microscope an access to scan in depth the image, equivalent in object space to :

$$\rho = 4 \lambda F^2 \left(\frac{1+g}{g}\right)^2 \tag{28}$$

In image plane holography and systems serving as relays, one is restricted to remain inside this scanning range. There is only one gaussian image plane. In the case of three dimensional objects, lensless holographic recording gives access to different object planes in best focus conditions, whereas in image plane holography the explorable depth of field is predetermined and the imaged plane is in the best focus only. The advantage in favour of lensless holography is counterset to some extent by the fact that whereas image plane holograms can be reconstructed using white light, thereby considerably smoothing speckles, the former of necessity have to be reconstructed using a coherent light source.

The possible variants of holographic microscopy are summarized in Table 1.

Table 1 - Assessment of resolution limit, depth of field and information read out in holographic microscopy

Schematics of the optical systems	Resolution limit	Depth of field	Information read out
1. Pre-Magnification			
	Set by the numerical aperture of the microscope objective.	Set by the depth of field of the microscope.	Information spread over several holograms as a hologram records one microscopic pixel.
	Set by the resolution of the relay lens.	- do -	- do -
2. Post-Magnification			
	Set by the resolution of the recording film and the resolution of the imaging lens.	Set by the depth of field of the imaging system during reconstruction.	A hologram records whole object information. Read out carried out by means of a scanning microscope.
	- do -	- do -	- do -

O : object; OB : object beam; M : microscopic objective; M.H. : microscopic hologram; R : reference;
C.R. : conjugate reference; R.L. : relay lens; A.I. : aerial image; L : imaging lens; H : hologram.

7.2   Optical system

The proposed optical system consists in harnessing holographic interferometry and holographic microscopy for a common cause - detection and minute inspection of cracks. The functioning of the system is based on the complementary
nature of the two techniques. Whereas holographic interferometry permits to detect the initiation of a crack and its global whereabout, it is unsuited to
furnish a detailed information on the geometrical features of the crack. On the
other hand, severally handicapped by its field of view holographic microscopy is
helpless in detecting the onset of cracking; although once this anomaly and its
approximate detection has been transmitted to it by holographic interferometry,
it becomes a powerful means of visualizing the crack process in all its finer
details (length, width, branching, microstructure details around cracks, etc.).

The set-up is shown in Fig.15. Following Section 6.1, the hologram recorded at P_1 can be recognized to be uniquely sensitive to the out-of-plane displacement components. An aerial image of the concrete block is simultaneously
relayed at a distance away from the object. The aerial image is followed by a
microscope whose output image is picked by a TV camera. The images transmitted
by the two TV cameras are made to appear simultaneously on the monitor by means
of an image mixing device.

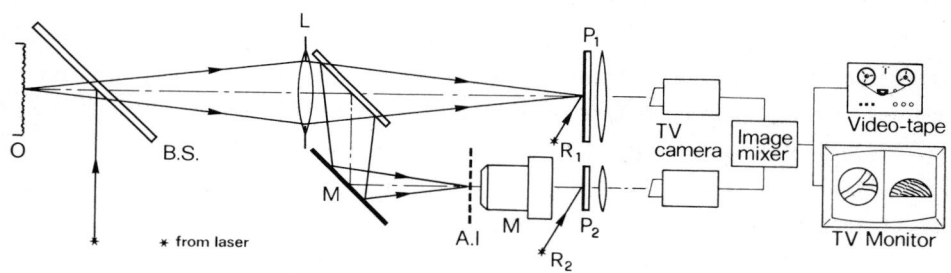

Fig.15 - Schematic of holographic microscopy used in conjunction
with holographic interferometry for the detection
and the study of crack propagation.

The set-up offers five distinct possibilities :
(i)   Possibility to visualize out-of-plane displacement fields in real-time at
       the output P_1. The displacement fringes release signals of alarm for re-
       gions susceptible of crack formation and the eventual onsetting of cracks.
(ii)   Possibility to follow the regions of interest such as hairline crack flaws
       in real-time, at the output P_2 , using optical microscope.
(iii) Possibility to record a hologram at each loading step at P_1 and preserve

it for examining it a posteriori by means of a scanning microscope (post magnification variant).

(iv) Possibility to record microscopic holograms at the output P_2. The holograms can then be reconstructed a posteriori. A rugged surface like concrete does not lend itself well to direct microscopic observation as the frontal distance of the instrument is very small. Moreover, the limited depth of field makes it difficult to focus on a surface spotted with persistent hills and valleys natural to the concrete. For this reason the image of concrete is transmitted at a distance to perform holographic microscopy in its pre-magnification variant. However, in this case the limits on resolution are dictated by the optical system preceding the image.

(v) Possibility to visualize out-of-plane displacement fields in the microscopic regions of interest (holographic interferometry microscopy) in real-time at P_2.

Feature (i) has been discussed in Section 6.1. Once an anomaly has been signalled by way of discontinuities in the fringes, the microscope is translated to the region of interest and at each loading step the perturbed zone is scanned (feature (ii)). This is equivalent to real-time image plane holographic microscopy in its post magnification variant. A fine longitudinal movement incorporated in the microscope permits to scan the object in depth as well. The zones of interest having thus been located, these regions are frozen by making successive holograms as the scanning microscope follows the crack path (feature (iv)). This is achieved by means of a holocamera of photothermoplastic film type. The recording time for a single hologram is about 40 seconds. The recorded holographic film can then be run at a later time and examined at leisure; focusing can be altered over a limited range permitting to focus on the two sides of the crack and thereby preserving information on the relative displacement direction of the crack lips.

Sometimes it may be interesting to perform holographic interferometry microscopy in the regions of strong anomalies and cracks (feature (v)). These regions are characterized by :

(i) High density fringes.

(ii) Decreasing fringe contrast due to the change in the microstructure in the region surrounding the cracks and relatively large differential movements.

Holographic interferometry microscopy permits to overcome some of these problems. The contrast of the displacement fringes around the crack tip can be improved considerably by adequately compensating the rigid-body movements. The

effect of change in microstructure on fringe contrast can be reduced if the
wavefronts emanating on two successive loading steps are compared rather than
the initial wavefront with a wavefront at any given time. The resolution of
fringes does not pose any problem and the sign of displacement in the micro
region can be determined by adjoining equispaced parallel fringes obtained by
introducing a slight tilt between the two object waves.

8   CONCLUSION

With the aim to detect the presence of a crack on the surface of the concrete as early as its formation, to pin-point it on the model surface, to follow
its propagation and finally to visualize the deformation fields at the neighborhood of the crack tips, three groups of methods and their main variants have
been presented : speckle photography, speckle interferometry and holographic
interferometry.

Rather than to describe the anatomy of each, we have primarily aimed to
assemble these methods around some fundamental properties common to each. In
addition, the first promising applications, showing the benefits that can be
drawn from the use of these methods in the crack analysis of concrete, have been
briefly passed in review.

The synoptic table (Table 2) summarizes the nature of the displacement
components accessible in each type of measurement system. The numerical values
on the range and sensitivity have been calculated for $\lambda = 0.5$ μm and are based on
the assumption that one-fifth of an interfringe can be resolved, which roughly
corresponds to a visual mode of analysis. However, more elaborate methods, as
heterodyning, allow a much better fringe interpolation and in consequence a
higher gain in sensitivity is obtained as the later is directly proportional to
the gain realized in fringe interpolation. Finally, the estimation on the model
size is based on the restrictions imposed by the system optics (for example,
collimated illumination) or by the power of the laser used (1 Watt, TEM_{00} mode
in our case). Moreover, from a practical point of view, the newly developed recording materials as the photothermoplastic plates, and the interferogram digitization device greatly facilitate the employment of these methods.

The other chapters of the book testify to the progress recorded in the
theoretical analysis of the cracking in concrete. By their good spatial resolution and high sensitivity in measurement of the deformation fields, the
methods of speckle metrology and holographic interferometry present themselves
as highly performing experimental tools, well adapted to the demands of these
new theoretical developments.

Table 2 - Synoptic representation of the described methods

	TYPE OF METHODS	MEASURABLE COMPONENTS	PARASITE COMPONENTS	ELIMINATION	RANGE	SENSITIVITY (based on 1/5 of a fringe)	MODEL SIZE
SPECKLE PHOTOGRAPHY	Focused Speckle Photography	u and v	w	possible *	5 - 500 μm	1 μm	1 m^2
	Defocused Speckle Photography	$\frac{\partial w}{\partial x}$ and $\frac{\partial w}{\partial y}$	u,v,w	possible *	$10^{-4} - 5.10^{-3}$ rd	2.10^{-5} rd	0.25 m^2
	White Light Speckle Photography	u and v	w	possible *	20 - 500 μm	5 μm	5 m^2
SPECKLE INTERFEROMETRY	Out-of-plane SI	w	u and v	yes **	0.25 - 10 μm	0.05 μm	0.25 m^2
	In-plane SI	u or v	-	-	0.5 - 20 μm	0.1 μm	0.25 m^2
	Shearing SI	$\left[\frac{\partial u}{\partial x} \text{ or } \frac{\partial w}{\partial y}\right.$ $\left.\frac{\partial u}{\partial x} \text{ or } \frac{\partial v}{\partial y}\right]$	vice versa	possible *	$2.10^{-5}-8.10^{-4}$ rd or 10^2-10^3 μstrain	4.10^{-6} rd or 20 μstrain	0.25 m^2
HOLOGRAPHIC INTERFEROMETRY	Out-of-plane HI	w	u,v	yes **,***	0.25 - 20 μm	0.05 μm	0.25-1 m^2
	In-plane HI	u or v	w	yes ***	0.5 - 20 μm	0.1 μm	0.25 m^2

*   Analysis of several simultaneously recorded specklegrams or interferograms.
**   Collimated illumination, normal to the object surface and the aperture of the observation system at infinity.
*** Compensation.

REFERENCES

1   P.F. Packman, in A.S. Kobayashi (Ed.), Experimental Techniques in Fracture
    Mechanics, vol. 2, Iowa State University Press, Ames, Iowa, 1975, ch. 2,
    p. 59.
2   W.N. Sharpe, D.R. Martin, in Proc. 6th Int. Conf. of Exp. Stress Anal.,
    Munich, Sept. 18-22, 1978, VDI-Verlag GmbH, Düsseldorf, 1978, pp. 195-201.
3   G. Marci, P.F. Packman, Int. Journal of Fracture, 16 (1980), pp.133-153.
4   D.E. Macha, D.M. Corbly, J. W. Jones, Exp. Mech., 19 (1979), pp.207-213.
5   W.N. Sharpe, Opt. Eng., 21 (1982), pp. 483-488.
6   C. Sok, J. Baron, D. François, Cem. Concr. Res., 9 (1979), pp. 641-648.
7   S. Lowenthal, H. Arsenault, J. Opt. Soc. Am., 60 (1970), pp. 1478-1483.
8   J.W. Goodman, in J.C. Dainty (Ed.), Laser Speckle and Related Phenomena,
    Springer Verlag, Berlin, 1975, ch. 2, p. 9.
9   M. Françon, Granularité Laser - Speckle, Masson, Paris, 1978.
10 J.C. Dainty (Ed.), Laser Speckle and Related Phenomena, Springer Verlag,
    Berlin, 1975.
11 R.K. Erf (Ed.), Speckle Metrology, Academic Press, New York, 1978.
12 J. Ch. Viénot, C. Froehly, J. Monneret, J. Pasteur, in E.R. Robertson and
    J.M. Harvey (Eds.), Proc. Symp. The Engineering Uses of Holography, Glasgow
    1968, Univ. Press, Cambridge, 1970, pp. 133-150.
13 J. Monneret, Opt. Commun., 2 (1970), pp. 159-162.
14 S. Walles, Opt. Acta, 17 (1970), pp.899-913.
15 K.A. Stetson, J. Opt. Soc. Am, 66 (1976), pp.1267-1270.
16 W. Schumann and M. Dubas, Holographic Interferometry, Springer Verlag,
    Berlin, 1979.
17 P. Jacquot and P.K. Rastogi, Appl. Opt., 18 (1979), pp. 2022-2032.
18 P. Jacquot, Tech. Digest Topical Meeting on Hologramm Interf. and Speckle
    Metrology, Cape Cod, June 2-4, 1980, OSA, 1980, pp. MA4-1, MA4-4.
19 E. Archbold, J.M. Burch, A.E. Ennos, Opt. Acta, 17 (1970), pp. 883-898.
20 E. Archbold, A.E. Ennos, Opt. Acta, 19 (1972), pp. 253-271.
21 R.P. Khetan, F.P. Chiang, Appl. Opt., 15 (1976), pp. 2205-2215.
22 K.A. Stetson, J. Opt. Soc. Amer., 67 (1977), pp. 1587-1590.
23 M.L. Roblin, G. Schalow, B. Chourabi, J. Opt., 8 (1977), pp. 149-158.
24 G.H. Kaufmann, A.E. Ennos ,B. Gale and D.J. Pugh, J. Phys. E : Sci. Instr.,
    13 (1980), pp. 579-584.
25 B. Ineichen, P. Eglin, R. Dändliker, Appl. Opt., 19 (1980), pp.2191-2195.
26 G.B. Smith, K.A. Stetson, Appl. Opt., 19 (1980), pp. 3031-3033.
27 J.A. Leendertz, J. Phys. E : Sci. Instr., 3 (1970), pp. 214-218.
28 J.N. Butters, J.A. Leendertz, J. Phys. E : Sci. Instr., 4 (1971), pp. 277-
    279.
29 C.M. Vest, Holographic Interferometry, Wiley, New York, 1979.
30 Y.I. Ostrovsky, M.M. Butusov, G.V. Ostrovskaya, Interferometry by Holo-
    graphy, Springer Verlag, Berlin, 1980.
31 R.J. Collier, C.B. Burckhardt, L.H. Lin, Optical Holography, Academic
    Press, New York, 1971.
32 H.J. Caulfield (Ed.), Handbook of Optical Holography, Academic Press,
    New York, 1979.
33 W.T. Welford, Opt. Commun., 1 (1969), pp. 123-125.
34 K.A. Stetson, J. Opt. Soc. Am., 64 (1974), pp. 1-10.
35 J. Leroy, Nouv. Rev. Opt., 6 (1975), pp. 329-337.
36 A.E. Ennos, J. Sci. Instr., 1 (1968), pp. 731-734.
37 J. Tsujiuchi, N. Takeya, K. Matsuda, Opt. Acta, 16 (1969), pp. 709-722.
38 S.K. Dhir, J.P. Sikora, Exp. Mech., 7 (1972), pp. 323-327.
39 M. Dubas, W. Schumann, Opt. Acta, 21 (1974), pp. 547-562.
40 M. Dubas, W. Schumann, Opt. Acta, 22 (1975), pp. 807-819.
41 J. Ebbeni, J.C. Charmet, Appl. Opt., 16 (1977), pp. 2543-2545.

154

42   E.B. Aleksandrov, A.M. Bonch-Bruevich, Sov. Phys.-Tech. Phys., 12 (1967), pp. 258-265.
43   J.W.C. Gates, Opt. Technol., 1 (1969), pp. 247-250.
44   N. Abramson, Appl. Opt., 11 (1972), pp. 1143-1147.
45   P.M. Boone, L.C. De Backer, Optik, 37 (1973), pp. 61-81.
46   V. Fossati-Bellani, A. Sona, Appl. Opt., 13 (1974), pp. 1337-1341.
47   J. Ebbeni, in Proc. of the 5th Int. Conf.of Exp. Stress Anal., Udine, May 1974, CISM, Udine, 1974, pp. 4.20-25.
48   R. Dändliker, B. Eliasson, B. Ineichen, F.M. Mottier, in E.R. Robertson (Ed.), Proc. Symp. the Engineering Uses of Coherent Optics, Glasgow 1975, Univ. Press, Cambridge, 1976, pp. 99-117.
49   C.A. Sciammarella, J.A. Gilbert, Exp. Mech., 16 (1976), pp. 215-220.
50   G. Cadoret, Annales de l'Institut Technique du Bâtiment et des Travaux Publics, Série : Essais et Mesures, no. 170, 373 (1979), pp. 26-74.
51   W. Schumann, M. Dubas, Optik, 46 (1976), pp. 377-392.
52   P.K. Rastogi, M. Spajer, J. Monneret, Opt. Lasers Eng., 2 (1981), pp.79-103.
53   M. Spajer, P.K. Rastogi, J. Monneret, Appl. Opt., 20 (1981), pp.3392-3402.
54   E. Archbold, A.E. Ennos, M.S. Virdee, in Proc. 1st Europ. Conf. on Opt. Appl. to Metrol., Strasbourg, Octobre 1977, SPIE vol. 136, Washington, 1978, pp. 258-264.
55   M.G. Pedretti, F.P. Chiang, J. Opt. Soc. Am., 68 (1978), pp. 1742-1748.
56   P. Jacquot, P.K. Rastogi, Opt. Lasers Eng., 2 (1981), pp. 33-55.
57   H.J. Tiziani, K. Leonhardt, J. Klenk, Opt. Commun., 34 (1980), pp.327-331.
58   W.J. Beranek, A.J.A. Bruisma, in Proc. 6th Int. Conf. of Exp. Stress Anal., Munich, Sept. 18-22, 1978, VDI-Verlag GmbH, Düsseldorf, 1978, pp.417-425.
59   L. Pflug, in Proc. of 7th Nat. Congress of the Italian Society for Stress Analysis, supplement, Cagliari, Sept. 1979, A.I.A.S. 1979, pp. 5-26.
60   L.C. De Backer, Nondestruct. Test., 8 (1975), pp. 177-180.
61   W.T. Evans, A.R. Luxmoore, Eng. Fract. Mech., 6 (1974), pp. 735-743.
62   A. Archbold, A.E. Ennos, Nondestruct. Test., 8 (1975), pp. 181-184.
63   A.R. Luxmoore, in A. Lagarde (Ed.), Proc. IUTAM Symp. Optical Methods in Mechanics of Solids, Poitiers, Sept. 1979, Sijthoff and Noordhoff, 1981, pp. 509-524.
64   M.A. Trafalian, C.E. Taylor, in A. Lagarde (Ed.), Proc. IUTAM Symp. Optical Methods in Mechanics of Solids, Poitiers, Sept. 1979, Sijthoff and Noordhoff, 1981, pp. 337-342.
65   D.A. Gregory, Opt. Laser Technol., 8 (1976), pp. 201-209.
66   F.P. Chiang, R.M. Juang, Appl. Opt., 15 (1976), pp. 2199-2204.
67   P.K. Rastogi, P. Jacquot, Opt. Eng., 21 (1982), pp. 411-426.
68   D.A. Gregory, in R.K. Erf (Ed.), Speckle Metrology, Academic Press, New York, 1978, ch. 8, p. 183.
69   K.A. Stetson, I.R. Harrison, in Proc. 6th Int. Conf. of Exp. Stress Anal., Munich, Sept. 18-22, 1978, VDI-Verlag GmbH, Düsseldorf, 1978, pp.149-154.
70   J.M. Burch, C. Forno, Opt. Eng., 14 (1975), pp. 178-185.
71   P.M. Boone, L.C. De Backer, Optik, 44 (1976), pp. 343-355.
72   F.P. Chiang, A. Asundi, Eng. Fract. Mech., 15 (1981), pp. 115-121.
73   J.N. Butters, in E.R. Robertson (Ed.), Proc. Symp. the Engineering Uses of Coherent Optics, Glasgow 1975, Univ. Press, Cambridge, 1976, pp. 155-169.
74   J.N. Butters, R. Jones, C. Wykes, in R.K. Erf (Ed.), Speckle Metrology, Academic Press, New York, 1978, ch. 6, p. 111.
75   D.E. Duffy, Appl. Opt., 11 (1972), pp. 1778-1781.
76   D.B. Barker, M.E. Fourney, Exp. Mech., 17 (1977), pp. 241-247.
77   Y.Y. Hung, Opt. Commun., 11 (1974), pp. 132-135.
78   Y.Y. Hung, I.M. Daniel, R.E. Rowlands, Appl. Opt., 14 (1975), pp. 618-622.
79   J. Brdicko, M.D. Olson, C.R. Hazell, Opt. Acta, 25 (1978), pp. 963-989.
80   A. Rosenberg, J. Politch, Opt. Commun., 26 (1978), pp. 301-304.
81   Y.Y. Hung, in R.K. Erf (Ed.), Speckle Metrology, Academic Press, New York, 1978, ch. 4, p. 51.

82   Y.Y. Hung, Opt. Eng., 21 (1982), pp. 391-395.
83   A. R. Luxmoore, Nondestruct. Test., 6 (1973), pp. 258-263.
84   M.F. Light, A.R. Luxmoore, Mag. Concr. Res., 24 (1972), pp. 167-172.
85   P. Stroeven, H.M. De Haas, in Proc. RILEM Symp. New Developments in Non-
      Destructive Testing on Non-Metallic Materials, Constanza, Romania, V2,
      1974, p. 19.
86   H.M. De Haas, P. Stroeven, in Proc. 10th Annual Conf. BSSM, Strain Measure-
      ment and Analysis as an Aid to Design, Warwick, 1974, Stevin Rep. 1-75-1,
      Delft, 1975.
87   C.M. Vest, in R.K. Erf (Ed.), Holographic Nondestructive Testing, Academic
      Press, New York, 1974, ch. 8.6, p. 289.
88   T.D. Dudderar, Exp. Mech., 9 (1969), pp. 281-285.
89   H. Spetzler, C.H. Scholz, C.P.J.Lu, Pure Appl. Geophys., 112 (1974), pp.
      571-582.
90   T.D. Dudderar, E.M. Doerries, Exp. Mech., 16 (1976), pp. 300-304.
91   T. Lefevre, P. Boone, M. De Caluwe, in Proc. 6th Int. Conf. of Exp. Stress
      Anal., Munich, Sept. 18-22, 1978, VDI-Verlag GmbH, Düsseldorf, 1978, pp.
      407-415.
92   C.A. Sciammarella, S.K. Chawla, in Proc. 6th Int. Conf. of Exp. Stress
      Anal., Munich, Sept. 18-22, 1978, VDI-Verlag GmbH, Düsseldorf, 1978, pp.
      775-779.
93   G. Cadoret, in Proc. 1st Europ. Conf. on Opt. Appl. to Metrol.,
      Strasbourg, Octobre 1977, SPIE vol. 136, Washington, 1978, pp. 114-126.
94   H. Fagot, F. Albe, P. Smigielski, J.L. Arnaud, in 4th Coll. Int. Meth.
      Nondestruct. Test., Grenoble 11-14 Sept. 1979, Université scientifique et
      médicale de Grenoble, 1979, pp. 171-175.
95   J. Monneret, P.K. Rastogi, M. Spajer, in Proc. 1st Europ. Conf. on Opt.
      Appl. to Metrol., Strasbourg, Octobre 1977, SPIE vol. 136, Washington,
      1978, pp. 82-89.
96   C.A. Sciammarella, S.K. Chawla, Exp. Mech., 18 (1978), pp. 373-381.
97   W.J. Beranek, A.J.A. Bruisma, Exp. Mech., 20 (1980), pp. 289-300.
98   C.A. Sciammarella, P.K. Rastogi, P. Jacquot, R. Narayanan, Exp. Mech., 22
      (1982), pp. 52-63.
99   C.A. Sciammarella, Opt. Eng., 21 (1982), pp. 447-457.
100   R. Dändliker, in E. Wolf (Ed.), Progress in Optics, Vol. XVII, North
      Holland, Amsterdam, 1980, ch. 1, p. 3.
101   R.W. Smith, T.H. Williams, Optik, 39 (1972), pp. 150-155.
102   M.B. Rhodes, R.F. Cournoyer, in SPIE vol. 104, Multidisciplinary Micro-
      scopy, 1977, pp. 21-28.
103   M.E. Cox, in SPIE vol. 104, Multidisciplinary Microscopy, 1977, pp.69-72.
104   M.E. Cox, K.J. Vahala, Appl. Opt., 17 (1978), pp. 1455-1457.
105   R.A. Briones, L.O. Heflinger, R.F. Wuerker, Appl. Opt., 17 (1978), pp.
      944-950.

Fracture Mechanics of Concrete,
edited by F.H. Wittmann, 1983
Elsevier Science Publishers B.V., Amsterdam — Printed in The Netherlands

Chapter 3.5

FORMATION AND PROPAGATION OF CRACKS AND ACOUSTIC EMISSION

by U. Diederichs, U. Schneider and M. Terrien

1. Introduction

1.1 Origin of Acoustic Emission

The method of acoustic emission analysis depends on the phenomenon, that solid materials emit acoustic waves, when they are non elastically stressed. With increasing stresses the main part of the introduced energy is stored in form of elastic deformations in the crystal lattice. The lattice strains increase until fixed limits are exceeded, thereafter a conversion into new states of equilibrium with lower energy takes place. The released energy propagates in form of elastic waves and can be detected in the environment as an ultrasonic pulse. Therefore the acoustic emission depends on irreversible changes of conditions, which will be described as follows [1 to 5].

Dislocation movement: The plastic deformation of crystalline materials under mechanical stresses is caused by the movement of lattice dislocations as shown in fig. 1. When reaching a critical shear strain the dislocation line starts moving and vanishes at last at a grain boundery - a permanent gliding of specific crystal areas has been performed.

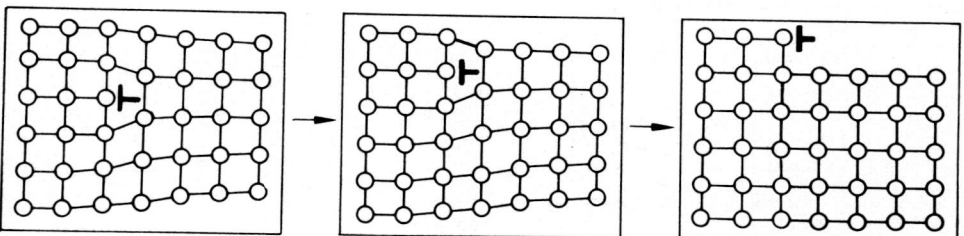

Fig. 1. Dislocation movement

Twinning: Under mechanical stresses crystalline materials may slide off in such a way, that the shifted part of the crystal reaches a homologeous position to the not shifted area [6].

Phase transformation: In some cases phase transformations are connected with acoustic emission; e.g. many freezing processes (liquid → solid) or the well known martensite formation during the hardening of steel (fcc- → bcc-structure).

Formation and propagation of cracks: When stress increases a zone of plastic deformation occurs in the environment of the crack tip. During the crack opening the state of equilibrium is disturbed. To compensate the disturbance a dilational wave and a shear wave are emitted from the crack tip, which can be detected as acoustic emission.

Friction: The deformation of flawed materials causes stick-slip oscillations between the crack borders, whereby acoustic emission occurs. This can be used for detection and localization of cracks. But in many cases it is difficult to examine the formation of cracks.

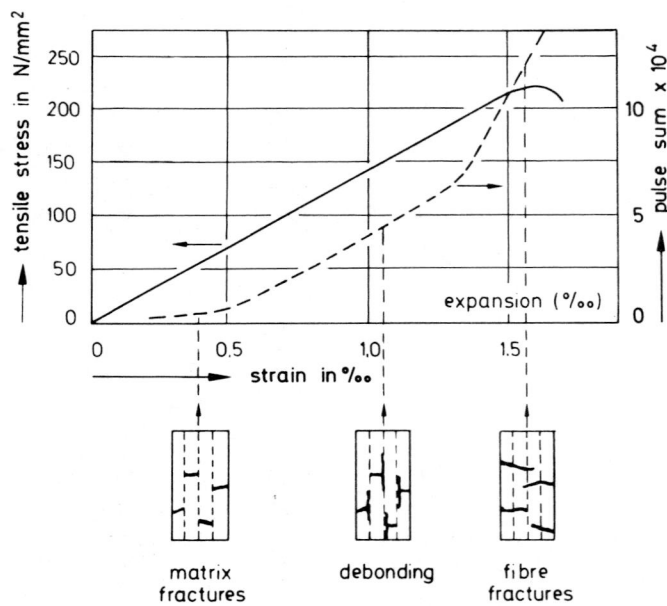

Fig. 2. Types of fracture and pulse sum of GFK-laminates [7]

Fibre fractures - matrix fractures - debonding: The acoustic emission of fibre reinforced materials under tensile stresses mainly depends on the various failure processes. It is possible to draw conclusions from the alterations in the pulse sum on the different kinds of fractures as shown in fig. 2. When mainly matrix fractures take place, numerous amplitudes of low magnitude are produced. In the case of fibre-fractures or debonding the emission of pulses with high amplitudes occur. This is true for glassfibre but not in general for carbon fibre.

Displacements in xero gels: Xero gels are coherent dispersive systems consisting of solid matter, which embed voids of microscopical and submicroscopical dimensions. These voids may be either interconnected to each other or enclosed, e.g. pumice, silicagel and cement paste. They can be described as a three-dimensional cross linked framework concerning their deformation behaviour. In case of external stresses relative displacements between the gelparticles occur which are connected with the spontaneous release of stress waves. These displacements are comparable to the dislocation movements of crystalline solids [8].

1.2 Propagation of Acoustic Emission Waves

With the propagation of acoustic waves it is normally subdivided a short-range field and a remote field. In the short-range field there are undisturbed waves in the centre of a crack. It occurs, if the material is homogeneous, isotropic and unlimited. In the remote field the waves are reflected and attenuated because of the finiteness of the material and its internal structure including dislocations, anisotropies and inhomogeneities. The waves are mainly transformed into natural motions of the solid matter. A smaller part of them is changed into surface waves. In the acoustic emission analysis the surface waves are very important, because their attenuation is only proportional to $1/\sqrt{r}$. In contrary to that the spherical waves are attenuated by $1/r$. The surface waves can be subdivided into RAYLEIGH- and LAMB-waves as a function of the relation between wave length and the thickness of the material (fig. 3). Caused by the complexity of the above factors it is very difficult to draw conclusions on the formation processes from the measurement of acoustic emissions [2, 3, 5].

1.3 Shape and Size of the Acoustic Emission Signals

The shape of signals can be subdivided into two basic categories (fig. 4): Continuous sound emission and burst signals. Continuous sound emission is characteristic of processes which occur in a more or less continuous manner, e.g. during the plastic deformation of ductile materials. In such cases, pulses show low amplitudes and fairly long durations connected with a relatively narrow frequency spectrum. Burst signals occur if the release processes are temporally separated i.e. not permanent. This is mainly to be expected during crack initiation and the fractures in brittle materials. The burst signals are decaying sinusoids and generally show a wide frequency spectrum [9].

160

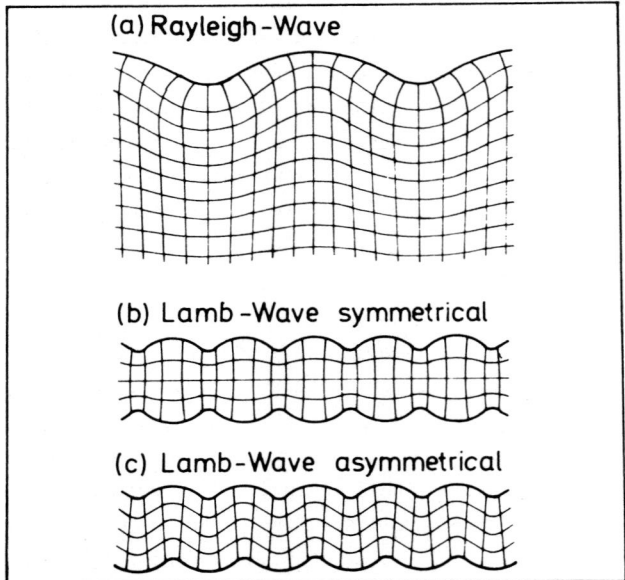

Fig. 3. Different types of surface waves [3]

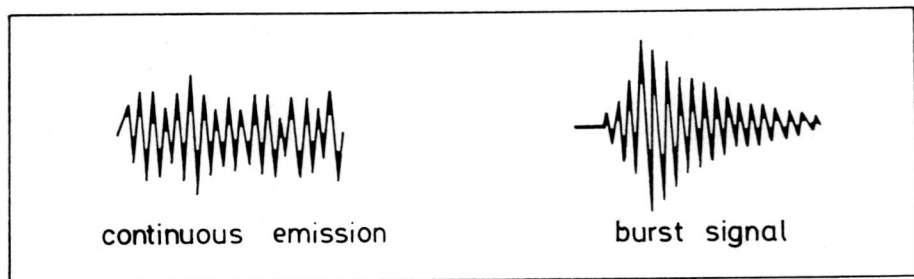

Fig. 4. Different types of acoustic emission signals [2]

The intensity of the signals is influenced by many parameters: Applied stress, strain rate, grain size, temperature, anisotropy, inhomogeneity and fracture behaviour, but also the transmission characteristics of the continuum, the coupling and the mechanical properties of the transducer are of importance [10]. Another basic influence is that the intensity of acoustic emission signals depends on the magnitude of energy originating from their various effects of formation. Some values of the relative intensity, normalized to the energy, which is released during the occurence of one single effect, are given below [2, 3, 5]:

- dislocations   1 ...   10
- phase transformations 5 ... 1000
- crack formation   20 ... 1000

2. Experimental

2.1 Transducer and Amplifier

For converting the mechanical vibrations into electric signals there are optical, capacitive and piezo-electric pick-up transducers. The range of frequencies are approximately between 50 kHz and 2 MHz [11]. This is the range most suitable for acoustic emission analysis because at higher frequencies the emission looses intensity and at lower frequencies the measurements may be superimposed by extraneous noises. In the acoustic emission analysis the piezo-electric transducers are by far the most common as they are very easy to handle and have a simple construction. Their principle of measurement is known from vibration techniques and various ultrasonic systems [12, 13]. It depends on the generation of surface charges at the piezo-electric crystal under mechanical stresses. Two different types of transducers are known, resonance and wideband transducer. Over the main part of the frequency range the wideband transducers show a very flat transmission frequency characteristic. The frequency spectrum of the electric signal is more or less in conformity with the acoustic waves, therefore it is possible to make suitable frequency analysis. In contrary to this resonance transducers have a significant maximum of sensitivity and therefore a very small frequency range [13]. They can be used well for quantitative measurements, if a high signal to noise ratio is necessary. For absolute frequency analysis they are not suitable.

The measuring points are mostly installed on the surface of the materials, i.e. a detection of the surface waves is aspired, which have higher intensities. The coupling of the transducers is done by contact agents like oil of high viscosity or artificial plastic resin in the same way as in ultrasonic systems. It must be payed attention to the thickness of the contact agent area [14].

The piezo-electric transducers yield electric signals up to 100 µV. It is necessary to lead the signal along a very short cable to a preamplifier with a low impedancy output. The preamplifier has to be installed as near as possible to the specimen, often it is directly integrated within the transducer. The main amplifier for a subsequent working should be tuned to the kind of transducer used. For a wideband transducer it is suitable to employ a wideband amplifier in connection with high and low pass filters, whose cut off frequencies can be changed. Using resonance transducers the frequency range of the main amplifier should be tuned to the resonance frequency, improving the signal-to-noise-ratio. In most cases an amplification up to 100 dB (e.g. 40 dB preamplifier, 0 ... 60 dB main amplifier) is sufficient [2, 3, 5, 15].

2.2 Evaluation of Acoustic Emission Signals

The measuring system used for single channel study of acoustic emission is illustrated in fig. 5, where the now mostly used evaluation methods are shown. In the following some of the important aspects concerning the practical application are compiled [1, 2, 3, 4, 5, 15, 16]:

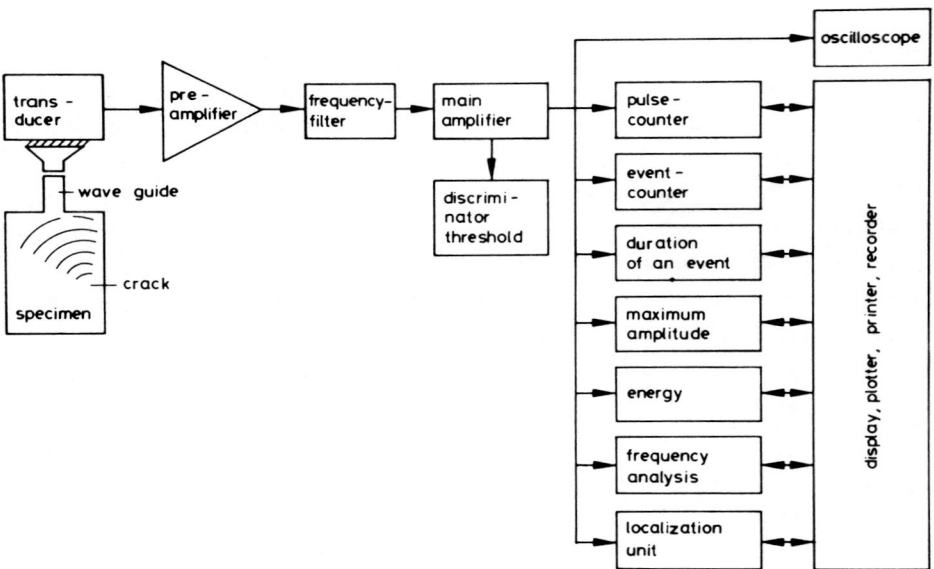

Fig. 5. Sketch of acoustic emission measuring system

Ring-down-counting: All those pulses, whose voltage exceeds a certain predetermined threshold value will be transformed into normalized square wave impulses and counted by a pulse counter. The total number of these pulses is called
"pulse sum". Another possibility is the counting of the pulses per a unit of
time. This measuring value is called "pulse rate". Fig. 6 illustrates the principle of these methods. In the upper part the temporal curve of a signal - one
single burst - can be seen.

The threshold value in this figure is exceeded four, two and one times, resp.,
depending on the received signal intensity. According to this square wave impul-

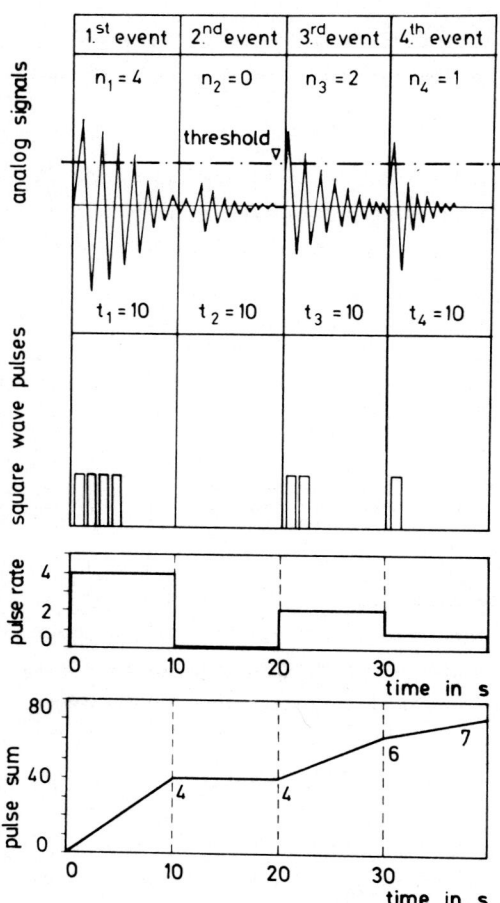

Fig. 6. Evaluation of acoustic signals [17]

ses are produced, which elevate the counter position by the corresponding number. Due to this evaluation, events of high intensity rise the counter position
to a greater extend than events of low intensity do.

Counting of events: For separating events it is necessary to count only one
pulse when the threshold value is exceeded for the first time. The length of the
internal produced square wave impulse must be enlarged until the counter position is elevated for one point per burst signal. Thereafter the counter will be
opened for the next pulse, if no pulse has been detected for an adjustable
"deadtime".

Duration of an event: With this method an event is defined by the identification of an interruption between the pulses. This technique allows the determination of the duration of one event in a very simple way. At the first exceeding
of the threshold value a digital clock will be started and when identificating
an interruption it will be stopped again.

Maximum amplitude: An additional kind of evaluation is the determination of
the maximum amplitudes. For analysing the experimental results it is possible
to plot a curve of a statistic distribution. A simple one has the following
form [5, 18]:

$$N(A_{max}) = (\frac{A_{min}}{A_{max}})^b$$

A_{max} : maximum amplitude of an event
A_{min} : lowest amplitude of all events
b : exponent, depending on the ratio of the numbers of high
           to low amplitudes, $b = 0.5 \ldots 2$

Energy: A physically very suitable measuring value for the estimation of alterations taking place in a material is the released energy. For its determination the following procedure can be used: The electric vibration per event will
be squared because the relation between transducer voltage U and stimulating
energy E is [19, 20]:

$$E \sim \int_{t} U^2(t^+) \, dt^+$$

The received signal drives a voltage-frequency-transformer while low voltages
induce low frequencies and high voltages the higher ones. The pulses at the out-
flow of the transformer can be counted, the sum is the measuring value of the
energy. Another more simple kind of detecting the energy, which has in the

opinion of some workers better dynamic properties, is the estimation of energy
with the aid of the area under the signal amplitudes according to the equation:

$$|f| = \int\limits_{t} |U(t^+)| \, dt^+$$

Frequency analysis: The measurement of the frequency spectrum of events
enables the identification and separation of individual sources of sounds,
which occur during a test, but it is a rather elaborated measuring technique.
The evaluation of the frequency spectrum is made often by Fourier's analysis
with the aid of computers. Therefore the events must be compiled in data log-
gers like magnetic disk memory, magnetic tape or perforated tape and can be
evaluated after finishing the measuring operation [2, 3, 5, 9].

Localisation of cracks: The method of crack localisation depends on the measurement of differences in transit time of acoustic waves, when their propagation velocity is known. The localisation of an acoustic emission burst in a linear object needs two transducers in the simplest case. Normally there are
three or more necessary. Using geometrical equations, transducer coordinates,
propagation velocity and transit time the sources can be calculated [2, 3, 4,
21].

Available systems: Many companies like DUNEGAN/ENDEVECO, GÄRTNER, KLETEK
PHYSICAL ACOUSTICS CORPORATION, BRÜEL & KJAER etc. offer complete measuring
units which are able to carry out all the described evaluation methods. Among
the single equipments there are individual differences in getting the various
measuring values. When determining the energies e.g., both methods mentioned
above could be found. Either the use of a voltage-frequency-transformer or the
measuring of the pulse area with the aid of several pick up circuits, which are
adjusted on different trigger levels.

The object of the acoustic emission analysis is to derive informations on
the internal structure and structural changes from the sound emission, which
occurs within a material in consequence of the manufacturing process and of
characteristic loading conditions, respectively. To achieve this, the received
acoustic signals must be suitably processed and described by certain numerical
values mentioned above, respectively. Further a correlation between one of
these characteristic values and the conditions of manufacturing or the loading
conditions concerned must be proved. In many cases, however, it is not necessary to have a detailed knowledge of the formation and propagation mechanisms.
In such cases it is sufficient to use only one of the characteristic values.

166

3. Common Application of the Acoustic Emission

Tensile Test: The acoustic emission with tensile tests has been examined systematically by KAISER for the first time [22]. In case of materials with definite yield points as to be found with most of the steels, a high intensity of
acoustic emission occurs at the elastic limit and in the region of LÜDERS deformations (fig. 7a). In the adjoined range, where strain hardening starts, the intensity decreases distinctly. This is in accordance with the observation that
the LÜDERS deformations are connected with many glidings which come to rest in
the strain-hardening range. Materials without a definite yield point show a continuous increase of acoustic emission, which indicates the increase of plastic
deformation processes (fig. 7b). Here the maximum in the acoustic emission curve
may indeed be used as a criterion for defining a quasi yield point [4, 6, 14].

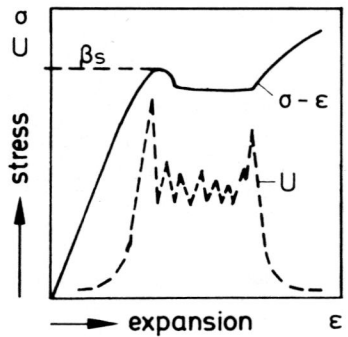 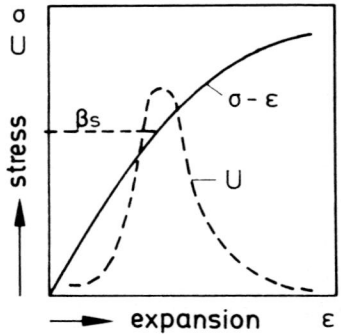

a)inhomogeneous deformation   b)homogeneous deformation

Fig. 7. Tensile tests of metals [23]
    (σ: stress; U: acoustic emission; ε: expansion; ß_s: yield point)

With repeated load cycles the so-called "KAISER-effect" can be observed. It
means, that a second load cycle up to the previous maximum stress does not cause
any acoustic emissions. Only if this maximum is exceeded acoustic emission
starts again. This behaviour can be used for the detection of preceeding stresses in steel members.

Compressive test: With compressive tests with concrete the acoustic emission
pulse sum increases in the range of creep strength. The "KAISER-effect" has

been identified at this material, too. Investigations on the relation between
acoustic emission and structure damages are just on the way [14, 15].

Crack detection: When permanent stresses are applied to materials, the existing initial cracks will grow until unstable crack propagation takes place.
The danger of crack propagation can be estimated if long term test data are at
hand (fig. 8). During a fatigue test it is possible to detect a damage much
earlier by observing the acoustic emission intensity [2, 3, 5, 24].

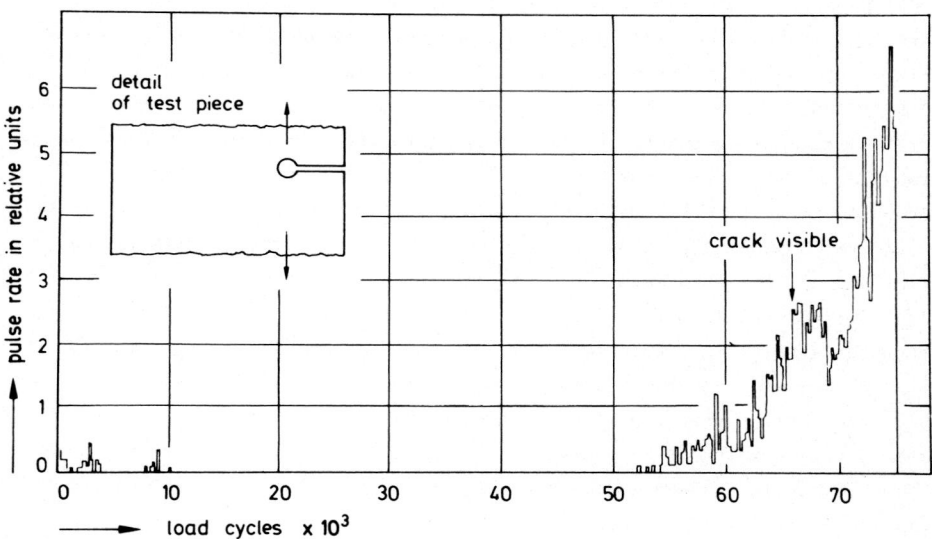

Fig. 8. Crack detection during a fatigue test [2]

Acoustic emission can also be used during the construction of structures,
e.g. by observing the welding or other kinds of thermal processes. For the monitoring of pressure vessels and pipe lines it is of good utility, too. There it
is favourable to localize areas with great acoustic emission, which then can be
examined by other methods without testing the whole structural member [25 , 26].

4. Historical Review of Acoustic Emission in Concrete

The initial studies of acoustic emissions were reported by OBERT in 1941 [27]
and by HODGSON in 1942 [28]. Both workers were interested in the prediction of
cracks in mines, which were generated during the rock excavation process. They
formulated the basic ideas of acoustic emission monitoring. Further they attempted to develop crack location techniques as to determine the zone of maximum

stress intensity in rocks, and performed laboratory tests in order to substantiate the results of their field studies.

In the early 1950s KAISER [22] studied acoustic emission or stress-wave emission technique in metals (zinc, steel, aluminium, copper and lead). His work has significant impact studying the rate of cracking and growth of fatigue cracks in rocks, glasses, ceramics, plastics and composites for more than two decades [29 to 45]. In comparison to this the application of acoustic emission for nondestructive testing of concrete is rather new.

In 1959 RÜSCH [46] made rough preliminary observations concerning acoustic signals produced by stressed concrete. The irreversibility of the signals (i.e., the KAISER-effect) was identified for stresses up to 70 - 85 percent of the ultimate load. In 1959 and 1960 almost concurrently L'HERMITE [47, 48] reported preliminary results on noises generated during the deformation of concrete. In 1965 ROBINSON [49] investigated the acoustic emissions generated in mortar and concrete with various amounts and sizes of aggregate. The emissions were mainly of two frequencies, namely, 2 kHz and 13 - 14 kHz. The emissions occured at loads well below those levels, where the ultrasonic velocity and Poisson's ratio are affected. Hence, the important conclusion was that acoustic emission detects earlier and smaller structural changes than other conventional methods do.

In 1970 WELLS [50] built an apparatus to record acoustic emissions from concrete under strain. He "listened" in a frequency regime from about 2 kHz to 20 kHz. Three- and four-inch cubes of both mortar and concrete were tested. The aims of the work, namely, to detect noises emitted from strained concrete and to record the waveform of the noises were achieved. However, only little work was done to obtain a reproducible "signature" of the acoustic emissions versus stresses for various specimens.

In 1970 GREEN [51] has done one of the most comprehensive acoustic emission studies with concrete. He used twelve 15 x 30 cm cylindrical concrete specimens; four cylinders each made from three different aggregate materials (limestone, graywacke and chert). Acoustic emissions were monitored under standard ASTM test procedures for compressive strength, static Young's modulus, Poisson's ratio and tensile splitting strength. Additionally, a model prestressed concrete pressure vessel was proof-loaded while acoustic emissions were monitored. Emissions as high as 100 kHz were recorded. GREEN's work is significant in so far as he proved clearly that acoustic emissions from concrete are an indicator of failure processes. Early warning of total compressive failure and preliminary correlation to material modulus was indicated. Resolution of gross cracking, onset of pressure vessel leakage and failure and prestressing rod failure was accomplished.

In the last decade hand in hand with the advent of modern micro electronics the application of acoustic emission analysis increased significantly. Two main fields in the application with concrete have been developed:

- Techniques for studying the initiation and propagation of cracks as well as the mode of concrete failure,
- Techniques for the localisation of cracks.

A lot of data and publications are available. Nevertheless many questions remained unsolved. Up to now no unambiguous relation between crack parameters and detected acoustic emission signals could be found, respectively. Further basic research is needed to improve this method to such an extent, that its potential for material tests can be fully utilized.

5. Application of Acoustic Emission in Plain Concrete
5.1 Uniaxial Short-term Compressive Tests

The acoustic emission of concrete during short-term compressive loading has been the subject of many investigations [15, 19, 46 to 65]. Nevertheless, the acoustic emission technique is still under consideration. This is due to the fact that many of the reported data are hard to interprete as the:

- test materials are different,
- test descriptions are incomplete,
- test procedures are different,
- test conditions are not comparable,
- acoustic emission techniques and the evaluations of the signals are different.

However, a lot of data, which have been published up to now, describe important aspects with respect to crack formation and propagation in concrete. The main features studied are:

- threshold values for characteristic emissions in concrete,
- irreversibility of crack formation (KAISER-effect),
- different types of loading,
- effect of different types of specimens,
- different aggregates (lightweight aggregate), size of coarse aggregate,
- load history.

In the following typical results and concluding remarks of the above mentioned topics are summarized.

General observations: Independent of the technique or evaluation method used results similar to those indicated in fig. 9 have been observed. The curve reveals three characteristic regions of acoustic emission or crack development. It appears that even only a small number of acoustic emission events with small amplitudes occur in region I. Distinct signals clearly distinguishable from interfering noises occur from about 15 - 25 per cent of the failure load onward.

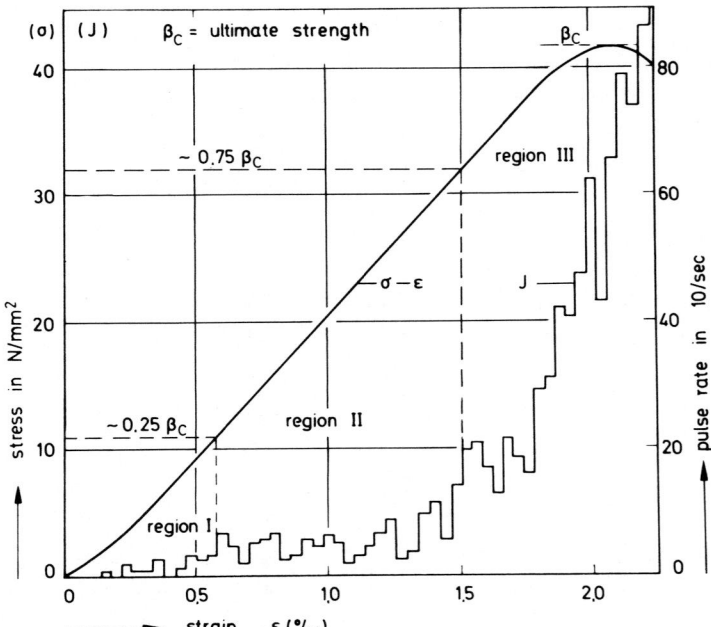

Fig. 9. Acoustic emission pulse rate from normal concrete under uniaxial compressive load [17]

After this there is for the time being no appreciable further increase in pulse
rate, mean amplitude or event rate, respectively. The formation of micro-cracks
or stable crack propagation is assumed to be the essential mechanism in this
region (region II). An abrupt increase in the pulse rate at about 75 - 90 per
cent of failure load indicates incipient destruction of the internal structure.
A high and dense pulse spectrum as well as a steep rise of the pulse rate indicate the occurence of unstable crack propagation leading to failure (region III).

From the literature reviewed it can be stated that with the aid of the acoustic emission analysis both the development and the type of structural damage
can be estimated. In cases where only the development of the structural damage
is of interest, normally counting methods (pulse rate, pulse sum, event rate or
event sum) are sufficient. If one needs a more precise idea of the type of structural damage a signal valuation is necessary (pulse area analysis, mean amplitude, mean amplitude analysis or frequency analysis). In most cases the development and type of structural damage can be valuated sufficiently accurate by the

three mean characteristics pulse sum, pulse area (or energy) and pulse amplitude.

Threshold values for characteristic emissions in concrete: With the aid of the three above mentioned characteristic values:

- pulse sum or event sum,

- signal amplitude,

- pulse area sum or pulse energy sum

it is possible to explain roughly the development of structural damage or crack development in concrete. There are three more or less marked regions (see fig. 10 - 15) which may be related to typical threshold values of concrete:

- Region I.

  Immediately after loading low energy emissions occur which are typical for friction and slip phenomena. The reasons are the initial compaction of voids formed during hydration and elasto-plastic deformations of the matrix (cement

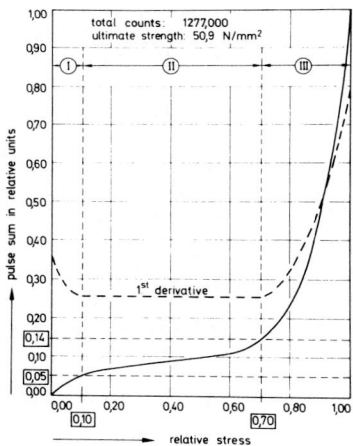

Fig. 10. Relative pulse sum of light-weight aggregate concrete versus stress (ultimate stress = 1) [19]

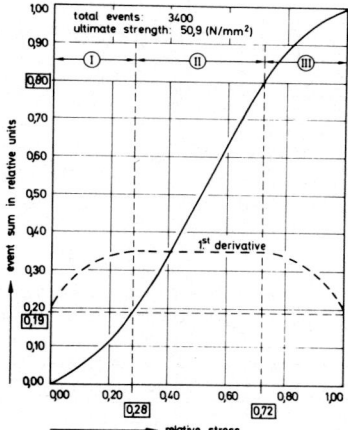

Fig. 11. Relative event sum of light-weight aggregate concrete versus stress (ultimate stress = 1) [19]

paste) accompanied by initial formation of microcracks. This region is characterized by a steep increase of the pulse sum curve connected with small amplitudes and small pulse areas. Region I comprises the stress intervall $0 < $ stress/ultimate strength $\leqq 0.15$.

- Region II

    During the course of increasing stresses the formation of micro-cracks increa-
    ses more or less intensively. The intensity depends on the compressive
    strength and the size of the coarse aggregates. The region II extends up to
    the so-called "limit of discontinuity" (σ_d). This threshold value separates
    the region of stable microcrack formation from the region of unstable micro-
    crack formation. It is characterized by a distinct raise of the pulse sum

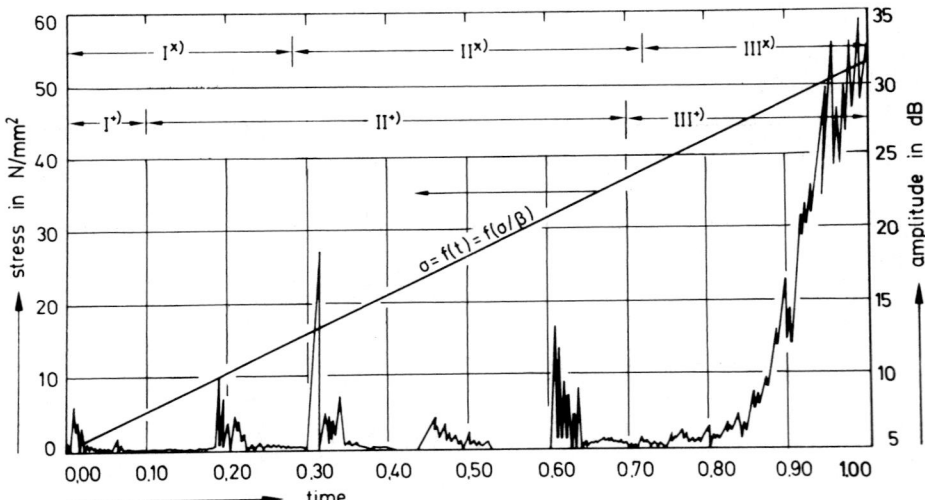

Fig. 12. Applied stress and measured peak amplitudes of the acoustic emission
signals of lightweight aggregate concrete versus loading (time until
fracture = 1, ultimate strength = 50.9 N/mm^2)
x) region I - III determined from the event sum curve,
+) region I - III determined from the pulse sum curve [19]

curve which is approximately proportional to the applied stress. Small signal
amplitudes and a more or less pronounced shift of the pulse area distributions
in direction to higher relative pulse areas are observed. The "limit of discontinuity" σ_d overlaps the so-called "microcrack limit" $\sigma_D \approx 0.7 - 0.8$ as
mentioned in [17, 19, 52, 53]. The actual value of σ_D, however, depends on the
type of concrete, the ultimate strength, the maximum size of the coarse aggregate, etc.

- Region IIIa

    Due to the accumulation of microcracks macrocracks develop. With the aid of pulse area or energy analysis also other kinds of structural damage are observable, for instance fracture of grains in lightweight aggregate concrete. Both the relative pulse sum curve and the relative pulse area sum curve indicate a nearly exponential raise. Simultaneously the signal amplitudes increase

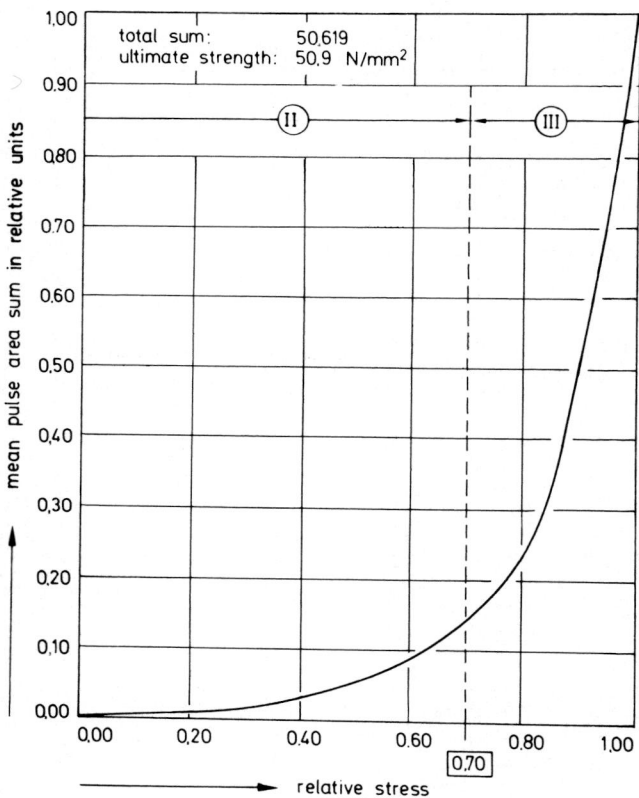

Fig. 13. Mean pulse area sum versus applied stress (ultimate stress = 1) of lightweight aggregate concrete [19]

significantly. The region is limited by the area of unstable crack propagation which can be determined by amplitude measurements (see fig. 12, raise of the signal amplitude > 20 dB). The "stability boundary" separates the areas of stable and unstable micro-cracking. The transition from smaller structural disintegrations to a total structural destruction is called "upper structural threshold" [17, 19, 52, 53]. Depending on the concrete type and mixture the related stress σ_{stab} is in the interval $0.85 < \sigma_{stab} \leqq 0.95$.

174

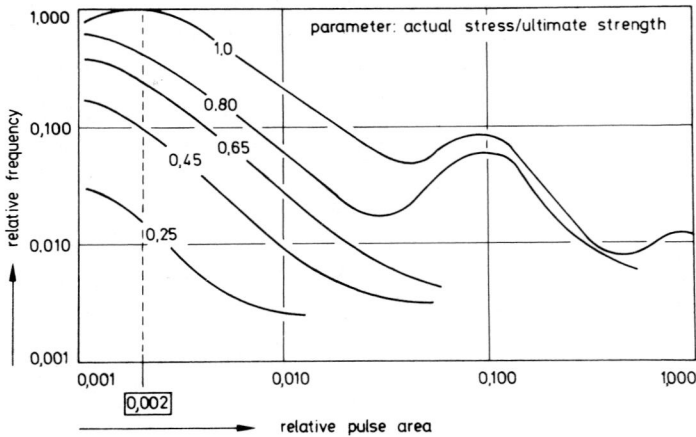

Fig. 14. Pulse area distributions as function of load level of lightweight
aggregate concrete (ultimate strength 50.9 N/mm^2) [19]

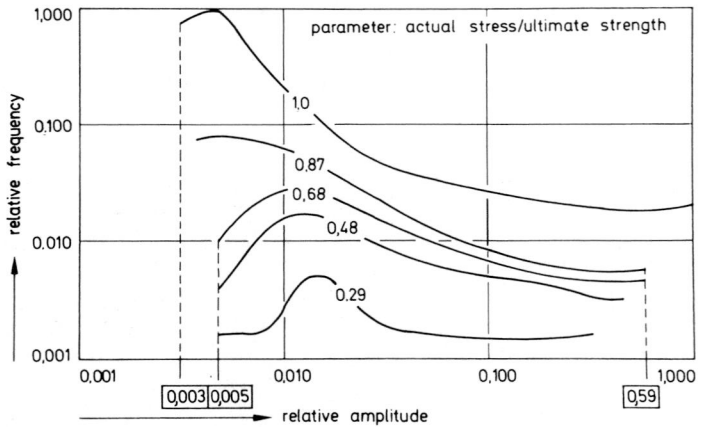

Fig. 15. Peak amplitude distribution as function of load level of lightweight
aggregate concrete (ultimate strength 50.9 N/mm^2) [19]

- Region IIIb

The stability boundary (σ_{stab}) subdivides the phase of stable from the phase
of unstable macrocracking (σ_{inst}). Approaching this boundary the specimen
shows more and more continuous cracks. They consist only of prismatic areas
which are more or less mechanically interconnected. If this critical value is
exceeded a rapid destruction of the specimen occurs. The related acoustic
emission signals show mostly maximum amplitudes and rates and the processing
of the signals is in almost all cases heavily influenced by the electronics'
capacity [19]. Therefore in this region mainly indifferent maximum signal
values in all analysis methods were recorded. With respect to the concrete
type, mix proportion, specimen geometry and load devices the related stress
σ_{inst} is in the range $0.9 < \sigma_{inst} < 1$ [52, 53].

By frequency analysis of the acoustic emission waves it is possible to trace
the changes of internal structure of concrete more detailed [17, 19, 51, 63, 64].
Fig. 16 (a, b and c) shows the stress (σ)-strain (ε) curve, the result of location search of local fracture and the relationship between power spectra of
acoustic emission waves and relative stress level (σ/σ_{max}), respectively. The
acoustic power spectra of recorded acoustic emission waves were determined by
applying the fast Fourier transformation method. The estimation of the location

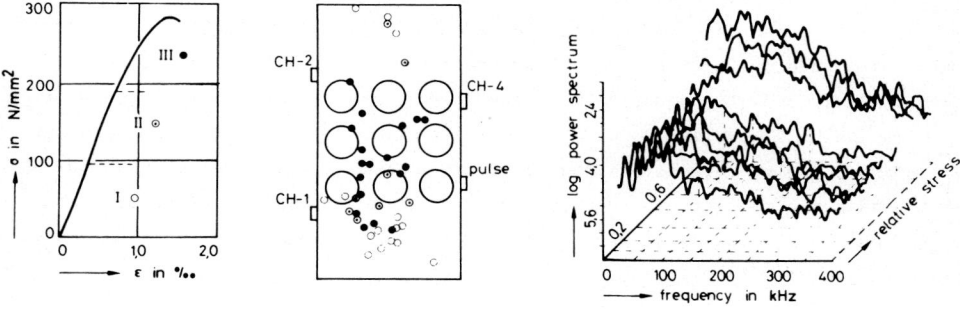

Fig. 16. Stress-strain curve, crack location and frequency spectrum of a model
concrete specimen containing nine model aggregates under compressive
load [64]

of the local fracture ensued as follows: The average of three source locations
was calculated from the time lags of the onset of acoustic emission waves picked
up by four sensors at the same time. Simultaneously the longitudinal wave velo-

city through the test specimen which is necessary to calculate the source location was measured by feeding a rectangle pulse into the specimen.

Fig. 16c shows that the low frequency components are prominent and the power spectra of the acoustic emission waves are almost constant in the stress region I and II. In region III the high frequency components increase suddenly. Nearby the maximum stress, the lower frequency components under 50 kHz also increase, and the power of acoustic emission increases suddenly. It is assumed that these tendencies in the frequency characteristics approximately correspond to the subsequent fracture process of concrete: the occurence of bond cracks and mortar cracks, propagation of these cracks, the interconnection to each other, and finally occurence of macroscopic cracks.

Frequency analysis of different workers [19, 63, 64, 65] show that the acoustic emission signals occur in the range of approximately 10 kHz to 30 MHz. However, interference noises and sound attenuation limit the practically utilizable range to between 50 kHz and 1.5 MHz. Above 800 kHz the signal amplitudes are reduced by more than 20 dB. The frequency range of higher intensities - that means of sufficient amplitudes - comprises 50 kHz to 300 kHz for normal concrete and 50 kHz to 400 kHz for lightweight aggregate concrete.

KAISER effect: The occurence of the KAISER effect in concrete depends essentially on the recovery time after loading and the previously attained stress maximum [15, 46, 56, 58]. Fig. 17 shows the influence of the previously applied stress level. The upper part of the figure indicates the pulse sum of a normal concrete specimen which was loaded to 36 per cent of its ultimate stress then unloaded and after that immediately reloaded to failure. Obviously no acoustic emission occurs during the second load cycle until previously applied stress level is exceeded. The lower part of the figure shows the pulse sum of a similar specimen which is preloaded to 75 % of its ultimate load. In this case acoustic emissions occur shortly before the previous maximum stress level is reached. As mentioned above this stress level separates two different mechanisms of crack formation. This value of stress level coincides with the transition from region II to region III (see fig. 10).

Investigations with longer periods between unloading and reloading have shown that the KAISER effect decreases with increasing recovery periods [58]. This is assumed due to the visco-elastic behaviour of the cement paste. The recovery processes which are nearly independent of the previous stress level make the pulse sum and the pulse area sum (energy sum) raise again. The influence of the rest period duration is already reduced after periods of 28 days to an imme-

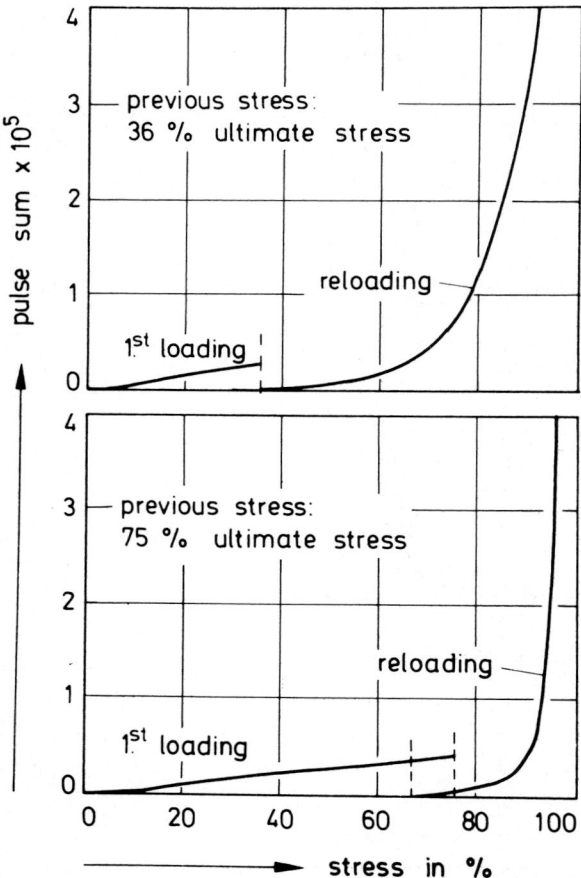

Fig. 17. Acoustic emission pulse sum of reloaded normal concrete demonstrating
the "KAISER effect" (stress in % of ultimate stress) [15]

asurable extent. However, depending on the rest periods principeally the KAISER
effect does occur. With increasing previous stress levels, however, the KAISER
effect is less marked.

The KAISER effect has been observed in thermal stressed concrete, too [15,
60], Fig. 18 shows such an example. The upper part shows the pulse sum of a
normal concrete specimen during the first heating cycle. Obviously from 350 $^{\circ}$C
onwards a distinct acoustic emission appears. In the second heating cycle there
is no acoustic emission until the previously applied maximum temperature is exceeded. However, in general the KAISER effect due to thermal stresses is less

178

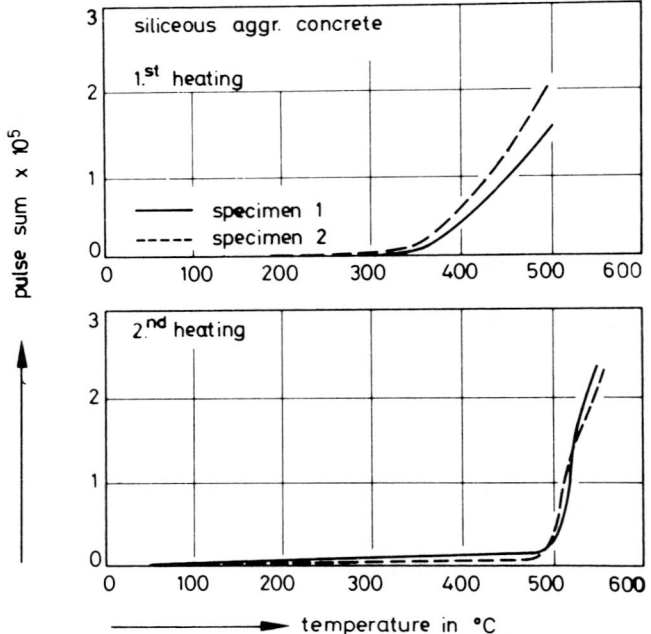

Fig. 18. Acoustic emission pulse sum of heated normal concrete demonstrating
the "thermal KAISER effect" [15]

distinct than the mechanical stress induced KAISER effect.

The KAISER effect is assumed to occur also after cyclic loading (fatigue
tests) and long-term compressive loading. Comprehensive studies on this problem
which allow to give unambigeous statements are lacking. Information about the
use of the KAISER effect for reliable supplementary determinations of maximum
stresses in concrete structures or members are not at hand.

Effects of loading devices: Deformation and strength of concrete specimens
are significantly influenced by the loading device used. This is due to the
restrained lateral deformations of the specimens caused by friction effects between the specimens and more or less stiff platens. As a result of this different modes of fracture occur which can be detected by the acoustic emission
technique [52]. This is shown in Fig. 19, which contains the pulse sum curves
versus applied stress of 10 cm concrete cubes tested with different loading
devices. It is clearly recognized that rigid steel platens more or less prevent
the development of cracks up to high stresses. Interlayers of lubricated alumi-

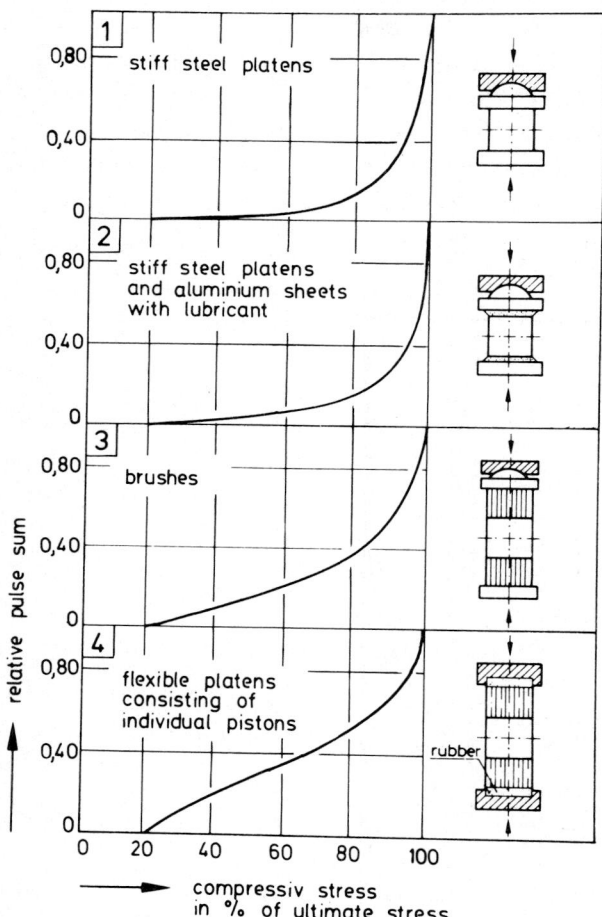

Fig. 19. Relative pulse sum versus compressive stress of 10 cm normal concrete cubes tested by means of different loading devices [52]

num sheets modify this result only to a small extent. When steel brushes are used the development of significant cracks starts earlier. Flexible platens finally show a steady increasing of fracture already from low stresses onwards. These results are supported by tests of prismatic concrete specimens with different lengths but the same cross section [53]. From these tests it can be seen that by the use of flexible platens the dimension of the specimen is of minor importance. On the other hand stiff platens indicate great differences - the more stout the specimens are the slighter is the initial slope of the pulse sum curve and the steeper is the increase near the ultimate stress.

Specimen size effects: It is well established in the testing of concrete cylinders that the size of the test cylinder itself can be a contributing factor
towards the ultimate strength. Attempts have been made to see whether the acoustic emission response follows similar lines [56]. Typical results are indicated
in Fig. 20. It shows the load versus acoustic emission response of 3-day old
concrete cylinders of different diameters (7.62; 15.24 and 22.86 cm; height/
diameter = 2). It can be seen that nearly over the whole stress level the smal-

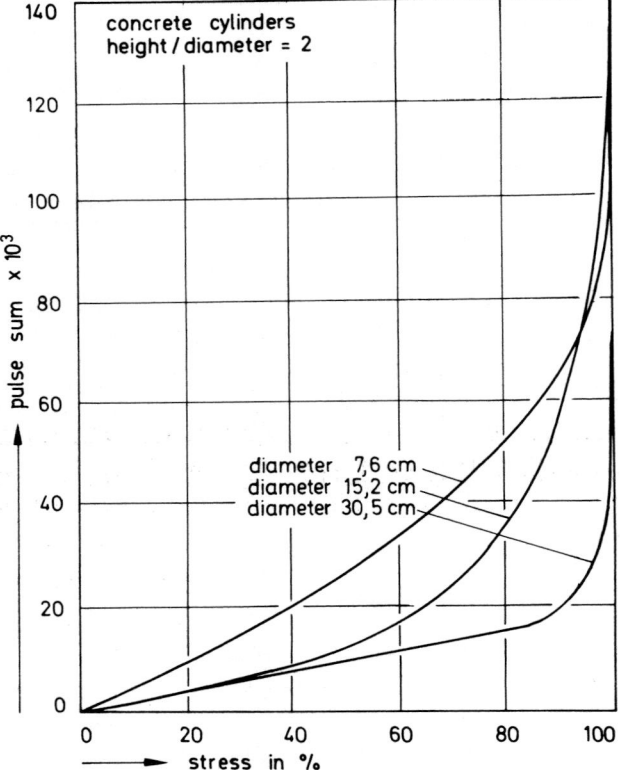

Fig. 20. Pulse sum versus stress (in % of ultimate stress) of three-day old
normal concrete showing the effect of specimen seize [56]

ler cylinder are more emittive than the larger ones. This is mainly due to the
fact that in small specimens the acoustic waves can reach the transducer in
form of surface waves to a greater extent. These surface waves are less attenu-

ated compared to spherical waves from the inner of the specimen. Therefore more
events with relative low original amplitudes can be detected. On the other hand
per event a greater number of pulses exceeds the discriminator threshold. Further in small specimens at relative low stresses already a great number of small
cracks are forming because for the redistribution of local peak stresses only a
relatively small volume exists. In case of larger specimens there is a larger volume for the visco-elastic stress redistributions, therefore at low stress levels
only a small number of cracks occur. In total, however, a large specimen contains in total a greater number of dislocations and additionally a greater energy content is stored during the loading, therefore a large specimen emits single
bursts already at a lower stress level, too.

Effects of aggregates: To determine aggregate effects a lot of investigations
with normal and lightweight aggregate concrete have been performed [19, 57] (see
Fig. 21). The available reports on normal concrete also involve investigations
on concretes with mineralogically different sorts of coarse aggregates, various
amounts and sizes of aggregates [29, 51, 55]. For a conclusive estimation of
the influence of the mineralogical character on the acoustic emission the reported data are insufficient yet.

However, it has been proved that the acoustic emission is significantly influenced by the size of aggregates. - A reduction of the diameter of the coarse
aggregate - transition from concrete to mortar - leads to a more homogeneous material. Therefore the peak stresses occuring in the contact zone are reduced.
As a result of this the part of the continuous emission is elevated and the total pulse sum of mortar is greater compared to concrete. Additionally, because
of the better homogeneity of mortar this material reacts with increasing loads
already at low stress levels by "gliding" and the formation of small cracks.
Therefore from each "gliding" or cracking process emissions of low energy content are emanating. The measured total energy sum is reduced compared to concrete by this effect.

However, the greatest differences occur between lightweight and normal concretes. This is due to the fact that the load bearing mechanisms in both concretes are of distinct difference. In normal concretes mainly the aggregates
bear the external loads. On the other hand the lightweight aggregates have lower
strengths and Young's modulus than the cement paste. In this case the stresses
concentrate in the paste. This results in an elevated total pulse sum of lightweight concrete compared to normal concrete. In lightweight concrete predominantly continuous emission occurs, too. This is due to the fact that the extreme
stresses in the cement paste matrix lead to displacements in the cement gel before a crack formation occurs.

182

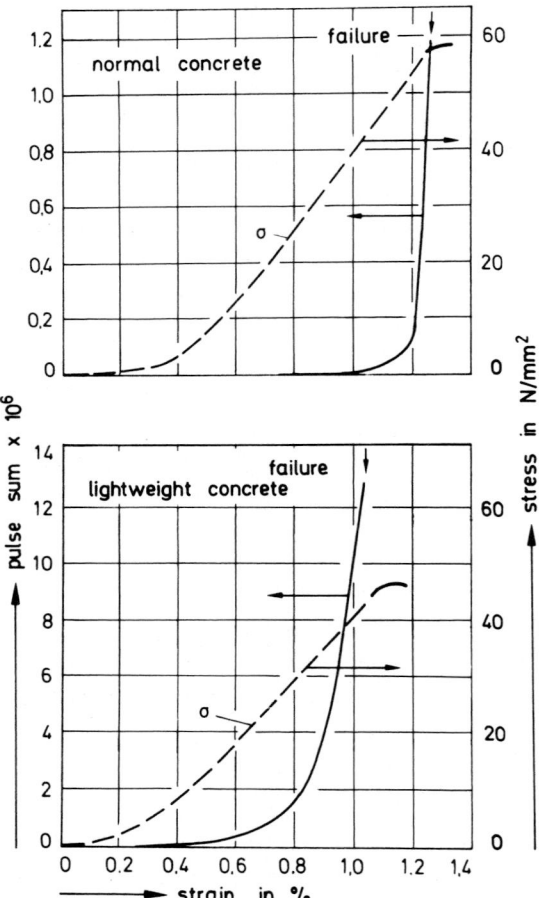

Fig. 21. Pulse sum and stress as function of strain showing the different fracture behaviour of normal and lightweight concrete under compressive
loading [57]

The lower tensile strength of the lightweight aggregate is a reason also of
higher emissivity at lower stress levels. Aggregate fracture in normal concrete
occurs only in the last stage before the complete specimen failure.

In accordance with detailed signal evaluations it can be stated that in case
of comparable ultimate strength lightweight aggregate concretes fail more
brittle than normal concretes [19].

Effects of ultimate strength: One must admit that the true effect of concrete strength or mix proportion, age and specimen size on acoustic emissions may actually be concealed by the characteristics of the source, the transmission and the reception of the elastic waves [19, 56]. Especially information concerning the changes in the amplitudes and frequencies of emitted signals and the attenuation of those signals and the variation of the various types of impulses (i.e. longitudinal, RAYLEIGH, LAMB etc.) as function of the strength (age, mix proportion, specimen size) must be analysed in more detail before the apparent effects can be ambigeously substantiated.

Although a lot of questions are still unsolved at least a qualitatively clear picture of the effect of concrete strength can be drawn. - Due to the adjustment of the Young's modulus and the strength of the cement matrix to the according values of the aggregates high strength concretes are more homogeneous in the bearing stress distribution. The differences between aggregates stress levels and matrix stress levels are diminished. As a result also the contact zone matrix/aggregates in the case of high strength concrete gets only relatively low stresses. This hold for mortars, too.

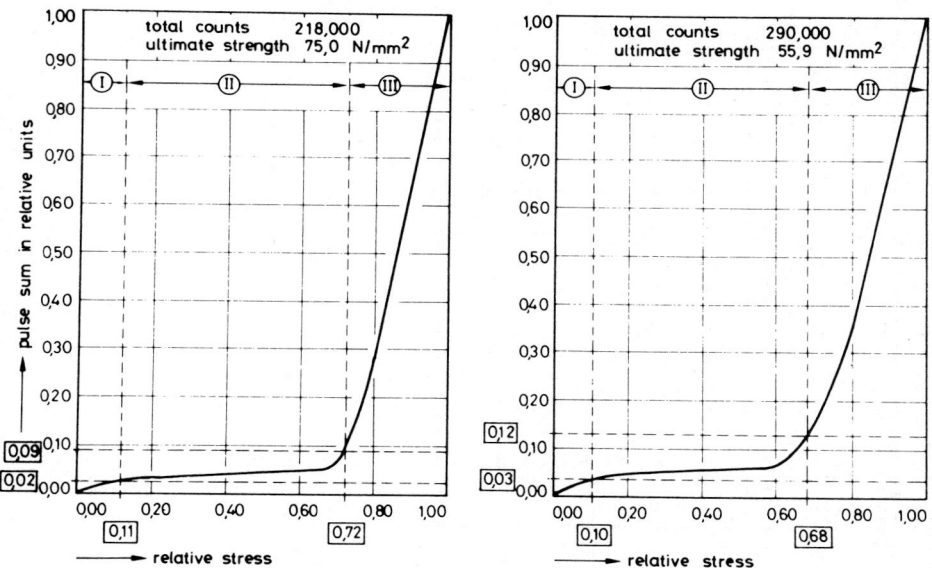

Fig. 22. Relative pulse sum versus stress (ultimate stress = 1) of normal concrete showing the effect of concrete strength on the emanating acoustic emission [19]

184

Figure 22 gives an impression of the effect of ultimate strength on the pulse sum curves. The initial slope of the high strength concrete is less than that of the other concrete. Additionally the total pulse sum of the high strength concrete exceeds that of the concrete with lower strength. Also the limits of the regions I and II are shifted to higher stresses for the high strength concrete.

Detailed valuations (pulse area analysis, frequency analysis) of signals confirm the well known fact that high strength concretes fail in a more brittle manner than low strength concretes do [19].

5.2 Uniaxial Long-term Compressive Tests

Occasionally it has been attempted to investigate the effects of constant sustained load with acoustic emission techniques with the aim to ascertain the suitability of the method for the monitoring of time-dependent processes [46, 56]. The example presented in fig. 23 shows the relationship between pulse rate, load duration and as an additional parameter load level.

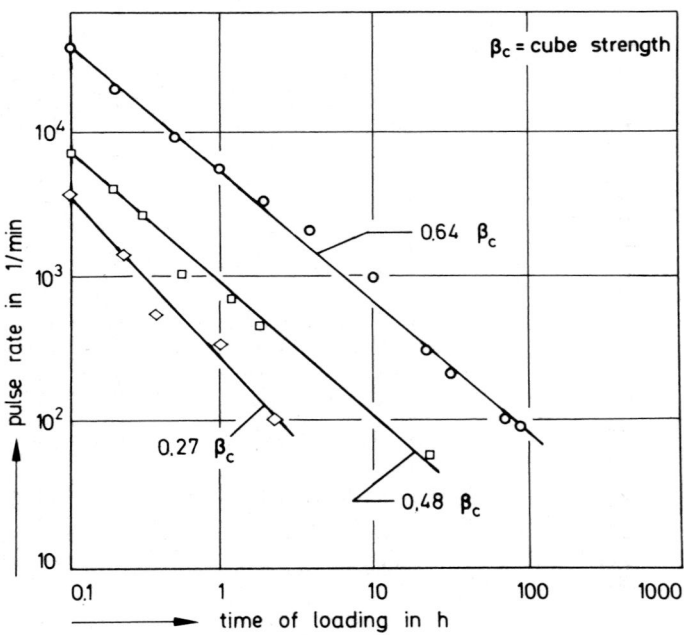

Fig. 23. Acoustic emission from normalweight concrete under sustained load [56]

It can be seen that the acoustic emissions increased rapidly as the load approaches the ultimate fracture load. Further it can be seen that the acoustic
emission is being generated still after 80 h of load, however, the pulse rate
is extremely reduced at this stage. This can be explained by an internal stress
redistribution which takes place due to time dependent visco-elastic deformation of the cement paste. This is connected with spontaneous reductions of peak
stresses by crack formation. These mechanisms are assumed to lead to the well
established residual strains after unloading, too.

5.3 Uniaxial Cyclic Compressive Tests

In [66 - 68] comprehensive tests to determine the effect of a load with various load levels and frequencies are reported. The aim was to answer the
question in what way the magnitude and frequency of pulsating loads contribute
to structural damage and how the damage develops as a function of time during
the service life of the structure. In this connection the results of acoustic
emission measurements and direct measurements of volume changes with cylindrical concrete specimens (5 cm in diameter and 10 cm high) were compared. Single-
stage tests (WÖHLER) and multi-stage tests (MINER) were performed.

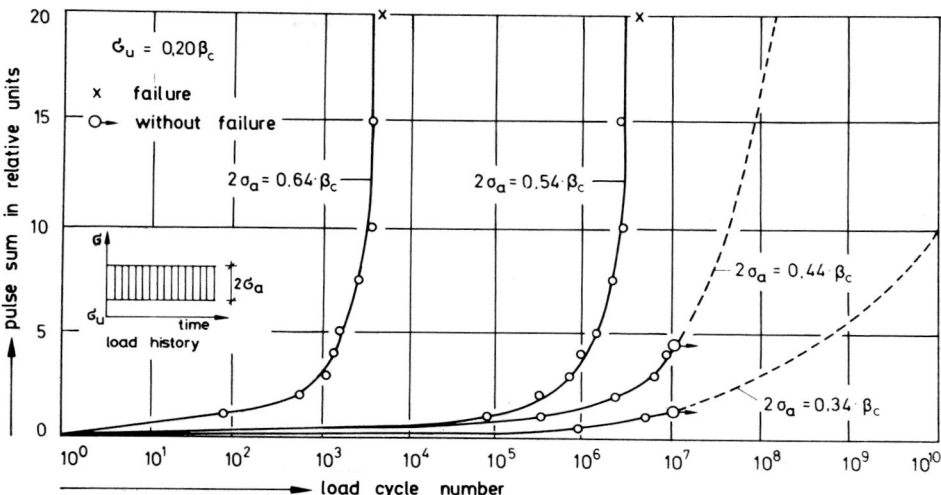

Fig. 24. Damage development under single-stage loading (WÖHLER test) indicated
by pulse sum versus load cycles [67]

Fig. 24 shows the pulse sum of the acoustic emissions during single-stage tests (WÖHLER). These tests were performed with constant sinusoidal fatigue loads. The lower stress limit, referred to the static compressive strength, was σ_u = 0.20 β_c the upper stress limit varied between 0.64 and 0.34. The load frequency was 1 Hz and 10 Hz. From Fig. 24 the following emerges: For the loads in the finite fatigue strength range, which leads to a failure of the specimen, it is possible to plot complete damage curves. The tests with loading in the quasi fatigue strength range were terminated at about 10^7 cycles. By extrapolating the curves beyond the range of investigation with particular respect to the curves obtained in the finite life range, failure load cycle numbers between 10^8 and 10^{12} are obtained. This would indicate that fatigue stress ranges below the quasi-fatigue strength deduced from WÖHLER lines, which would produce failure of the specimens, too.

Plots of the relative pulse sum (referred to the instant of failure) against the specific number of load cycles (ultimate load cycle = 1) reveal three characteristic regions of damage development. To each region certain stages of internal cracking are generally believed to correspond. It appears that even a small number of load cycles will bring about certain changes, e.g., in consequence of micro-cracking (region I). After this there is at first no further increase of damage (region II). It seems as if a notable increase begins only in the last quarter of the service life (region III). A high and dense pulse spectrum as well as a steep rise of the pulse sum are indicated. Finally the occurence of unstable crack propagation leads to failure. This characteristic behaviour is clearly manifested by volume change measurements [66, 67].

From the results of two-stage tests it likely appears that the damage also depends on the combination of the load collective.

5.4 Splitting Tensile Tests

Connected with investigations on the relationship between fracture mode and acoustic emission characteristics splitting tensile tests on mortar cylinders were performed as well. Especially source location measurements and frequency analysis were conducted [64, 65].

The source location tests performed show that at relative low stress levels local fracture occurs nearby the plane along the axis of load. At higher stress levels the locations gradually tend to shift from the outer regions to the center.

Fig. 25 indicates the mean power spectra of a specimen under splitting tension at different relative stress levels ($\sigma/\sigma_{ultimate}$). In the range of $\sigma/\sigma_{max} \geq$ 0.4 the peak value nearly 20 kHz of the power spectra gradually decreases with the increase of the relative stress level. The high frequency components of 150 - 200 kHz become prominent in this region.

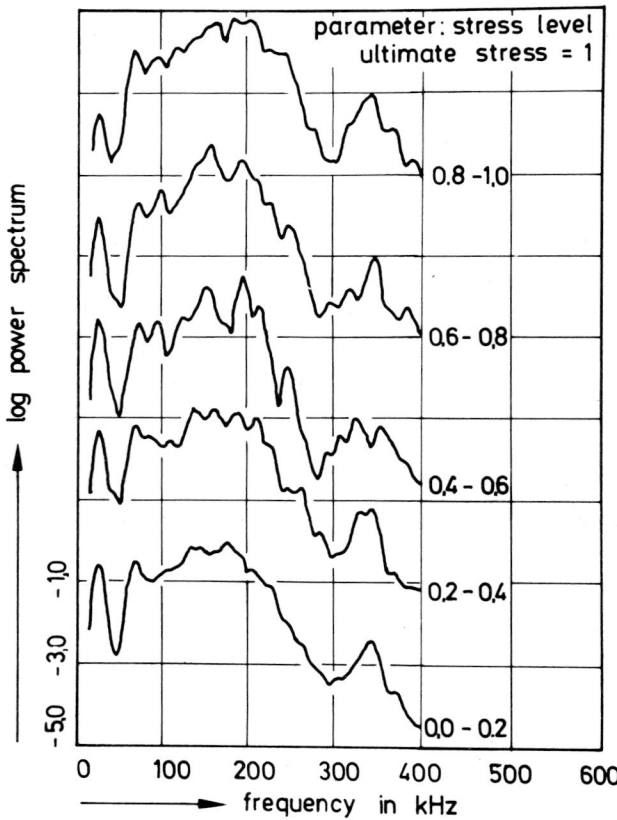

Fig. 25. Acoustic emission power spectra of mortar cylinders under splitting tensile loads [65]

A comparison of these results with results obtained from shear compression tests as well as flexural tests and compressive tests shows that the high frequency components of the acoustic emission waves tend to increase with the increase of stress level, independent of the fracture mode of specimens [51, 64, 65].

5.5 Tensile Tests

On the occurence of acoustic emission in specimens made of mortar and concrete under tensile stresses a detailed report is given in [69]. The cylindrical
specimens were strain-controlled loaded. If an acoustic emission event was detected the strain was kept constant until the spectrum analysis of the received
signal was completed. This takes e.g. 2 min. During this hold time, however,
further signals were received, stored and also processed.

Simultaneously to this sound localisations were performed by using two adap-
ters. The measurements showed that several emission centres are simultaneously
active, that means that crack formation occurs at numerous places in the specimen. The existence of a great number of crack centres is assumed to be a stabi-
lizing factor because the cracks hinder themselves during their propagation.

The analysis of the acoustic emission spectra indicated the existence of
three typical kinds of spectra. These are: the "low frequency spectrum" with
maximum frequencies of 100 kHz (important frequencies are in between 30-50 kHz),
the "middle frequency spectrum" with maximum frequencies of 200 kHz containing
only 30 - 50 kHz frequencies of minor amplitudes, and the "high frequency spectrum" containing frequencies of relevant amplitudes up to 300 kHz, respectively.
The "low frequency spectrum" occurs at low stress levels and is due to the
starting of micro crack formation at the aggregate matrix interface. An increase
of the applied stress promotes the micro crack development and the cracks become
wider during their propagation through the mortar. These processes are believed
to produce the "middle" and the "high frequency spectra", respectively. The tensile tests show similar patterns. As already mentioned in connection with compressive and tensile splitting tests: increasing stresses lead to an increase
of the detected frequencies.

5.6 Bending Tests

McCabe et al. [56] performed a series of tests with concrete beams (10 x 15 x
76 cm^3) under three point loading. Fig. 26 shows the results. The transducers
were mounted on the compressive face and on the tension face of the beams. Tension was observed to be the governing failure mode in the unreinforced beams. It
should be noted that the acoustic emission detected at the tension face was considerably higher than that detected at the compression face at all stress levels
up to tensile failure. These findings are supported by investigations in [60],
where similar beams (20 x 20 x 50 cm^3) were tested under three point loading.

Additionally it was found, that the acoustic emission before failure strongly
depends on the stress state. In the case of compressive loading the foregoing

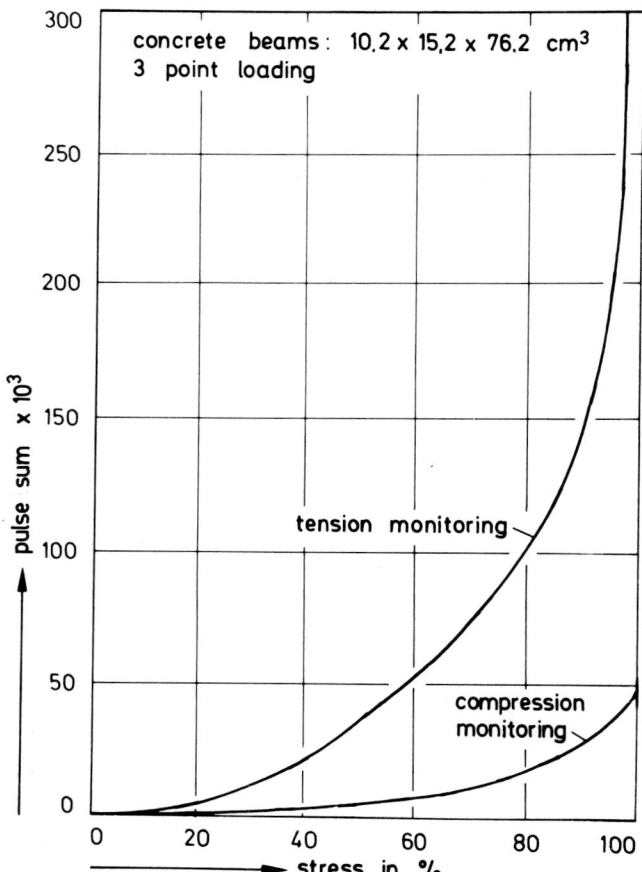

Fig. 26. Load versus pulse sum of concrete beams under three point loading with transducer mounted either at the compression face or at the tension face [56]

warning that the failure point is reached is more distinct than in the case of bending. This is assumed due to the lack of frictional noises during the opening of tensile cracks and the different energy content of the acoustic emissions caused by the different ultimate strengths in compression and tension, respectively [56].

Tests with hold time periods during the loading show that in accordance with compressive tests in bending tests stable states are indicated by decreasing acoustic emission activity during the hold time periods [60]. Frequency characteristics of acoustic emission waves under flexure show that the component of high frequencies are prominent in the power spectra, nearly independent of the stress level, and as mentioned above, the energy content of the acoustic emissions is relatively low. Additionally, the low frequency components of about 25 kHz are less than those in the power spectra of other failure modes [64, 65].

Bending tests (four point loading) performed with unreinforced beams made from pure cement paste lead to somewhat different results: after the formation and detection of initial cracks unstable cracking led to failure of the specimen without any warning [70].

5.7 Shear Tests

To some extent shear tests have been performed with mortar specimens by using the methods of crack localization and frequency analysis [64, 65]. The purpose of these acoustic emission tests was to examine the acoustic emission characteristics and to trace the fracture processes. The specimens used were of prismatic shape (10 x 10 x 30 cm^3) and the angles of the shear plane were 0° (uniaxial compression), 45° and 60°, respectively.

The crack localization measurements have shown that independent of the shear plane angle the cracks are located mainly in the central portion of the specimen even if high stress levels are applied. In specimens whose shear plane angles were 45° and 60°, resp., the cracks are distributed preferentially in the circumference of the shear plane and the specimen axis. In specimens under uniaxial compression the generated cracks are irregularly distributed in the central portion of the specimens at middle and higher stresses. The authors concluded from their results that the tension fracture occurs at the same time as the shear compression fracture in the specimen does.

The frequency analysis show that the low frequency components below 50 kHz decrease and the high frequency components increase gradually with increasing stress levels, resp. But the middle frequency components are prominent throughout all stress levels, this seems to be typical under both splitting tension and shear compression. That means, there are no ambiguous characteristics in the power spectra, which allow to distinguish between these two modes of failure.

5.8 Triaxial Compressive Tests

Acoustic emission measurements during triaxial compressive loading of model concrete specimens (36 mm in diameter and 72 mm in length) made with 5 mm glass balls and cement paste (400 kg/m^3) have been reported in [71]. The tests were carried out in a triaxial cell. The samples were loaded using the classical two

step method: first of all, increasing the hydrostatic pressure (P) to the
desired lateral pressure; secondly increasing axial stress (σ_1), while keeping
the lateral pressure constant, whereby the axial strain rate was kept constant
during the axial loading. Simultaneously the acoustic emission was detected by
one transducer mounted on the loading piston.

Fig. 27 represents the pulse sum (dashed line) versus the deviatoric stress
(σ_1 - P) for different levels of the lateral pressure (P) and, curves of constant "acoustic emission rates" (N^*) defined as the first derivate $N^* = \Delta N/\Delta\sigma$.
From these curves and the comparison with lateral strain versus axial strain
curves the authors deduce three characteristical domains in the σ_1-P-plane:

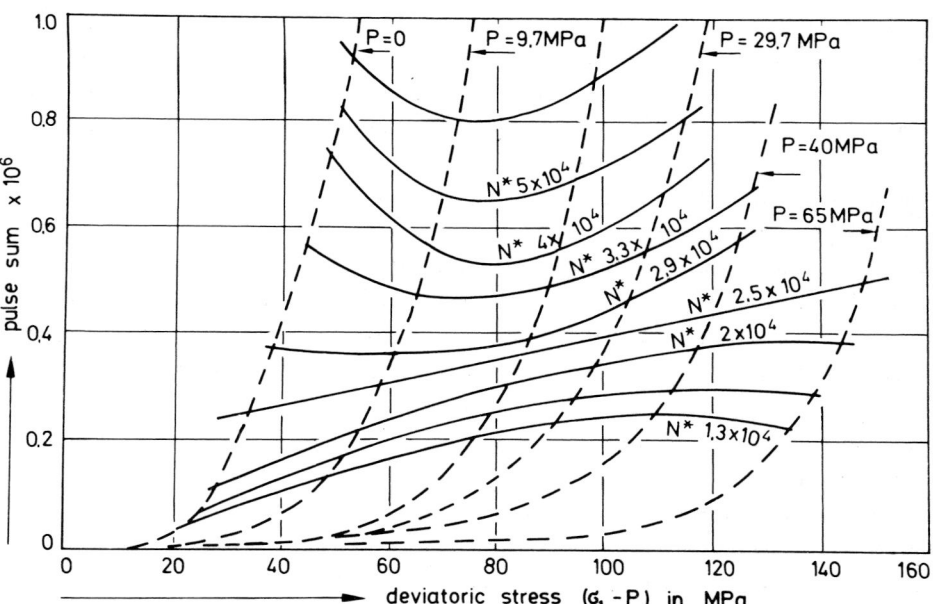

Fig. 27. Acoustic emission pulse sum versus deviatoric stress and curves of
          constant "emission rates" ($N^* = \Delta N/\Delta\sigma$) of model concrete during
          triaxial loading [71]

The "elastic domain", the "non-microcracking domain" and the "domain of
microcracking". The "elastic domain" is defined as being the limit of linearity of the stress strain curves. The domains of microcracking and non-microcracking are separated by the straight line corresponding to $N^* = 2.5 \cdot 10^4$

counts per MPa (see fig. 27). By triaxial creep tests the author can show that creep becomes stable if σ_1-P-loading combinations corresponding to the non-microcracking domain of the σ_1-P-plane are applied. A delayed failure can be observed if σ_1-P-loading combinations corresponding to the microcracking domain are applied. The acoustic emissions in the non-microcracking domain are assumed due to sliding and friction of original crack surface.

5.9 Thermal Tests

Thermal induced acoustic emissions have been studied in the high as well as in the low temperature region. To investigate the influence of elevated temperatures on the crack formation and crack propagation tests were performed on concrete cylinders (14 cm in diameter, 14 cm long) [60]. The specimens were heated with a constant heating rate of 5 K/h to 200 °C. When the temperature had been reached it was maintained for 10 ... 100 h. After this hold time period the specimens were cooled with a constant cooling rate of 5 K/h to ambient temperature. Some specimens were heated again to investigate the "thermal KAISER effect".

The observed acoustic emission of the heated specimens indicated an increase in the pulse rate in conformity with the temperature rise; during the hold time period the pulse rate decreased markedly and during the cooling a gradual increase of the pulse rate was observed. Further it has been proved that the thermal induced acoustic emission shows a very low intensity in these experiments compared to compressive tests. The "thermal KAISER effect" could be observed as expected.

Under consideration of the investigation of the fire behaviour of materials Schneider et al. [15, 72] extended the temperature region in their tests up to 900 °C. They used cylindrical specimens (8 cm in diameter, 30 cm long) made of various mortars, cement pastes and concretes, resp. The specimens were heated in a special furnace with a heating rate of 2 K/min up to 900 °C. Simultaneously the thermal strains, the surface temperatures of the specimens and the event rates were measured. In these experiments quartz-glass rods (70 cm in length) were used as waveguides, which were glued on the specimen surface with a thermal resistens resin.

A typical test result is represented in fig. 28. It shows the thermal strain and the event rate versus the surface temperature of a normal concrete specimen made of siliceous aggregate and Portland cement. Four temperature regions with a distinct crack behaviour can be distinguished:

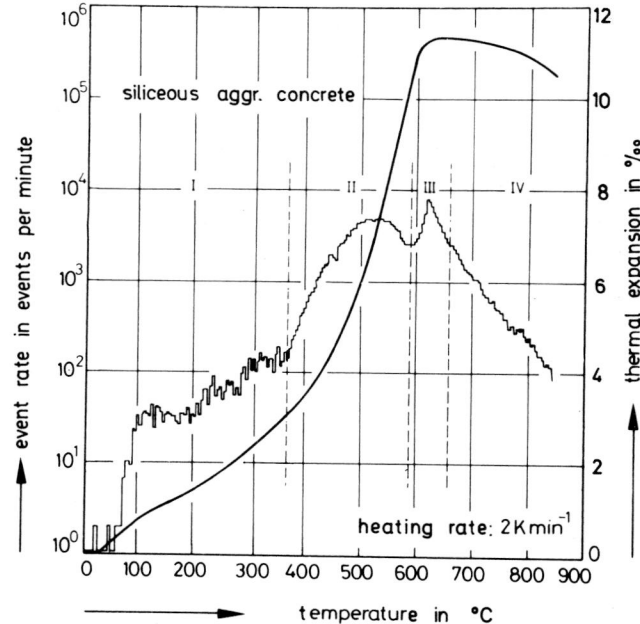

Fig. 28. Event rate and thermal expansion as function of temperature of
          siliceous aggregate concrete [15]

region I: First of all due to the evaporation of the pore water shrinkage of
             the cement paste occurs, whereby marked internal stress intensities
             are induced. Similar to shrinkage stresses due to the common drying
             process these thermal induced shrinkage stresses lead to micro-
             cracking.

region II: At temperatures up to nearly 600 °C the high thermal expansion of
             the siliceous aggregate causes large deformation of the cement stone
             matrix, which results in cracks. Furthermore the decomposition of
             the portlandite takes place. Both phenomena cause event rates up to
             5.000 per minute. Known from experience, in these region siliceous
             aggregate concrete shows the steepest losses in strength.

region III: Again a sharp increase of the event rate occurs if the temperature
             approaches nearly 600 °C. This is assumed to be due to the micro-

194

cracking caused by the spontaneous volume increase corresponding to the quartz inversion at 573 °C.

region IV: Above approximately 650 °C a further degradation of the material occurs due to the deterioration of the cement paste, which simultaneously shows a volume decrease. The event rate gradually decays because the concrete gets more and more the characteristics of a high deteriorated debris in which no internal stresses exist. Additionally ultrasonic attenuation measurements have shown that due to the strong degradation of the matrix the ultrasonic attenuation rises significantly. Naturally this leads to a decreasing of the detectable events.

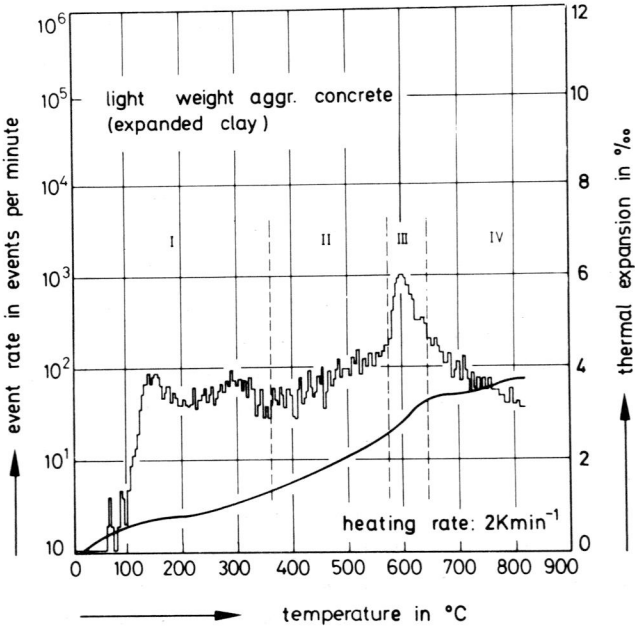

Fig. 29. Event rate and thermal expansion as function of temperature of lightweight aggregate concrete (expanded clay) [15]

Fig. 29 shows the event rate of similar specimen made with lightweight aggregate concrete. Remarkable in comparison to fig. 28 are the low thermal strains
of the lightweight concrete and the relatively low event rates in the temperature region 350 to 800 $^{\circ}$C. Due to the content of siliceous fines in the concrete mix, at about 600 $^{\circ}$C a peak in the acoustic emission activity (event rate)
occurs, which indicates the quartz inversion.

From additional measurements with mortars and cement paste it follows that
thermal loaded concretes may show differences in their crack behaviour, where
the main determining factor is the type and content of the coarse aggregate
used.

Concerning low temperature tests with concrete only few results are available
[60, 73], but the performed tests give evidence that freezing of pore water
leads to intensified internal stresses. If the local strength of the material
is exceeded, this leads to crack formations. The detection of the emitted sounds
is very difficult. This is due to freezing noises of condensed water vapor from
the ambiency, which occurs preferentially on the cold ends of the transducer or
wave guide. Beyond it the amplitudes of the acoustic emission signals are far
below those amplitudes which are commonly detected during high temperature and
mechanical loading. A well elaborated measurement technique, however, would improve the acoustic emission technique for investigations of frost damage of concrete.

5.10 Special Applications

In [74] the conversion process of high aluminum cement paste (HAC) (monocal-
cium aluminate decahydrate, dicalcium aluminate octahydrate and alumina gel →
tricalcium aluminate hexahydrate and gibbsite) was examined by acoustic emission
analysis. As it is well known the conversion proceeds very slowly at ambient
temperatures. It is accelerated at elevated temperatures. If HAC concrete members are manufactured with a high water/cement-ratio and served under elevated
temperature conditions (for example 60 $^{\circ}$C) the conversion process, which is
accompanied by the evolution of free water, may produce a significant weakening
of the members.

In [72] tests were performed with HAC specimen (20 x 10 x 7.5 cm^3) immersed
for 72 h in a stirred water bath at 60 $^{\circ}$C. The measurements show that no acoustic
emission occurs during the conversion of HAC-cement paste if the water/cement
ratio was less than 0.35. In the case of higher water/cement ratios the conversion was accompanied by a high level of acoustic emission activity. This is
assumed to be due to the liberated water, which leads to a hydrostatic pressure,
whereby microcracking will be enhanced. It is concluded that this process could
result in a significant decrease in strength, whereby eventually a failure of
HAC concrete load bearing members may occur.

The latter problem has been studied in detail by determining the characteristic strength of concrete beams in situ by means of acoustic emission monitoring [75].

Meiser and Tressler [76] used the acoustic emission analysis for the examination of the fracture behavior of aluminious cement composites. They found out a difference between the fracture mechanism of low density aluminous cement composites and high density composites. They showed either a failure mechanism caused by cumulative damage or a failure caused by the propagation of existing cracks.

Chhuy Sok and Baron |77, 78] investigated unreinforced notched concrete specimens of large size (double cantilever beam type, 178 cm in length) with special reference to fracture mechanics. They determined the onset of crack formation at the crack tip by one-dimensional source location measurements. It was found out that an extended damage zone occurs during the loading in front of the root long before any cracks become visible. The length of this damage zone has been reported to be more than 20 cm. The authors concluded that the use of small specimens for true fracture mechanics investigations lead to unreliable parameters. Therefore the specimens should have a minimum extension of about 50 cm in length.

6. Application of Acoustic Emission with Bond Effects

The published works concern three main research fields: bond between aggregate and mortar matrix [64, 65], fibre reinforced concrete [70], bond between embedded steel bars and concrete [79 - 81]. The latter concerns mainly the fire problem.

For investigating the bond between aggregate and mortar, Tanigawa et al. [64, 65] performed shear compression, splitting tensile and bending tests with two phase specimens. The specimens were made by cutting large aggregate pieces of sufficient size in a form of trapezoidal prismens, half cylinders, or half beams, resp. The aggregate pieces were moulded together with cement mortar paste as to get the test specimens (prismens 10 x 10 x 30 cm^3, cylinders 15 cm in diameter and 10 cm long, beams 10 x 10 x 40 cm^3). Source location measurements and frequency analysis were accomplished.

It has been found out, that the low frequency components under 50 kHz are small and almost constant throughout all stress levels in the specimens at shear compression bond tests in contrast to those at shear compressing tests. This difference is assumed to result from the difference between the fracture mechanism of shear compression and that of shear compression bond failure. Further it has been found out, that the fracture in the two phase specimens only occur at the interface between matrix and aggregate.

Bond specimens, which failed under splitting tension, showed a similar acoustic emission spectra than pure mortar specimens under splitting tension (see chapter 5.4). The results of source locations indicate that most of the crack locations are distributed nearby the plane along the axis of load, but shifted a little into the matrix half.

In the frequency spectra of bond specimens, which failed by bending the frequency components of about 150 kHz, are prominent at maximum, though the number of observed acoustic emission waves was very low. Unfortunately, results of source location measurements are not at hand.

The acoustic emission of fibre reinforced concrete was investigated by Faninger et al [70]. They performed bending tests with carbon fibre and fibre glass reinforced normal concrete and aerated concrete with carbon fibre reinforcement, resp. The specimens (beams 10 x 10 x 100 cm) were four point loaded. During the loading up to failure the deflection and the pulse sum were recorded.

The results show that unlike to the unreinforced material the fibre reinforced concrete indicates, as to be expected, distinct differences in the deflection behavior and the corresponding acoustic emission. Cement stone specimens without reinforcement fail spontaneously after the occurence of the first initial crack by unstable crack propagation without any warning by acoustic emission. In glass fibre and carbon fibre reinforced specimens an acoustic emission occurs after a critical deflection or bending stress level is exceeded. At this stage inter fibre fracture with partial fibre fracture is assumed to be the governing mode of fracture. Further loading leads to progressive deflection accompanied by a sharp increase of the signal amplitudes and the pulse rate, which now indicates that mainly fibre fracture occurs. A further load increase leads to unstable cracking and subsequent failure of the specimen whereby the pulse rate shows again a steep increase. With the aid of acoustic emission it is possible to dinstinguish fracture effects, which occur in the fibres of in adjacent areas by the evaluation of amplitudes of the acoustic emission signals.

Acoustic emission studies of bond effects between embedded steel bars and concrete are reported in [79 - 81]. Royles et al. [79] examined the influence of elevated temperature on residual bond between concrete and steel over the range of temperatures from 20 to 800 $^{\circ}$C. Cylindrical pull-out specimens were employed. Bond stress-slip relations together with associated acoustic emission measurement were performed, where the signals were picked up from the steel/ concrete interface. The tests were carried out in the cold state after heating to the desired temperature and subsequent cooling.

198

From figures in [79] representing the acoustic emission event sum in relation
to slip and bond stress it can be seen that the bond stress - event sum relation has a similar form as the bond stress - slip relation. The authors assume
that the emission is caused by a breakdown of adhesion and cracking along the
interface and associated local crashing under the ribs.

In [80,81] similar experiments are reported, however, the tests were performed in the hot state. Acoustic emission was examined also during heating.
Fig. 30 shows typical relations between both the acoustic emission pulse sum
and the difference between thermal expansion of steel and surrounding concrete

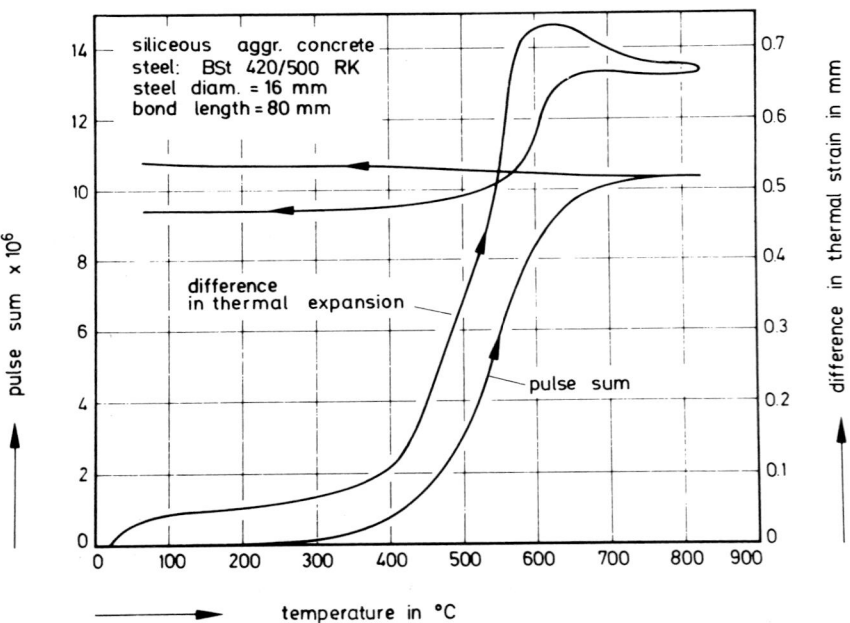

Fig.   30.   Curves of acoustic emission pulse sum and difference in thermal
            expansion of concrete and embedded steel during a heating -
            cooling cycle [81]

during a heating/cooling cycle (heating and cooling rate 1 K/min). It can be
seen that the event sum during the temperature raise is very similar to the
difference between the thermal strains of steel and the surrounding concrete.
This means, that thermal incompatibility is the main reason for the decrease
of the bond strength. During cooling the acoustic emission is relatively small

because due to the high deterioration of the concrete, further internal stresses cannot develop.

7. Application of Acoustic Emission with Concrete Members and Structures

Most tests on concrete members and structures have been performed in connection with non-destructive testing and in-service monitoring of primary containments for nuclear reactors [51, 59, 60, 82 - 90], which are either thick-walled steel pressure vessels (pressurized water reactors) or massive prestressed concrete structures (gas-cooled reactors). The ability to detect, localize and analyze cracks in these structures is important for ensuring structural integrity throughout their 30 to 40 year design life. The investigations mostly have included reactor noise measurements, leak detection with controlled leaks of primary water and the development of a modified plant startup technique, which greatly assists monitoring of crack growth signals during pressurization.

In urban buildings, bridges, arches etc. the acoustic emission technique, e.g. for the detection of prestressing tendons failure, has not been applied yet. This seems to be due to the fundamental difficulty in applying the extremely large loads, which are necessary to get sufficient deflections and stress levels. On the other hand the acoustic emission as a technique for monitoring structural integrity has been successfully applied in concrete beams and flat slabs, and it was shown that the technique is able to identify, locate and indicate the severity of crack propagation in concrete members as well as to detect the onset of failure [75, 82, 84, 91].

8. Summary

The report summarizes our knowledge on acoustic emission techniques with concrete and concrete members. It gives a survey on origin, propagation, shape and size of acoustic emission signals. The experimental techniques including detection and processing of the received signals are illustrated as well as the evaluation methods. Furthermore the common use of the acoustic emission technique in testing materials like metals, glasses, ceramics, plastics with reference to crack formation and propagation is mentioned.

After a brief historical review of the application of acoustic emission analysis with concrete the results of acoustic emission tests with plain concrete, bond effects and concrete members and structures are reported and discussed. It was found that with this method one is able to trace crack formations due to different kinds of loading as uniaxial compression, splitting tension, tension, bending shear etc. and thermal incompatibilities. It is possible to separate different modes of fracture by means of detailed signal valuations, e.g. amplitude evaluations and frequency analysis.

The acoustic emission is a nondestructive method, which allows to measure in situ. A special advantage of this method relates to the fact that it detects crack events in real time. Further advantages are that crack formation in concrete can be observed in its initial state, even before visible cracks can be observed. By means of the "KAISER effect" it is possible to estimate previous stress levels or temperature actions. Besides that it is possible to differentiate stable and unstable crack development under a sustained load. Also phase transformations like freezing and HAC-conversion may be studied. The method enables long-term monitoring of concrete members and structures with the aim to identify, locate and indicate the severity of crack propagation as well as the onset of failure.

Although the acoustic emission method requires a rather elaborate signal evaluation it is a useful tool in the field of non-destructive testing of concrete. Up to now the acoustic emission analysis has not yet been fully strained to the utmost limits of its capacity. An unambigeous theory connecting the acoustic emission signals with crack parameters is lacking.

9. REFERENCES

[1] J. Becht, J. Eisenblätter and P. Jax, Werkstoffprüfung mit der Schall-
     emissionsanalyse (SEA), Z. f. Werkstofftechnik/J. of Materials Technol.,
     4 (1973), Nr. 6, pp. 306-314.
[2] T. Licht, Acoustic Emission, Brüel & Kjaer, Technical Review, No. 2, 1979.
[3] J. Kolerus, Schallemissionsanalyse, Teil 1: Schallemission: Entstehung,
     Ausbreitung und Anwendung, Teil 2: Verfahren und Geräte, Technische Mes-
     sen, 47. Jg. (1980), Heft 11 und 12.
[4] J.R. Mitchell, Fundamentals of Acoustic Emission and Application as an NDT
     Tool for FRP, 34th Annual Technical Conference, (1979), Reinforced Pla-
     stics/Composites Institute, The Society of the Plastics Industry, Inc.,
     Section 3-F, pp. 1-3.
[5] H. Schillalies, Grundlagen und Anwendungen der Schallemissionsanalyse SEA
     bei nichtmetallischen anorganischen und spröden Werkstoffen, Handbuch der
     Keramik, Verl. Schmid GmbH, Freiburg i.Brg. (1979), (Gruppe III T).
[6] P. Jax, Schallemission bei plastischer Verformung von Metallen, Ber. zum
     Symp. d. Deut. Ges. f. Metallkunde: Schallemission, München (1974),
     pp. 59-117.
[7] H.-A. Crostack and E.Roeder, Anwendung der Schallemissionsanalyse zur Be-
     stimmung von Vorgängen der Rißbildung und Rißausbreitung, Der Maschinen-
     schaden 48, H.1 (1975), p.17.
[8] F. Wittmann, Bestimmung physikalischer Eigenschaften des Zementsteins,
     Deutscher Ausschuß für Stahlbeton, Heft 232, Verlag Wilhelm Ernst & Sohn,
     Berlin (1974).
[9] R.W.B.Stephens, A.A.Pollock,Waveforms and Frequency Spectra of Acoustic
     Emissions, The J. of the Acoustical Soc. of Amer. 50 (1971), pp. 904-910.
[10] S.H. Carpenter and C.R. Heiple, Acoustic Emission Generated by Dislocation
     Mechanisms during the Deformation of Metals, Symp. on Acoustic Emission
     during a Joint Meeting of the American and Japanese Acoustical Societies,
     Honolulu, Hawaii (1978).
[11] J.D. Blacic and R.L. Hagman, Wide-band Optical-mechanical System for Mea-
     suring Acoustic Emissions at High Temperature and Pressure,
     Rev. Sci. Instrum. 48 (1977), No. 7, pp. 729-732.
[12] H. Borchers and H.-M. Tensi, Eine verbesserte piezoelektrische Methode zur
     Untersuchung von Vorgängen in Metallen bei mechanischer Beanspruchung und
     bei Phasenänderung, Zeitschrift f. Metallkunde, 51 (1960), Heft 4.
[13] W. Brockmann and T. Fischer, Die Schallemissionsanalyse als Prüfverfahren
     für Metall-Klebeverbindungen, Materialprüf. 19 (1977), Nr. 10, pp. 430-436.
[14] J. Eisenblätter, Schallemissionsanalyse. Ein neues zerstörungsfreies Prüf-
     verfahren, Ingenieur Digest, 11. Jg., Heft 10 (1972), pp. 62-67.
[15] U. Schneider, W. Rosenberger and U. Diederichs, Untersuchung der Rißkine-
     tik in Beton durch Schallemissionsanalyse, Mitteilungsblatt für die amtli-
     che Materialprüfung in Niedersachsen, Vol. 20/21 (1980/81), Bad Harzburg,
     (1981).
[16] H. Schmidt, Udbredelse af Akustisk Emission i Beton, Danmarks Tekniske
     Hojskole, Afdelingen for Baerende Konstruktioner, Lyngby. R 93. Lyngby,
     (1978), 123 S.
[17] H. Weigler und D. Klausen, Die Schallemissionsanalyse, Betonwerk + Fertig-
     teil-Technik 45 (1979), H. 12, pp. 709-716.
[18] A.G. Evans and M. Linzer, Acoustic Emission in Brittle Materials, Ann.
     Rev. Mater. Sci. (1977), Nr. 7, pp. 179-208.
[19] N.Feineis, Anwendung der Schallemissionsanalyse (SEA) als zerstörungsfrei-
     es Prüfverfahren für Beton, Dissertation TH Darmstadt (1982).
[20] K.G. Schmitt-Thomas, H.M. Tensi and H. Zeitler, Möglichkeiten der Schall-
     emissionsmessungen bei werkstoffkundlichen Untersuchungen. Ber. zum Symp.
     der Deutschen Gesellschaft für Metallkunde: Schallemission, München (1974),
     pp. 1-23.
[21] A. Tobias, Acoustic-emission Source Location in Two Dimensions by an Array
     of Three Sensors, Non-destructive Testing 9 (1976), pp. 9-12.

[22] J. Kaiser, Untersuchungen über das Auftreten von Geräuschen beim Zugver-
      such. Dissertation TH München (1950).
[23] J. Eisenblätter und G. Fanninger, Zur Anwendung der Schallemissionsanalyse
      in Forschung und Technik. Metall H. 1 (1977), S. 51 und H. 2 (1977),
      S. 139.
[24] G. Fischer und W. Pompe, Schallemissionsmessungen an Glaskeramiken und
      Keramiken unter Dauerstandbedingungen, Ber. Dt. Keram. Ges., 55 (1978),
      pp. 431-436.
[25] J. Ruge und V. Jürgens, Einsatz der Schallemissionsanalyse zur Beurteilung
      der Qualität von Schmelzschweißverbindungen an niedriglegierten, hochfe-
      sten Baustählen, Schweißtechnik, 25 (1975), Nr. 9, pp. 410-411.
[26] J. Ruge und V. Jürgens, Einsatz der Schallemission zur Beurteilung von
      Schweißverbindungen an niedriglegierten Baustählen, Ber. zum Symp. der
      Deutschen Ges. f. Metallkunde: Schallemission, München (1974), pp.310-321.
[27] L. Obert, Use of Subaudible Noises for Prediction of Rockbursts, U.S.Bu-
      reau of Mines, Report Investigation 3555 (1941).
[28] E.A. Hodgson, Bulletin Seismological Society of America, Vol. 32, No. 249
      (1942).
[29] P. Jax und D. Hums, Feststellung und Überwachung beginnender Rißbildung
      und des Rißwachstums in keramischen Werkstoffen mit Hilfe der Schallemis-
      sionsanalyse, Keramische Zeitschrift, 26 (1974), Nr. 11, pp. 625-630.
[30] P. Jax, Acoustic Emission as an Aid to Detecting and Monitoring Crack
      Formation and Crack Growth in Ceramic Materials, Proc. of the 3rd Meeting
      of the European Working Group on Acoustic Emission, Ispra (1974), pp.
      55-86.
[31] D. Hums und P. Jax, Ermittlung beginnender Rißbildung in keramischen Werk-
      stoffen bei thermischer Belastung mittels Schallemissionsanalyse (SEA),
      Science of Ceramics, 7 (1973), pp. 281-295.
[32] B.J. Dalgleish et al., The Temperature Dependence of the Fracture Tough-
      ness and Acoustic Emission of Polycrystalline Alumina, J. of Materials
      Science 14 (1979), pp. 2605-2615.
[33] M.J. Noone, R.L. Mehan, R.C. Bradt, D.P.H. Hasselmann und F.F. Lange (Ed.),
      Observation of crack propagation in polycrystalline ceramics and its re-
      lationship to acoustic emissions, Conf. University Park, PA., USA,
      11-13 July 1973. Symposium on Fracture Mechanics of Ceramics. Vol. I,
      New York USA (1974).
[34] Kh.-G. Schmitt-Thomas, H.M. Tensi und H. Zeitler, Schallemissionsmessungen
      an einer Legierung vom Typ AlCuMg₂ mit unterschiedlichen Aushärtungszu-
      ständen, Aluminium 51 (1975), Nr. 8, pp. 520-524.
[35] G.C. Robinson, C.R. Reese und E.A. La Roche jr., Determination of Thermal-
      ly Induced Acoustic Emission of Ceramics, Ceramic Bulletin 53 (1974),
      No. 6, pp. 482-485.
[36] J.H. Speake and G.J. Curtis, Characterization of the Fracture Processes in
      CFRP Using Spectral Analysis of the Acoustic Emission Arising from the
      Application of Stress. Carbon Fibres, Conf. London, 18-20 Febr., (1974).
[37] J. Holt and I.G. Palmer, Interpretation of the Acoustic Emission Signals
      from the Deformation of Low Alloy Steels, Ber. zum Symp. d. Deutschen Ge-
      sellschaft für Metallkunde: Schallemission, München (1974), pp. 24-44.
[38] J.P. Weiss, Zerstörungsfreie Werkstoffprüfung mit Schallemission, Sonder-
      druck aus: Zeitschrift für Schweißtechnik, Nr. 10 (1978), pp. 221-228.
[39] M. Mirabile, Acoustic Emission Energy and Mechanisms of Plastic Deformati-
      on and Fracture, Ber. zum Symp. d. Deutschen Gesellschaft für Metallkunde:
      Schallemission, München (1974), pp. 189-221.
[40] F.J. Esper and H.M. Wiedenmann, Thermal Shock Resistance of Ceramics De-
      termined by Acoustic Emission Analysis, Ber. Dt. Keram. Ges. 55 (1978),
      Nr. 12, pp. 507-510.
[41] L.J. Graham, G.A. Alers, R.C. Bradt, D.P.H. Hasselmann and F.F. Lange,
      Microstructural Aspects of Acoustic Emission Generation in Ceramics. Conf.
      University Park, PA., USA, July 1973. Symposium on Fracture Mechanics of
      Ceramics., Vol. I., New York (1974), USA.

[42] A.S. Tetelman, A.G. Evans, R.C. Bradt, D.P.H. Hasselmann and F.F. Lange,
      Failure Prediction in Brittle Materials Using Fracture Mechanics and
      Acoustic Emission. Conf. University Park, PA., USA, 11-13 July 1973. Sym-
      posium on Fracture Mechanics of Ceramics., Vol. II. New York (1974), USA.

[43] J.J. Schuldies, The Acoustic Emission Response of Mechanically Stressed
      Ceramics. Mater. Eval. (USA). Co: MAEVAD. Vol. 31, No. 10, pp. 209-213,
      Oct. (1973).

[44] J. Eisenblätter, Industrielle Anwendung der Schallemissionsanalyse, Ber.
      zum Symp. d. Deutschen Ges. f. Metallkunde: Schallemission, München (1974),
      pp. 222-253.

[45] H. Ahlborn, J. Becht and H.-J. Schwalbe, Untersuchungen zur Rißzähigkeit
      verschiedener GFK-Laminate unter Einsatz der Schallemissionsanalyse (SEA),
      Kunststoffberater 6 (1975), pp. 300-306.

[46] H. Rüsch, Physikalische Fragen der Betonprüfung, Zement-Kalk-Gips, Heft 1
      (1959), pp. 1-9.

[47] R.G.L'Hermite, What do we know about the Plastic Deformation and Creep of
      Concrete?, RILEM Bulletin, No. 1, März (1959), pp. 21-51.

[48] R.G. L'Hermite, Volume Changes of Concrete. Proceedings, Fourth Intern.
      Symposium on Chemistry of Cement, V. II, National Bureau Standard Washing-
      ton, D.C. (1960), pp. 659-694, NBS Monograph No. 43.

[49] G.S. Robinson, Methods of Detecting the Formation and Propagation of Mi-
      crocracks in Concrete, Internat. Conf. on the Structure of Concrete: Ses-
      sion C (1965), pp. 131-145.

[50] D. Wells, An Acoustic Apparatus to Record Emissions from Concrete under
      Strain, Nuclear Engineering and Design, V. 12 (1970), pp. 80-88.

[51] A.T. Green, Stress Wave Emission and Fracture of Prestressed Concrete
      Reactor Vessel Materials, Second Interamerican Conference on Materials
      Technology, American Society of Mechanical Engineers, V. 1, Aug. (1970),
      pp. 635-649.

[52] G. Schickert, Acoustic Emission Technique Applied to Tests with Concrete
      Cubes. 2nd Intern. RILEM Symposium on "New Developments in Non-destruc-
      tive Testing of Non-metallic Materials", Constanta/Rumänien (1974).

[53] G. Schickert, Schwellenwerte beim Beton-Druckversuch, Forschungsbericht
      70, Bundesanstalt für Materialprüfung (BAM), Berlin (1980).

[54] V.M. Malhotra, Testing Hardened Concrete: Nondestructive Methods. Am.
      Concr. Inst. Monogr. No. 9 (1975), pp. 1-188.

[55] D.C. Spooner and J.W. Dougill, A Quantitative Assessment of Damage
      Sustained in Concrete During Compressive Loading, Magazine of Concrete
      Research, 27, No. 92 (1975), pp. 151-160.

[56] W.M. McCabe, R.M. Körner and A.E. Lord jr., Acoustic Emission Behavior of
      Concrete Laboratory Specimens, ACI Journal, 73 (1976), pp. 367-371.

[57] P. Jax and H. Gaar, Verfolgung von Bruchvorgängen im Mikro- und Makrobe-
      reich an Gläsern und Glaskeramik mit Hilfe der Schallemissionsanalyse
      (SEA), Glastechn. Ber., 50 (1977), No. 9, pp. 229-236.

[58] J. Nielsen and D.F. Griffin, Acoustic Emission of Plain Concrete, J. of
      Testing and Evaluation, 5, No. 6 (1977), pp. 476-483.

[59] H. Hick, Entwicklung der Schallemissionsmeßtechnik am Physikinstitut
      (1977 - 1981), Österreichisches Forschungszentrum Seibersdorf GmbH.,
      OEFZS, Ber. No. A0265, PH-330/81, Oktober (1981).

[60] H. Hick, T. Schmeskal, R. Schwara, S. Zagiczek, C. Werbik, J. Němet, L.
      Weißbacher, E. Aschenbrenner und R. Recker, Untersuchung zur sicherheits-
      technischen Überwachung sowie zur Fehlerauffindung mittels Schallemissi-
      onsanalyse für Bauwerke aus Spannbeton. Österreichisches Forschungszentrum
      Seibersdorf GmbH., OEFZS, Ber. No. A0274, PH-334/81, Dezember (1981).

[61] K. Okada, W. Koyanagi und K. Rokugo, Energy transformation in the fracture
      process of concrete. Memoirs of the Faculty of Engineering, Kyoto Univer-
      sity, Vol. XXXIX, part 3, July (1977), Kyoto, Japan, pp. 389-402.

[62] Y. Niwa, M. Ohtsu and H. Shiomi, Waveform analysis of acoustic emission in
      concrete. Memoirs of the Faculty of Engineering, Kyoto University, Vol.
      XLIII, Part 4, October (1981).

[63] D.G.Fertis, Concrete Material Response by Acoustic Spectra Analysis. ASCE J. Struct. Div., V. 102, N. 2, (Febr. 1976), pp. 387-400.

[64] Y. Tanigawa, K. Yamada and S.-I. Kiriyama, Frequency characteristics of acoustic emission waves of concrete, Transactions of the Japan Concrete Institute, Vol. 2, (1980), pp. 155-162.

[65] Y. Tanigawa, K. Yamada and S.-I. Kiriyama, Relationship between fracture mode and acoustic emission characteristics of mortar, The 24th Japan Congress on Materials Research - Non-metallic materials, (March 1981), pp. 241-247.

[66] D. Klausen, Festigkeit und Schädigung von Beton bei häufig wiederholter Beanspruchung. Dissertation TH Darmstadt (1978).

[67] H. Weigler und D. Klausen, Ermüdungsverhalten von Beton - Auswirkung einer Beanspruchung im Dauerfestigkeitsbereich, Betonwerk + Fertigteil-Technik, Heft 4 (1979), pp. 214-220.

[68] H. Weigler, Beton bei häufig wiederholter Beanspruchung, Contribution to "Techn.-wiss. Zementtagung", Hannover (1980).

[69] M. Terrien, Emission acoustique et "compartiment méanique post-critique" d'un béton sollicité en traction, Bull. Liaison L.P.C. no. 105, Janvier/Février (1980), Réf. 2398.

[70] G. Faninger, K.H. Grünthaler and H.J. Schwalbe, Faserverstärkter Beton und Schallemission, Betonwerk + Fertigteil-Technik, Heft 2 (1977), pp. 82-86.

[71] J. Bergues and M. Terrien, Study of Concrete's Cracking under Multiaxial Stresses, Advances in Fracture Research, Ed.: D. Francois et al. Pergamon Press, Oxford and New York (1980).

[72] U. Schneider, U. Diederichs, W. Rosenberger and R. Weiß, Hochtemperatur-verhalten von Festbeton, Sonderforschungsbereich 148 "Brandverhalten von Bauteilen", Arbeitsbericht 1978 - 1980, Teil II, B3-1 bis B3-142, TU Braunschweig (1980).

[73] E.Y. Anderson, Acoustic Emission from Frost Damage in Concrete. Proc. of the Institute of Acoustics.

[74] R.W. Parkinson and C.T. Peters, Acoustic Emission Activity during Accele-rated Conversion of high Alumina Cement Paste, J. of Materials and Science 12 (1977), pp. 848-850.

[75] M. Arrington, B.M. Evans, Acoustic Emission Testing of High Alumina Cement Concrete, NDT International 10, April (1977), pp. 81-87.

[76] M.D. Meiser and R.E. Tressler, Failure Modes in Low Density Aluminous Cement Bonded Composites, Cement and Concrete Research, Vol. 12, (1982), pp. 279-288.

[77] M. Chhuy Sok, J. Baron and D. François, Mecanique de la rupture Appliquée au Béton hydraulique, Cement and Concrete Research 9, (1979), pp. 641-648.

[78] M. Chhuy Sok, Etude de la Propagation d'une Fissure dans un Béton non armé. Bul. Liaison L.C.P., No. 98, Nov-Déc. (1978), Réf. 2240.

[79] R. Royles, P.D. Morley and M.R. Khan, The Behaviour of Reinforced Concrete at Elevated Temperatures with Particular Reference to Bond Strength. In: Bartos, P. (Editor): Bond in Concrete, Applied Science Publishers, London, (1982).

[80] U. Diederichs and U. Schneider, Changes in Bond Behaviour due to elevated Temperatures, in: Bartos, P. (Editor): Bond in Concrete, Applied Science Publishers, London, (1982).

[81] U. Diederichs, Untersuchungen über den Verbund zwischen Stahl und Beton bei hohen Temperaturen. Diss. TU Braunschweig (1982).

[82] D.J. Naus, Grouted and Nongrouted Tendons for Prestressed Concrete Pressu-re Vessels. Transactions of the 5th International Conference on "Structu-ral Mechanics in Reactor Technology", Paper H 3/6, Berlin, 13.-17.Aug. (1979).

[83] D.J. Naus, Acoustic emission monitoring of steel and concrete structural elements with particular reference to primary nuclear containment struc-tures. International Conference on Acoustic Emission, Anaheim, Ca., USA, 10. Sept. (1979), p. 38.

[84] M.C. Reymond, Acoustic Emission in reinforced Concrete flat Slabs, Meeting on stress wave emission. Lyon, France. 17. März (1975). Journess d'études sur l'émission acoustique. pp. 387-398, Lyon, Institut National des Sciences Appliquées (1975).

[85] B. Audenard, M.A. Mayer and R.Schuttler, Système de localisation d'émission acoustique - Précision et représentation de la mesure. 4. Colloque Int. sur les Méthodes de contrôle Non-Déstructive, Grenoble,11.-14.Sept.(1979).

[86] J.B. Vetrano, W.D. Jolly and P.H. Hutton, Continuous Monitoring of Nuclear Reactor Pressure Vessels by Acoustic Emission Technique, Inst. Mech. Engrs. C 58/72, pp. 221-226.

[87] E. Fontana, G. Grugni, C. Panzani, B. Pirovano, G. Possa and F. Tonolini, Acoustic Emission Monitoring During Hydrotest of a Thin Wall Pressure Vessel, Proc. of the 3. Meeting of the European Working Group on Acoustic Emission, Ispra (1974), pp. 126-140.

[88] R. Gopal, Experiences with Acoustic Emission Monitoring in Nuclear Power Plants, Int. Mech. E. Conf., Publ. n. 1979-4, Period. Insp. of Pressurized Components, London, Mai 8-10 1979, Publ. by Mech. Eng. Publ. Ltd. London (1979).

[89] A.G. Glover, J. Holt and J.A. Williams, The Examination of a Mild Steel Pressure Vessel under Creep Loading using Acoustic Emission, Ber. zum Symp. d. Deut. Ges. f. Metallkunde: Schallemission, München (1974), pp. 118-132.

[90] H. Jonas, Erste Akustische Leckdetektionsmessungen an Liner-Modellen, Inst. f. Nukleare Sicherheitsforschung d. Kernforschungsanalage Jülich, Aktennotiz SEA 11/79 (1979).

[91] M.C. Reymond, P. Acker, M. Ben Kirane, A.M. Paillere, F. Saba, A. Sakellarion, and A. Raharinaivo, Essai de caractérisation de l'endommagement du béton par analyse de l'émission acoustique. Revue d'acoustique, No. 59, pp. 256 ff, (1981).

Fracture Mechanics of Concrete,
edited by F.H. Wittmann, 1983
Elsevier Science Publishers B.V., Amsterdam — Printed in The Netherlands

Chapter 3.6

DETECTION of CRACKS by MERCURY PENETRATION MEASUREMENTS

by U. Schneider and U. Diederichs

1. Introduction

Some of the methods illustrated in the chapters 3.1 - 3.5, namely, microscopic observation, X-ray technique, photoelastic methods, Speckle metrology, holographic interferometry and acoustic emission have the disadvantage that the information obtained is related to very small, probably not representative sections of the total structure. A more complete information necessitates an outspending of time for serial investigations. In so far porometry methods are advantageous as the total volume of a specimen can be investigated by integrated values for the crack development. Up to a certain extent one gets quantitative informations about crack widths, crack surface areas and crack volumes. By some porometry methods calibration tests with uncracked specimens are necessary to separate the crack parameters from the inherent porosity parameters of the plain material.

From the numerous methods feasible, as there are water suction, permeability, diffusion, helium pycnometer and sorption experiments, which all have their special merits with respect to special investigation of the porous structures as cement paste, mortar and concrete, resp., the mercury penetration method has an exceptional position. The method enables to measure quantitatively distribution functions of crack widths, crack volumina and crack surface areas, which is principally accomplished by sorption measurements, too. However, the application of sorption measurements is limited to maximum crack widths of approximately 300 $\overset{\circ}{A}$. This is due to the fact that according to the Kelvin-Thompson equation for radii higher than 300 $\overset{\circ}{A}$ the capillary condensation takes place approximately at 0.99 of the saturation vapor pressure. A value which is extraordinarily difficult to handle in experiments because very low atmospheric pressure variations lead to a total condensation of the gas on plain surface areas.

In comparison to sorption methods for mercury penetration the range of application lies approximately between 1.3 nm and 150 μm [1]. This range is assumed to be sufficient for the detection of almost the entire region of crack width which is relevant in connection with concrete and cement paste investigations.

In the following a brief description of the mercury penetration method is given. The measuring and valuation problems are shortly discussed. Furtheron typical results available from literature and some recent test results are presented. The tests were performed with special reference to fracture mechanics. The potential and the limits of the method are being discussed.

2. Experimental

2.1 Principle

The principle of the mercury-penetration-technique - usually called mercury-porosimetry - is based on the fact that the mercury behaves towards most substances like a nonwetting liquid. Consequently, it does not penetrate into the openings and cracks of these substances without the application of external pressure. If a sample is filled into a vessel, which is tapered with a capillary tube, evacuated, filled with mercury and then put under increasing pressure, the mercury penetrates into the cracks resp. pores, and its level in the capillary decreases. If the decrease of the level is registered in dependence on the pressure, a porosimetric curve may be obtained, indicating the volume, which is penetrated into the cracks resp. pores of the sample at a given pressure.

The general equation for the capillary pressure across a curved liquid - vapor interface -   is the Laplace equation [2]:

$$\Delta P = \sigma \; (\frac{1}{R_1} + \frac{1}{R_2}) \tag{1}$$

where σ is the surface tension of the liquid and $(\frac{1}{R_1} + \frac{1}{R_2})$ is the mean curvature at any point on the curved surface. The radii, R_1 and R_2, are functions of the geometry of the space, in which the curved surface exists. Because the values of R_1 and R_2 are usually not necessary to be known it is possible to substitude them with the aid of geometrical relations and the contact angles ϑ between mercury and the sample. In the case of cylindrical assumed pores the radii R_1 and R_2 become equal and equation (1) occurs to

$$\Delta P = \sigma \; \frac{2 \cos \vartheta}{R} \tag{2}$$

using: $R = R_{1/2} \cdot \cos \vartheta$, where R is the radius of the pore.

It is very rare, that pores in real solids are even circular cylinders, so equ. (2) is rarely exact. Square pores may be approached by elliptical cylinders,

$$\Delta P = \sigma \cos \vartheta \; (\frac{1}{r_1} + \frac{1}{r_2}) \tag{3}$$

where $2 \, r_1$ and $2 \, r_2$ are the minor and major axis of the ellipse.

For slit pores - which means plain cracks -, r_2 is nearly infinite, and

$$\Delta P = \sigma \; \frac{\cos \vartheta}{r_1} = \sigma \; \frac{2 \cos \vartheta}{d} \tag{4}$$

where d is the width of the slit.

Using equation (4) it is possible to get the width of cracks by measuring the employed pressure. The measurement firstly yields an integral crack width resp. pore size distribution, whereby pore volumina are plotted versus the crack width.

If cylindrical pore shape is assumed, it is also possible to calculate a surface area distribution from the V - P plots.

Nevertheless one has to keep in mind that the pore structure of a sample consists of an irregular network with constrictions, cavities, junctions, etc. This leads to mistakes in the measuring results for example, the volume of the cylindrical pore drawn in fig. 1 is equal to that of the "real" pore, but the measured pore diameter and the calculated surface area will deviate from the actual values.

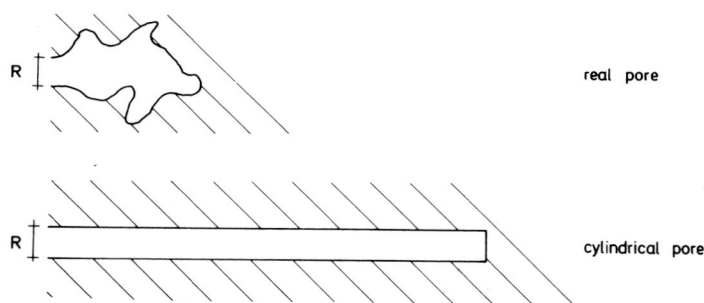

real pore

cylindrical pore

Fig. 1. Shape and size of a real pore in comparison to an ideal cylindrical one

2.2 Equipments

A mercury porosimeter in principle consists of three parts disregarded the variety of modifications and perfections:

(a) the container for the sample, the dilatometer;

(b) the source of the pressure, the pressure measuring unit;

(c) the equipment for monitoring the progress of the penetration of mercury into the sample.

Depending on the value of the pressure used, the porosimeters fall into two groups, viz. low and high pressure instruments. Low pressure porosimeters operate from vacuum to atmospheric pressure, high pressure porosimeters above 1 bar up to 6000 bar. With low pressure porosimeters a pressure very near to zero cannot be obtained generally, because of the disadvantage that a mercury column is above the sample when the measurement starts. The height of the column and in accordance to it the lowest possible pressure depends on the different equipments used.

Investigations with thermal and mechanical stressed concrete performed by the authors were made with the aid of two porosimeters, a low pressure porosimeter with a measuring range of 0.13 bar to 8 bar and a high pressure porosimeter with a measuring range of 1 bar to 2000 bar.

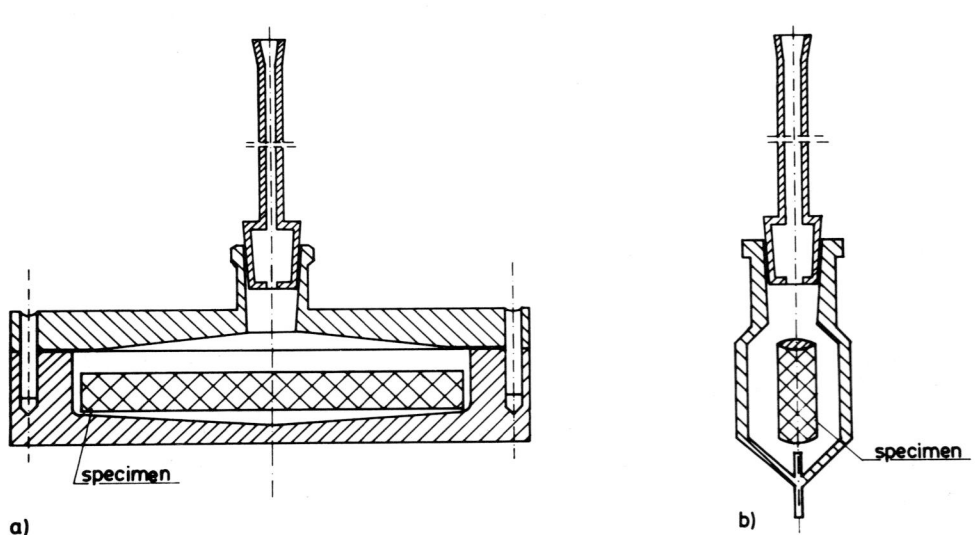

a)   b)

Fig. 2. Cross sections of the two used specimen containers

The pressure range of 0.13 bar to 8 bar is connected to the width of cracks resp. pore radii of approximately 60 μm to 1 μm. To investigate cracks of this size it is necessary to use specimens of sufficient large dimensions. Therefore a special specimen container shown in fig. 2a has been developed. With it concrete specimens of 70 x 70 x 10 mm^3 size can be investigated.

In the high pressure porosimeter with the range 1 bar to 2000 bar, according to smaller pores of approximately 7.5 μm to 4 μm, the specimen container shown in fig. 2b was applied. The specimens here were bore kernels, 10 mm in diameter and 35 - 40 mm long.

When the measurements were carried out, corrections had to be made concerning the compressibility of the dilatometer, the specimen container and the mercury.

These effects can easily be eliminated by blank runs.

2.3 Evaluation and Interpretation

The equations (4) and (5) give relations between the applied pressure p and the crack width d resp. pore radius R:

$$R = d = 2 \, \sigma \, \frac{\cos \vartheta}{P} \qquad (4)$$

in which σ is the surface tension of mercury (σ = 480 dyn/cm) and ϑ the contact angle. Concerning the contact angle of mercury with concrete many authors report values between 117° and 142° [1, 3, 4, 5, 6]. These differences are caused by the fact that concrete is an inhomogeneous matter consisting of many mineralogi- cally different materials, which in addition may suffer alterations during a stressing, e.g. thermal exposure. Therefore the authors choose the fix value of ϑ= $141,3^{\circ}$ for the contact angle, which leads to the simple relation:

$$R = d = \frac{7.5}{P \, [bar]} \quad [\mu m] \qquad (5)$$

The measurements gives the cumulative distribution of the pore volume by plotting the integrated pore volume versus the pore radii. In order to obtain a better understanding of the pore structure the cumulative distribution should be accompanied by the differential distribution of the pore volume versus pore radius, enabling direct evidences about the abundance of pore siz es resp. crack widths. However, one must be aware of the great sensitivity against small changes in the measured volumes, which is inherent in any numerical differentiation. Therefore, a numerical fit of the cumulative distribution data using overlapping logarithmic curves was carried out. Each section of the fit is logarithmically differentiated in order to get the ordinate of the differential distribution. In the following the plotting dV/d log R versus log R is used. Due to this type of plotting areas of equal sizes below the distribution curves correspond to equal pore volumina.

3. Crack Detection in Concrete

3.1 Shrinkage induced Cracks

Besides external mechanical or thermal stresses shrinkage of the cement paste may lead to crack formation. With dried specimen of pure cement paste and at very fresh concretes or mortars, resp., often cracks are already visible with the naked eye.

Without discussing the shrinkage and crack mechanisms in detail a typical re- sult of a mercury penetration measurement will be briefly discussed in the fol-

lowing. On fig. 3 a SEM micrograph of a pure cement paste specimen made with
blast furnace slag cement (w/c = 0.3) which was dried at ambient temperature
for 15 h under vacuum is shown. In the middle part of the figure there are
clearly recognizable inclusions of portlandite crystals and a number of cellu-
lar-type arranged cracks showing crack widths of about 1 µm.

Fig. 3. SEM micrograph of a hardened pure cement paste specimen dried at ambient
temperature for 15 h under vacuum distinctly showing shrinkage induced
cracks

The result of a mercury penetration measurement performed on a similarly
prepared cement paste specimen (core: diam. 10 mm, 40 mm length) is shown on
fig. 4. Here, the existence of cracks with corresponding crack widths is indicated by the maximum of the curve at approximately 1.5 µm.

This example clearly shows, that the penetration method in principal is
suitable for the detection of crack formations due to shrinkage. However, this
method includes some difficulties, namely, the impossibility to carry out blank
runs with undried specimens as to separate the initially present voids from the
true shrinkage cracks. - Studies on this method to detect and estimate shrinkage
induced cracks to our knowledge are published only in connection with investigations of the thermal behaviour of concrete [7, 8, 9]. But in this case shrinkage
is superimposed by thermal incompatibilities and thermal degradation of the
cement paste and aggregate, resp. Therefore an unambigeous separation of the
single effects is hardly possible.

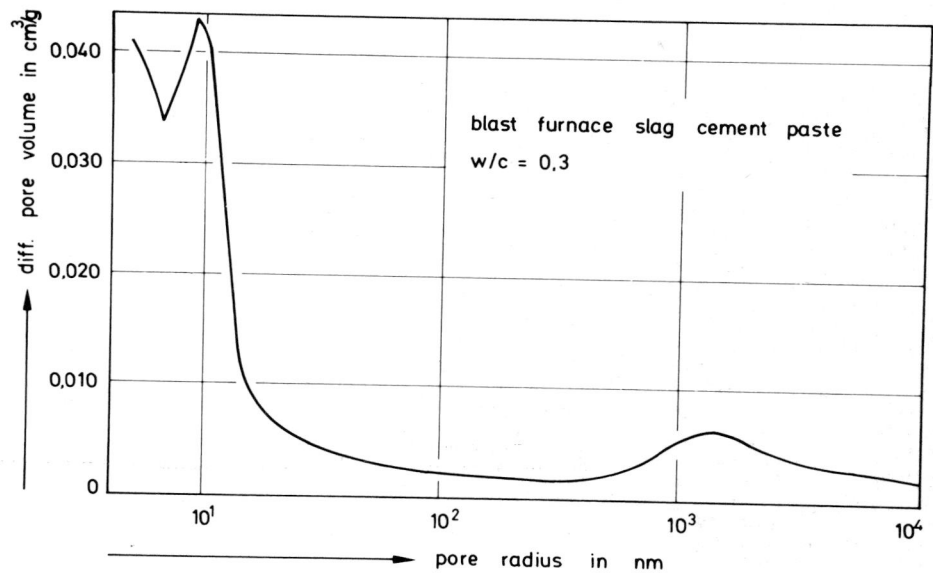

Fig. 4. Differential pore size distribution of a hardened pure cement paste
specimen dried at ambient temperature for 15 h under vacuum prior to
the mercury penetration test

3.2 Mechanically induced Cracks

Mercury penetration measurements with mechanical stressed concrete specimens
have been performed, too. The specimens of 300 x 120 x 70 mm^3 size had been excentrically loaded. The maximum expansions resp. shortenings along the cross
section of the specimens, which occured during the dynamical loading, are qualitatively shown in fig. 5.

The loaded specimen can be subdivided into three main sections:

I   Section of tensile loading,

II   Nearly unloaded section

III Section of compressive loading.

Eight pieces of 70 x 70 x 10 mm^3 size have been sawed out of the middle of
the loaded specimens for performing the penetration measurements. The positions
of the various pieces, named 2 to 9, is indicated in fig. 5 connected with the
limit of the maximum load.

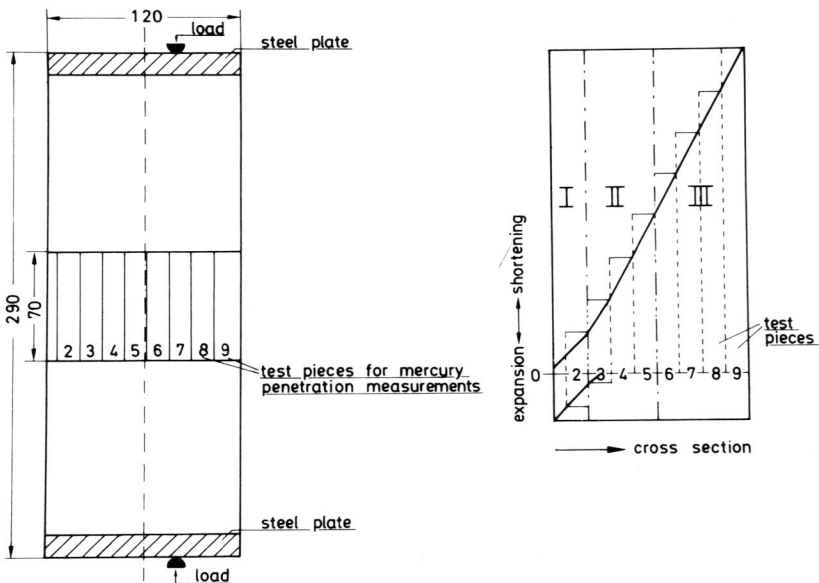

Fig. 5. Size and expansion-shortening distribution along the cross section of
an excentrically loaded specimen

The mercury penetration measurements were carried out in the low pressure
range, as to detect cracks between 1 μm and 60 μm. The equipment described in
part 2.2 has been used. For comparison a series of unloaded concrete specimens
of the same size has been tested, too. The results of the penetration measurements are shown in the figures 6 and 7. Fig. 6 represents the crack volume of
cracks greater than 1 μm along the cross section. In fig. 7 the differential
crack width distribution versus the crack width is plotted in dependence of the
sample position in the cross section of the whole loaded specimen.

Fig. 6 shows, that the distribution of the crack volume along the cross section correlates with the employed stresses. The three above mentioned different
loaded ranges are clearly indicated. It can be assumped, that the damage in this
concrete specimen occured, when a pore volume of approximately 6 cm^3/kg was exceeded.

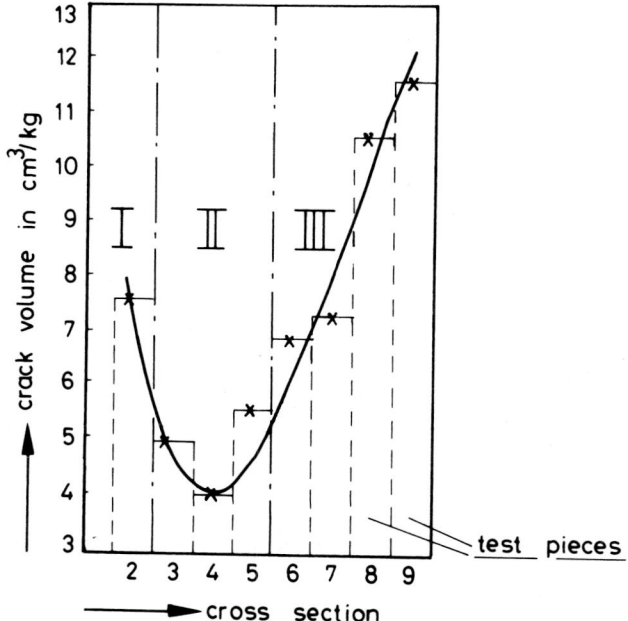

Fig. 6. Crack volume along the cross section of an excentrically loaded
specimen

A more distinct result about the kind of damage and its distribution can be
obtained from fig. 7. The ranges I, II, III can easily be observed, but furthermore an evidence about the abundance of crack widths is given. The tensile
stressed part of the concrete specimen (sample 2) shows a distinct increase
of crack development and two main crack widths at 1,8 μm and 4 μm. The difference in comparison to unloaded specimens is accentuated in fig. 7 by darker
and lighter shaded areas. The curves of the following three samples (3, 4, 5)
do not differ very much from the curve of unloaded concrete, i. e. the nearly
undamaged section can be recognized. The last four samples (6, 7, 8, 9) are
showing an increase of the crack distribution in the range up to 5 μm.

The increasing compression correlates with the increasing sample number and
can be observed at the curves. The mostly stressed sample (9) shows a great maximum at 11 μm; it must be assumed that many cracks of this width have been
occured.

216

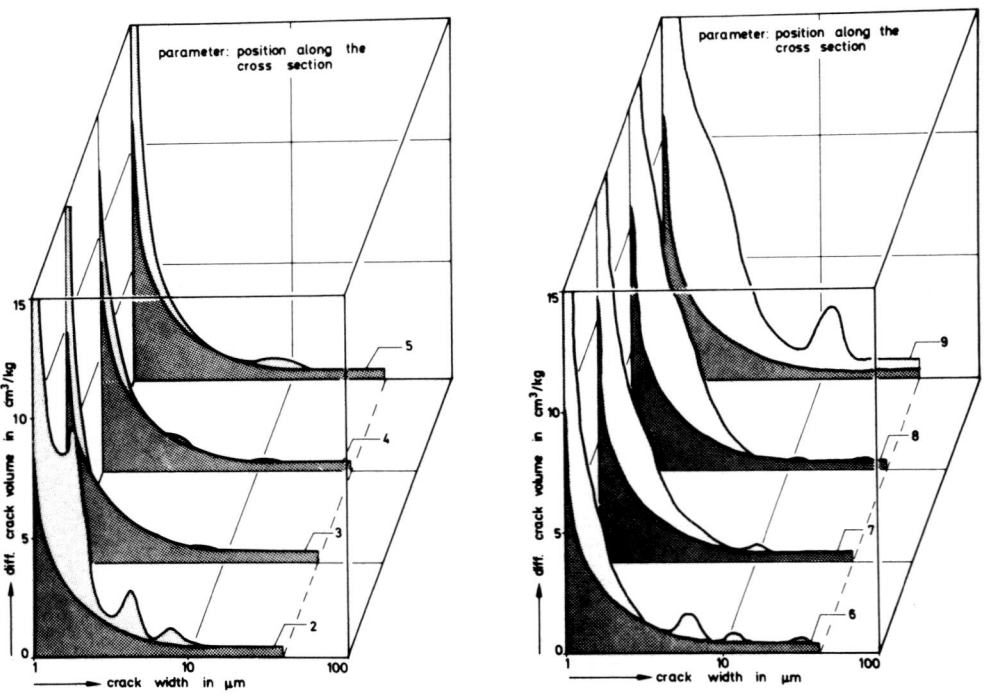

Fig. 7. Differential crack width distribution versus crack width in dependence
on the position in the cross section of an excentrically loaded specimen

3.3 Thermal induced Cracks

The crack development caused by high temperatures (up to 900 °C) as well as
by extremely low temperatures (down to - 170 °C) has been studied by means of
the mercury penetration method to a considerable extent [7 - 13]. In this connection it should be mentioned that in the case of low temperatures the formation and growth of cracks is caused by thermal incompatibilities. In the case
of high temperatures this damage mechanism is superimposed by shrinkage and dehydration reactions.

In [11] cylindrical specimen (diam. 12 mm, length 40 mm; mix proportions:
cement : sand : water = 1 : 3 : 0,5 by weight) were used, which were drilled

from mortar cylinders of 50 mm diameter 90 days after casting. The specimens
were water saturated prior to the temperature treatment. The loss of moisture
during temperature cycling was prevented by a plastic sealing. Cooling was performed at a constant rate of 1 K/min to - 170 °C. Upon reaching - 170 °C this
temperature was maintained for 45 minutes. Then a constant rate heating up to
+ 20 °C followed (see fig. 8). This type of temperature cycle was repeated

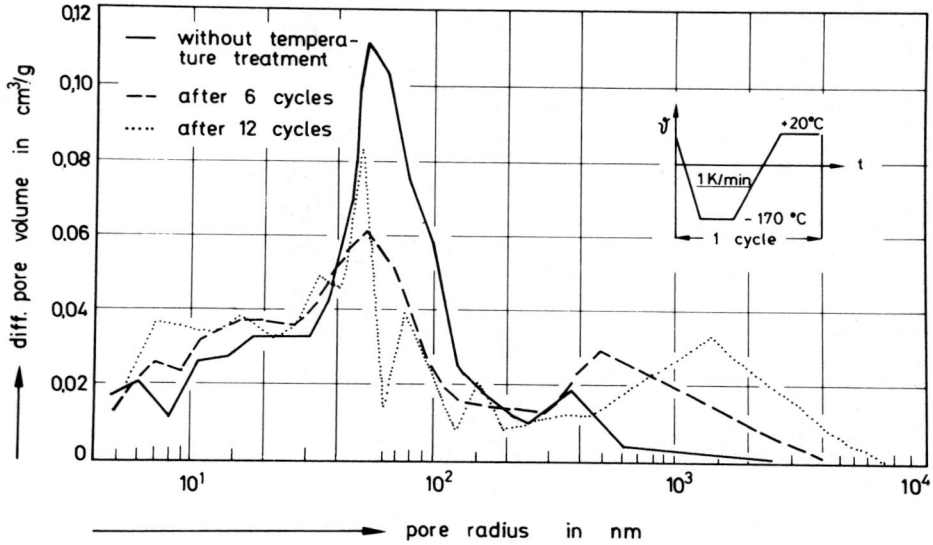

Fig. 8. Differential pore size distribution versus pore radius or Portland
cement mortar specimens prior and after low temperature cycling [11]

1, 4, 6 and 12 times. Since the specimens were water saturated they had to be
oven-dried at 105 °C to weight constancy prior to the mercury penetration measurements to permit unimpeded mercury intrusion.

Fig. 8 shows the differential pore size distribution versus pore radius for
a Portland cement mortar exposed to several cycles of extremely low temperatures in comparison with a mortar specimen without low temperature treatment.
The results of the latter are represented by the full line. It can be seen
that the Portland cement mortar without a cyclic temperature treatment shows
a first maximum of the differential pore size distribution between 50 and
80 μm and a second maximum between 300 and 500 μm, which can be considered as
the crack volume. In course of the temperature cycling the original first maxi-

mum of the differential pore size distribution decreases. Depending on the number of cycles the second maximum shifts towards larger pore radii (resp. crack widths). Simultaneously the area below the envelope of the second maximum increases, which is identical with the increase of the crack volume. This is caused by internal pressure resulting from the intricate processes of phase transition mainly of water (see also [9, 10, 11]).

Heating of cement mortar leads to crack formation, too. Besides thermal incompatibilities increasing temperatures activate a series of reactions in the hardened cement paste. These reactions commence with the complete desiccation of the pore system leading to shrinkage, thereafter follows with increasing temperatures the decomposition of hydration products and destruction of the gel structure [7, 8, 12 - 14]. The influences on the pore structure and crack development show fig. 9 and 10. The pore size distributions of mortar specimens (12 mm in diam., 40 mm long) made with Portland cement (mix proportion by weight 1 : 3 : 0,5 cement : sand : water) are indicated.The test specimens were unsealed heated in a tubular oven at a constant heating rate of 10 K/min up to the desired temperatures and then cooled slowly to ambient temperature.

From the fig. 9 and 10 it can be clearly seen that the distribution functions show almost independently of the maximum temperature two main maxima. Obviously the cement mortar contains different penetrating pore systems, which is indicated by the differently shaded areas beneath the envelope of the pore size distribution. The darker shaded region is mainly assumed to represent the pore system of the hardened cement paste matrix and partially the pore system of the interfacial zone between the bulk cement stone and the aggregate. The distribution function of this pore system indicates no remarkable changes in its shape and its location up to temperatures of nearly 450 $^\circ$C. At 450 $^\circ$C a superimposed pore spectrum occurs - indicated in fig. 9 by the cross dotted area -, which arises from the dehydration of portlandite. In fig. 10 this pore spectrum is not further separated from the cement stone porosity by a special dotting for the sake of clearity.

Only onwards about 550 $^\circ$C the pore size distribution of the cement stone matrix shows alterations due to further dehydration reactions of the different CSH-phases accompanied with a shift of the distribution function towards larger pore radii and an increase of the area beneath the respective distribution functions.

The second pore system indicated in fig. 9 und 10 by the lighther shaded areas is assumed to represent besides the enlarged porosity of the interfacial zone between aggregate and bulk cement stone mainly pure cracks. The corresponding pore size distribution function shows from 20 $^\circ$C onwards with increasing maximum temperatures a strong shift of its peak location towards larger pore

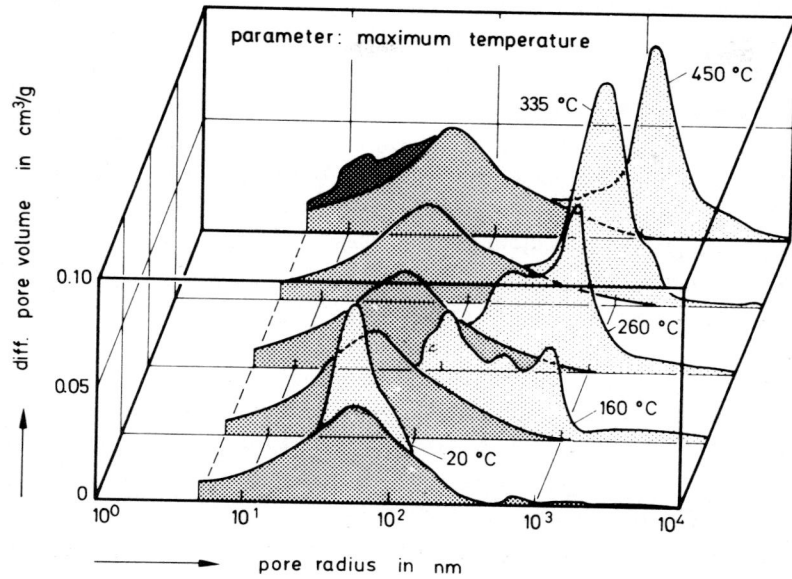

Fig. 9. Pore size distribution of mortar specimens made with Portland cement after thermal exposure [12]

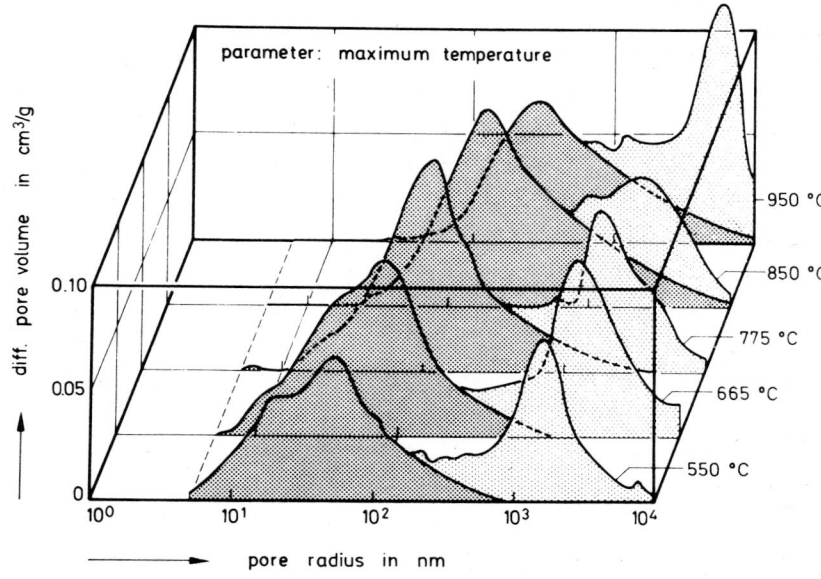

Fig. 10. Pore size distribution of mortar specimens made with Portland cement after thermal exposure [12]

radii resp. crack widths. Simultaneously the areas beneath the distribution
functions increase monotonously i.e. the total crack volume is increasing, too.

An excellent impression from the shift of the peak maxima of the pore size
distribution due to thermal exposure is given in fig. 11. It shows the peak locations versus maximum temperature for the above discussed Portland cement mortar and additionally for a blast furnace slag cement mortar, resp. Clearly it
can be seen that for both mortars the location of the first peak does hardly
alter up to the onset of dehydration reactions of CSH phases at approximately
600 °C. On the other hand, the location of the second peak which corresponds to
cracks shifts nearly independent of the type of cement used with the temperature
increase. Whereby the "main crack width" already in the temperature region from

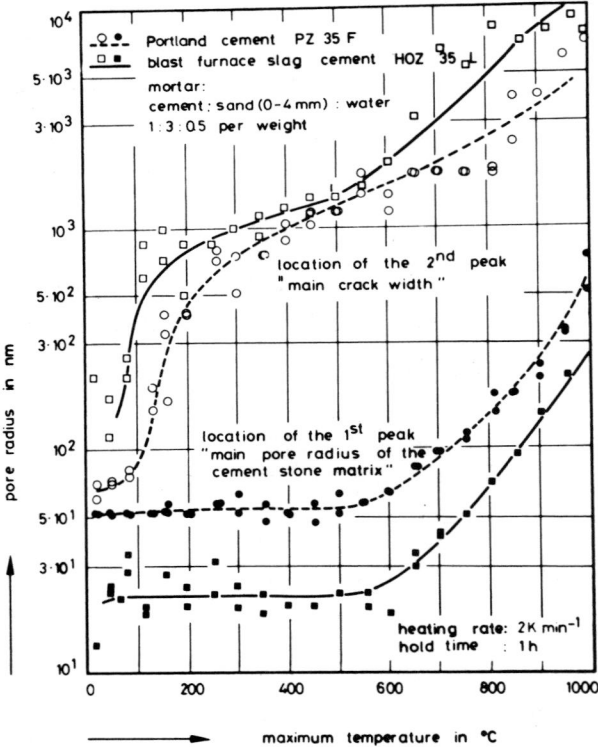

Fig. 11. Location of the maxima of the differential pore size distribution of
thermally exposed mortar specimen versus maximum temperature [12]

20 ^{O}C to approximately 200 ^{O}C increases by one order of magnitude. The governing mechanisms are assumed to be internal shrinkage of the hardened cement paste – although an overall thermal expansion of the cement paste will be measured – and differences in the thermal strains between the matrix and the aggregate.

As wellknown in course of a further temperature increase the cement paste undergoes significant shrinkage – Portland cement paste about – 18 % at 800 ^{O}C e.g. – whereas the siliceous sand undergoes a significant expansion. Due to this thermal incompatibilities cracks develop especially in the relatively weak inter-facial zone between the aggregate and the cement stone matrix. The latter effect has been observed many times by SEM inspections of the authors with temperature deteriorated mortar and concrete specimens. These observations support the re-sults, which were obtained by the mercury penetration method.

4. Summary

The report gives a brief introduction to the mercury penetration method and il-lustrates the evaluation and interpretation of the obtained measurement results with respect to crack measurements. It is concluded that this method in compari-son to other crack detection methods, commonly used porometry methods included, bears different advantages. In contrast to other methods the mercury penetration method conceives the total specimen volume. Furthermore one gets quantitative results on crack widths and crack volumes. Experimental results indicate further that the method is suitable for getting information about cracks resulting from internally or externally induced stresses namely, shrinkage, mechanically and thermally applied stresses. In our opinion this method fits the requirements as well for high as for low temperature deterioration investigations with mortar and concrete.

Investigations on shrinkage induced and mechanical induced cracks by the mer-cury penetration method after our best knowledge have not been published before, however, according to our experiments it can be stated, that the method also gives for mechanical and shrinkage induced cracks reliable and interesting re-sults.

5. References

[1] J. van Brakel, S. Modrý and M. Svata, Mercury Porosimetry: State of the Art, Powder Technology, 29 (1981), pp. 1 – 12.
[2] H. Vogel, Gerthsen·Kneser·Vogel – Physik, Ein Lehrbuch zum Gebrauch neben Vorlesungen, 12. völlig neubearbeitete und erweiterte Auflage, Springer-Verlag, Berlin-Heidelberg-New York (1974).
[3] A. Auskern and W. Horn, Capillary Porosity in Hardened Cement Paste. Journal of Testing and Evaluation, Vol. 1, No. 1 (1973), pp. 74 – 79.

[4] L.C. Drake, Pore-Size Distribution in Porous Materials - Application of
     High Pressure Mercury Porosimeter to Cracking Catalysts. Industrial and
     Engineering Chemistry, Vol. 41, No. 4 (1949), pp. 780 - 785.
[5] S. Diamond and W.L. Dolch, Generalized Log-Normal Distribution of Pore
     Sizes in Hydrated Cementpaste, Journal of Colloid and Interface Science,
     Vol. 38, No. 1 (1972).
[6] B. Kroone and D.N. Crook, Studies of Pore Size Distributions in Mortars.
     Magazine of Concrete Research; Vol. 13, No. 39 (1961), pp. 127-132,
     a. Vol. 14, No. 40 (1962), pp. 43-46.
[7] U. Schneider, R. Weiß and U. Diederichs, Ursachen und Auswirkungen der Ent-
     festigung von Beton bei hoher Temperatur. Mitteilungsblatt für die amtliche
     Materialprüfung in Niedersachsen, Nr. 18/19, Clausthal (1979), pp. 50-57.
[8] U. Schneider, U. Diederichs, W. Rosenberger and R. Weiß, Hochtemperatur-
     verhalten von Festbeton, Sonderforschungsbereich 148 "Brandverhalten von
     Bauteilen", Arbeitsbericht 1978-1980, Teil II, Braunschweig (1980), pp.
     3-145.
[9] G. Wiedemann, Zum Einfluß tiefer Temperaturen auf Festigkeit und Verformung
     von Beton, Dissertation, TU Braunschweig (1982).
[10] F.S.Rostásy, U. Schneider and G. Wiedemann, Behavior of Mortar and Concrete
     at Extremely Low Temperatures, Cement and Concrete Research, Vol. 9 (1979),
     pp. 365-376.
[11] F.S. Rostásy, R. Weiß and G. Wiedemann, Changes of Pore Structure of Cement
     Mortars due to Temperature, Cement and Concrete Research, Vol. 10 (1980),
     pp. 157-164.
[12] U. Diederichs, K. Hinrichsmeyer, W. Rosenberger and U. Schneider, Nachweis
     von Strukturveränderungen beim Erhitzen von Zementstein und Mörtel durch
     Quecksilberporosimetrie und N_2-Sorption, Workshop über die Charakterisierung
     von Festkörpern durch Physisorption, Chemisorption und Quecksilberpenetra-
     tion im Max-Planck-Institut für Chemie und im Institut für Anorganische
     Chemie und Analytische Chemie, Universität Mainz, Mainz, 11. - 13. Mai
     (1982).
[13] U. Schneider, U. Diederichs and K. Hinrichsmeyer, Nachweis von Strukturver-
     änderungen beim Erhitzen von Zementstein und Mörtel durch Quecksilberporo-
     simetrie (to be published in tiz (1983)).
[14] R. Weiß and U. Schneider, N_2-Sorptionsmessungen zur Bestimmung der spezifi-
     schen Oberfläche und der Porenverteilung von erhitztem Normalbeton, Cement
     and Concrete Research, Vol. 6 (1976), pp. 613-622.

Fracture Mechanics of Concrete,
edited by F.H. Wittmann, 1983
Elsevier Science Publishers B.V., Amsterdam — Printed in The Netherlands

Chapter 4.1

ANALYSIS OF ONE SINGLE CRACK

by A. Hillerborg

1. SOME GENERAL CONSIDERATIONS

Before discussing methods of crack analysis it is necessary to make clear the intention of the analysis. The main intention must be the possibity to predict the formation and behaviour of a crack in a structure, including the influence of the crack on its safety and serviceability. For concrete we have the very special condition that cracks of a nearly unlimited length are accepted in the service state, provided that they are no too wide and that they do not cause failure. From this point of view concrete differs very much from metals, where only limited crack lengths are accepted, but where crack widths as a rule are of secondary importance.

In concrete most cracks start from an uncracked surface and grow through a large portion of the depth of the specimen. Both formation and growth are influenced by stresses from imposed deformations, shrinkage, temperature etc. Thus fracture mechanics, when applied to concrete, should be able to analyse the following:

1. The formation of a crack in a specimen which is not notched or precracked.
2. The growth of a crack to a size of the same order as that of the specimen.
3. The influence of imposed deformations on crack formation and crack growth.

Most work in fracture mechanics so far deals with metals. Some of the analytical methods used for metals are not suitable for concrete, as the fracture behaviour of the two materials is very different.

In metals fracture is mostly preceeded by yielding, giving a typical stress-elongation relation in tension according to Fig. 1. This has the following two consequences

1. The stresses in the fracture zone in front of a crack tip remain constant or show a slight increase with an increasing deformation until local fracture occurs and the crack advances, see Fig. 2.

2. The yielding is accompanied by a tendency to contraction perpendicular to the stress direction. If this contraction is not prevented (plane stress conditions), great yield deformations can take place with a corresponding high energy absorption. If the contraction is prevented (plane strain conditions) the amount of yield deformation and thus also the energy absorption, is substantially smaller. Near the surface of a specimen there is a zone where pla-

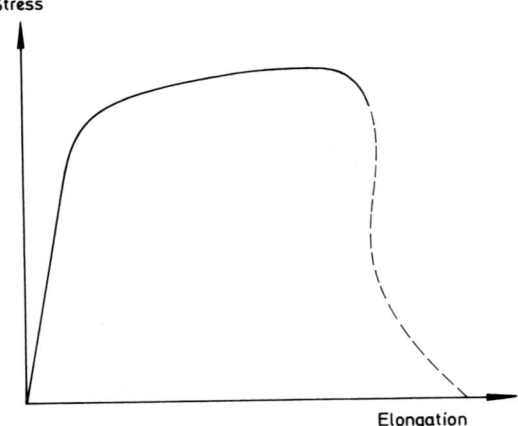

Fig. 1. Typical stress-elongation curve for a metal.

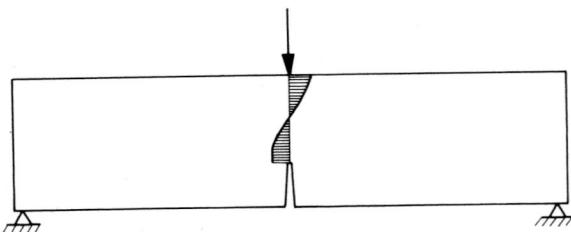

Fig. 2. Typical stress distribution in a loaded metal beam with a crack. The position of the crack tip is well-defined and the stress is high close to the crack tip.

ne stress conditions prevail, while the interior of a thick specimen is main-ly under plane strain conditions. Therefore the width of the specimen (the length of the crack front) is of importance for the average energy absorption per unit increase in crack area. As the crack advances, the zone with plane stress may increase in width, which may lead to an increase in resistance against crack growth. One way for taking this phenomenon into account is R-curve (resistance-curve) analysis. The difference between plane stress and plane strain has to be taken into account in design and in tests. Test pie-ces are mostly given such sizes, that plane strain conditions prevail, i e they are wider than a minimum size.

For concrete a typical stress-elongation relation from a tension test is shown in Fig. 3 (ref. 1, 2). The test is performed in a very stiff testing machine, which makes it possible to follow the descending branch.

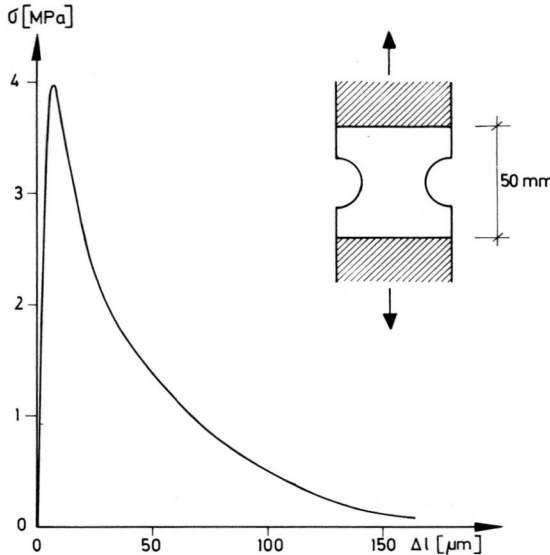

Fig. 3. Typical stress-elongation curve for concrete
in a tension test according to the figure. Gauge length
40 mm. Smallest cross section 20 x 30 mm^2.

If we compare Fig. 3 to Fig. 1, we find that the fracture behaviour of concrete is very different from that of metals. This is due to the fact that the fracture of concrete is preceeded by micro-cracking instead of yielding.

One consequence of this is that the stresses decrease with increasing microcracking and thus with increasing deformation. Some of the microcracks grow to
form visible cracks while the specimen still takes a substantial load. The fact
that a crack is visible thus is no proof that the stress has fallen to zero.

Due to the microcracking the stresses in front of a crack tip may have a typical distribution according to Fig. 4. In this case we have no well defined crack
tip, but rather a fracture zone, within which the cracking increases and the
stresses decrease as the deformation increases. We cannot identify any well defined crack tip, but some points of interest may be noted:

1. The point where the stress has its maximum. On further deformation the stress
    decreases due to increasing microcracking. This may be looked upon as the
    first sign of fracture and this point may serve as a limit for the fracture
    zone.

2. The points where the crack becomes visible with or without a microscope.

3. The point where the stress transfer ends. This point may serve as a limit
    between the fracture zone and what may be termed the "real crack", i e that

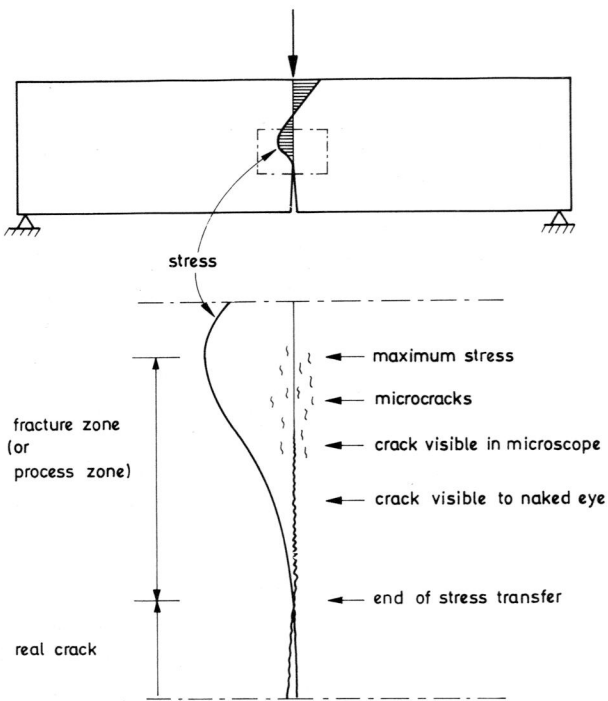

stress

fracture zone
(or
process zone)

real crack

← maximum stress

← microcracks

← crack visible in microscope

← crack visible to naked eye

← end of stress transfer

.Fig. 4. A loaded concrete beam with a crack and a fracture
zone. No well-defined crack tip exists.

part of the crack, over which no stresses are transferred. This point may be
looked upon as the most natural definition of the crack tip in fracture mechanics, but for concrete it is not possible to observe this point in a test.

From the above it is evident that <u>no well-defined crack tip exists</u> for concrete (like it does for metals), as there is a gradual disruption with increasing
deformation and cracking within the fracture zone.

A consequence of this is that fracture mechanics formulas based on the study
of the stability or growth of a crack tip are never strictly applicable to concrete. Sometimes such formulas may however be used as reasonable approximations,
viz if the fracture zone is small compared to the size of the specimen. The motivation for this is that the most likely position of the point corresponding to
the theoretical crack tip is within the fracture zone and thus its position gets
better defined the smaller is the fracture zone.

It has been demonstrated in tests (ref. 3, 4) and by means of calculations
(see below) that the length of the fracture zone for concrete is of the order of
50-100 mm or more. The fracture zone thus cannot be regarded as small in comparison with the size of the specimen unless the specimen has a depth of several me-

ters.

The microcracking of concrete is not accompanied by any substantial contraction corresponding to that which occurs when metals yield. For this reason no essential difference exists between plane strain and plane stress conditions for concrete. For concrete we thus do not need to bother about the difference between plane strain and plane stress, which causes a lot of complexity for metals. One consequence of this is that the width of the specimen (the length of the crack front) cannot be expected to be of any major importance for concrete, like it is for metals. This conclusion has also been confirmed by tests, see (ref. 28).

Cracks may form in different ways, depending on the type of stress acting. Thus a difference can be made between an opening mode (mode I), where tensile stresses act perpendicular to the crack plane, and two different shear modes (II and III). This report deals primarily with the opening mode, which is by far the most essential one fore concrete.

A review of the application of fracture mechanics to concrete has recently been given by Swamy (ref. 5).

2. THE FICTITIOUS CRACK MODEL, FCM

The first model to be described and discussed is the fictitious crack model, FCM (ref. 6, 7), because this model may be used as a basis for the discussion of other models. A model of essentially the same character, but with a different formulation for the computer program, has been used by Bazant and Oh (ref. 29).

The fundamental idea of the FCM is best demonstrated by means of a tension test, Fig. 5. The test is assumed to be deformation-controlled and stable, so that it is possible to follow the descending branch of the stress-deformation curve all the way down to zero load.

We assume that the deformation is measured along two equal gauge lengths A and B with the results shown in the diagram.

As the specimen is assumed to be homogenous and to have a constant area, the curves A and B coincide until the maximum load is reached. On further deformation a fracture zone forms somewhere in the specimen. This fracture zone has a limited width in the direction of the stress. As the fracture zone develops the force will decrease due to the formation of microcracks and the corresponding weakening of the material. The decreasing load results in a decrease in deformation everywhere outside the fracture zone, corresponding to the unloading curve in the stress-strain diagram. No more fracture zone can form, as the load decreases.

In Fig. 5 it is assumed that the whole fracture zone falls within gauge length A. The deformations within gauge length B can then be described by means of a stress-strain curve, including an unloading branch.

The deformation within gauge length A includes also the deformation of the fracture zone. The additional deformation due to the fracture zone is the diffe-

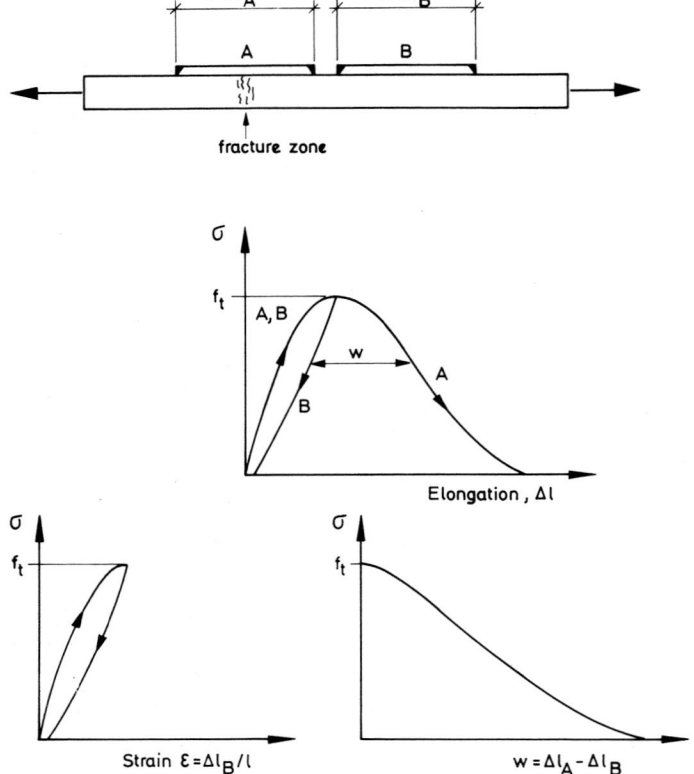

Fig. 5. The principles for division of the deformation properties into a σ-ε-diagram and a σ-w-diagram, where w is the additional deformation due to formation of a fracture zone.

rence w between the descending branches of curves A and B. We thus introduce the notation

w = additional deformation due to the fracture zone

We can now describe the deformation properties of the test piece by means of two diagrams:

1. The stress-strain-(σ-ε) diagram, including the unloading branch
2. The stress-deformation-(σ-w) diagram for the fracture zone.

By means of these two diagrams we can calculate the deformation $\Delta\ell$ of any gauge length ℓ_0, where the gauge end is not situated within a fracture zone.

If there is no fracture zone within the gauge length, the deformation is

$$\Delta\ell = \varepsilon\ell_0 \tag{1}$$

If a fracture zone is situated within the gauge length, the deformation is

$$\Delta \ell = \varepsilon \ell_o + w \qquad (2)$$

It has to be noted that w is a length contrary to ε, which is a strain.

For a non-yielding material like concrete the σ-w curve is probably independent of the size of the specimen, although this should be better confirmed by
tests.

The width of the fracture zone does not enter into the equations above. We
thus have a certain freedom to make a suitable assumption regarding the width of
the fracture zone. The simplest possible <u>assumption</u> then is, <u>that the original</u>
<u>width of the fracture zone is zero</u>. The total width of the fracture zone then
equals w. Another possible assumption is that the width of the fracture zone has
a certain value w_c and that the strain is equal within this width. This assump-
is used in (ref. 29), with w_c proportional to the aggregate size.

According to the assumption of zero original width, the fracture zone may be
described as a tied crack with width w, i e a crack which can transfer a stress
σ according to the σ-w curve when its width is w.

As the fracture zone in reality has a certain width, the tied crack which is
introduced as a simplified description is no real crack. It has therefore been
called a <u>fictitious</u> crack and this term has been used to characterize the model.

The application of the fictitious crack model, FCM, to the description of the
tensile test is shown in Fig. 6.

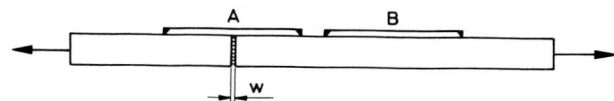

Fig. 6. The simplified description of the fracture zone as a
        "fictitious crack" with width w.

During a tensile test to complete separation, energy is absorbed inside and
outside the fracture zone. With the FCM <u>the energy absorbed in the fictitious</u>
<u>crack</u> is

$$A \int_0^{w_\ell} \sigma \, dw = AG_F \qquad (3)$$

where A = cross sectional area; w_ℓ = w-value for $\sigma = 0$; G_F = area below the σ-w
curve, Fig. 7.

G_F thus is the absorbed energy per unit crack area for the complete separation
of the crack surfaces. The crack area in question is the projected area, not the

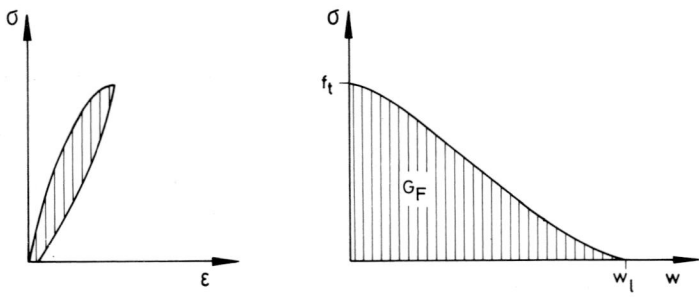

Fig. 7. Energy dissipation related to the σ-ε- and the σ-w-diagrams.
The values of the shaded areas represent the energy dissipation per
unit material volume and per unit crack area respectively.

total area of the irregular crack surface. The notation G_F is identical to 2γ, γ_p
or $(2\gamma+\gamma_p)$, used in some publications, although these notations may also sometimes have differing definitions.

The energy absorption outside the fictitious crack is determined in the usual
way as the volume of the specimen times the area below the σ-ε curve, Fig. 7. For
a purely elastic material this energy absorption is zero.

The FCM can be applied not only to the tension test, but also to more complicated stress situations. Of primary interest is its application to the analysis
of the stability and growth of a crack.

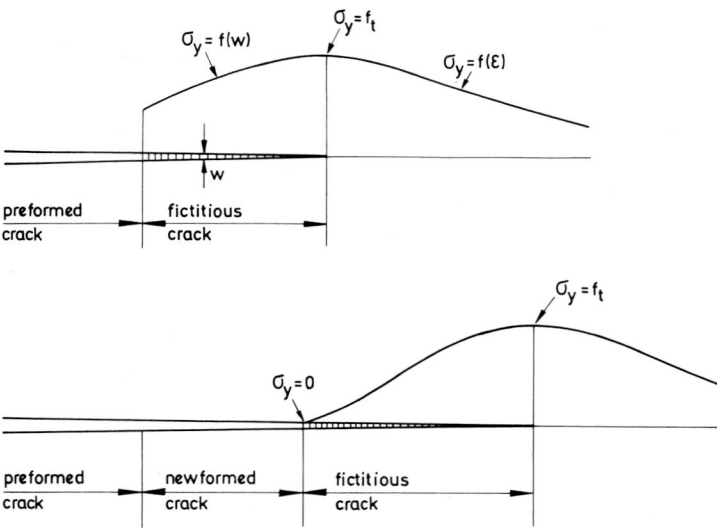

Fig. 8. Stress distribution in front of a crack tip before and
after growth of the real crack.

Fig. 8 shows the stress distribution in front of a notch or a crack tip in a beam under the action of a growing imposed deformation (or load). A fracture zone has developed and this fracture zone is described as a fictitious crack (cf Fig. 4, which gives the corresponding description of the reality).

Within the fictitious crack the relation between the stress σ and the crack width w is given by the σ-w curve. Everywhere outside the fictitious crack the σ-ε curve for the material is valid.

As the deformation is increased, the stresses in front of the fictitious crack tip increase. No stress is assumed to be higher than the tensile strength f_t. As soon as a stress has reached f_t, any increase in deformation causes the development of a fictitious crack at that point. Thus the stress at the fictitious crack tip is f_t as long as the fictitious crack grows.

The σ-ε and σ-w curves to be used in this application of the FCM are not necessarily the same as in the tension test. At the fictitious crack tip we have a three-dimensional state of stress, which may influence the tensile strength f_t. For metals it is quite evident that this influence may be of great importance and this is one reason why it is necessary for metals to make a distinction between plane strain and plane stress.

For concrete the difference between plane strain and plane stress is probably insignificant, as has been pointed out earlier.

There is reason to believe that for concrete the same σ-ε and σ-w curves can be used in the case of crack growth as in the case of a tension test (see ref. 2, 8). This however has to be confirmed by means of further tests. If, however, great compressive stresses act perpendicular to the tensile stress, this will influence the tensile strength and the σ-ε and σ-w curves.

The FCM has a very general applicability. It can be used to analyse the formation and growth of fracture zones and cracks, whether the fracture starts from a crack, a notch, an irregularity or a plain surface. It can also be used where shrinkage or temperature strains act and for non-isotropic materials.

On the other hand it is hardly ever possible to find analytical solutions based on the FCM. Thus numerical methods have to be used. All calculations so far have been performed by means of the finite element method, FEM (ref. 2, 6, 7, 8). With this method it is easy to follow the growth of fictitious and real cracks, which coincide with the sides of elements. The elements are just separated by distances w and forces corresponding to σ from the σ-w curve are introduced across the crack. It is also possible to analyse cracks which do not follow the original element mesh, but this is more complicated (ref. 7).

In FEM calculations it is very timeconsuming and expensive to use non-linear σ-ε curves. It is however relatively inexpensive to use stepwise linear σ-w curves. Thus a non-linear σ-w curve can be taken into account more easily than a non linear σ-ε curve.

The simplest possible assumptions regarding σ-ε and σ-w curves to be used in FEM calculations are according to Fig. 9, i e straight line approximations for both curves. Most calculations performed so far have been based on these assumptions.

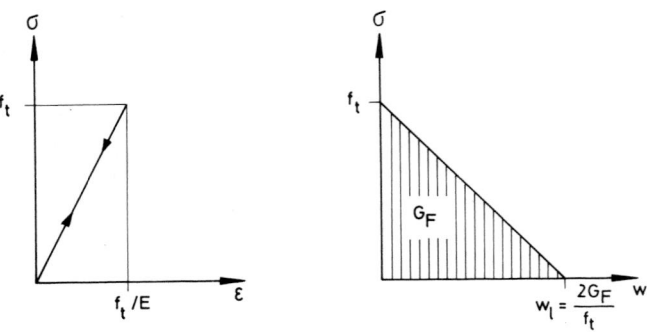

Fig. 9. Simple approximate assumption for use in numerical calculations.

With the shapes of the σ-ε and σ-w curves given in Fig. 9, the material properties are fully defined by f_t the tensile strength; E the modulus of elasticity; G_F the fracture energy.

It is also suitable to combine these values into a characteristic length

$$\ell_{ch} = \frac{EG_F}{f_t^2} \qquad (4)$$

The charateristic length ℓ_{ch} is a pure <u>material property</u>, which has <u>no direct physical correspondance</u>. It thus must not be mistaken for "critical crack length", "length of fracture zone" or any other real physical length. On the other hand these lengths are often approximately proportional to ℓ_{ch}. Thus the length of the fracture zone at crack growth often is of the order 0.3-0.5 ℓ_{ch}.

With the assumption of Fig. 9 the accuracy of a FEM calculation depends to a great degree on the relation between the length of the fracture zone and the size of the finite element mesh. As the length of the fracture zone is approximately proportional to ℓ_{ch}, the accuracy depends on the relation between ℓ_{ch} and the size of the elements. Next to the crack path the distance between node points should not be chosen greater than about 0.1-0.2 ℓ_{ch}. The number of elements thus increases with a decrease in ℓ_{ch} and with an increase in the size of the structure.

It must be noted that the result of a calculation does not only depend on ℓ_{ch}, but also on E, G_F and f_t and on the shapes of the σ-ε and σ-w curves, which may sometimes play an important role, e g for fibre reinforced concrete (ref. 1, 2, 9).

The σ-w curve has to be determined by means of a tension test, which gives a stable complete descending branch. This requires a very stiff test rig and special test arrangements (ref. 1, 2, 10, 11, 30).

The value of G_F can easily be determined by means of a bend test on a notched beam (ref. 2, 12).

3. LINEAR ELASTIC FRACTURE MECHANICS, LEFM

LEFM is based on the assumption that the material is fully elastic everywhere and that no fracture zone exists. The distribution of the stress σ_y perpendicular to the crack plane in front of a crack tip (cf Fig. 10) can then be expressed by the equation (ref. 13).

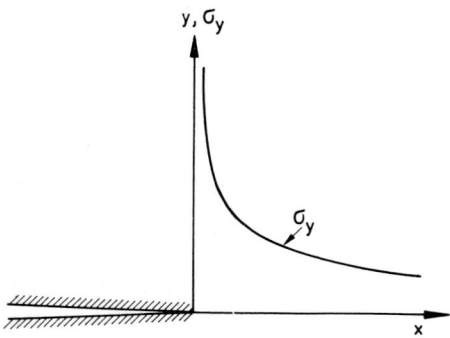

Fig. 10. Stress distribution in front of a crack tip according to the theory of elasticity.

$$\sigma_y = \frac{K}{\sqrt{2\pi x}} + \cdots \cdots \cdots \qquad (5)$$

where the dots indicate terms which are small in comparison with the first one when x approaches zero.

K in Eq (5) is the <u>stress intensity factor</u> and it depends on the load acting, the crack length and the shape of the specimen. It is usually expressed in the form

$$K = Y\sigma \sqrt{a} \qquad (6)$$

where σ is the stress which should act if the crack were not present; a is the length of an edge crack or half the length of an interior crack; Y is a coefficient, which depends on the type of load, the shape of the specimen and the crack length. Often the value of Y is about 1.7-2.0, but for long cracks it may take on much higher values.

If the material can carry the stresses, the crack will not grow. According to Eq. (5) the stress at the crack tip approaches infinity. The criterion of crack growth thus cannot be based on a stress value, which is compared to a strength value as in ordinary design calculations.

Instead of a stress criterion it has been found suitable to use a criterion based on the stress intensity factor K, which is a finite value. The critical stress intensity factor K_c is the value of K, for which the crack starts growing. K_c is assumed to be a material property.

When a crack propagates, this is accompanied by a release of potential energy. The energy release rate G is the energy release per unit growth of crack area. The energy release rate can be used as an alternative crack growth criterion in LEFM (ref. 14, 15, 16, 17). The critical energy release rate G_c is the value of G, for which the crack starts growing. G_c is assumed to be a material property.

A crack will start growing as soon as a stress is applied if there is no mechanism which can absorb energy at crack growth. This energy absorption must, according to LEFM, take place at the crack tip, as all the material is supposed to be elastic, thus taking part in the energy release but not in the energy absorption. G_c can be interpreted as the energy absorption rate at the crack tip, i e the energy absorbed at the crack tip per unit growth of crack area.

The following relation is valid between K and G (and thus also between K_c and G_c) (ref. 13)

$$K = \sqrt{EG} \text{ for plane stress} \tag{7}$$

$$K = \sqrt{EG/(1-\nu^2)} \text{ for plane strain} \tag{8}$$

where E is the modulus of elasticity; ν is Poissons ratio

For concrete, with ν about 0.2, the error in K will be less than 2 percent if we always use

$$K = \sqrt{EG} \tag{9}$$

$$K_c = \sqrt{EG_c} \tag{10}$$

This simplified relation will thus always be used below.

For simple cases, the values of K and G can be calculated analytically. For more complicated cases K and G can be computed by means of FEM or other numerical techniques (ref. 18).

LEFM can never describe the behaviour of a real material correctly - it is always an idealization, and thus an approximation, as it assumes infinite stresses (and thus infinite strength) and no fracture zone (or a fracture zone of zero length).

In a real material there is always a finite stress and a fracture zone of finite length in front of a crack with tensile stresses perpendicular to the crack plane.

Compared to a real material, LEFM can be looked upon as the limiting case when the length of the fracture zone approaches zero for a linear elastic material.

With LEFM it is often possible to find general analytical solutions, which are much easier to handle than numerical solutions by means of FCM. LEFM should therefore be preferred in all situations, where this approximation can be accepted. Comparisons with results from FCM-calculations can show how great errors that can be expected when LEFM is used (see below).

In the practical application of LEFM a calculated value of K or G is compared to the material property K_c or G_c. It is then of course essential than K_c and G_c are correctly determined by means of tests.

According to definition K_c is the value of K when the crack starts growing. According to the earlier discussion there does not exist any welldefined crack tip for concrete and thus it is not possible to determine the load for which a crack starts in a test. Normally it is therefore assumed that this load coincides with the maximum load, which should be the case according to LEFM. As there does not seem to be any better alternative definition, the following definition has to be accepted where the evaluation of concrete tests is concerned: K_c and G_c are the values of K and G corresponding to the maximum loads in the tests.

It should be noted that other definitions are used in metal testing (ref. 19), where the shape of the load-deflection curve is taken into account. This is due to the fact that for metals there is a great difference in energy absorption between plane strain and plane stress conditions (see above). For concrete this problem does not occur and thus there is no reason to use any other value than the maximum load.

For the determination of K and G it is necessary to know the crack length. As no welldefined crack tip exists, the only welldefined crack length in a test is the notch depth in a notched beam. Crack lengths observed on the surface of a beam are as a rule too uncertain to be used for the evaluation. Sometimes crack lengths are indirectly determined from compliance measurements. As the crack tip in reality is not welldefined, the cracklengths determined from compliance measurements must also be regarded as unreliable.

In order to study the degree of approximation involved in the application of LEFM, The the three-point bend test of a notched beam has been studied by means of FCM (ref. 2, 7, 8). The maximum load F_{max} has been calculated and inserted into the formula (ref. 20).

$$K_c = \frac{3F_{max}\ell}{2bd^2} \sqrt{a}\left(1.93 - 3.07\left(\frac{a}{d}\right) + 14.53\left(\frac{a}{d}\right)^2 - 25.11\left(\frac{a}{d}\right)^3 + 25.8\left(\frac{a}{d}\right)^4\right) \tag{11}$$

From K_c it is possible to calculate G_c according to Eq. (10).

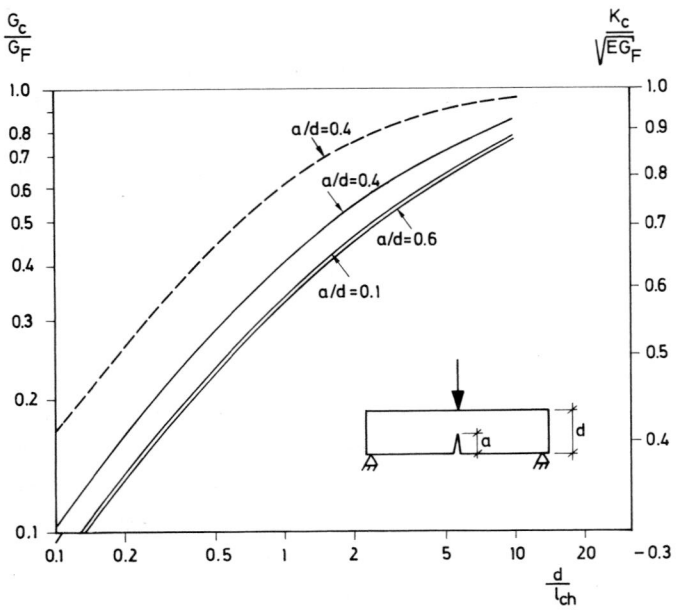

Fig. 11. Theoretical variation with beam depth d of G_c or K_c, determined by means of a three-point bend test and application of LEFM formulas (from ref. 2).
———————— σ-w curve according to Fig. 9.
- - - - - - - - - σ-w curve according to Dugdale (Fig. 14).

Results of such calculations are given in Fig. 11 (ref. 2) for assumed material properties according to Fig. 9 and also with a σ-w curve according to Dugdale, i e $\sigma=f_t$ until $w=G_F/f_t$ and $\sigma=0$ for higher w-values, (Fig. 14). From Fig. 11 it is evident that the value of K_c or G_c from a test depends very much on the ratio d/ℓ_{ch} if d/ℓ_{ch} is small. Thus K_c and G_c in these cases are not material properties.

With increasing d/ℓ_{ch}, K_c and G_c approach constant values which may be regarded as material properties. Thus if K_c and G_c are regarded as material parameters, these have to be determined from tests with a minimum value of d/ℓ_{ch}.

From Fig. 11 it can be seen that G_c approaches G_F as d/ℓ_{ch} increases. Thus $\underline{G_c}$ can be treated as a material property only when it approaches G_F.

This also means that as a pure material property G_c equals G_F.

A distinction must be made between G_c as a pure material property and G_c as determined from a test. If the test piece is too small, G_c can take on any value below G_F.

It is quite natural that G_c approaches G_F when d/ℓ_{ch} grows, because the length of the fracture zone is approximately proportional to ℓ_{ch}. If the length of the fracture zone approaches zero, the conditions approach those assumed in LEFM. The energy absorption per unit growth in crack area (G_c) then approaches G_F, as that unit crack area will absorb all the energy associated with the deformation of the fracture zone until complete separation.

With $G_c \rightarrow G_F$ we can write

$$K_c \rightarrow \sqrt{EG_F} \tag{12}$$

and

$$\left(\frac{K_c}{f_t}\right)^2 \rightarrow \ell_{ch} \tag{13}$$

For metals it is often prescribed (ref. 19) that for a tree-point bend test $a/(K_c/\sigma_y)^2 \geq 2.5$ and $a/d \approx 0.5$. With the yield strength $\sigma_y \approx f_t$, $(K_c/\sigma_y)^2 \approx \ell_{ch}$, this means $d/\ell_{ch} > 5$ for metals. For metals the σ-w curve is of the Dugdale type, and thus according to Fig. 11 this limitation will ensure that the K_c- and G_c-values will not be too dependent on the beam depth, but may be regarded as material properties, with $G_c \rightarrow G_F$.

For concrete, whose σ-w curve is more like the straight line of Fig. 9, it is seen from Fig. 11 that the limit of d/ℓ_{ch} has to be put higher than for metals, i e $d/\ell_{ch} > 10$-15, in order to get a comparable validity of the results. As ℓ_{ch} for concrete is of the order 0.2-0.4 m, beam depths of 2-6 m would be reuired, which is not reasonable for laboratory tests.

An alternative way of determining K_c and G_c might be the use of small specimen and a correction of the results by means of Fig. 11 or some other diagram. As ℓ_{ch} depends on G_F, which is not known, Fig. 11 is not suitable for this purpose. Fig. 12 shows a rearrangement of Fig. 11, in which $d/(EG_c/f_t^2)$ is given on the abscisse axis, where G_c is the value obtained in the test. In principle it should be possible to calculate the material property $G_F = G_c/(G_c/G_F)$ from Fig. 12 if G_c, E, f_t, and the shape of the σ-w curve are known. In practice, however, the G_F-value will be very uncertain if the depth is too small, because the value of G_c/G_F will depend too much on uncertainties in E, f_t and in the shape of the σ-w curve. Thus for instance if G_c/G_F has been found to be 0.3, an uncertainty in f_t of $\pm$ 20 % makes G_c/G_F vary between about 0.17 and 0.44, i e with about $\pm$ 45 %. The uncertainty in the shape of the σ-w curve may mean even more. The conclusion is that K_c- and G_c-values determined on small specimens are very uncertain also if they are corrected according to the curves in Fig. 12. Values of d/ℓ_{ch} smaller than about 5 should not be used for concrete or cement paste even if the results are corrected by means of Fig. 12. This corresponds to beam depths of about 1-2 m for concrete.

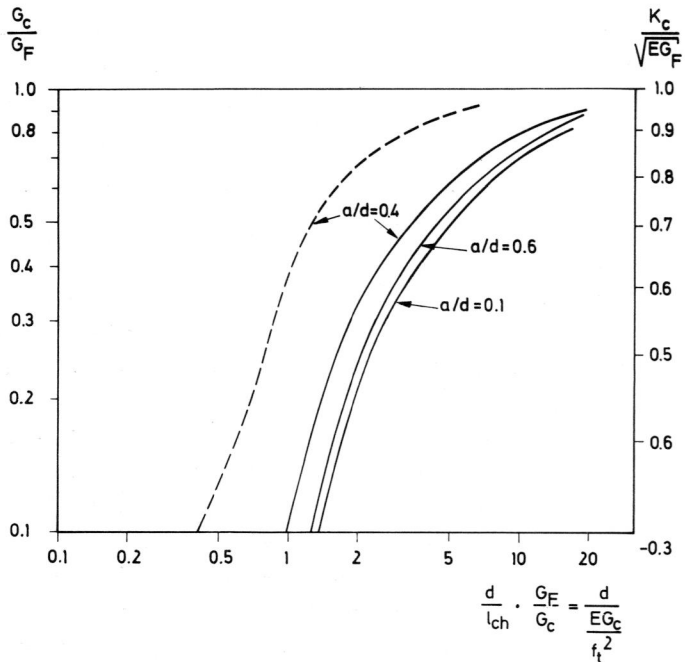

$\dfrac{G_c}{G_F}$

$\dfrac{K_c}{\sqrt{EG_F}}$

a/d=0.4

a/d=0.6

a/d=0.1

$$\frac{d}{l_{ch}} \cdot \frac{G_F}{G_c} = \frac{d}{\dfrac{EG_c}{f_t^2}}$$

Fig. 12. Rearrangement of Fig. 11 with respect to value on abscisse axis.

It must be noted that the curves of Figs. 11 and 12 are calculated for a three-point bend test on a beam of a constant width. They are thus not directly applicable to other test specimens. A specimen which is thinner along the expected crack path (a "waisted" specimen) will get longer fracture zones and thus an increased demand on the specimen size, approximately proportional to the relation between the width of the specimen and the width of the "waist".

Fig. 11 also gives an indication of the degree of accuracy to be expected when LEFM is applied in the analysis of fracture phenomena. The accuracy will evident- ly be rather poor if the crack length or the ligament size d-a is smaller than l_{ch}, which corresponds to crack lengths less than about 200-400 mm for concrete and 5-15 mm for cement paste. The accuracy can be increased by using a reduced G_c- or K_c-value estimated e g from Fig. 11 or by using an "effective" crack length according to section 9 below.

LEFM may be applied by means of numerical methods, e g FEM. The fact that the stresses approach infinity at the crack tip introduces special problems, which can be solved in different ways, (ref. 18).

One recent effective method of application of FEM to the analysis of cracks in concrete is the Blunt Crack Band Model according to Bazant et al (ref. 26, 27).

It must be noted that all methods, where the ascending branch of the stress-strain diagram is assumed to be linear and where the length of the fracture zone is not taken into account, belong to LEFM, and thus have the limited applicability mentioned above.

4. THE J-INTEGRAL

The J-integral (ref. 12) is a path-independent contour integral, which is taken outside the fracture zone from a stress-free point at one crack face to a stress-free point at the other crack face.

For an <u>elastic</u> (linear or nonlinear) material, J is equal to the energy release rate at crack propagation, i e equivalent to G in LEFM. When J reaches a <u>critical value</u> J_c, the crack advances. Just like G_c, J_c can be expected to equal G_F if the material has ideal elastic properties.

As no material is purely elastic, the J-approach is an approximation of reality, just like LEFM. The J-approach however is valid not only for linear elastic but also for non-linear elastic materials. The main property of an elastic material is that the stress-strain diagram is unique, i e there is no difference between loading and unloading curves. This assumption is formally fulfilled for all materials as long as no unloading occurs (we neglect here creep, shrinkage etc). Thus the J-approach may be expected to give correct results as long as no unloading of material takes place.

For many metals the stress is nearly constant within the fracture zone (Dugdale model) and thus the amount of unloading of material before crack advance is small and at the same time the crack tip is well defined. Under these circumstances the J-approach can be expected to give correct results, as the assumptions are rather well satisfied.

For concrete the J-approach may not be expected to give correct results because

1. Unloading of material takes place within the fracture zone due to the descending σ-w curve. The size of the fracture zone is often appreciable and thus the amount of unloading of material is also appreciable.

2. The position of the crack tip is not well defined.

The error introduced by these facts depends on the relative length of the fracture zone, which in its turn is approximately proportional to ℓ_{ch}. As an example of this dependence Fig. 13 shows the theoretical variation (according to FCM-analysis, ref. 2) of J_c/G_F with d/ℓ_{ch} for a three-point bend test. J_c has been evaluated as the difference in area below the load-deflection curves up to the maximum load for the beams with different notch depths, divided by the difference in crack area.

A comparison between Figs. 11 and 13 shows that the determination of J_c by means of this method can be made on smaller beams than the determination of G_c

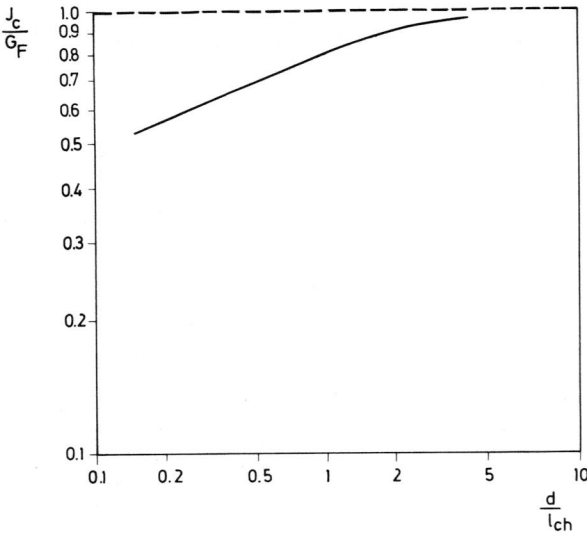

Fig. 13. Theoretical variation of J_c with beam depth d, determined by means of three-point bend tests on beams with two different notch depths a/d = 0.2 and a/d = 0.3 (ref. 2).
————— σ-w curve according to Fig. 9.
--------- σ-w curve according to Dugdale (Fig. 14).

based on LEFM. On the other hand the method for J_c-determination is more compli-cated than that for G_c-determination, and also more complicated than that for G_F-determination.

If the crack tip is well defined, it is possible to determine J_c by studying the change in the load-deflection curve with crack growth. As the crack tip is as a matter of fact never well defined for concrete, it is not possible to get reliable values by means of this method.

The application of the J-approach to structures requires the determination of the J-integral. For linear elastic materials J is identical to G according to LEFM. For non-linear materials the J-integral has to be calculated by means of numerical methods, e g FEM.

In the application, the J-approach has the same disadvantages as LEFM.

The J-approach does not seem to offer any advantages with regard to the prac-tical application of fracture mechanics to concrete structures.

5. R-CURVE ANALYSIS

For yielding metals the crack growth resistance (R) increases with the advan-vance of the crack, as the zone with plane stress conditions widens (see sektion) 1 above).

For metals it may therefore be of interest to study how some measure of crack growth resistance (F_c, K_c, J_c etc) varies with the change in crack length of a test piece. This is called R-curve analysis (ref. 19).

The corresponding phenomenon does not occur for concrete, and thus the reason to apply R-curve analysis to metals is not valid for concrete.

R-curve analysis is based on the observation of crack growth. If it is not possible to observe a well-defined crack tip, the proper basis for the analysis is lacking. This is the case for concrete, as a crack may well be visible although it still transfers stress.

It can be demonstrated that the result of an R-curve analysis, based on the growth of the visible concrete crack in fibre-reinforced concrete, depends on the size of the specimen (ref. 1, 2). The R-curve thus in that case is not a pure material property and it cannot be used for design purposes for specimens of a size that differs from that of the tested specimen. R-curve analysis therefore seems to be of a very limited practical value for concrete, including fibre reinforced concrete.

6. THE DUGDALE "STRIP-YIELD" MODEL

The Dugdale model (ref. 22) is a special case of FCM, where the σ-w curve is assumed to have the shape in Fig. 14, i e a constant yield stress, followed by a sudden rupture.

This model may be suitable for many metals, but it does not give a good description of the properties of concrete.

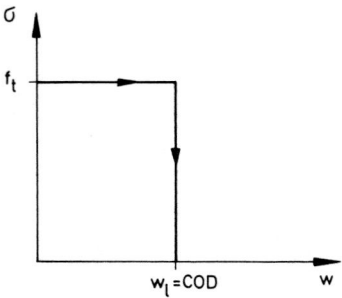

Fig. 14. σ-w curve according to the Dugdale assumption for yielding materials.

7. THE COD-APPROACH

COD (Crack Opening Displacement) is a measure of the width of the blunted crack tip when the crack starts growing, Fig. 14. It is only meaningful if there exists as well-defined crack tip, which gets blunted before it starts advancing. This is not the case for concrete. Thus the COD-appraoch does not seem to be applicable to concrete.

8. THE BARENBLATT MODEL

In the Barenblatt model (ref. 23) a cohesive force is assumed to act across a fictitious crack ahead of the real crack tip. It is in this respect very like FCM. However Barenblatt described these cohesive forces as atomic forces, acting across very small distances. Thus the width of the "fictitious crack" seems to be of the order 10^{-9} m in the Barenblatt model, whereas widths of the order 10^{-6} - 10^{-3} m are used in FCM. The Barenblatt model never seems to have been applied in design.

9. LEFM CORRECTED FOR "EFFECTIVE" CRACK LENGTH

The assumption of LEFM that the stresses may approach infinity is unrealistic, as a plastic or fracture zone will always develop where the strains get too high.

The influence of the limited stresses in the fracture zone may be approximately taken into account by "cutting off" the peak in the stress curve. In order to retain equilibrium it is then necessary to make corrections to the stress curve.

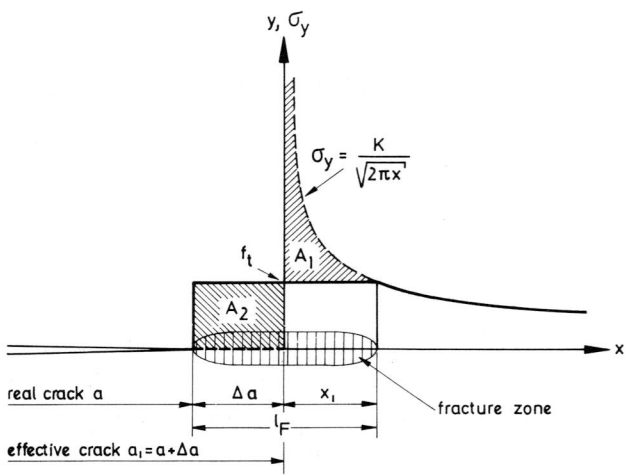

Fig. 15. Effective crack length, fracture zone length, and corresponding stress distribution.

A simple way of doing this is the following (ref. 19).

Fig. 15 shows a LEFM stress curve for an "effective" crack length $a_1 = a + \Delta a$, greater than the real crack length a. The curve has been cut off at $\sigma = f_t$ and the stress has been assumed to equal f_t between the crack tip and this point. This stress distribution may be reasonable for a yielding material with the yield stress equal to f_t (Dugdale model).

According to LEFM the cut-off point is determined from

$$f_t = \frac{K}{\sqrt{2\pi x_1}} \tag{14}$$

giving

$$x_1 = \frac{1}{2\pi}(\frac{K}{f_t})^2 \tag{15}$$

It can easily be demonstrated that the area A_1, which has been cut away from the stress block, is

$$A_1 = x_1 f_t \tag{16}$$

This area has to be compensated by the area A_2 of the stress block between the real and the effective crack tips

$$A_2 = A_1$$

$$\Delta a f_t = x_1 f_t$$

$$\Delta a = x_1 = \frac{1}{2\pi}(\frac{K}{f_t})^2 \tag{17}$$

The crack growth will start when $K=K_c$. At that moment we thus have

$$\Delta a = x_1 = \frac{1}{2\pi}(\frac{K_c}{f_t})^2 \approx \frac{\ell_{ch}}{2\pi} \tag{18}$$

The corresponding length ℓ_F of the fracture zone $\ell_F = a + x_1$ is

$$\ell_F \approx \frac{\ell_{ch}}{\pi} \tag{19}$$

These values of Δa and ℓ_F are of course only valid as long as the size of the specimen is large compared to ℓ_F, i e ℓ_F comprises only a small part of the ligament in front of the crack. Even if this assumption is fulfilled, however, the given values of Δa and ℓ_F must be regarded as lower limits, at least for the fundamental case of a crack in an infinite plate. The following three arguments support this statement:

1. The analytical application of the Dugdale model (ref. 24) gives the relation

$$\frac{a}{a+\ell_F} = \cos\frac{\pi\sigma}{2f_t} \tag{20}$$

An evaluation of this expression gives more than 23 % higher values of ℓ_F than (19).

2. When only the first term of the LEFM solution is taken into account, the stresses in front of the crack tip are understimated, see Fig. 16. This also leads to an underestimation of ℓ_F.

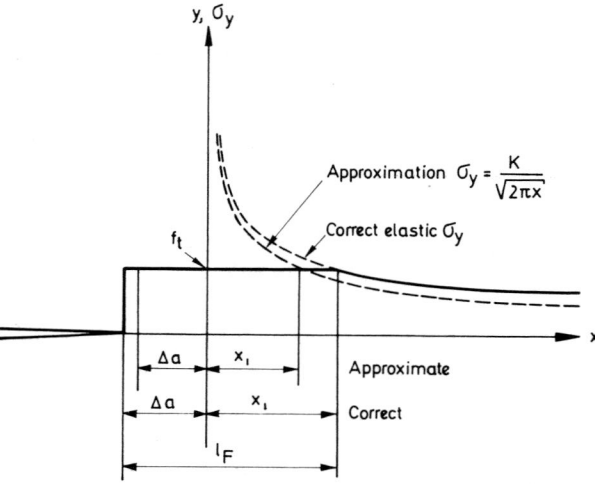

Fig. 16. With a correct elastic σ_y-distribution the length of the fracture zone is greater than according to Fig. 15.

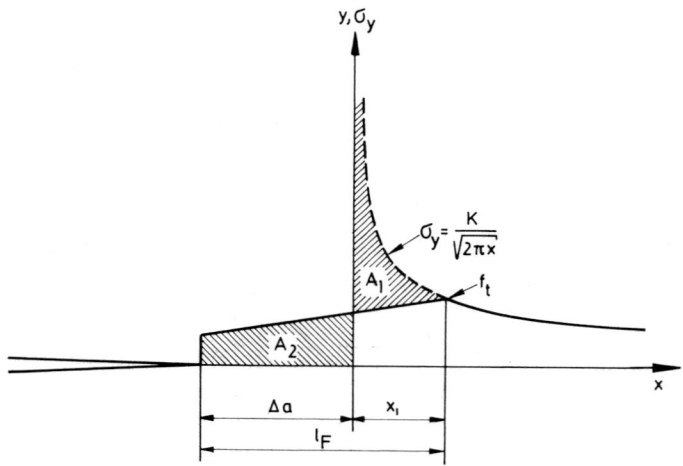

·Fig. 17. With a descending σ-w curve the distance Δa increases compared to that of Fig. 15.

3. In concrete we do not have yield with a constant stress within the fracture zone, but the stress decreases e g according to Fig. 17. The equality between the areas A_1 and A_2 then requires an increase in Δa and thus in ℓ_F.

The use of an "effective" crack length $a_1 = a + \Delta a$ is thus a very uncertain method for taking the length of the fracture zone into account. In spite of this it may be recommended to use an effective crack length $a_1 = a + \Delta a$ instead of a in analyses of test results and in design based on LEFM. The correction will presumably always have a positive effect on the accuracy. The value of Δa for plain concrete can be chosen as about $0.2 \, \ell_{ch}$ at crack advance if no more correct value is known. If Δa exceeds $0.2a$ or 0.2 times the ligament, the result will be uncertain.

10. NOTCH SENSITIVITY

A material or a structure can be more or less sensitive to the influence of cracks, notches or other irregularities.

A notch on the tension side of a bent beam according to Fig. 18 will of course always reduce the moment capacity, as the net section is reduced. The formal net strength f_{net} will depend on the notch depth according to the equation

$$f_{net} = \frac{6M_{max}}{b(d-a)^2}$$

(21)

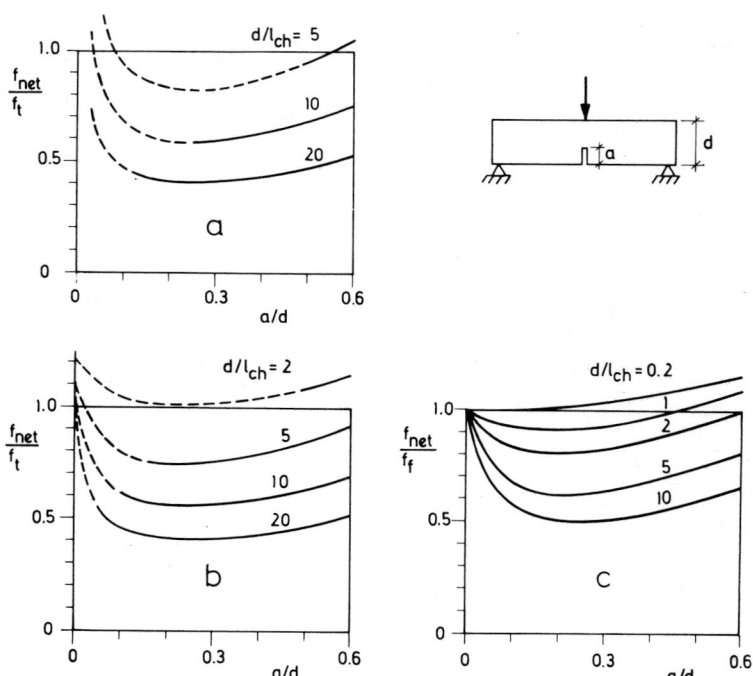

Fig. 18. Theoretical variation in net stress at failure with beam depth and notch depth
a) LEFM /25/
b) LEFM with correction of "effective" notch length $a + 0.2\ell_{ch}$
c) FCM /2/
Broken lines indicate areas outside the assumed validity. Notice the difference between f_t = tensile strength and f_f = flexural stength.

A structure is said to be notch sensitive if f_{net} at failure decreases when the notch depth a increases from zero. If on the other hand f_{net} is nearly independent of a, the structure is notch insensitive.

The theoretical variation of f_{net} with the notch depth can be calculated in different ways. Thus Ziegeldorf et al (ref. 25) have demonstrated how this variation can be calculated by means of LEFM. The result is shown in Fig. 18a. This approach can be expected to give unreliable results for small values of d/ℓ_{ch} and for small values of a/d.

A more reliable result, which is applicable for all values of d/ℓ_{ch} and a/d can be reached by means of FCM. Results of such calculations are shown in Fig. 18c. In this diagram the net stress at failure has been compared to the flexural strength of the unnotched beam, whereas in Fig. 18a (and b) the net stress at failure has been compared to the tensile strength, as LEFM does not give the flexural strength.

In Fig. 18b a corrected "effective" notch length $a+0.2 \ell_{ch}$ has been used together with LEFM according to section 9 above.

From Fig. 18 it is quite clear that "notch sensitivity" is not a pure material property, but a size- and material-dependent property of a structure. For a beam depth of 50 mm cement paste($\ell_{ch} \approx 10$ mm) will seem notch sensitive, whereas mortar ($\ell_{ch} \approx 150$ mm) will not seem notch sensitive. On the other hand cement paste will not seem notch sensitive for a beam depth of 5 mm, whereas mortar will seem notch sensitive for a beam depth of 500 mm.

Notch sensitivety thus cannot be used as a pure material property, but it can be used as a relative quality when different materials are compared. For such a comparison it is however better to use ℓ_{ch}, as a decrease in ℓ_{ch} corresponds to an increase in notch sensitivity and ℓ_{ch} is easier to quantify.

11. TYPICAL MATERIAL PROPERTIES

The applicability of different methods of analysis depends mainly on the relation between the size of the specimen and the characteristic length ℓ_{ch} of the material. The shape of the σ-w curve is also of importance.

LEFM may be applied as long as both the length of the crack and the length of the ligament is greater than approximately 2-5 ℓ_{ch}.

LEFM with a corrected "effective" crack length may be applied as long as the length of the crack and the length of the ligament is greater than approximately ℓ_{ch}.

FCM may be applied indenpendently of the size of the crack or the ligament, thus down to zero crack length. The computer work involved however increases with increasing size and with decreasing ℓ_{ch}, as the distance between nodes in FCM-FEM calculations should not be chosen greater than 0.1-0.2 ℓ_{ch}.

Thus the relation between the size (depth, crack length, ligament etc) of the structure and the characteristic length of the material determines the suitable choice of method of analysis. LEFM is most suitable where this relation is high, whereas FCM is most suitable where this relation is low. The choice of method thus depends on the value of ℓ_{ch}.

For cement paste and concrete the following values of ℓ_{ch} may be regarded as typical (ref. 2, 7, 12), but great deviations from these values may be found, especially where fibre reinforcement is present.

Cement paste	5-15 mm
Mortar	100-200 mm
Concrete	200-400 mm
Glass fibre reinforced mortar	0.5-3 m
Steel fibre reinforced concrete	2-20 m

From these figures it is evident that LEFM has a limited applicability to ordinary concrete and that it cannot be applied to fibre reinforced concrete.

For the application of FCM the shape of the σ-w curve is of interest. Typical such curves are shown in Fig. 19 (ref. 1, 2). Great variations from these curves may be expected, especially where fibres are present.

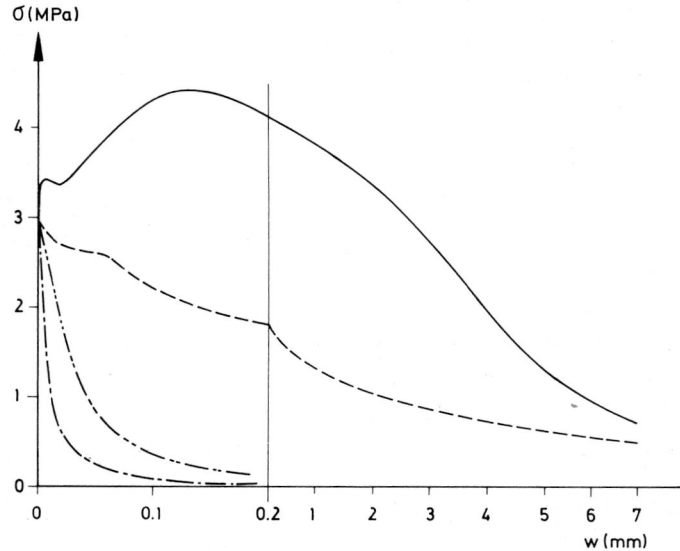

Fig. 19. Examples of σ-w curves for plain and fibre-reinforced mortar and concrete.

—·—·—·— plain mortar
— ·· — ·· —···—plain concrete
——————— about 1 % indented steel fibres d/ℓ=0.3/30 mm
-------- about 1 % plain steel fibres d/ℓ=0.3/30 mm

248

12. SUMMARY AND CONCLUSIONS

1. The fracture behaviour of concrete differs so much from that of metals, that many fracture mechanics methods used for metals are unsuitable for concrete.

2. No well defined crack tip exists for concrete, only a fracture zone where microcracks gradually change into more or less visible cracks, while the concrete still carries some stress. This lack of a well defined crack tip limits the applicability to concrete of some methods which are suitable for metals.

3. For concrete no essential difference exists between plane strain and plane stress conditions. This significantly simplifies the application of fracture mechanics to concrete as compared to metals.

4. Linear elastic fracture mechanics (LEFM) may be applied to the study of cracks if the size of the specimen and of the crack are large enough in relation to the characteristic length ℓ_{ch} of the material. This poses a severe limitation to the applicability of LEFM to concrete.

5. The applicability of LEFM can be extended by the introduction of an "effective" crack length of the order $a+0.2\ell_{ch}$.

6. The fictitious crack model (or other similar models) can be applied irrespective of specimen size and also to other types of fracture than those which start from an existing crack or notch. Its practical limitation lies in the computer work, which increases with increasing specimen size.

REFERENCES

1 Petersson, P-E: Fracture Mechanical Calculations and Tests for Fibre-reinforced Cementitions Materials, Proc from Advances in Cement Matrix Composites, Mat Res Soc, Annual Meeting, Boston 1980.
2 Petersson, P-E: Report TVBM-1006, University of Lund 1981, 174 pp.
3 Entov, V M; Yagust, V I: Mechanics of Solids, V10, No 4 (1975), 87-95.
4 Sok, C, Baron, J,Francois, D: Cement and Concrete Res 1979, 641-648.
5 Swamy, R N: Fracture Mechanics Applied to Concrete. Developments in Concrete Technology - 1, London 1979, 221-281.
6 Hillerborg, A, Modéer, M, Petersson, P-E: Cement and Concrete Res 1976, 773-782.
7 Modéer, M: Report TVBM-1001, University of Lund 1979, 120 pp.
8 Petersson, P-E: Cement and Concrete Res 1980, 79-89.
9 Hillerborg, A: Cement Composites 1980, 177-184.
10 Hughes, B P, Chapman, G P: RILEM Bulletin 1966, p 99.
11 Evans, P H, Marathe, M S: RILEM Mat a Struct 1968, No 1.
12 Petersson, P-E: Cement and Concrete Res 1980, 91-101.
13 Irwin, G R: J Appl Mech 1957, p 361.
14 Griffith, A A: Phil Trans Roy Soc, 1921, p 163.
15 Griffith, A A: Proc 1 st Int Congr Appl Mech, Delft 1924, p 55
16 Irwin, G R: Fracturing of Metals, ASM Cleveland, 1948.
17 Orowan, E: Rep Progr Phys 1949, p 185.
18 Hellen, T K: Numerical Methods in Fracture Mechanics. Developments in Fracture Mechanics - 1, London 1979, 145-181.
19 Knott, J F: Fundamentals of Fracture Mechanics, London 1973.

20 Brown, W F M, Srawley, J E: ASTM STP 410, 1967.
21 Rice, J R: Journ Appl Mech 1968, 379-386.
22 Dugdale, D S: Journ Mech Phys Solids 1960, 100-104.
23 Barenblatt, G I: Advances in Applied Mechanics 1962, p 55.
24 Burdekin, F M; Stone, D E W: Journ Strain Anal 1966, p 145.
25 Ziegeldorf, S, Müller, H S; Hilsdorf, H K: Cement and Concr Res 1980, 589-599.
26 Bažant, Z P, Cedolin, L: J Eng Mech Div ASCE 1979, 297-315.
27 Bažant, Z P, Cedolin, L: J Eng Mech Div ASCE 1980, 1287-1306.
28 Mindess, S, Nadeau, J S: Cement and Concrete Res 1976, 529-534.
29 Bâzant, Z P, Oh, B H: Concrete Fracture via Stress-strain Relations, Northwestern Univ Report 81-10/665c, p 50.
30 Terrien, M: Bull Liaison Labo. P et Ch, 105, jan-févr 1980, 65-72.

LIST OF SYMBOLS

A   cross sectional area

E   Young's modulus

F   load

G   energy release rate

G_c critical value of G for crack growth

G_F fracture energy at total separation

K   stress intensity factor

K_c critical value of K for crack growth

M   bending moment

Y   stress intensity factor coefficient

a   crack length, half length of internal crack

f_f flexural strength

f_t tensile strength

f_{net} formal bending strength of net section

ℓ_o gauge length

ℓ_{ch} characteristic length $= EG_F/f_t^2$

w   additional deformation due to the fracture zone (=width of the "fictitious crack")

w_ℓ value of w when the stress transfer ceases

x   coordinate in crack direction

y   coordinate perpendicular to crack

$\Delta\ell$ elongation of gauge length ℓ_o

ε strain

σ stress without regard to cracks

σ_y stress in y-direction

Fracture Mechanics of Concrete,
edited by F.H. Wittmann, 1983
Elsevier Science Publishers B.V., Amsterdam — Printed in The Netherlands

Chapter 4.2

CRACK PROPAGATION IN A COMPOSITE MATERIAL

by Y. ZAITSEV

1.   INTRODUCTION

There have been many previous studies concerning the fracture behaviour of concrete. Most of them, however, are primarily based on purely phenomenological observations and on continuum mechanics theory. Thus, they refer to homogeneous materials ignoring the presence of the unhomogeneous structure of concrete. Such an approach is not sufficient if one wants to explain the different behaviour of hardened cement paste as well as lightweight, normal and high-strength concrete under high load.

For descriptions of behaviour of concrete under high load, the methods of fracture mechanics can be very useful. However, the application of classical fracture mechanics to concrete has been rather limited. One significant reason for this is the complicated composite structure of concrete. In a composite structure failure is usually not caused by the extension of one critical crack. In fact far below the ultimate load cracks begin to propagate and are arrested again by high strength aggregate particles. Thus failure really means a gradual degradation of the structure.

By means of computerized simulation methods it is possible to generate simplified model structures which statistically described the distribution of pores and inhomogeneities (inclusions) in the volume of concrete. The stress field in such a simulated stochastic structure can be calculated as a function of the applied load. If the fracture mechanics parameters of the different components can be estimated it is possible to study crack propagation within the composite structure by means of "computer experiments".

In this section the heterogeneous structure of concrete is described in terms of a multi-level system according to the chapter by Wittmann "Structure of hardened cement paste and concrete". The following assumptions are made :

1) A solid body consisting of a matrix with inclusions (aggregate particles), defects of the first kind (pores) and of the second kind (cracks) on different levels of structure is considered.

2) Inclusions may have different shapes (circular or polygonal) and are considered only on the meso-level.

3) Defects of the first kind may have different shapes (circular, ellyptical, polygonal) and are considered only on the micro-level.

4) Defects of the second kind may have different shapes (straight, curve-shaped) and are considered on micro- and meso-levels.

5) Spacial three-dimensional states of stress may be replaced by two-dimensional states of stress.

6) Material between inclusions and defects is homogeneous and istropic.

7) Dimensions of inclusions and defects are small as compared to dimensions of the body.

8) Strains are small.

9) For short-time loading, the effect of creep may be neglected, i.e. the behaviour of the main part of material (except zones in the vicinity of crack tips) is elastic.

10) For sustained or slowly variable loading the "elastic" solutions may be used if instead of the $1/E$-value some operator $1/\hat{E}$ will be inserted.

On the basis of these assumptions, conditions for crack propagation in two-dimensional model structures with randomly distributed pores and inclusions are studied analytically. By using the derived formulae crack propagation and failure of hardened cement paste in lightweight, normal and high-strength concrete are discussed. Computer experiments for various loads and conditions are described. It is shown that crack propagation and final degradation of composite structures can be simulated realistically. It is also shown that experimental results are in reasonable agreement with theoretical predictions. Finally, some practical applications of the theory are mentioned.

2. CRACK PROPAGATION, STRENGTH OF HARDENED CEMENT PASTE AND CONCRETE IN TENSION

2.1 Hardened cement paste

According to the chapter by Wittmann, "Structure of hardened cement paste and concrete", the structure of hardened cement paste will be treated on the micro-level of the hierarchic system of structure of materials. The main defects of the structure of unloaded cement paste may be divided into two groups:

- defects of the first kind - pores
- defects of the second kind - pre-existing cracks, which are results of shrinkage and other processes in ageing of cement paste.

The stress concentration, caused by defects of the first kind (pores), highly affects the fracture initiation. It is known that stress concentration around holes and cavities depends on their shapes.

In porous materials such as hardened cement paste, many different pore shapes can be observed. If a load is applied, high tensile stress occurs in certain regions of hardened cement paste, which is, according to the assumption 6.9 (paragraph 1), assumed to be homogeneous, elastic and isotropic. Under such assumptions, the stress distribution around pores may be found with the help of methods of elasticity theory. Taking into account also assumptions 5 (see above), we shall reduce this problem to a two-dimensional case. Stress distribution around holes of simple shapes (circular, ellyptical, etc) within the range of plane elasticity theory is well known. Stress distribution around square and triangular holes which are randomly orientated with respect to the direction of the applied load has been studied by Zaitsev (ref. 1, 2). Stress concentration around holes of various shape was treated also by Savin (ref. 3).

If one assumes the different pore shapes and the orientation of all pores with respect to load direction to be randomly distributed, the probability of the occurence of stress of a given level in the material can be calculated using Monte-Carlo-Methods. If in addition one knows the pore shape distribution, a "random pore" can be mathematically simulated.

By a microscopic analysis of pore shape frequency in hardened cement paste (ref. 4) the following pore shape distribution has been observed (table I).

TABLE 1

Observed frequency of different pore shapes in hardened cement paste.

Pore Shape	Sample I Number	%	Sample II Number	%	Sample III Number	%
Circular	17	32.7	17	26.6	35	42.2
Square	9	17.3	7	10.9	3	3.6
Triangular	7	13.5	6	9.4	9	10.8
Ellyptic (1:2)	9	17.3	24	37.5	16	19.3
Ellyptic (1:3)	8	15.4	10	15.6	19	22.9
Rectangular (1:2 to 1:3)	2	3.8	--	----	1	1.2
Sum	52	100	64	100	83	100

From references 1-3 it is known that coefficients of stress concentration near triangular, square and rectangular holes do not significantly differ from corresponding coefficients for ellyptical holes having a proper main axes relation a:b. Therefore it is reasonable to analyse the stress concentration around a random pore using a simplified pore distribution and replacing triangular pores by elliptical (a:b = 1:5), square pores by elliptical (a:b = 1:3.5), rectangular pores by elliptical (a:b = 1:3.5). Corresponding frequencies of different pore shapes are given in Table II.

TABLE II
Simplified frenquencies of different pore shapes in hardened cement paste.

Pore Shape	Sample I Number	%	Sample II Number	%	Sample III Number	%
Circular (1:1)	17	32.7	17	26.6	35	42.2
Ellyptic (1:2)	9	17.3	24	37.5	16	19.3
Ellyptic (1:3)	8	15.4	10	15.6	19	22.9
Ellyptic (1:3.5)	11	21.1	7	10.9	4	4.8
Ellyptic (1:5)	7	13.5	6	9.4	9	10.8
Sum	52	100	64	100	83	100

The angle between the maximum axis of a hole and the load direction is indicated by α (see Fig. 1). The angle α is considered to have a uniform distribution between 0 and 2π. Using Monte-Carlo-Methods, it is possible to calculate the probability of the occurence of stress of a given level. Stress is considered to be a random function of the angle θ, which defines the position of a point on the pore outline (see Fig. 1). Some realizations of the computed random function $S(\theta)$ are shown in figure 2a, 2b. Results of the statistical analysis of the random function $S(\theta)$ simulated according to the frequency of different pore shapes (see tables I, II) are given in figure 3. Curve 1 corresponds to the observed distribution (according to table I), curve 2 to the simplified one, curve 3 to simplified approximations:

$$M_S(\theta) = 1-2 \cos 2\theta$$
$$D_S(\theta) = 1.4 (1.01-\cos 2\theta)$$

Let us now analyse crack propagation in hardened cement paste subjected to uniaxial tension. According to the simplifying assumptions made above (see "Introduction") we shall investigate the problem of crack propagation in an elastic plate containing ellyptical holes; each hole contains two pre-existing cracks, which are orientated in the most dangerous way; i.e. coplanar with the major axis of the ellypse, which is perpendicular to the direction of the applied external load (see Fig. 4). This problem can be treated with help of a simplified method of Panasjuk (ref. 5). According to this method, stress in the vici-

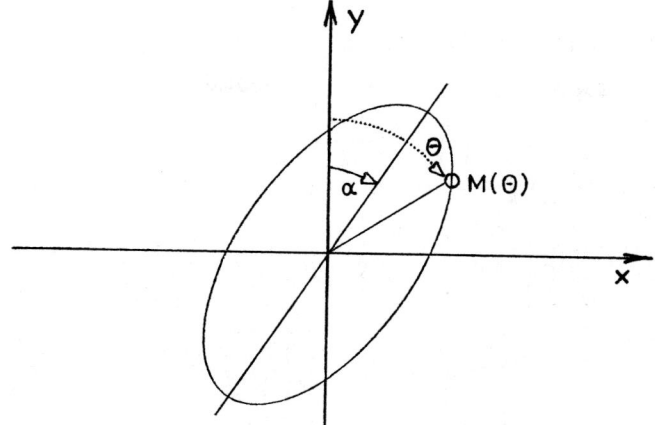

Fig. 1. The definition of the angles α and θ.

nity of the crack tip can be regarded as a sum :

$$\sigma_y(x, 0) = \sigma_0(x, 0) + \sigma_1(x, 0); \quad (x < a, \; x > b)$$

where $\sigma_0(x, 0)$ is the stress in a plate with an ellyptical hole (without cracks) caused by tensile stress $\sigma_y = q$; $\sigma_1(x, 0)$ is the stress in a plate with a crack along Ox-axis, $-a \leq x \leq b$, when both sides of the crack are subjected to the pressure $p_n(x) = \sigma_0(x, 0)$, $-a \leq x \leq -d$ and $d \leq x \leq b$.

Stress $\sigma_0(x, 0)$ in a plate with an ellyptical hole (without cracks) can be expressed in the ellyptical coordinates u, v($x = $ ch $u \cdot$ sin v; $y = $ ch $u \cdot$ cos v). On the crack line we have :

$$y = 0; \; v = \pi/2, \; x = \text{ch } u, \; \text{and} \; \sigma_0(x, 0) = qF \{u(x)\} \;;$$

$$F\{U(x)\} = 1 + \frac{\text{ch } u}{2\text{sh}^2 u} \{e^{u_0}(e^{2u_0} - 3) \times (1 + \frac{\text{cth } u}{2}) e^{-2u} + \text{ch } u_0 \text{ cth } u\},$$

where U_0 is the U-value on the boundary of the hole, which can be found from the following conditions :

$$\text{cth}_0 = d/d_1; \; \text{ch } u_0 = \frac{d/d_1}{\sqrt{d^2/d_1{}^2 - 1}}; \; \text{sh } u_0 = \frac{1}{\sqrt{d^2/d_1{}^2 - 1}}$$

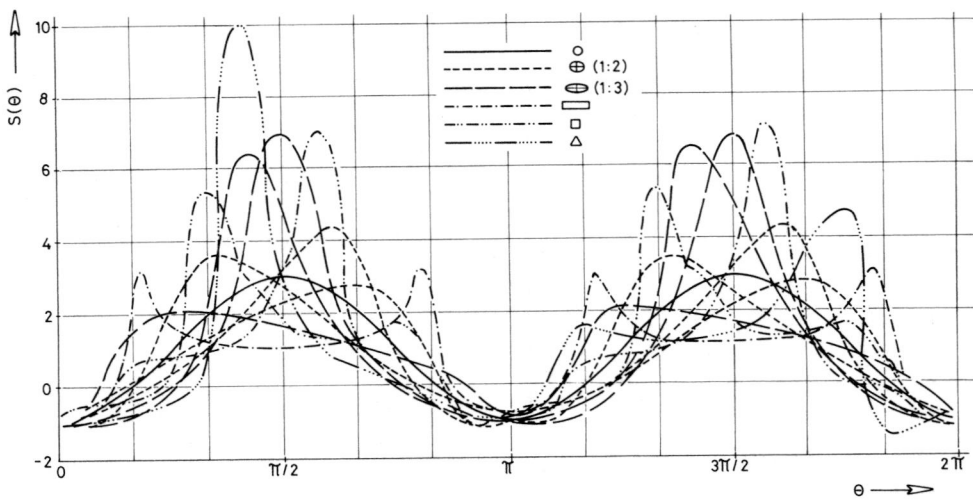

Fig. 2a. Simulation of stress concentration around a "random" pore. Observed pore shape distribution.

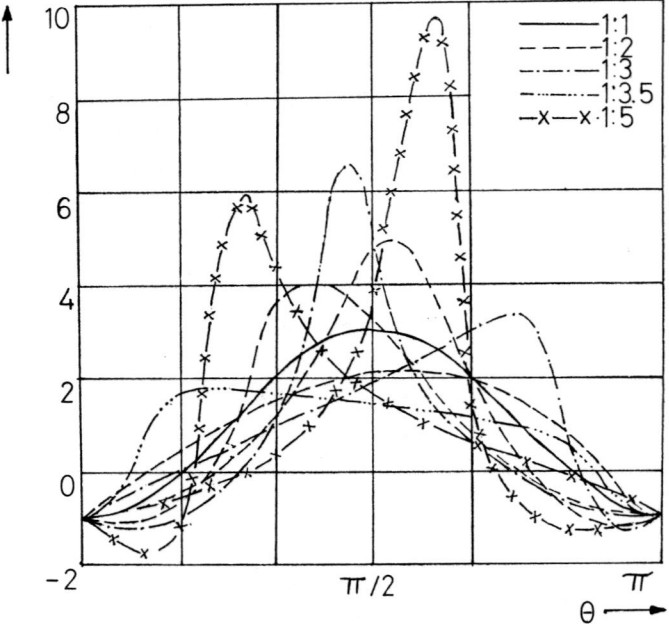

Fig. 2b. Simulation of stress concentration around a "random" pore. Simplified pore shape distribution.

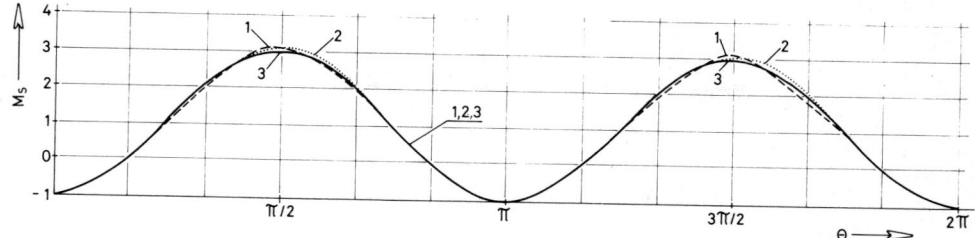

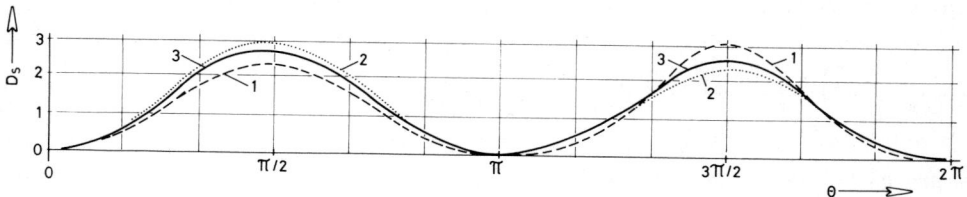

Fig. 3. Mean value M_S and dispersion D_S for stress around a "random pore" :
1) by observed pore shape distribution;
2) by simplified pore shape distribution;
3) by approximation of M_S and D_S.

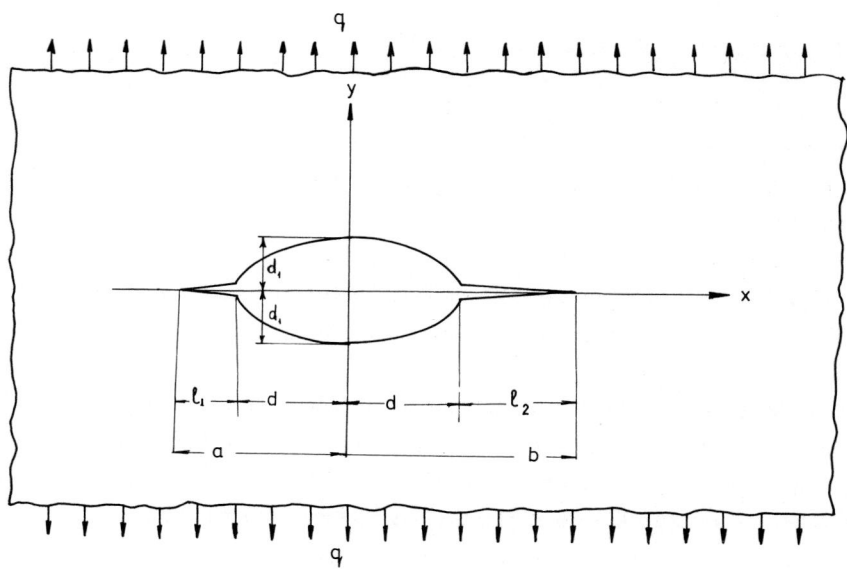

Fig. 4. Schematic representation of a pore and coplanar cracks in hardened cement paste subjected to uniaxial tension.

Stress $\sigma_1(x, 0)$ according to /5/ :

$$\sigma_1(x, 0) = \frac{1}{\pi \sqrt{(x-b)(x+a)}} \int_{-a}^{b} \frac{p_n(\xi) \sqrt{(b-\xi)(a+\xi)}}{/x-\xi/} \, d\xi;$$

$$x \leq -a; \quad x \geq b; \quad a \leq b;$$

$$p_n(\xi) = \begin{array}{ll} \sigma_0(\xi, 0) & -a \leq \xi \leq -d; \\ 0 & -d < \xi < d; \\ \sigma_0(\xi, 0) & d \leq \xi \leq b \end{array}$$

The ultimate load value $q = q_\star$ can be found from the expression :

$$\frac{1}{\sqrt{b+a}} \int_{-a}^{b} \frac{p_n^\star(\xi)\sqrt{(b-\xi)(a+\xi)}}{b-\xi} \, d\xi = \sqrt{\frac{\pi}{2}} \cdot K_{IC}$$

where $p_n^\star(\xi)$ is defined as the value of $p_n(\xi)$ when $q = q_\star$ (i.e. in the critical stage). From the expressions given above one can obtain :

$$q_\star = \sqrt{\frac{\pi}{2}} K_{IC} \sqrt{a+b} \Big/ f(a, b)$$

where

$$f(a, b) = \int_{-a}^{-d} F\{u(t)\} \sqrt{\frac{a+t}{b-t}} \, dt + \int_{d}^{b} F\{u(t)\} \sqrt{\frac{a+t}{b-t}} \, dt$$

Integral evaluation can be done numerically by computer. Figure 5 gives results of such calculation for various values of d/d_1 (curve 1). It was assumed that both cracks have the same length $\ell_1 = \ell_2 = \ell$. The results are given in terms of related crack length $\lambda = \ell/d$ and related external tensile load $q_\star^0 = q_\star \sqrt{2\pi d}/K_{IC}$. Setting $d = d_1$ (a circular hole), we obtain the lower line 1, and by assuming absence of pores (Griffith's crack of the length $2(\ell+d)$), we obtain the upper line 3.

It may be seen from figure 5 that the influence of pore shape is significant only when the crack length is small as compared to pore dimensions. If the crack length has the same order of magnitude as the pore dimensions ($\lambda = 1$) or is greater, it is possible to neglect the presence of a pore and use the Griffith's solution for a crack of length $2(\ell+d)$.

The crack propagation described above is unstable. This means that by monotonically increasing the external load, the pre-existing cracks at first do not grow. However,

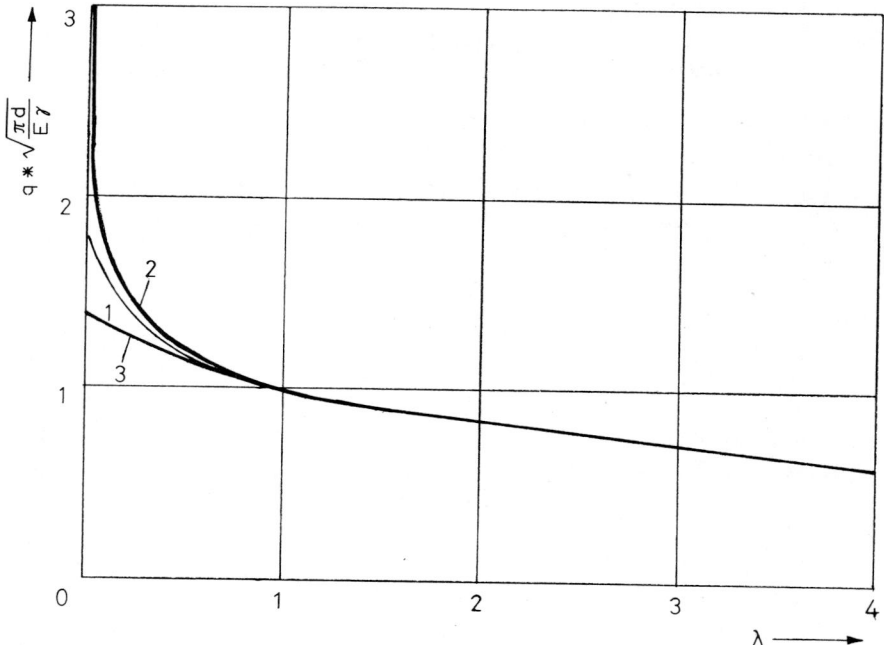

Fig. 5. Relationship between related external load and related crack length :
1) by ellyptical pores
2) by a round (circular) pore
c) by absence of pore (Griffith's crack).

when the external load has reached the critical value for one of the most dangerous cracks, this crack will propagate spontaneously, which leads to the failure
of the sample of material.

2.2   Concrete

According to the chapter by Wittmann "Structure of hardened cement paste and
concrete", the structure of concrete will be treated on the meso-level of the
hierarchic system of the structure of materials. The main defects of the structure
of unloaded concrete are pre-existing cracks. These cracks are the result of sedimentation (bleeding), shrinkage and other processes in fresh and hardening concrete (see table 2 in the chapter by Wittmann).   Most of these cracks are interfacial
(bond) cracks between hardened cement paste and coarse aggregates (see chapter
by Mindess, "The application of fracture mechanics to cement and concrete: a
historical review", paragraph 2).

Let us now analyse crack propagation in concrete subjected to uniaxial tension. According to the simplifying assumptions made above we shall investigate the problem of crack propagation in an elastic plate of relative thickness containing circular inclusions (coarse aggregate particles). Each inclusion has one pre-existing bond-crack; the dimension of the most dangerous crack in the matrix (hardened cement paste), defining its strength, is small as compared to the dimension of the inclusions. We shall assume also that the value of K_{IC} for inclusions is greater than the value of K_{IC} for the matrix, and the value of K_{IC} for the matrix is greater than the value of K_{IC} for the interface, i.e.

$$K_{IC}^{INCL} > K_{IC}^{M} > K_{IC}^{IF}$$

With these assumptions, this problem has been solved by Cherepanov. He has found the following relationship between the external tensile load q_* and the angle θ, which defines the dimension of the crack :

$$q_* = K_{IC}^{IF} \; F(\theta) \; \sqrt{R};$$

$$F(\theta) = \frac{4(3-\cos \theta)/\sqrt{\pi}}{\sqrt{\sin \theta (44+12\cos \theta+12 \cos^2 \theta-4\cos 4\theta+\sin^4 \theta)}}$$

Line 1 on figure 6 shows this relationship; on the y-axis values of the related load $q_{IF} = q_* \sqrt{\frac{2R}{\pi}}/K_{IC}^{IF}$ are given. When $\theta<\theta_0 (\theta_0 = \pi/4)$ a crack propagates in an unstable fashion (descending part of the curve), but when $\theta>\theta_0$ the crack becomes stable (ascending part of the curve). The crack can propagate in a stable fashion only before the external load has reached a definite critical value, corresponding to the appearance of matrix branches of the crack according to figure 6 (below), which will again lead (by $\theta=\theta_2$) to an unstable fashion of crack propagation. The precise value of θ_2 depends on the value of K_{IC}^{IF}/K_{IC}^{M}: the greater this value is, the less the angle θ_2 is. By $K_{IC}^{IF}/K_{IC}^{M} \cong 0.6$ (which is near to the experimental results of Alexander, Hillermaier and Hilsdorf, Ziegeldorf and Hilsdorf, etc) $\theta_2 = \pi/2$. This situation is shown in figure 6, where line 2 gives the values of the related load for the interfacial part of the crack $q_M = q_* \sqrt{\frac{2R}{\pi}}/K_{IC}^{M}$ (i.e. $q_M^0 = 0.6 \; q_{IF}^0$), and line 3 gives the values of the related load for the crack in the matrix. The last values have been found in an approximative way as for a Griffith's crack of the length $2(\ell+R)$. The load value q_0 (beginning at the stable stage of crack propagation along the interface) is equal to approximately 80%

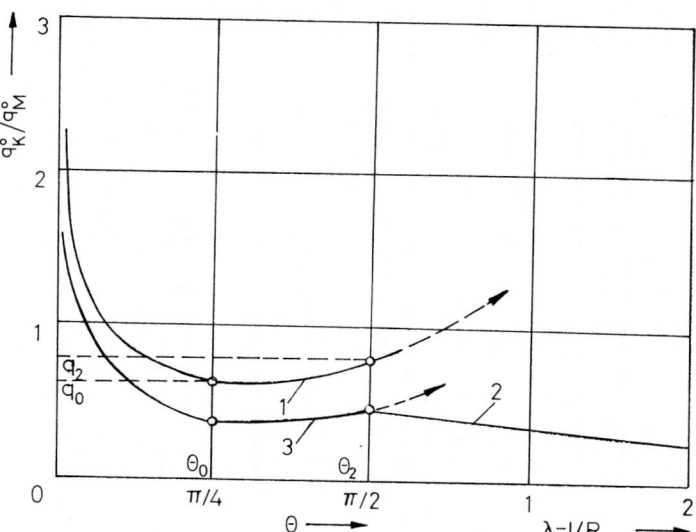

Fig. 6. Relationship between related external load and related length of an interfacial crack.
1) interfacial crack (load is related to K_{IC}^{IF})
2) interfacial crack (load is related to K_{IC}^{M})
3) interfacial crack and additional matrix (load is related to K_{IC}^{M}).

of the value of q_2^0, which corresponds to the beginning of the unstable stage of crack propagation in the matrix. The following stages of crack propagation in concrete can be analysed in an analogous way.

Now we can simulate the structure of concrete and the crack propagation using Monte-Carlo-Methods. Typical examples of one of the computer realisations are shown in figure 7. For the simulation, 50 circular inclusions have been produced, and each of them is assumed to have one pre-existing interfacial crack. It was also supposed that there are pre-existing cracks in the matrix (according to paragraph 2.1). It was assumed that the centres of cracks in the matrix and the centres of aggregate particles are uniformly distributed on the area of the sample. The length of pre-existing matrix and interfacial cracks as well as the diameter of aggregate particles are assumed to have a Gaussian distribution.

Taking the opening of cracks into consideration, the contribution of cracks to the longitudinal deformation ε can be found, if we assume linear superposi-

262

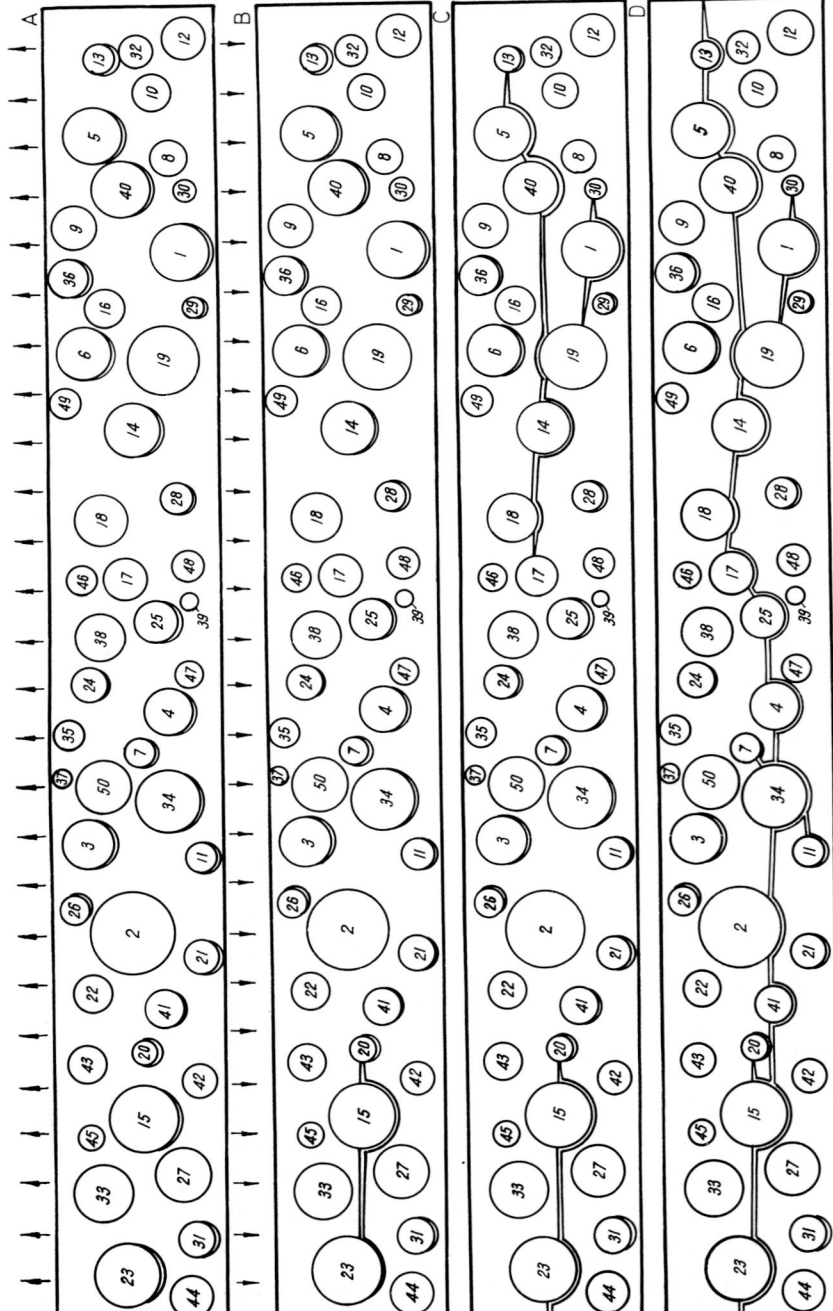

Fig. 7. Crack pattern for different load levels in normal concrete subjected to uniaxial tension. Cracks originating from interfaces run around aggregates.

tion of the two components

$$\varepsilon_1 = \varepsilon_1' + \varepsilon_1'',$$

where ε_1' is the contribution of the unfractured material and ε_1'' is the inelastic component caused by the displacement of crack edges. According to reference 5, the half width v of a crack of the length 2l in a material with modulus of elasticity E subjected to uniaxial tensile stress p can be found as follows :

$$v(x, 0) = \frac{2p}{E} \sqrt{l^2 - x^2},$$

whereby the maximal half-width is equal to :

$$v_{max} = 2pl/E$$

The component $\varepsilon_1' = \sigma/E$, while the component ε_1'' can be found if we assume that deformation caused by the displacement of the crack edges is "smeared" on the volume of a sample. Thus for one crack we have :

$$\varepsilon_2'' = \frac{2 \int_0^{2l} v(\xi) \, d\xi}{bh}$$

and for all n cracks, propagating in a sample of concrete :

$$\varepsilon_2'' = 2 \sum_{i=1}^{n} \int_0^{2l_i} v(\xi) d\xi = \frac{4p}{E} \sum_{i=1}^{n} \int_0^{2l_i} \sqrt{l_i^2 - \xi^2} \, d\xi$$

Figure 8 shows stress-strain curves for concrete according to the results of simulation of crack propagation. On the X-axis, values of the related strain $\varepsilon/\varepsilon_{el}^*$ are given ($\varepsilon_{el}^* = Q^*/E$, Q^* is the fracture stress), and on the Y-axis values of related load $q_*^0 = q_* \sqrt{2\bar{R}/\pi}/K_{IC}^M$, where $\bar{R}$ is mean value of the diameter of aggregate particles.

The simulated σ-ε curves for concrete in tension may be divided into the three regions shown in figure 8. In the region 1, which corresponds to the crack pattern in figure 7(a), the external load is less than the q_0-value (see Fig. 6) for each of the pre-existing cracks. This means that cracks do not pro-

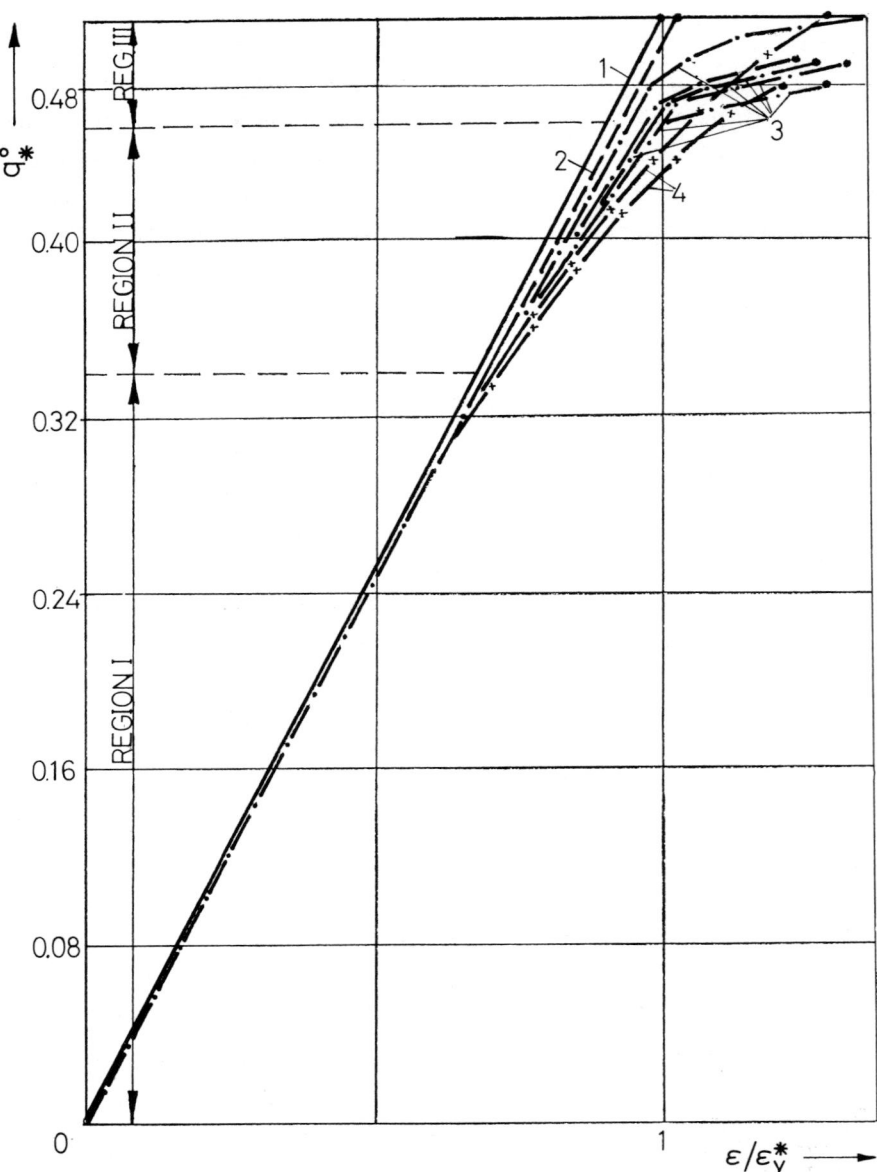

Fig. 8. Stress-strain curves for concrete according results of simulation of crack propagation.
1) Linear for a homogeneous material without cracks
2) Linear for a material with pre-existing cracks, which do not propagate
3) non-linear taking into account crack propagation
4) experimental curves for concrete in tension.

pagate and there is only very little extension (widening) of these cracks. The σ-ε diagram (line 2 in Fig. 8) is linear, but it has a little bit greater declination as compared to the line 1, which corresponds to a material without cracks; this difference is due to the crack extension (widening) mentioned above.

In the region 2, cracks according to figure 7(a) begin to propagate in a stable fashion, which corresponds to external load values greater than the q_0-value (see Fig. 6) for the most dangerous cracks. The corresponding σ-ε curves become slightly non-linear (see lines 3 in Fig. 8, each line corresponds to one realization of the Monte-Carlo-Method).

In region 3 (see Fig. 7(b,c)) cracks begin to propagate through the matrix between the aggregate particles. The σ-ε curves become significantly non-linear (see line 3 in Fig. 8, upper parts). Finally, failure of the sample occurs (see Fig. 7, d). It was found that region 1 corresponds to $\sigma < 0.65\ \sigma_{ult}$; region 2 to $0.65\ \sigma_{ult} < \sigma < 0.9\ \sigma_{ult}$ and region 3 to $\sigma > 0.9\ \sigma_{ult}$. It was also found that, by assumed parameters of crack redistribution, the critical load for pre-existing matrix cracks (according to paragraph 2.1) was higher than σ_{ult} for concrete, thus the failure of concrete in this case only depends on the propagation of pre-existing interfacial cracks.

Figure 9 shows the mean value (from 20 realizations) of the summarized length of cracks (related to the area of the sample) as a function of the related strain $\varepsilon/\varepsilon_\star$. Lines 1 and 2 correspond to interfacial and matrix cracks. Lines 3 and 4 are given according to experimental results (ref. 6). All results of simulation of crack propagation in concrete described above are in reasonable agreement with existing experimental results.

Finally we shall try to find the relationship between K_{IC} for components of concrete structure on the meso-level (matrix, interface) and tensile strength R_B on the macro-level (both levels are determined according to the chapter by Wittmann). From the expression for the related load $q_\star^0 = q_\star\ \sqrt{\frac{2\bar{R}}{\pi}}/K_{IC}^M$, which is given in figure 8, we get (assuming that by $q_\star^0 = 0.5$, see Fig. 8, $q_\star = R_B$) :

$$K_{IC}^M = R_B \sqrt{\frac{8\bar{R}}{\pi}}$$

For $R_B = 2 \div 4$ MPa and $\bar{R} = 1$ cm or 2 cm, we get $K_{IC}^M = 0.32 \div 0.64$ MPa $\cdot$ m$^{\frac{1}{2}}$ and $K_{IC}^M = 0.45 \div 0.90$ MPa $\cdot$ m$^{\frac{1}{2}}$ respectively. We have also assumed that $K_{IC}^{IF} = 0.6\ K_{IC}^M$, i.e. K_{IC}^{IF}-values lay between 0.19 and 0.38 MPa $\cdot$ m$^{\frac{1}{2}}$ for $\bar{R} = 1$ cm and between 0.27

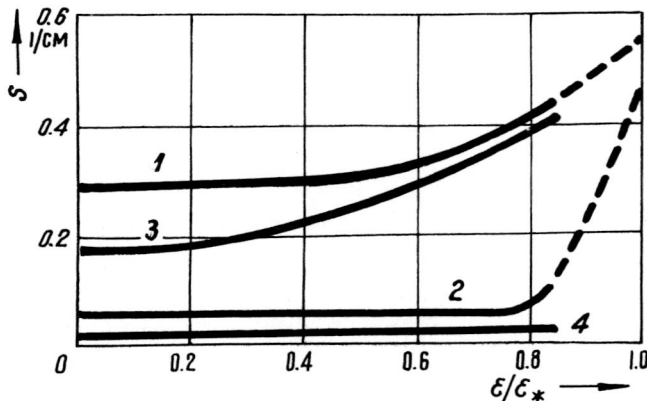

Fig. 9. Summarized length of cracks as a function of related strain $\varepsilon/\varepsilon_*$.
1) simulated (interface cracks)
2) simulated (matrix cracks)
3) experimental (ref. 6) interface cracks
4) experimental (ref. 6) matrix cracks

and 0.54 MPa $\cdot$ m$^{\frac{1}{2}}$ for $\bar{R} = 2$ cm. These values are within the range of existing experimental results for K_{IC}^{M} and K_{IC}^{IF}.

3. CRACK PROPAGATION AND STRENGTH OF HARDENED CEMENT PASTE AND CONCRETE IN COMPRESSION

3.1   Hardened cement paste

It was mentioned above (paragraph 2.1) that the structure of hardened cement paste (micro-level) has defects of the first kind (pores) and of the second kind (pre-existing cracks). It was found (see paragraph 2.1) that around
pores the concentration of stress of the same sign as the external stress depends significantly on the shape of the pores; this was the reason to take into
account the real shape of the pores by external tension. On the other hand the
concentration of stress of the opposite sign as compared to the external stress
practically does not depend on the shape of pores. Thus, by external compressive stress, it is possible not to take the real shape of pores into consideration and assume all the pores to be circular.

Under such assumptions we shall now analyze the problem of crack propagation
in an elastic plate of thickness 1, containing circular holes; each hole contains two coplanar pre-existing cracks, which are orientated in the most dangerous direction - parallel to the direction of the applied external load (see
Fig. 10). According to (ref. 5) the stress in the vicinity of crack's tip can

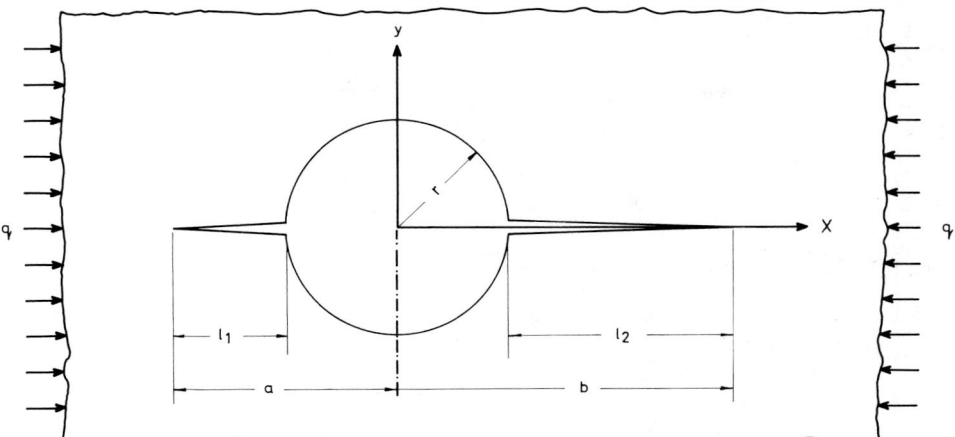

Fig. 10. Schematic representation of a pore and coplanar cracks in hardened cement paste subjected to uniaxial compression.

be regarded as a sum

$$\sigma_y(x,\ 0) = \sigma_0(x,\ 0) + \sigma_1(x,\ 0)$$

where $\sigma_0(x,\ 0)$ and $\sigma_1(x,\ 0)$ have the same meaning as in paragraph 2.1.

Stress $\sigma_0(x,\ 0)$ in a plate with a circular hole (without cracks) is well known :

$$\sigma_0(x,\ 0) = q\ (\frac{r^2}{2x^2} - \frac{3r^4}{2x^4})$$

Stress $\sigma_1(x,\ 1)$ according to (ref. 5) :

$$\sigma_1(x,\ 0) = \frac{1}{\pi\ \sqrt{(x-b)(x+a)}} \int_{-a}^{b} \frac{Pn(\xi)\ \sqrt{(b-\xi)(a+\xi)}}{/x-\xi/}\ d\xi$$

$x \leq -a;\ x \geq b;\ a \leq b$

$$Pn(\xi) = \begin{array}{ll} \sigma_0(\xi,0) & \text{by } -a \leq \xi \leq -r \\ 0 & \text{by } -r < \xi < r \\ \sigma_0(\xi,0) & \text{by } r \leq \xi \leq b \end{array}$$

The critical load value $q=q_*$ can be found in the same way as in paragraph 2.1. It is equal to :

$$q_* = \sqrt{\frac{\pi}{2}} K_{IC} \sqrt{a+b}/f(a, b)$$

where

$$f(a, b) = A_+ \sqrt{(a+r)(b-r)} - A_- \sqrt{(a-r)(b+r)} -$$

$$- B \ell n \frac{\{\sqrt{ab} + \sqrt{(a-r)(b+r)}\}^2 + r^2}{\{\sqrt{ab} + \sqrt{(a+r)(b-r)}\}^2 + r^2}$$

$$A^\pm = - \left[\frac{r^2}{b^2}\left(\frac{5}{8} + \frac{1}{8}\frac{b}{a}\right) \pm \frac{r^3}{b^3}\left(\frac{15}{16} + \frac{1}{4}\frac{b}{a} + \frac{3}{13}\frac{b^2}{a^2}\right)\right]$$

$$B = \frac{r^2 \sqrt{ab}}{32a^3 b^4} \left[8a^2 b^2(a+b) - 3r^2\left(7a^3 - 3a^2b + 5ab^2 - b^3\right)\right]$$

If both cracks have the same length ($\ell_1 = \ell_2 = \ell$), we get the following expression for the critical load value :

$$q_* = - \sqrt{\frac{\pi}{2r}} \sqrt{\frac{(1+\lambda)^7}{2\{(1+\lambda)^2-1\}}} K_{IC},$$

where λ is the related crack length, $\lambda = \ell/r$.

Figure 11 shows the relationship between related load value $q_*^0 = |q_*|\sqrt{\frac{2r}{\pi}}/K_{IC}$ and related crack length $\lambda = \ell/r$. Line 2 corresponds to the expression for q_* obtained above under assumption that $\ell_1 = \ell_2 = \ell$. Line 1 is given to comparison and corresponds to the case of only one crack (i.e. $\ell_2 = 0$, $\ell_1 = \ell \neq 0$, $a = r$). We can see from figure 11, that by $\lambda < \lambda_0$ ($\lambda_0 = \sqrt{7/5}-1 \cong 0.18$) crack propagation is unstable, but by $\lambda > \lambda_0$ it becomes stable.

By increasing the load, cracks begin to interact and beyond a certain load the crack propagation can become unstable again. In order to analyse this stage of crack propagation we shall now investigate the following problem. An infinite plate of thickness 1 has two equal circular holes and four coplanar cracks as shown in figure 12.

This problem can be treated in a similar way as the one solved above. The solution of this problem was given in (ref. 7). The relationship between rela-

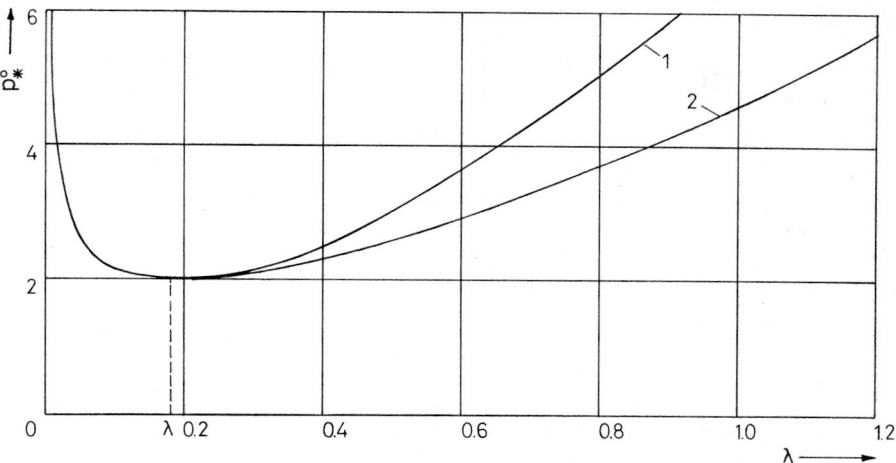

Fig. 11. Relationship between related external load and related crack length.
1) circular pore and one crack
2) circular pore and two coplanar cracks of the same length.

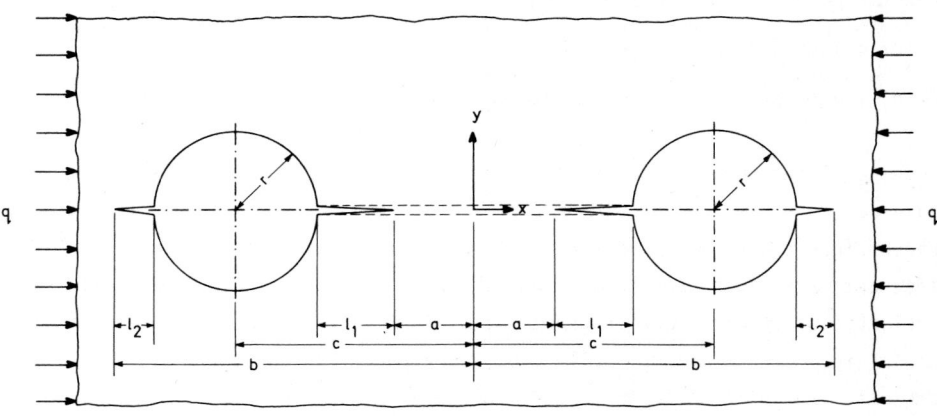

Fig. 12. Schematic representation of two pores and coplanar cracks in hardened
cement paste subjected to uniaxial compression.

ted external load $q_*^0 = /q_*/ \sqrt{\dfrac{2r}{\pi}} \; K_{IC}$ and related crack length $\lambda = \ell_i/r$ or $\Psi = \ell_i/$
(c-r) according to reference 7 is given in figure 13. Curves 2 and 3 in figure
13 show that unstable crack propagation (by $\lambda < \lambda_0$) will change (by $\lambda > \lambda_0$) to
stable crack propagation, in accordance with the relationship for cracks near
an isolated hole (without interaction of cracks) - see curves 1a, 1b. The

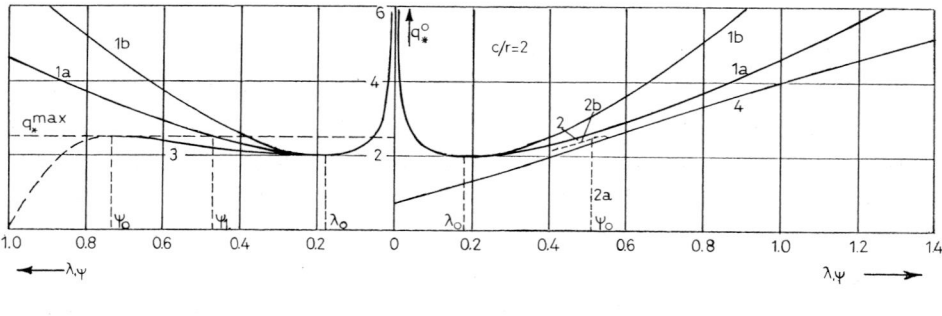

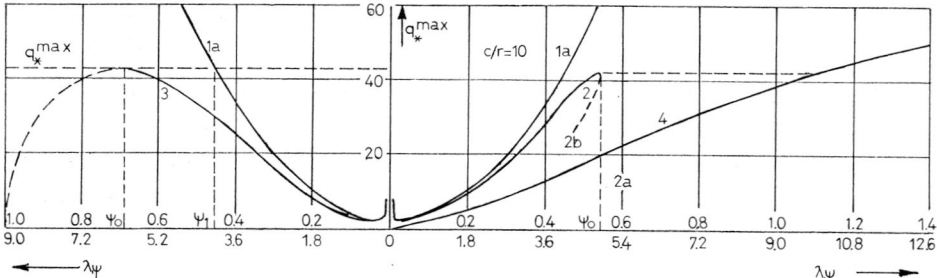

Fig. 13. Relationship between related external load and related crack length near two pores : a) $c/r = 2$; b) $c/r = 10$.
1) Relationship for two cracks (1a) and one (1b) isolated crack (without interaction) according to figure 11.
2) Relationship for interacting external cracks.
3) Relationship for interacting internal cracks.

presence of neighbouring holes causes the unstable fashion of crack propagation again (by $\psi = \psi_0$ and $q_*^0 = q_*^{max}$, see curves 2 on Fig. 13); curves 2a correspond to irreversible cracks and curves 2b - to reversible cracks. The calculations for different c/r-values have shown that the related crack length (at the end of the length of stable crack propagation) ψ_2 is nearly constant ($\psi_2 = 0.69 \div 0.73$ for internal cracks and $\psi_2 = 0.51 \div 0.55$ for external cracks). The related crack length ψ_1, which corresponds to q_*^{max} on the line 1a, can also be regarded as a constant ($\psi_1 = 0.46 \div 0.47$).

The process of crack propagation discussed above will lead to the joining of two internal cracks and formation of a new crack, crossing both holes (see dotted lines on Fig. 12). The problem of propagation of such a crack was analysed in reference 7. The results, which were obtained in a similar way, are given in figure 13 (line 4). A comparison of curves 2 and 4 shows that if the load has reached the value q_*^{max} (i.e. cracks have entered into the zone of unstable propagation), both interacting cracks join at once and the λ-value "jumps" from curve 2 to curve 4 (wave line).

The following stage of crack propagation, when three, four and more coplanar cracks interact, is a process having a statistical nature, because this interaction depends significantly on the distribution of distance between pores. This process can be simulated using Monte-Carlo-Methods. The results of the simulation described in reference 7 show a satisfactory agreement with existing experimental results.

In the model described above it is assumed that all the cracks are coplanar. It is, however, well known that real hardened cement paste has cracks oriented at random. Therefore a more complicated model has been analyzed. It was assumed that all pores are statistically uniformly distributed on the area in figure 18, which represents the specimen size. Each pore has two pre-existing cracks, the length of which is uniformly distributed within the range $0 < l/r < 2$. The angle of crack orientation α_i with respect to the external load is also uniformly distributed within the limits of 0 and 2π. The interaction of cracks was taken into account by the assumption that two cracks ($i = 1; 2$) will interact and coalesce if α_1 and α_2 are both below $\pi/6$ or if α_1 is below $\pi/6$ while α_2 is above $5\pi/6$. The random structure has been created by a computer program. If crack propagation is studied by Monte-Carlo Methods, the results are similar to those obtained by using the one-dimensional model with spelling cracks. The results of crack simulation and mode of fracture of such a model will be discussed below, in paragraph 3.2.

For both model structures, the coplanar cracks and the two-dimensional crack arrangement, the equations describing crack propagation as a function of increasing load can be written in general form as follows :

$$q = \frac{K_{IC}}{\sqrt{2r}} \; g(\lambda)$$

where $g = g(\lambda)$ is a function of the related crack length λ.

Simulation of crack propagation as described above makes it possible to estimate the effect of pore size distribution on the fracture mechanism, mechanical strength and strain behaviour of hardened cement paste. In particular, in reference 8 it was found that by increasing the mean size of pores while keeping the quantity constant, the ultimate load will decrease. Increasing the maximum size of pores but keeping a constant mean size and a constant quantity of pores will decrease the ultimate load too.

3.2   Concrete

As in paragraph 2.2, the structure of concrete will be treated on the meso-level of the hierarchic system of the structure of materials. As in paragraph 2.2, we shall investigate a problem of crack propagation in an elastic plate of thickness 1, containing inclusions (coarse aggregate particles). Inclusions having in this case random polygonial shape are randomly distributed in a matrix. The size and shape distribution and the volume content of the inclusions can be varied in order to simulate different concrete mix proportions. Each inclusion has one pre-existing bond crack; the dimension of the most dangerous crack in the matrix (hardened cement paste), defining its strength, is small as compared to the dimension of the inclusions.

We begin with the simplest case of a randomly inclined crack in an elastic plate loaded at infinity. In figure 14 an initial crack having a length of 2l, and an inclination of α with respect to the direction of external compressive load $q(q < o)$ is shown. This crack might propagate along the same inclined line as a shear crack (Mode II). It can be shown, however, that in our example at the tips of the initial crack, two branching cracks of Mode I (splitting or opening cracks) are created (ref. 9). By introducing simplifying assumptions the crack length ℓ_2 according to reference 10 can be expressed as follows :

$$\frac{P}{\sqrt{\pi \ell_2}} = K_{IC}$$

where $P = T \sin \alpha$ and T is the resulting force of shear stress $\tau_{\zeta\eta}$, causing sliding of two opposite sides of the inclined crack. Taking the coefficient of friction ρ into consideration, T can be expressed as follows :

$$T = 2\ell_1 \tau_{\zeta\eta} = 2\ell_1 q (\sin \alpha \cos \alpha - \rho \sin^2\alpha)$$

From these equations we can obtain (crack of Mode I in matrix) :

$$q_I^M = -\frac{\sqrt{\pi \ell_2}}{2\ell_1} \frac{K_{IC}^M}{A(\alpha, \rho)}$$

where $A(\alpha, \rho) = \sin^2\alpha \cos \alpha - \rho \sin^3\alpha$. From the lower equation one can see that the propagation of such a crack is stable, i.e. ℓ_2 will steadily increase as q increases. A similar stable propagation of such cracks has been observed experimentally (ref. 9).

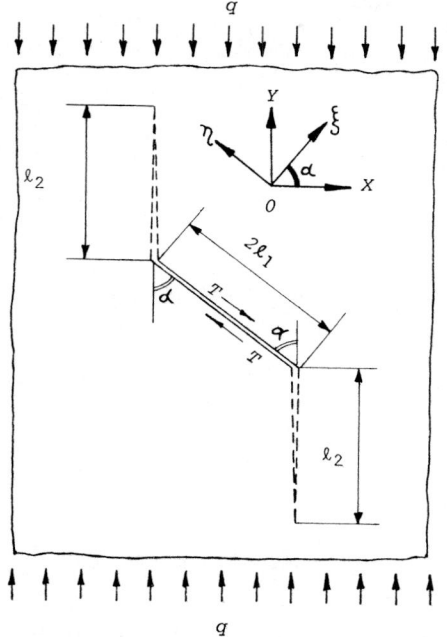

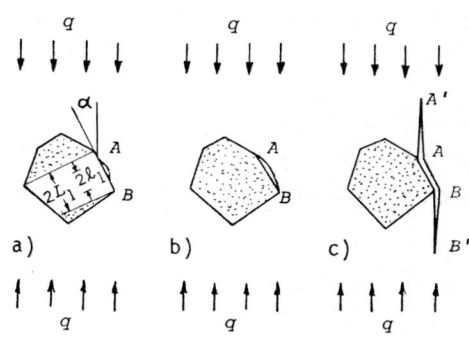

Fig. 14. Schematic representation of the development of branching cracks and definition of symbols used in corresponding equations.

Fig. 15. An initial crack with length $2\ell_1$ (a) grows in an unstable fashion along an interface AB (B) and finally stable branching cracks AA' and BB' are created as the load is increased.

After analysing this simplest case, we shall now consider a homogeneous matrix with one polygonial inclusion, representing an aggregate particle in an infinite matrix. An initial interfacial crack with length $2\ell_1$ is assumed to be located along one side AB (see Fig. 15). This problem can be treated in a similar way as the one with an inclined crack in a homogeneous matrix, taking into consideration, however, concentration of shear and normal stress in the interface. This can be done by introducing coefficients of stress concentration k_τ n k_σ. It can be shown that the initial shear crack spreads (Mode II) in an unstable fashion as soon as the critical load q_{II}^{IF} (Mode II, interface) is reached:

$$q_{II}^{IF} = -\frac{k_{IIC}^{IF}}{\sqrt{\pi\ell_1}\, D_{IF}(\alpha, \rho)}$$

where $D_{IF} = k_\tau \sin\alpha \cos\alpha - k_\sigma\rho \sin^2\alpha$. The shear crack reaches the length $2L_1$ (see Fig. 11b) and stops, because further crack propagation in the same inclined direction would take place through the matrix, where $K_{IIC}^M \gg K_{IIC}^{IF}$.

If, however, the external load is increased to a higher critical value q_I^M :

$$q_I^M = -\frac{\sqrt{3/4}}{\sqrt{\pi\ell_1}} \frac{K_{IC}^M}{D_{IF}(\alpha,\ \rho)}$$

branching cracks in the matrix will develop (see Fig. 15c). The actual crack length in the matrix can be given as a function of load analogously with the case of a homogeneous material treated alone :

$$q = -\frac{\sqrt{\pi\ell_2}}{2L_1} \frac{K_{IC}^M}{A_{IF}(\alpha,\ \rho)}$$

where $A_{IF}(\alpha,\ \rho) = D_{IF}(\alpha,\ \rho) \sin\alpha$, and ℓ_2 corresponds to the distance AA' as shown in figure 15c.

In figure 16 the situation as shown in figure 15c is repeated, but now it is assumed that the branching crack AA' meets a second inclusion as it propagates. Further crack growth will either take place through the inclusion maintaining the same direction, or the crack has to follow the interface MN as a crack of Mode I, or the crack has to follow the same interface MN, but as a crack of Mode II (see Fig. 16c). To know which path will be followed, three critical load values must be compared :

$$q_I^{INCL} = -\frac{\sqrt{\pi\ell_2}}{2L_1} \frac{K_{IC}^{INCL}}{A_{IF}(\alpha,\ \rho)}$$

$$q_I^{INT} = -\frac{2K_{IC}^{IF}\sqrt{\pi\ell_2/L_1}}{A_{IF}(\alpha,\ \rho)\ \{3\cos\frac{\beta}{2} + \cos\frac{3\beta}{2}\} - 3B_{IF}(\alpha,\ \rho)\ \{\sin\frac{\beta}{2} + \sin\frac{3\beta}{2}\}}$$

$$q_{II}^{INT} = -\frac{2K_{IC}^{IF}\sqrt{\pi\ell_2/L_1}}{A_{IF}(\alpha,\ \rho)\ \{\sin\frac{\beta}{2} + \sin\frac{3\beta}{2}\} + B_{IF}(\alpha,\ \rho)\ \{\cos\frac{\beta}{2} + 3\cos\frac{3\beta}{2}\}}$$

The index INCL denotes that values are valid for the crack propagation through the second inclusion; L_2 corresponds to the distance AA' as shown in

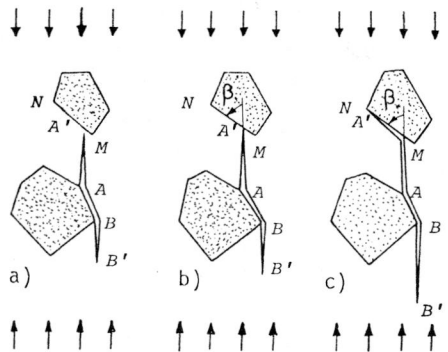

Fig. 16. A crack path as shown in figure 15, in the vicinity of the second inclusion (a); the crack meets the second inclusion (b); finally the crack will propagate along the interface MN (c).

figure 16b and $B_{IF}(\alpha, \rho)$ is equal to :

$$B_{IF}(\alpha, \rho) = D_{IF}(\alpha, \rho) \cos \alpha$$

From the expressions for q_I^{INCL}, q_I^{INT}, q_{II}^{INT} it is evident that the further crack path depends on the relationship between interface and inclusion characteristics K_{IC}^{INCL}, K_{IC}^{IF}, K_{IIC}^{IF} as well as on the geometry of the crack path A'ABB', (see expressions for $A_{IF}(\alpha, \rho)$, $B_{IF}(\alpha, \rho)$) and the inclination of the interface MN.

For further discussion of these theoretical predictions we shall look into three different cases :
- normal concrete (where $K_{IC}^{IF} \ll K_{IC}^{INCL}$)
- high strength concrete (where $K_{IC}^{IF} \approx K_{IC}^{INCL}$)
- lightweight concrete (where $K_{IC}^{IF} \gg K_{IC}^{INCL}$)

These three cases shall be dealt with separately.

In the case of normal concrete a crack will propagate mainly along the interface MN, because the critical q_I^{INCL}-value will be too high as a result of high K_{IC}^{INCL}-values. Whether crack propagation according to Mode I or to Mode II will take place depends significantly on the sign of β (see Fig. 16). If $\beta < 0$ a crack of Mode I is to be expected but if $\beta > 0$ (as in Fig. 16), a crack of Mode II is more probable.

It must be noted that shear cracks (Mode II) are facilitated by shear components of external pressure, whereas the presence of normal confining components
of external compression makes the formation of opening cracks in the interface
(Mode I by $\beta < 0$) less likely as compared to the formation of shear cracks (Mode
II by $\beta > 0$). This theoretical prediction has been verificated experimentally-
see figures 23-25 (ref. 11). As a consequence, in a material with randomly distributed inclusions (where the probability of occurence of positive and negative values of β are equal) new interfacial cracks will propagate mainly according to Mode II by $\beta > 0$. This means that a resulting crack running through the
total specimen will contain some interfacial parts, which deviate largely in
the same direction ($\beta > 0!$) as the external load direction (see also Fig. 19).
Thus, the resulting crack will be slightly inclined and not exactly parallel to
the external load direction.

In the case of high-strength concrete a crack as shown in figure 16 will
propagate either according to the mechanisms described above (i.e. along the
interface), or it will penetrate into the inclusion. With increasing values of
β crack propagation along an interface occurs at higher loads.

Above a critical value β_* of the angle of inclination of the interface side
β, cracks will choose their way through inclusions. The critical value, β_*, is
practically independent of α and/or ρ and depends only on the ratio χ_1 of K_{IC}^{INCL}
and K_{IC}^{IF}. In particular, by $\chi_1 = K_{IC}^{INCL}/K_{IC}^{IF} = 1$ we obtain $\beta_* = \pi/3$. Crack propagation in the inclusion also depends on the ratio $\chi_1 = K_{IC}^{INCL}/K_{IC}^{IF}$. If $\chi_1 < 1$, cracks
grow more rapidly and if $\chi_1 > 1$ they grow more slowly as compared to the homogeneous matrix, but the whole crack pattern in high strength concrete must be
expected to be similar to the one in a homogeneous matrix. Thus, the probability of crack deviation from the direction of the external load is much less as
compared to the case of normal strength concrete.

In the case of lightweight concrete, where the interface strength is much
higher than the matrix strength, fracture surfaces run across the matrix and
aggregate particles. For this case the model of crack initiation at pores according to paragraph 3.1 can be used. It has been shown that such cracks propagate
in lightweight concrete in a stable fashion analogous to the case of hardened
cement paste discussed above. After a crack has reached an inclusion it propagates into the inclusion. This propagation becomes instable because of the
small K_{IC}^{INCL}-values as compared to corresponding K_{IC}^{M}-values. When the crack,
after having passed through the inclusion, reaches the matrix again, it stops.
After further increase of the external load (depending on $\chi_1 = K_{IC}^{INCL}/K_{IC}^{M}$-value)
the crack will propagate through the matrix again as if it were a homogeneous

material.

After describing all essential elements of crack propagation in a two-phase material, we can simulate the structure of concrete and the crack propagation using Monte-Carlo-Methods. Typical examples of computer realizations of the structure of normal and lightweight concrete are shown in figure 17. In the case of normal concrete 30 polygonal inclusions have been produced. Each particle is supposed to have one interfacial crack. The structure can also be used to study crack propagation in high strength concrete. In this case the geometrical arrangement is maintained, but the fracture mechanics parameters are modified. The random structure of lightweight concrete is simulated by 20 round inclusions and small pores spread over the matrix. A typical example of a computer simulation of the porous structure of hardened cement paste is shown in figure 18.

As the load in the computer experiment is increased, first the most critical cracks will propagate. Further increase of the load produces a characteristic crack pattern and finally one crack will run through the total specimen. This is defined to be failure of the material.

Different load levels and corresponding crack patterns are found by means of a computer experiment as shown in figures 19-22. A discussion of this result is published (ref. 12). Cracks in normal concrete do not penetrate aggregates, thus they contain some interfacial parts which mostly have an angle of inclination of the same sign, and the resulting "over-all" crack will be slightly inclined. To the author's knowledge, this often observed behaviour is now theoretically explained for the first time.

Computer experiments provide a solid basis for further systematic investigations. Differing material structures as well as failure under a multiaxial state of stress can be studied in a similar way. Time-dependent processes such as failure under high sustained load can be included in this type of investigation too.

4.   CRACK PROPAGATION AND STRENGTH UNDER A MULTIAXIAL STATE OF STRESS

4.1   Bi-axial compression

Statistical analysis of stress concentration around a "random pore" (see paragraph 2.1) can be extended to the bi-axial state of stress, in particular, to the bi-axial compression. In reference 4 it has been shown that the mathematical treatment which has been outlined briefly in paragraph 2.1 may be extended to three-dimensional pores.

In figure 26 the number of crossings under bi-axial state of stress is given as function of the axial stress ratio. The parameter is the applied load. The

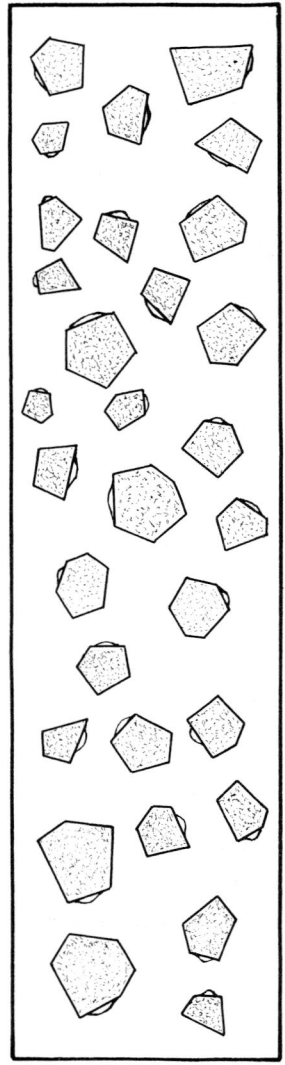 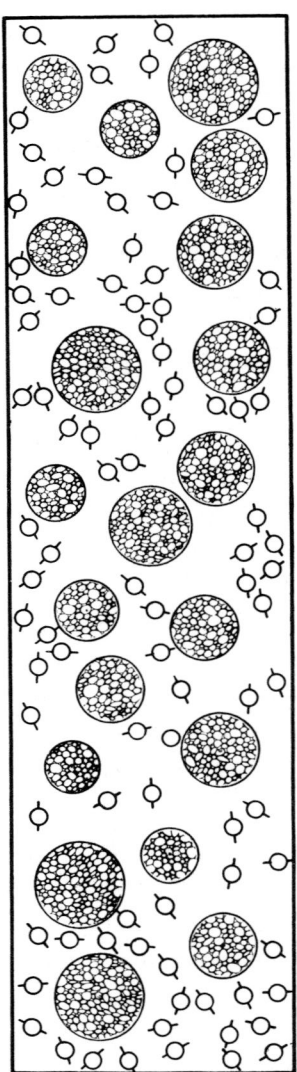

a   b

Fig. 17. Typical computer realization of the random structure.
a) Normal concrete.
b) Lightweight concrete.

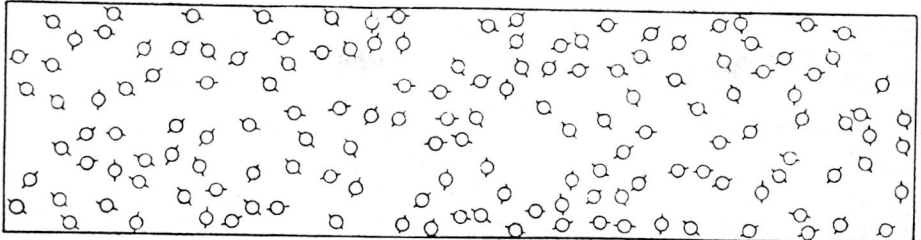

Fig. 18. Typical computer realization of the random structure of hardened cement paste.

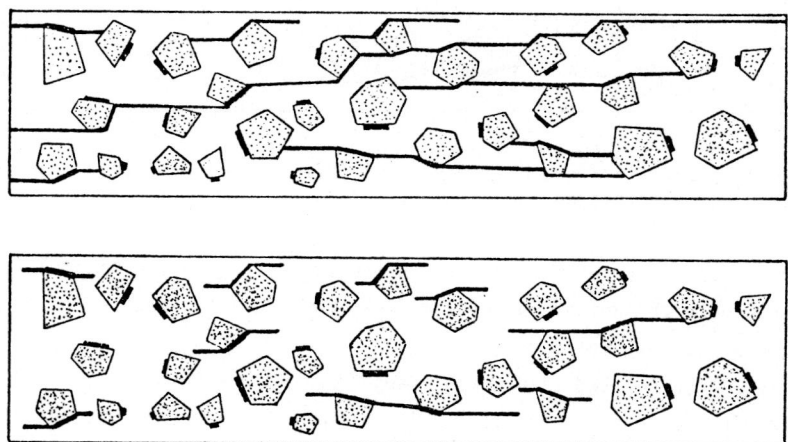

Fig. 19. Crack pattern for two different load level in normal concrete. Cracks are running around aggregate particles.

curves (see Fig. 26) are calculated under the assumption that the external load is increased in steps of two percent, the lowest curve corresponding to the lowest level.

Before the statistical stress analysis can be interpreted in terms of strength, it is necessary to introduce a failure criterion. A certain number of crossings was chosen, that is to say, a certain number of local defects in the matrix, as the failure criterion. The actual number depends on the material and on influences such as temperature or moisture content. An equal number of crossings under varying ratios σ_z/σ_y are represented by a horizontal line (see Fig. 26). It can be seen that starting from $\sigma_z/\sigma_y = 0$ the load must be increased

280

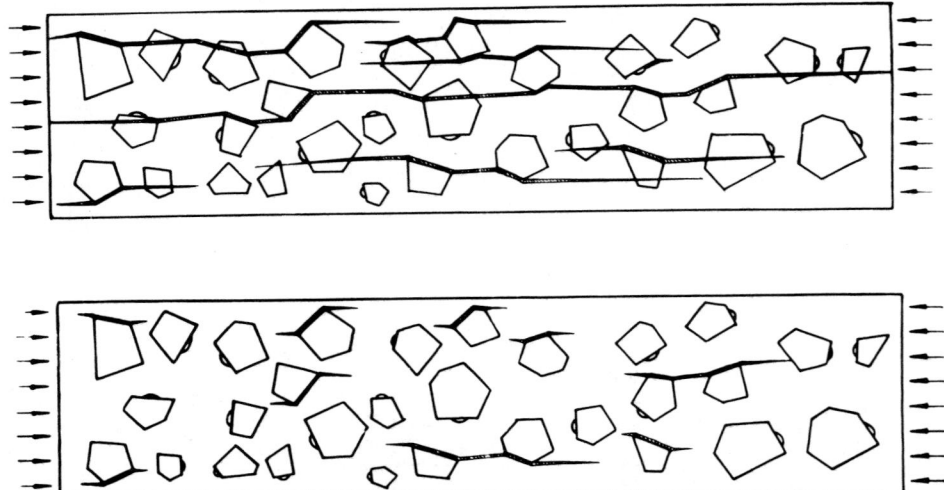

Fig. 20. Crack pattern for two different load levels in high strength concrete. Some cracks penetrate through aggregate particles.

to cause an equal number of microcracks. At about $\sigma_z/\sigma_y = 0.5$ the highest load is necessary and with further increase of σ_z/σ_y a moderate decrease of the corresponding load can be observed.

In figure 27 the related load which leads to microcracks in 2.8%, 3.8% and 5.3% of the pores present in the matrix is shown in the commonly used diagram. The more pronounced the increase of strength under bi-axial load is, the greater is the number of microcracks necessary to reach the failure criterion. Some experimental results are shown (see Fig. 28). Because of the considerable scatter of experimental results no certain check of the theory is possible. The actual scatter is even wider than shown (see Fig. 28) if data from additional authors are represented. The large scatter can be explained partly by differences in experimental procedures.

One aspect of the theory presented in this paper, however, can be checked. In figure 29 results of experiments carried out with concrete, hardened cement paste, and plaster of Paris are shown. Only a few microcracks are necessary in a brittle material to cause failure. Thus hardened cement paste and plaster of Paris in accordance with the discussion of figure 26 do not show a significant increase of strength under a bi-axial state of stress. In concrete a great number of microcracks are arrested as soon as they interfere with the aggregate. Only when

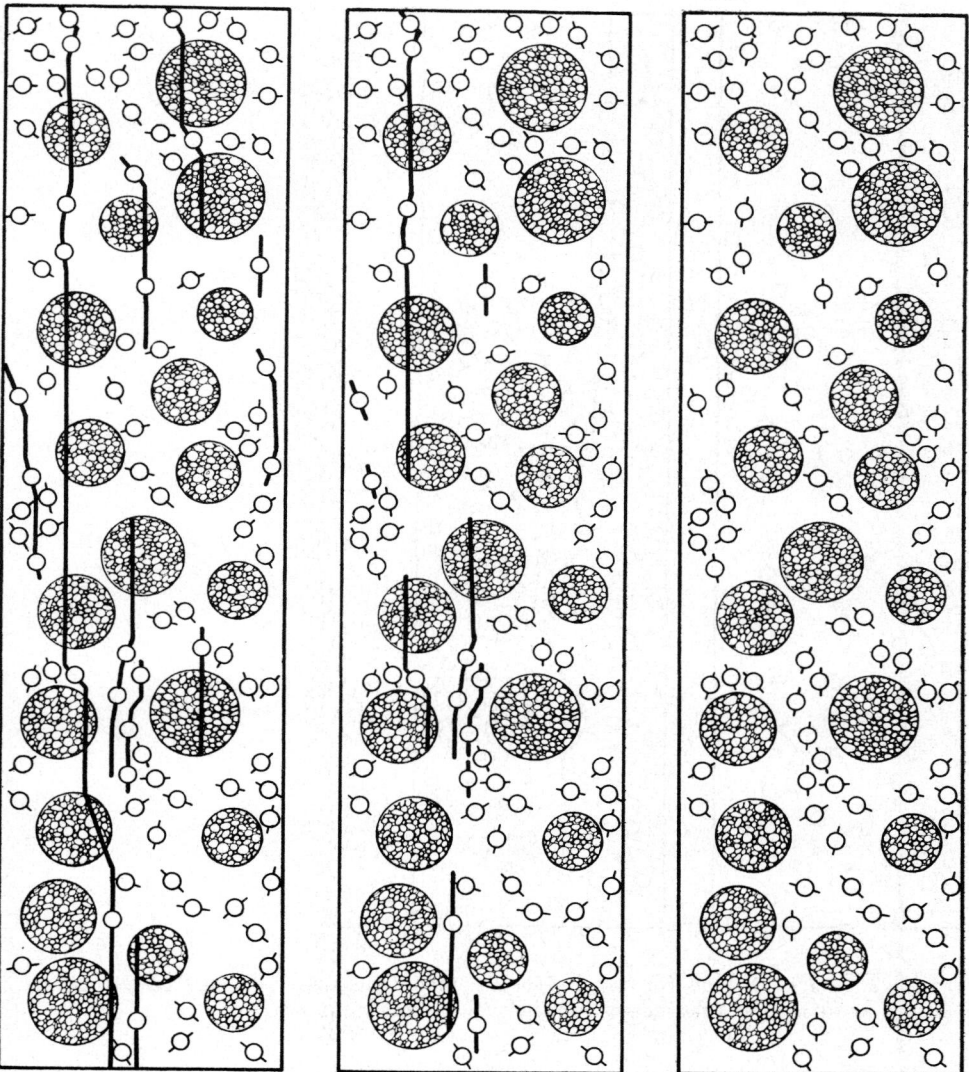

Fig. 21. Crack pattern for two different load levels in lightweight concrete. Cracks originating from pores run through lightweight aggregate particles.

the load is increased are new microcracks formed and eventually failure occurs. As a consequence, strength of concrete is more severely influenced by a bi-axial state of stress.

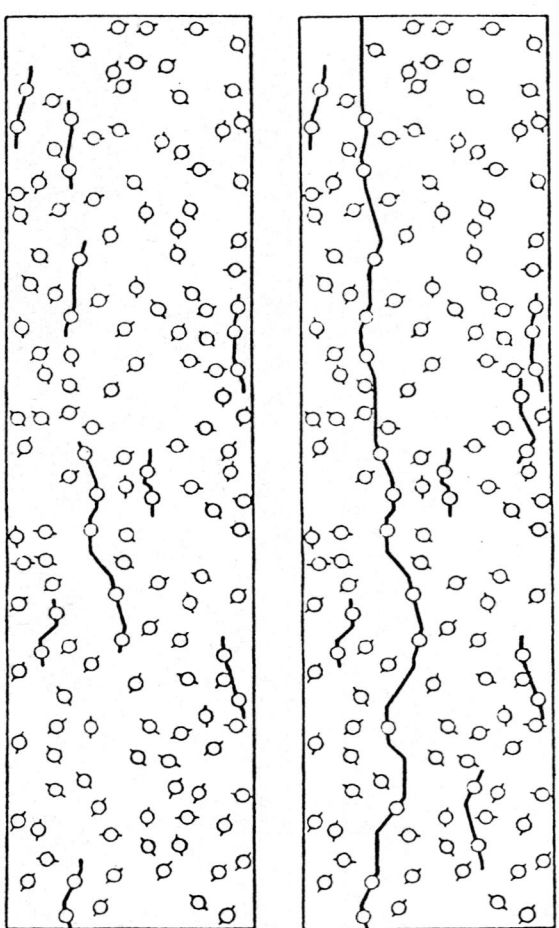

Fig. 22. Crack pattern for two different load levels in hardened cement paste. Cracks originating from pores run through homogeneous matrix.

4.2   Tri-axial compression

The strength of concrete subjected to tri-axial compression (axial compression σ_3 with confining lateral pressure $\sigma_1 = \sigma_2$ $\sigma_{1,2,3} < 0$) can be described with help of the coefficient of efficiency of confining lateral pressure :

$$K = \frac{|\sigma_3^*| - R_{cyl}}{|\sigma_2^*|}$$

where R_{cyl} is the uniaxial strength of a companion cylindrical specimen, and

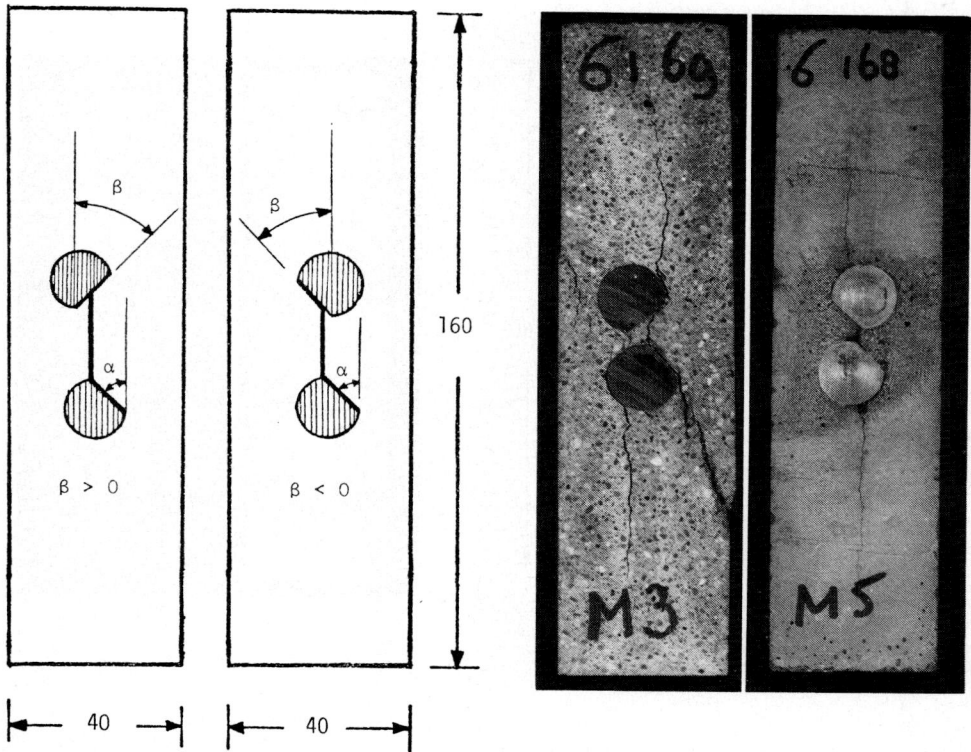

Fig. 23. Geometrical arrangement and photographs of mortar specimens with two inclusions. The inclinations of the two artificial cracks have opposite (left) and equal (right) signs.

$\sigma_2^*(\sigma_3^*)$ are the values of $\sigma_2(\sigma_3)$ by the failure of a specimen subjected to tri-axial compression. Experimental values of K usually have a significant scatter, especially with a low confining pressure and a theoretical approach to the problem of tri-axial strength of concrete is extremely desirable.

The model of concrete used in this paragraph consists of polygonial aggregate inclusions randomly distributed in a homogeneous matrix (mortar). This model takes into account pre-existing randomly inclined interface cracks between inclusions and matrix caused by shrinkage of hardened cement paste (see Fig. 30a), and is similar with the model described in paragraph 3.2.

Behaviour of a specimen having the shape of a cylinder and loaded with axial pressure σ_3 and confining pressure $\sigma_1 = \sigma_2$ will be analysed. This problem can be approximately reduced to a plane problem when the behaviour of a plate of thickness "1", cut out from the cylinder and loaded with axial stress $\sigma_y = q$

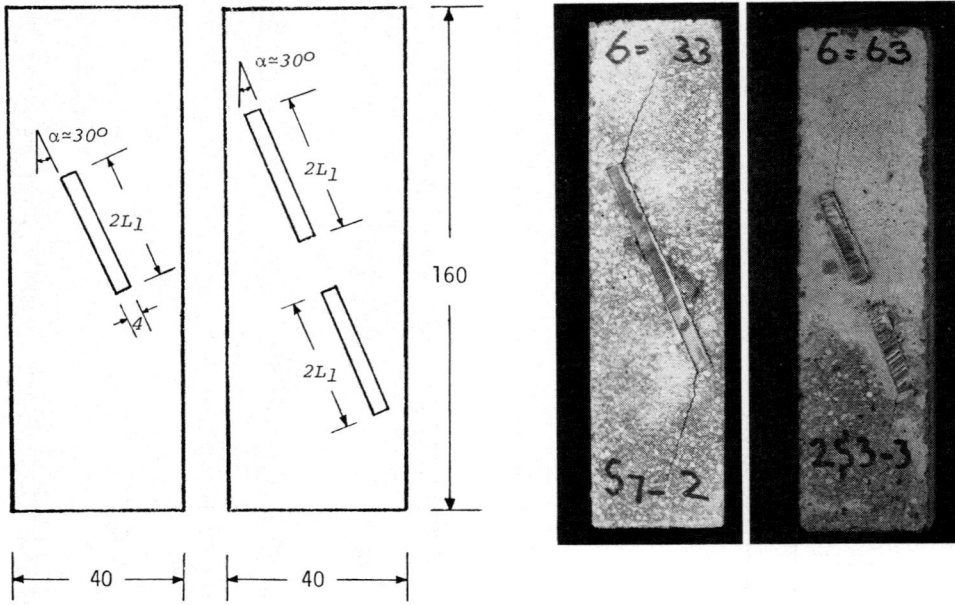

Fig. 24. Geometrical arrangement of one and two inclined flat steel plates and photographs of two failed specimens.

and confined pressure $\sigma_x = \eta_q (\eta = \text{const})$ is analysed.

It can be shown with help of methods of fracture mechanics that initial bond cracks between matrix and inclusions will propagate according to Mode II. An analogous result has been obtained with help of the finite elements method of Palaniswamy and Shah (ref. 14). The value of external load required for propagation of an initial crack of the length of $2l_1$ inclined by the angle of α to the y-axis is equal to :

$$q_{II}^{IF} = -\frac{K_{IIC}^{IF}}{\sqrt{\pi l_1} D(\alpha, \rho, \eta)}$$

where K_{IIC}^{IF} is the critical value of the stress intensity factor for interfacial cracks (Mode II), ρ is the coefficient of friction between matrix- and aggregate materials,

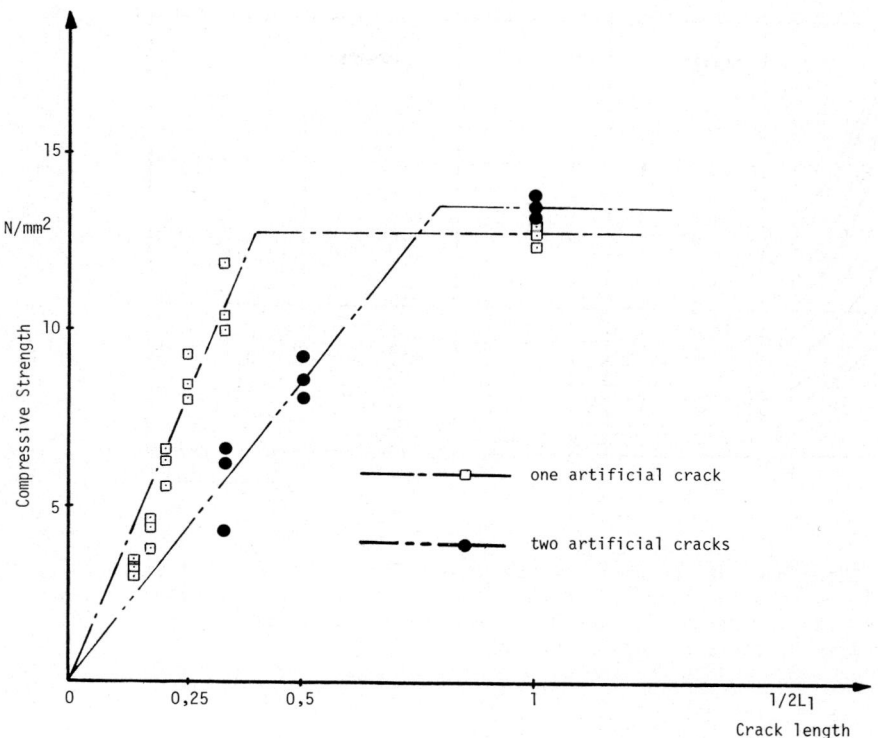

Fig. 25. Compressive strength of mortar prisms having one and two artificial cracks respectively as a function of the reciprocal crack length.

$$D(\alpha, \rho, \eta) = K_\tau (1 - \eta) \sin \alpha_2 \cos \alpha -$$

$$- \rho(k_\sigma' \sin^2\alpha + \eta k_\sigma'' \cdot \cos^2\alpha)$$

and k_τ, k_σ', k_σ'' are stress concentration factors on the inclusion boundary for respectively shear stress, normal stress from axial pressure, and normal stress from confining pressure. For typical values of ρ for concrete ($\rho = 0.7$-0.9) the propagation of initial cracks is most likely on the contact surfaces inclined by the angle of $\alpha \cong 25^0$ to the sample axis, independent from the confining pressure. At the same time, the value of q_{II}^{IF} depends significantly on the value of the confined pressure.

The next stage of crack propagation is connected with crack penetration into the matrix. Cracks start to propagate approximately parallel to the y-axis as cracks

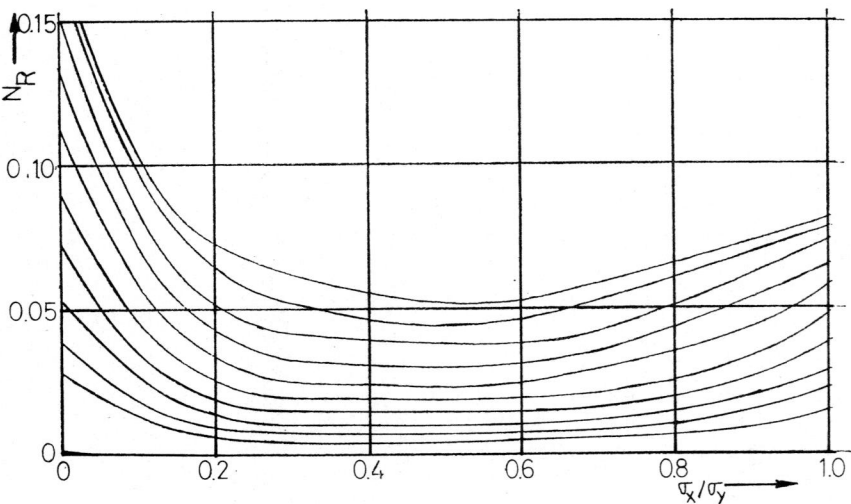

Fig. 26. The number of crossings of the stress field around a "random pore" with a given level as a function of the ratio σ_z/σ_y. The load is increased in steps of 2% from the lowest to the highest curve. The number of crossings is related to the number of pores within the material.

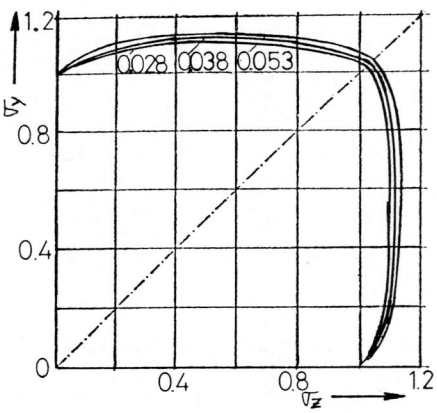

Fig. 27. Lines of equal number of microcracks (crossings) in the σ_y - σ_z - diagram. If it is assumed that 2.8%, 3.8% or 5.3% of the total number of pores are damaged by microcracks, the corresponding lines indicate strength of a porous material under a bi-axial state of stress.

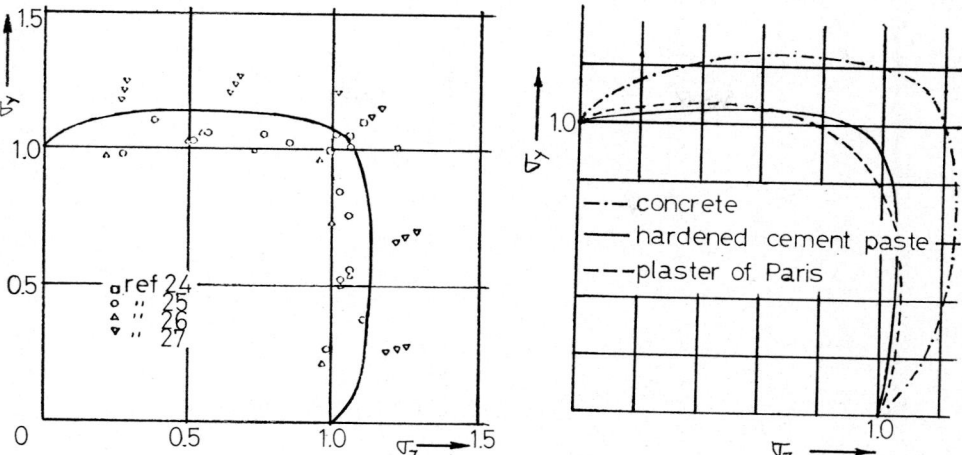

Fig. 28. Some results of experiments to determine compressive strength of concrete under a bi-axial state of stress. The line corresponding to 5.3% of pores damaged by microcracks from fig. 27 is shown once more.

Fig. 29. The compressive strength of concrete hardened cement paste and plaster of Paris under bi-axial load as determined (ref. 13).

of Mode I (splitting cracks) by the value of axial pressure :

$$q_I^M = - \frac{K_{IC}^M \sqrt{3/4}}{\sqrt{\pi L_1} \; D(\alpha, \, \rho, \, \eta)}$$

whereby K_{IC}^M is the critical value of the stress intensity factor for cracks in matrix (Mode I), $2L_1$ is the length of the side of an inclusion, where the initial crack is situated. Predominant propagation of longitudinal (splitting) cracks at the initial stages of loading of a concrete sample is verified experimentally by Berg and Solomenzev (ref. 15). Further stages of crack propagation can be described as follows :

$$q = \frac{K_{IC}^M \sqrt{\pi L_1}}{\frac{2}{\pi} \sqrt{\frac{L_1}{\ell_2}} \; A(\alpha, \, \rho, \, \eta) - \eta \sqrt{\frac{\ell_1}{L_1}}}$$

where ℓ_2 is the length of the longitudinal part of the crack,

$$A(\alpha, \, \rho, \, \eta) = D(\alpha, \, \rho, \, \eta) \cdot \sin \alpha$$

288

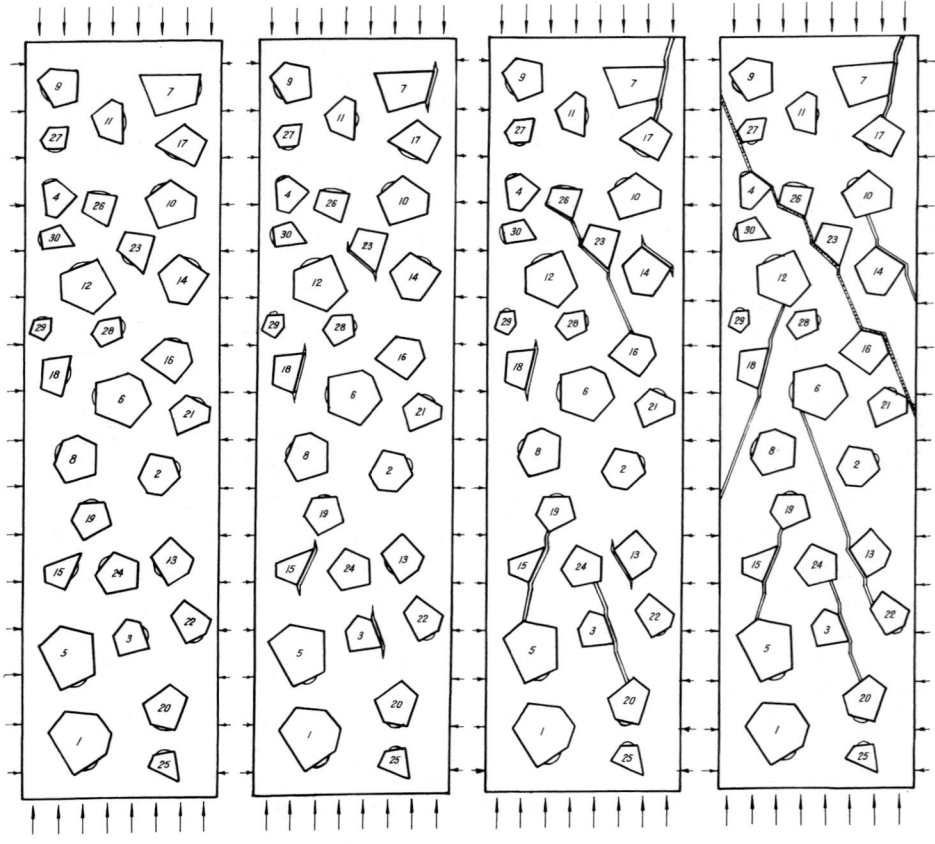

Fig. 30. One of the realizations of the Monte-Carlo-Method showing consequent stages of crack propagation in concrete subjected to tri-axial compression.

Figure 31 shows the relationship between axial pressure q_*^0 and λ, where 2λ is the length of a crack ($2\ell_1$ or $2L_1 + 2\ell_2$). It was assumed that $\alpha = 30^0$; $K_{IIC}^{IF}/K_{IC}^{M} = 0.675$; $\eta = 0$; $\eta = 0-0.2$. Line 1 gives values $q_*^0 = |q_{II}^{IF}| \sqrt{L_1}/K_{IC}^{M}$, corresponding to the propagation of initial bond cracks. For the case $\eta = 0.2$ line 2 is also shown. It gives $q_*^0 = |q_I^M| \sqrt{L_1}/K_{IC}^{M}$, corresponding to penetration of an inclined crack into the matrix. Initial bond cracks by load increase at first do not grow (vertical lines by $\lambda = \lambda_1$ on Fig. 31). By load q_1, the crack grows in an unstable manner and reaches the length of $\lambda = 1$ (i.e. the crack occupies the whole side of contact between matrix and inclusion).

At the load value of q_2, the crack "jumps" on one of lines 3 corresponding to stable crack propagation in the matrix. As can be seen from figure 31, confining pressure (especially by $\eta = 0.2$) leads to a significant increase of q_1- and q_2- values, i.e. stresses, required for initial crack propagation and penetration of a crack into the matrix. Even more pronounced is the influence of confined pressure on the propagation of longitudinal cracks in the matrix : by $\eta = 0.1$ the length of these cracks (line 3) is significantly less as compared to uniaxial tension ($\eta = 0$), and for the case $\eta = 0.2$ cracks formed in the matrix practically do not grow. Thus, the only way for fracture of concrete with the system of cracks described above is the propagation of shear cracks (Mode II). This stage of crack propagation can be described in an analogous way. The interaction of cracks can be approximately estimated according to Panasjuk (ref. 5).

Figure 30 (b-d) gives consequent stages of crack propagation (including fracture of a sample) as described above for one of the realizations of a Monte-Carlo-Method.

Figure 32 (line "T", i.e. "Theory") shows the relationship between the K-coefficient of efficiency of lateral pressure as described above and related confined pressure $|\sigma_2|/R_{cyl}$ where R_{cyl} is the uni-axial strength of concrete.

As can be seen from figure 32, the value of K increases significantly with low values of lateral pressure. It can be explained by the significant influence of confining lateral pressure on propagation of longitudinal splitting cracks.

By the increase of lateral pressure the importance of splitting cracks (Mode I) in the general picture of fracture will decrease, and final fracture will be caused by shear cracks.

Figure 32 also gives experimental values of K, obtained in various investigations; figures near lines give uniaxial strength of concrete R_{cyl} in MPa. Experimental data confirms the significant increase of K by low $|\sigma_2|/R_{cyl}$ values, found with the help of the method described above. On the other hand, these experimental data indicate the influence of uni-axial strength R_{cyl} on K-values, which was

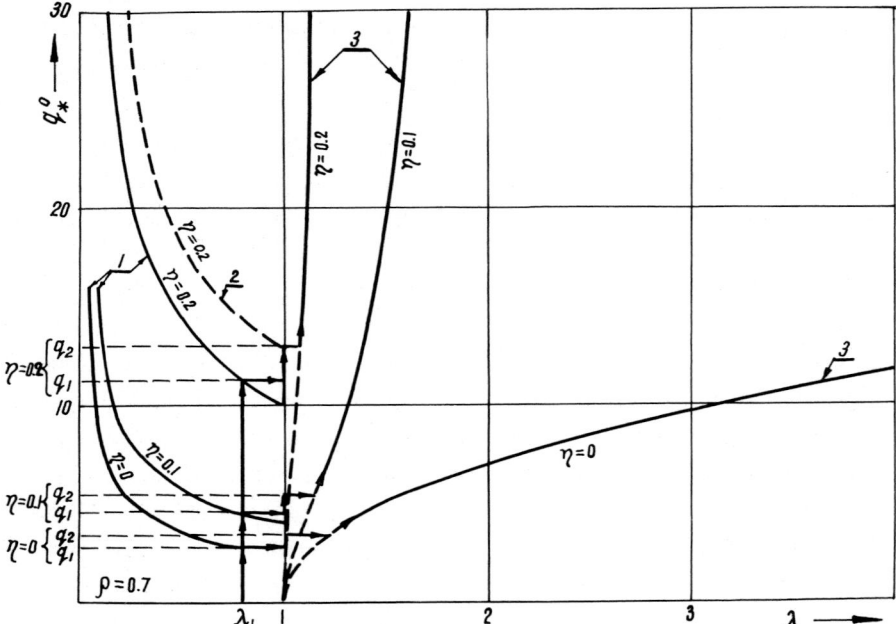

Fig. 31. Relationship between related axial stress q_*^0 related crack length λ.

also predicted by Monte-Carlo-Method. Details of this investigation are given in reference 16.

5. CRACK PROPAGATION AND STRENGTH UNDER SUSTAINED LOAD

5.1 Constant sustained load

If the material is viscoelastic, cracks can propagate due to the fact that creep occurs in the immediate vicinity of the crack tips. In reference 7 it has been shown that formulas which are derived under the assumption that a material reacts in an ideal way, can be extended so that they can be applied to describe the behaviour of a viscoelastic material as well. In this case the elastic modulus E has to be replaced by a time dependent operator $\tilde{E}$ where :

$$\frac{1}{\tilde{E}} = \frac{1}{E} + \int_{\tau_1}^{t} K(t, \tau) \frac{\sigma(\tau)}{\sigma(t)} \frac{1}{E(\tau)} \, d\tau$$

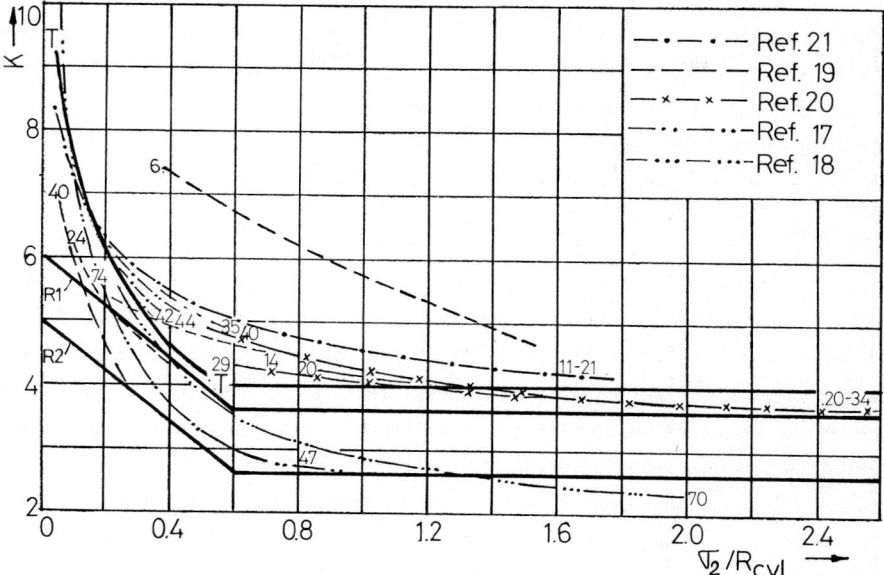

Fig. 32. Relationship between related value of confined pressure σ_2/R_{cyl} and coefficient of efficiency of confined pressure K. Lines 1-5 give experimental data (details are given in ref. 16).

In this equation k(t, τ) has the following meaning :

$$K(t, \tau) = - E(\tau) \frac{\partial}{\partial \tau} \left[\frac{1}{E(\tau)} + K(t, \tau) \right]$$

Now it is possible to calculate the time t when a specimen fails under a high sustained load, the load being applied at time τ. The basic assumption here is that the specimen will fail as soon as the total crack length increases due to creep of the material in the crack tips and reaches a value that equals the critical crack length of the short time experiment. The related strength under sustained load is then given by the following expression :

$$\eta(t, \tau_1) = \frac{m(t, \tau_1) \, R_K(t)}{R_K(\tau_1)} \sqrt{\frac{E(\tau_1)}{E(t)} \frac{1}{1+\phi(t, \tau_1)}}$$

$R_K(t)$ and $R_K(\tau_1)$ being the short time strength of a companion specimen at time t when the sample fails and at time τ_1 when the load has been applied, respectively. $E(t)$ and $E(\tau)$ represent the elastic modulus at the indicated age.

Creep in the material near the crack tips not only increases the crack length but reduces the stress concentration at the same time which leads to an increase in strength and $m(t, \tau)$ takes this effect into consideration.

The same concept was extended to the bi-axial state of stress. Finally the ultimate load is found to be :

$$q = \sqrt{\frac{\pi E \gamma}{R}} \; g(\lambda)$$

where $g(\lambda)$ is a function of the related crack length. By introducing time dependent material parameters such as creep into this equation, one finds a final result identical with a uniaxial state of stress. That means that one can expect the strength to decrease under high bi-axial stress in the same way as detected in experiments using a uniaxial state of stress.

In figure 33 the comparison of the proposed theory with experiments carried out with concrete is shown. The drawn line has been calculated with the aid of the equation given above. Within the usual range of scatter which is characteristic of this type of experiment, fair agreement between theoretical prediction and experimental results has been found.

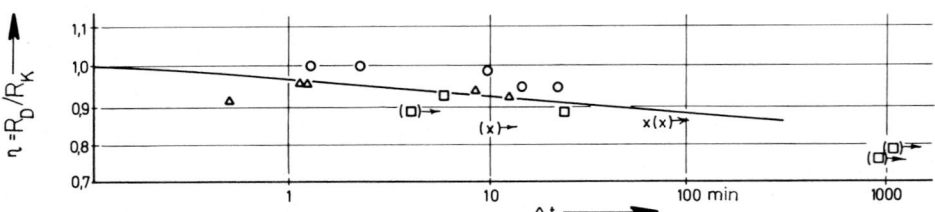

Fig. 33. Decrease of strength under high sustained load as function of the required time until failure.

5.2   Variable sustained load

The application of a formerly developed fracture mechanics method to a more complex case, i.e. a variable sustained load was considered. As a result, the following equation was obtained, which describes the change of total length of cracks, S, with the help of introducing the related intensity of stress σ (applied at the moment $\tau_1 = t_0$) $q_0 = \sigma(t_0)/\sqrt{\pi E(t_0)\gamma_s(t_0)}$:

$$S(t, t_0) = F\left\{q_0\left[\frac{R(t_0)}{m(t, \tau)R(t)} \sqrt{\frac{E(t)}{\hat{E}}} \frac{\sigma(t)}{\sigma(t_0)}\right]\right\}$$

where $1/\hat{E}$ is an operator assumed in a general form according to reference 7 which takes into consideration the creep strain of concrete; $R(t)$ and $E(t)$ represent the short term strength and the elastic modulus of concrete.

In this equation factors $R(t_0)/m(t,t_0)R(t)$ and $\sqrt{E(t)}/\hat{E}$ take into account (as in the case of constant load) the effect of the two opposing processes - hardening and loosening of the material structure.

The multiplier $\sigma(t)/\sigma(t_0)$ reflects the effect of load change at time t compared to the initial load at time t_0.

Let us introduce a function :

$$M(t, t_0) = \frac{\sigma(t)}{m(t, \tau_1)R(t)} \sqrt{\frac{E(t)}{\hat{E}}}$$

The initial level of loading being denoted by $\rho = \sigma(\tau_1)/R(\tau_1)$, we obtain :

$$S(t, \tau_1) = F\{q_0 M(t, \tau_1)/\rho\}$$

Let us extend the assumption made above - the equality of critical total length of cracks S* under short-term (S_s^*) and constant sustained load (S_c^*) - for the case of time-variable sustained load (S_v^*).

If the initial intensity of such a load is ρq_0 then the following equality should be satisfied at the moment of failure under time-variable sustained load (similar to the case of short-term and constant sustained load) :

$$S_v^*(t, \tau_1) = F\{q_0/\rho\}$$

i.e. in accordance with the expression for $S(t, t_0)$ for the moment of failure :

$$M(t, \tau_1) = 1$$

The change of the function $M(t, \tau_1)$ for the case of constant sustained load is shown in figure 34, where n_* is the related long-term strength. It can be shown that the values of $M(t, \tau_1)$ are always in the range 0 to 1, whereby $M(t, \tau_1) = 0$ for unloaded material ($\rho = 0$) and $M(t, \tau_1) = 1$ for the case of failure. Therefore the function $M(t, \tau_1)$ may be defined as the "measure of material destruction".

294

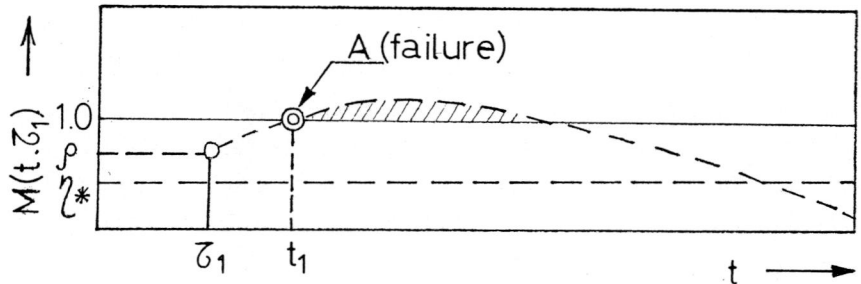

Fig. 34. Criterion of failure of concrete under constant sustained load (measure of destruction).

The concept of the measure of destruction being introduced, the time to failure under time-variable load may be determined.

The following equality should be satisfied at the moment of failure according to equations given above :

$$\frac{\sigma(t)}{m(t, \tau_1)R(t)} \sqrt{\frac{E(t)}{\tilde{E}}} = 1$$

The time to failure may be determined from this equation if all the time-dependent functions are known.

6.   SOME PRACTICAL APPLICATIONS

6.1   Prediction of long-term strength of concrete

The long-term strength of concrete has not been investigated so widely as compared to its creep performance. For practical use an approximated limit value of the related long-term strength $\eta^* = 0.8$ is commonly recommended. On the other hand, some investigations have shown that this value is affected by the concrete quality. Theoretical investigations (see paragraph 5) have shown that the related strength under sustained load is given by the following expression :

$$\eta(t, \tau_1) = \frac{m(t, \tau_1)R(\tau_1)}{R(t)} \sqrt{\frac{E(t)}{E(\tau_1)}} \cdot \frac{1}{1+E(\tau_1)C_0(t, \tau_1)}$$

$R(t)$ and $R(\tau_1)$ being the short time strength of a companion specimen at time t when the sample fails and at time τ_1, when the load has been applied. $E(t)$ and $E(\tau_1)$ represent the elastic modulus at the indicated age. $m(t, \tau_1)$ takes into consideration the increase of short-time strength of concrete under sustained load as compared with short-time strength of an unloaded specimen (m-effect). The variables R and E in the right hand part can be expressed as follows (ref. 22) :

$$R(\tau) = \{1 - \frac{2 - \tau/t_1}{1+R(28)/R_1} \lg \frac{t_2}{\tau}\} R(28)$$

$$E(\tau) = 5300 R(\tau)/ \{8.5p+R(\tau)/R_1\}$$

where $t_1 = 500d$, $t_2 = 28d$, $R_1 = 10 MP_a$, $R(28) = R(\tau_1=28d)$ and $P =$ specific mass content of cement paste in concrete. C_o-values can be obtained according to reference 22. As a first approximation (and for extra safety) we assume that $m(t, \tau_1) = 1$. This leads to a lower estimation for the long-term strength of concrete. The long-term strength curves were calculated (stress level versus time of failure). As theory predicts sustained load strength shows an increase after a certain duration of load ($t = t_*$). In physical terms this means that by $t > t_*$ no long term failure can occur any more under constant sustained load and the minimum values n_* of the sustained load strength. Figure 35 shows these values as a function of the short term strength of concrete R(28).

Lines "A" and "B" are given for comparison and represent empirical relations for middle and old age concrete. These values can be used for prediction of long term strength of concrete with respect to its quality and age of loading.

6.2   Creep fracture of concrete in prestressed concrete members during manufacture

Evaluation of the strength of prestressed concrete members during manufacture (at the moment of prestress transfer) is one of the practical problems encountered when designing prestressed concrete structures. Failure of concrete in this case may take place under the prestress force.

Consider a prestressed concrete member, subjected to axial precompression forces. The total length of cracks and the time to failure are given by the equations of paragraph 5.2. The change of stresses $\sigma(t)$ is mainly defined by prestress losses resulting from concrete creep. These losses under high stresses depend in their turn on the initial stress level.

Values of stress losses, as well as short term strength R(t), modulus of elasticity E(t) and specific linear creep strain $C_o(t, \tau_1)$ may be obtained using the expressions given in reference 22. Thus all the necessary data are available for evaluation of prestressed concrete strength. Figure 36 shows changes of the mea-

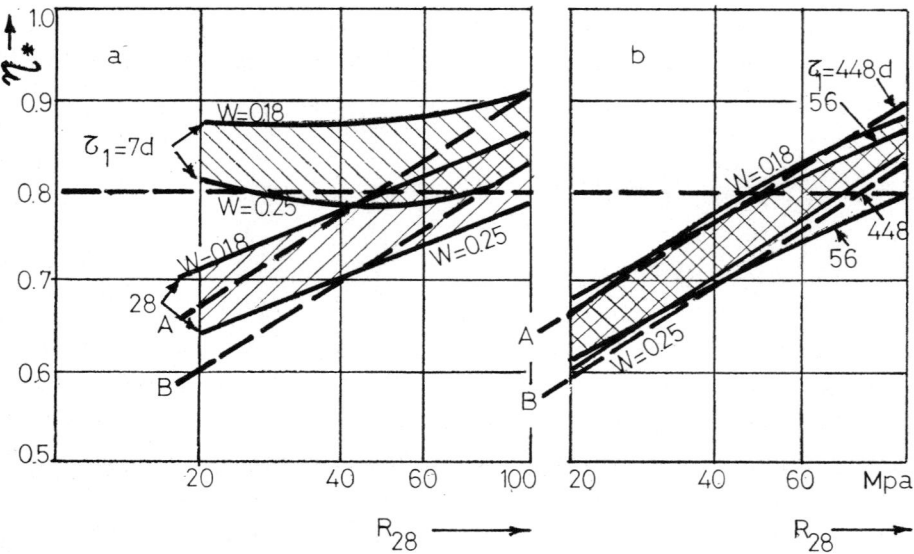

Fig. 35. Minimum related strength of concrete under sustained load
for young and medium age concrete (a), old concrete (b).

sure of destruction $M(t, \tau_1)$ calculated according to the procedure mentioned
above. Various prestress levels of axially prestressed members were taken into
consideration.

It can be seen from figure 36 that the main features of the function $M(t, t_0)$
essentially depend on the percentage area of prestressed reinforcement μ_{pr}. When
the value of μ_{pr} is high ($\mu_{pr} > 0.02$), stress losses resulting from concrete
creep are significant, while the process of destruction under rapidly decreasing
load is slightly in evidence. Accordingly, right from the start, the measure of
destruction $M(t, \tau_1)$ is decreasing irrespective of the prestress level.

When the value of μ_{pr} is low, stress loss proceeds more slowly and material
loosening processes are of great importance. Under such conditions, the prestress
level being high enough, failure of concrete takes place (see the top curve when
$\rho = 0.98$ in figure 36). The same may be obtained for the case of an eccentric pre-
compressed member. The danger of destructive processes in prestressed members
with a low percentage area of reinforcement under a high prestress level is confirmed experimentally too.

The results of analyses carried out by the procedure mentioned above for concrete with a range of cubic strength (R = 20 : 50 MPa) and using various values of
specific (volumetric) water content in the concrete mix w (w = 0.18 and w = 0.25)

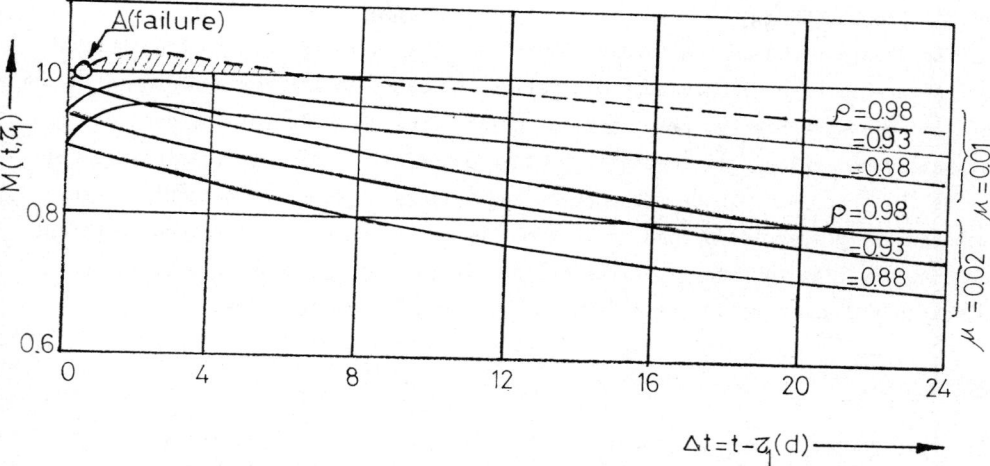

Fig. 36. Criterion of failure of prestressed concrete members subjected to the prestressing force.

are given in figure 37. The lines in this figure reflect the higher boundary of safe (in regard to long term concrete strength) prestress level ρ_{max}, related to the short term concrete strength at the moment of prestress transfer. Within these boundaries the load bearing capacity of concrete cannot be exhausted, i.e. $M(t, \tau_1) < 1$. Results obtained above have been used in the new Building Code.

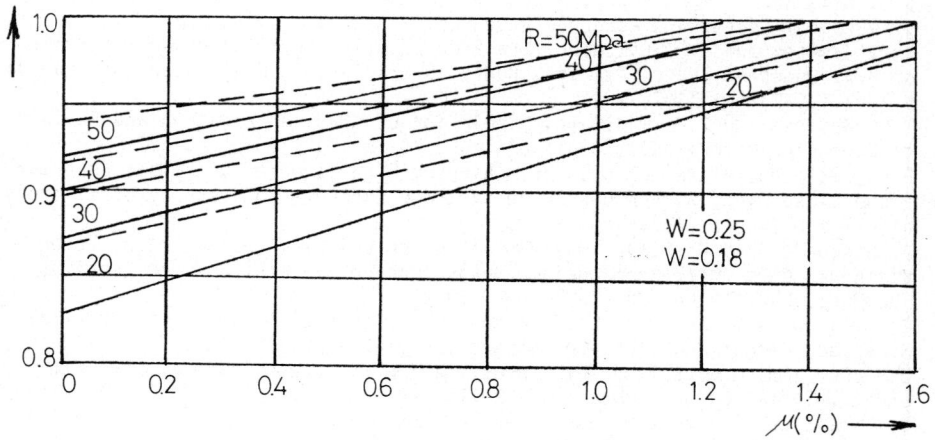

Fig. 37. Higher boundary of a prestress level, calculated as described.

7. CONCLUSIONS

The theory based on fracture mechanism methods, with respect to the statistical nature of crack propagation can deliver valuable information about the influence of inclusions, porosity, pre-existing cracks and other elements of concrete structure on mechanical properties of this material. It is also possible to describe the process of crack propagation and failure of concrete not only under monotonically increasing load (uni- or multi-axial) but under sustained (constant or slowly variable) load. This theory can, in the future, provide a basis for research and the creation of materials with predicted properties.

REFERENCES

1 Y.V. Zaitsev, Stress Concentration around Square Holes, Izvestija Vusov, Stroitestvo i Architectura, No 5 (1970) pp. 47-49 (in Russian).
2 Y.V. Zaitsev, Stress Concentration around Triangular Holes, Izvestija Vusov, Stroitestvo i Architectura, No 5 (1971) pp. 59-61 (in Russian).
3 G.N. Savin, Stress Distribution around Holes, Kiev, "Naukova Dumka" (1968) pp. 888 (in Russian with English summary).
4 Y.V. Zaitsev and F.H. Wittmann, Eine theoretische Studie des Verhaltens von Beton, Materialprüfung, 16 (1974) pp. 170-174.
5 V.V. Panasjuk, Stress Distribution around Cracks in Plates and Shells, Kiev, "Naukova Dumka" (1976) pp. 444 (in Russian).
6 K.T. Krishnaswamy, Mechanism of Failure and Microcracking of Plain Concrete under Uniaxial Tension, Indian Concrete Journal, 45, no 5 (1971).
7 Y.V. Zaitsev and F.H. Wittmann, Verformung und Bruchvorgang poröser Baustoffe unter kurzzeitiger Belastung und unter Dauerlast, Deutscher Ausschuss für Stahlbeton, Heft 232 (1974) pp. 65-145.
8 Y.V. Zaitsev, Influence of Structure on Fracture Mechanism of Hardened Cement Paste, Proc. Int. Congr. Chemistry of Cement, Paris (June 30 - July 4, 1980) pp. VI 176 - VI 180.
9 S.S. Dias and H.K. Hilsdorf, Fracture Mechanisms of Concrete under Compressive Loads, Cement and Concrete Research, 3 (1973) pp. 363-388.
10 Y.V. Zaitsev and F.H. Wittmann, Crack Propagation in a Two-Phase Material such as Concrete, Proc. 4th Int. Conf. on Fracture (ICF-4), Waterloo, Canada, Vol. 3 (1977) pp. 1197-1204.
11 Y.V. Zaitsev and F.H. Wittmann, Simulation of Crack Propagation and Failure of Concrete, Matériaux et Constructions, 14, No 83 (1981) pp. 357-365.
12 F.H. Wittmann, Micromechanics of Achieving High Strength and other Superior Properties, Proc. Workshop on High Strength Concrete, University of Illinois at Chicago Circle (1979) pp. 8-30.
13 A. Stegbauer and D. Linse, Das Verhalten von Leichtbeton, Gasbeton, Zementstein und Gips unter zweiachsiger Beanspruchung, Bericht aus dem Institut für Massivbau, Techn. Univ. München (1972).
14 R.G. Palaniswamy and S.P. Shah, A Model for Concrete Subjected to Triaxial Stresses, Cement and Concrete Research, 5 (1975) pp. 273-284.
15 D.Y. Berg and G.G. Solomenzev, Investigation of Stress State of Concrete under Triaxial Compression, Trudy ZNIIS, 70 (1969) pp. 106-123 (in Russian).
16 Y.V. Zaitsev, Fracture Mechanism and Strength of Concrete under Triaxial Compression, Proc. 5th Int. Conf. Fracture (ICF-5) Cannes, France (March 29 - April 3, 1981) pp. 2281-2282.

17 O.Ya. Berg and G.G. Solomenzev, Investigation of Stress State of Concrete under Triaxial Compression. In: Trudy ZNIIS, Vol. 70 (1969) pp. 106-123 (in Russian).

18 J. Bergues, P. Habib and P. Morlier, Critère de Rupture des Bétons Soumis à des Sollicitations Triaxiales, Cahiers Groupe Français, Rhéologie, Vol. 2, No 5 (1971) pp. 347-354.

19 A.A. Gvozdev, Calculation of Bearing Capacity of Structures with Method of Limit Equilibrium, Moscow, "Gosstrojizdat" (1949) (in Russian).

20 N.W. Krahl and others, The Behaviour of Plain Mortar and Concrete under Triaxial Stress, ASTM, Proc., Vol. 65 (1965) pp. 695-711.

21 Yu.N. Malashkin and others, Investigation of Strength and Process of Deformation of Concrete with Respect on Kind of State of Stress. In: Trudy Koordinationnych Soveshchanij po Gidrotechnike, Vol. 99 (1975) pp. 21-25 (in Russian).

22 E.N. Scerbakov and Y.V. Zaitsev, Proc. 2nd Int Conf. Mechanical Behaviour of Materials, Boston (August 16-20, 1976) pp. 460-464.

23 Y.V. Zaitsev and E.N. Scerbakov, Proc. 4th Int. Conf. Fracture (ICF-4), Waterloo (June 19-24, 1977) pp. 1219-1222.

24 A. Föppl, Mitteilung aus dem Mech. Techn. Laboratorium der Königl. Techn. Hochschule München, Hefte 27 u. 28 (1900).

25 N.N. Davidenko und V.A. Jarkov, Sprödbruch unter zweiachsiger Druckspannung. Zurnal Techniceskoj Fisiki, 25 (1955) pp. 2200-2202.

26 O.Ya. Berg und N.V. Smirvov, Uber die Festigkeit des Betons bei zweiachsigem Druck, Beton i Zelezobeton, No 11, 37 (1965) (in Russisch).

27 H.B. Kupfer, H. Hilsdorf and H. Rüsch, Behaviour of Concrete under Diaxial Stresses, Proceedings ACI, 66, (1969) pp. 656.

Fracture Mechanics of Concrete,
edited by F.H. Wittmann, 1983
Elsevier Science Publishers B.V., Amsterdam — Printed in The Netherlands

Chapter 4.3.

A STOCHASTIC THEORY FOR FRACTURE OF CONCRETE

by H. MIHASHI

1. INTRODUCTION

Most of the theoretical studies on fracture mechanics are based on continuum mechanics theory and these refer to homogeneous materials ignoring the presence of the heterogeneous microstructure of the solid. In a real material, however, there are many flaws of various sizes, shapes and orientations. Considering this randomness of the internal structure, a probabilistic concept seems to be essential to study the fracture phenomena of real materials.

Up to now, several attempts have been made to consider such effects. Weibull (ref. 1) proposed an empirical formula to relate the probability of failure to stress using a statistical method. He is the first to apply the weakest link concept to fracture phenomena of solids. Freudenthal (ref.2) conjuncted this asymptotic distribution function based on the weakest link model with the Griffith crack instability criterion to discuss more generally the scatter of fracture phenomena of brittle materials. Jayatilaka and Trustrum (ref. 3) developed a general expression for the failure probability of a brittle material based on the properties of the flaw size distribution and the stress necessary to propagate from an inclined crack. A model to explain the failure of brittle materials in compression was also proposed (ref. 4).

On the other hand, Hirata (ref. 5) pointed out that the fracture phenomena in solids might be usually interpreted as a kind of Markoff process and he first applied the theory of stochastic process to the brittle fracture of glass. Yokobori (ref. 6) has developed a nucleation theory and extended the theory of stochastic process to various types of failure of polycrystalline metals (ref. 7). An application of the stochastic process theory was carried out by Hori (ref. 8) to analyze the statistical aspects of flexural strength of portland cement mortar under a sustained load. The statistical distribution of strength was discussed by Nishimatsu (ref. 9) considering the specific rate of crack initiation to fracture as a random variable caused by the randomness of the shape and orientation of the Griffith crack.

Concrete materials, however, are extremely heterogeneous and include not only flaws but various kinds of structural defects. A new concept to describe the stochastic nature of failure of concrete materials has been published by Mihashi and Izumi (ref. 10). Since this theory is based on physically relevant probability models, it provides a realistic basis for a mathematical formulation of the variability of strength of concrete. Moreover, the influence of temperature, of the size of specimen and of the rate of loading on the strength of concrete are also explained uniformly. This theory has recently been extended for the fatigue of concrete materials (ref. 11), that is fracture of concrete under sustained loads and repeated loads. In the following sections, the outline of the stochastic theory is described and the essential theoretical predictions are compared with published experimental findings.

2. FUNDAMENTAL CONCEPTS OF THE STOCHASTIC THEORY FOR FRACTURE OF CONCRETE

It may be incorrect that all kinds of cracking equally affect the failure process. Some of them are very stable and the others quickly propagate. Before the final stage of failure process, most of the cracking is limited to around some larger defects. It is well known that the failure process is highly affected by some larger material defects. Around larger material defects, the highly stressed region may be wider and the possibility that some weaker defects are acted upon by higher tensile stress may increase.

In the present theory, the following experimental findings are considered to build up the theoretical model. 1). The broken parts in normal concrete systems are the cement paste phase and cement-aggregate interface. They include a lot of fine cracks, flaws and voids. 2). Aggregates and voids play a role to cause local stress concentration and sometimes even to change the sort of stress such as local tensile stress under compressive load. 3). In the tensile stress field, micro-cracks easily grow into mesolevel cracks in order to release the locally stored strain energy. 4). Until fracture of the specimen occurs, a series of typical states of crack propagation are followed (ref. 12).

Fundamental concepts of the stochastic theory are represented by the following three characteristics. [1]. 'Statistical Approach' is taken to describe the random properties of local strength caused by 'Geometrical Random'. [2]. 'Stochastic Process Theory' is introduced to present the random properties of the rate of cracking based on the kinematic and thermodynamic properties of a solid. [3]. 'Changeable Elements' are used to represent the imperfectly brittle failure process of concrete materials. The details and background of these concepts are shown in the following sections.

2.1 'Statistical Approach' to describe the random local strength

Since a concrete system is quite heterogeneous, the local stress is extremely random. Only when a local region is considered, the stress σ_{local} seems to be constant. Among various kinds of stress, tensile principal stress of the local region may dominantly affect the local fracture of hardened cement paste phase. For the purpose of simplicity, only Mode I, which is popular in Fracture

Mechanics, is considered hereafter (Fig. 1).

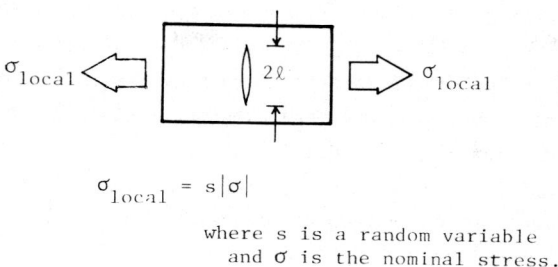

$$\sigma_{local} = s|\sigma|$$

where s is a random variable
and σ is the nominal stress.

Fig. 1. System with a crack

On the basis of such consideration, the local dominant principal stress may be described by $\sigma_1 = s|\sigma|$, where σ is the nominal stress and 's' is a microstress disturbance coefficient caused by material defects. s is a kind of random variable. When only a notch or a pre-existing large crack is considered as the mesolevel material defect, the coefficient s may be a function of the stress intensity factor K.

According to the element of Fracture Mechanics or to Griffith Theory, the strength of a system with a crack which has the length of 2a is described as follows:

$$f_t \propto \sqrt{\frac{E\gamma}{a}} \tag{2.1}$$

This equation means a system with a larger value of $E\gamma$ and a smaller value of a gives higher strength. When the strength of a solid is compared with that of another solid, the equivalent crack length $2a/E\gamma$ may be useful to evaluate the difference. Since the local stress acting on a microcrack is exaggerated by s times in a heterogeneous solid, stress intensity factor around the equivalent crack can be written as follows:

$$K \propto \sqrt{\pi a/E\gamma}\ s\sigma \tag{2.2}$$

Hereafter the equivalent crack length is described with $2c^2$ as follows:

$$2c^2 = 2s^2 a/E\gamma \tag{2.3}$$

According to some experimental findings by means of SEM (ref. 13), there are a lot of weak zones in the cement paste phase which has been normally considered as a continuous and rather homogeneous solid. Even under tensile load, the crack initiation is caused not from the crack tip but from the region closed to the crack tip (ref. 14). It means that such a weak zone acts as a crack which is

304

similar to one proposed by Griffith. Although there are many cracks with
various shapes and sizes, the larger microcracks are much more sensitive to
cracking than smaller ones. Therefore it may be useful to notice the population
of microcracks with the largest length. The mean length of those microcracks (
that is, the expected value of the largest microcrack length) is given by $2\bar{a}$.
For the purpose of the simplicity, only the expected largest value $2\bar{a}$ will be
used. Normally a specimen has its own value of $E\gamma$ which may be evaluated as a
constant of cement paste, but s is randomly changeable in the specimen.
Therefore the equivalent crack length becomes a random variable and the
following equation is obtained.

$$c = s\sqrt{\frac{\bar{a}}{E\gamma}} \tag{2.4}$$

In other words, all kinds of mechanical factors which may affect the fracture
toughness of a microcrack are included in the equivalent crack length. The
reason why the equivalent crack length is called 'fracture factor', is that the
dimension is not the same as that of a real crack length. The probability density
function of the fracture factor : $f_c(c)$ is described as follows:

$$f_c(c) = \sqrt{\frac{E\gamma}{\bar{a}}} f_s(s) \tag{2.5}$$

where $f_s(s)$ is the probability density function of microstress disturbance
coefficient caused by material defects (Fig. 2).

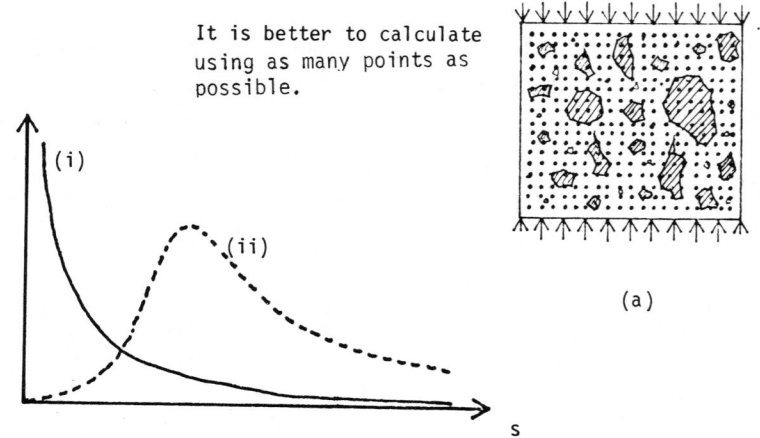

It is better to calculate
using as many points as
possible.

(i)

(ii)

(a)

(b). Probability Density Function of 's'.

Fig. 2. Evaluation of $f_s(s)$

At this moment, F.E.M. analysis and Speckle Pattern Technique seem to be available to get the information about the probability density function $f_s(s)$, which shows the statistical property of random local stress. On the each point of Fig. 2.a, the principal stress can be calculated, in which tensile principal stress is important for local fracture. If it is obvious that cracking is limited only in the region of the aggregates, it is not necessary to evaluate the local stress inside of the aggregates.

2.2 'Stochastic Process Theory' to present the random properties of the rate of cracking

Concrete is a highly heterogeneous material not only on the mesolevel but also on microscopic level. Even though the effects of aggregates and voids are taken into consideration, the cement paste phase itself is quite heterogeneous and the mechanical properties are affected by the internal structures. It is surely necessary to pay attention to the sensitivity of the internal micro-structures, which is extremely different from those of metallic materials.

Fracture of concrete is usually caused by a series of cracking which is essentially random. Since the growth of a crack seems to be the result of the nucleation and the linking of mesolevel cracks, the nucleation process may be considered as a rate-determining process. At first, a unit of a brittle material with a crack is considered (Fig. 3). Most of previous studies such as those by Weibull (ref. 1) and Freudenthal (ref. 2) might name 'sample' or 'element' in stead of 'unit'.

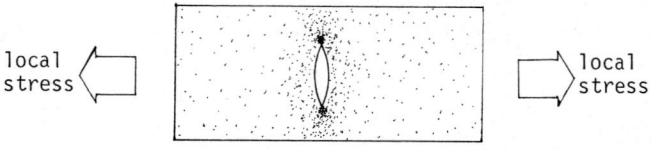

local stress   local stress

Regions around the crack tip
are activated because of the
highly concentrated stress.

Fig. 3. Unit with a microcrack

According to the theory of rate process, the rate of crack initiation on the atomic level is given by the following equation:

$$r = \frac{kT}{h} \exp[-\{U_0 - \phi(\sigma_\ell)\}/kT] \tag{2.6}$$

It is assumed that $\phi(\sigma_\ell) >> kT$, where $\phi(\sigma_\ell)$ is an increasing function of σ_ℓ and

σ_ℓ is the local stress at the area of interest. U_0 equals activation energy in the non-stressed material, k is equal to Boltzmann constant, h equals Plank constant and T means absolute temperature. The transition aspect is described in Fig. 4. The solid (initial state : 0) is easily changed into another stable

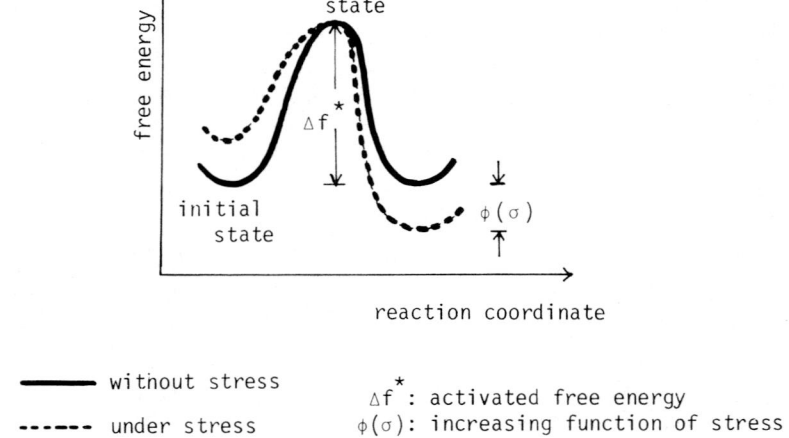

Fig. 4. Transition aspect

state (fractured state : 1) through the activated state under a stressed condition. The activation energy $U(\sigma)$ is dependent on the type of the objective phenomena. In the case of brittle fracture under a static load, the function $U(\sigma)$ is given by equation (2.7) according to the theory of heterogeneous nucleation (ref. 7).

$$U(\sigma) = U_b - \frac{1}{n_b} \ln(q\sigma) \qquad (2.7)$$

where U_b and n_b are material constants, q is the local stress concentration coefficient that means $q\sigma$ is the local stress around a crack tip.

$$q\sigma = \alpha_b K$$

where α_b is a constant and K is the stress intensity factor. The same function is given for creep fracture, too. Hence the equation (2.6) can be rewritten as follows:

$$r = \frac{kT}{h} \exp\left(-U_b/kT\right)\left(q\sigma\right)^{\frac{1}{n_b kT}} \qquad (2.8)$$

Since a crack is most likely to nucleate in the stress concentrated zone, the rate of real crack nucleation: I may be assumed to be proportional to the number of molecules in the vicinity of the tip of a pre-existing crack. That means the rate is proportional to the crack length. Then the following equation is obtained.

$$I = VZr \qquad (2.9)$$

where Z is the number of molecules in the stress concentrated zone and V is the effective volume but not the total volume of the specimen (ref. 7).

$\mu(t)dt$ which means the probability of fracture initiation between a given time t and $t+dt$ may be represented by equation (2.10), as $\mu(t)$ is proportional to the rate of real crack nucleation: I.

$$\mu(t)dt = Z'A(c\sigma)^\beta \bar{a} \, dt \qquad (2.10)$$

where A and β are material constants dependent on the absolute temperature and Z' is a material constant. The mean value $\bar{\mu}(t)$ for a large number of n is then given by the following equation.

$$\bar{\mu}(t) = \int_0^\infty \mu(t) \, g(\mu) \, d\mu = \int_0^\infty \mu(t) \, f_c(c) \, dc \qquad (2.11)$$

where $g(\mu)$ is a density function of μ. In the case when only one specific crack length: $2\bar{a}$ is considered, the following equation is obtained.

$$\bar{\mu}(t) = \bar{Z}A(E\gamma)^{-\beta/2} \bar{a}^{\beta/2+1} \sigma(t)^\beta \int_0^\infty s^\beta f_s(s) \, ds \qquad (2.12)$$

because

$$f_c(c) \, dc \propto f_s(s) \, ds$$

When an element is composed of n units, the rate of crack initiation which leads to the fracture of the element is given by the following equation:

$$\bar{p}(t) = n\bar{\mu}(t)$$

$$= L\sigma(t)^\beta \qquad (2.13)$$

where

$$L = n\bar{Z}A(E\gamma)^{-\beta/2} \bar{a}^{\beta/2+1} R \qquad (2.13.1)$$

and

$$R = \int_0^\infty s^\beta f_s(s) \, ds \qquad (2.13.2)$$

Failure processes of brittle materials may be regarded as a kind of Markoff process, because the probability of cracking is influenced only by the state of the previous instant but it is independent of the character of the motion preceding the time instant t. Markoff process is characterized by a transition probability: $\bar{p}_{01}(t)$ representing the probability that the state of the element is changed from state 0 (non-fracture) to state 1 (fractured) (Fig. 5).

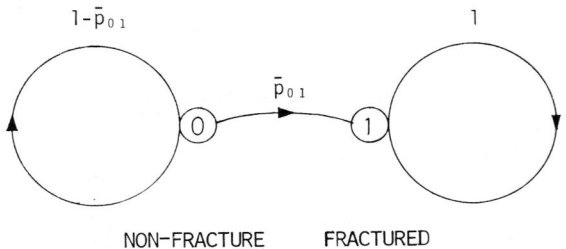

Fig. 5. Transition line graph of a stochastic process

Since the fracture of brittle materials is caused by cracking, the transition probability may be equal to the rate of crack initiation: $\bar{p}(t)$ as shown in equation (2.13). According to the theory of stochastic process, the differential of the survival probability: $P_0(t)$, is represented by equation (2.14).

$$\frac{dP_0(t)}{dt} = -\bar{p}_{01}(t)\, P_0(t) \tag{2.14}$$

where $P_0(t)$ means the probability that no fracture occurs before the time t in any units. The random variable of the survival probability is not always t but it is usually dependent on the loading condition; for example, σ under a monotonically increasing load, t under a sustained load and the number of times: N under a repeated load (Fig. 6).

2.3 'Changeable Elements' to represent the imperfectly brittle failure process

Failure process is composed of a series of local brittle fracture (that is cracking) and the regions where the activated energy is highly concentrated are named 'phase' in the present theory (Fig. 7). The number of units n might be connected with the number of some kinds of rather large microcracks. A large number of n may describe a very porous solid. It is also possible to change the number of n in accordance with the state of elements. For example, the internal microstructures of interface (bond-layer) may be different from that of other matrix regions.

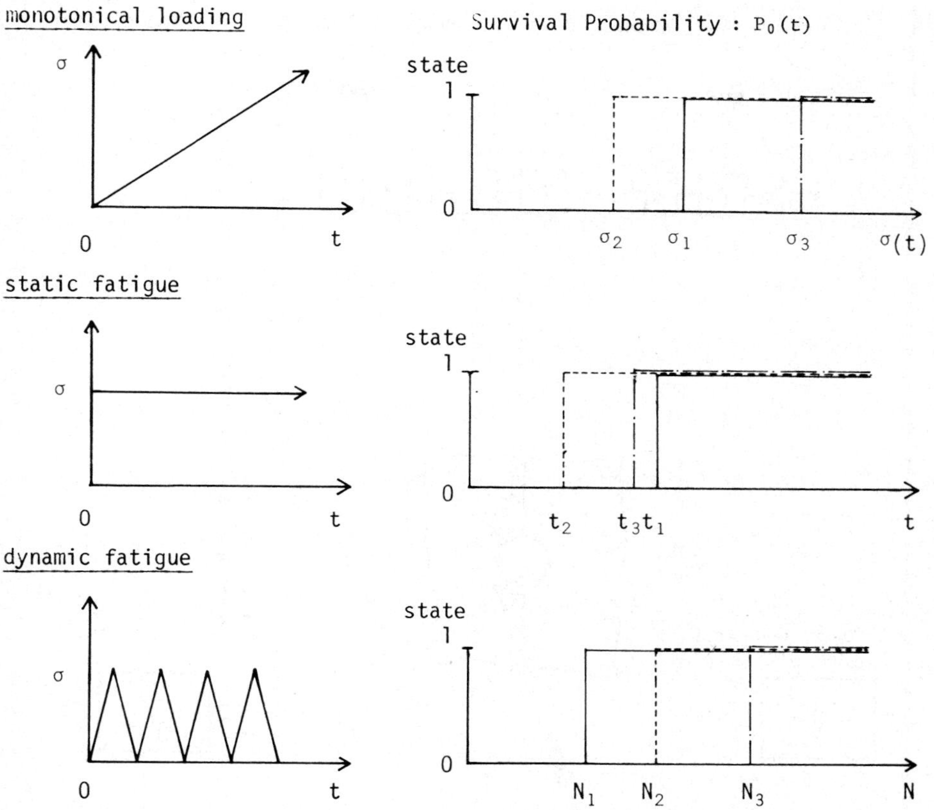

monotonical loading
Survival Probability : $P_0(t)$
σ
state
1
0
0
t
σ_2 σ_1 σ_3 $\sigma(t)$
static fatigue
σ
state
1
0
0
t
t_2 $t_3 t_1$ t
dynamic fatigue
σ
state
1
0
0
t
N_1 N_2 N_3 N

Fig. 6. Loading condition, random variable and transition of element's state

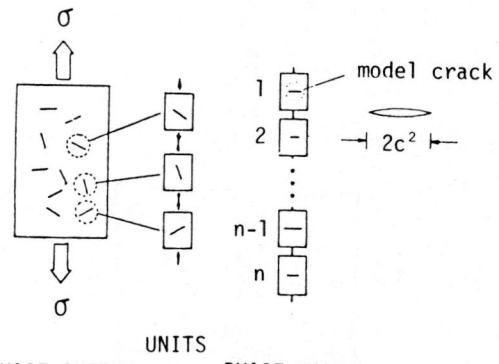

σ
σ
model crack
1
2
n-1
n
$2c^2$
UNITS
PHASE SYSTEM
PHASE MODEL

Fig. 7. Phase system and model

The concepts of the linking models are shown in Figs. 8 to 11. The element

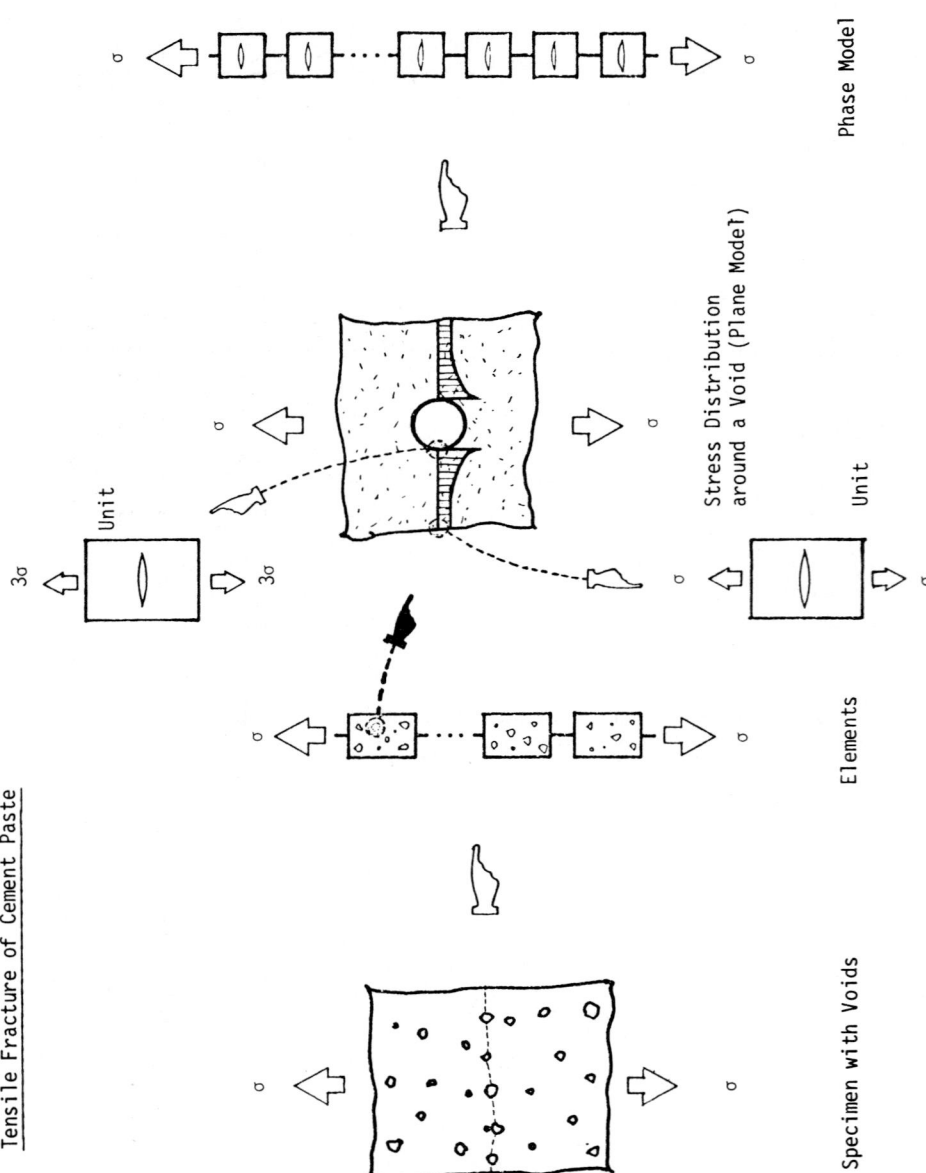

Fig. 8. Tensile fracture of cement paste

311

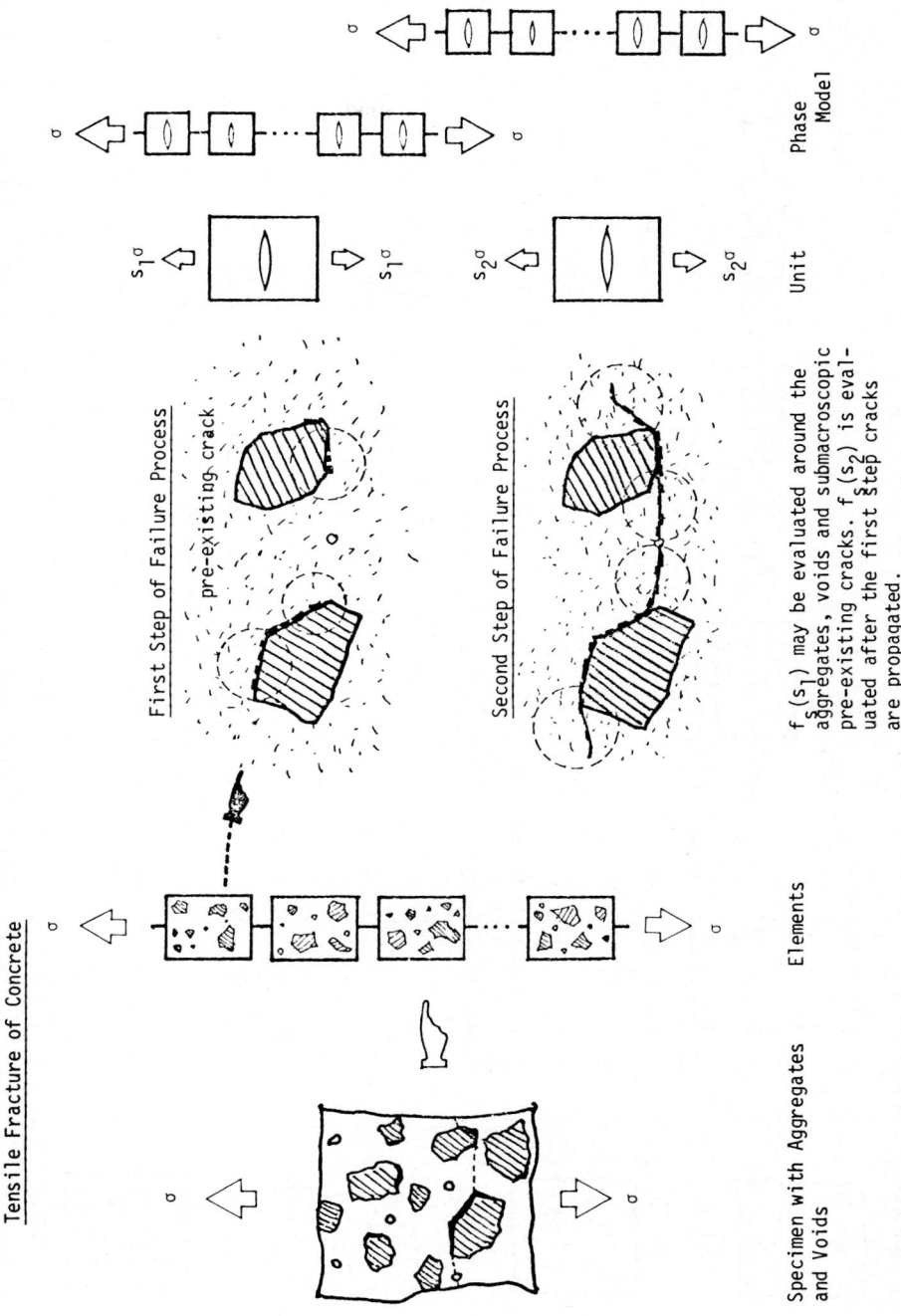

Fig. 9. Tensile fracture of concrete

312

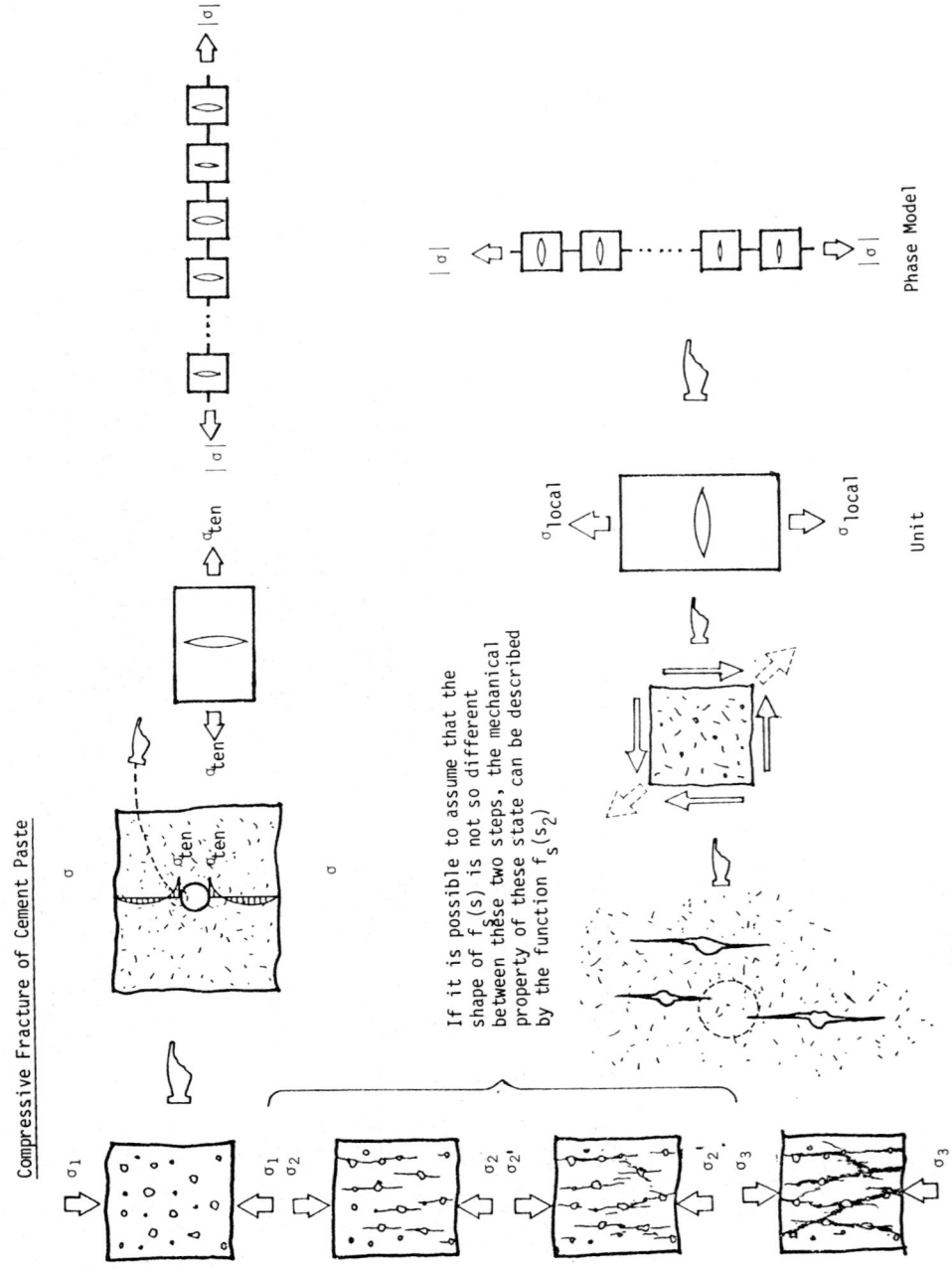

If it is possible to assume that the shape of $f_s(s)$ is not so different between these two steps, the mechanical property of these state can be described by the function $f_s(s_2)$

Fig. 10. Compressive fracture of cement paste

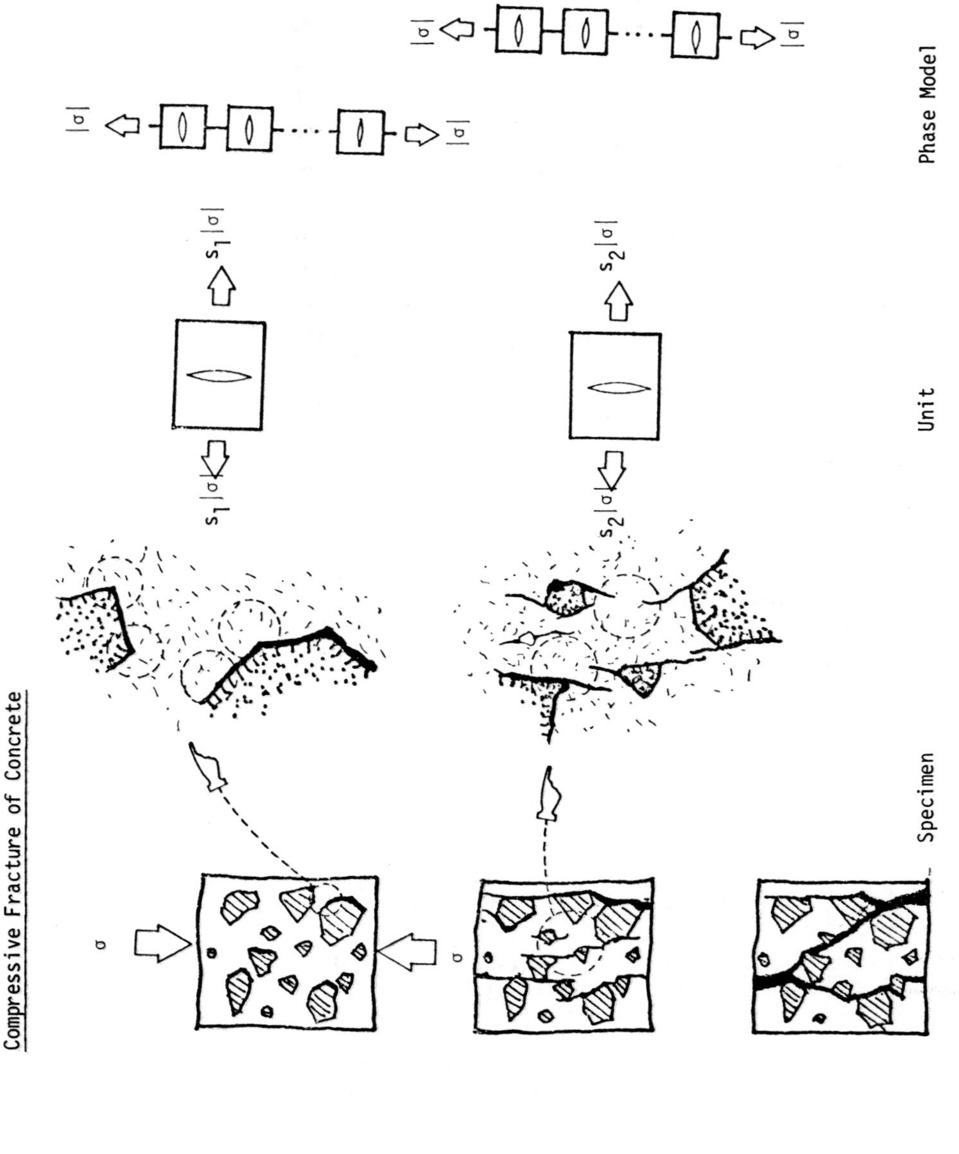

Fig. 11. Compressive fracture of concrete

314

number m means a size parameter of the system (specimen). The state of an element
is not always unique, but sometimes does it change from the initial state to
some saturated state with cracking. The number of states may be connected with
the failure process. Finally a linking crack runs through the specimen. This
final cracking process is modelled by the fracture of a single element. In the
case of compressive fracture (Fig. 10 and 11), it is not so easy to divide the
specimen into elements visually as those in tensile fracture, because the
cracking direction is changed as the load is increasing. However, it is still
possible to pick up some characteristic state in the failure process and the
mechanism by which the change of state is carried out can be described in the
same way as that in tensile fracture (ref. 12). The characteristic failure
process of concrete materials may be described by two different types of models
(Fig. 12). If the failure process is modelled by two states (i.e. non-fractured

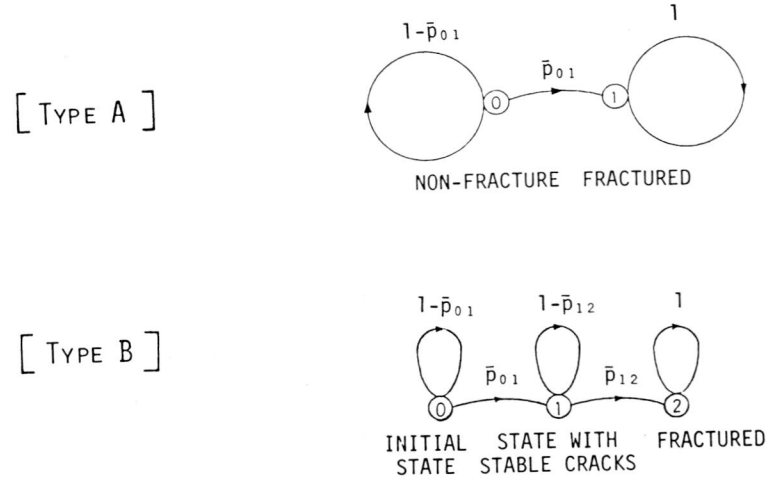

Fig. 12. Transition line graph to describe the change
of element's state

state and fractured state), Type A model is available even for compressive
fracture. However, it is well known that such a simple model as Type A may be
normally available only for tensile, flexural and shearing fracture, which shows
rather brittle fracture phenomena (Fig. 13). Sometimes the stable crack is
existing even under a tensile or bending load. When the first step of failure
process does not mean the fracture of the specimen, Type B should be used at
least. The mechanical properties of these elements are changed from the initial
state to the second state. The linking models of both types are shown in Fig. 14.

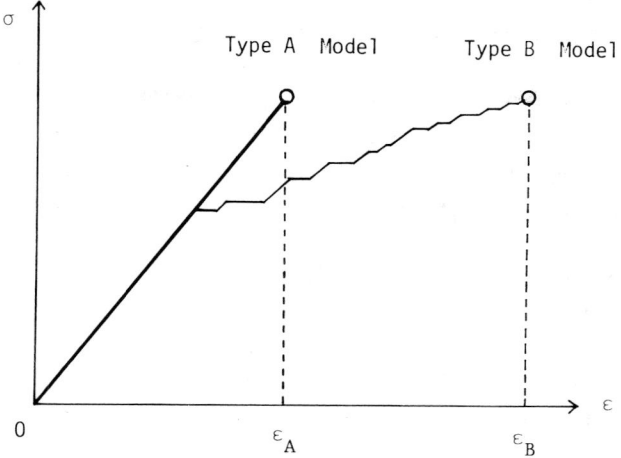

Fig. 13. Stress-strain behaviour modelled with Type A and Type B

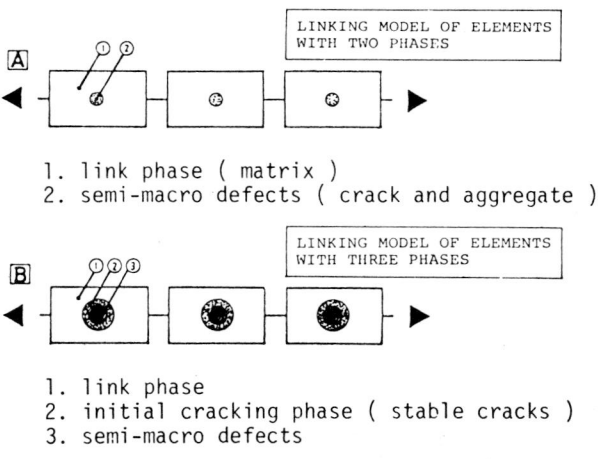

1. link phase (matrix)
2. semi-macro defects (crack and aggregate)

1. link phase
2. initial cracking phase (stable cracks)
3. semi-macro defects

Fig. 14. Linking model of elements

It is also possible to make much more complex models than Type B only by
increasing the number of states (refs. 10 and 11). Although such complex model
can cover also Type A and Type B, the equation becomes very complex and it is
not easy to define the coefficients rationally.

The mechanical property of each state is given by the function $f_c(c)$. Since
the internal microstructure of cement paste phase may not be changed after the
state of the element is transited by mesolevel cracking, the change of the

mechanical property of the element may be described by the change of function $f_s(s)$. The change of the shape of $f_s(s)$ may be understood in accordance with Fig. 15. As the results of these assumptions, the transition probability is

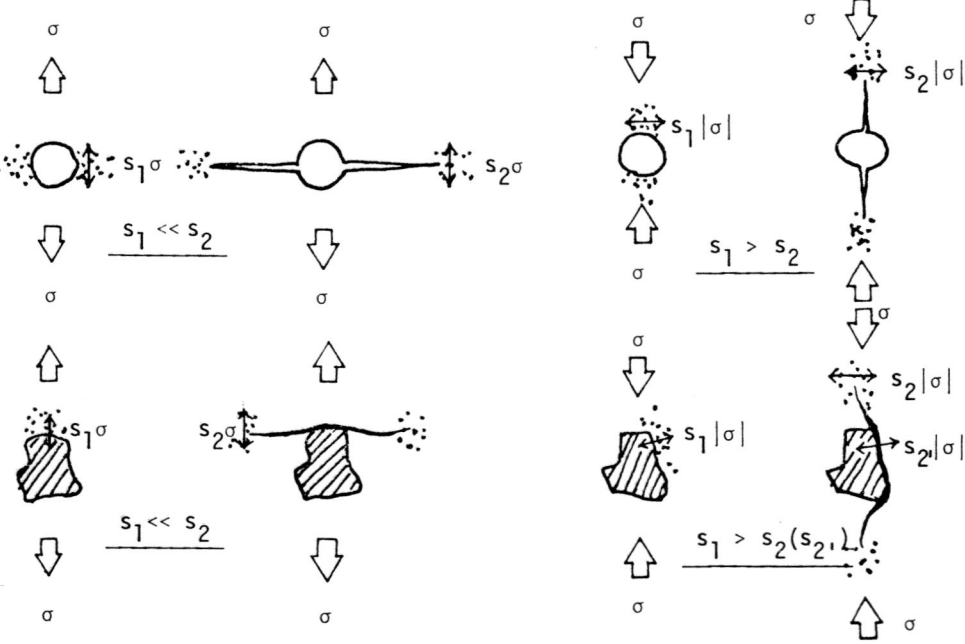

Fig. 15. Mechanical concept to change the shape of $f_s(s)$

described as follows:

Type A : $\bar{p}_{01}(t) = L\sigma(t)^\beta$, where $L = n\bar{Z}A(E\gamma)^{-\beta/2}\bar{a}^{\beta/2+1}R$; $R = \int_0^\infty s^\beta f_s(s)ds$

Type B : $\bar{p}_{ij}(t) = L_i|\sigma(t)|^\beta$, $L_i = n\bar{Z}A(E\gamma)^{-\beta/2}\bar{a}^{\beta/2+1}R_i$; $R_i = \int_0^\infty s^\beta f_s(s_i)ds$

How to evaluate m, n and $\bar{a}$ quantitatively, and how to connect these parameters with other parameters such like w/c, pore size distribution curve and others are now under investigation.

3. STOCHASTIC THEORY FOR FRACTURE OF CONCRETE
3.1 Strength under a monotonically increasing load
3.1.1 Fracture of a material of Type A.   From equation (2.14), one obtains the following equation.

$$\frac{1}{\bar{p}_0(t)} d\bar{p}_0(t) = -\bar{p}_{01}(t)dt \tag{3.1}$$

where $\bar{P}_0(t)$ is the survival probability of an element and $\bar{p}_{01}(t)$ is the mean value of transition probabilities from the state 0 to the state 1. Therefore the survival probability representing that no fracture occurs from the element before a given time t is presented by equation (3.2).

$$\bar{P}_0(t) = \exp\{-\int_0^t \bar{p}_{01}(t)dt\} \tag{3.2}$$

The survival probability of a specimen composed of m elements under a uniformly distributed stress condition is given by equation (3.3).

$$P(t) = [\exp\{-\int_0^t L\sigma(t)^\beta dt\}]^m = \exp\{-\int_0^t mL\sigma(t)^\beta dt\} \tag{3.3}$$

When the stress is monotonically increasing, the stress at a certain time t is given by equation (3.4).

$$\sigma(t) = \dot{\sigma}t \tag{3.4}$$

where $\dot{\sigma}$ is the rate of loading. In such a case, equation (3.3) is rewritten as follows:

$$P(\sigma) = \exp\{-\frac{mL}{(\beta+1)\dot{\sigma}} \sigma^{\beta+1}\} \tag{3.3.1}$$

Since the *strength* usually means the stress level on which the specimen is fractured, the cumulative distribution function of strength that is the probability of fracture is given by equation (3.5).

$$F(\sigma) = 1 - P(\sigma) = 1 - \exp\{-\frac{mL}{(\beta+1)\dot{\sigma}} \sigma^{\beta+1}\} \tag{3.5}$$

Then the probability density function of strength is obtained as follows:

$$q(\sigma) = \frac{mL}{\dot{\sigma}}\sigma^\beta \exp\{-\frac{mL}{(\beta+1)\dot{\sigma}} \sigma^{\beta+1}\} \tag{3.6}$$

The peak value of the function $q(\sigma)$ is expressed as follows:

$$\bar{\sigma} = (\frac{\beta\dot{\sigma}}{mL})^{\frac{1}{\beta+1}} \tag{3.7}$$

Hence the mean value of strength $\bar{f}$ is given by equation (3.8).

$$\bar{f} = \int_0^\infty \sigma\, q(\sigma)d\sigma = \{\frac{(\beta+1)\dot{\sigma}}{mL}\}^{1/(\beta+1)} \Gamma(\frac{\beta+2}{\beta+1}) \tag{3.8}$$

where Γ is the Gamma Function. Equation (3.7) indicates that large value of L (heterogeneity, etc.) and m (the volume of the specimen) decrease the value of $\bar{f}$. It should be noticed, moreover, that equation (3.8) leads to explanations of kinetic effects of the rate of loading and temperature which could not be explained by usual statistical theories (refs. 1, 2, 3 and 4). The variance $V^2(\sigma)$ is obtained as follows:

$$V^2(\sigma) = \{\frac{(\beta+1)\dot\sigma}{mL}\}^{\frac{2}{(\beta+1)}}\{\Gamma(\frac{\beta+3}{\beta+1}) - \Gamma^2(\frac{\beta+2}{\beta+1})\} \qquad (3.9)$$

This predicts that large values of L and m decrease the scattering of strength, too.

In the case of three-point bending tests (Fig. 16), the stress distribution is not uniform and the transition probability of an element locating on a certain point: $X(x,y)$ is obtained as follows:

$$\bar{P}_{01}(x,y,t) = L\sigma_0(t)^\beta(\frac{2}{dh})^\beta y^\beta(d-|x|)^\beta \qquad (3.10)$$

where d is a half length of the span, h is the height of the beam and $\sigma_0(t)$ is the stress at the outer fiber in the center of the span. When the load is

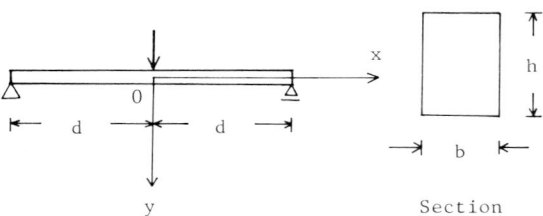

Fig. 16. Dimension of bending test

monotonically increased, the survival probability and the mean value of flexural strength are given by equations (3.11) and (3.12), respectively.

$$P(\sigma_0) = \exp\{-\frac{m_f bdhL}{(\beta+1)^3\dot\sigma_0}\sigma_0^{\beta+1}\} \qquad (3.11)$$

$$\bar{f} = \{\frac{3(\beta+1)^3\dot\sigma_0}{m_f bdh(\beta+3)L}\}^{\frac{1}{\beta+1}}\Gamma(\frac{\beta+2}{\beta+1}) \qquad (3.12)$$

where m_f is a size parameter for the three-point bending test.

3.1.2 Fracture of a material of Type B. Generally speaking, it is more suitable to describe the fracture phenomena of concrete as successive events. In such cases, let $\bar{P}_i(t)$ be the probability that the state of an element has been transformed into the state i of the successive stochastic processes at a given time t. Then the following differential equations are obtained.

$$d\mathring{P}_0(t)/dt = -\bar{p}_{01}(t)\,\mathring{P}_0(t) \tag{3.13}$$

$$d\mathring{P}_i(t)/dt = -\bar{p}_{i,i+1}(t)\,\mathring{P}_i(t) + \bar{p}_{i-1,i}(t)\,\mathring{P}_{i-1}(t) \tag{3.14}$$

where $\bar{p}_{ij}(t)$ indicates the probability that an element makes a transition from state i to state j in a unit interval at a certain time t. The transition probability $\bar{p}_{ij}(t)$ may be described as follows:

$$\bar{p}_{ij}(t) = L_i|\sigma(t)|^\beta \tag{3.15}$$

where

$$L_i = n\bar{Z}A(E\gamma)^{-\beta/2}\bar{a}^{\beta/2+1}R_i \tag{3.15.1}$$

Hence the following equations are obtained for i=0 and i=1, respectively.

$$\bar{p}_{01}(t) = L_0|\sigma(t)|^\beta \quad ; \quad \bar{p}_{12}(t) = L_1|\sigma(t)|^\beta \tag{3.15.2}$$

From equation (3.13) under the initial condition: $\mathring{P}_0(0)=1$, $\mathring{P}_i(0)=0$ ($i\neq0$), one can obtain the following equation.

$$\overset{e}{P}_i(t) = \exp\{-\int_0^t \bar{p}_{i,i+1}(t)\,dt\}\,u_i(t) \tag{3.16}$$

where

$$u_i(t) = \int_0^t \{\bar{p}_{\ell i}(t)\exp[\int_0^t \{\bar{p}_{i,i+1}(t) - \bar{p}_{\ell i}(t)\}dt]u_\ell(t)\}dt \tag{3.17}$$

where $\ell=i-1$ and the initial conditions concerning $u_i(t)$ are $u_0(0)=1$; $u_i(0)=0$; $i\geq1$. Under a monotonically increasing stress, equation (3.16) is rewritten as follows:

$$\mathring{P}_i(\sigma) = \exp\{-\frac{L_i\sigma^{\beta+1}}{(\beta+1)\dot{\sigma}}\}\int_0^\sigma [\frac{L_\ell\sigma^\beta}{\dot{\sigma}}\exp\{\frac{L_i-L_\ell}{(\beta+1)\dot{\sigma}}\sigma^{\beta+1}\}\,u_\ell(\sigma)]d\sigma \tag{3.18}$$

Now the survival probability of the element on a given stress level is written as follows:

$$\mathring{P}(\sigma) = \sum_{i=0}^{1} \mathring{P}_i(\sigma) \tag{3.19}$$

The survival probability of the specimen can be obtained by equation (3.20).

$$P(\sigma) = \{\hat{P}(\sigma)\}^m \tag{3.20}$$

Therefore the following equation is obtained.

$$P(\sigma) = [\sum_{i=0}^{1} \prod_{k=0}^{1} \frac{L_k}{(L_i - L_k)} \exp\{-\frac{L_i}{(\beta+1)\dot{\sigma}} \sigma^{\beta+1}\}]^m \tag{3.21}$$

where $k \neq i$. Then the mean value of strength is given by equation (3.22).

$$\bar{f} = \{\frac{(\beta+1)\dot{\sigma}}{mL_0}\}^{\frac{1}{\beta+1}} [\frac{L_1}{mL_0} \Gamma(\frac{2\beta+3}{\beta+1}) + \frac{\{(2m-1)L_0 - L_1\}L_1}{2m^2 L_0^2} \Gamma(\frac{3\beta+4}{\beta+1})$$

$$+ \frac{\{m^2 L_0^2 + (1-3m)L_0 L_1 + L_1^2\}}{6m^2 L_0^2} \Gamma(\frac{4\beta+5}{\beta+1})]$$

$$= \{\frac{(\beta+1)\dot{\sigma}}{mL_0}\}^{\frac{1}{\beta+1}} f_1(\lambda, m, \beta) \tag{3.22}$$

where $\lambda = L_1/L_0$ and $f_1(\lambda, m, \beta)$ is a function which is influenced only by λ, m and β.

3.2 Strength under a sustained load

3.2.1 Fracture of a material of Type A. When the load is sustained on a certain level, equation (3.3) which gives the survival probability is rewritten as follows:

$$P(t) = \exp(-\int_0^t mL\sigma^\beta dt) = \exp(-mL\sigma^\beta t) \tag{3.23}$$

The mean value of life time: $\bar{t}$ is described by equation (3.24).

$$\bar{t} = \frac{1}{mL\sigma^\beta} \Gamma(2) \tag{3.24}$$

When the stress level is described by the stress ratio: η to a certain level: σ_0 (for example, the strength of the same concrete obtained by a standard test), equation (3.24) is rewritten by equation (3.25).

$$\bar{t} = \frac{1}{mL\eta^\beta} \frac{\Gamma(2)}{\sigma_0^\beta} \tag{3.25}$$

Then the following equation is obtained :

$$\ln \bar{t} = -\beta \ln \eta + f_2(L, m, \beta, \sigma_0) \tag{3.26}$$

where $\eta = \sigma/\sigma_0$ and $f_2(L, m, \beta, \sigma_0)$ is a function which is influenced only by L, m, β and σ_0.

3.2.2 <u>Fracture of a material of Type B.</u> Since the failure process of concrete is not completely brittle especially under compressive loads, the probability $\overset{\beta}{P}_i(t)$ that the state of an element has been transformed into the stage i of the successive stochastic processes at a given time t, should be considered. One can obtain the survival probability on an element: $\overset{\beta}{P}(t)$ through some mathematical procedure from equations (3.14) and (3.15) as follows:

$$\overset{\beta}{P}(t) = \overset{\beta}{P}_0(t) + \overset{\beta}{P}_1(t)$$

$$= \frac{L_1}{(L_1-L_0)} \exp(-L_0|\sigma|^\beta t)\{1 - \frac{L_0}{L_1}\exp(L_0-L_1)|\sigma|^\beta t\} \tag{3.27}$$

The survival probability of a specimen with m elements can be described by the following relation.

$$P(t) = \{\overset{\beta}{P}(t)\}^m$$

As the results of that, the following equations are obtained (ref. 11).

$$P(t) = \frac{1}{(L_1-L_0)^m}\{L_1 \exp(-L_0|\dot\sigma|^\beta t)-L_0 \exp(-L_1|\sigma|^\beta t)\}^m \tag{3.28}$$

$$\bar{t} = \frac{1}{m^2 L_0 \sigma^\beta}[\lambda\Gamma(3)+\frac{\{(2m-1)\lambda-\lambda^2\}}{2m} \dot\Gamma(4)+\frac{\{\lambda^2+(1-3m)\lambda+m^2\}}{6m} \Gamma(5)]$$

$$= \frac{1}{m^2 L_0 \sigma^\beta} f_3(m, \lambda) \tag{3.29}$$

3.2.3 <u>Life time under a sustained load considering ageing effects.</u> Young's modulus E, surface energy γ and pre-existing crack length: $2\bar{a}$ may vary with the age of the specimens. They may be described as functions of the age τ as follows:

$$E(\tau) = E_1 \cdot \tau^{\alpha 1}, \quad \gamma(\tau) = \gamma_1 \cdot \tau^{\alpha 2}, \quad \bar{a}(\tau) = \bar{a}_1/\tau^{\alpha 3} \tag{3.30}$$

Under a sustained load, the survival probability may be represented by equation (3.31).

$$P(t,\tau_0) = \exp\{\frac{m\bar{L}C_1\sigma^\beta}{(1-\alpha)} \tau_0^{1-\alpha}\}\exp\{\frac{-m\bar{L}C_1\sigma^\beta}{(1-\alpha)} (t+\tau_0)^{1-\alpha}\} \tag{3.31}$$

where

$$L = \bar{L} L(t) \quad \text{and} \quad L(t) = C_1\tau^{-\alpha}$$

τ_0 is the age of the specimen at which the load has been applied. When a is unity, this equation should be changed (ref. 11). According to equation (3.31), the mean value of life time is given by equation (3.32).

$$\bar{t}(\tau_0,\sigma) = \exp\{\frac{m\bar{L}C_1\sigma^\beta}{(1-\alpha)}\tau_0^{1-\alpha}\}[\{\frac{(1-\alpha)}{m\bar{L}C_1\sigma^\beta}\}^{\frac{1}{1-\alpha}}\Gamma(\frac{2-\alpha}{1-\alpha},\frac{m\bar{L}C_1\sigma^\beta}{1-\alpha}\tau_0^{1-\alpha})$$

$$-\tau_0\Gamma(1,\frac{m\bar{L}C_1\sigma^\beta}{1-\alpha}\tau_0^{1-\alpha})] \quad (3.32)$$

where $\Gamma(b, z)$ is the Incomplete Gamma Function.

In the case of the loading condition shown in Fig. 17, the survival probability should be evaluated with the consideration to $P(\sigma)$; where $P(\sigma)$ is

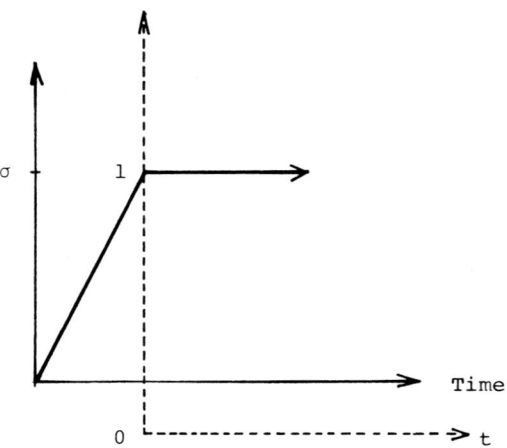

Fig. 17. Loading condition of static fatigue

the survival probability on the point 1. Hence the following equations are obtained.

$$P(t) = \exp(-mL\sigma^\beta t) \cdot P(\sigma)$$
$$= \exp(-mL\sigma^\beta t)\ \exp\{-\frac{mL}{(\beta+1)\dot{\sigma}}\sigma^{\beta+1}\} \quad (3.33)$$

$$\bar{t} = \frac{1}{mL\sigma^\beta}\ \exp\{-\frac{mL}{(\beta+1)\dot{\sigma}}\sigma^{\beta+1}\} \quad (3.34)$$

Because of the same reason, equation (3.32) may be rewritten as follows after the introduction of $\bar{\sigma}$.

$$\bar{t}= \frac{1}{m\bar{L}C_1}(\frac{m\bar{L}C_1}{\beta\dot{\sigma}})^{\frac{\beta}{\beta+1}}\{\tau_0^{\frac{\alpha}{\beta+1}}\frac{1}{\eta\beta}+\frac{\alpha}{m\bar{L}C_1}(\frac{m\bar{L}C_1}{\beta\dot{\sigma}})^{\frac{\beta}{\beta+1}}\frac{1}{\tau_0^{1-\frac{2\alpha}{\beta+1}}}(\frac{1}{\eta\beta})^2\}$$

$$\cdot\exp(\frac{-\beta}{\beta+1}\eta^{\beta+1})\ f_4(\lambda,\ m) \quad (3.35)$$

where $\eta=\sigma/\bar{\sigma}$ and $\bar{\sigma}$ is the peak value of the probability density function $q(\sigma)$ as shown in equation (3.7).

3.3   Strength under a repeated load

3.3.1   Fracture of a material of Type A.   To begin with the description of
fatigue of concrete, the simplest model (Type A) will be used to study the
failure process. A rectangular pulse loading history is being assumed. Then the
survival probability after one cycle of loading as shown in Fig. 18.(a), is

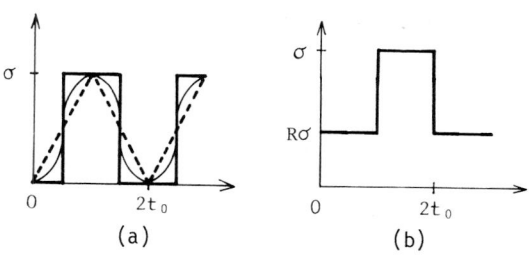

Fig. 18. Cyclic loading condition

given by equation (3.36).

$$P(1) = \exp \left\{ - \int_0^{t_0} mL\sigma^\beta \, dt \right\}$$

(3.36)

After N cycles of loading, the following equation expresses the survival
probability.

$$P(N) = \exp(-mLNt_0\sigma^\beta)$$

(3.37)

The mean value of fatigue life is then described as follows:

$$\bar{N}_1 = \frac{1}{mL\, t_0\sigma^\beta} \cdot \Gamma(2)$$

$$= 2f/mL\sigma^\beta$$

(3.38)

where

$$f = 1/2t_0$$

is the frequency of the cyclic loading history. If $\bar{\sigma}$ shown in equation (3.7) is
introduced as the static strength, equation (3.38) will be rewritten as follows:

$$\ln \bar{N}_1 = -\beta \ln \eta + \ln \left\{ \frac{2f}{mL} \left(\frac{mL}{\beta\bar{\sigma}} \right)^{\frac{\beta}{\beta+1}} \right\}$$

(3.39)

where $\eta = \sigma/\bar{\sigma}$. If it is supposed that the transition probability is dependent
only on the stress, the survival probability after N cycles of triangular pulse
load history as shown in Fig. 18.(a) is described by equation (3.40).

$$P(N) = \exp\left(-\frac{2NmL\sigma^{\beta}}{\beta+1}\,t_0\right) \tag{3.40}$$

The mean value of fatigue life is given by equation (3.41).

$$\bar{N}_2 = \frac{(\beta+1)}{mL\,\sigma^{\beta}}\,f \tag{3.41}$$

Sine-wave loading history is frequently used for experimental studies. In such a case, the survival probability after one cycle is given by the following equation.

$$P(1) = \exp\left\{-\int_0^{1/f} mL\cdot(\sigma_m + \sigma_a\cdot\sin 2\pi ft)^{\beta}\cdot dt\right\} \tag{3.42}$$

where σ_m is the mean stress level and σ_a is the amplitude. Unfortunately it is very difficult to find an analytical solution of this equation. However the upper limit and the lower limit of N_3, the fatigue life under sine-wave cyclic load, may be estimated as follows:

$$\bar{N}_1 \leq \bar{N}_3 \leq \bar{N}_2 \tag{3.43}$$

3.3.2 Fracture of the material of Type B.   According to the model of Type B, the following survival probability and the mean value of fatigue life are obtained if rectangular pulse loading history is assumed (ref. 11).

$$P(N) = \frac{L_1^{m}}{(L_1-L_0)^m}\cdot\exp(-mL_0\sigma^{\beta}\,t_0 N)\cdot\left[1-\frac{L_0}{L_1}\exp\{L_0-L_1)\cdot\sigma^{\beta}\cdot t_0 N\}\right]^{m} \tag{3.44}$$

$$\bar{N} = \frac{1}{m^2\sigma^{\beta}\cdot L_0 t_0}[\frac{L_1}{L_0}\cdot\Gamma(3)+\frac{L_1\cdot(2mL_0-L_1-L_0)}{2mL_0^2}\cdot\Gamma(4)+\frac{\{L_1^2+(1-3m)\cdot L_1 L_0+m^2 L_0^2\}}{6mL_0^2}\cdot\Gamma(5)] \tag{3.45}$$

The corresponding equations for survival probability $P(N)$ and the mean fatigue life $\bar{N}$ under triangular loading history are found to be as follows:

$$P(N) = \frac{L_1^{m}}{(L_1-L_0)^m}\exp\left(-\frac{2mL_0\sigma^{\beta}t_0}{\beta+1}\,N\right)\left[1-\frac{L_0}{L_1}\exp\{\frac{2(L_0-L_1)}{\beta+1}\sigma^{\beta}t_0 N\}\right]^{m} \tag{3.46}$$

$$\bar{N} = \frac{\beta+1}{2m^2 L_0 t_0 \sigma^\beta} \left[\frac{L_1}{L_0} \Gamma(3) + \frac{L_1(2mL_0 - L_1 - L_0)}{2mL_0^2} \Gamma(4) + \frac{\{L_1^2 + (1-3m)L_0 L_1 + m^2 L_0^2\}}{6mL_0^2} \Gamma(5) \right] \qquad (3.47)$$

From equations (3.39), (3.41), (3.45) and (3.47), the general relation between the upper stress level η and the mean value of fatigue life $\bar{N}$ for all kinds of constant amplitude loading condition is described as follows:

$$\ln \bar{N} = -\beta \cdot \ln \eta + \text{const.} \qquad (3.48)$$

where the constant is a function dependent on the chosen loading history, L, m and β.

3.3.3 Influence of time-dependent deformation on fatigue life.

Under repeated loading conditions, an influence of time-dependent deformation on fatigue life cannot be excluded. Possible mechanisms causing the change of internal structure include the followings:

(1). Increase of the radius of microcrack tip caused by microscopic creep deformation.

(2). Stress redistribution caused by the accumulation of stable microcracks. The survival probability of an element under rectangular loading conditions as shown in Fig. 18.(b), is given by equation (3.49).

$$\overset{e}{P}(N) = \exp\left\{- \int_0^N L \cdot (R^\beta + 1) \cdot \sigma^\beta \cdot t_0 \cdot dN\right\} \qquad (3.49)$$

If the change of internal structure under a certain sustained load is considered, the following relation can be assumed.

$$\overset{e}{P}(N) = \exp\left\{- \int_0^N \frac{L \cdot (R^\beta + 1) \cdot \sigma^\beta \cdot t_0}{a \cdot (bS^\alpha N + 1)^\delta} \, dN\right\} \qquad (3.50)$$

where

$$S = \int_0^{2t_0} \{\sigma(t)\}^{\alpha 1} \cdot dt = (R^{\alpha 1} + 1) \sigma^{\alpha 1} \cdot t_0 \qquad (3.51)$$

and α may be a kind of material parameter affected by temperature, humidity and the frequency of repeated load. Hence the survival probability of the specimen is obtained.

$$P(N) = \exp\left\{\frac{mL \cdot (R^\beta + 1) \cdot \sigma^\beta \cdot t_0}{a \cdot b \cdot S^\alpha \cdot (1-\delta)}\right\} \cdot \exp\left\{- \frac{mL(R^\beta + 1) \cdot \sigma^\beta t_0}{ab \cdot S^\alpha \cdot (1-\delta)} (bS^\alpha N + 1)^{1-\delta}\right\} \qquad (3.52)$$

The mean value of fatigue life (the number of load repetitions up to fracture) may be predicted by the following equation:

$$\bar{N} = \frac{2fa}{mL(R^{\beta}+1)\sigma_u^{\beta}} \left\{ 1 + \frac{ab\delta(R^{\alpha_1}+1)^{\alpha}(2f)^{1-\alpha}}{mL(R^{\beta}+1)\sigma_u^{\beta-\alpha_2}} \right.$$

$$\left. + \frac{a^2b^2 \cdot \delta \cdot (2\delta-1) \cdot f^{2(1-\alpha)} \cdot (R^{\alpha_1}+1)^{2\alpha}}{m^2 \cdot L^2 \cdot (R^{\beta}+1)^2 \cdot \sigma_u^{2(\beta-\alpha_2)}} \right\} \qquad (3.53)$$

where f is frequency, R is the ratio of σ_1/σ_u with σ_u and σ_1 being upper and lower bound of the applied stress respectively, m is a size parameter of the specimen, β is a kind of material parameter, a, b, α_1 and δ are parameters to introduce the effect of time-dependent deformation, α_2 is the product of α_1 and α, L is a parameter representing the heterogeneity of internal structure. Equation (3.53) will give a non-linear relationship between $\ln\eta$ and $\ln\bar{N}$, in which the value of $\bar{N}$ extremely increases in the region of the small value of η.

4. COMPARISON WITH PUBLISHED EXPERIMENTAL RESULTS

4.1 Stress rate effect

From equations (3.8), (3.12) and (3.22), the influence of rate of loading may be described as follows:

$$\frac{\bar{f}}{\bar{f}_0} = \left(\frac{\dot{\sigma}}{\dot{\sigma}_0} \right)^{\frac{1}{\beta+1}} \qquad (4.1)$$

$\bar{f}$ and $\bar{f}_0$ stand for strength under a high rate of loading (dynamic) and a low rate of loading (static) respectively. The corresponding rates of loading are described by $\dot{\sigma}$ and $\dot{\sigma}_0$. In Figs. 19 and 20, some experimental results presented by Mihashi and Wittmann (ref. 11) and Takeda and Tachikawa (ref.15), are shown.

TABLE 1. Experimental program to study the influence of rate of loading (ref. 11).

Group	Material	Loading Condition	Different Rates of Loading	Number of Tests
I	Mortar(W/C=0.45)	Bending	5	148
II	Mortar(,, =0.65)	Bending	5	149
III	Mortar(,, =0.45)	Compression	6	188
IV	Mortar(,, =0.65)	Compression	6	190
V	Concrete	Compression	2	60

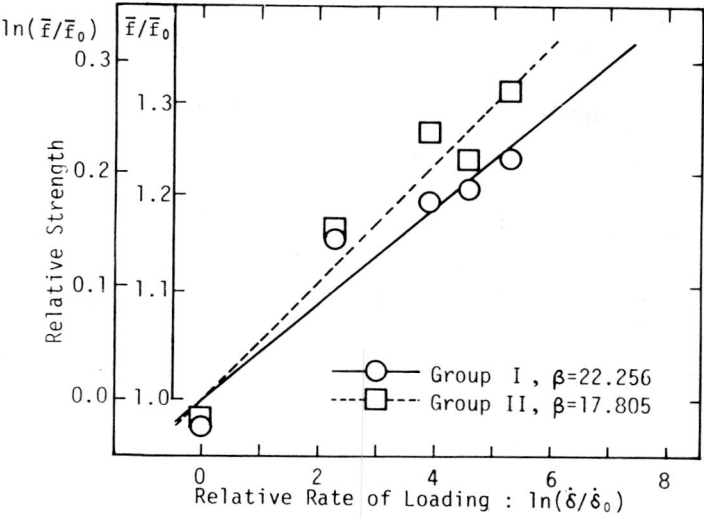

Fig. 19.a. Relation between the related rate of loading and the related mean value of flexural strength (ref. 11)

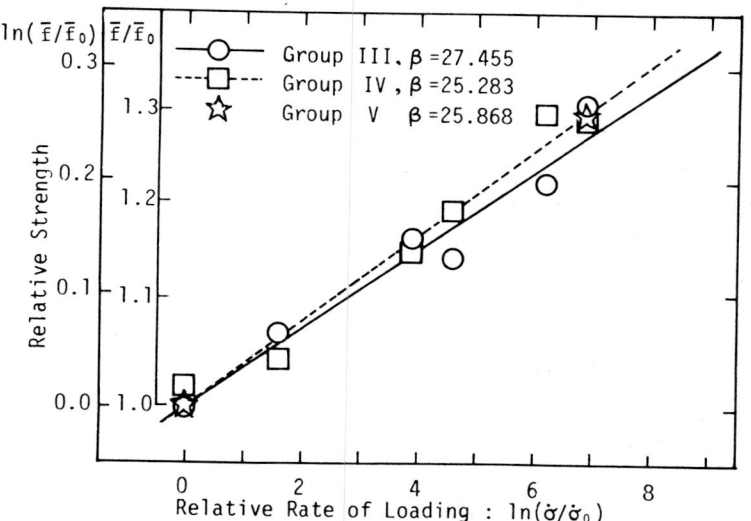

Fig. 19.b. Relation between the related rate of loading and the related mean value of compressive strength (ref. 11)

It deserves to be noted that equation (4.1) describes the dependence of strength under high rate of loading satisfactorily in a very wide range.

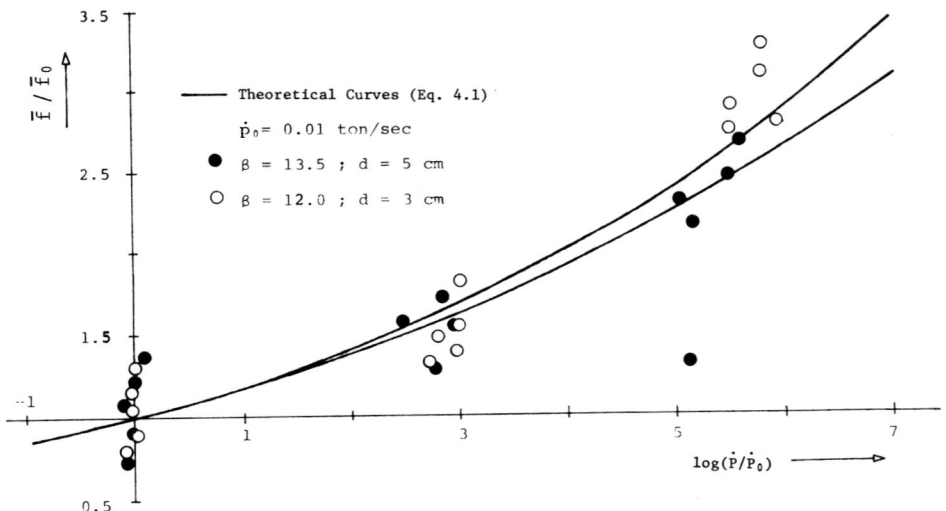

Fig. 20. Related flexural strength of concrete slab as
function of related rate of loading (after
J. Takeda et al. (ref. 15))

4.2   Environmental temperature effect

The failure of solids may be generally considered to be caused by the
nucleation processes dependent on stress. In such processes, thermal agitation
has an effect equivalent to enhancing the locally concentrated tensile stress.
From this standpoint, the constant A in equation (2.12) is described as follows
(ref. 16):

$$A = T \exp(-U_0/kT) \, c_1{}^{\beta} \tag{4.2}$$

where $\beta = {}^1/n_b kt$; k is the Boltzmann constant; T is absolute temperature; U_0 is
the activated free energy for microcracking; n_b and c_1 are material constants.
The temperature effect on strength is shown from equations (3.8) and (3.22) as
follows:

$$\ln \Lambda_{temp} = \ln\{{}^{\sigma}T/\sigma_{T_0}\} = (\lambda_T - 1) \ln(\dot{\sigma}/C_k T_0{}^2)^{n_b kT} + f_5(\lambda_T) \tag{4.3}$$

where T_0 is the standard temperature, $f_5(\lambda_T)$ is a function of λ_T and $\lambda_T = {}^T/T_0$;
C_k is equal to $n_b kmn\bar{Z}R$. In a narrow range of temperature variation, the
quantity of $f_5(\lambda_T)$ may be negligible and equation (4.3) is rewritten as follows:

$$\ln \Lambda_{temp} = C(\lambda_T - 1) \tag{4.4}$$

Some results comparing the theoretical relationship with experimental data are
shown in Fig. 21. These show that strength is inversely proportional to the

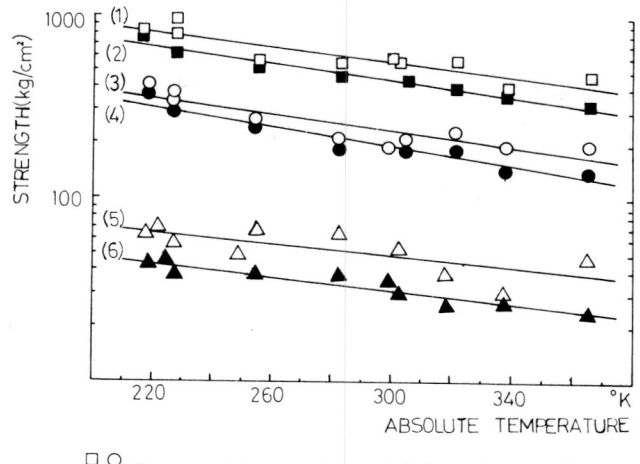

Fig. 21. Relationship between strength of concrete and
         absolute temperature (after J.C. Saemann and
         G.W. Washa (ref. 17))

temperature ascent. The same tendency may be observed for flexural strength
(ref. 16).

4.3 Size of specimen

It is well known that a larger specimen has a lower strength. From equation
(3.8), the following equation expressing such an effect in cases of brittle
fracture is obtained.

$$\Lambda_s^{I} = \frac{\bar{f}}{\bar{f}_0} = (^{m_0}/m)^{\frac{1}{\beta+1}} \tag{4.5}$$

This expression is just the same as that proposed by Weibull (ref. 1). On the
other hand, another equation is obtained in cases of imperfectly brittle
fracture (Type B) from equation (3.22) as follows:

$$\Lambda_s^{II} = (\frac{m_0}{m})^{\frac{1}{\beta+1}} \frac{[\frac{\lambda}{m}\Gamma(\frac{2\beta+3}{\beta+1})+\frac{\{(2m-1)-\lambda\}\lambda}{2m^2}\Gamma(\frac{3\beta+4}{\beta+1})+\frac{m^2+(1-3m)\lambda+\lambda^2}{6m^2}\Gamma(\frac{4\beta+5}{\beta+1})]}{[\frac{\lambda}{m_0}\Gamma(\frac{2\beta+3}{\beta+1})+\frac{\{(2m_0-1)-\lambda\}\lambda}{2m_0^2}\Gamma(\frac{3\beta+4}{\beta+1})+\frac{m_0^2+(1-3m_0)\lambda+\lambda^2}{6m_0^2}\Gamma(\frac{4\beta+5}{\beta+1})]} \tag{4.6}$$

Equation (4.6) predicts that equation (4.5) does not always express the effect
of size on strength if the microstructure of specimen is heterogeneous and that
the property of size effect in cases of tensile fracture may be different from
that in cases of compressive fracture (Figs. 22 and 23). Theoretical curves

obtained from equation (4.6) are also shown in Figs. 22 and 23. These
theoretical curves are in good agreement with experimental results.

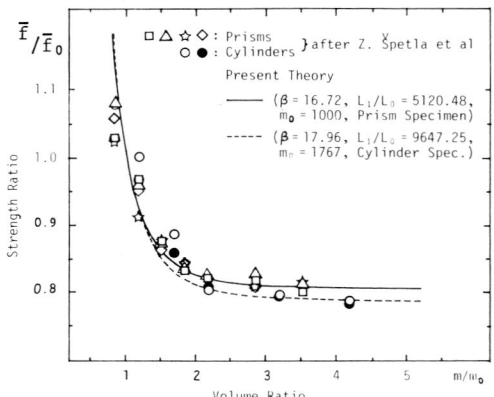

Fig. 22. Size effect on tensile strength of concrete (ref. 18)

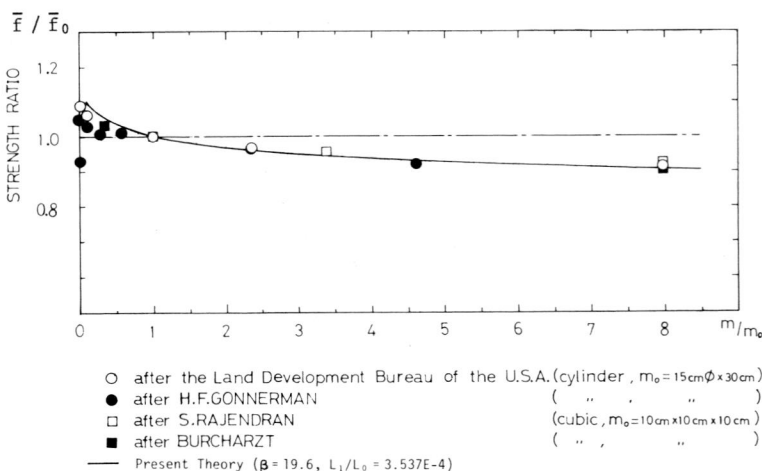

Fig. 23. Size effect on compressive strength
of concrete (refs. 19 and 20)

4.4   Variability of strength

One of the strong points of the present theory is that it provides a
realistic basis for a mathematical formulation of the variability of strength of
concrete. Some experimental results of the probability of fracture of mortar
prisms under bending loads and of normal concrete prisms under compressive loads
are shown in Figs. 24 and 25 respectively (ref. 11). Solid lines show the

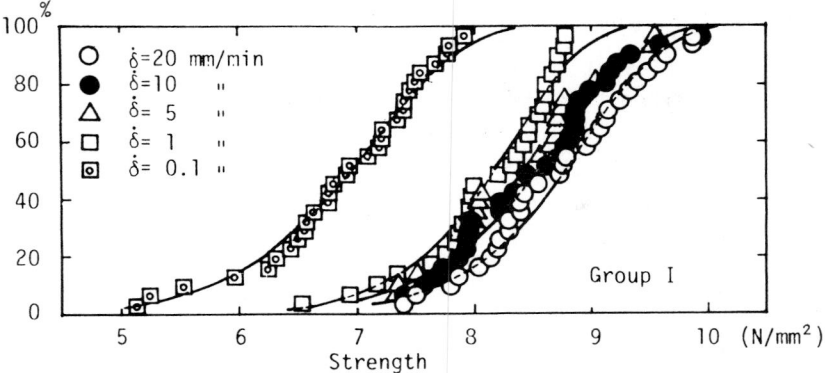

Fig. 24. Probability of fracture of mortar under
bending load (ref. 11)

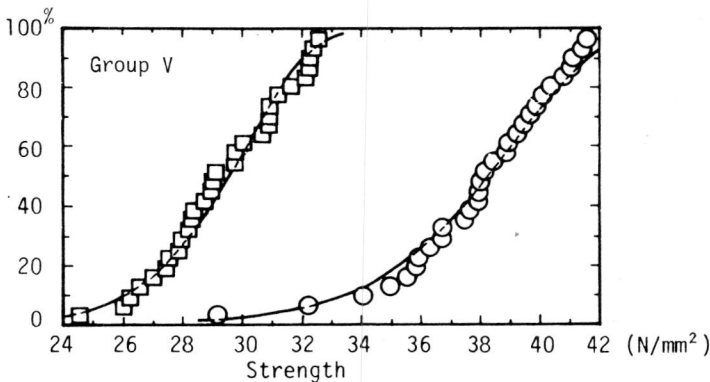

Fig. 25. Probability of fracture of concrete under
compressive load (ref. 11)

probability of fracture $F(\sigma)$ calculated with the help of equation (3.5).

In the case of a material of Type A, the survival probability: equation (3.4)
is rewritten as follows:

$$\ln\{-\ln P(\sigma)\} = \ln\{mL/(\beta+1)\dot{\sigma}\} + (\beta+1)\ln\sigma \qquad (4.7)$$

Equation (4.7) shows that the relationship between $\ln\{-\ln P(\sigma)\}$ and $\ln\sigma$ is
linear and that the slope of that line is $(\beta+1)$ (Figs. 26 and 27) (ref. 10).
In the case of a material of Type B, however, those relationships become
nonlinear.

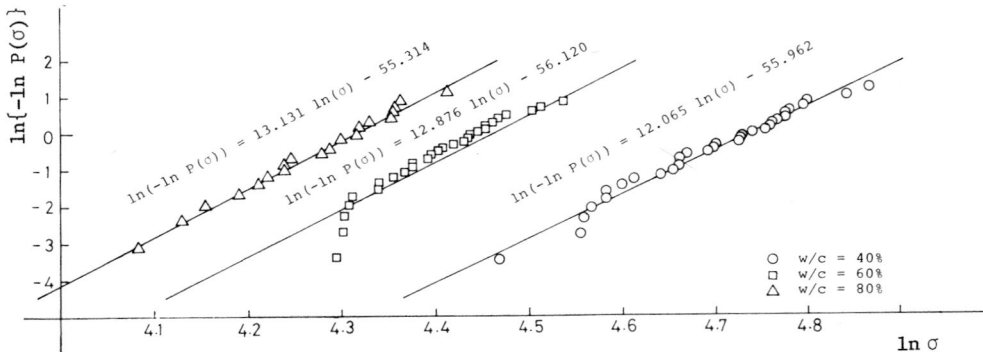

Fig. 26. Relationship between survival probability
and flexural stress of cement paste (ref. 10)

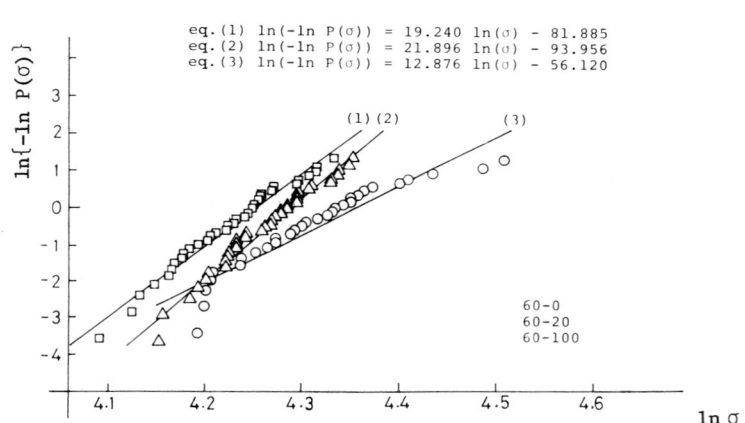

Fig. 27. Relationship between survival probability and
flexural stress of mortar (w/c=60%; s/c=0, 20
and 100%) (ref. 10)

The present theory predicts that the coefficient of variability is not influenced by the rate of loading (see equations (3.8) and (3.9)). Within the range of accuracy, this theoretical prediction is verified by an experimental study (ref. 11) as shown in TABLE 2. This result is especially interesting from the view point of a realistic reliability assessment of a concrete structure.

Since the coefficient of variability is not influenced by rate of loading, it is reasonable to expect that the distribution function of the strength which is normalized by the mean value for each group, will be the same. Fig. 28 gives the histograms for normalized strength and the theoretical density function in accordance with equation (3.6).

TABLE 2. Mean value, standard deviation and coefficient of variation
of strength of concrete under some rates of loading (ref. 11)

Rate of Loading	Mean Value of Strength (N/mm^2)	Standard Deviation	Coefficient of Variation
Group I : $\ln \sigma = 0.043 \ln \dot{\delta} + 2.050$; $\beta = 22.2$			
20.0 (mm/min)	8.72	0.677	0.078
10.0	8.48	0.713	0.084
5.0	8.39	0.685	0.082
1.0	8.11	0.585	0.072
0.1	6.87	0.729	0.106
Group II : $\ln \sigma = 0.053 \ln \dot{\delta} + 1.980$; $\beta = 17.8$			
20.0 (mm/min)	8.44	0.716	0.085
10.0	7.93	0.820	0.103
5.0	8.13	0.629	0.077
1.0	7.45	0.542	0.073
0.1	6.28	0.465	0.074
Group III : $\ln \sigma = 0.035 \ln \dot{\sigma} + 3.675$; $\beta = 27.4$			
50.505 (N/mm^2 sec)	46.51	5.075	0.109
25.253	43.37	6.348	0.146
5.173	40.63	5.993	0.147
2.586	41.30	5.900	0.143
0.259	37.98	5.295	0.139
0.052	35.52	3.905	0.110
Group IV : $\ln \sigma = 0.038 \ln \dot{\sigma} + 3.361$; $\beta = 25.2$			
50.505 (N/mm^2 sec)	33.15	4.669	0.141
25.253	33.38	3.577	0.107
5.173	30.63	3.036	0.099
2.586	29.55	3.014	0.102
0.259	26.86	3.906	0.145
0.052	26.22	3.689	0.141
Group V : $\ln \sigma = 0.037 \ln \dot{\sigma} + 3.271$; $\beta = 25.9$			
17.341 (N/mm^2 sec)	37.86	2.805	0.074
0.018	29.31	2.151	0.073

334

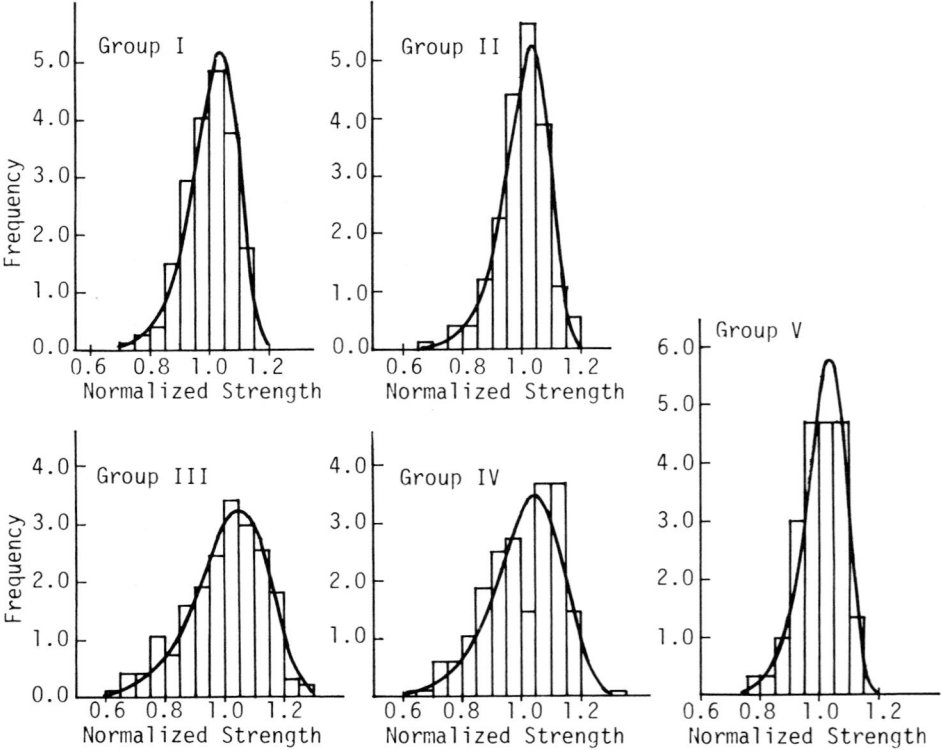

Fig. 28. Histograms of normalized strength and theoretical curves
of probability density functions (ref. 11)

4.5   Ageing effect on the life time under a sustained load

If the load is kept constant on a level slightly below the critical load,
the survival probability decreases as a function of the duration of load: t.
Fig. 29 shows the comparison of the present theory (that is equation (3.35))
with experimental data obtained by Wittmann and Zaitsev (ref. 21). These solid
lines are drawn according to the following equation.

$$\bar{t}=0.03126\{\tau_0^{0.2344}\cdot(1/\eta)^{21.278}+\frac{66.484}{\tau_0^{0.5311}}\cdot(1/\eta)^{42.556}\}\exp(-0.9551\cdot\eta^{22.278})\qquad(4.8)$$

4.6   Dynamic fatigue of concrete

From equation (3.52), the survival probability after N cycles of loading may
be described as follows:

$$\ln(-\ln P(N) + Y_c)\approx(1-\delta)\ln N + X_c$$

$$(4.9)$$

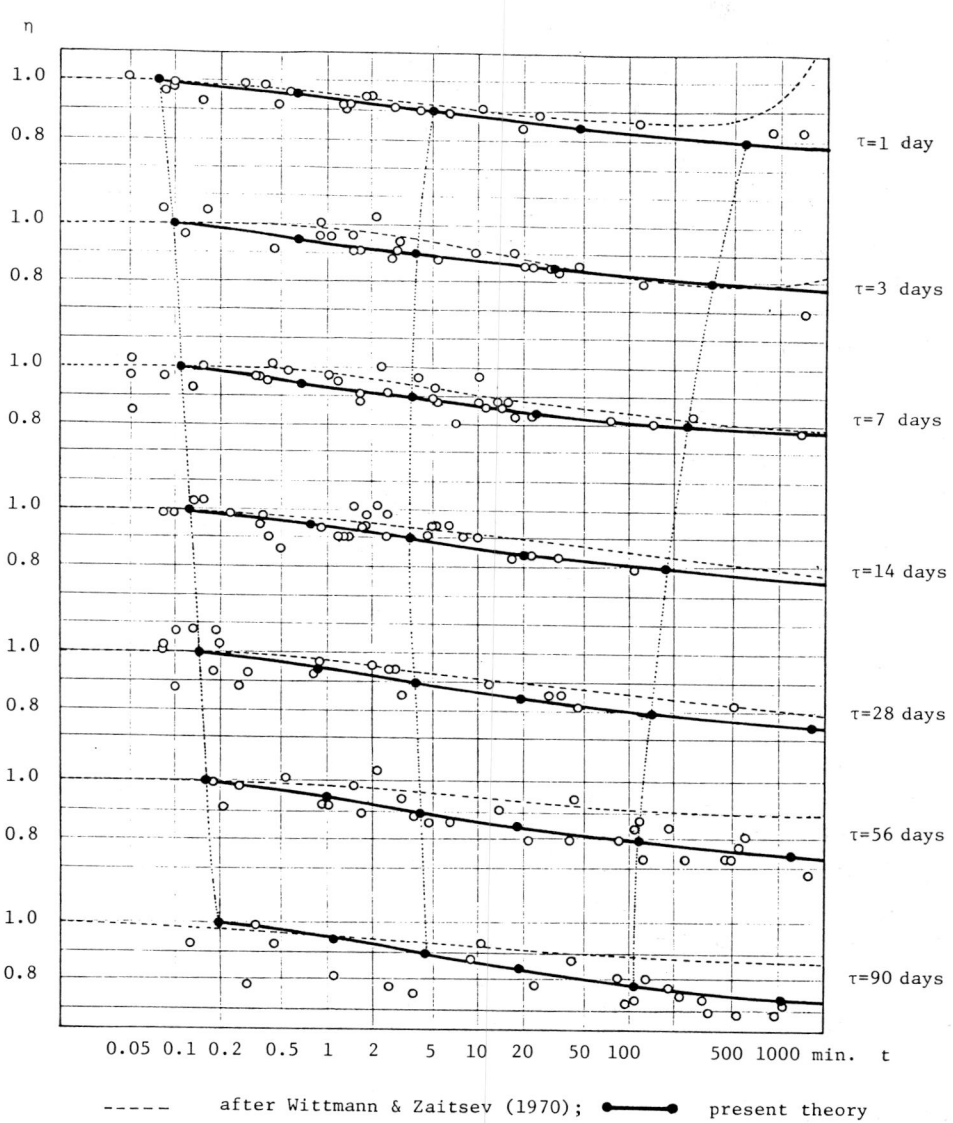

335

η

τ=1 day

τ=3 days

τ=7 days

τ=14 days

τ=28 days

τ=56 days

τ=90 days

0.05 0.1 0.2   0.5   1   2   5   10   20   50 100   500 1000 min.   t

- - - - -   after Wittmann & Zaitsev (1970);   ●——●   present theory

Fig. 29. Comparison of equation (4.8) with experimental
data presented by Wittmann and Zaitsev (ref. 21)

where P(N) is the survival probability, δ is a parameter to introduce the effect of time-dependent deformation, X_c and Y_c are constants essencially affected by the stress level and temperature. In Fig. 30, experimental results are compared with equation (4.9).

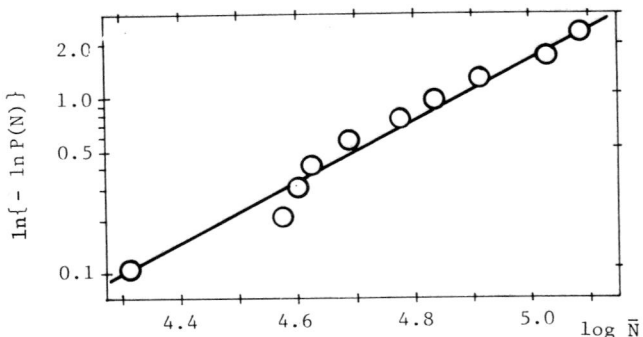

Fig. 30. Comparison of equation (4.9) with experimental
data presented by Leeuwen and Siemes (ref. 22)

According to equation (3.48), the mean value of fatigue life (the number of load repetitions up to fracture): $\bar{N}$ may be predicted by the following equation if the effect of time-dependent deformation is negligible.

$$\ln \bar{N} = -\beta \ln \eta + f_6 (f, L, m, \beta)$$ (4.10)

where f is the frequency. Fig. 31 shows the comparison of equation (4.10) with

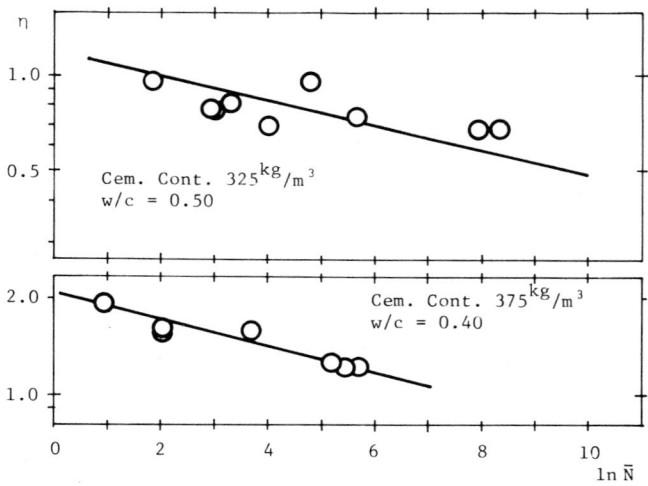

Fig. 31. Fatigue life under impact tensile load
presented by Reinhardt (ref. 23)

the impact fatigue under direct tensile repeated load. A good agreement
between the experimental results and the theory is observed. Fatigue of
concrete under compressive repeated load is also described so satisfactorily by
equation (4.10) as shown in Fig. 32.

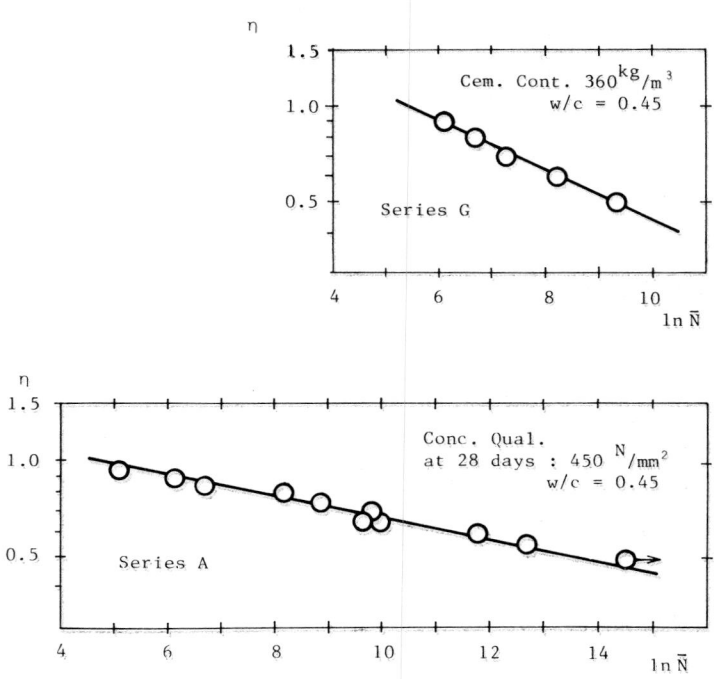

Fig. 32. Fatigue life under compressive load presented
by Leeuwen and Siemes (ref. 22)

When such effect is enhanced, the relation between the mean value of fatigue
life and applied maximum stress may be described by the following non-linear
equation which is obtained from equation (3.53).

$$\ln \bar{N} = -\beta \ln \eta + f_7(f, L, m, \beta, \eta, R) \tag{4.11}$$

The fatigue life may increase extremely in the range of lower values of η. The
influence of the lower bound of applied stress on the dynamic fatigue of
concrete is presented in Fig. 33. The solid lines are drawn with respect to
equation (3.53). This figure suggests that a repeated loading condition with a
higher value of the lower bound: R makes the fatigue life longer by means of
some changes of the internal microstructure caused by the same mechanism as the
time-dependent deformation.

338

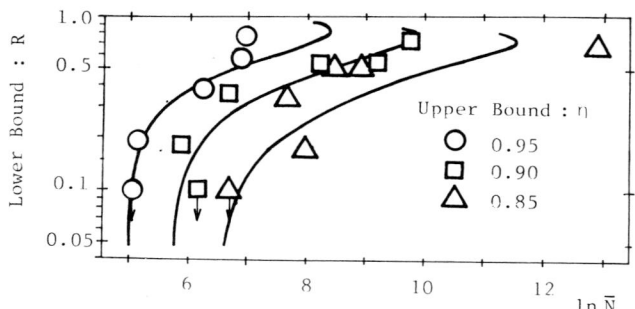

Fig. 33. Influence of the lower bound of the applied stress
on fatigue life comparing equation (4.10) with
experimental data presented by Leeuwen and Siemes
(ref. 22)

5. CONCLUSIONS

The stochastic theory for fracture of concrete satisfactorily describes the
experimental findings of strength properties which are influenced by loading,
environmental temperature, size of the specimen and age of the specimen. It is
also possible to describe the failure process not only under a monotonically
increasing load but under time dependent loading conditions such as a sustained
load and as a repeated load.

Moreover, the present theory provides a realistic basis for a mathematical
formulation of the variability of aggregative materials such as concrete.

REFERENCES

1. W. Weibull, Proceedings of the Royal Swedish Institute of Engineering
   Research, 151 (1939) pp. 151
2. A.M. Freudenthal, in H. Liebowitz (Ed.), Fracture, II, Academic Press, 1968,
   pp. 591-619.
3. A. de S. Jayatilaka and K. Trustrum, Journal of Materials Science, 12, 7
   (1977) 1426-1430
4. A. de S. Jayatilaka and K. Trustrum, Journal of Materials Science, 13, 2
   (1978) 455-457
5. M. Hirata, Science of Machine, 1, 5 (1949) 231-234, (in Japanese)
6. T. Yokobori, Journal of the Physical Society of Japan, 7, 1 (1952) 44-47
7. T. Yokobori, An Interdisciplinary Approach to Fracture and Strength of
   Solids, Iwanami Book Co., 1974, Second Ed., 334 pp., (in Japanese)
8. M. Hori, Journal of the Physical Society of Japan, 14, 10 (1959) 1444-1452
9. Y. Nishimatsu, Proceedings of the 12th Japan Congress on Material Research
   (1969) 240-243
10. H. Mihashi and M. Izumi, Cement and Concrete Research, 7 (1977) 411-421
11. H. Mihashi and F.H. Wittmann, Heron, 25, 3 (1980) 1-54
12. For example, H. Mihashi, T. Sasaki and M. Izumi, Proceedings of the Third
    International Conference on Mechanical Behaviour of Materials, 3, Cambridge
    (1979) 97-107

13. S. Mindess and S. Diamond, Cement and Concrete Research, 10 (1980) 509-519
14. D.D. Higgins and J.E. Bailey, Proceedings of the Conference on Hydraulic Cement Pastes: their structure and properties, Scheffield (1976) 283-296
15. J. Takeda, H. Tachikawa and K. Fujimoto, Proceedings of the Second International Conference on Mechanical Behaviour of Materials, Boston (1976) 1468-1472
16. M. Izumi and H. Mihashi, Transactions of Architectural Institute of Japan, 287 (1980) 1-13 (in Japanese)
17. J.C. Saemann and G.W. Washa, Journal of American Concrete Institute, 54,20 (1957) 385-395
18. Z. Špetla and V. Kadleček, Bulletin RILEM, 33 (1966)
19. Y. Kondo and S. Ban (Eds.), Concrete Handbook, Sakurai Book Company, 1957, 745 pp. (in Japanese)
20. S. Rajendran, Bulletin RILEM, 26 (1965)
21. F.H. Wittmann and J. Zaitsev, Schriftenreihe Deutscher Ausschuss für Stahlbeton, Heft 232 (1973)
22. J. van Leeuwen and A.J. Siemes, Heron, 24, 1 (1979) 1-34

Fracture Mechanics of Concrete,
edited by F.H. Wittmann, 1983
Elsevier Science Publishers B.V., Amsterdam — Printed in The Netherlands

Chapter 4.4

DAMAGE THEORY APPLIED TO CONCRETE
By M. Lorrain and K.E. Loland

1 Introduction

Introduced in 1958 by KACHANOV (ref. 1), the damage theory has been used in 1968 by RABOTNOV (ref. 2), in 1974 by BROBERG (ref. 3), and in 1978 by LEMAITRE and CHABOCHE (ref. 4), to describe the progressive deterioration of materials under monotonic or repeated loading. The main purpose of this research was the analysis of the creep-behaviour of metals.

In 1977, JANSON and HULT (ref. 5) suggested the term "Continuous Damage Mechanics" (CDM) to designate methods of rupture analyses involving damage concepts. They also proposed to combine CDM and Fracture Mechanics (FM).

The multitude of attempts to analyse the behaviour of concrete with the aid of this concept which is new to this kind of material, reveals the interest shown in the damage theory (see Fig. 1), the main characteristic of which being that a mathematical parameter replaces other physical functions such as the variation of volume, for instance.

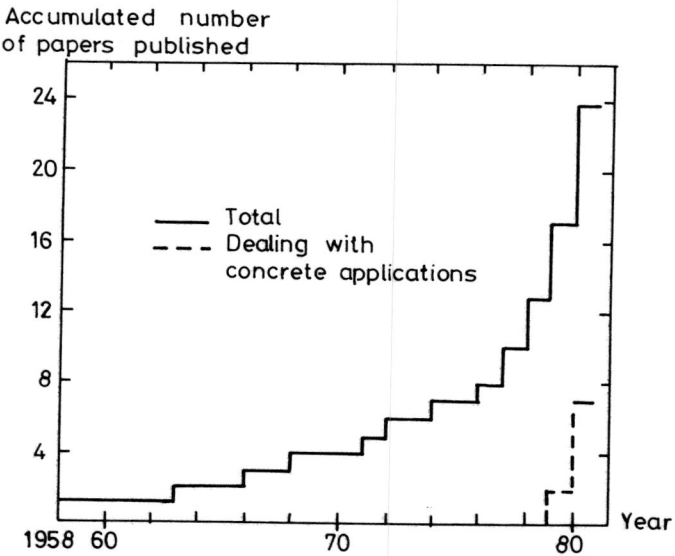

Fig. 1 - Publications dealing with damage theory.

2 The damage concept

2.1 Definition. Damage corresponds to irreversible degradation of the cohesion of the material under internal and/or external straining. This may lead to
failure of an elementary volume.

The main difference between damage and fracture mechanics (as indicated in
Fig. 2) might be stated after JANSON and HULT (ref. 5) in the following
terms :

- in damage mechanics, the strength of a loaded structure is determined by
the deterioration of the material caused by loading. This deterioration -or
damage- may be described in terms of a continuous defect field ;

- in fracture mechanics, the strength of a loaded structure is determined
by the severity of a single defect such as a sharp crack. The medium around
the crack is assumed to be mechanically intact.

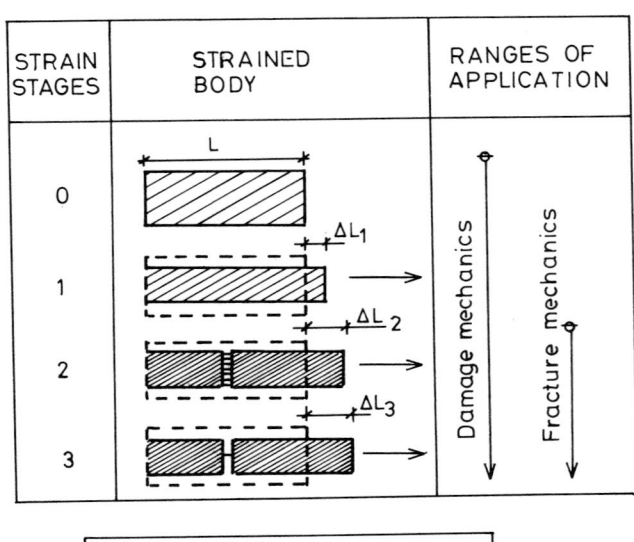

Fig. 2 - Sketch showing ranges of application for damage and fracture
         mechanics.

A more realistic assessment of the behaviour of a loaded structure may be
obtained by combining these two approaches. This may be done by studying the
deterioration of a loaded structure, where a macrodefect -or damaged fracture
zone- is included in a matrix with randomly distributed microdefects

-or damage- following JANSON and HULT (ref. 5), LØLAND and GJØRV (ref. 6) and LØLAND (ref. 7).

2.2 <u>Damage characterization with a scalar parameter</u> (uniaxial states of stress). Under uniaxial states of stress, the damage theory can be characterized by a scalar parameter, D, which denotes the concentration of microcracks (microvoids, microdefects) existing in an elementary volume of the material :

- $D = 0$ corresponds to the intact material without deterioration ; it is the reference state.
- $D = 1$ corresponds to the failure of an elementary volume of the material.

$1 - D$, in the uniaxial state, can be expressed as the fractional area of undamaged material.

$$1 - D = \frac{A_n}{A} \tag{1}$$

where : A_n = undamaged (net) area of the cross-section ;

A = total -or nominal- area of the cross-section.

Hence :

$$D = \frac{A - A_n}{A} \tag{2}$$

where $(A - A_n)$ is the damaged area in the cross-section (ref. 5), (ref. 7), (ref. 8).

An effective net stress σ_n acting on the undamaged material can also be defined as :

$$\sigma_n = \frac{\sigma}{1 - D} \tag{3}$$

where σ_n is the nominal stress ($\sigma = \frac{F}{A}$ for uniaxial tension).

Then σ_n is the nominal stress calibrated by A_n :

$$\frac{F}{A_n} = \frac{F}{A (1 - D)} = \frac{\sigma}{1 - D} = \sigma_n \tag{4}$$

For $D = 1$, $\sigma_n \rightarrow \infty$. However, failure of the elementary volume occurs at $\sigma_n < \infty$. At this stage, $\sigma = f$ = nominal failure stress of the material and D equals a characteristic value $Dc < 1$:

$$f_n = \frac{f}{1 - D_c} \tag{5}$$

$$\text{or} : D_c = \frac{f_n - f}{f_n} \tag{6}$$

where f_n is the local cohesive strength of the material.

2.3 Thermodynamic interpretation of the damage parameter

LEMAITRE and CHABOCHE (ref. 4) consider the damage parameter D as one of the hidden thermodynamic variables, which are responsible for irreversible processes. These variables are called "hidden" because they do not appear in the thermostatic analysis of a continuum ; they remain constant during reversible thermostatic transformations (see GERMAIN (ref. 9)).

So, D depends directly on the geometry and on the distribution of microdefects like microvoids and microcracks, but this dependence is expressed in terms of functional relationships between undamaged material subject to nominal stress and damaged material subject to effective stress.

Therefore, a difference exists between the more physical definition of A_n as net resistant area and the formal definition of A_n as an equivalent resistant area after LEMAITRE and CHABOCHE. In contrast, the definition of σ_n -effective net stress- does not change for both interpretations of A_n (eq. (3)).

In a uniaxial state of stress -assuming a linear damage accumulation- we can write the above mentioned equivalence of behaviour of undamaged and damaged material as follows :

$$\sigma_n = E_n \cdot \varepsilon$$

Using eq. (3) this yields :

$$\sigma = E_n (1 - D)\varepsilon$$

$$\text{or} : \sigma = E\varepsilon$$

$$\text{where} : E = E_n (1 - D) \tag{7}$$

and : E_n is the modulus of elasticity of the undamaged material.

The damage parameter, D, describes the variation of the elastic modulus, E.

Note : Both approaches of the damage concept show that the damage parameter, D, indicates the severity of a situation with reference to failure of the elementary volume.

2.4 Damage parameter measurement (uniaxial stress-state)

The damage parameter, D, can be measured during loading cycles. The stress-strain relation for decreasing load may be described as follows :

$$\frac{d\sigma}{d\varepsilon} = \frac{d}{d\varepsilon} \left[E_n (1 - D)\varepsilon \right]$$

$$= \frac{dE_n}{d\varepsilon} (1 - D)\varepsilon + E_n (1 - D) - E_n\varepsilon \frac{dD}{d\varepsilon}$$

Since damage is assumed to be irreversible, we may write for the unloading case :

$$\frac{dD}{d\varepsilon} = 0$$

Furthermore we assume that the net modulus of elasticity is a material property which yields :

$$\frac{dE_n}{d\varepsilon} = 0$$

Hence : $\frac{d\sigma}{d\varepsilon} = E_n (1 - D)$

and : $D = 1 - \frac{1}{E_n} \cdot \frac{d\sigma}{d\varepsilon}$ (8)

If E^* is the mean slope of an unloading branch of the $\sigma - \varepsilon$ diagram eq. (8) can be written as :

$$D = 1 - \frac{E^*}{E}$$ (9)

where : E is the initial value of YOUNG's modulus being measured at the beginning of the first loading cycle.

This relation allows to calculate the damage, D, at any stress or strain level from the slope of the unloading branch (Fig. 3). The method has been successfully used for ductile metals.

346

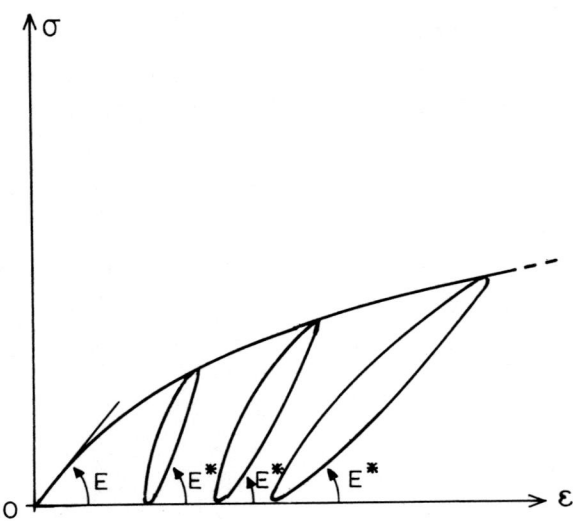

Fig. 3 - Cyclic loading - indicating variations in elastic properties.

3 Application of damage theory to concrete

3.1 <u>Applicability</u>. The damage concept was initially used to analyse and describe the behaviour of metals under sustained and repeated loads.

The concept may, however, be applicable even better to concrete since :
- in concrete the process of cracking (damage development) is continuous and begins at very low applied stresses or strains ;
- concrete is brittle and little or no yielding takes place ;
- the direction of crack development equals more or less the direction of the maximum principal stress, tension being positive.

Further, the concept allows to take into account initial damage in the form of defects (cracks or voids) preceeding the mechanical deterioration caused by external loads.

Several laboratories have undertaken research programs on the application of damage theory to concrete, especially :
- a Norwegian group from Trondheim University led by LØLAND ;
- a French group from Paris VI University led by LEMAITRE (MAZARS et al.).

A comparison of their points of view regarding the modeling of the behaviour of concrete in tension and compression by damage theory is undoubtedly of high interest.

3.2 Application of damage theory to the modeling of the behaviour of concrete under uniaxial tension.

MAZARS's model

In describing the stress-strain curve of concrete under uniaxial tension, MAZARS (ref. 10) distinguishes :
- before peak load : $\varepsilon < \varepsilon_c$, where
    - net stress : $\sigma_n = E\varepsilon \ (= \sigma_{eff})$
    - damage : $D = 0$
    - nominal stress : $\sigma = E.\varepsilon$ $\qquad\qquad\qquad\qquad\qquad$ (10)
- beyond peak load : $\varepsilon > \varepsilon_c$, where
    - net stress : $\sigma_n = \dfrac{\sigma}{1 - D}$

$$- \text{damage} : D = 1 - \frac{\varepsilon_c \ (1 - A_1)}{\varepsilon} - A_1 \exp\ (B_1\ (\varepsilon_c - \varepsilon)) \qquad\qquad (11)$$

- nominal stress : $\sigma = E\ (1 - D)\ \varepsilon$

where : ε_c is the strain at maximum nominal stress ;

A_1, B_1 are constants which can be determined experimentally.

These relations are the ones given initially for metals by RABOTNOV (ref. 2) and KACHANOV (ref. 1). The experimental $\sigma - \varepsilon$ curves which are taken as reference (Fig. 4), have been obtained by TERRIEN (ref. 11). Fig. 5 shows the variation of σ, σ_n and D with reference to strain, ε, after MAZARS.

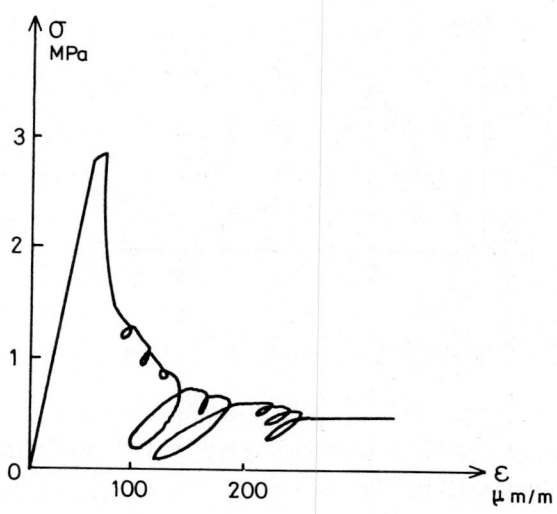

Fig. 4 - Experimental stress-strain curves in uniaxial tension according to TERRIEN.

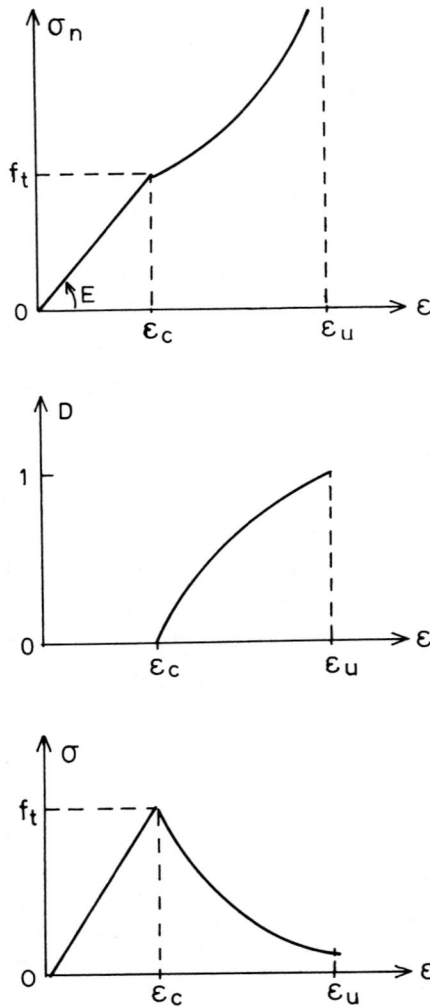

Fig. 5 - Mathematical modeling according to MAZARS.

LØLAND's model

Also LØLAND (ref. 6), (ref. 7) differentiates between two phases in modeling the σ - ε curve in uniaxial tension :

- before peak load : $\varepsilon \leq \varepsilon_c$, where :
  - net stress : $\sigma_n = E_n \varepsilon$
  - damage : $D = D_o + (1 - D_o) C_1 n_1^{\nu_1}$ (12)
  - nominal stress : $\sigma = \sigma_n (1 - D)$

- beyond peak load : $\varepsilon > \varepsilon_c$; where
    - net stress : $\sigma_n = E_n \varepsilon_c = $ const.
    - damage : $D = D_0 + (1 - D_0)(C_2 + C_3 (\eta - 1))$ (13)
    - nominal stress : $\sigma = \sigma_n (1 - D)$

where : $\eta = \dfrac{\varepsilon}{\varepsilon_c}$ relative strain,

      D_0 = initial damage, existing before loading.

The constants C_1, C_2, C_3 and ν_1 can be determined experimentally as follows :

$C_1 = 1 - \lambda_t = C_2$

$C_3 = \lambda_t / (\eta_u - 1)$ (14)

$\nu_1 = \lambda_t / (1 - \lambda_t)$

where : $\lambda_t = f_t / (E_i \cdot \varepsilon_c)$, denotes linearity   (15)

and : $\eta_u = \dfrac{\varepsilon_u}{\varepsilon_c} = $ ultimate relative strain,

      E_i = initial modulus of elasticity,

      f_t = maximum nominal stress = tensile strength.

This formulation (Fig. 6) makes it possible to determine the $\sigma - \varepsilon$ curve very easily based only on two material parameters λ_t and ε_u, the latter being calculated by means of a fracture mechanics analysis as described in 3.2.

$$\varepsilon_u = \frac{2G_D (1 - D(\varepsilon_c))}{f_t \cdot L} + \varepsilon_c - \frac{f_t}{E_i}$$ (16)

where : G_D = damage energy - equal the energy necessary to break a notched
             concrete specimen under three - point bending ;
      L = length of the strained element.

This formulation also takes into account the well known fact that the descending branch of the $\sigma - \varepsilon$ curve is size-dependant. LØLAND (ref. 8) denotes this model the Continuous Damage model (CDm).

Note : LØLAND's model fits very well the experimental $\sigma - \varepsilon$ curves as far as the ascending branch is concerned (Fig. 7). The linear approximation of the descending branch, however, seems to be too simple for analysing crack instability, but for any other purpose it might be fair.

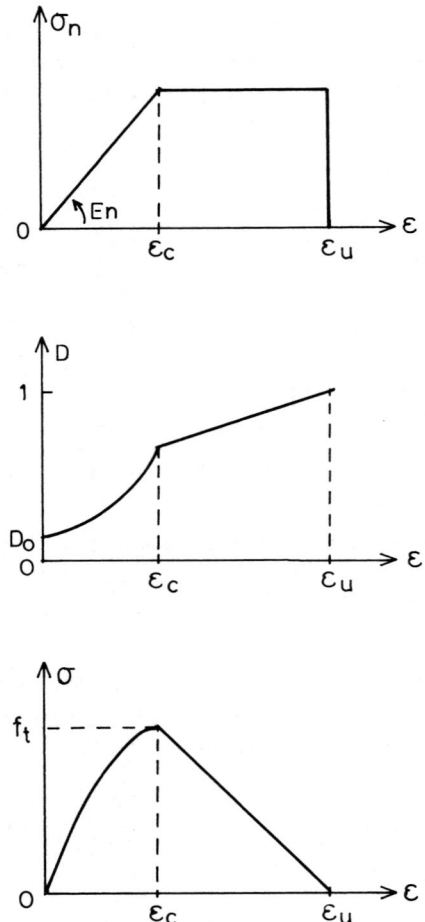

Fig. 6 - Mathematical modeling according to LØLAND.

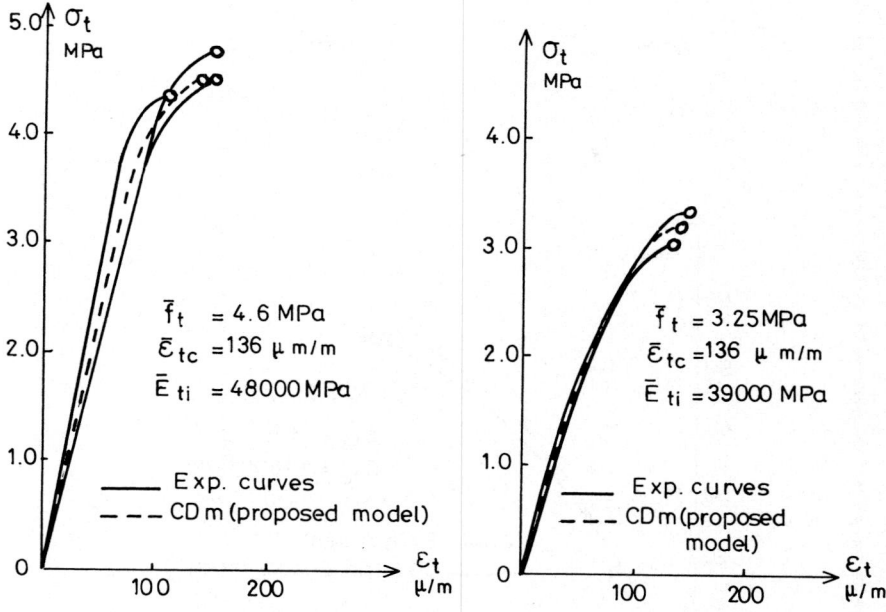

Fig. 7 - Experimental and CDm-calculated stress-strain curves for concrete
in uniaxial tension according to LØLAND.

3.3 <u>Application of damage theory to the modeling of the behaviour of concrete under uniaxial compression</u>. The previous proposals might be extended to
the uniaxial compression state of stress, assuming crack development in the
direction parallel to the loading axis. This is successfully done by LØLAND
(ref. 7), but in spite of good correlation between experimental and calculated
damage (Fig. 8) he states that the phenomenological basis is not as realistic for uniaxial compression as it is for uniaxial tension. LØLAND uses the
same mathematical formulation for uniaxial compression and tension.

Several authors have proposed different mathematical formulations to predict
the stress-strain curve in uniaxial compression on a pure empirical basis.
Some of these, however, connect the changes in elastic modulus, energy dissipation, etc... during loading to damage creation in the material (e. g. SPOONER
and DOUGILL (ref. 12), COOK and CHINDAPRASIRT (ref. 13), RODUKO (ref. 14) and
others). However, the phenomenological basis is relatively weak for all of
these. In spite of this, the damage concept seems very useful to predict changes in material properties.

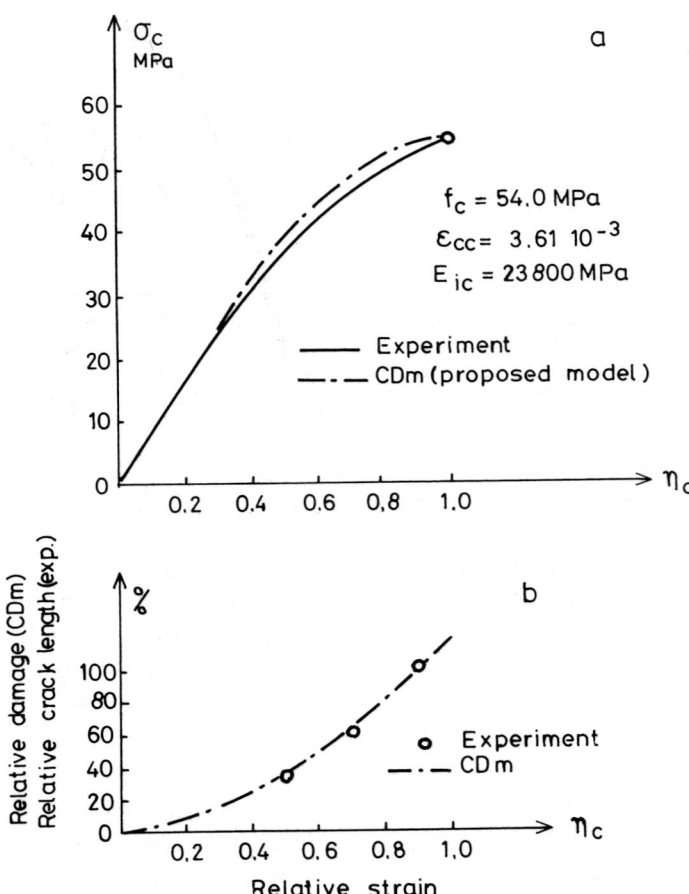

Fig. 8 - Stress-strain and damage-strain curves for concrete in uniaxial
compression according to LØLAND ($\eta_c = \varepsilon/\varepsilon_c$).

3.4   Application of damage theory to the modeling of the behaviour of concrete under multiaxial stresses.

For the multiaxial state of stress, one has to incorporate important variations of POISSON's ratio, and also to generalize the unidimensional damage
concept in a multidimensional way.

This requires transformation of the scalar damage parameter, D, into a tensorial damage parameter, $[d]$. This approach is followed by BENOUNICHE (ref. 15)
who proposes an anisotropic three dimensional formulation originally suggested
by CORDEBOIS (ref. 16) and modified to remain consistent with the uniaxial tension formulation.

Multiaxial formulation of damage.

The damage is considered now as a tensorial parameter $[dij]$. The previous
equation (3) becomes :

$$\Sigma_n = \left[I - d \right]^{-1/2} \Sigma \left[I - d \right]^{-1/2}$$

(17)

where : Σ_n, Σ = tensors of stress

$\qquad\qquad$ I   = unitary tensor

$\qquad\qquad$ $d = [dij]$

Tensorial damage parameter for concrete under uniaxial compression :
BENOUNICHE's model.

BENOUNICHE uses a tensorial damage parameter because of the important variation of POISSON's ratio due to the development of cracks in a direction parallel
to the loading axis. This tensorial damage parameter $[dij]$ takes into account
the assumption of axial symmetry and isotropy for concrete under uniaxial
compression :

$$[dij] = \begin{bmatrix} D_1 & 0 & 0 \\ 0 & D_2 & 0 \\ 0 & 0 & D_2 \end{bmatrix}$$

(18)

The elastic characteristics of the damaged material can be calculated from
the equations (17) and (18), and we obtain ;

$$E = E_n (1 - D_1)^2$$

(19)

$$\nu = \nu_n \frac{1 - D_1}{1 - D_2}$$

(20)

where :

E_n = YOUNG's modulus, undamaged material
ν_n = POISSON's ratio, undamaged material
D_1, D_2 = longitudinal and transversal damage parameters

The longitudinal and transversal damage parameters can be written as follows :

$$D_1 = 1 - \left[\frac{E}{E_n} \right]^{1/2} \tag{21}$$

$$D_2 = 1 - \frac{\nu_n}{\nu} \left[\frac{E}{E_n} \right]^{1/2} \tag{22}$$

D_1 and D_2 can be calculated during loading cycles from the initial E_n and ν_n values of YOUNG's modulus and POISSON's ratio of the undamaged material and from the values E and ν measured when unloading.

Note : BENOUNICHE's model fits well with the experimental $\sigma - \varepsilon$ curves (see (ref. 15)) as well on the ascending part as on the descending branch.

4   Combination of damage theory and fracture mechanics

4.1   <u>JANSON and HULT's combination for tensile failure</u>. JANSON and HULT (ref. 5) proposed several combinations of damage theory and fracture mechanics in general without mentioning any specific material. Particularly they proposed to calculate the limit tensile stress (or fracture stress) from the energy condition of fracture mechanics.

$$U = W \tag{23}$$

where : U = available deformation energy ;
and   : W = energy needed to create new crack surfaces.
W is given by fracture concepts as :

$$W = 2 \mu A \tag{24}$$

where : μ = specific surface energy.

U is calculated assuming that stored strain energy in close vicinity of the fracture surfaces will be available to supply the work needed to create new surfaces in fracturing. The well known expression for the ideal fracture strength of flawless, crystalline, linearly elastic materials emerges by assuming a width of 2b (where b is the lattic spacing) on each side of the separation supply the needed energy. Hence :

$$U = 4b A \overline{U} \tag{25}$$

where : $\overline{U} = \frac{1}{2} f_t \varepsilon_c = \frac{f_t^2}{2E}$ = strain energy density.

Using eq. (23), we obtain for materials with no damage creation ;

$$f_t = \sqrt{\frac{\mu E}{b}} \tag{26}$$

For "real" materials with damage creation -assuming strain to be completely
reversible and damage to be completely irreversible- the strain energy density
$\overline{U}$, can be written as (Fig. 9) :

$$\overline{U} = \frac{f_t^2}{2E \, (1 - D_c)}$$
(27)

where : D_c = "damage capacity" corresponding to ε_c.

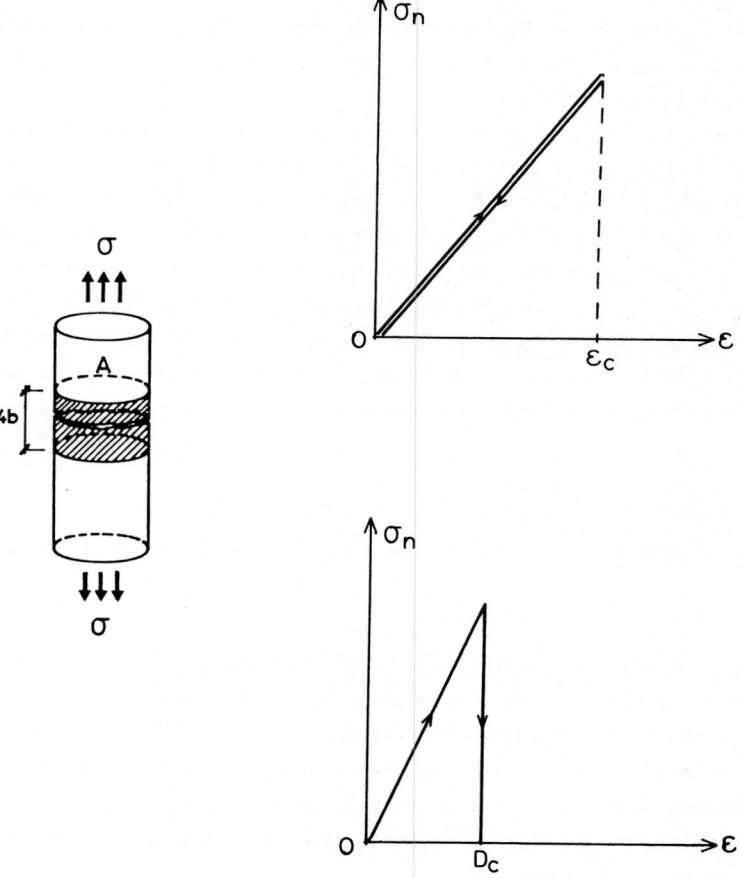

Fig. 9 - Idealized σ_n - ε curve and σ_n - D curve according to JANSON and HULT.

Substituting eq. (27) and (23), the following fracture stress emerges for materials where damage is developed :

$$f_t = \sqrt{\frac{\mu E \ (1 - D_c)}{b}} \tag{28}$$

4.2 LØLAND's combination for the tension and compression stress states.
LØLAND (ref. 7) proposes a combination of the Continuous Damage model (CDm) and ordinary fracture mechanics to describe the failure mechanism and the descending branch of the $\sigma - \varepsilon$ relationship under uniaxial tension and compression.
Uniaxial tension.

A modified version of the Fictitious Crack Model (FCM) (ref. 17)- described elsewhere in this report- iş used to characterize the cracking process ahead of a macrocrack in the material.

The most essential point here is that fracture energy is consumed during damage creation. This means that instead of relaxation of stresses as the fracture zone (or fictitious crack as in FCM) opens, the area which transmits stresses continuously decreases . This corresponds to growing damage. The damage energy, G_D, denotes the energy required to create one unit of damage in the fracture zone. G_D is determined experimentally through three-point bending of notched specimens as for G_C in the FCM. Using linear accumulation of damage in the fracture zone -related to the fracture zone extension- the ultimate fracture zone extension, w_u, can be written as :

$$w_u = \frac{2G_D}{f_t} \tag{29}$$

The ultimate strain for a concrete specimen in uniaxial tension can then be written as :

$$\varepsilon_u = \varepsilon_c - \varepsilon_r + \varepsilon_f \tag{30}$$

The various strain components are shown in Fig. 10.

This yields equation (16) for the ultimate tensile strain, ε_u. Thus it is possible to estimate the complete stress-strain curve for uniaxial tension. LØLAND points out, however, that it is very difficult to experimentally determine the descending branch of the stress-strain curve in a stable manner for ordinary concrete with acceptable aggregate size, and he doubts it would be possible to do this satisfactorily.

LØLAND also proposes to use either ultimate fracture zone extension, w_u, or ultimate tensile strain, ε_{tu}^{100} -for a specified specimen length of 100 mm- as toughness parameters for concrete. In this context, it is interesting to note the lack of correlation between G_c and f_t (Fig. 11), while there is a

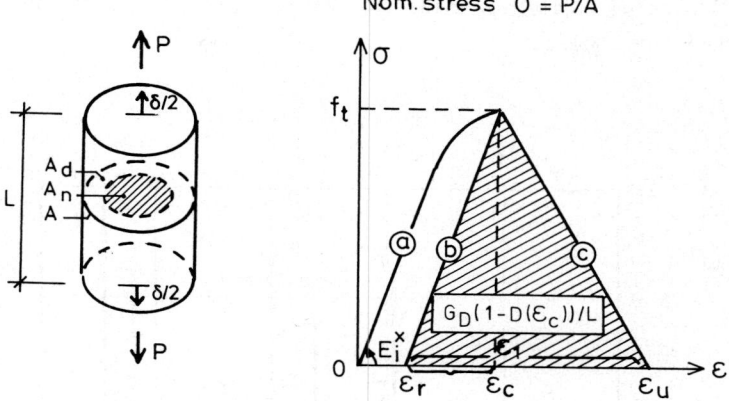

Fig. 10 - CDm-description of the nominal stress-strain relationship for an uniaxial loaded concrete specimen under constant strain increase according to LØLAND. Damage energy, G_D, and some strain values are shown.

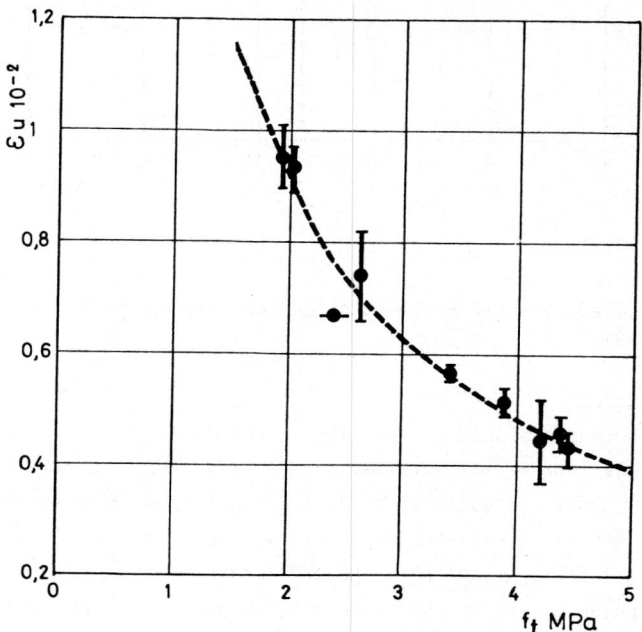

Fig. 11 - Ductility, ε_{tu}^{100}, related to tensile strength according to LØLAND and GJØRV.

satisfactory correlation between f_t and ε_{tu}^{100} (denoted "ductility") as shown in Fig. 12. Consequently, the correlation between G_c and ε_{tu}^{100} is rather poor because ε_{tu}^{100} includes damage outside the fracture zone (or fictitious crack) while G_c does not.

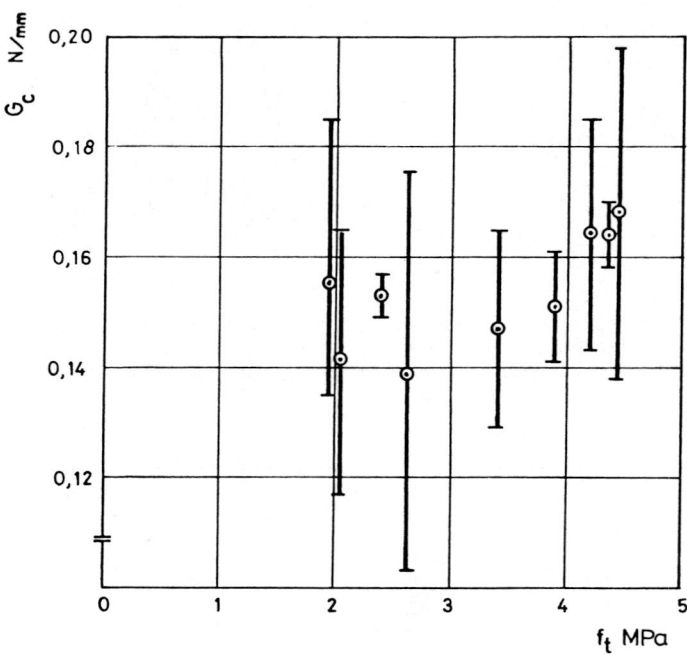

Fig. 12 - G_c related to tensile strength for the same specimens shown in Fig. 11.

Uniaxial compression.

  LØLAND also includes fracture mechanics in the mathematical modeling of the stress-strain relationship for uniaxial compression. This, however, is done on a more empirical basis. In spite of this, the description fits very well the ascending experimental stress-strain curves, as shown in Fig. 13. He also proposes empirical formulations for repeated loads.

  The same formulae as for the uniaxial tension are used, but different cons-tants are now determined from compression tests. The ultimate tensile strain is calculated using the aforementioned damage energy, G_D, which gives the following formula :

$$\frac{\varepsilon_u}{\varepsilon_c} = 1 + \frac{1.67}{f_c}\left[\frac{\alpha G_D}{\varepsilon_c} - \frac{f_c}{\lambda_c}\left[0,5 - \frac{(1 - \lambda_c)^2}{2 - \lambda_c}\right]\right] \tag{31}$$

where : α = empirical constant which is very close to 1.0 ;

$\lambda_c = \dfrac{f_c}{E_i\,\varepsilon_c}$, denoted "linearity".

In compression, ε_u denotes the strain value where the stress on the descending branch of the σ - ε curve has reached a level of 0.2 f_c.

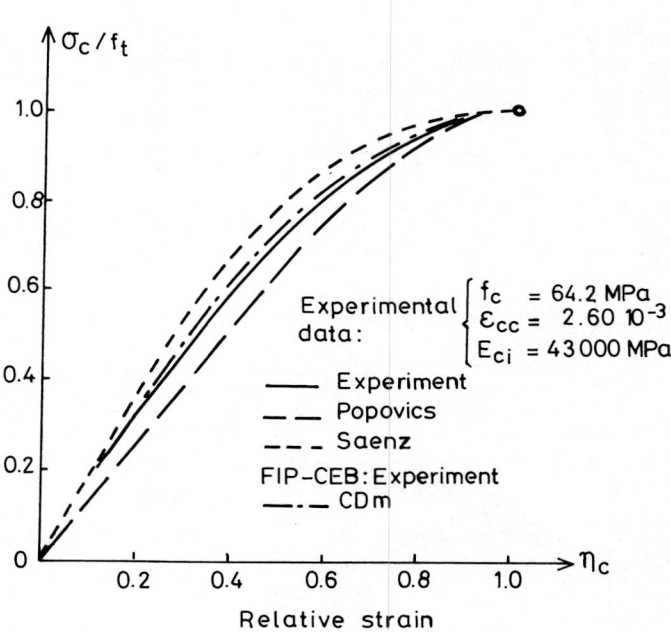

Fig. 13 - Typical uniaxial stress-strain curve for high quality concrete in compression compared to Continuous Damage model (CDm) and different others analytical expressions.

LØLAND's formulation provides a good fit with experimental σ - ε curves as e. g. can be seen in Fig. 14. Fig. 15 shows a formulation for repeated loads proposed by LØLAND, which also fits the actual behaviour very well.

360

This, however, is based on more empirical considerations.

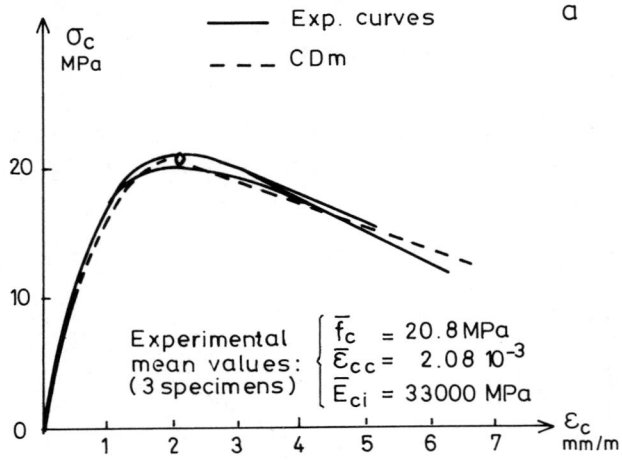

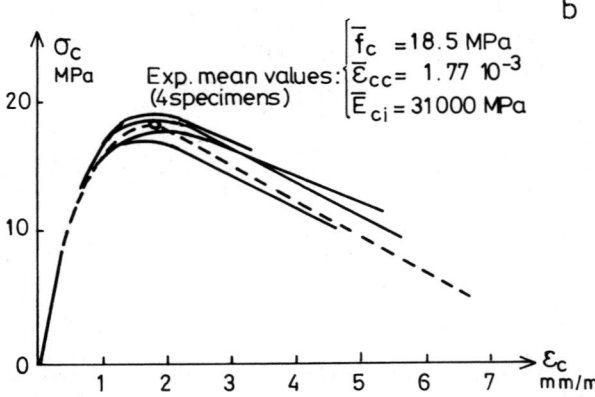

Fig. 14 - Stress-strain curves for concrete in uniaxial compression related to CDm description.

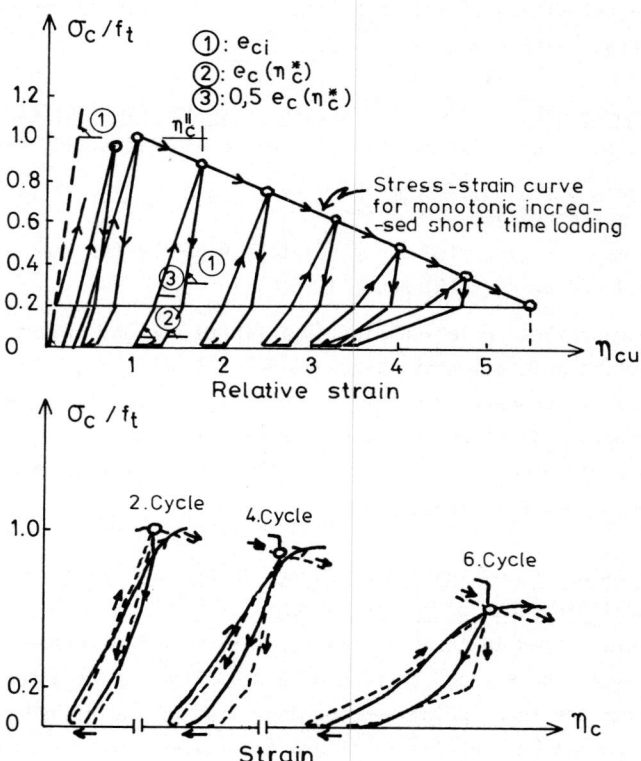

Fig. 15 - CDm modeling of repeated compressive loadings.

4.3 <u>MAZARS's criterion for failure of concrete structures in flexure</u>.
MAZARS (ref. 10) combines tensile behaviour relationships defined in section 2
with a generalized cracking criterion.
<u>Generalized tensile strain</u>.

A combination of strains, $\bar{\varepsilon}$, is defined as follows :

$$\bar{\varepsilon} = \sum_{i=1}^{3} \left[<\varepsilon_{ii}>^2 \right]^{1/2} \tag{32}$$

where : $<\varepsilon_{ii}> = \varepsilon_{ii}$, principal strain, if $\varepsilon_{ii} > 0$ (tensile strain)
$<\varepsilon_{ii}> = 0$, if $\varepsilon_{ii} < 0$.

This formulation is used instead of the uniaxial tensile strain, ε, to represent the state of strain which is actually triaxial as far as concrete structures are concerned. It might be called MAZARS's generalized tensile strain. Notice, however, that this does not directly to physical concepts. It is connected more to the definition of a generalized tensile strain criterion :

$\bar{\varepsilon} > \bar{\varepsilon}_f$: crack extension.

Damage theory and generalized tensile strain criterion combination.

MAZARS replaces ε (uniaxial tensile strain) in equations (10) and (11) (Section 2) by $\bar{\varepsilon}$. Hence :

$\bar{\varepsilon} < \varepsilon_c$: elastic region

$\bar{\varepsilon} > \varepsilon_c$: damaged region. The $D - \bar{\varepsilon}$ relationships are the same as previously defined by eq. (10) and (11).

Consideration of these relationships in a finite element program can be useful to study the initiation and propagation of cracks in a concrete structure. It suffices to compare at every step of the study, $\bar{\varepsilon}$ to ε_c, as follows :

$\bar{\varepsilon} < \varepsilon_c$: the program continues in the linear elastic zone ;

$\bar{\varepsilon} > \varepsilon_c$: the program properly modifies the stiffness matrix according to the damage theory relationships.

Application of MAZARS's combination of damage theory and fracture mechanics to describe the real behaviour of concrete beams.

The beams were tested in three and four point-bending, respectively. Their span was 1.40 m and their cross section 0.22 x 0.15 m^2. The reinforcement in some of the beams for three-point bending consisted of two ribbed bars $\emptyset$ 12 mm. The strain in the extreme tension fiber and the maximum deflection were measured.

The results show -after adjustment of ε_c- a satisfactory correlation between experiment and finite element calculation, except for the reinforced beam (Fig. 16, 17, 18).

It is worth noticing that an adjustment of ε_c has been necessary to obtain satisfactory correlation between theory and experiment. The value of ε_c varies noticeably between two tests, e.g :

$\varepsilon_c = 115.10^{-6}$ for the reinforced beam,

$\varepsilon_c = 70.10^{-6}$ for the plain beam.

These contradictions show clearly the difficulties arising for this particular formulation when the behaviour of damaged concrete in structures should be described with only one damage parameter.

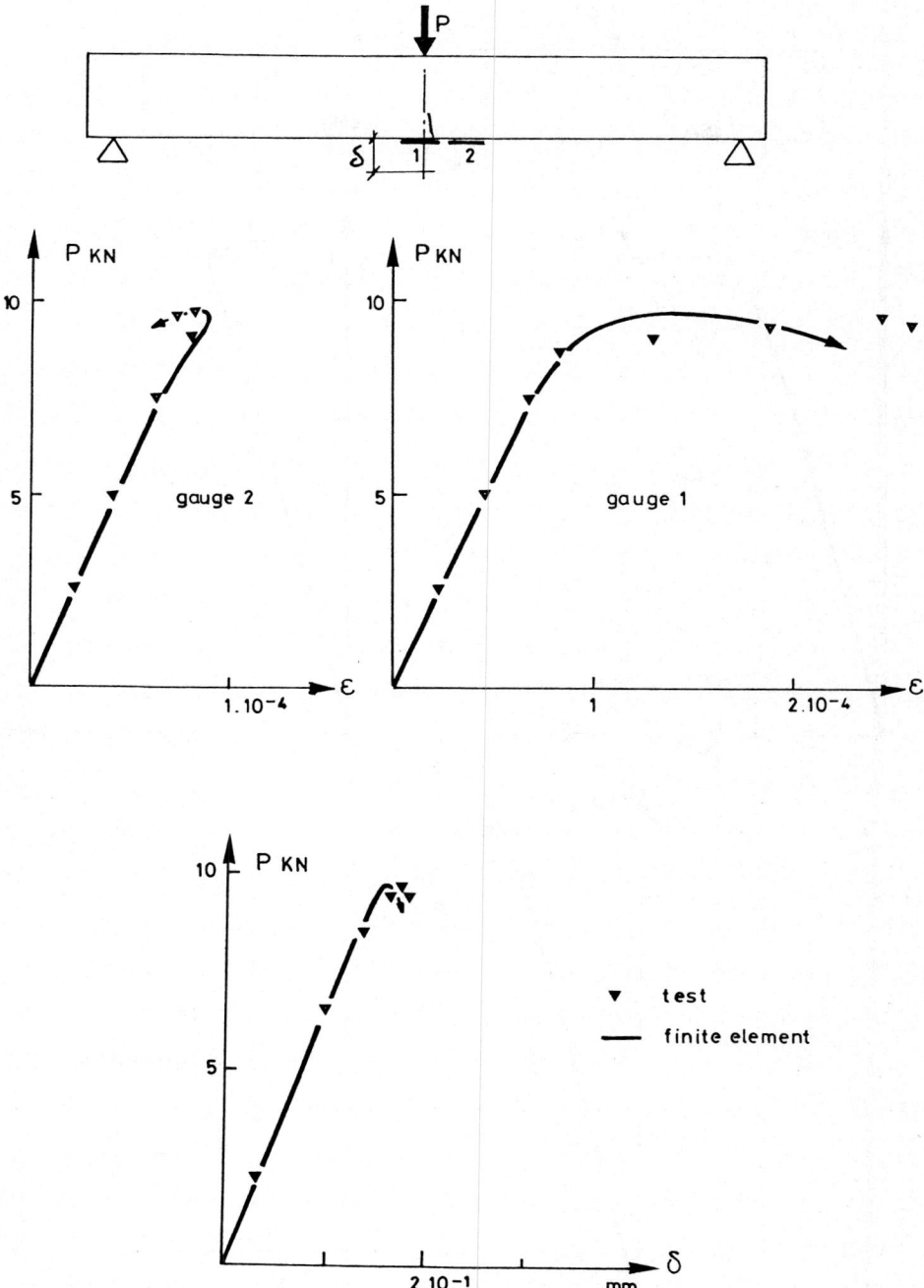

Fig. 16 - Experimental and calculated P - δ and P - ε curves for an unreinforced concrete specimen under three-point bending according to MAZARS.

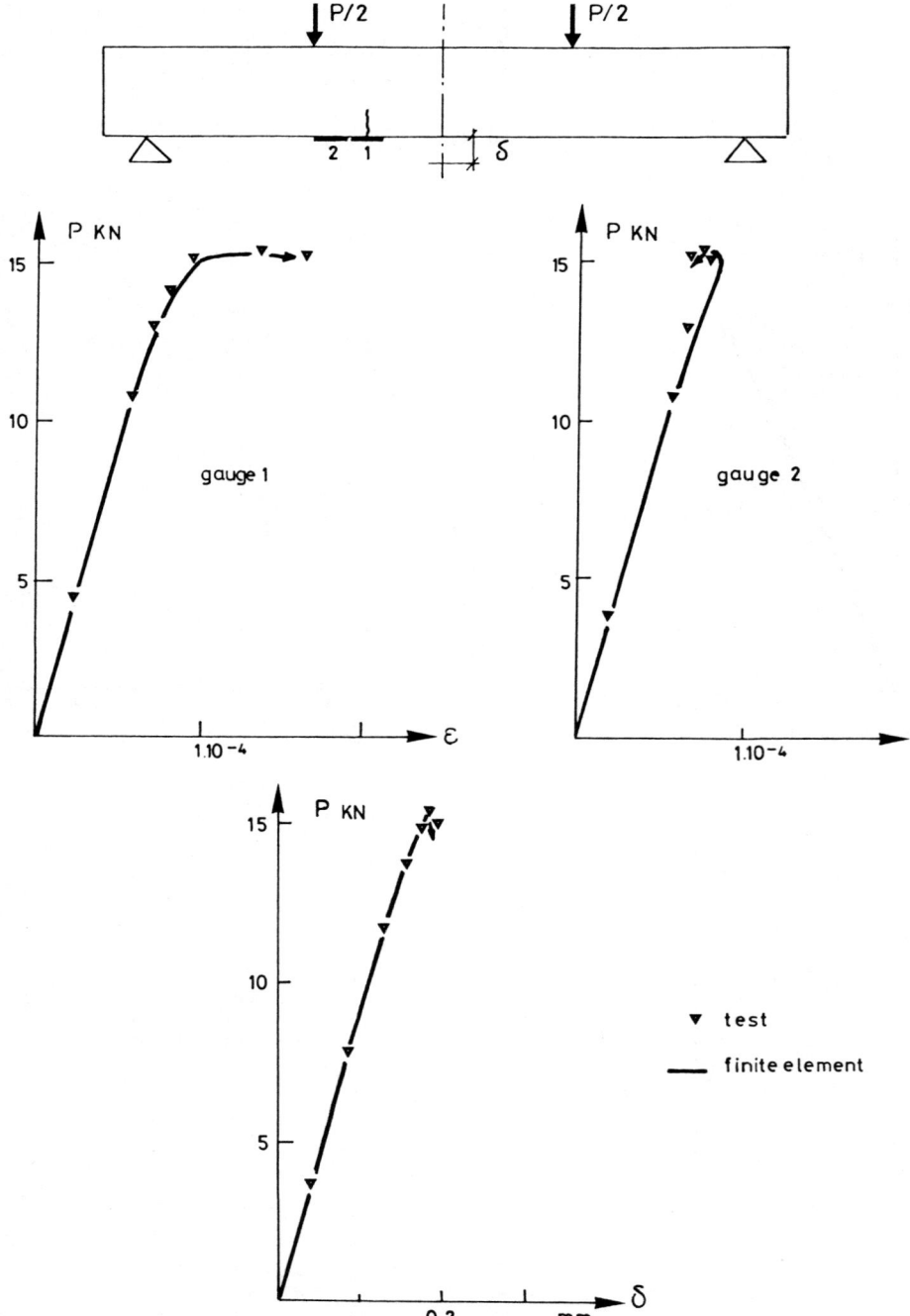

Fig. 17 - Experimental and calculated P - δ and P - ε curves for an unreinforced concrete specimen under four-point bending according to MAZARS.

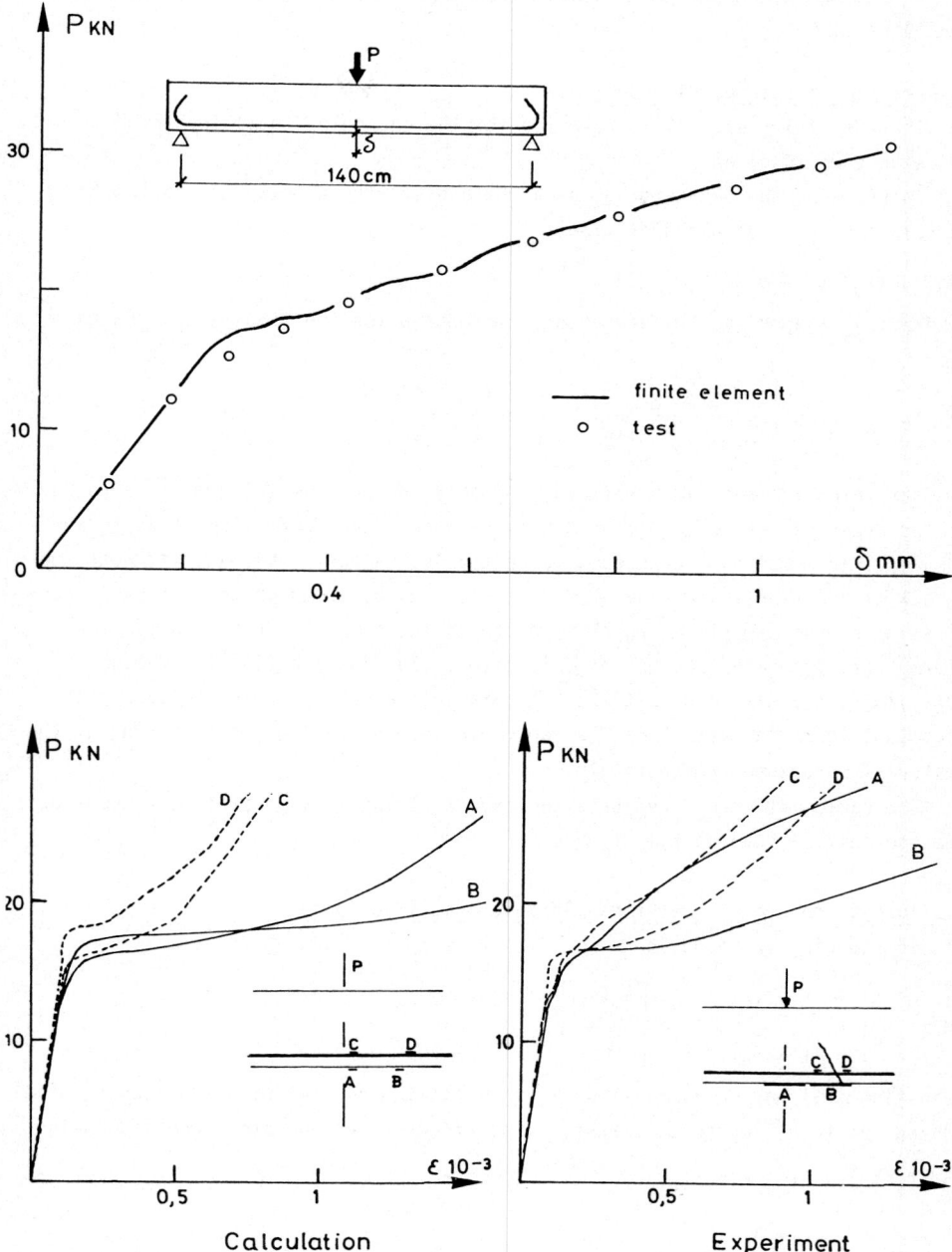

Fig. 18 - Experimental and calculated P - δ and P - ε curves for a reinforced concrete specimen under three-point bending according to MAZARS.

4.4 ZAITSEV and SCERBAKOV's combination for the failure of concrete under sustained load.

Generalized fracture criterion.

In (ref. 18) (ref. 19), a relation for the change of total length of cracks, S, with time is given.

Introducing the parameter q_o as a measure of stress intensity caused by an external stress σ applied at time t_o

$$q_o = \sigma(t_o) \ (\pi \ E \ (t_o) \ \mu_s \ (t_o))^{-1/2}$$

where : μ_s = specific surface energy, we obtain for the final crack length :

$$S \ (t, \ t_o) = Fq \left[\frac{R(t_o)}{m(t, \ t_o).R(t)} \quad \sqrt{\frac{E(t)}{\tilde{E}} \ \frac{\sigma(t)}{\sigma(t_o)}} \right] \qquad (33)$$

According to (ref. 18) (ref. 19) $(\tilde{E})^{-1}$ is an operator which takes into account creep of concrete. R(t) and E(t) represent the short term strength and the elastic modulus of concrete, respectively. Creep in the material near the crack tips not only increases the crack length, but reduces the stress concentration at the same time, which leads to an increase in strength R(t). The term $m(t, \ t_o)$ takes account of this effect. In equation (33) the ratios $R(t_o)/m(t, \ t_o) \ R(t)$ and $(E(t)/\tilde{E})^{1/2}$ take into account (as in the case of a constant load) the effect of the two processes - hardening and softening of the material structure, respectively.

The ratio $\sigma(t)/\sigma(t_o)$ reflects the effect of load change at time t compared to the initial load at time t_o.

Failure of concrete under time-variable sustained load.

Introducing the function :

$$M(t, \ t_o) = \frac{\sigma(t)}{m(t, \ t_o) \ R(t)} \left[\frac{E(t)}{\tilde{E}} \right]^{1/2} \qquad (34)$$

equation (33) may be simplified. We then obtain the following equation for the change of total length of cracks, S, as affected by concrete creep and ageing :

$$S \ (t, \ t_o) = F(q_o.M(t, \ t_o)/\rho) \qquad (35)$$

where : $\rho = \dfrac{\sigma(t_o)}{R(t_o)}$

In (ref. 18) (ref. 19), the equality of critical total length of cracks under short-term loads, S_s^*, and under constant sustained load, S_c^*, is proposed, we may apply this criterion to the case of a time-variable sustained load : S_v^*. If the initial intensity of such a load is ρq_o, then the following equality should be satisfied at the moment of failure under time-variable sustained load (similar to the case of short-term and constant sustained load) :

$$S_v^* (t, t_o) = F (q_o/\rho) \tag{36}$$

i. e. in accordance with eq. (35) for the moment of failure :

$$M (t, t_o) = 1 \tag{37}$$

The change of the function $M (t, t_o)$ for the case of a constant sustained load is shown in Fig. 19, where n_* is the relative long-term strength. It can be shown that the values of $M (t, t_o)$ are always in the range of 0 to 1, whereby $M (t, t_o) = 0$ for an unloaded material ($\rho = 0$) and $M (t, t_o) = 1$ for failure. Therefore, the function $M (t, t_o)$ may be defined as a "measure of material damage".

With this concept, the time to failure under time-variable load may be determined, as also shown schematically by LØLAND (ref. 7).

According to equations (34) and (37), the following equality should be satisfied at the moment of failure.

$$\frac{\sigma (t)}{m (t, t_o) R (t)} \left(\frac{E(t)}{\tilde{E}}\right)^{1/2} = 1 \tag{38}$$

The time to failure may be determined form equation (36) if all the time-dependent empirical functions are known.

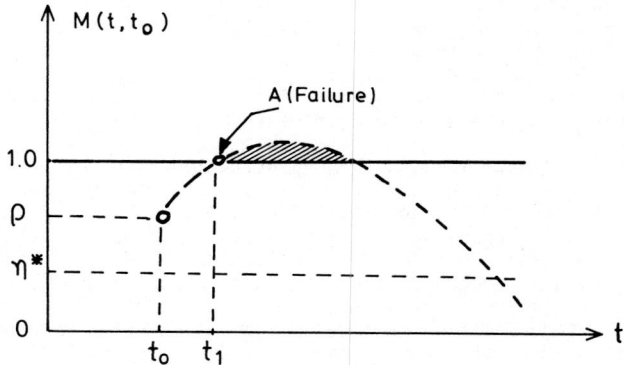

Fig. 19 - Criterion of failure of concrete under constant sustained load after ZAITSEV and SCERBAKOV. (M (t, t_o) = measure of destruction).

5 - Conclusions

Damage theory undoubtedly allows an appreciation of the behaviour of concrete beyond "the elastic limit". Further it makes it possible to create a phenomenological platform for the mathematical modeling. This is essential for the understanding of the different processes contributing to the nonlinearity in the behaviour of concrete.

However, the complex phenomena, particularly in the cracking process of concrete structures under multiaxial states of stress, are difficult to describe with the aid of one damage parameter only. Hence, we must not be surprised by some unsatisfactory correlations between theory and experiments.

REFERENCES

1   Kachanov, L. M., "On the Creep Fracture Time", Izv. AN SSSR, Otd, Tekhn. Nauk, No. 8, 1958, pp. 26-31 (in Russian).
2   Rabotnov, I. N., "On the Equations of State for Creep", Progress in Applied Mechanics - The Prager Anniversary Volume, 1963, pp. 307-315.
3   Broberg, H., "Damage Measures in Creep Deformation and Rupture", Swedish Solid Mech. Rep., 1974.
4   Lemaître, J. et CHABOCHE, J. L. "Aspect Phénoménologique de la Rupture par Endommagement", Journal de Mécanique Appliquée, Vol. 2, No. 3, 1978, pp. 317-365.
5   JANSON, J. et Hult, J., "Fracture Mechanics and Damage Mechanics : A Combined Approach", J. de Mécanique Appliquée, Vol. 1, 1977, pp. 69-84.
6   Loland, K. E. and Gjorv, O. E., "Ductility of Concrete and Tensile Behaviour", Presented at the Annual Convention, ASCE, Florida, Oct. 1980.
7   Loland, K. E., "Mathematical Modelling of Deformational and Fracture Properties of Concrete based on Damage Mechanical Principles - Application on Concrete with and without Addition of Silica Fume", Dr. Ing. Thesis, Univ. of Trondheim, Norway, Sept. 1981.
8   Loland, K. E., "Continuous Damage Model for Load-Response Estimation of Concrete", Cement and Concrete Research, Vol. 10, 1980, pp. 395-402.
9   Germain, P., "Mécanique des Milieux Continus" Tome I, Théorie Générale, Paris, 1973, Masson.
10 Mazars, J., "Mechanical Damage and Fracture of Concrete Structures" 5th Int. Conf. on Fracture, Cannes 1980.
11 Terrien, M., "Etude des Bruits Emis par le Béton au cours de la Fissuration du Mortier, du Granulat ou de leur interface". Rapport de Contrat - 1978 Ecole Polytechnique - Laboratoire de Mécanique des Solides - 91128 Palaiseau France.
12 Spooner, D. C. and Dougill, J. W., "A Quantitative Assessment of Damage Sustained in Concrete during Compressive Loading", Mag. of Concr. Res., Vol. 27, No. 92, Sept. 1975, pp. 151-160.
13 Cook, D. J. and Chindaprasirt, P., "A Mathematical Model for the Prediction of Damage in Concrete" Submitted for publication in Cem. and Concr. Res.
14 Rokugo, K., "Energy Approach to the Failure of Concrete and Concrete Members", Ph. D. thesis, Univ. of Kyoto, Japan, Nov. 1979.
15 Bénouniche, S., "Modélisation de l'Endommagement du Béton Hydraulique par Microfissuration en Compression", Thèse 3ème Cycle, Univ. Paris VI, Déc. 1979.
16 Cordebois, J. P. et Sidoroff, F., "Elasticité Anisotrope induite par l'Endommagement", Euromech 115 Anisotropie, Grenoble, 1979.

17 Hillerborg, A., Modéer, M. and Petersson, P. E., "Analysis of Crack Forma-
    tion and Crack Growth by means of Fracture Mechanics and Finite Elements",
    Cem. and Concr. Res., Vol. 6, No. 6, Nov. 1976, pp. 773-781.
18 Zaitsev, Y. and Wittmann, F. H., "Fracture of Porous Viscoelastic Materials",
    3rd Int. Conf. on Fracture, Munich, 1973, Dusseldorf, 1973, IX-323.
19 Zaitsev, Y. and Scerbakiv, E. N., "Creep Fracture of Concrete in Prestres-
    sed Concrete Members during Manufacture", 4th Int. Conf. on Fracture,
    Waterloo, 1977, 3-1219.

DAMAGE MECHANICS REFERENCES, NOT REFERRED TO IN THE TEXT.

Bostrom, P. O., "Damage Induced Instability in Beam Bending", Int. J. Non Linear
Mech., Vol. 11, 1976, pp. 303-311.

Dufailly, J. , "Modélisation Mécanique et Identification de l'Endommagement
Plastique des Métaux", Thèse 3ème Cycle, Univ. Paris VI, 1980.

Janson, J., "Dugdale-Crack in a Material with Continuous Damage Formation",
Eng. Fract. Mech., Vol. 9, 1977, pp. 891-899.

Janson, J., "A Continuous Damage Approach to the Fatique Process", Eng. Fract.
Mech., Vol. 10, 1978, pp. 651-657.

Janson, J., "Damage Model of Crack Growth and Instability", Eng. Fract. Mech.,
Vol. 10, 1978, pp. 795-806.

Kachanov, M. L., "Deformation of Medium with Cracks", Izvestia VN11G, Leningrad,
USSR, Vol. 99, 1972, pp. 195-210 (in Russian).

Kachanov, M. L. "Continuum Model of Medium with Cracks", Journ. of the Eng.
Mech. Div., ASCE, Vol. 106, No. EM5, Oct. 1980, pp. 1039-1051.

Krajcinovic, D., "Distributed Damaged Theory of Beams in Pure Bending", J. Appl.
Mech., Vol. 46, 1979, pp. 592-596.

Krajcinovic, D., Srinivasan, M. G., Fonseka, G. U., Valentin, R. A., "Progressive Damage of Dynamically Loaded Brittle Rod", Presented at the Annual
Convention, ASCE, Florida, 1980.

Lemaître, J. et Mazars, J., "Modélisation du Comportement et de la Rupture du
Béton", 4ème Symposium Franco-Polonais de Mécanique, Marseille, 1980.

Odquist, F. K. G., "Mathematical Theory of Creep and Creep Rupture", Clarendon
Press, Oxford, 1966.

Rabotnov, Y. N., "Creep Rupture", Proc. XII Int. Congr. Appl. Mech. Stanford-
Springer, 1968.

Srinivasan, M. G., Krajcinovic, D., Fonseka, G. U. and Valentin, R. A., "The
Distributed Damage Theory and its Application in Dynamically Loaded Structures",
Abstracts, 16th Annual Meeting, Society of Engineering Science, Inc., Sept. 1979
Northwestern University.

Vakulenko, A. R. and Kachanov, M. L., "Continuum Model of Medium with Cracks",
Mechanika Tverdogo Tela (Mechanics of Solids), No. 4, Moscow, USSR, 1971,
pp. 159-166 (in Russian).

Fracture Mechanics of Concrete,
edited by F.H. Wittmann, 1983
Elsevier Science Publishers B.V., Amsterdam — Printed in The Netherlands

Chapter 5.1
FRACTURE MECHANICS PARAMETERS OF HARDENED CEMENT PASTE, AGGREGATES
AND INTERFACES
by. S. Ziegeldorf

1.   INTRODUCTION
    One potential approach to a better understanding of the fract-
ure of hardened cement paste, mortar and concrete is given by
fracture mechanics. The most important fracture mechanics parameter K_{IC} is a critical value of the stress intensity factor,
which can be regarded as a one-parameter description of the stress
and displacement field in the region of the crack tip. The evaluation of K_{IC} and of most other fracture mechanics parameters pre-
sumes a linear elastic, homogeneous and isotropic material.
Furthermore, the propagation of only a single crack is assumed.
All these assumptions are incorrect for concrete and even hardened
cement paste. Nevertheless, several previous investigators reported a surprising agreement between experiment and theory.
    This report presents a review of relevant information regarding
experiments on the components of concrete, i.e. on hardened cement
paste, aggregates and interfaces. Many tests, specimens, techniques and analysis procedures have been proposed for these materials. The objective of this report is
1) to review the different test methods most widely used;
2) to review the experimental results received by different investigators;
3) to summarize the data with regard to applicability of LEFM to
   the components of concrete.

2.   FRACTURE MECHANICS METHODS AS APPLIED TO HARDENED CEMENT
      PASTE
    Hardened cement paste can be regarded to be a model material
concerning the application of linear elastic fracture mechanics to
cementitious materials. Contrary to concrete, hardened cement
paste is homogeneous and exhibits rather brittle behavior in
tension as well as in compression. Thus the most important pre-
requisite for the applicability of linear elastic fracture mechanics to this material seems to be fulfilled.

The following fracture mechanics parameters have been evaluated on hardened cement paste in the past:

- K_{IC} (fracture toughness)
- G_{IC} (critical value of energy release rate)
- J_C (critical value of J-Integral)
- γ (fracture surface energy).

The COD- and R-curve approach have till now been applied to concrete (see the following paper by R.N.Swamy), however not to hardened cement paste.

Using the relationship (applicable for plane strain and linear elastic materials)

$$G \cdot E = (1 - \nu^2) \cdot K^2 \tag{1}$$

values of G_{IC} can be calculated from K_{IC}-values.

2.1 Methods of Evaluating K_{IC}

2.1.1 General remarks. Different types of specimen have been used for the evaluation of K_{IC} and G_{IC} for hardened cement paste. The wide variety of specimen forms is partly due to the fact that - unlike with metallic materials (ASTM E 399) - no standard speci- mens are recommended for the determination of fracture mechanics parameters of cementitious materials. The same holds true for the experimental procedure in such a test. It is noteworthy that a wide discrepancy of results is observed in experiments on hardened cement paste in which different experimental techniques were used. Thus generally caution is recommended when comparing fracture chanics parameters obtained by using different specimen forms. According to Kasperkiewicz (ref. 1) it has till now not been checked how various non-elastic effects influence the results of particular tests. For example the effect of creep of cement paste at the crack tip may be different in tests with different specimen forms. It has to be accepted furtheron that the microcrack form- ation ahead of the crack tip is strongly influenced by the experi- mental conditions.

In the following the most important specimen geometries and the relevant methods of evaluation are described.

2.1.2 Notched beam (ref. 2-16). The most commonly used specimen type for K_{IC}- or G_{IC}-evaluation is the notched beam loaded in 3-point- or 4-point-bending (Fig. 1). The advantages of the notched

beam are its economic fabrication and its relatively simple testing procedure. The notches are fabricated by inserting thin foils
into the specimen mould or by a saw cut into the hardened specimen.

When performing a test a gradually increased load is applied to
the notched beam until a stress level is reached which results in
crack propagation. Dependent on the notch depth and the stiffness
of the material and of the loading frame the resultant load-dis-
placement-diagrams exhibit catastrophic, semi-stable or stable
fracture (Fig. 2)(ref.2). Prediction methods for the occurrence of
catastrophic and stable fracture, resp. have been proposed by several authors (ref. 16-18).

Assuming a beam of cross-section b x W and of crack length a
and denoting the critical nominal stress at the crack tip by σ_{nc}
the fracture toughness can be given by:

$$K_{IC} = \sigma_{nc} \sqrt{\pi \cdot a} \, Y(a/W) \tag{2}$$

By using the relationship

$$\sigma_{nc} = \frac{6M_c}{b(W-a)^2} \tag{3}$$

with M_c = critical bending moment.

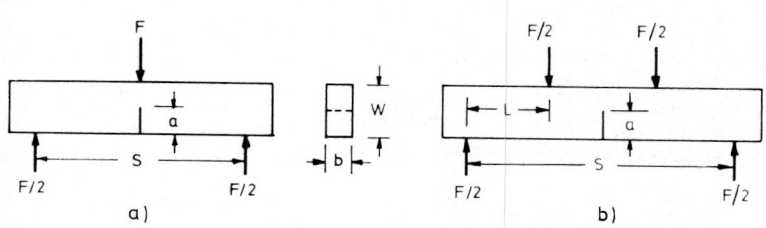

Fig. 1: Notched beam loaded in a) 3-point b) 4-point bending

Eq. 2 can be rewritten and one obtains

$$K_{IC} = \frac{6M_c}{b(W-a)^2} \sqrt{\pi \cdot a} \, Y(a/W) \tag{4}$$

For the polynominal Y(a/W) different expressions have been proposed in evaluating experiments on cementitious materials. The

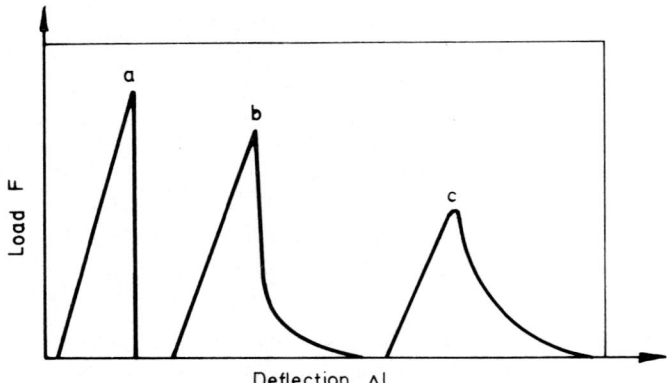

Fig. 2: Typical load-deflection diagrams of hardened cement paste
         representing
         a) catastrophic fracture
         b) semi-stable fracture
         c) stable fracture

most important ones are given in the following:

a) Relations due to Brown and Srawley (ref. 19)

$$Y_1(a/W) = \frac{1}{\sqrt{\pi}}(1 - \frac{a}{W})^2 [1,99 - 2,47(a/W) + 12,97(a/W)^2 -$$
$$- 23,17(a/W)^3 + 24,80(a/W)^4]$$

$$Y_2(a/W) = \frac{1}{\sqrt{\pi}}(1 - \frac{a}{W})^2 [1,96 - 2,75(a/W) + 13,66(a/W)^2 - \qquad (5)$$
$$- 23,98(a/W)^3 + 25,25(a/W)^4]$$

$$Y_3(a/W) = \frac{1}{\sqrt{\pi}}(1 - \frac{a}{W})^2 [1,93 - 3,07(a/W) + 14,53(a/W)^2 -$$
$$- 25,11(a/W)^3 + 25,80(a/W)^4]$$

for pure bending ($Y_1(a/W)$), three-point bending with
s/W = 8($Y_2(a/W)$) and three-point bending with s/W = 4 ($Y_3(a/W)$),
respectively.

b) Relation due to Lott and Kesler (ref. 20)

$$Y_4(a/W) = \frac{1}{\pi}(1-a/W)^2 \cdot \sqrt{2 [10,08(a/W)^2 - 1,225(a/W) + 0,1917] W/a} \quad (6)$$

for three-point-bending.

c) Relation due to Winnie and Wundt (ref. 21)

$$Y_5(a/W) = (1 - a/W)^2 \qquad\qquad (7)$$

both for three- and four-point-bending.

A graphical representation of the different formulas is given in Fig. 3. From this Fig. it follows that the polynomials exhibit some degree of variation. Certainly the relation $Y_4(a/W)$ should not be used in a wide crack length regime. It seems evident that

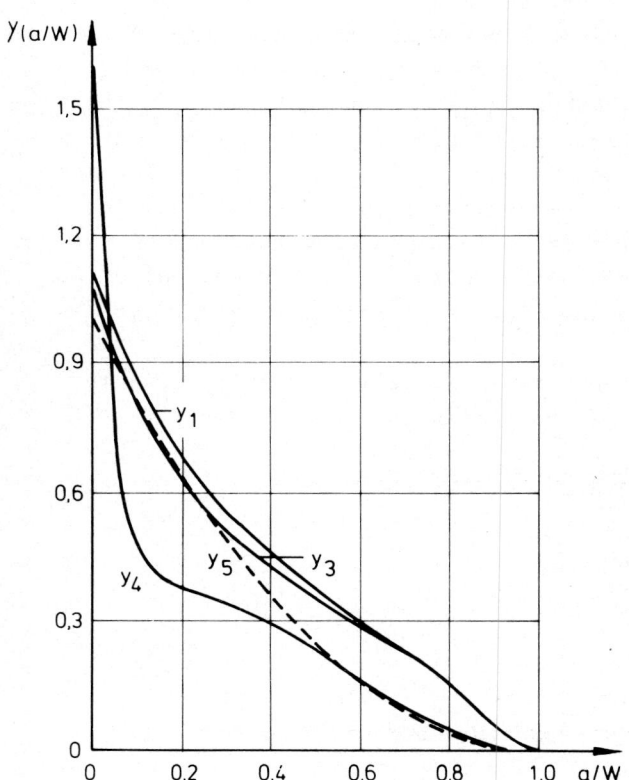

Fig. 3: Graphical representation of the Y(a/W)-functions (ref.1)

a recalculation of experimental data seems necessary comparing
K_{IC}-values evaluated with different polynomials. In order to facilitate the comparability of results obtained in bending experiments on cementitious materials it is proposed to use in future
only the relations due to Brown and Srawley (eq.5). These relations have the advantage

- to consider three- and four-point bending
- to be valid in a large crack length regime ($0 < a/W < 0,6$) with
  high accuracy ($\leq 0,2\%$) (ref. 19)
- to be widely used in experiments on metallic and ceramic materials.

2.1.3 Double cantilever beam (DCB). The DCB-technique has been
used by several investigators because it has the advantage that
crack growth can be analysed over a range of crack lengths
(ref. 5, 7).

The principle of the technique is explained in Fig. 4. The specimen has the form of a long prism with two crack-guiding grooves
cut symmetrically along its length. At one end there is a notch
acting as a crack starter joining the grooves at one end. The
crack is propagated by mechanically separating the ends of the two
cantilevers. As with the notched beam, there are several effectively equivalent derivations for the critical strain intensity
factor K_{IC}.

In experiments on cementitious materials a relation due to
Swanson and Gross (ref. 22) has often been used:

$$K_{IC} = \frac{2\sqrt{3}\, F_c \cdot a}{b \cdot W \cdot h\ h} \sqrt{1 + 1,32(h/a) + 0,532(h/a)^2} \tag{8}$$

with F_c - critical load
        a - crack length
        b - beam width
        W - web width
        h - beam height

Recently Mai (ref. 23) proposed a simpler equation

$$K_{IC} = \frac{2\sqrt{3}\, F_c \cdot a}{\sqrt{b \cdot W \cdot h}\ h} (1 + 0,7(h/a)) \tag{9}$$

As the difference between eq.8 and 9, resp., is smaller than
1% it is recommended to use eq. 9 in future experiments.

When the web width is not uniform but made so that

$$C \cdot W = a^2 (0,7 + a/h) \qquad C = const. \qquad (10)$$

then eq. 9 becomes

$$K_{IC} = \frac{2\sqrt{3} \, F_c}{\sqrt{b \cdot h} \, h} \sqrt{C} \qquad (11)$$

so that the effect of the increasing crack length is exactly balanced by the increasing web width and K_{IC} can be measured without determining a. For such specimens F_c is independent of crack length provided there is no change in fracture toughness with crack length.

2.1.4 Double torsion beam (DTB) (ref. 8, 9). As with the double cantilever beam test the double torsion test has the advantage that the stress intensity factor at the notch depends only on the load and on the specimen dimensions and not on the crack length. In the experiment the test specimen (Fig.5) is loaded by torsional forces, and the stress intensity factor is given by the following formula (ref. 24):

$$K_I = F \cdot W_m \cdot \sqrt{\frac{3(1+\nu)}{Wt^3 \, t_n}} \qquad (12)$$

The parameters are defined in Fig.5.

2.1.5 Compact tension specimen (CTS) (ref. 25, 26). Because of its compactness the CT-specimen is the most widely used form in fracture toughness testing of metallic materials. It can be regarded as a DCB of small length and without side grooves.

The fracture toughness can be calculated from the following formula

$$K_{IC} = \frac{F_c \sqrt{a}}{W \, b} \cdot F(a/W) \qquad (13)$$

with F_c - critical force.

An expression for the correction polynomial $F(a/W)$ is given by ASTM standard E 399 for $0,45 < a/W < 0,55$:

$$F(a/W) = 29,6 - 185,5(a/W) + 655,7(a/W)^2 - 1017(a/W)^3 + 638(a/W)^4 \qquad (14)$$

Hillemeier (ref.28) enlarged the validity range of eq.14 by developing three correction polynomials, each valid for a part of the crack length regime $0,125 < a/W < 0,925$.

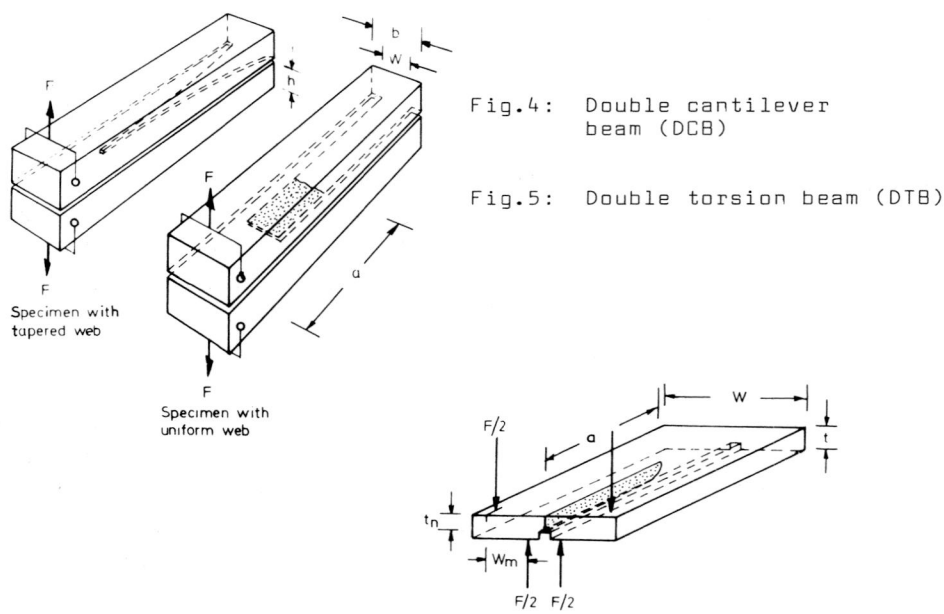

Fig.4: Double cantilever beam (DCB)

Fig.5: Double torsion beam (DTB)

Instead of pulling the two cantilever beams directly apart which always resulted in catastrophic crack growth, Hillemeier and Hilsdorf (ref.26) used a loading system including a wedge and needle bearings (Fig.7). By this method the stiffness of the loading system is increased, so that controlled crack growth can be achieved.

2.1.6 Plate specimen. In an early investigation Kesler, Naus and Lott (ref.27) determined the fracture toughness of cement paste, mortar and concrete by testing plate specimens (Fig.8) containing two cracks and a loading hole located at the center of the specimen. An expression for the stress intensity factor for this specimen (for $h \to \infty$) is given by the following expression (ref.29):

$$K_{IC} = \frac{F_c}{b \cdot \left[\pi\, a \, \frac{W}{2\pi a} \cdot \sin \frac{2\pi a}{W} \right]^{1/2}} \tag{15}$$

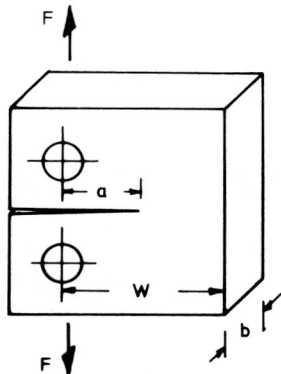

Fig.6: CT-specimen

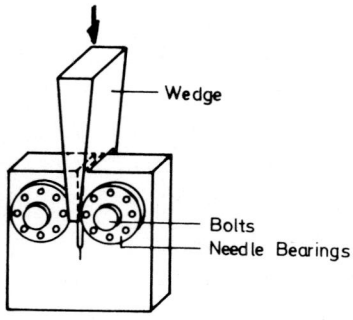

Fig.7: Wedge loaded compact tension specimen

The meaning of the parameters included is explained in Fig.8.

2.1.7 Other specimen forms. There are various other specimen
forms used to determine K_{IC}-values of cementitious materials, e.g.
- circumferentially notched bar under bending (CNRBB) or under
  eccentric compression (CNRBEL) (ref. 30, 31)
- single edge notch prism under eccentric compression or tension
  (SENEL) (ref. 32, 33)

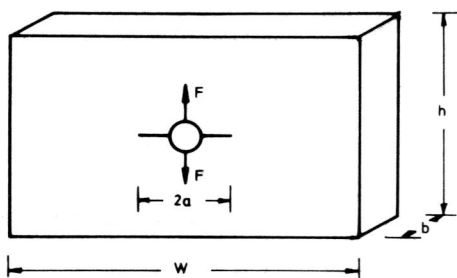

Fig. 8:   Plate specimen

- compression of a cylinder or a prism containing a central vertical notch (CVNCC or CVNPC) (ref. 33-35)
- central notch plate in tension (CNPT) (ref.33).

However, as these specimen forms have not been used in connection with hardened cement paste, they are not furtheron treated
here.

2.1.8 Compliance method.   There is one method to determine the
fracture toughness of a material which differs from the others
just mentioned, in so far that it is applicable to any forms of
specimen.   This method was developed by Irwin and Kies (ref.36)
and consists of measuring the change of compliance dC/da as a
function of crack length and the critical load:

$$G_c = \frac{1}{2b^2} \; F_c^2 \cdot \frac{dC}{da} \tag{16}$$

In order to determine dC/da the compliance function $C=C(a/W)$
has to be evaluated experimentally.   This can be done ba measuring
the compliance of specimens which have different crack length but
which are otherwise identical (Fig.9).

The critical value of energy release rate G_c can be determined
from eq.16 by inserting the critical load F_c at onset of rapid
crack propagation and the pertinent dC/da-value.

There is another procedure which, too, is called compliance
method, but which is not directly related to the determination of
a critical toughness value.   In this latter method the relation
between compliance and crack length is used to determine non-
destructively the crack length at any state of crack growth. This

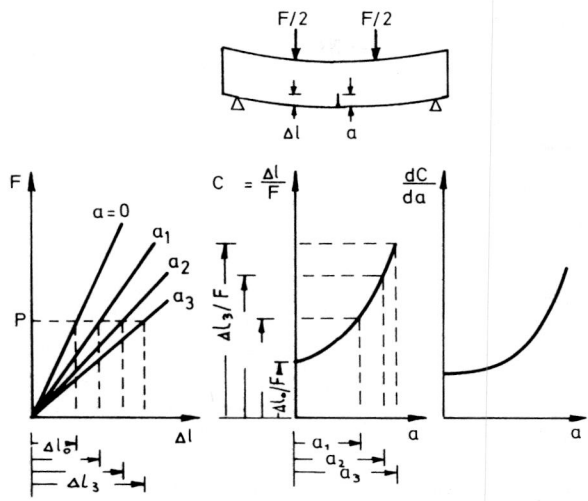

Fig. 9:   Experimental determination of dC/da

method therefore is often applied in combination with the load
relaxation method (described below).

   In determining a crack length by using the compliance method
one has to take consideration, however, that the compliance of a
sawn-notch-specimen can be quite different from the compliance of
a natural-crack-specimen. This is because in the second case
there is considerable interaction between the two sides of the
crack path, giving an apparently higher stiffness for the same
crack length (ref.37).

2.1.9 Load relaxation method. In a normal fracture toughness experiment, the specimen is loaded until rapid crack propagation is
initiated. Thus each specimen gives only one K_{IC}-value. A better
result can be obtained by avoiding rapid crack propagation.

   Brown (ref.5) in testing DCB- and 4-point-bending specimens,
proposed a method in which the load is taken off just before the
onset of rapid crack propagation. Loading and unloading is repeated several times, each time using a specimen with a slightly
greater crack length (Fig.10).

382

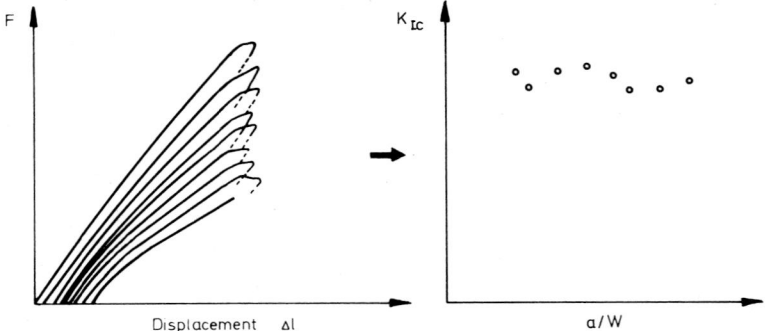

Fig.10: Load relaxation method

As the crack length of the specimen at each loading cycle can
be evaluated using the compliance method, the fracture toughness
can be measured as a function of crack length. Another important
advantage of this method is the possibility of obtaining slow crack
growth corrected K_{IC}-values. This can be done as follows: K_{IC}- or
G_{IC}-values are computed by inserting into the respective formulas
the critical load of the nth cycle and the crack length of the
(n+1)th cycle.

The method of loading the same specimen several times has been
used in the past on DCB- and 4-point-bending specimens (ref.5), on
double torsion specimens (ref.9) and on CT-specimens (ref.26).

A method similar to the one just described consists in the load
relaxation method used by Nadeau, Mindess and Hay (ref.9) on
double torsion specimens. The load is increased until a crack
forms at the tip of the machined notch. This is signalled by a
load drop as the specimen suddenly becomes more compliant. The
cross-head is stopped and the load continues to fall slowly as the
crack grows at a decreasing rate. Each point on the resulting load-
time curve corresponds to a different stress intensity and crack
velocity. In this way the relation between stress intensity factor
and crack velocity can be established. K_{IC}-values can be obtained
by loading the specimen quickly until fast crack growth occurs.

2.2   Methods of Evaluating J_{IC}

Fracture Mechanics can be divided into 2 categories: Linear
Elastic Fracture Mechanics (LEFM) and Elastic-Plastic Fracture
Mechanics. However, while Linear Elastic Fracture Mechanics (K_{IC},
G_{IC}) is rather well known and well developed, the latter is still
evolving.

Recently the J-Integral, which was first described by Rice
(ref. 38) has been proposed as a failure-criterion for metals with
elastic-plastic behavior. The J-integral is a path-independent
line integral. It results by integration of deformation energy
density along a closed path around the crack under consideration.
Rice showed, that for both linear- and nonlinear-elastic materials
J is equivalent to the energy necessary to increase the cracked
surface by an infinitesimal amount. For the linear-elastic materials, J is equal to the strain energy release rate G. In this case
the following relations are valid at the critical stress state
(for plane strain):

$$J_{IC} = G_{IC} = K^2_{IC} \frac{1-\nu^2}{E} \qquad (17)$$

Some assumptions and limitations have to be respected, when the
integral is used as a failure criterion. The most important limitation is - in the case of application to cementitious materials -
that unloading is not permitted, i.e. that J_{IC} applies only to onset of crack propagation. However, since the same can be stated
for K_{IC}, this limitation does not rule out J_C as a fracture criterion (ref. 13).

A possible advantage of the J-integral approach is, that this
concept may be applied to specimens with dimensions which do not
meet the requirements put forward by LEFM. Thus Begley and Landes
(ref. 39) reported rather constant values for medium-strength alloy steel specimens of different geometries.

The J-integral is defined as the change in potential energy
when the crack is extended by an infinitesimal amount "da" (see
Fig.11):

$$J = - \frac{1}{b} \frac{dU}{da} \qquad (18)$$

Based on eq. 18, various methods have been proposed in order to
evaluate experimentally a critical value of J. All methods use
the area under the load deflection curve, the deflection being
measured along the load line. There is an "exact" method due to

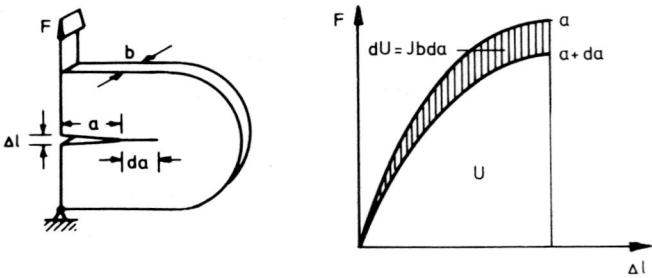

Fig. 11: Definition of J-integral

Begley and Landes (ref.39) which, however, is very time-consuming, as a great number of specimens with different notch depths have to be tested. Rice et al. (ref.40) developed a simplified method to evaluate J. He showed that for deeply notched specimens ($a/W \geq 0,6$) J_{IC}-integral values can be calculated from the load vs. load-point-displacement curve of a notched and unnotched specimen. In performing this procedure the following relation can be used:

$$J_{IC} = \frac{2}{b(W-a)} (A_T - A_U) \tag{19}$$

with A_T, A_U - strain energy (see Fig.12)

b   - specimen thickness

W-a   - uncracked ligament

In many cases it is even sufficient to consider only the cracked specimen so that eq.19 as well as the experimental procedure are still more simplified (ref.41):

$$J_{IC} = \frac{2A_T}{b(W-a)} \tag{20}$$

2.3   Methods of Evaluating the Fracture Surface Energy

According to the Griffith theory the tensile strength of a material depends on the surface energy γ (for plane strain):

$$B_t = \sqrt{\frac{2E \cdot \gamma}{\pi \cdot a(1-\nu)}} \tag{21}$$

The surface energy of a linear elastic material can be measured in a slow bend test. In this test a notched beam specimen is slowly loaded in flexure in a suitable, stiff testing machine, such

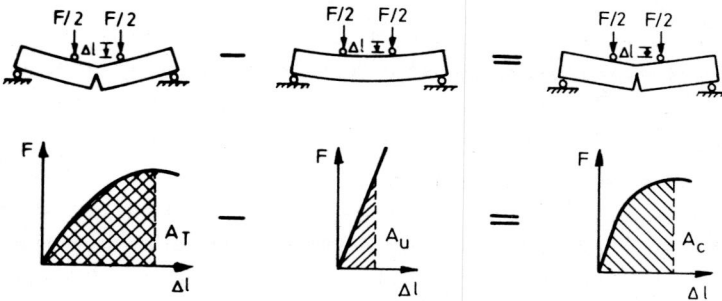

Fig. 12: J_{IC}-evaluation using single specimen formula (ref.41)

that the complete load-deflection curve until final failure can be
obtained.   Assuming that all energy expended is changed into surface energy - which is quite natural for a linear elastic material
- the fracture surface energy is then obtained by dividing the total work to fracture U (area under the load-deflection curve) by
the fracture surface area 2A

$$\gamma = \frac{U}{2A}$$ (22)

Cementitious materials, however, are not linear elastic. Furthermore the true area of the fracture surface may be considerably
larger that the cross-sectional area of the member. Microscopic
studies by Moavenzadeh et al. (ref. 2) and Higgins et al.(ref.11)
lead to the assumption of the following approximate ratios of true
and apparent fracture surface area:
- hardened cement paste: (1 - 2) : 1
- mortar   : (5 - 10) : 1
- concrete   : (15 - 20) : 1

Studies of fracture surface also indicate that for concretes
made of normal weight aggregates having a strength up to approximately 30 N/mm² the cracks normally propagate only in the matrix
or in the interfaces. For light weight aggregate concretes or for
concretes with a higher strength also crack propagation through
aggregate particles may occur.

The total energy used to fracture a member consists of various
components (ref.42):

386

$$\text{tot } U = U_k + U_f + U_{pl} + U_s \tag{23}$$

In the case of unstable crack growth (on a macroscopic or on a microscopic scale) a certain amount of kinetic energy U_k will be used up and transformed into heat. Friction between individual particles at the crack surfaces will use up the energy U_f. True plastic deformations may occur within the solid particles thus leading to the term U_{pl}. Finally the term U_s describes the amount of energy which has been used for the formation of new surfaces. The terms U_k, U_f, U_{pl} and U_s are difficult to separate. Provided that macroscopic unstable crack growth is minimized they may be expressed by the term U_{eff}, the effective surface energy. On the basis of these simplifications an effective fracture energy may be defined as follows:

$$\gamma_{eff} = \frac{U_{eff}}{2A_o} \tag{24}$$

with A_o - cross-sectional area of the fractured member.

Often it is assumed that the effective fracture energy is a materials characteristic. With cementitious materials, however, one has to be aware that especially the effect of crack branching poses severe problems in the application of this mode, as it cannot be accepted without further ado that the amount of crack branching remains constant all over a given fracture path. In the case of increasing branching - as often observed in concrete specimens - the local effective fracture energy

$$\gamma_{eff} = \frac{d}{d2A} U_{eff} \tag{25}$$

will increase with increasing crack length.

2.4 Basic Problems in Applying Fracture Mechanics Methods to Cementitious Materials

It has been pointed out repeatedly in this report that the application of fracture mechanics, and especially of LEFM to cementitious materials is not unproblematic. For instance
- LEFM normally is based on the analysis of a single crack in a homogeneous, isotropic medium. Cementitious materials, however, are inhomogeneous and - because of crack branching - exhibit multiple cracking.
- Furtheron LEFM as well as the J-integral concept are no more applicable when local cracking occurs in the material before

rapid crack propagation begins. In cementitious materials, such cracking is a well known fact.

The application of LEFM to cementitious materials was justified in the past mainly by two reasons:

a) Crack branching can be treated analogous to the plastic zone at the crack tip of metallic materials, i.e. the applicability of LEFM is justified when crack branching is confined to a small scale compared to the dimension of the specimen and of the crack.

b) In many experiments on metallic and ceramic materials it had been observed that the application of LEFM led to reasonable results even if not all basic requirements of LEFM had been fulfilled.

The neglecting of such basic requirements is, nevertheless, not unproblematic and a thorough analysis of such procedure is necessary. Furthermore there are other problems involved in the application of LEFM to cementitious materials which play a minor role when performing experiments on brittle metallic or ceramic materials. In the following some of these problems are discussed in detail.

2.4.1 Specimen size. If one wishes to get valid K_{IC}-values in experiments on metallic or ceramic specimens, one has to observe the following equations (according to ASTM E 399 standard):

$$a,b \geqslant 2,5 \cdot \left[\frac{K_{IC}}{\sigma_y}\right]^2 \qquad W \geqslant 5 \cdot \left[\frac{K_{IC}}{\sigma_y}\right]^2 \qquad (26)$$

with a - crack length
      b - specimen thickness
      σ_y - yield strength
      W - specimen width

Cementitious materials do not exhibit yielding like metallic materials. Because of creep and crack formation stress-strain diagrams of cementitious materials deviate at relatively small stresses from linear elastic behavior. The application of eq.26 to cementitious materials by inserting the tensile strength B_t in place of σ_y therefore will lead to invalid results. Modeer (ref. 16), in applying the "fictitious crack model" to hardened cement paste, mortar and concrete, proposed the following minimum value for the specimen width W

$$W \geq 10 \left[\frac{K_{IC}}{\beta_t}\right]^2 \qquad (27)$$

which leads to the results as given in Table 1. In this table medium values of K_{IC} and β_t have been taken from the literature.

TABLE 1 : Minimum Specimen Width

material	K_{IC} MN/m$^{3/2}$	β_t MN/m^2	W_{min} mm
hardened cement paste	0,4	6	44
mortar	0,6	5	144
concrete	1,0	4	625

The W_{min}-value for concrete corresponds to another estimation recently published by Hilsdorf (ref.43). In most fracture toughness experiments on cementitious materials specimen widths smaller than the W_{min}-values given in Table 1 have been used.

2.4.2 Crack length. When performing fracture toughness experiments on cementitious materials the exact determination of the crack length at onset of rapid crack propagation poses severe problems. One reason for this difficulty is given in the effect of slow stable crack growth. Because of this effect the effective crack length at the onset of rapid fracture may be substantially greater than the initial crack length. As the evaluation of the amount of slow crack growth is difficult to perform, this effect has not been considered in most fracture toughness experiments on cementitious materials known from the literature.

Another effect which renders difficult crack determination is due to microcrack formation at the crack tip. It has been shown that numerous microcracks develop in hardened cement paste as well as in concrete prior to crack propagation. These microcracks are not arranged in a circular sphere comparable to the yield zone of metals. The microcracked zone rather is concentrated on the region of the ligament and has a (longitudinal) extension of about 100 µm in hardened cement paste (ref.11) and of about 100-200 mm in concrete (ref.32, 44). As the microcracks reduce the stresses in the cracked zone similar to the stress reduction caused by yielding in metals it is clear that the extension of the microcracked zone has to be taken into account when determining the crack length.

   It has already been mentioned that in cementitious materials
there is considerable interaction between the two sides of a crack
thus reducing the effective crack length. Furthermore the amount
of crack branching in the neighbourhood of a natural crack will
vary which, too, influences the effective crack length. Therefore
the behavior of a specimen with a crack produced by a saw cut or
by inserting a metal foil will differ from a specimen with a natural crack of identical crack length.

   In summary it can be stated that the determination of a "true"
crack length in cementitious materials is not unproblematic.

2.4.3 Critical load. According to ASTM E 399 the critical load
for evaluating a valid K_{IC}-value is determined by a secant line
through the origin of the test record with slope $(F/\Delta l) =
0,95(F/\Delta l)_0$, where $(F/\Delta l)_0$ is the slope of the tangent to the
initial linear part of the record (see Fig. 13a). In evaluating
experiments on cementitious materials this procedure is not appropriate, as load-displacement diagrams of these materials
(Fig. 13b)
a) generally are rather nonlinear
b) often exhibit a marked curvature even at the origin due to
    crack closure effects.

   Instead of this the maximum load is utilized mostly, leading
to substantially higher K_{IC}-values. Thus the procedure becomes
rather simple, but it should be checked more thoroughly if it is
in agreement with the requirements of LEFM.

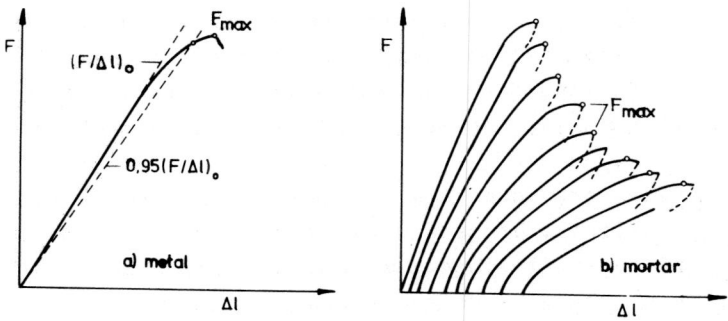

Fig. 13: Determination of Critical Load

3.   RESULTS OF FRACTURE MECHANICS EXPERIMENTS ON HARDENED CEMENT
      PASTE

Taking consideration of the fact that the conditions for the
applicability of LEFM normally are not fulfilled in experiments on
concrete and mortar specimens some authors previously used the expressions "apparent" (ref. 11, 27), "effective" (ref. 3, 8) or
"pseudo" (ref. 45) fracture toughness. However, hardened cement
paste being a rather homogeneous and brittle material, the expression "fracture toughness" is further used in this paper in
connection with this material.

3.1   Notch Sensitivity of Hardened Cement Paste

Frequently, the applicability of fracture mechanics to concrete,
mortar and hardened cement paste has been associated with the existence of notch sensitivity of such materials. In this context a
material is considered notch sensitive if the net failure stress
at the notched cross-section decreases with increasing crack
length. In calculating the net stress, stress concentrations at
the crack tip are not taken into account.

In Fig. 14 some typical results of studies on the notch sensitivity of hardened cement paste, mortar and concrete are presented.

The most important features of these studies may be summarized
as follows:
- Hardened cement paste is a strongly notch sensitive material.
   The net failure stress of notched specimens decreases to a minimum of approximately 20 to 50% of the strength of unnotched
   specimens.
- Results of several authors indicate concrete and mortar to be
   notch insensitive (ref. 4, 27, 46). Other authors observed
   some notch sensitivity in experiments with these materials
   (ref. 12, 15). In any case mortars and concrete are considerably less notch sensitive than hardened cement paste. Furthermore the notch sensitivity decreases with increasing aggregate
   size.
- In most cases where notch sensitivity has been observed the net
   failure stress at first decreases with increasing relative
   notch depth, then passes through a minimum and increases with a
   further increase of relative notch depth.

From these data several authors concluded that linear elastic
fracture mechanics may be applied to hardened cement paste, however not to concrete.

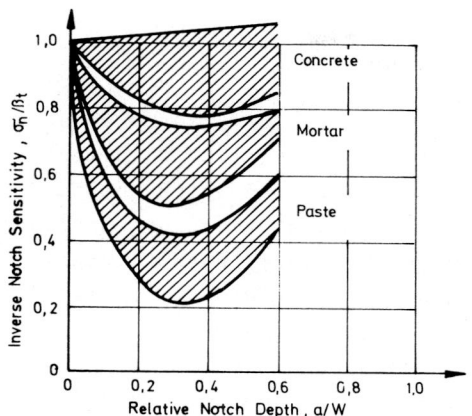

Fig. 14: Experimental results of notch sensitivity as function
          of relative notch depth

Recently (ref. 47), a simple model law for the notch sensitivity of brittle materials has been deduced under the assumption that
linear elastic fracture mechanics is applicable to the specimen in
question. According to this model law the following equation for
the notch sensitivity expressed as the ratio of net failure stress
σ_n to modulus of rupture β_t of an unnotched specimen is valid
(in the case of four-point-bending):

$$\frac{\sigma_n}{\beta_t} = \frac{K_{IC}}{\beta_t} \frac{1}{\sqrt{a} \; (1 - \frac{a}{W})^2 F(\frac{a}{W})} \leq 1 \tag{28}$$

with   a   - crack length
        W   - specimen width
        F(a/W) - correction polynominal

Equation 28 indicates that notch sensitivity of a specimen is a
necessary but not a sufficient condition for the applicability of
linear elastic fracture mechanics.
    The equation predicts furtheron
- a maximum of notch sensitivity at a relative crack length of
   a/W = 0.246 (four-point-bending);
- an increase of notch sensitivity with increasing specimen size;
- a decrease of notch sensitivity with increasing ratio of K_{IC}/β_t.

In summing up it can be stated that the model law is able to predict qualitatively the notch sensitivity behavior of hardened cement paste, mortar and concrete.

3.2   Effect of Specimen Dimensions

It is obvious that for either K_{IC}, G_{IC}, J_{IC} or γ to be a valid
fracture criterion, the values must be independent of specimen
dimensions, i.e. the validity of the fracture mechanics approach
can only be established if it can be shown that tests on specimens
of significantly different geometry but of otherwise sufficient
dimensions (according to the conditions put forward by fracture
mechanics) yield identical values.   Reviewing the experiments on
cementitious materials it seems that this condition is neither
fulfilled in the test results on concretes and mortars nor in
test results on hardened cement pastes (hcp).

Higgins and Bailey (ref. 11) using notched hcp-beams observed
that the calculated values of K_{IC}-values increase with increasing
specimen size.   They showed that K_{IC} appears to tend towards a
constant limiting value (for large specimens with deep notches)
of about 0,79 MN/m$^{3/2}$ for a w/c-ratio of 0,3 (Fig. 15).

Higgins and Bailey state: "It is not possible to define any definite minimum size since the size at which it is no longer

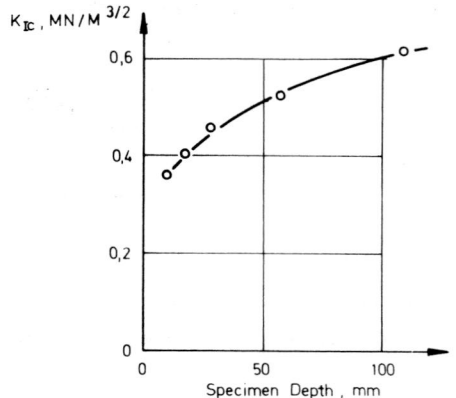

Fig. 15:   Influence of specimen size on fracture toughness of hcp
           (ref. 11)

possible to detect an increase in this factor will be determined
by the scatter in the experimental results and so the more carefully the experiments are performed, the larger the required minimum size will be. The largest beams used in the present study
(600 mm long by 110 mm deep) were still too small for the experimental scatter to mask the change in the calculated stress intensity factor at failure with size".

The results of Higgins and Bailey regarding experiments on
hardened cement paste are confirmed by the results of earlier experiments of Kaplan (ref. 46) and of Moavenzadeh et al. (ref. 2)
on mortar and concrete. Both authors observed significantly lower
G_{IC}-values for small beams than for large beams. Kaplan suggested
that this discrepancy might be due to the effect of slow crack
propagation and also due to shear.

In a recently published paper, Modeer (ref. 16) proposed a similar dependence of fracture toughness on specimen dimension.
Combining a variety of the Dugdale model ("Fictitious Crack Model")
and a finite element analysis in a theoretical study on cementitious materials, he developed a relation according to Fig. 16.

The characteristic length is a material parameter:

$$l_{ch} = (K_{IC}/\beta_t)^2 \tag{29}$$

According to Modeer (ref. 16) l_{ch}-values for concrete are
200 - 300 mm, for mortar 100 - 200 mm, and for hardened cement
paste 5 - 10 mm. Fig. 16 indicates that the value W/l_{ch} should

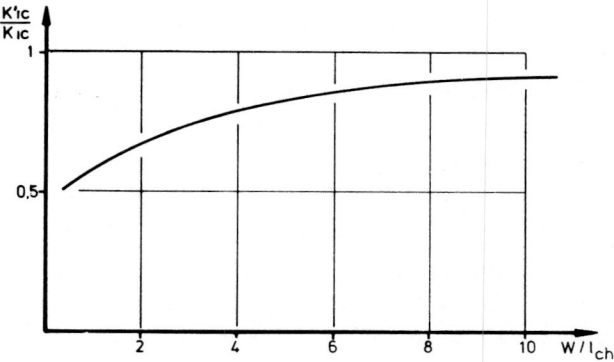

Fig. 16:   Apparent fracture toughness over fracture toughness vs.
           specimen depth over characteristic length (ref. 16)

exceed 10 to reduce the difference between "apparent" and "true"
fracture toughness sufficiently. This demand would require minimal dimensions of correct test specimens (W) of 50 - 100 mm for
hardened cement paste, in good accordance with the above mentioned experimental results of Higgins and Bailey.

3.3   Effect of Crack Length

    The effect of crack length on fracture toughness is comparable
to the effect of specimen dimensions on fracture toughness: in
both cases the condition of applicability of linear elastic fracture mechanics claims independence of fracture toughness of the
respective variable. The experimental results on the effect of
crack length on fracture toughness are, however, rather contradictory:

-   The results of Brown (ref. 5) and Brown and Pomeroy (ref. 7) on
    flexure and double cantilever beams indicate that there is
    little change of fracture toughness with crack growth, whilst
    when aggregates are present the toughness increases with crack
    length.

-   Higgins and Bailey's (ref. 11) experiments on flexure beams
    with variable overall depth, too, have yielded a constant K_{IC}-
    value over a wide range of crack depths.

-   In 1977, Hillemeier and Hilsdorf (ref. 26) conducted an analy-
    tical and experimental investigation on hardened cement paste
    using compact-tension specimens. Their results showed that
    K_{IC}-values decrease with increasing crack length, finally ap-
    proaching a constant value. The constant K_{IC}-value was inde-
    pendent of both the initial crack length and the method by
    which the crack was prepared.

-   Mindess at el. (ref. 13) compared J_{IC}- with K_{IC}- and G_{IC}- va-
    lues for hardened cement paste, plain concrete and different
    volume fractions of steel and glass fiber reinforced concretes
    in the form of notched beams tested in four-point-bending.
    All their fracture parameters decreased with increasing notch
    depth in agreement with Hillemeier and Hilsdorf's results.

-   Shah and McGarry (ref. 4) reported increasing "toughness"-va-
    lues (toughness being defined as area under the load-deflection
    curve in a three-point-bending test) with increasing notch
    depth. However, since the failure in their paste experiments
    seemed to be catastrophic, the validity of these experiments is
    questionable.

The decrease of fracture toughness with increasing crack
length is explained by Hillemeier and Hilsdorf (ref. 26) "by the
presence of numerous microcracks which surround the tip of the
initial notch. For multiple crack growth, a higher energy rate is
required than for the propagation of a single crack. Consequently
K_{IC} is large. Eventually one major crack is formed. The propagation of this crack can then be described by a characteristic value of K_{IC} which is independent of the initial crack length".

This point of view is supported by recent experiments of Ziegeldorf and Hilsdorf (ref. 48) on compact-tension specimens of
hardened cement paste modified with a polymer dispersion. In
these experiments independence of fracture toughness of crack
length was observed. The explanation of these apparently contradictory results is the shrinkage behavior of both materials:
According to relevant measurements (ref. 43), the shrinkage of
unmodified cement paste is more pronounced than that of the modified paste. It is imaginable, therefore, that multiple microcracks surrounding the initial crack tip develop in the pure hardened cement paste but not in the modified one.

3.4   Effect of Water-Cement-Ratio

The effect of water-cement-ratio on fracture toughness and
surface energy of hardened cement paste was investigated by several authors (ref. 2, 3, 7). Their results are summarized in
Fig. 17.

In general there is a pronounced decrease of K_{IC} and γ with increasing w/c-ratio. This is analogous to the decrease in strength
with increasing w/c-ratio. Fig. 17 indicates, however, that the
effect of w/c-ratio is particularly marked for w/c-ratios < 0,5.

3.5   Effect of Curing Period

The effect of curing period has been investigated by ref. 2, 3,
7, 11, 14, 26 . The results are summarized in Fig. 18.

Most investigations indicate that up to about 7 - 14 days wet
curing there is a large increase in K_{IC} and γ with time, but thereafter, within the time-scales used (i.e. up to 100 days) there is
relatively little change. Though the matrix strength continues to
increase after 7 - 14 days, the same cannot be stated for the
fracture toughness and surface energy. The lack of increase of
toughness with decreasing porosity might be due to the fact that
the hardened cement paste becomes increasingly more brittle with

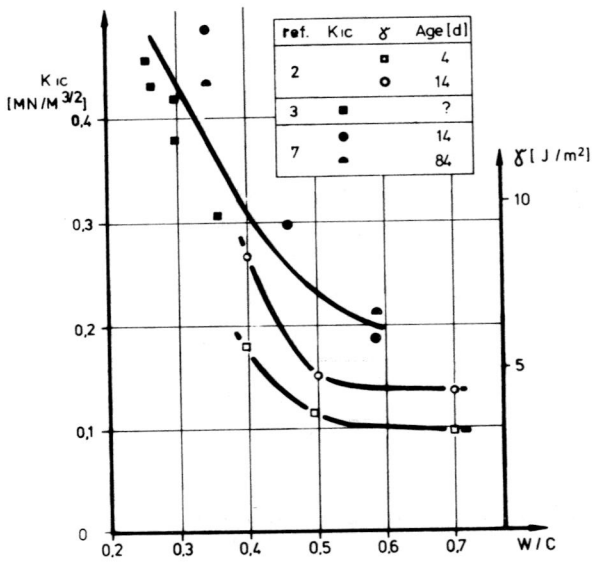

Fig. 17:   Effect of w/c-ratio on fracture toughness and surface
           energy

increasing curing time.   The observation shows that the toughness
and surface energy are not merely dependent on the capillary porosity.

3.6   Effect of Loading Rate

Brown and Pomeroy (ref. 7) used DCB-specimens to investigate
the effect of crack growth rate on fracture toughness.   Experiments with crack growth rates of about 0,5, 35 and 1100 mm/min
have shown that the toughness increases smoothly with crack growth
rate for all specimen ages: at 14 days the toughness at the
fastest speed was about 25% larger than that at the slowest. These
results are qualitatively confirmed by 3-point-bending experiments
of Higgins and Bailey (ref. 11) and by Nadeau, Mindess and Hay
(ref. 9).   The latter used the load relaxation method on a double
torsion specimen and   -   obtaining crack velocities between 0,05
and 500 mm/min   -   found a linear relationship between log

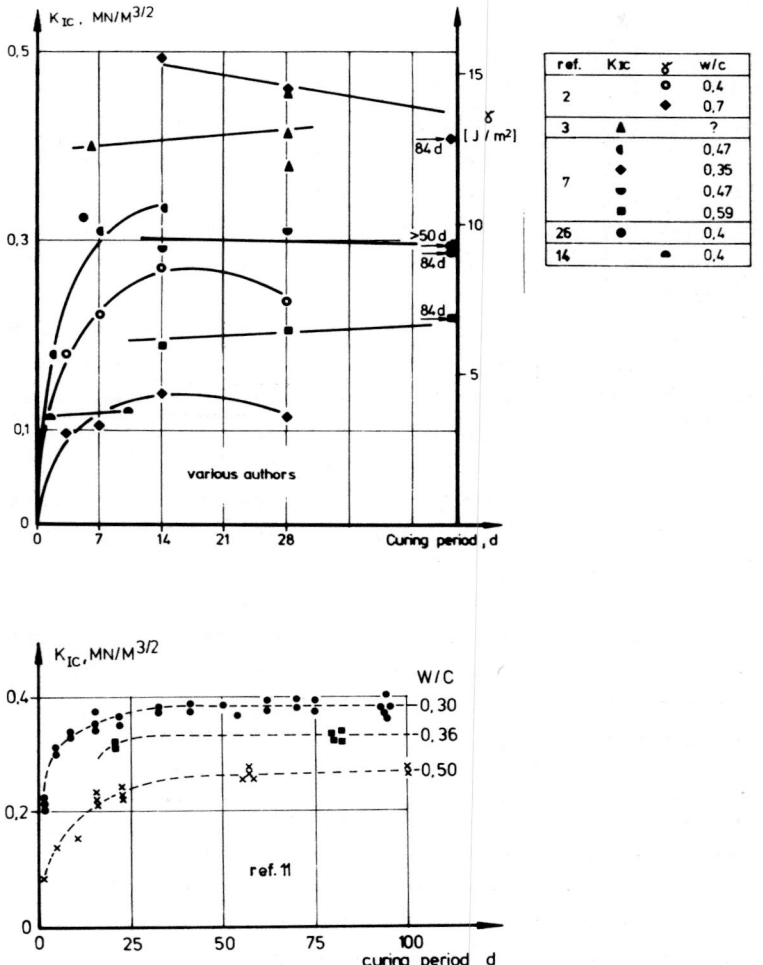

Fig. 18:   Effect of curing period on fracture toughness and surface energy

(velocity) and log K_{IC} with a slope of 35.

3.7 Effects of Notch Width and Crack Propagation

The results of experiments on the dependence of fracture toughness on notch width are somewhat contradictory.

- Brown (ref. 5) reports in agreement with measurements by Welch
  and Haisman (ref. 49) on concrete that his experiments on hardened cement paste specimens also show that there is no significant difference between the toughness for sawn and cast notches.
  He explains this behavior by slow crack growth before the maximum load is reached, which ensures that the cracks are naturally
  sharp when K_{IC} is determined. A similar opinion is shared by
  Nadeau, Mindess and Hay (ref. 9). They state that the K_{IC}-
  differences between the results for machined notch and a sharp
  crack in the same specimen are < 10%.
- However, in a parallel paper by Mindess, Nadeau and Hay (ref. 8),
  the same authors came to another conclusion. The K_{IC}-values
  measured on notched beams (with relatively large notches) are
  about 36% larger than those measured on double torsion specimens with a sharp crack, i.e. one that had been extended from
  a machined notch. They attributed the difference to the greater difficulty of starting a crack from the tip of the machined
  notch.
- This latter opinion is supported by Hillemeier and Hilsdorf
  (ref. 26). They measured a characteristic fracture toughness
  on CT-specimens using the load relaxation method only after the
  crack had propagated a certain distance and stated that the decrease of fracture toughness with increasing crack length within a given test might be caused by the presence of numerous microcracks which surround the tip of the initial notch. For
  multiple crack growth, a higher energy release rate is required
  than for the propagation of a single crack. Consequently K_{IC} is
  large.

A detailed investigation on the effect of notch width on fracture toughness has been performed by Higgins and Bailey (ref. 11).
Fig. 19 shows the result of their measurements.

For wide slits, the fracture toughness decreases as the slit
width decreases. However, for slit widths below about 0,5 mm no
significant change in fracture toughness with slit width was detected.

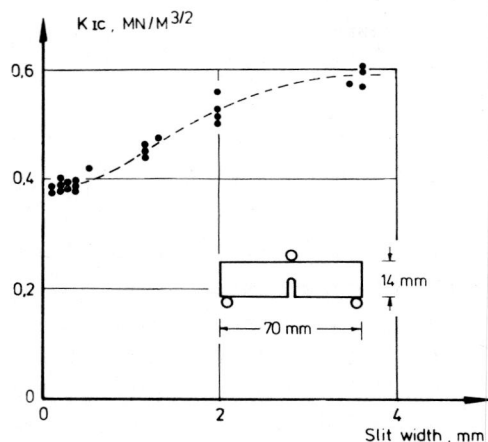

Fig. 19:   Effect of notch width on fracture toughness (ref.11)

3.8   Effect of Air Content

The matrix strength is reduced when the air content of hardened cement paste is increased.   The same tendency might be expected for the fracture toughness.   This is confirmed by Naus and Lott (ref. 2), who found that there was a 23,4%-decrease in K_{IC} when the air content was increased from 2% to 8%.   Hillemeier and Hilsdorf (ref. 26), too, observed a similar marked decrease of K_{IC} with the addition of entrained air or hollow polymer particles (8% particles $\rightarrow \Delta K_{IC}$ = 17%).

3.9   Effect of Moisture Content

There are only few investigations known to the author on the effect of moisture content on the toughness of hardened cement paste and these two are contradictory.   Cooper and Figg (ref.6) observed an increase of surface energy values of about 20% when the specimens were dried.   Mindess, Nadeau and Hay (ref. 9) on the other hand measured smaller fracture toughness values for dried specimens than for wet ones.   A conclusion cannot be drawn from their results, however, since the w/c-ratios of the dry and wet specimens were different and since crack formation in the dry specimens was not accounted for.

3.10 Comparison of Fracture Parameters

3.10.1 Comparison of K_{IC} and J_{IC}. For linear-elastic materials
the following equations are valid:

$$(1-\nu^2)K^2_{C}/E = G_c = J_c = 2\gamma \tag{30}$$

Thus it is possible to calculate K_{IC}-values from G_{IC}-, J_{IC}- and
γ-values and vice versa.

Mindess, Lawrence and Kesler (ref. 13) measured J_{IC}- and G_{IC}-
values for hardened cement paste using 4-point-bending prisms.
They observed J_{IC}- and G_{IC}-values which agreed very closely with
each other and state the J_{IC}-criterion for fracture to be valid.

A similar investigation was carried out by Harder (ref. 25) on
CT-specimens. He evaluated J_{IC}-values according to three different procedures and compared them with the respective K_{IC}-values
which were determined corresponding to ASTM E 399. His results
indicate that the agreement between K_{IC}-values according to ASTM
E 399 and respective K_{IC}-values calculated from J_{IC}-values is rather good, especially for relative notch depths greater than 0,5
(Fig. 20).

3.10.2 Comparison of γ and G_{IC}. Kaplan (ref. 46) has shown that
the energy requirement for crack propagation in hardened cement
paste is an order of magnitude larger than the surface energy of
the new crack surface, i.e. the surface energy of tobermorite.
Since hardened cement paste does not exhibit ductile properties,
Glucklich (ref. 50) suggested that the increased energy requirement for crack propagation in hardened cement paste is caused by
formation of a microcracking region near the crack tip.

According to Moavenzadeh and Kuguel's (ref. 2) measurements γ
as obtained by $\gamma = G_{IC}/2$ was about 50% lower than the respective
value obtained by the slow bend test. This is in accord with Kaplan's observations, however, using quantitative microscopy Moavenzadeh et al. did not observe any side cracking in fractured specimens or hardened cement paste but conceded that in a very high magnification such microcracking might be observed. On the other hand
they observed that the neat cement paste fractured in an unstable
manner. Thus, the surface energy calculated by the slow bend test
was high since it contained excess energy not related to the formation of new surfaces.

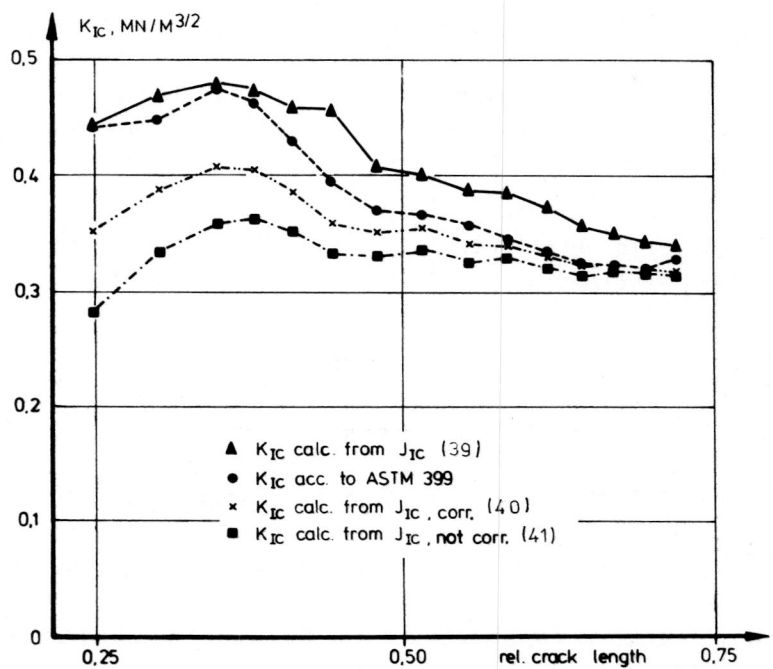

Fig. 20: Comparison of K_{IC}-values evaluated using different
          methods (ref. 25)

4.   FRACTURE MECHANICS EXPERIMENTS ON AGGREGATES

There are only a few investigations on fracture mechanics parameters of aggregates which are linked to crack propagation of
concrete. However, there are several investigations on fracture
mechanics parameters of rocks performed in the field of rock mechanics.

Generally it is observed that the fracture toughness of aggregates is distinctly larger than the corresponding values of hardened cement paste or the cement paste-aggregate interface (see
next section). For limestone, the measured values range from
0,70 MN/M$^{3/2}$ to 2,0 MN/M$^{3/2}$ (ref. 26, 51-57). For sandstone values

between 1,80 MN/M$^{3/2}$ and 3,29 MN/M$^{3/2}$ (ref. 58) and for granite a value of 0,58 MN/M$^{3/2}$ is given (ref. 59) which, however, is too small because of an incorrect evaluation of K_{IC}-values (In the fracture tests the fracture load exceeds the critical load for crack propagation by a factor > 3, whereas ASTM E 399 permits max. 1.1). Only for chalk a K_{IC}-value comparable to that of hardened cement paste is reported (0,25 MN/M$^{3/2}$) (ref. 53). In the following the influence of several parameters on K_{IC}-values of aggregate materials is characterized.

Orientation. Rock material with fine, globular grains shows a very low anisotropy. In relatively coarse grained limestone and sandstone, however, Hoagland et al. (ref. 51) observed a strong influence of orientation on K_{IC}-values. With both materials the easy crack propagation plane coincides with the bedding plane ($K_{IC,min}$). $K_{IC,max}$ is measured in a plane perpendicular to the bedding plane and is about 4 x $K_{IC,min}$ (ref. 51).

Crack length. Schmidt (ref. 52) observed increasing fracture toughness with increasing crack length in experiments on Indiana limestone. It is interesting to note that - similar to observations on hardened cement paste - his fracture toughness values approach a limiting value for the longest crack. This can be explained by the fact that the criterion for crack length a and specimen thickness b ($a, b > 2,5(K_{IC}/\sigma_s)^2$; σ_s - 0,2 percent-offset yield strength in tension) has not been satisfied for any of the tests reported in the paper.

Specimen size. Schmidt (ref. 52) also reports increasing fracture toughness values with increasing specimen size, such that for aggregates a specimen size dependence comparable to that given in Fig. 14 for hardened cement paste can be expected.

Porosity. Generally decreasing K_{IC}-values with increasing porosity are observed (ref. 53-56). This dependence was surprising to the authors, who initially had expected that the pores would cause a decrease of the stress intensity factor and therefore, an increase of fracture toughness.

5.   FRACTURE TOUGHNESS EXPERIMENTS ON INTERFACES

A crack in a hardened cement paste-aggregate interface is located in a bimaterial specimen, so that the conditions of homogeneity and isotropy are not fulfilled. It can be shown, however, that the stress field in the neighbourhood of the crack tip is characterized by two stress intensity factors (K_I and K_{II}) and by a $1/\sqrt{r}$-dependence of the stresses (ref. 60):

$$\sigma(r,\theta) = \frac{K_I}{\sqrt{r}} \, f_1(\theta, (\begin{smallmatrix}sin\\cos\end{smallmatrix})(\varepsilon \ln r)) + \frac{K_{II}}{\sqrt{r}} \, f_2(\theta, (\begin{smallmatrix}sin\\cos\end{smallmatrix}) \, (\varepsilon \ln r)) \qquad (31)$$

with ε - bimaterial constant

Thus the concept of stress characterization by K_c-factors is still valid. In the case of the hardened cement paste-aggregate CT-specimen (see Fig. 20) the K_{IC}-factor plays the dominant role ($K_{II}/K_I \sim 0,1$), such that the concept of characterizing the toughness of the contact zone by indicating its K_{IC}-factor seems justified.

As is true with fracture mechanics parameters of aggregates there are only few experiments concerning fracture mechanics parameters of interfaces in cementitious materials. First experimental results were published by Hillemeier and Hilsdorf (ref. 26) using the modified CT-specimn as shown in Fig. 21a.

A typical result is given in Fig. 21b. Starting with a value of K_{IC} = 0.42 MN/M$^{3/2}$ the fracture toughness decreases very rapidly after a crack growth of a few millimeters and approaches a value of approximately 0,12 MN/M$^{3/2}$ which is considerably smaller than the fracture toughness of hardened cement paste.

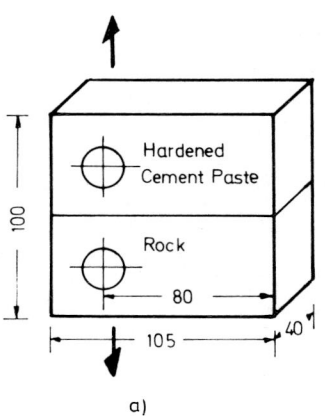

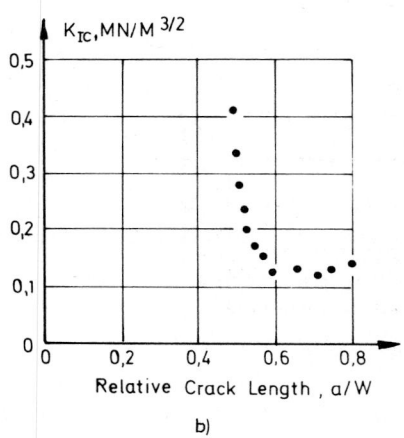

a)

b)

Fig. 21: a) Modified CT-specimen
b) K_{IC} of hardened cement paste-quartz-interface

A disadvantage of the specimen shown in Fig. 20 is the fact
that the two halves are of different compliance. This often led
to deviation of the crack path from the interface (ref. 28).
An improvement of the specimen geometry was achieved by using the
asymmetric CT-specimen (Fig. 22) (ref. 48) which was dimensioned
so that the compliance values of both parts of the specimen are
equal. With this specimen the authors succeeded in separating
the specimen exactly in the interface. The following K_{IC}-values
were evaluated (ref. 48):

- K_{IC}(paste-limestone)   : 0.16 $\pm$ 0.04 MN/M$^{3/2}$
- K_{IC}(paste-quartzite)   : 0.21 $\pm$ 0.04 MN/M$^{3/2}$
- K_{IC}(modified paste-limestone) : 0.37 $\pm$ 0,06 MN/M$^{3/2}$
- K_{IC}(modified paste-quartzite) : 0.31 $\pm$ 0.04 MN/M$^{3/2}$

The modified cement paste contained 25 W.-% dispersion (Poly-
vinylpropionate) which resulted in largely diminished shrinkage
properties. This was taken as an explanation

a) for the large values of the modified paste-interface-K_{IC}-values;
b) for the indepencence of these K_{IC}-values from crack length –
   contrary to the behavior of pure cement paste – as to be seen
   in Fig. 22.

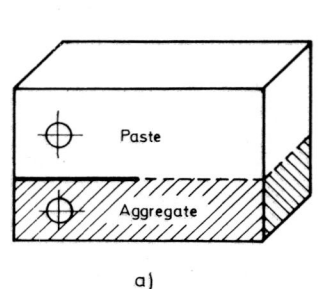

a)

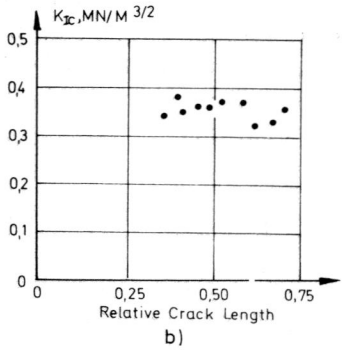

b)

Fig. 22: a) Asymmetric CT-specimen
        b) K_{IC} of modified hardened cement paste-quartz interface

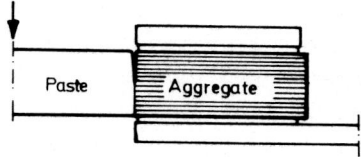

Fig. 23:   Cantilever method for paste-aggregate interfaces

Modeer (ref. 16) used the cantilever method (Fig. 23) to measure
G_{IC} of aggregate-hardened cement paste interfaces. The G_{IC}-values
reported for the interfaces between hardened cement paste and
aggregates with a rough surface when the adhesive capacity has
been at its maximum (G_{IC}(interface) = 17 N/m) were almost twice
that of the paste (G_{IC}(paste) = 9.0 N/m). Modeer explains this
discrepancy by the fact that the surface energy of the tobermorite
gel is 0.5 N/m and the surface energy of $Ca(OH)_2$, which is the
main part of the interface, is twice this value. However, the results of these tests remain astonishing since a crack always pur-
sues a path of least resistance. Thus, the interface being too
strong, the crack path would traverse the hardened cement paste
only.

6.   APPLICABILITY OF LEFM TO THE COMPONENTS OF CONCRETE

6.1   Aggregates

   As aggregates normally are brittle materials, the applicability
of LEFM to these materials seems unquestionable. However, it must
be kept in mind that the dimensions of aggregate particles present
in concrete generally do not satisfy the demands of LEFM.

6.2   Interfaces

   The applicability of LEFM to specimens with interfaces is theoretically not yet well established. In such specimens a mode II
crack surface movement in addition to the mode I movement does
take place. Because of this and other problems the fracture mechanics parameters for interfaces mentioned herein should be considered qualitatively rather than quantitatively.
   There is another effect which seems to be still more important
and which must not be neglected when discussing the results of experiments on specimens with aggregate-hardened cement paste inter-

faces. It is well known that hydrating cement paste specimens expand when stored in water but shrink when stored in a sealed condition. The shrinkage is even more pronounced when the specimen is not sealed so that loss of humidity occurs. In any case there will be relative volume change of the cement paste part relative to the aggregate part of an interface specimen leading to residual stresses and eventually to cracks. It must be recognized, therefore, that the bond strength as measured on specimens with an aggregate-hardened cement paste interface is no more a material parameter but a variable depending on the shrinkage characteristics of the cement, on the storage conditions and on the size and shape of the specimen (ref. 61). Thus the fracture mechanics parameters for interfaces mentioned in this report should be considered qualitatively rather than quantitatively.

6.3   Hardened Cement Paste

The validity of LEFM parameters as criteria for fracture can only be established if it can be shown that tests on specimens of significantly different dimensions yield identical values. This demand implies the independence of fracture parameters of crack length. Higgins and Bailey (ref. 11) and other authors (ref. 2,19) have shown that the "apparent" fracture toughness increases with increasing specimen size thus suggesting that LEFM is not readily applicable to hardened cement paste. As hardened cement paste is a rather brittle material, this statement seems surprising.

There are different reasons given in the literature to explain the observed behavior of hardened cement paste.

a) Using a staining technique, Kaplan (ref. 46) has shown that slow crack growth is initiated before the crack extends catastrophically. This finding was only recently confirmed by Mindess, Nadeau and Hay (ref. 8), who quantitatively determined the $V-K_I$-plot for hardened cement paste, i.e. a plot of crack velocity versus stress intensity factor.

b) Higgins and Bailey (ref. 11) assume that the zone of stress perturbation around the notch tip is not small in comparison with the dimensions and geometry of laboratory scale specimens. This assumption is supported by the observation that their K_{IC}-values tend towards a constant limiting value for large specimens with large notches.

c) Lott and Kesler (ref. 20) suggested that in cement paste a zone

of microcracks forms near the crack tip. This microcracking
increases the resistance of the material to crack propagation.
As the crack grows the zone of microcracking increases in size
until it reaches a limiting size when fast crack propagation
occurs.

The problem of slow crack growth can be avoided by using specimens
like DCB or double torsion specimen for which K_{IC} is independent
of crack length. It is encouraging, that both Brown (ref. 5) and
Brown and Pomeroy (ref. 7) on DCB specimens, as Nadeau, Mindess
and Hay (ref. 8, 9) on double torsion specimens, measured K_{IC}-
values which were independent of crack length.

The effect of microcracking on K_{IC} is difficult to account for
quantitatively. It would be important to know now if the microcracked zone has always the same size when fast crack propagation
occurs.

In summarizing this report it seems clear that LEFM may be applied to hardened cement paste, aggregate and interface specimens
of sufficient dimensions provided that the special properties of
these materials are taken into account.

8.   REFERENCES

1 Kasperkiewicz, J.: Personal Note.
2 Moavenzadeh, F., Kuguel, R., Keat, L.B.: Res. Rep. R.68-5;
   Mass.Inst.Techn., 1968.
3 Naus, D.J., Lott, J.L.: ACI Journal, 481 (1969), pp. 481-489.
4 Shah, S.P., McGarry, F.J.: Journal of the Eng. Mech. Div. ASCE
   (1971), pp. 1663-1667.
5 Brown, J.H.: Mag. Concr. Res. 24 (1972), pp. 185-196.
6 Cooper, G.A., Figg, J.: Trans. Brit. Cer. Soc. 71, 1 (1972),
   pp. 1-4.
7 Brown, J.H., Pomeroy, C.D.: Cem. Concr. Res. 3 (1973),
   pp. 475-480.
8 Mindess, S., Nadeau, J.S., Hay, J.M.: Cem. Concr. Res. 4 (1974),
   pp. 953-965.
9 Nadeau, J.S., Mindess, S., Hay, J.M.: J. Am. Cer. Soc. 57 (1974),
   pp. 51-54.
10 Auskern, A., Horn, W.: Cem. Concr. Res. 4 (1974), pp. 785-795.
11 Higgins, D.D., Bailey, J.E.: J. Mat. Sc. 11 (1976), pp.1995-2003.
12 Gjørv, O.E.. Sørensen, S.J., Arnesen, A.: Cem. Concr. Res.7
   (1977), pp. 333-344.
13 Mindess, S., Lawrence, F.V., Kesler, C.E.: Cem. Concr. Res. 7
   (1977), pp. 731-742.
14 Watson, K.L.: Cem. Concr. Res. 8 (1978), pp. 651-656.
15 Strange, P.C., Bryant, A.H.: J. Mat. Sc. 14 (1979),
   pp. 1863-1868.
16 Modeer, M.: Report TVBM-1001; University of Lund 1979.
17 Cooper, G.A.: J. of Mat. Sc. 12 (1977), pp. 277-289.
18 Petersson, P.E.: Cem. and Concr. Res. 10 (1980), pp. 78-89.

19 Brown, W.F.M., jr., Srawley, J.E.: ASTM STP 410 (1967).
20 Lott, G.L., Kesler, C.E.: T.A.M. Report No. 648 (1964).
21 Winnie, D.H., Wundt, B.M.: Trans. ASME 80 (1958), pp.1643-1655.
22 Swanson, E.D., Gross, G.E.: J. Appl. Phys. 40 (1969),
   pp. 4684-4685.
23 Mai, Y.W.: J. Mater. Sci. 14 (1979), pp. 2091-2102.
24 Williams, D.P., Evans, A.G.: J. Test. and Ev. (1973), pp.264.
25 Harder, D.: Diplomarbeit, Universität Karlsruhe, 1977.
26 Hillemeier, B., Hilsdorf, H.K.: Cem. Concr. Res. 7 (1977),
   pp.523-536.
27 Kesler, C.E., Naus, D.J., Lott, J.L.: Int. Conf. Mech. Beh. Mat.
   Kyoto, 1971.
28 Hillemeier, B.: Dissertation, Universität Karlsruhe, 1976.
29 Irwin, G.R.: J. Appl. Mech. 24 (1957), pp. 361-364.
30 Barr, B., Bear, T.: Concrete, April (1976), pp. 25-27.
31 Barr, B., Bear, T.: Concrete, April (1977), pp. 30-32.
32 Entov, V.M., Yagust, V.J.: Mech. of Sol. 10 (1975), pp. 87-95.
33 Mazars, J.: Proc. of Int. Conf, on Fract., Waterloo (1977),
   pp. 1205-1209.
34 Kitagawa, H., Suyama, M.: Prox. 19th Jap. Congr. Mat. Res.,
   Tokyo, pp. 156-159.
35 Khrapkov, A.A., Trapesnikov, L.P., Geinats, G.S., Pashchenko,
   V.J., Pak, A.P.: Proc. of Int. Conf. on Fract., Waterloo (1977),
   pp. 1211-1217.
36 Irwin, G.R., Kies, J.A.: Welding Journal, 33 (1954), pp. 193.
37 Mindess, S.: Personal Note.
38 Rice, J.R.: J. Appl. Mech. 35 (1968), pp. 379-386.
39 Begley, J.A., Landes, J.D.: ASTM STP 514 (1972), pp. 2-23.
40 Rice, J.R., Paris, P.C., Merkle, J.C.: ASTM STP 536 (1973),
   pp. 231-245.
41 Landes, J.D., Begley, J.A.: ASTM STP 560 (1974), pp. 170-186.
42 Hilsdorf, H.K., Ziegeldorf, S.: NATO ARI Symp. on Recycling of
   Concrete, Paris (1980).
43 Hilsdorf, H.K.: Forschungsbeiträge für die Baupraxis, W. Ernst
   u. Sohn ed., Berlin, München, Düsseldorf (1979), pp. 59-73.
44 Sok, C., Baron, J., Francois. D.: Cem. Concr. Res. 9 (1979),
   pp. 641-648.
45 Lott, J., Kesler, C.E.: Highw. Res. Bd. Special Report 90
   (1966), pp. 204-218.
46 Kaplan, M.F.: J. ACI 58 (1961), pp. 591-610.
47 Ziegeldorf, S., Müller, H.S., Hilsdorf, H.K.: Cem. Concr. Res.
   10 (1980), pp. 589-599.
48 Ziegeldorf, S.: 10. Forschungskoll. des DAfStb, Karlsruhe (1979),
   pp. 43-47.
49 Welch, G.B., Haisman, B.: Mat. Constr. 2 (1969), pp. 171-177.
50 Glucklich, J.: J. Eng. Mech. Div., Proc. ASCE 89 (1963),
   pp. 127-138.
51 Hoagland, R.G., Hahn, G.T., Rosenfield, A.R.: Rock Mechanics 5
   (1973), pp. 77-106.
52 Schmidt, R.A.: Exp. Mech., May (1976), pp. 161-167.
53 Henry, J.P., Paquet, J.: Bull. Soc. géol. Fr. 6 (1976),
   pp. 1573-1582.
54 Henry, J.P., Paquet, J., Tancrez, J.P.: Int. J. Rock.Mech. Min.
   Sci. 14 (1977), pp. 85-91.
55 Henry, J.P., Paquet, J.: C-R. Acad. Sc. Paris, 284 (1977),
   pp. 571-514.
56 Henry, J.P., Paquet, J.: Mech. Res. Comm. 4 (3) (1977),
   pp. 193-198.
57 Carpinteri, A.: Nota Tecnica N. 47, Univ. di Bologna, 1980.

58 Erismann, T., Heuberger, H., Preuss, E.: Tschermaks Min. Petr.
   Mitt. 24 (1977), pp. 67-119.
59 Peng, S., Johnson, A.M.: Int. J. Rock Mech. Min. Sci., 9 (1972),
   pp. 37-86.
60 Corten, H.T.: in "Fracture - an Advanced Treatise" H.Liebowitz
   ed., Academic Press (1972), pp. 676-771.
61 Ziegeldorf, S., Hilsdorf, H.K.: Cem. Concr. Res. 10 (1980),
   pp. 723-724.

Fracture Mechanics of Concrete,
edited by F.H. Wittmann, 1983
Elsevier Science Publishers B.V., Amsterdam — Printed in The Netherlands

5.2 LINEAR ELASTIC FRACTURE MECHANICS PARAMETERS OF CONCRETE

by R. N. SWAMY

1. INTRODUCTION

Materials such as cement pastes, mortars and concretes, and all cement-based composites are essentially discontinuous, anisotropic, heterogeneous and multi-phase systems. Such materials contain interfacial bond microcracks and other inherent flaws arising from volume changes and other effects during fabrication of the materials. It is these bond microcracks and interfacial discontinuities which create through their geometry the nuclei for potential crack propagation and fracture.

When such materials are subjected to external loading, the initial response is largely linearly elastic (Fig. 1) and the existing microcracks progressively increase in number and grow. The process of progressive discrete microcracking results in non-linear stress-strain behaviour, and a quasi-ductile mode of failure in contrast to the ideal brittle or Griffith material in which the onset of crack growth is synonymous with fracture. The strength, stiffness and mode of failure of such materials are also affected by the size of the specimen.

There has been extensive argument about the applicability of the conventional Griffith concept to concrete materials, and whether concrete fracture follows the weakest link concept or the classical bundle concepts (1). There is a considerable volume of experimental evidence to show that there are three basic characteristics which differentiate concrete materials from other materials. Firstly, many tests show that cracking in concrete materials is irregular, tortuous and rough so that the newly formed surface area is many times larger than the effective fracture area.

Secondly, the heterogeneity of the material ensures that cracking itself is a heterogeneous process. Even the cement paste matrix, which is probably the nearest to an elastic material in the concrete system, is itself not homogeneous, and suffers some degree of microcracking prior to fracture. Five stages of crack growth can generally be therefore distinguished in the concrete system: initiation of the crack, slow stable crack growth, crack arrest, a critical crack condition and unstable crack propagation. Slow satellite cracking is thus an essential feature of the fracture behaviour of concrete, and there is substantial stress transfer across the interfacial

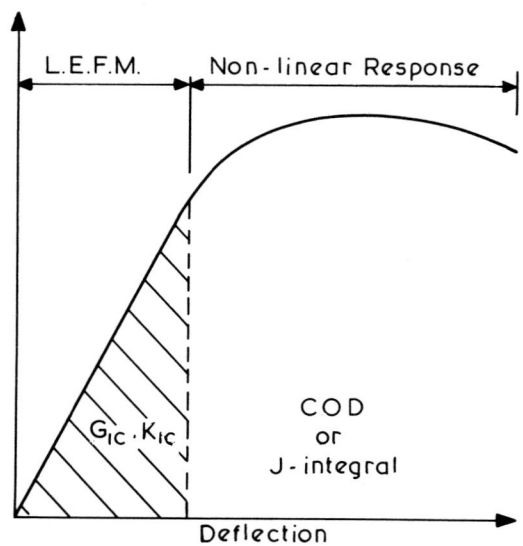

Fig. 1 Typical load-deflection curve of cement-based materials

microcracks.

Lastly, because of the enormous fracture surface area and the heterogenous
nature of the fracture process, the energy dissipating mechanism in concrete
materials is not merely confined to the surface energy (2, Fig. 2). There is
as yet no experimental or theoretical quantification of the various energy
dissipating mechanisms involved in concrete fracture. However, several
applications of linear elastic and non-linear (3) fracture mechanics have
been made to cement paste, mortar and concrete. The results obtained by
different investigators show a wide range of experimental values for the
critical release rate of elastic strain energy, the stress intensity factor,
and other fracture parameters, and probably confirm that forms of energy
dissipation other than surface energy are involved and not accounted for in
the calculated values.

In this section, the available evidence on the application of linear
elastic fracture mechanics to concrete materials is reported and discussed.

2. LINEAR ELASTIC AND NON-LINEAR FRACTURE MECHANICS PARAMETERS

The fracture toughness parameters of concrete materials may be obtained
either by applying linear elastic fracture mechanics (LEFM) or non-linear
fracture analysis. If the material is assumed to be fully elastic or if

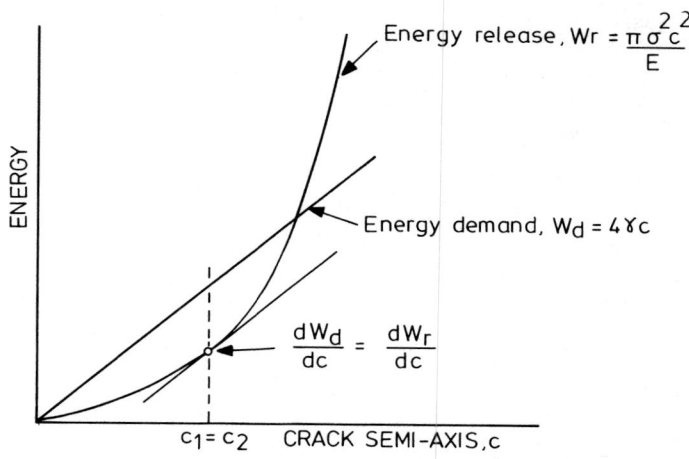

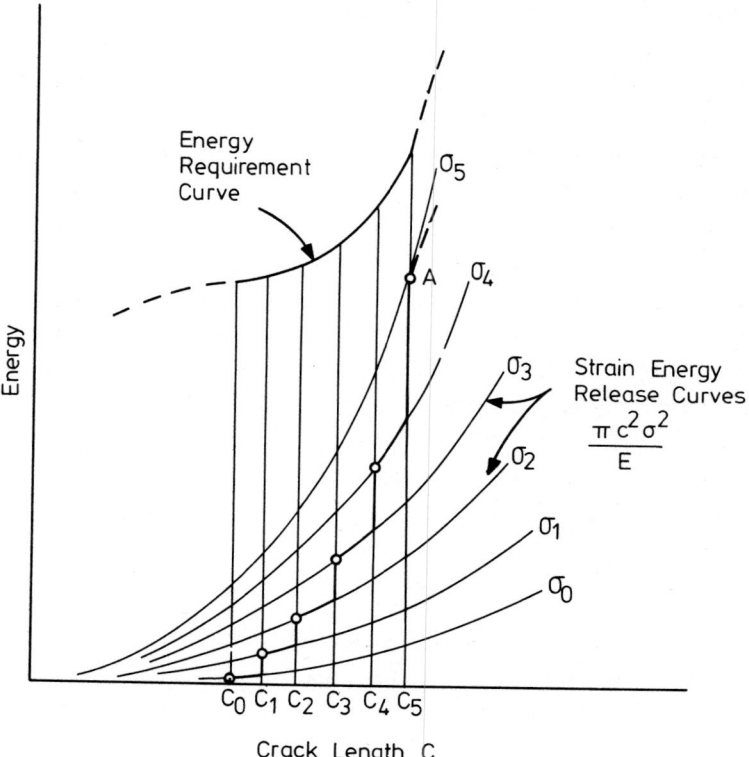

Fig. 2 (a) The ideal Griffith Material (b) Material showing stable crack growth

consideration is confined to the linear part of the load-deflection curve
(Fig. 1), the fracture toughness of a material may be determined by LEFM
provided the non-linear effects are sufficiently small to be neglected and the
test specimen geometry is sufficiently large to ensure plane strain conditions.
There is some argument that so far as concrete is concerned there is no
essential difference between plane strain and plane stress conditions, but the
evidence of test data is not conclusive in this respect.

The two fracture parameters measured by LEFM techniques are the critical
strain energy release rate G_{lc} and the critical stress intensity factor K_{lc}.
The experimental data appear to show that both these parameters are strongly
dependent on the specimen geometry, the type of test and the method of
measurement. For simple cases the values of G_c and K_c can be calculated
analytically; in complex cases, finite element methods or other numerical
techniques are required to evaluate G_c and K_c.

Two major reasons viz: (i) that concrete itself is not a perfectly brittle
material, and (ii) that aggregate or fibre inclusions impart to the matrix
considerable apparent ductility and large amounts of inelastic deformation
prior to fracture, often invalidate the application of LEFM to concrete
materials and fibre cement composites. In these materials it is most likely
that both crack opening and crack propagation occur more or less at the same
time. Separation of notch surfaces without extension of the notch cannot occur
in concrete materials since plastic yielding at the notch tip is limited and
the tensile strength of cementitious materials is only a fraction of their
compressive strength. So non-linear fracture toughness methods developed for
metallic materials cannot be strictly applied to concrete systems, although
many of the existing linear elastic criteria have been extended into the
elasto-plastic region. The one advantage of these non-linear methods is that
they recognise energy dissipation systems other than the creation of fracture
surfaces, and in this respect, they are superior to LEFM methods.

The most common techniques of elasto-plastic fracture mechanics which have
been applied to concrete materials and fibre cement composites are (3)

    i)   the critical crack opening displacement method (COD) (3-5)

   ii)   the J-integral approach (5-7)

  iii)   the R-curve analysis, (8-12) and,

   iv)   the fictitious crack model (13-15).

A number of investigations has been reported using these techniques. The
experimental data so far reported again show some contradictions and uncertainties in the application of these techniques as well to cementitious materials.
Only a limited amount of experimental results has been reported using R-curve
analysis, and it is not clear yet whether its application to concrete

materials is valid. The fictitious crack model has been largely developed by
Hillerborg and his collaborators, and experimental evidence on its validity is
also at the moment limited.

3. FRACTURE TOUGHNESS MEASUREMENTS IN CONCRETE MATERIALS

The purpose of this chapter is to describe the various experimental methods
of determining the linear elastic fracture mechanics parameters of concrete.
It is not intended to provide here a complete list of all the investigations
carried out or reproduce all the data obtained; rather, the aim of this
chapter is to identify the test methods and techniques applied to determine
fracture toughness parameters, and compare these, to a limited extent, with
data obtained from other non-elastic test methods. For a complete list of
published research on the cracking and fracture of concrete, attention is
directed to chapter 8 which provides a detailed annotated bibiliography of
papers published between 1928 and 1981.

3.1 Measurements of Critical Strain Energy Release Rate

It has been shown earlier in this publication that the basic fracture
mechanics theory can be developed in terms of an energy criterion or a stress
intensity factor. The strain energy release rate is given by

$$G = \pi c \sigma^2 / E \tag{1}$$

G would depend on the material, the specimen geometry and the state of
growth of the crack. The critical strain energy release rate, G_c, may then be
interpreted as a material parameter and can be measured in a laboratory with
sharply notched test specimens. Only the applied force and crack length need
to be measured to compute G. The applied load at the instability point is
easily obtained from an autographic load-displacement record. The appropriate
crack length is more difficult to measure. The critical crack length c_c should
include the equivalent length of slow crack growth in addition to the notch
depth. This length of slow crack growth may be found by a compliance method,
or by an ink staining technique or high speed photography.

A more reliable method of measuring G_c is by determining the compliance
of the specimen at different stages in the growth of the crack. This eliminates
the need for the difficult experimental measurements of critical stress and
critical crack length at the onset of fast crack propagation. It can be shown
that

$$G = \tfrac{1}{2} P^2 (dC)/(dA) \tag{2}$$

where P is the applied load and C, the specimen compliance. Measurements of

G_c have been made for glass, glassy polymers, adhesives, ceramics and concrete.

3.2 Fracture Toughness in Compression

Unlike those in a tensile stress field, cracks in a compression stress field are stable for a major part of the loading cycle and contribute to progressive crack propagation. There is, however, little experimental data on the determination of fracture toughness parameters in compression. Glucklich (16) has shown that corresponding to the classical Griffith equation for uniformly stressed infinite plate under plain strain conditions,

$$\sigma_{comp} = \left(\frac{8 \; E \; \gamma}{\pi \; b (1-\nu^2)} \right)^{\frac{1}{2}} \tag{3}$$

where 2b is the minor axis of the crack. Glucklich has also suggested two different expressions for the critical strain energy release rate G_c (17, 18).

$$G_c = \frac{\pi \; b \; \sigma^2}{2 \; E} \tag{4}$$

and,

$$G_c = \frac{\pi \; c \; \sigma^2 \; \sin^2 \phi \; \cos^2 \phi \; (1-\nu^2)}{E} \tag{5}$$

where ϕ is the inclination of the crack to the line of action of the applied stress.

Table 1 shows values of G_c in compression obtained by Knox on perspex plates (19) and Desayi on mortar and concrete prisms (20). These data show large variations.

3.3 Fracture Toughness in Tension

Values of G_c have been computed by various investigators from flexural and direct tension tests. Representative data from some of these studies (21-33) are summarised in Table 2, which also includes values of G_c for other construction materials for comparison purposes. These studies show that G_c is influenced by a number of parameters such as size and geometry of test specimen, quality of the cement paste or mortar matrix, type and volume of aggregate, nature of aggregate-matrix bond, allowance for crack growth, the value of Young's modulus and the type of loading. The higher the value of G_c, the more difficult it becomes to achieve brittle fracture.

In G_c measurements, it is important to consider the slow crack propagation which occurs prior to rapid crack propagation (21), and this implies that the velocity of crack propagation must also be taken into account (2). Neglecting slow crack growth can lead to considerable underestimation in the computed values of G_c (25). The moisture conditions of testing are also significant and

TABLE 1   Strain Energy Release Rate in Compression

Investigator	Test Geometry	Material System	Remarks	Experimental Results G_c, kN/m
Knox (19)	Perspex plates, 152 x 152 x 19mm	Concrete	Crack growth considered Crack angle $20^{\circ} - 35^{\circ}$	14.2 - 352
Desayi (20)	Prisms, 152 x 152 x 305mm	Mortar Concrete	No allowance for crack growth	0.1352 - 0.3470* 0.0427 - 0.1732** 0.1350 - 0.4740* 0.0536 - 0.2317**

* based on equation (4)

** based on equation (5)

TABLE 2   Strain Energy Release Rate in Tension

Investigator	Test Geometry	Material System	Remarks	Experimental Results G_c, kN/m
Kaplan (21)	Notched beam, 4-point bending 76 x 102 x 406mm and 152 x 152 x 508mm	Mortar	No correction for slow crack growth	0.0173 - 0.0285
		Concrete		0.0140 - 0.0294
	Notched beam, 3-point bending 76 x 102 x 406mm 152 x 152 x 508mm	Mortar		0.0109 - 0.0268
		Concrete		0.0067 - 0.0275
Romualdi and Batson (22)	Plate specimen with centre crack 610 x 813 x 63mm	Mortar	No correction for slow crack growth	0.0053 - 0.0123
Glucklich (23)	Notched beam, 4-point bending 51 x 102 x 1067mm	Mortar	Correction for slow crack growth	unnotched : 0.0201 notched : 0.0193
Lott and Kesler* (24)	Notched beam, 4-point bending 102 x 102 x 305mm	Mortar Concrete	No correction for slow crack growth	0.0035 0.0051

* Values of G_c derived from K_{1c}

1	2	3	4	5
Welch and Haisman (25)	Notched beam, 4-point bending 102 x 102 x 508mm	Paste	Correction for slow crack growth + E (dynamic)	0.0245
		Mortar		0.0385
		Concrete		0.0187 – 0.0363
		Concrete	no crack growth + E (dynamic)	0.0063 – 0.0154
Moavenzadeh and Kuguel (26)	Notched beam, 3-point bending 25 x 25 x 305mm	Paste	No correction for slow crack growth	0.0035 – 0.0050
	Cracked (line notch) beam, 3-point bending 25 x 25 x 305mm	Mortar		0.0245 – 0.0043
		Concrete		0.0084 – 0.0096
Naus and Lott (27)	Cracked (line notch) beam 4-point bending 102 x 102 x 305mm	Concrete	No correction for slow crack growth	0.0072
Okada and Koyanagi (28)	Notched beam, 4-point bending 47 x 100 x 388mm	Paste	No correction for slow crack growth	0.0078
		Mortar		0.0098 – 0.0157
		Concrete		0.0118 – 0.0147

1	2	3	4	5
Glucklich and Korin (29)	Notched beam, 4-point bending 12 x 25 x 280mm	Mortar	Corrected for slow crack growth Variable moisture content	0.0098 - 0.0490
Mindess, Lawrence and Kesler (30)	Notched beam, 4-point bending 76 x 76 x 381mm	Paste	No correction for crack growth	0.0088 - 0.0154
		Concrete		0.0172 - 0.0175
		Steel fibre concrete (0.23 - 2.0%)		0.0114 - 0.0434
		Glass fibre concrete (1%)		0.0194 - 0.0219
Knox (19)	Notched beam, bending	Concrete		0.0175
Barr and Bear (31)	Circumferentially notched round bar, 4-point bending (CNRBB)	Mortar		0.0648 - 0.1174
Mazars (32)	Plate specimen with centre crack and single-edge crack 600 x 340 x 80mm Notched beam, 3-point bending 150 x 100 x 1000mm	Concrete	Variable crack length Finite element analysis	0.0102

1	2	3	4	5
George (33)	Notched beam, 3-point loading 76 x 76 x 286mm	Soil cement	No correction for slow crack growth	0.0035 - 0.0166
Hahn, Kanninen and Rosenfield (34)		Plain carbon steels ductile fracture		≈ 500 - 900
		High strength steels ductile fracture		5 - 130
		Low to medium strength steels brittle fracture		0.6 - 60
		Epoxy resin		0.22
		Douglas Fir, parallel to grain		0.03
		Glass		0.002 - 0.008
		Polymethyl Methacrylate		0.5

tests by Glucklich and Korin (29) show that drying of cement mortar specimens from saturation to almost oven dryness increases G_c gradually, after an initial decrease.

3.4 Fracture Energy Measurements

When applying fracture toughness measurements care must be taken to distinguish measurements which give "work of fracture" by observing the total energy required to fracture a notched specimen, and measurements which yield G_c. In a perfectly brittle material, measurements of the surface energy γ from "suitable experiments" would correspond to determination of G_c. In materials which are not perfectly brittle, measurements of γ yield values greater than the true surface energy. Equivalence with G_c can then be assumed only if the effective γ is measured at the onset of rapid crack propagation. In any case computations of fracture energy and fracture toughness from areas of load-displacement curves can be grossly misleading and may contain significant errors because of the testing machine stiffness characteristics on the energy absorbed during fracture. Further, machine interaction can produce extraneous fracture patterns and mask the true behaviour of a material. Considerable discrepancies can thus occur in the measured and computed values of γ if this care is not exercised (21, 26, 35). (Table 3).

There is considerable evidence from the results of tests on brittle ceramic materials and composites that fracture energies computed from notched beam tests, work-of-fracture method and double cantilever beam tests show considerable variation depending upon the type of test. Further, fracture energies derived from work of fracture method and notched beam tests themselves show considerable scatter, and definite dependence on the notch depth - specimen thickness ratio from about 0.2 to 0.9. It therefore appears that all such data can only be used for qualitative comparison and cannot be treated as absolute values. The work of fracture method is strain rate dependent and invariably gives fracture energies totally unrelated to fracture energies obtained from fracture toughness type of tests.

It should also be added that surface energy measurements are influenced by (i) intrusion of impurities on the fractured surfaces, (ii) presence of moisture (iii) the geometry of the cleavage surface, and (iv) temperature.

Using the fracture energy reported by Cooper and Figg (36) and assuming plane stress conditions, the expected flaw sizes, for net tensile fracture stress of 5 N/mm^2 in hardened cement paste with a Young's modulus of 14 kN/mm^2, likely to be present in cement paste which would propagate in an unstable manner are 7 mm (dry) and 6 mm (wet). However, these estimates of flaw size should be treated with caution with due regard to their applicability.

TABLE 3   Fracture Energies of Various Materials

Investigator	Material	Test Conditions	Experimental Fracture Surface Energy J/m^2
Brunauer, Kantro and Weise (35)	Tobermorite	Heat of dissolution method	0.39
Moavenzadeh and Kugel (26)	Hydrated cement paste (28 days)	25 x 25 x 300 mm; W/C = 0.5	2.35-4.22
	Mortar (7 days)	Three-point bending test	2.14-4.75
	Concrete (28days)		3.50-6.25
Cooper and Figg (36)	Hydrated cement paste	25 x 50 x 200 mm; W/C = 0.5 Slow bend test	14.90 (dry) 12.40 (wet)
Kaplan (21)	Mortar 1 : 2.4	75 x 100 x 400 mm, and 150 x 150 x 500 mm 3- and 4-point bend tests	19.30 (measured) 1.61 (calculated)

1	2	3	4
Harris, Varlow and Ellis (37)	Mortar 1: 3.0	Notch beam, 3-point bending Dry, variable age Wet, variable age	20 – 23 25 – 40
Auskern and Horn (38)	Paste Polymer impregnated Paste	13 x 13 x 76, three point bending test	8.7 – 9.5 66.0 – 84.0
Beaumont and Aleszka (39)	Fibre concrete Polymer impregnated fibre concrete	25 x 25 x 125 mm, three point bending	1.75 20.00
	Silica glass Polymethyl Methacrylate Mild steel	Dry, variable age Wet, variable age	5 – 10 350 – 600 2970 – 3200 3800 – 4050

Although the energy approach has generally been abandoned in favour of the stress intensity factor approach, G_c, defined as fracture toughness, has recently been analytically related to compressive strength and the flexural moment of the modulus of rupture test (40). The values of G_c computed from existing direct tension and flexural tests show reasonable agreement between the results. This approach has been used to evaluate the shear at diagonal tension cracking in both flexurally cracked and flexurally uncracked reinforced beams, and the torque at diagonal tensional cracking, in a rein- forced concrete beam subjected to torsion.

The fictitious crack model also appears to be capable of application to concrete structural members but much more work needs to be done before the validity of these models can be critically assessed. Whilst both the energy approach and the fictitious model appear to be able to predict failure loads, they seem to be unable to explain logically the failure mechanisms involved, and this is an important requirement from a designer's point of view. Future work needs to be directed towards this goal to provide a physical basis to the energy models.

4. MEASUREMENTS OF STRESS INTENSITY FACTOR

One of the limitations of direct measurements of G_c is that it does not involve a knowledge of the stress distribution in the vicinity of the crack. Further, the phenomena of surface energy and plastic work accompanying fracture development are not yet fully understood, so that the stress intensity approach is now generally preferred to the energy approach. The K_c parameter is often a more useful representation of a material for engineering applications, and G_c is more useful when fracture mechanics is applied to material development.

The basis for the application of linear elastic stress field fracture mechanics is that a value of K_c or K_{lc} determined from the unstable fracture of suitable test specimens will be identical to the K value at fracture instability for a crack in a structure in service. The K value as a function of load and specimen geometry for test specimens (called K calibration) may be determined by theoretical elastic stress analysis (41) or by experimental compliance measurements.

The experimental methods of fracture toughness determination are well established (42, 43). The general procedure is to adopt a test specimen, of suitable geometry and dimensions, into which a crack of suitable size is introduced. The specimen is then loaded at a slow rate so that the crack is incrementally extended, and some means provided to measure crack extension.

Since K_{lc} is considered to be a material property, there are minimum dimension requirements to guard against general yield of the specimen. The specimen dimensions are determined by the ratio of toughness to the yield

strength of the material (Fig. 3). The required degree of sharpness of the
notch or slit is also important. In general, a natural crack should be
initiated at the root of the notch or slit. The largest source of error in K_c
(and G_c) measurements is the uncertainty about the exact crack length at the
moment of instability. Several methods of measuring crack length have been
used, such as penetrating liquids (ink staining), cinematography, optical
microscopy, electric potential, acoustic emission, and displacement gauges.
More recently, ultrasonics have been used for detecting the onset of crack
extension and for monitoring continued crack growth in fracture toughness tests.

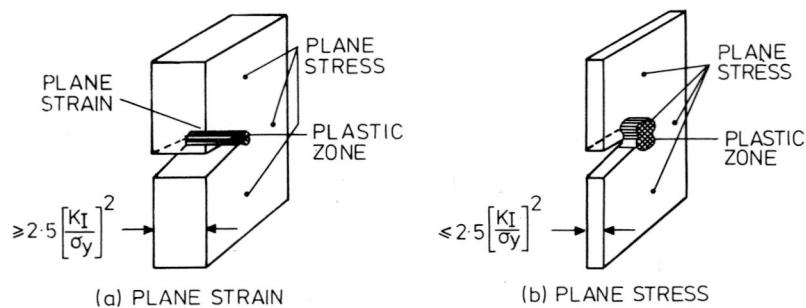

Fig. 3   State of stress in a cracked plate (a) Plane strain (b) Plane stress

Although a wide range of specimen geometries has been adopted for fracture
mechanics measurements, in concrete fracture toughness testing, there are no
standard specimen dimensions or methods of testing, and many of the reported
tests do not conform to the requirements of K_{lc} determination.

In spite of the lack of standard specifications for specimen geometry and
test methods, a considerable amount of research has been carried out to
determine the fracture parameters K_{lc} (and G_{lc}) of hardened cement·paste,
mortar, concrete and rock. The test methods used to determine these parameters
include 3-point bending and 4-point bending of notched beams, plate specimens
with centre crack, double cantilever beam (DCB), compact tension (CT)
specimen, double torsion plate specimen and double-edge-cracked tests
(Figs. 4-9). The influence of a wide range of parameters on fracture toughness
has been studied, and some representative test results are shown in Tables
4 and 5.

The circumferentially notched round bar tests have been used to evaluate
the fracture toughness of rocks and mortars primarily because of the relatively
small dimensions of the test specimens (31, 56, 57). In these tests the
specimens are either subjected to four-point bending (the CNRBB test)
(Fig. 9a) or to an eccentric longitudinal load (the CNRBEL test) (Fig. 9b).
The bend specimen and the eccentrically loaded specimen can be seen to be
similar to the standard bend and compact tension specimens used to evaluate

the fracture toughness of high-strength metals.

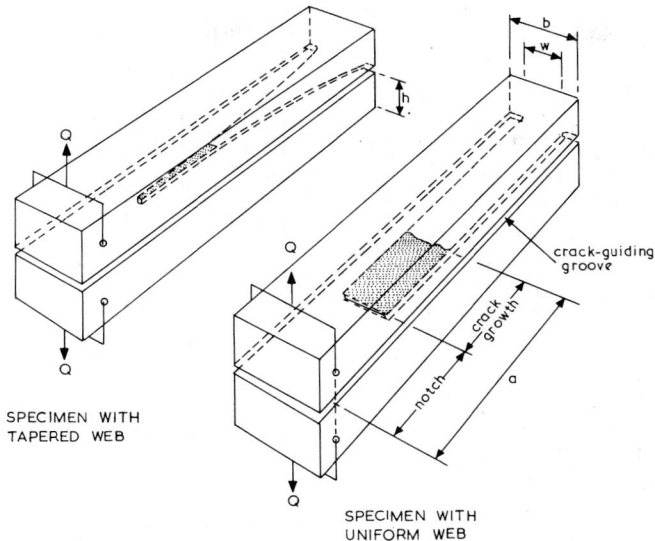

Fig. 4   The double cantilever beam test for fracture
         toughness.   General arrangement of DCB specimens

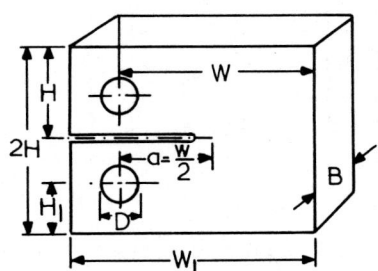

Fig. 5   Compact tension (CT) specimen.   Wedge opening loading (WOL).
         $W = 2.0B$, $D = 0.5B$, $a = 1.0B$, $W_l = 2.5B$, $H = 1.2B$, and $H_l = 0.65B$

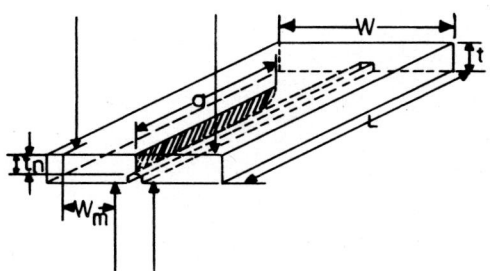

Fig. 6   Double torsion test: plate specimen

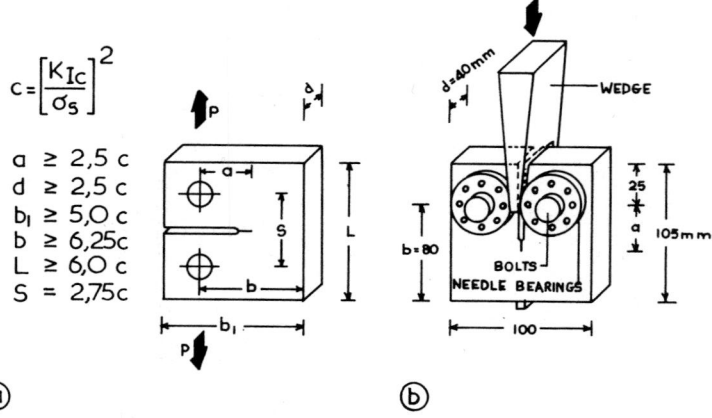

Fig. 7   Compact tension (CT) specimen.   (a) Specimen according to ASTM, Plane strain Fracture Toughness of Metallic Materials, Designation E 99/72.   (b) Wedge loaded specimen

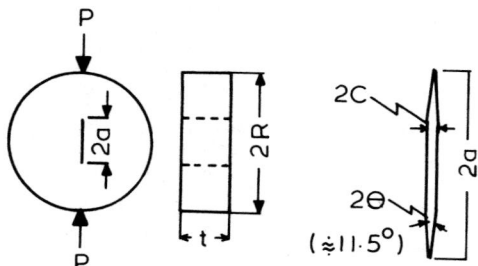

Fig. 8   Diametral compression test specimen

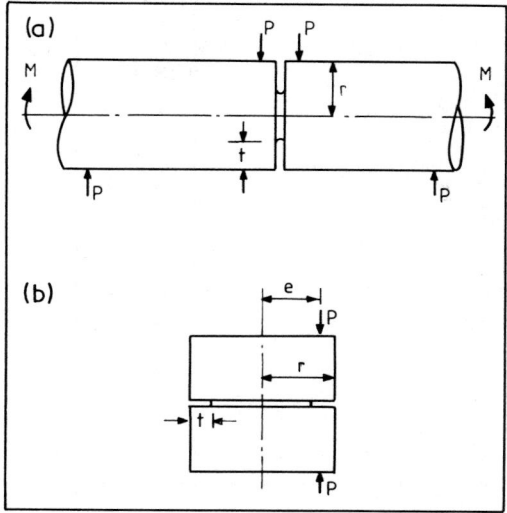

Fig. 9 (a) Circumferentially notched round bar subjected to four-point
bending (CNRBB) (b) Circumferentially notched bar under eccentric
loading (CNRBEL).

The fracture toughness values are given by (57, Fig. 9)
for the CNRBB specimen :

$$K_1/M = 3/\{2\sqrt{[\pi r^5(1 - t/r)^5(0.89 + 0.11\ r/t)]}\} \qquad (6)$$

for the CNRBEL specimen:

$$K_1/P = 3e/\{2\sqrt{[\pi r^5(1 - t/r)^5(0.89 + 0.11\ r/t)]}\}$$
$$- 1/\{2\sqrt{[\pi r^3(1 - t/r)^3(0.89 + 0.2\ r/t)]}\} \qquad (7)$$

In the CNRBEL test, the points of application of the load are taken as the
inside edges of the steel load distribution bars, since the load is concentrated to the inside edges as deformation of the specimens proceeds.

From the reported test data, the following conclusions can be drawn.
Factors influencing mix design also generally affect fracture toughness.
Water-cement ratio, air content, curing regime, fine aggregate content, and
the type and volume fraction of coarse aggregate all influence fracture toughness. The two major factors affecting fracture toughness appear to be the
matrix strength and the type and volume concentration of coarse aggregate
inclusions which act as crack arresters.

TABLE 4 Critical Stress Intensity Factors from Fracture Toughness Tests

Investigator	Test Geometry	Material System	Remarks	Experimental Results K_{lc}, $MN/m^{3/2}$
Kaplan* (21)	Notched beam, 4-point bending 76 x 102 x 406 mm and 152 x 152 x 508 mm	Mortar		0.64-0.78
		Concrete	No correction for slow crack growth	0.55-0.92
	Notched beam, 3-point bending 76 x 102 x 406 mm and 152 x 152 x 508 mm	Mortar		0.67-0.89
		Concrete		0.57-1.15
Romualdi and Batson* (22)	Plate specimen with centre crack 610 x 813 x 63 mm	Mortar	No correction for slow crack growth	0.34-0.52
Lott and Kesler (24)	Notched beam, 4-point bending 102 x 102 x 305 mm	Mortar	No correction for slow crack growth	0.29-0.33
		Concrete		0.34-0.40

* Values of K_{lc} computed from G_c

1	2	3	4	5
Welch and Haisman (25)	Notched beam, 4-point bending 102 x 102 x 508 mm	Paste Mortar Concrete Concrete	Correction for slow crack growth + E (dynamic) No crack growth + E (dynamic)	0.85 1.08 0.78-1.28 0.45-0.83
Moavenzadeh and Kuguel (26)	Notched beam, 3-point bending 25 x 25 x 305 mm Cracked (line notch) beam, 3-point bending 25 x 25 x 305 mm	Paste Mortar Concrete (0.5 W/C; age of test 3-28 days)	No correction for slow crack growth	0.13-0.17 0.14-0.15 0.23-0.26
Naus and Lott (27)	51 x 51 x 356 mm (paste and mortar) Cracked (line notch) beam, 4-point bending 102 x 102 x 305 mm	Paste Mortar Concrete	No correction for slow crack growth	0.31-0.45 0.21-0.57 0.37-0.77

1	2	3	4	5
Kesler, Naus and Lott (44)	Plate specimen with through-thickness flaw, wedge loaded 51 x 305 x W (457-914) mm	Paste Mortar Concrete	With and without correction for slow crack growth	0.08-0.26 0.30-1.26 0.34-1.43
Harris, Varlow and Ellis (37)	Notched beam, 3-point bending	Mortar	Dry, variable age Wet, variable age No correction for slow crack growth	0.40-0.29 0.41-0.44
Walsh (45)	Notched beam, 3-point bending 76 mm x 1.5d x 5d (d = 152 and 254 mm)	Concrete		0.54-1.07
Okada and Koyanagi (28)	Notched beam, 4-point bending 47 x 100 x 388 mm	Paste Mortar Concrete	No correction for slow crack growth	0.22 0.29-0.39 0.33-0.46

1	2	3	4	5
Brown and Pomeroy (46)	Cracked (line notch) beam, 4-point bending, 38 x 38 x 250 mm	Paste		0.30
	Double cantilever beam (DCB)	Mortar and Concrete	Crack growth considered	0.45-0.95
	50 x 100 x 350 mm (Fig. 4)	Paste		0.30-0.45
		Mortar		0.35-1.10
Carmichael and Jerram (47)	Compact tension (CT) specimen 25.4 mm thick (wedge opening loading) (Fig. 5)	Mortar		0.32-0.48
Nadeau, Mindess and Hay (48)	Double torsion plate specimen 229 x 76 x 13 mm (Fig. 6)	Paste	Machined notch sharp crack	0.34 0.29
	Notched beam, 3-point bending 13 x 13 x 38 mm	Paste		0.32
Mindess, Nadeau and Hay (49)	Double torsion plate specimen, 229 x 76 x 10 mm	Paste	Variable curing conditions	0.31-0.37
	Notched beam, 3-point loading 10 x 10 x 37 mm	Paste		0.33-0.48

1	2	3	4	5
Mindess and Nadeau (50)	Notched beam, 3-point bending; B x 51 x 203 mm (B = 45-254 mm)	Mortar; Concrete	Variable width of crack front	0.47; 0.76
Higgins and Bailey (51)	Cracked beam (line notch), 3-point bending, 25 x 14 x 90 mm; Cracked specimens, direct tension (double-edge-cracked specimen)	Paste; Paste	No correction for slow crack growth (variable notch width); Variable specimen depth No correction for slow crack growth	0.36-0.46; 0.40-0.65; 0.32-0.65; 0.39-0.49
Kitagawa and Suyama (52)	Notched beam, 3-point bending 40 x 80 x 320 (span) mm, 30 x 60 x 240 (span) mm, and 20 x 40 x 160 (span) mm	Mortar; Concrete	No correction for slow crack growth	0.27; 0.34
Hillemeier and Hilsdorf (53)	Wedge loaded compact-tension specimen, 100 x 105 x 40 mm	Paste; Aggregate; Aggregate-Matrix interface; Paste	Correction for slow crack growth; Marble; Quartz; Polymer modified matrix	0.49-0.31; 2.0; 3.3; 0.10-0.12; 0.24

1	2	3	4	5
Khrapkov, Trapesnikov, Geinats, Pashchencko and Pak (54)	Notched cylinders, 100 x 150 (dia.) mm	Concrete	Variable age	0.15-0.61
	Notched cylinders, 400 x 400 (dia.) mm	Concrete	Variable aggregate size	0.56-1.01
Mindess, Lawrence and Kesler (30)	Notched beams, 4-point bending 76 x 76 x 381 mm	Paste	No correction for	0.50-0.66
		Concrete	slow crack growth	0.87-0.88
George (33)	Notched beam, 3-point bending 76 x 76 x 286 mm	Soil cement	No correction for slow crack growth	0.09-0.14
Hahn, Kanninen and Rosenfield (34)		Plain carbon steel, ductile fracture		≈ 300-400
		High strength steels, ductile fracture		30-150
		Low to medium strength steels brittle fracture		10-100
		Epoxy resin		0.80
		Douglas Fir, parallel to grain		0.30
		Douglas Fir, normal to grain		0.40
		Glass		0.3 - 0.6

TABLE 5   Critical Stress Intensity Factors for Fracture Toughness Tests

Investigator	Test Geometry	Material System	Remarks	Experimental Results K_{lc}, $MN/m^{3/2}$
Kitagawa, Kim and Suyama (55)	Notched ring, diametral compression test 100 x 33 mm and 150x50mm	Mortar Concrete	No correction for crack growth	0.27-0.29 0.30-0.36
Barr and Bear (31, 56)	Circumferentially notched round bar, 4-point bending (CNRBB) 41x150mm and 54x150mm	Mortar	No correction for	0.51-0.63 0.74-1.00
	Circumferentially notched round bar, eccentric loading (CNRBEL), 54 mm dia. (Fig. 9)	Mortar	Variable notch depth	0.71-1.02 0.51-0.86

1	2	3	4	5
bear and Barr (57)	CNRBB test: ext. dia. 53.6 mm dia. at notch 33.4 mm	Mortar (0.6-2.4 mm grain size)	No correction for slow crack growth	0.74-0.86
	CNRBEL test	Mortar		0.71-0.87
	CNRBEL test	Slate (rock)	specimens drilled perpendicular to bedding plane	0.20-0.26
			specimens drilled parallel to bedding plane	0.07-0.15
	CNRBB/CNRBEL test	Dolomite (rock)	weathered rock	0.79-0.93
		Dolomite	fracture of calcite veins	1.09-1.53
		Dolomite	fracture of dolomite matrix	1.71-2.02
Barr, Evans and Dowers (63)	Split-cube test (Fig. 12) 100 mm cubes	Concrete 10 mm aggregate	No correction for slow crack growth	0.56-0.64

The fracture tougness values of aggregates (e.g. marble and quartz) are about an order of magnitude larger than those of the cement paste matrix (53). On the other hand in testing rock samples, the existence of cleavage planes and of anisotropy should be recognised. Slate, for example, has well defined cleavage planes, and tests (57) have shown that the K_{lc} value for specimens drilled parallel to the bedding planes is about half (or even less) than that for specimens drilled perpendicular to the bedding planes (Table 5). Dolomite on the other hand is relatively isotropic with no visible bedding planes. Cores taken in three mutually perpendicular directions and subjected to CNRBB and CNRBEL tests (57), however, showed large scatter. Analysis of data showed that the scatter related to failures in the various mineral components in the dolomite, and that when the test data were grouped on the basis of failure, the results were remarkably consistent. In testing rock samples therefore, their multiphase nature, the existence of cleavage planes, and similar special features need to be considered.

Addition of suitable aggregate can therefore increase fracture toughness, but it also results in a progressive increase in toughness with crack growth (Fig. 10, 46), although crack initiation probably occurs at a K_{lc} value corresponding to that for the paste. The higher the proportion of aggregate, the larger is generally the increase in toughness. Volume fraction, rather than grading, of aggregate appears to be more important. There is some evidence that K_{lc} varies approximately linearly with aggregate volume (46), and there is also some evidence that toughness increases with maximum size of the coarse aggregate particles (54). When considering different types of coarse aggregates, the aggregate-matrix bond may be the more influential factor affecting fracture toughness, since the fracture toughness of aggregate-matrix interfaces is considerably lower than that of the hardened cement paste (53, 58). However, with aggregates having cleavage planes, their fracture toughness can be as low as the aggregate-matrix interface, and the cement paste could also have a value of the same order of magnitude.

Many tests have ignored the effect of crack growth on K_{lc} (Table 4). Hardened cement paste is generally brittle, and it is difficult to restrict and control crack growth, particularly in flexural tests. K_{lc} would therefore be expected to be independent of crack growth in cement paste. Both notched beam and DCB tests show that this is so; wedge-loaded compact tension tests, however, show that fracture toughness decreases with crack growth and eventually stabilises to a constant value when one major crack develops to fracture (46, 53).

Khrapkov et al (54) also found that K_{lc} at unstable crack growth is less than that at stable crack growth. For concrete, they obtained values of 0.75-0.85 MN/m$^{3/2}$ at the former stage compared to a value of 1.90 MN/m$^{3/2}$

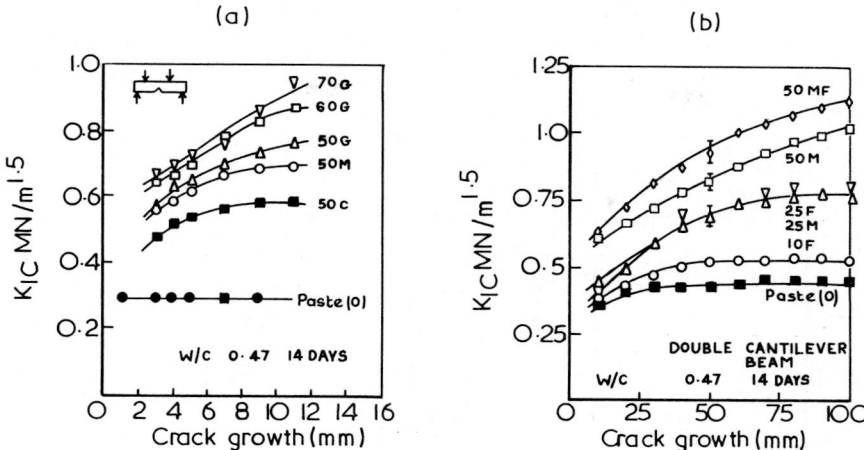

Fig. 10   Variation of fracture toughness with crack growth of cement paste,
mortar and concrete, (a) Notched beam, 4-point bending test.
(b) Double cantilever beam test. Aggregate sizes: F = fine
(52-100 mesh), M = medium (14-25 mesh), MF = mixture of equal
proportions of M and F, C = coarse (10-5mm) and G = graded
(10-100 mesh) (46).

at stable crack growth. The explanation probably lies in that at stable
crack growth there is extensive satellite microcracking at the pseudoplastic
zone of the crack tip. A higher energy is required for multiple crack growth.
As cracking proceeds and unstable crack growth is reached, not all the cracks
propagate at the same rate and final fracture is initiated by only one major
crack and hence the K_{lc} value is reduced. It may well be that these contradictions are also the result of the different types of tests used, and can
only be clarified by further research.

The addition of polymer dispersions and other admixtures also appears to
reduce the fracture toughness of the hardened cement paste matrix (53).

For hardened cement pastes, although fracture toughness increases in the
early ages, it remains practically constant after about 28 days. It would
then appear that fracture toughness of cement paste is dependent not merely on
capillary porosity but also on the structure and composition of the solid
constituents (46, 51). However, with concrete, fracture toughness is reported
to increase with age (54).

The crack geometry appears to have a clear effect on the fracture toughness parameters. The degree of sharpness, or lack of it, will influence the
ease or difficulty with which the crack is initiated at the crack tip. Tests
have been reported, however, that the type of notch has no significant effect
if slow crack growth is considered (25). The width of the crack or slit

i.e. the size of the flaw relative to the size of the microcracking zone has a significant influence on fracture toughness (Fig. 11, 44, 51). There is thus a distinct size effect on fracture toughness and many tests show that K_{lc} varies with specimen size (21, 44, 45, 51, 52).

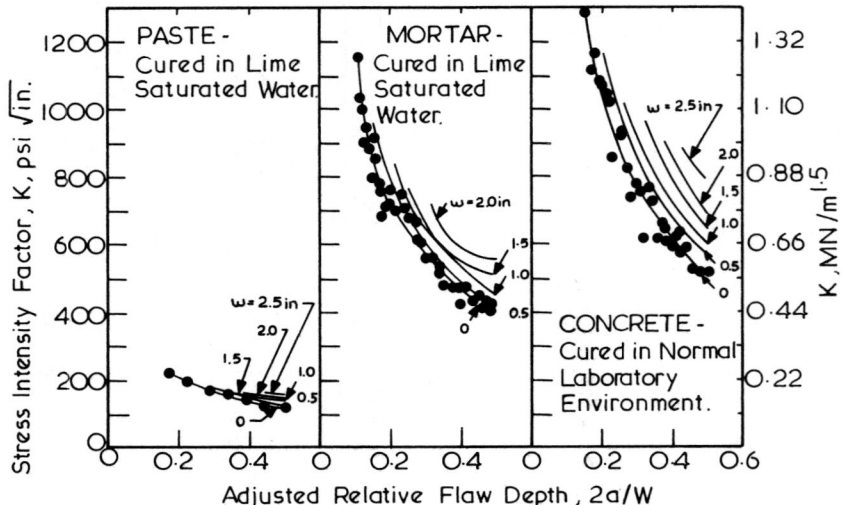

Fig. 11 Variation of stress intensity factor with relative flaw depth in relation to assumed microcracking zone size. (Plate specimens, crack length 2a, W = width of specimen, w = increase in total crack growth) (44)

Minor differences in fracture toughness have also been observed between sawn and cast-in notches, but these appear to be not so significant. Fracture toughness appears to be independent of the length of the crack front in notched beam tests (50); in plate tests with a central flaw, on the other hand, the length of the crack front appears to influence fracture toughness (44, 52).

Fracture toughness also appears to be relatively insensitive to rate of loading (31, 51).

Associated with the specimen size and crack geometry is the question of notch sensitiveness of cementitious systems. Notch sensitiveness depends very much on the type of test; notch beam tests show that hardened cement paste is notch sensitive (51, 53, 59), whereas notch plate tests show the opposite. These tests show that as the notch depth increases, the strength of the notched specimen is reduced by a factor of four; considering cementitious systems alone, the hardened paste is clearly notch-sensitive. However, if compared with other materials such as glass, where even the smallest of flaws could reduce strength by factors of the order of 100, the hardened cement paste could be considered to be relatively notch insensitive.

Although it is well established that size effects in mortar and concrete influence mean strength, scatter of results and mode of failure (60), in terms of fracture toughness mortar and concrete may be considered to be relatively notch insensitive compared to the hardened cement paste (21, 27, 59) probably because of the inelastic deformation or microcracking at the root of the notch. However, here again the perspective should not be lost. For laboratory size specimens, concrete may be relatively notch insensitive; but for larger systems like cladding panels and structural members, the material may be very much more susceptible to crack propagation (45, 60, 61).

The two tests on circumferentially notched round specimens (CNRBB and CNRBEL) (31, 56, 57) appear to give consistent and reproducible results, and the variability of data also appear to be low compared with other traditional tests (Table 5). The major advantages of these tests appears to be that they are suitable for testing in-situ samples, economic of material, and simple to test; the plane of crack propagation is determined by the circumferential notch, and the nominal crack area is relatively constant. For rock specimens, the tests are sufficiently sensitive to separate the fracture toughness values of the various mineral constituents in a rock.

4.1   The Split-cube Test

Since the circumferentially notched round bar tests are limited in application to rocks and mortars, the split-cube test has been developed by Barr and his colleagues for use with concrete mixes (62). In this test, double-edge notched standard cube specimens are subjected to eccentric loading (the DENCEL test) (Fig. 12). The loads are applied through two 6 mm square steel bars, of length equal to the side of the cube. The bars are positioned parallel to each other, at the edges of the test specimen as shown. The point of application of the load is taken at the edge of the steel bars nearest the notch-root since, as deformation proceeds, the load is concentrated in these edges.

Finite element solutions have been obtained for the DENCEL test specimens for both 100 mm and 150 mm cubes. Plane strain conditions were assumed, and several notch depths were considered giving a range of crack length/specimen width (a/d) ratios. For 100 mm cubes, the finite element solution gives the following expression

$$K_1 = \frac{P}{Bd^{\frac{1}{2}}} \left[18.3 \left(\frac{a}{d}\right)^{\frac{1}{2}} - 430 \left(\frac{a}{d}\right)^{3/2} + 3445 \left(\frac{a}{d}\right)^{5/2} - 11076 \left(\frac{a}{d}\right)^{7/2} + 12967 \left(\frac{a}{d}\right)^{9/2} \right]$$

$$(8)$$

where K_1 = stress intensity factor in opening mode

P = load

B = width of specimen (100 mm)

d = depth and width of specimen (100 mm)

a = depth of slot.

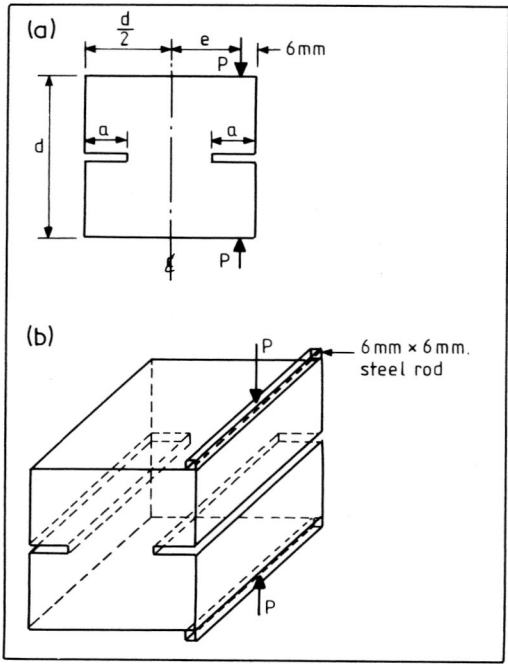

Fig. 12 Geometry and loading details of the split-cube (DENCEL) test

Some typical fracture toughness data obtained from split-cube tests (DENCEL specimens) are shown in Table 5 (63). The split-cube geometry appears to give reproducible results with low coefficients of variation (64). The advantage of the DENCEL test is that specimens can be manufactured easily (and at site) with standard moulds and the test carried out using a simple portable hydraulic bench mounted rig. Provided standard notch depths are introduced, the fracture toughness also appears to be proportional to the load at failure.

4.2 The J-integral

The parameter known as the J-integral, as an extension to nonlinear fracture mechanics, has been studied in detail as a quantitative measure of

toughness for plain concrete and fibre reinforced cement composites (3). The
evaluation of the J-integral is based on the method suggested by Rice et al
utilizing load-displacement diagrams of notched beams under 4-point bending
and applying the maximum load failure criterion (65). The concept was first
applied to concrete materials by Mindess, Lawrence and Kesler (30) and has
since been extensively studied by Halvorsen (6, 7) by Shah and Velazco (66)
and others (67).

Table 6 gives some representative data on the J-integral for comparison with
data on K_{lc} in Tables 4 and 5. Halvorsen's extensive study reveals some
interesting information as to the relationship between fracture toughness and
compressive strength, and in particular, the role of the concrete constituents
on crack resistance. He tested the paste and mortar phases of three different
concretes containing three different aggregates but having the same paste and
aggregate contents by volume. The paste phase, although substantially stronger
than the concretes, had about one third of the typical fracture toughness
value observed for the concretes with coarse aggregates (Table 6). The mortar
phase had also higher compressive strength but had the same order of fracture
toughness of the resulting coarse aggregate concrete. The concretes containing
the same paste and aggregate volume but different types of aggregates showed
that fracture toughness generally increased with compressive strength,
although not necessarily in proportion. These differences in fracture toughness related to the failure mode of the aggregates and their interfaces, and
thus generally to the tortuosity of the fracture path and the resulting energy
absorption in each case. Halvorsen further found that when the mix proportions
of the three concretes were adjusted to give comparable strengths, their
fracture toughness values were also similar (Table 6).

Halvorsen's data (7), although derived from an extension of the linear
elastic fracture mechanics, perhaps identify the basic mechanisms of crack
resistance in concrete composite materials. The variable fracture toughness
values often observed from apparently similar concrete materials are really
due to the differing fracture processes and fracture paths arising from the
relative stiffnesses and toughnesses of the matrix and the aggregates. Whilst
the nature of the aggregate and its interface with the matrix has a decisive
influence on toughness, the presence or absence of the fine or coarse
aggregates has a much more significant effect on the fracture properties of
the material. Toughness is also influenced by the strength or by the same
factors that influence strength. All these point out that the energy dissipation mechanisms in heterogeneous materials such as concrete composites are
not merely related to new fracture surfaces but other factors as well (1, 2).

The J-integral toughness parameter is subject to significant variations as
it is influenced by the shape of the curve, the maximum load and deformation.

TABLE 6 The J-integral Fracture Toughness

Investigator	Test Geometry	Material System	Remarks	Experimental Results J_{lc}, kN/m
Mindess, Lawrence and Kesler (30)	Notched beams, 4-point bending, 76 x 76 x 381 mm	Paste	No correction for	0.0110-0.0152
		Concrete	slow crack growth	0.0400-0.0422
Halvorsen (7)	75 x 75 x 380 mm beams span 255mm: notch depth 40 and 50 mm	Paste phase		0.011
		Mortar phase		0.031
		Concrete (same agg. paste volume but diff. strengths)	Florida limestone Illinois limestone Illinois gravel	0.029 0.035 0.037
		Concrete	Similar strengths: three diff. aggs.	0.029-0.032
		Concrete (low st)	Gravel & limestone	0.0255
		Concrete (high st)	Gravel & limestone	0.032-0.037

Nevertheless tests seem to indicate the J_{lc} as a promising fracture criterion for concrete materials in spite of the fact that it may well be underestimated (68).

5. FRACTURE TOUGHNESS OF FIBRE CEMENT COMPOSITES

The incorporation of short, discrete fibres in a relatively brittle cement matrix is known to improve the tensile strength properties and stiffness characteristics of the unreinforced matrix; but probably the most important influence of the fibre reinforcement is to delay and control the tensile cracking of the material. The uncontrolled and unstable tensile crack propagation, an inherent property of all cement-based matrices, is thus transformed by the fibres into a slow controlled crack growth. The crack controlling property of the fibre reinforcement has three major effects -

(1) it delays the onset of tensile cracking and increases the tensile strain capability of the composite,

(2) it imparts a well defined post-cracking behaviour to the composite, and the post-maximum load ductility increases, and

(3) it enhances the crack resistance and energy absorption characteristics of the composite material.

Toughness and ductility are thus the most important properties of fibre cement composites (Fig. 13, 1, 69, 70). The fracture toughness of fibre cement composites has also therefore been quite extensively studied and both linear elastic and non-linear fracture mechanics concepts have been applied (3, 6, 10, 12, 22, 71, 72).

The fracture toughness tests on fibre cement composites have yielded data on strain energy release rate G_{lc} (5), fracture energy (73), crack opening displacement (4, 5), stress intensity factor K_{ic} and the J-integral. A wide range of test techniques has been adopted to evaluate these fracture parameters - notched beam tests, single-edge-notched specimens, double cantilever beams, compact-tension tests, double torsion tests, circumferentially notched round bar tests, and split cube tests. Typical representative data on the critical stress intensity factor and the J-integral are presented in Tables 7 (5, 10, 30, 37, 63, 74-77) and Table 8 (5, 6, 30) respectively (Fig. 14).

As in the case of unreinforced concrete materials, the data on fibre cement composites also show considerable variations, differences and uncertainties among the different test techniques, and the validity of some of the test methods is not yet clear. The results, however, emphasize the superiority of fibre cement composites in terms of crack resistance and toughness. Crack arrest and slow crack growth appear to be inherent characteristics of fibre composites, although not all test methods and test geometries (76) appear to exhibit this property. Satellite and multiple cracking enable an increase

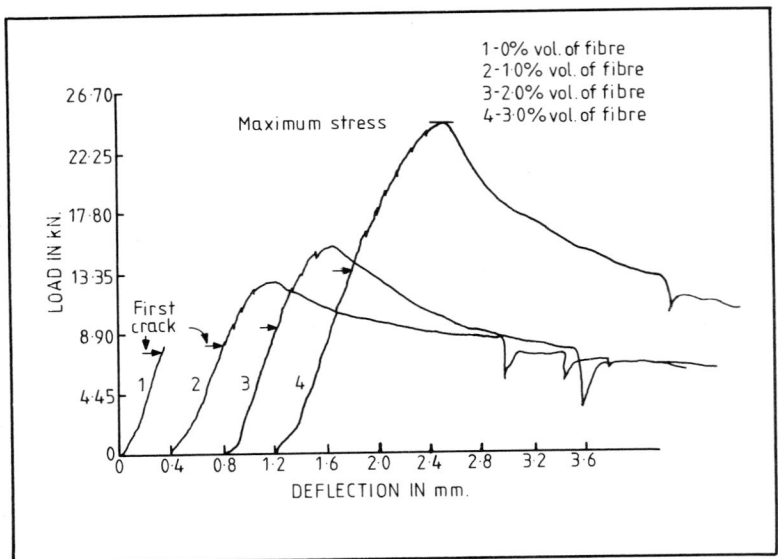

Fig. 13   Load-deflection behaviour of steel fibre reinforced mortar beams

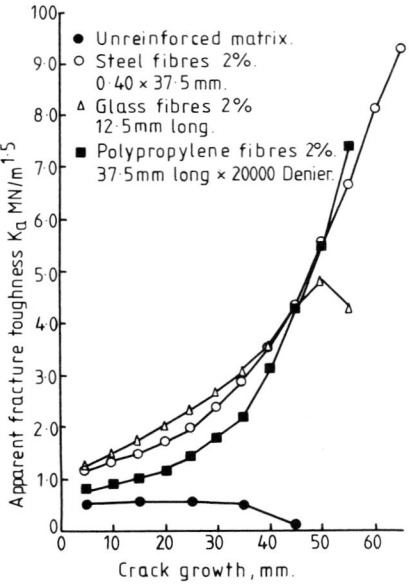

Fig. 14   Variation of fracture toughness with crack growth

TABLE 7   Fracture Toughness of Fibre Cement Composites

Investigator	Test Geometry	Material System	Remarks	Experimental Results K_{lc}, $MN/m^{3/2}$
Mindess, Lawrence and Kesler (30)	Notched beams, 4-point bending 76 x 76 x 381 mm	Steel fibre concrete (0.23% - 2.0%)	No correction for slow crack growth	0.68 - 1.32
		Glass fibre concrete (1.0%)		0.81-0.86
Harris, Varlow and Ellis (37)	Notched beam, 3-point bending	Fibre (mild steel) concrete	Dry, variable age	0.61-0.63
			Wet, variable age	0.74-0.83
			No correction for slow crack growth	
Javan and Dury (74)	CNRBB test 100 x 200 mm cylinders steel fibres: 0.25 x 25 mm Polypropylene fibres: 0.01 x 50 mm x 12000 denier	Plain concrete	No correction for slow crack growth	0.05-0.13
		steel fibre concrete (0.5%)		0.18-0.30
		steel fibre concrete (1%)	K_{lc} increases with notch depth	0.32-0.48
		Polypropylene concrete (½%)		0.10-0.20
		Polypropylene concrete (1%)		0.15-0.38

1	2	3	4	5
Mai, Foote, and Cotterell (10)	DCB 190 x 76 x 6 mm	asbestos-cement 10% fibres	No correction for slow crack growth	2.52 ± 0.27
		asbestos-cement 20% fibres		3.66 ± 0.24
	single-edge-notched specimen 190 x 50 x 6mm a/W = 0.10 to 0.50	asbestos-cement 10% fibres	K values increasing with crack extension for DCB test	2.23 ± 0.34
		asbestos-cement 20% fibres		3.39 ± 0.63
	Three-point bend test 190 x 26 x 6 mm, span = 100 mm, a/W = 0.10 - 0.60	asbestos-cement 10% fibres	K values geometry dependent for three-point bend test	2.34 ± 0.18
		asbestos-cement 20% fibres		3.44 ± 0.29
	Three-point bend test 190 x 50 x 6 mm, span = 153 mm, a/W = 0.10 - 0.60	asbestos-cement 10% fibres		3.60 ± 0.30
		asbestos-cement 20% fibres		4.90 ± 0.32
Brandt (5)	Notched beams, 4-point bending 50 x 50 x 300 mm, variable notch depth	steel fibre concrete 0.4 x 40 mm 2%	No correction for slow crack growth	0.30-0.80

1	2	3	4	5
Swamy (75)	Notched beams, 4-point bending 100 x 100 x 50 mm	steel fibre concrete (2%, 0.40 x 37.5 mm)	Cured in internal environment	1.0 - 9.5
		Polypropylene fibre concrete (2%, 50 mm fibrillated)	Corrected for slow crack growth	0.8 - 7.5
		Glass fibre concrete (2%, 12.5 mm)	K_{lc} increasing with crack growth	1.0 - 5.0
Barr, Evans and Dowers (63)	Split-cube test: 100 mm cube	Polypropylene fibre concrete 50 mm, 12000 denier 0.1-0.3%	No correction for	0.49 - 0.65
	Compact tension specimen	"		0.61 - 0.69

1	2	3	4	5
Yamm and Mindess (76)	Double torsion test 1220 x 406 x 51 mm	Concrete 10 mm aggregate Glass fibres 25 mm long 0.25-2% Steel fibres 12.5 and 25 mm 0.25-2% Steel fibres 0.6 x 50 mm	No correction for slow crack growth Straight Deformed	1.37-2.66 0.89-2.64 1.86-3.69
Barr and Liu (77)	Compact tension specimen 50 x 50 mm, 100 x 100 mm and 150 x 150 mm, 6-10 mm thick	Mortar Glass fibres 3-5%	No correction for slow crack growth Variable crack length-specimen width ratio	0.022-0.044

TABLE 8   The J-integral Fracture Toughness of Fibre Cement Composites

Investigator	Test Geometry	Material System	Remarks	Experimental Results J_{lc}, kN/m
Mindess, Lawrence and Kesler (30)	Notched beams, 4-point bending, 76 x 76 x 381 mm	Fibre concrete steel: 0.23 - 2%	No correction for slow crack growth	0.0305-0.8298
		Glass, 1%		0.1354-0.2389
Halvorsen (6)	Notched beam tests 4-point bending 75 x 75 x 380 mm	Steel fibre concrete length 12-50 mm dia. 0.15-0.58 mm aspect ratio 35-100 volume 0.25-2%	No correction for slow crack growth	0.035-3.84
	75 x 75 x 380, third-point	1.5% fibres	size effect	0.315
	75 x 75 x 380, centre-point	"		0.330
	150 x 150 x 535, third-point	"	455 mm span	0.865
	150 x 150 x 760, third-point	"	705 mm span	0.625
Brandt (5)	Notched beams, 4-point bending 50 x 50 x 300 mm	steel fibre concrete 0.4 x 40 mm 2%		0.030-0.060

of fracture toughness with crack growth, although at the unstable crack
propagation stage, there is only one major crack propagating to failure, and
there is a reduction in the crack resistance of the material (75). Slow crack
growth and non-linear crack behaviour should obviously be considered in any
correct evaluation of fracture resistance (78), but current test methods are
not independent of test geometry, specimen size and notch depth.

6   QUASI-STATIC FRACTURE TOUGHNESS TESTING

In materials where cracks run in a slow, controllable manner, so that
negligible generation of kinetic energy occurs during cracking, Gurney and
Hunt (79) have suggested a quasi-static method of fracture toughness determination. The quantity used in this method as a measure of fracture toughness
is the local work required to spread the crack expressed per unit of nominal
crack area. This quantity is represented by R* and is equal to G_c and K_c^2/E.
The method was initially applied to linear elastic materials and extended to
cracking where irreversible deformation is confined to a boundary layer near
the crack surface.

The method was originally developed for materials like polymethylmeth-
acrylate (perspex) in which cracks can be made to propagate slowly. The
specimen type used by Gurney and Hunt and the typical load-displacement graph
obtained autographically are shown in Fig. 15. The area OA_1A_2 then represents
the energy required to increase the crack area from A_1 to A_2 and the mean value
of R is given by

$$R = \frac{\text{Area } OA_1A_2}{A_2 - A_1} \qquad (9)$$

It can be shown that

$$R = \tfrac{1}{2} P^2 \,(dC/dA) = G \qquad (10)$$

so that by using the compliance technique, the need to measure the crack area
can be eliminated.

This method has also been extended to quasi-static cracking of materials
with high fracture toughness and low yield stress i.e. in situations where
general yielding precedes crack propagation. The size parameter which controls
the cracking-yielding transition is ER/σ_y^2 where E is the Young's modulus, R
fracture toughness and σ_y, yield stress. A simple and versatile laboratory-

---

*   It has been suggested that strictly speaking R measures fracture resistance
    rather than fracture toughness K_{1c} or G_c which relate to the onset of
    unstable crack propagation under essentially plane strain conditions.

size test rig which enables the large structure fracture mode of materials with high values of ER/σ_y^2 (> 50 mm) to be determined using small scale test pieces in conventional testing machines has recently been developed (80).

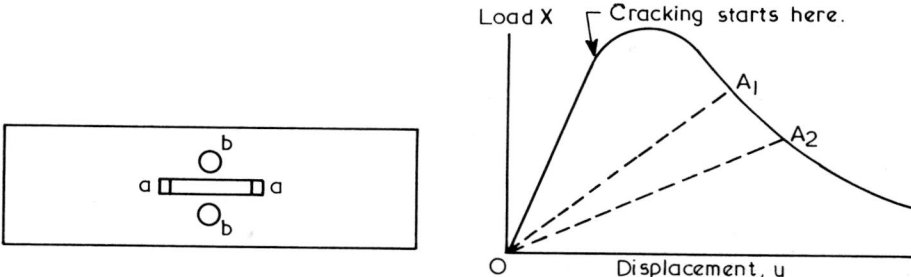

Fig. 15   (a) Perspex test specimen for quasi-static testing, slot aa, loading holes bb.   (b) Typical load-displacement diagram. A_1 and A_2 correspond to known crack areas.

The irreversible work area method of determining fracture toughness has been found particularly applicable to fibre reinforced composites and environmental fracture toughness testing in which both the environmental medium (liquids or gases) and temperature can be varied. Improved specimen geometries have also been developed such as the double-ended specimen for glass fibre reinforced plastics and the single-ended or double cantilever specimen for environmental testing (81). The technique can, however, be readily used for most materials which, in a suitable specimen form, exhibit stable crack growth when deformed in a stiff testing machine under normal laboratory conditions. Thus the method has been used to determine the fracture toughness of glass fibre reinforced cement using crack line loaded single-edge-crack specimens (82) and the work of fracture of asbestos cements using grooved double cantilever beam specimens.

7. TOUGHNESS INDEX

Many investigators have used the area obtained in a load-deflection test to assess the post-cracking toughness of materials. The "Toughness Index", as evaluated by the area under the load-deflection curve is then considered as a measure of the energy absorption capability of the material during fracture in the non-linear portion of the curve. The ACI Committee 544 (83-85) has defined the toughness index as the measure of the amount of energy required to deflect a fibre concrete beam, used in the modulus of rupture test, by a given amount (1.90 mm) compared to the energy required to bring the fibre beam to the point of first crack.

A modification to the toughness index has been proposed by Barr and his co-workers (77, 86). In this modification the test specimen is loaded until the load-deflection graph extends to twice the deflection at first crack. The proposed definition of the toughness index is shown in Fig. 16, and as defined, it varies from 0.25 for plain concrete to a maximum of unity for very tough fibre reinforced concrete materials. Since the denominator of the definition is given in terms of the area to twice the deflection at first crack, the maximum value of the index is unity compared to four if the ACI definition is used. The new definition does not impose any restrictions on specimen type.

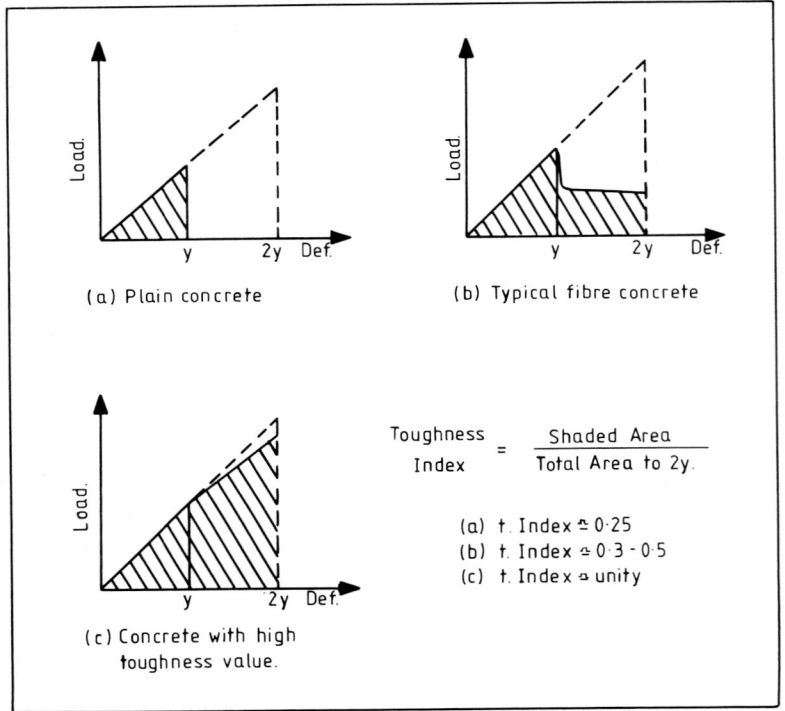

Fig. 16   Definition of toughness index

Barr and his colleagues (77, 86) have determined fracture toughness index results for polypropylene fibre concrete, steel fibre concrete and glass reinforced concrete (GRC). For polypropylene fibre concrete both compact tension (CTS) and compact compression (DENCEL or split-cube test) test specimens were used. Two types of notches were introduced into the compact tension specimens – firstly by means of a masonry saw and secondly by an inbuilt notch produced during casting. The compact compression test was used

for steel fibre concrete and the compact tension test for GRC. Typical data
obtained by Barr and his colleagues are shown in Tables 9 and 10.

These data show that the new fracture toughness index as defined by Barr
and his colleagues is a reasonably valid criterion. The results for polypropylene fibre concrete show that the index values are similar for different
loading systems and different types of notches, or when similar notches are
introduced into different test geometries. The results for GRC show good
correlation (Table 10) with the coefficient of variation within 2% for all
notch depths and specimen sizes tested. These data suggest that the new
toughness index may provide a criterion independent of both specimen size and
notch depth - specimen width ratio.

Obviously a lot more data, particularly on other test geometries, need to
be obtained to test the validity of the toughness index criterion. The
results obtained so far appear to indicate the index criterion as a promising
fracture toughness parameter.

8 CONCLUSIONS

The data presented in this chapter clearly show that the application of
fracture mechanics to concrete materials and concrete structures is very much
in its infancy, and that much of this application owes to developments that
have taken place in other material systems. Experience with metals shows that
fracture mechanics concepts can be powerful tools in predicting critical crack
lengths and flaw sizes, in determining the safe period of operation of a
structure under a given applied stress and in estimating the life of a
structure. Although application of fracture mechanics concepts to concrete
lags behind in almost all these aspects, there are welcome signs of the use of
these concepts to cracking in concrete structures under various stress
systems, and to fibre composites (40, 47). Khrapkov et al (54) have reported
the application of fracture mechanics to the analysis of massive concrete dams
operating in cold temperatures.

There is no doubt that there is the need to know more about the fracture
process in concrete and how this ultimately relates to mechanical behaviour and
strength both in laboratory tests and in service. The state of knowledge of
fracture processes in concrete is clearly not so advanced as it is for other
engineering materials like metals, glass, ceramics and rocks. The state of
knowledge appears to lack reliable quantitativeness, and many questions remain
unanswered. Much of the basic fracture mechanics concepts, with respect to
concrete materials at any rate, have remained to be of a purely theoretical
nature because in real engineering materials, the critical strain energy
release rate lacks a direct physical meaning and surface energies cannot be
easily measured. The problem is made more difficult by the lack of rational

456

TABLE 9   Fracture Toughness Index Results for Fibre Cement Composites (77, 86)

Type of fibre	Fibre content % by weight	Type of specimen	Type of notch	Toughness index	Remarks
Polypropylene	0.10	DENCEL	saw	0.325	
		CTS	saw	0.324	
		CTS	inbuilt	0.372	
	0.15	DENCEL	saw	0.333	Max. agg. size 10 mm
		CTS	saw	0.352	Fibres 12000 denier 50 mm
		CTS	inbuilt	0.426	Single size strand
	0.20	DENCEL	saw	0.406	DENCEL – 30 mm notch
		CTS	saw	0.378	CTS – 40 mm notch
		CTS	inbuilt	0.431	
	0.25	DENCEL	saw	0.376	
		CTS	saw	0.400	
		CTS	inbuilt	0.430	
STEEL	0.20	DENCEL	saw	0.340	
	0.40			0.360	
	1.0			0.490	Max. agg. size 10 mm
	2.0			0.600	30 mm notch depth
	3.0			0.730	
	4.0			0.840	

TABLE 10   Toughness Index Results for GRC (77, 86)

	Notch depth/Specimen width			Average toughness index	Remarks
	a/W = 0.40	a/W = 0.50	a/W = 0.60		
Fracture toughness	0.926	0.900	0.907	0.910	100 x 100 mm CTS
Number of specimens	32	34	37	–	Fibres 3-5%
Coefft. of Variation %	2.00	1.10	1.90	–	Specimen thickness 6-10 mm
Fracture toughness	0.946	0.927	0.929	0.930	150 x 150 mm CTS
Number of specimens	29	30	30	–	Fibres 3-5%
Coefft. of Variation %	0.80	1.40	1.00		

and logical mathematical models to relate fracture processes to other physical
data such as bond strength and effective flaw size. However, vast experience
has been accumulated in fracture mechanics testing and measurements for metal
systems, and more recently for ceramic materials, polymers and concrete composites, and this fund of information must be applied to the development of
concepts and techniques suitable for concrete structures.

In the present state of development of fracture mechanics as applied to
concrete, it is not clear as to which fracture toughness parameter - the energy
concept, the stress intensity factor approach, the J-integral or any other
non-linear parameter - will prove a valid and appropriate fracture criterion
for concrete materials. There are protagonists for each of the parameter
that has been evaluated, but what is important is to develop a parameter that
will rationally and logically explain the physical process of crack arrest
and fracture observed in concrete materials. It is no cynical or negative
attitude to say that only few of these parameters can claim to such a
relationship that will be attractive to designers. It is equally clear that
much more fundamental work needs to be undertaken, not merely in devising new
test methods to evaluate a fracture toughness parameter, but also to establish
a practical relationship between fracture toughness and service life of a
concrete structure.

9 REFERENCES

1 R. N. Swamy and C. V. S. Kameswara Rao, Proceedings, Conference on Fibre
   Reinforced Materials: Design and Engineering Applications, Institution of
   Civil Engineers, London, 1977, pp 87-96.
2 R. N. Swamy, Proceedings, International Conference on the Structure of
   Concrete, London, 1965, Cement and Concrete Association, London, 1968,
   pp 212-214.
3 R. N. Swamy in F. D. Lydon (Ed), Developments in Concrete Technology - 1,
   Applied Science Publishers Ltd, London, 1979, pp 221-281.
4 K. Nishioka, S. Yamakawa, K. Hirakawa, and S. Akihama, Proceedings, RILEM
   Symposium 1978, Testing and Test Methods of Fibre Cement Composites, The
   Construction Press, Lancaster, pp 87-98.
5 A. Brandt, The Inter. J. of Cement Composites, 2 (1980) 35-42.
6 G. T. Halvorsen, The Inter. J. of Cement Composites, 2 (1980) 13-22.
7 G. T. Halvorsen, The Inter. J. of Cement Composites, 2 (1980) 143-148.
8 Y. W. Mai, Journal of Materials Science, 14 (1979) 2091-2102.
9 J. C. Lenain and A. R. Bunsell, Journal of Materials Science, 14 (1979)
   321-332.
10 Y. W. Mai, R. M. Foote and B. Cotterell, The Inter. J. of Cement Composites,
   2 (1980) 23-34.
11 R. Andonian, Y. W. Mai and B. Cotterell, The Inter. J. of Cement Composites,
   1 (1979) 151-158.
12 G. Velazco, K. Visalvanich and S. P. Shah, Cement and Concrete Research,
   10 (1980) 41-51.
13 A. Hillerborg, M. Modeer and P. E. Petersson, Cement and Concrete Research,
   6 (1976) 773-782.
14 M. Modeer, Report TVBM-1001, University of Lund, 1979.
15 A. Hillerborg, The Inter. J. of Cement Composites, 2 (1980) 177-184.

16 J. Glucklich, Theoretical and Applied Mechanics Report No. 215, University of Illinois, 1962, pp 25.
17 J. Glucklich, J. Eng. Mech. Div., Proceedings, Am. Soc. Civ. Engrs, 89 (1963) 127-138.
18 J. Glucklich, Proceedings, Inter. Conf. on the Structure of Concrete, London 1965, Cement and Concrete Association, London 1968, pp 176-189.
19 W. R. A. Knox, in Fracture Toughness of High Strength Materials: Theory and Practice, Publication 20, Iron and Steel Institute, London, 1970, pp 158-162.
20 P. Desayi, RILEM Materials and Structures, 10 (1977) 139-144.
21 M. F. Kaplan, J. Am. Conc. Inst., 58 (1961) 591-610.
22 J. P. Romualdi and G. B. Batson, J. Eng. Mech. Div., Proceedings, Am. Soc. Civ. Engrs., 89 (1963) 147-168.
23 J. Glucklich, Proceedings, First Inter. Conf. on Fracture, 2 (1965) 1343-1382
24 J. L. Lott and C. E. Kesler, Symposium on Structure of Portland Cement Paste and Concrete, Highway Res. Board, Special Report 90, 1966, 204-218.
25 G. B. Welch and B. Haisman, RILEM Materials and Structures, 2 (1969) 171-177
26 F. Moavenzadeh and R. Kuguel, J. of Materials, 4 (1969) 497-519.
27 D. J. Naus and J.L. Lott, J. Am. Conc. Inst., 66 (1969) 481-489.
28 K. Okada and W. Koyanagi, Proceedings, Inter. Conf. on Mech. Behaviour of Materials, Kyoto, 1971, 4 (1972) 72-83.
29 J. Glucklich and U. Korin, J. Am. Ceramic Soc., 58 (1975) 517-521.
30 S. Mindess, F. V. Lawrence and C. E. Kesler, Cement and Concrete Res., 7 (1977) 731-742.
31 B. Barr and T. J. Bear, Concrete, 10 (1976) 25-27.
32 J. Mazars, Proceedings, Fourth Inter. Conf. on Fracture, 3B (1977) 1205-1209.
33 K. P. George, J. Soil Mech and Foundations, Proceedings, Am. Soc. Civ. Engrs. 96 (1970) 991-1010.
34 G. J. Hahn, M. F. Kanninen and A. R. Rosenfield, Annual Review of Materials Science, 2 (1972) 381-399.
35 S. Brunauer, D. L. Kantro and C. H. Weise, Canadian J. of Chemistry, 37 (1959) 714-716.
36 G. A. Cooper and J. Figg, J. British Ceramic Soc., 71 (1972) 1-4.
37 B. Harris, J. Varlow and C. D. Ellis, Cement and Concrete Res., 2(1972) 447-461.
38 A. Auskern and W. Horn, Cement and Concrete Res., 4 (1974) 785-795.
39 P. W. R. Beaumont and J. C. Aleszka, J. of Materials Science, 13 (1978) 1749-1760.
40 N. M. Hawkins, A. N. Wyss and A. H.Mattock, J. Struct. Div., Proceedings Am. Soc. Civ. Engrs., 103 (1977) 1015-1030.
41 P. C. Paris and G. C. Sih, In Fracture Toughness Testing, ASTM STP 381, Am. Soc. Test. Materials, Philadelphia, 1965, pp 30-39.
42 W. F. Brown and J. E. Srawley, In Plane Strain Crack Toughness Testing of High Strength Metallic Materials, ASTM STP 410, Am. Soc. Testing Materials, Philadelphia, 1966, pp 13.
43 J. E. Srawley and W. F. Brown, In Fracture Toughness Testing, ASTM STP 381, Am. Soc. Testing Materials, Philadelphia, 1965, pp 133-139.
44 C. E. Kesler, D. J. Naus and J. L. Lott, Proceedings Inter. Conf. on Mechanical Behaviour of Materials, Kyoto, 1971, 4 (1972) 113-124.
45 P. F. Walsh, Indian Concrete Journal, 46 (1972) 469-470: also Eng. Fract. Mech. 4 (1972) 533-541; J. Eng. Mech. Div., Proceedings Am. Soc. Civ. Engrs., 98 (1972) 1611-1614.
46 J. H. Brown and C. D. Pomeroy, Cement and Concrete Res., 3 (1973) 475-480.
47 G. D. T. Carmichael and K. Jerram, Cement and Concrete Res., 3 (1973) 459-467.
48 J. S. Nadeau, S. Mindess and J. M. Hay, J. Am. Cer. Soc., 57 (1974) 51-54.
49 S. Mindess, J. S. Nadeau and J. M. Hay, Cement and Concrete Res., 4 (1974) 953-965.
50 S. Mindess and J. S. Nadeau, Cement and Concrete Res., 6 (1976) 529-534.
51 D. D. Higgins and J. E. Bailey, J. of Materials Science, 11 (1976) 1995-2003.

52 K. Kitagawa and M. Suyama, Proceedings 19th Japan Congress on Materials Res.
    1976, pp 156-159.
53 B. Hillemeier and H. K. Hilsdorf, Cement and Concrete Res., 7 (1977) 523-536.
54 A. A. Khrapkov, L. P. Trapesnikov, G. S. Geinats, V. I. Pashchenko and
    A. P. Pak, Proceedings Fourth International Conference on Fracture, 3B
    (1977) 1211-1217.
55 K. Kitagawa, S. Kim and M. Suyama, Proceedings, 19th Japan Congress on
    Materials Res., 1976, pp 160-163.
56 B. Barr and T. J. Bear, Concrete, 11 (1977) 30-32.
57 T. J. Bear and B. Barr, Int. J. of Fracture, 13 (1977) 92-96.
58 R. N. Swamy, Proceedings, the Southampton 1969 Civil Engineering Materials
    Conference, Wiley-Interscience, 1971, pp 301-315.
59 S. P. Shah and F. J. McGarry, J. Eng. Mech. Div., Proceedings Am. Soc. Civ.
    Engrs., 97 (1971) 1663-1676.
60 R. N. Swamy and C. V. S. Kameswara Rao, Cement and Concrete Res., 3 (1973)
    413-427.
61 C. V. S. Kameswara Rao and R. N. Swamy, Cement and Concrete Res., 4 (1974)
    669-681.
62 B. Barr, W. T. Evans, R. C. Dowers and B. B. Sabir, The fracture toughness
    of concrete in Numerical Methods in Fracture Mechanics, Eds. D. R. J. Owen
    and A. R. Luxmoore, Pineridge Press, 1980, pp 737-749.
63 B. I. G. Barr, W. T. Evans and R.C. Dowers, Inter. J. of Cement Composites
    and Lightweight Concrete, 3 (1981) 115-122.
64 R. C. Dowers, The fracture toughness of concretes, Thesis for M. Phil degree,
    The Polytechnic of Wales, Aug. 1980, pp 187.
65 J. R. Rice, P. C. Paris and J. G. Merkle, In Flow Growth and Fracture
    Toughness Testing, ASTM STP 536, Am. Soc. Test. Mat., 1973, pp 231-245.
66 S. P. Shah and C. Velazco, Proceedings, Third Engineering Mechanics Division
    Speciality Conference, Am. Soc. Civ. Engrs., 1979.
67 R. A. Schmidt and T. J. Lutz, in Fracture Mechanics Applied to Brittle
    Materials, S. W. Freiman, Editor, ASTM STP 678, Am. Soc. Test. Mat., 1978,
    pp 166-182.
68 C. G. Chipperfield, J. of Testing and Evaluation, 6 (1978) 253-258.
69 R. N. Swamy, RILEM Materials and Structures, 8 (1975) 235-254.
70 R. N. Swamy and P. S. Mangat, Cement and Concrete Res. 5 (1975) 37-53.
71 S. W. Freiman, Editor, Fracture Mechanics Applied to Brittle Materials,
    ASTM STP 678, Am. Soc. Test. Mat., 1978, pp 220.
72 W. F. Chen and E. C. Ting, Editors, Fracture in Concrete, Am. Soc. Civ. Engrs.
    1980, pp 105.
73 T. Ohigashi, Proceedings, RILEM Symposium 1978, Testing and Test Methods of
    Fibre Cement Composites, The Construction Press, Lancaster, pp 67-78.
74 L. Javan and B. L. Drury, Concrete, 13 (1979) 31-33.
75 R. N. Swamy, Inter. J. of Cement Composites, 2 (1980)43-53.
76 A. S. T. Yam and S. Mindess, Inter. J. of Cement Composites and Lightweight
    Concrete, 4 (1982) 83-98.
77 B. I. G. Barr and K. Liu, Inter. J. of Cement Composites and Lightweight
    Concrete, 4 (1982) 163-171.
78 J. H. Brown, Magazine of Conc. Res., 25 (1973) 31-38.
79 C. Gurney and J. Hunt, Proceedings Roy. Soc. London, Series A, 229 (1967)
    508-524.
80 C. Gurney, Y. W. Mai and R. C. Owen, Proceedings Roy. Soc. London, Series A,
    340 (1974) 213-218.
81 D. G. Ashwell and M. G. Hancock, Proceedings Conference on Fibre Reinforced
    Materials, Design and Engineering Applications, Institution of Civil
    Engineers, London, 1977, pp 97-103.
82 W. A. Patterson and H. C. Chan, Composites, 6 (1975) 102-104.
83 American Concrete Institute, Committee 544, Measurement of Properties of
    Fibre Reinforced Concrete, J. Am. Conc Inst., 75 (1978) 283-289.
84 C. H. Henager, Proceedings RILEM Symp. 1978, Testing and Test Methods of
    Fibre Cement Composites, The Construction Press, Lancaster, 1978, pp 78-86.

85 C. H. Henager, Proceedings Concrete International 1980, Fibrous Concrete,
   The Construction Press, Lancaster, 1980, pp 16-28.
86 B. I. G. Barr, K. Liu and R. C. Dowers, Inter. J. of Cement Composites and
   Lightweight Concrete, 4 (1982) 221-227.

Fracture Mechanics of Concrete,
edited by F.H. Wittmann, 1983
Elsevier Science Publishers B.V., Amsterdam — Printed in The Netherlands

Chapter 5.3

NONLINEAR FRACTURE MECHANICS PARAMETERS

by M. WECHARATANA and S.P. SHAH

1.   INTRODUCTION

Fracture toughness of metallic materials in Mode I crack propagation under plane-strain conditions (K_{IC}) is usually determined by testing either compact tension specimens or notched-beam specimens (see Fig. 1). The method of testing and the dimension of these two types of specimens are specified by relevant ASTM Standard specifications (ref. 1). Among other things, these dimensions are specified to assure that :

1) the slow crack growth prior to attaining peak load is small compared to the initial notch length so that almost linear load-deflection relationship is obtained, and

2) the zone of nonlinearity around the crack tip (radius of plastic zone) is small compared to the length of the notch and that of the uncracked ligament.

If these restrictions are met then one can obtain specimen-independent fracture toughness (K_{IC}) using the concepts of linear elastic fracture mechanics (LEFM) from the test results.

Portland cement concrete is characterized by strain softening and nonlinear behavior and it is not clear that the testing specifications developed for metals can be applied to concrete. In addition, it has been shown that the nonlinear behavior of concrete is related to its heterogeneity; the larger the grain size and the larger the volume fraction of inclusions, the more nonlinear and thougher is the observed behavior (ref. 2). Since fairly large grain size are used in making concrete (23 mm), the extension of the formulae developed for metals would lead to extremely large size specimens.

An alternate approach is to modify the concepts of linear elastic fracture mechanics to include the effects of slow crack growth and crack-tip nonlinearity (process zone) in analyzing the results of fracture toughness tests. This is attempted in this chapter.

2.   SLOW CRACK GROWTH

To limit the extent of slow crack growth ASTM-E399 specifies that the relationship between the measured load (P) and the crack mouth displacement (Δ)

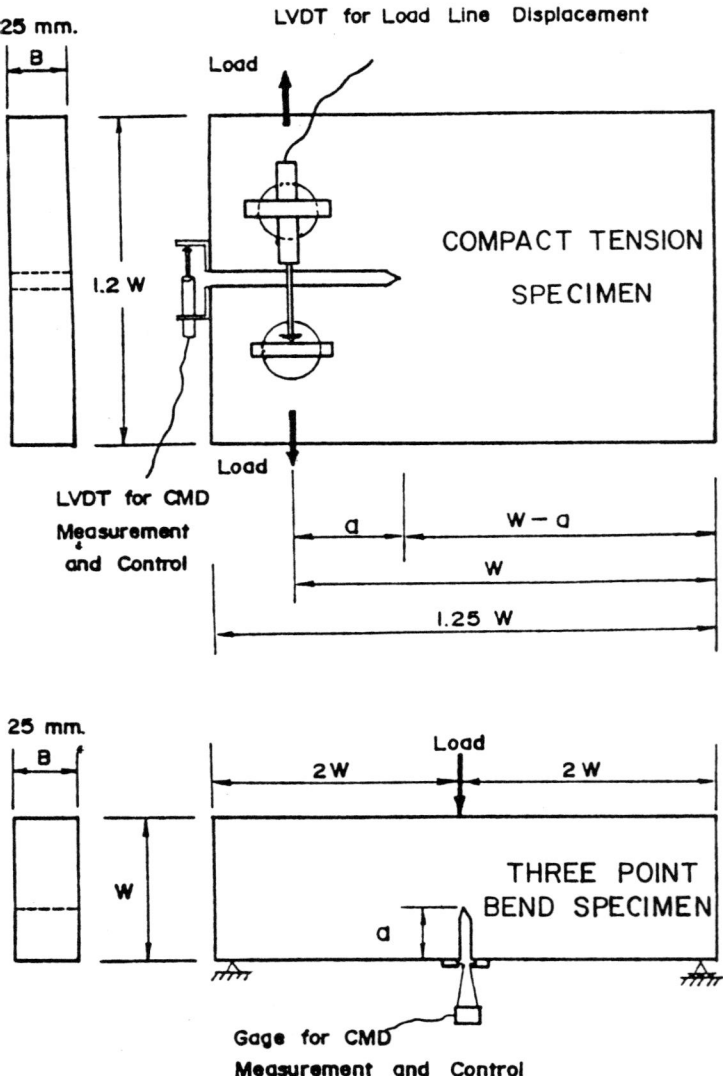

Fig. 1. ASTM - Standard Test Specimens.

should be almost linear (see Fig. 2). This is stipulated by specifying that the ratio of the maximum load (P_{max}) and P_5 (see Fig. 2, for definition) cannot be greater than 1.10 for a valid fracture toughness (K_{IC}) test. In figure 3 is shown a load vs. crack mouth displacement record of a portland cement mortar specimen (ref. 3). It can be seen that P_{max}/P_5 is much greater than 1.10. A substantial slow crack growth was observed during loading with the aid of a microscope (ref. 3). These results are for mortar specimens made with small grain size inclusions (sand particles). The extent of slow crack growth is likely to be even larger for concrete specimens which are made with larger size coarse aggregates.

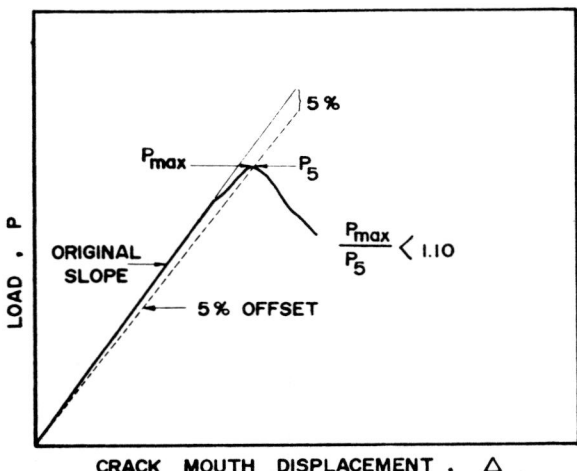

Fig. 2. ASTM Restriction for slow crack growth.

2.1 Resistance curves

If the slow crack growth is accurately measured then one can calculate load and strain energy release rate at each crack extension and obtain a relationship between the crack growth and the strain energy release rate. Such plots are called R-curves or resistance curves (see Fig. 4). The resistance to crack extension as measured by the corresponding strain energy release rate (G_R) at first increases due to the slow crack growth and then reaches a steady state value (G_{ss}). The concept of R-curves to characterize the resistance during slow crack growth for rocks, ceramics, asbestos cement, mortar and concrete has been used by many investigations (ref. 4-7).

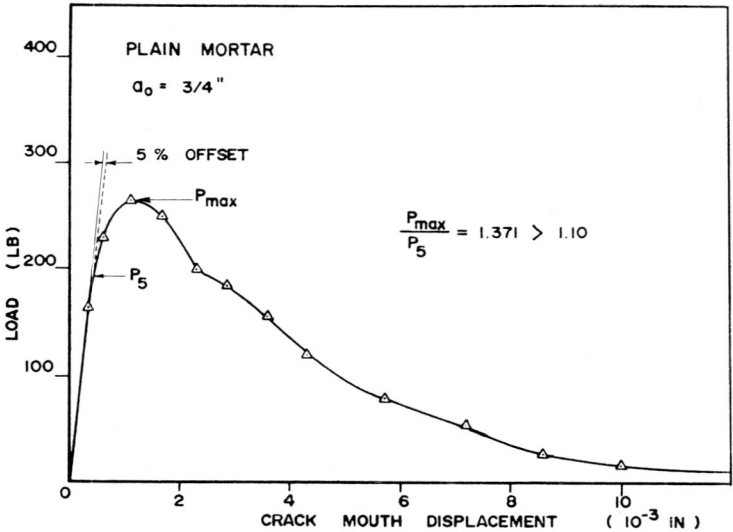

Fig. 3. Slow crack growth in mortar (ref. 3).

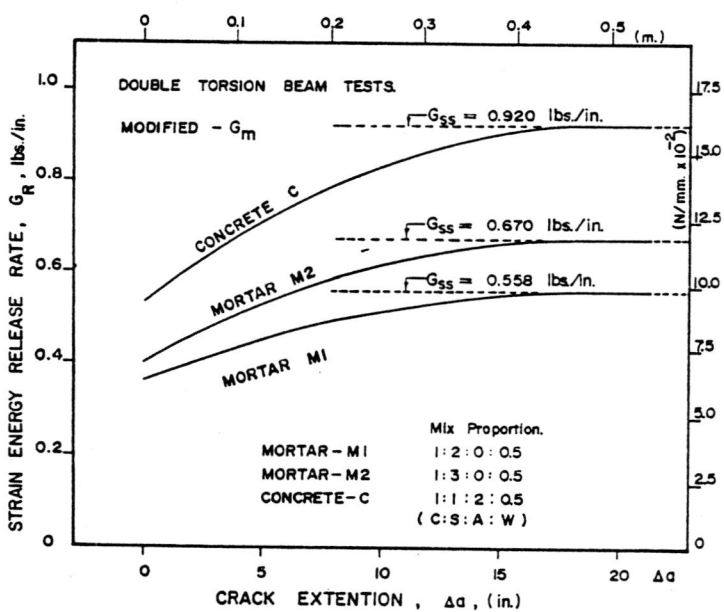

Fig. 4. Examples of R-curves (ref. 12).

2.2. Measurement of slow crack growth

To accurately measure the slow crack growth the specimens should be loaded at a constant rate of crack mouth displacement and the load stopped at frequent interval for measurement of crack growth. Even with the stable loading arrangement, the measurement of crack growth in concrete can be difficult because of the opaque nature of the material, and the tortuousity of the crack growth.

An alternate method to obtain the extent of slow crack growth is by socalled compliance calibration technique. In this method, compliance of a previous artificially notched specimen is compared with the compliance of a specimen under test. Several investigators have used this approach to estimate the slow crack growth (ref. 4, 5, 7-10). However, with this method it is assumed that the compliance of a specimen with an artificial crack is identical to that of a specimen with a real crack. This is not the case in mortar and concrete because of the irregular and rough nature of real cracks (ref. 3, 10). Resulting from aggregate-interlock these cracks have lower compliance than the artificially cast smooth cracks. Because of these difficulties, in the investigations carried out by the authors, the crack growth is measured with an internally illuminated microscope which is fitted with a micrometer. This optical arrangement permits measurement of a crack with a resolution of 1.3×10^{-3} mm.

2.3 Measurement of fracture resistance

From the measurement of load and crack growth the resistance to slow crack growth can be calculated using linear elastic fracture mechanics in terms of either stress intensity factor (K_R) or the strain energy release rate (G_R). This would be valid if the nonlinear effects around the crack tip resulting from microcracking ahead of the crack and the aggregate interlocking behind the crack were negligible. These effects, however, are substantial for mortar and concrete.

The significant extent of nonlinearity can be seen in the load vs. loadline deflection curves plotted for a double cantilever and a double torsion concrete specimen (see Figs. 5 and 6). The dimensions of the specimens tested are given in figures 7 and 8. These specimens were loaded at a constant rate of crack opening displacement. The load was periodically stopped to measure crack growth (microscopically). The specimens were then unloaded and again loaded.

For the case of linearly elastic, brittle materials, the strain energy release rate can be calculated from :

$$G_R = \frac{P^2}{2} \cdot \frac{dC}{dA} \qquad (1)$$

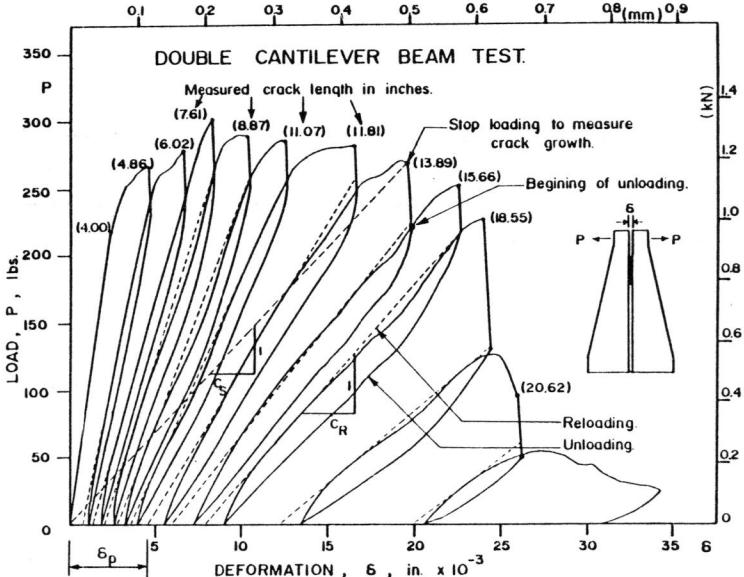

Fig. 5. Results from double cantilever beam tests (ref. 13).

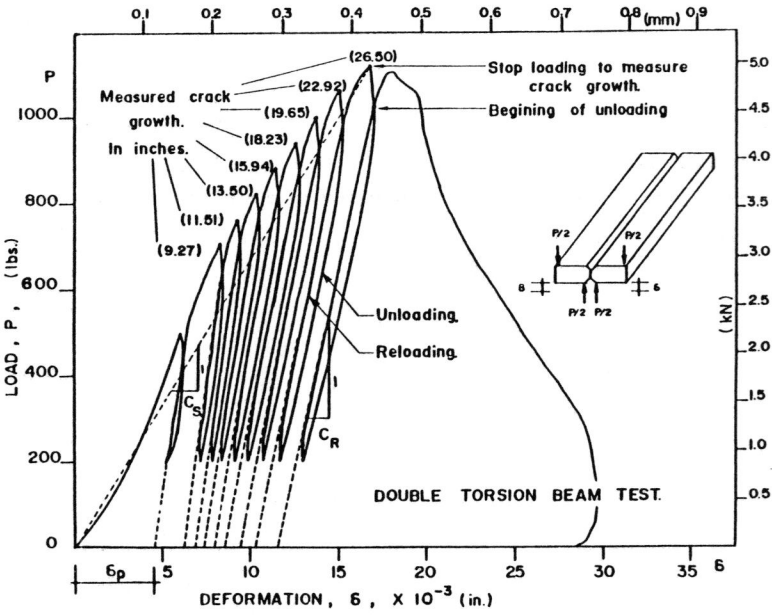

Fig. 6. Results from double torsion beam tests (ref. 13). Note definitions of C_S, C_R and δ_p.

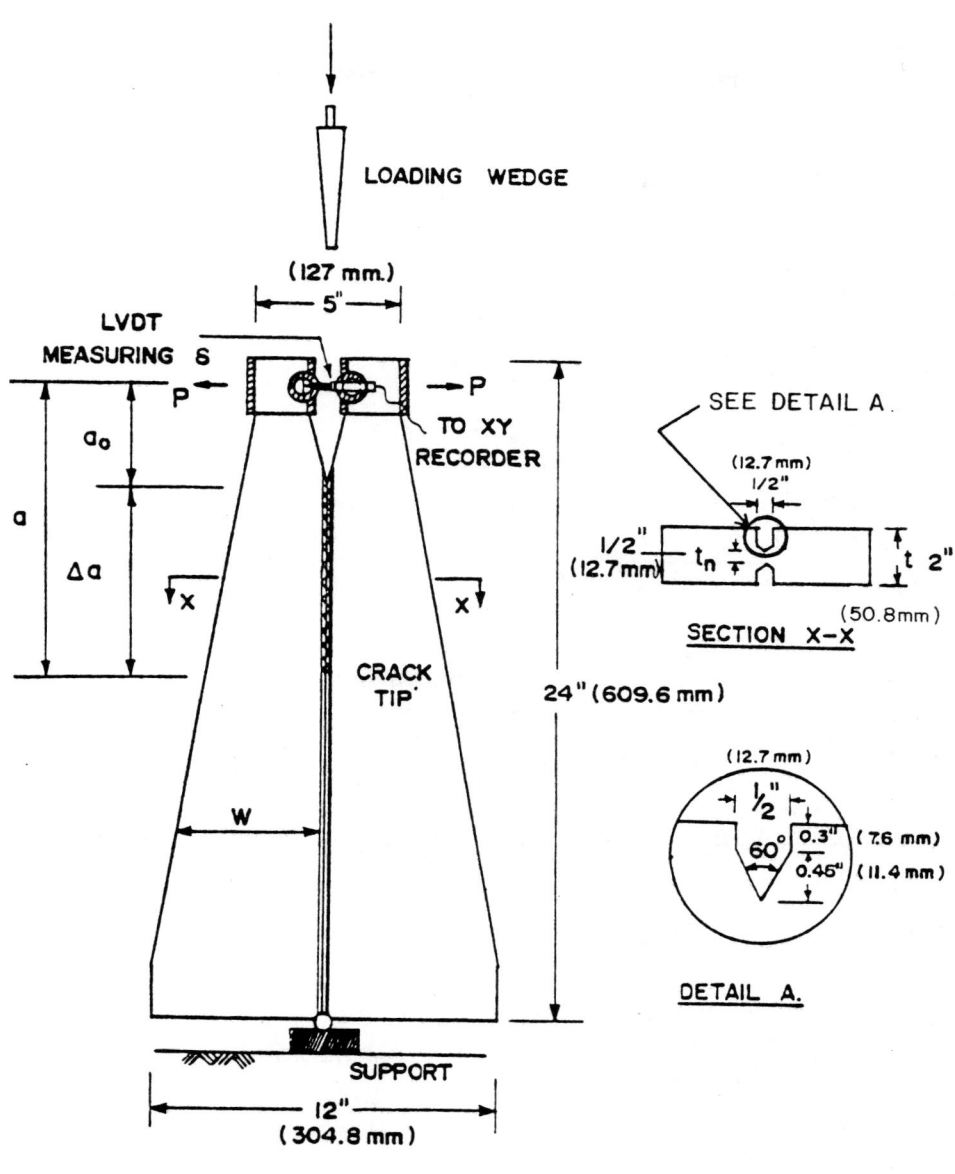

DOUBLE CANTILEVER BEAM.

Fig. 7. Details of testing of DCB specimens.

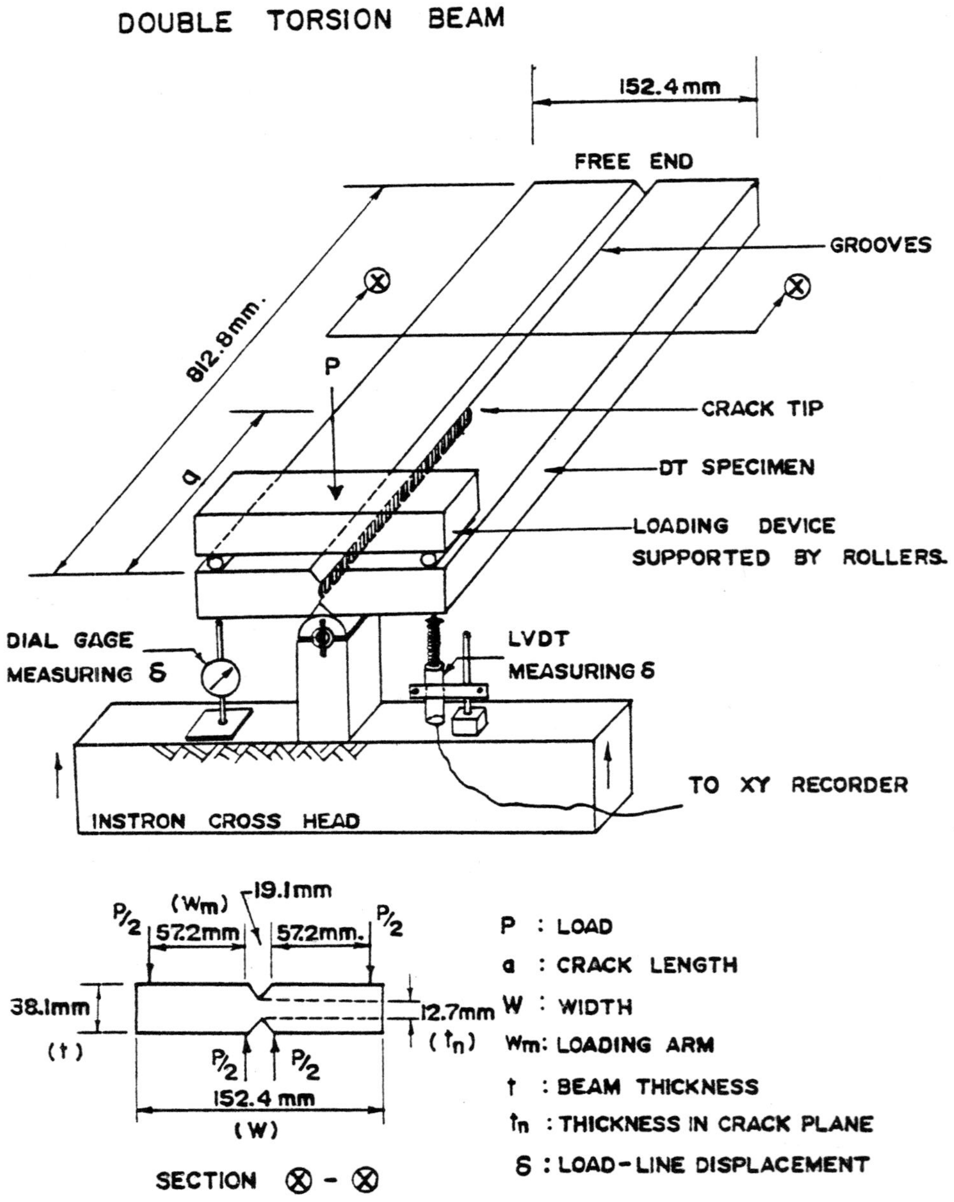

DOUBLE TORSION BEAM

Fig. 8. Details of testing of DT specimens.

where dC/dA is the rate of change of compliance with respect to crack growth.
To obtain G_R using equation (1) during slow crack growth from the test results,
one can calculate the changes in the values of secant compliances (C_S in Figs.
5 and 6) with the measured crack growth. This procedure cannot be very accurate
for concrete since it ignores the effects of observed nonlinear deformation
(δ_p) as shown in figures 5 and 6.

Many investigators use reloading (or unloading) compliance (C_R in Figs. 5
and 6) to calculate the value of G_R in equation (1) (ref. 4, 8, 11). However, it
was shown by the authors (ref. 12, 13), that the use of reloading compliances to
calculate G_R underestimates the fracture resistance for cementitious composites
since that method ignores the inelastic energy absorbed during the crack growth.
To include the inelastic energy during crack growth, the authors have proposed
modified definition of G_R as follows :

$$\text{(Modified) } G_R = \frac{P_1 P_2}{2} \frac{dC_R}{dA} + \frac{(P_1 + P_2)}{P_1 P_2} \frac{d\delta_p}{dA} \qquad (2)$$

where $\frac{d\delta_p}{dA}$ is the rate of change of permanent deformation with the crack growth
and P_1, P_2 are two neighboring consecutive loads. Attempts to modify the linear
elastic-fracture-mechanics definition of G_R has also been reported by Hodgkinson
and Williams for polymeric materials (ref. 14). The various compliance methods
to calculate G_I are shown in figure 9.

R-curves calculated using equation (2) for mortar and concrete are shown in
figure 4. It can be seen that the higher the maximum size of aggregates or volume fraction of aggregates, the higher is the value of fracture toughness as
measured by R-curves. It should be noted that the conventional compressive
strength of concrete specimens was lower than that of mortar specimens, but the
values of the fracture toughness as measured by R-curves are higher for concrete
specimens. On the other hand, the reported values of fracture toughness, calculated using LEFM, generally show direct relationship with the corresponding
strength values.

3.   PROCESS ZONE

To assure that the size of the test specimen is such that the dimensions of
the process zone (plastic zone in metals) is relatively small so that the concepts of LEFM can be applied while calculating K_{IC}, ASTM-E399 recommends that :

$$a, B, \text{ and } W-a \geq 2.5 \ \{K_{IC}/\sigma_y\}^2. \qquad (3)$$

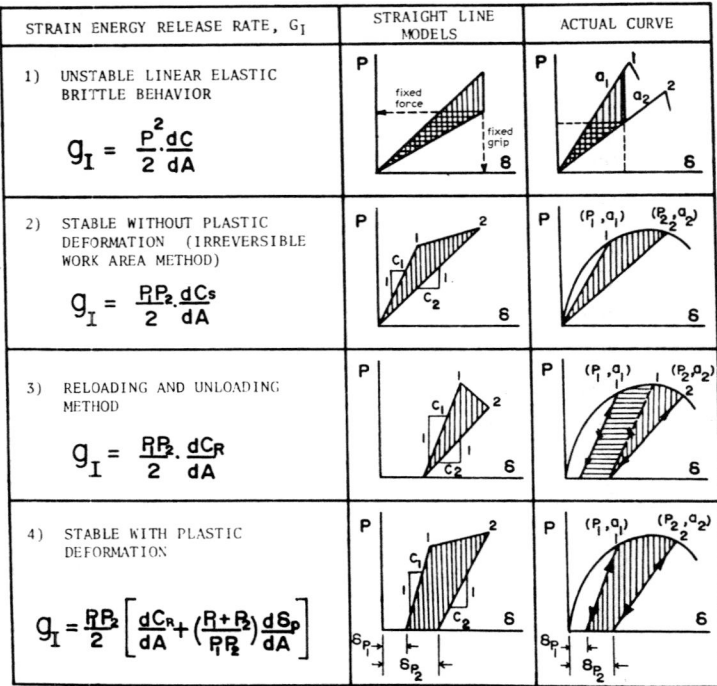

Fig. 9. Various methods of defining strain energy release rate.

where a, b and W are the length of the initial notch, the thickness of the specimen and the depth of the specimen (see Fig. 1) and σ_y is the yield strength
of the material. This formula is based on the assumption that the size of the
plastic zone is proportional to the quantity $(K_{IC}/\sigma_y)^2$.

Attempts have been made to use the same concept for non yielding material
such as rocks and concrete. For example, Schmidt and Lutz (ref. 6) showed that
if the uniaxial tensile strength f_t is substituted for the yield strength σ_y in
equation (3) then the steady-state value of fracture toughness was obtained for
rock specimens designed according to equation (3). The grain size of the rock
specimens used by Schmidt and Lutz were 0.75 mm. For that size, they found that
a crack length of 100 mm or greater is needed to obtain a fracture toughness
value independent of crack length. For a much larger grain size in concrete,
extremely large specimens would be necessary to obtain a steady state value of
fracture toughness using equation (3).

A similar conclusion was also reached by Hillerborg, Modeer and Petersson
(ref. 15). They report that to have a valid test based on LEFM, notched beams
should have a depth of 10 times a quantity which they term the characteristic

length, l_{ch}, which is given by :

$$l_{ch} = (K_{IC}/f_t)^2 \tag{4}$$

If their conclusion is valid then one needs a depth of beams equal to 2 to 3
meter to obtain a steady-state value of fracture toughness for concrete.

The process zone in mortar and concrete is related to the aggregate interlock
behind the crack tip. Thus, it seems likely that the size of this nonlinear zone
should depend among other things, on size of the aggregate, the strength of the
interface between the aggregates and matrix, and the geometry of the crack front.
These factors are not included in equation (3) or (4). To consider these effects,
a theoretical model was developed by the authors (ref. 16) to evaluate the
length of the process zone. Unlike many previous models (ref. 4, 7, 17) this
model is not restricted to small-scale nonlinear zone.

The details of the theoretical model as well as detailed comparison with the
experimental data are given in reference 18. A brief summary of the model is
presented here.

4.   THEORETICAL MODEL

To include the effect of large-scale crack-tip nonlinearity, a crack of a
given (traction free) length of a is replaced by an (elastic) effective crack
of length a_{eff} (= $a + l_p$) where l_p is the length of the process zone (see
Fig. 10). The stresses at the tip of the effective crack are assumed to be cal-
culable by using LEFM. A closing pressure is assumed to be distributed on the
length l_p of the effective crack. If the process zone is assumed to be localized
in a band of narrow width, then the closing stress distribution can be obtained
from the uniaxial tensile stress-displacement relationship. The crack initiation
is assumed to occur when the crack-front displacement at the tip of the trac-
tion-free crack equals the critical crack-displacement value, η_{max}. This value
is obtained from the descending part of the uniaxial stress-displacement relationship (that is, when the tensile stress is close to zero).

Using this model, R-curves were theoretically calculated for the double cantilever and double-torsion specimens (ref. 16, 18). For calculation of the
theoretical fracture toughness value (G_R) for a given crack length, a, the
first step was to assume a value of l_p and a crack profile. The closing pressure
distribution was then determined from the assumed value of l_p and the crack geometry. Then the value of the crack-tip-displacement (resulting from the applied
load P and the closing pressure) was calculated using elastic-beam theory. If

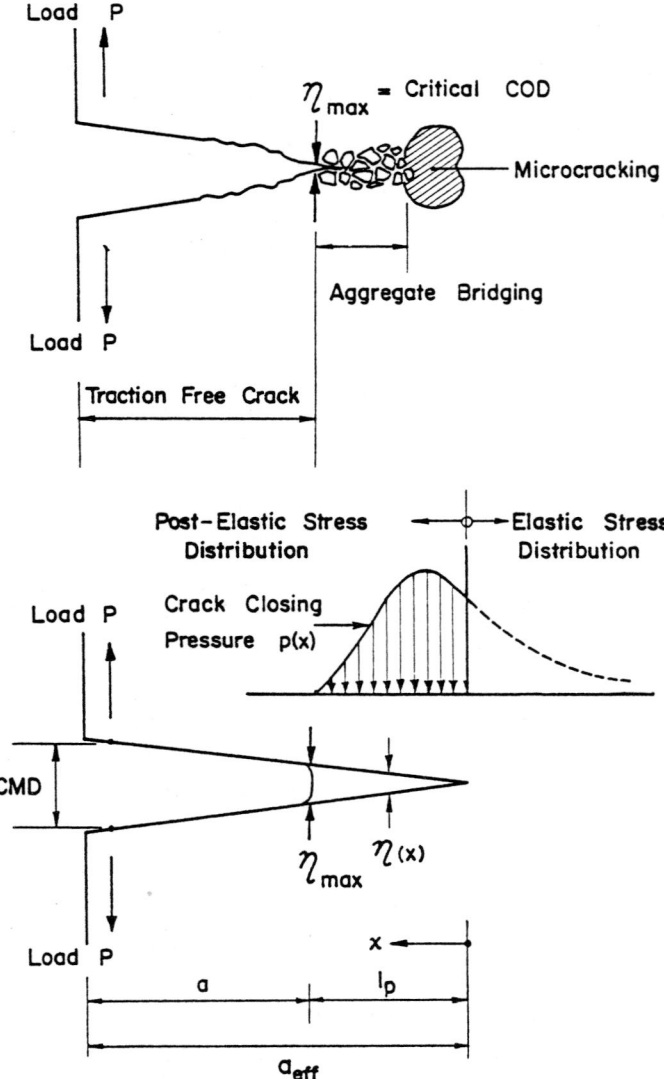

Fig. 10. The concept of effective crack length.

this value was not equal to η_{max}, then another value of ℓ_p was assumed. For the value of ℓ_p which corresponded to the critical crack-tip displacement, η_{max}, the load-line displacement was checked with the experimental data. If the theoretical and experimental values did not match, then the crack profile was changed and the entire operation was repeated. Finally, the values of strain energy release rate was calculated. Note that to calculate R-curves using this approach, only two materials functions are needed : elastic values and the tensile stress vs. crack-displacement relationship. Such tensile stress vs. crack width relationship have been recently available (ref. 19). A comparison of the theoretical and the experimentally measured values is shown in figure 11.

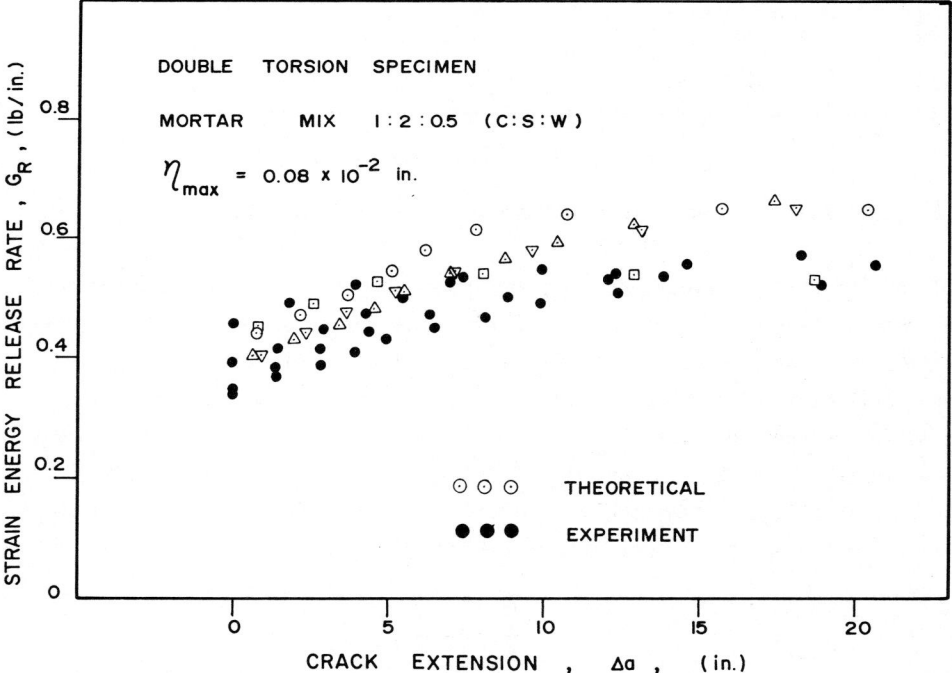

Fig. 11. Theoretical and experimental R-curves.

It was observed that the theoretically calculated length of the process zone remained essentially constant during the slow crack growth (see Fig. 12). Similar observation has been made for asbestos-cement compact tension specimens where the length of the process zone was measured using acoustic emission technique (ref. 4). The observed length of process zone was about 30 mm which remained constant during the crack growth. The length of process zone as well as

476

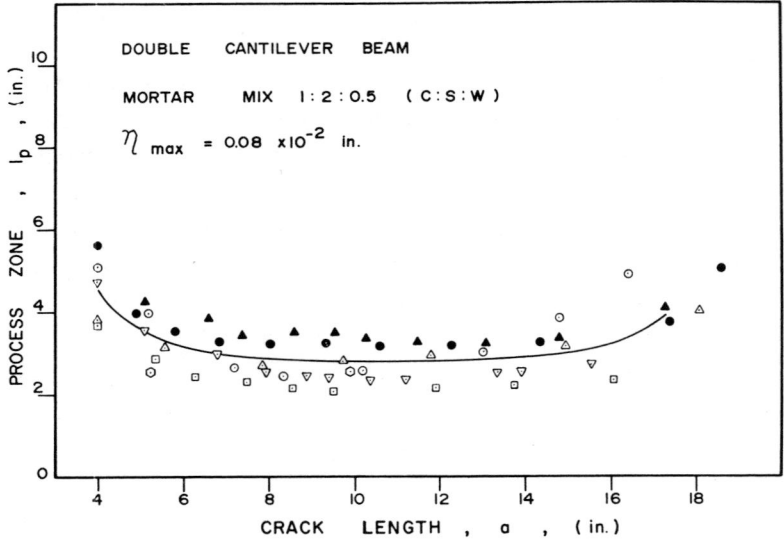

Fig. 12. Predicted process zone and crack growth relationship.

well as the critical crack opening displacement, η_{max}, have been measured for the compact-tension specimens of polymeric materials by using optical interference microscopy (see Fig. 13) (ref. 20). Note that the length of the process zone for PMMA is of the order of 40×10^{-3} mm; 30 mm for asbestos cement and 75 mm for mortar specimens. These large differences in the size of the process zone, must, among other things, be related to size of the microstructure of these materials. Note that the concept of a critical crack opening displacement has been noted for glassy polymers (see Fig. 13).

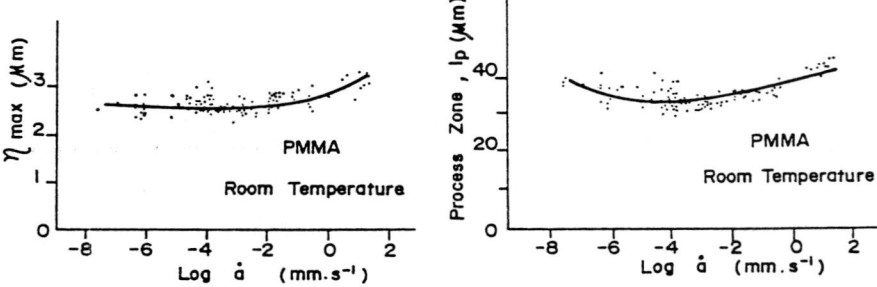

Fig. 13. Experimental relationships between process zone, critical crack opening displacement and crack growth.

The length of the process zone is likely to be a function of the specimen geometry. It was observed that the theoretical length of the process zone was only 25 mm for the double torsion specimens whereas it was 75 mm for the double cantilever specimen of the identical material. This should be expected since the crack opening displacement for a given crack extension, is different for the double torsion specimen as compared to that of the double cantilever specimen (ref. 13).

The length of the process zone for concrete can be relatively large, and can be dependent on the geometry of the specimen (see Fig. 14). Thus, if the fracture resistance (G_R) is plotted against the change in the visible, traction free crack length (Δa), then the R-curves may be the specimen-geometry dependent. However, if G_R is plotted against the change in the effective crack length $\Delta(a + \ell_p)$, then R-curves should be specimen-geometry independent. This is shown in figure 15 for the double cantilever and the double torsion mortar specimens.

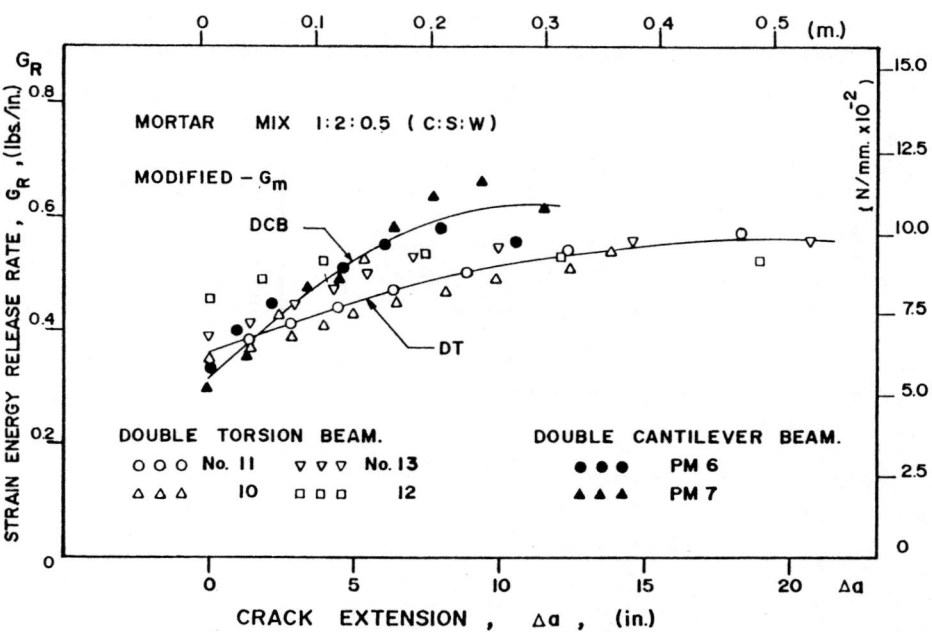

Fig. 14. The relation between G_R and Δa.

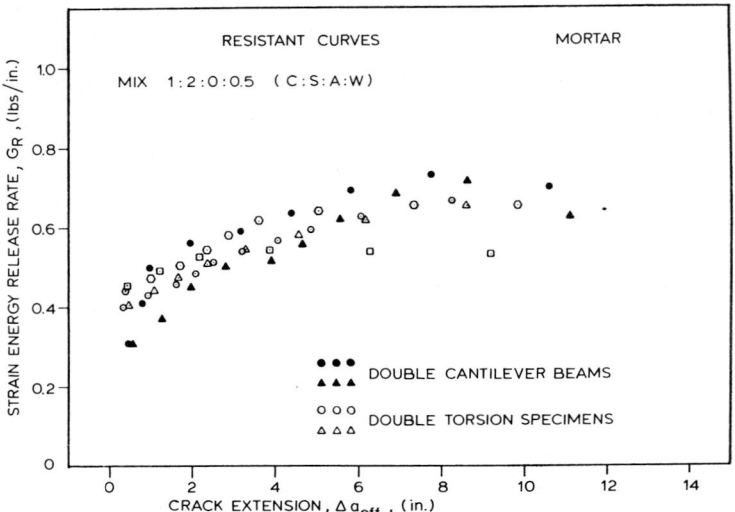

Fig. 15. The relation between G_R and Δ_{eff}.

For beam specimens, the opening displacement of a crack will depend on the dimensions of the beam as well as on the method of loading (3-point vs. 4-point) (see Fig. 16). A theoretical estimate of the process zone was obtained for the

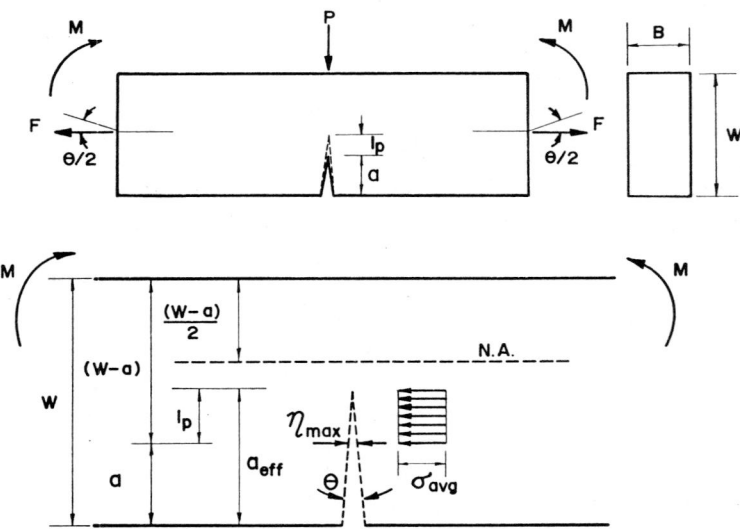

Fig. 16. Concept of effective elastic crack length for notched beams.

beams neglecting the effects of shear deformations (ref. 16, 18). Based on the theoretical calculations of the process zone, and the R-curves obtained form the experiments on the double cantilever (or double torsion) specimens, it was possible to predict the values of the peak load of beams tested by other investigators (ref. 16, 18). For example, in figure 17 the predicted values of the experimental results obtained by Walsh (ref. 21) are given.

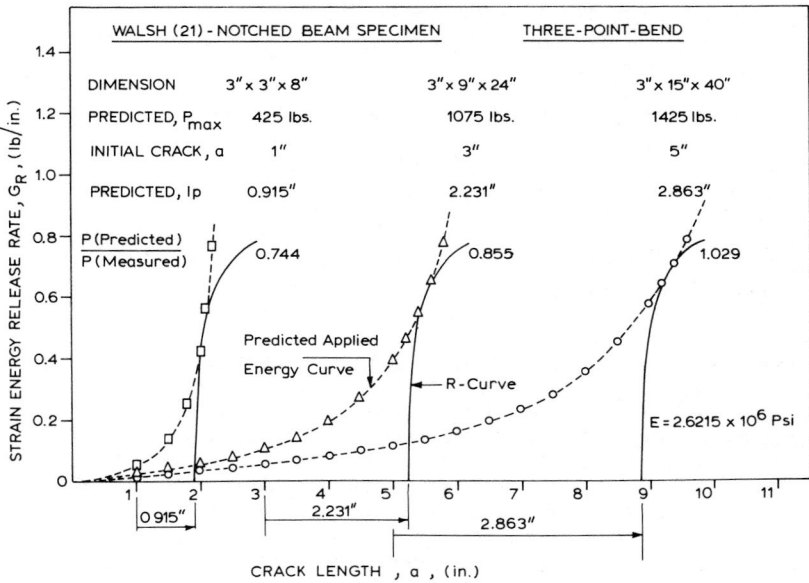

Fig. 17. Prediction of the test results from Walsh (ref. 21).

This means that R-curves can be useful to characterize the resistance of cementitious materials during slow crack growth provided that the ordinate of the R-curve, G_R is modified to include the inelastic energy, and the abscissa is modified to include the effect of process zone on specimen geometry.

REFERENCES

1 Standard Method of Test for Plane-Strain Fracture Toughness of Metallic Materials, ASTM Designation E399.74, Part 10, ASTM Annual Standard.
2 S.P. Shah and F.J. McGarry, Griffith Fracture Criteria and Concrete, Proceedings of ASCE, Journal of Eng. Mech. Div., Vol. 47, No EM6 (1971) pp. 1633-1676.
3 G. Velazco, K. Visalvanich and S.P. Shah, Fracture Behavior and Analysis of Fiber Reinforced Concrete Beams, Cement and Concrete Research, Vol. 10 (Jan. 1980) pp. 41-51.

4 J.C. Lenian and A.R. Bunsell, The Resistance to Crack Growth of Asbestos Cement, Journal of Materials Science, Vol. 14 (1979) pp. 321-332.

5 Y.W. Mai, Strength and Fracture Properties of Asbestos Cement Mortar Composites, Journal of Materials Science, Vol. 14, No 9 (Sept. 1979) pp. 2091-2102.

6 R.A. Schmidt and T.J. Lutz, K_{IC} and J_{IC} of Westerly Granite - Effects of Thickness and In-Plane Dimensions, in Fracture Mechanics Applied to Brittle Materials, ASTM-STP 678, ASTM, Philadelphia (1978) pp. 166-182.

7 R.M. Foote, B. Cottrell and Y.W. Mai, Crack Growth Resistance Curve for a Cement Composite, in Advances in Cement-Matrix Composites, Proceedings, Symposium L, Materials Research Society, Annual Meeting, Boston, Massachussets (Nov. 17-18, 1980) pp. 95-106.

8 J.H. Brown, Measuring the Fracture Toughness of Cement Paste and Mortar, Magazine of Concrete Research, Vol. 24, No 81 (Dec. 1972) pp. 185-196.

9 S.E. Schwartz, K.K. Hu and G.L. Jones, Compliance Monitoring of Crack Growth in Concrete, Proceedings of ASCE, Journal of Eng. Mech. Div., Vol. 104, No EM4 (August 1978) pp. 789-800.

10 D.J. Cook and G.D. Crookham, Fracture Thoughness Measurements of Polymer Concrete, Magazine of Concrete Research, Vol. 30, No 105 (Dec. 1978) pp. 205-214.

11 B. Hillemeir and H.K. Hilsdorf, Fracture Mechanics Studies on Concrete Composites, Cement and Concrete Research, Vol. 7 (1971) pp. 523-536.

12 M. Wecharatana and S.P. Shah, Double Torsion Tests for Studying Slow Crack Growth of Portland Cement Mortar, Cement and Concrete Research, Vol. 10 (1980) pp. 833-844.

13 M. Wecharatana and S.P. Shah, Slow Crack Growth in Cement Composites, Proceedings of ASCE, Journal of Structural Division (June 1982).

14 J.M. Hodgkinson and J.G. Williams, J and G Analysis of the Tearing of Highly Ductile Polymer, Journal of Materials Science, Vol. 16 (1981) pp. 50-56.

15 A. Hillerborg, M. Modeer and P.E. Petersson, Analysis of Crack Formation and Crack Growth in Concrete by Means of Fracture Mechanics and Finite Elements, Cement and Concrete Research, Vol. 6 (1976) pp. 773-782.

16 M. Wecharatana and S.P. Shah, Prediction of Nonlinear Fracture Process Zone in Concrete, to be published in the Engineering Mechanics Division, Journal of the American Society of Civil Engineers.

17 D.S. Dugdale, Yielding of Steel Sheets Containing Slits, Journal of Mechanics, Physics and Solids, Vol. 8 (1960) pp. 100-104.

18 M. Wecharatana, Fracture Resistance in Cementitious Composites, Ph.D. Dissertation, Department of Materials Engineering, University of Illinois at Chicago Circle (March 1982).

19 P.E. Petersson, Fracture Mechanical Calculations and Tests for Fiber Reinforced Cementitious Materials, in Advances in Cement-Matrix Composites, Proceedings, Symposium L, Materials Research Society, Annual Meeting, Boston, Massachussets (Nov. 17-18, 1980) pp. 95-106.

20 M.G. Schinker and W. Doll, Interference Optical Measurements of Large Deformations at the Tip of a Running Crack in a Glassy Thermoplastic, in Mechanical Properties of Materials at High Rates of Strain, Ed. J. Harding, Inst. Phys. Conf. Ser., No 47, Chapter 2 (1979) pp. 224-232.

21 P.F. Walsh, Fracture of Plain Concrete, Indian Concrete Journal, Vol. 46 (Nov. 1971) pp. 469-470, 477.

Fracture Mechanics of Concrete,
edited by F.H. Wittmann, 1983
Elsevier Science Publishers B.V., Amsterdam — Printed in The Netherlands

Chapter 6

\ THE FRACTURE OF FIBRE REINFORCED AND POLYMER IMPREGNATED CONCRETES: A REVIEW

by Sidney MINDESS

1   INTRODUCTION

As has already been pointed out by a number of authors (e.g. ref. 1), the
application of fracture mechanics to hardened cement paste, mortar, and plain
concrete is fraught with uncertainty. The simple linear elastic fracture
models which have been proposed, even for these relatively brittle materials,
do not appear to provide an adequate failure criterion. The linear elastic
fracture parameters K_c and G_c are one-parameter descriptions of the stress and
displacement fields in the region of a crack tip; their calculation assumes a
homogeneous, linearly elastic material. For hardened cement pastes, and even
more for concrete, this assumption is at best an approximation. Therefore,
many investigators have concluded that linear elastic fracture mechanics (LEFM)
cannot be applied to these cementitious materials, though this conclusion is
still the subject of considerable controversy.

Turning now to fibre reinforced concrete (frc), polymer concrete (PC) and
polymer impregnated concrete (PIC), the picture becomes much more confusing.
Not only is there an additional phase introduced into an already complex system, but there is also a fundamental change in the behaviour of these materials
under load. Fibre additions greatly improve the "toughness" of the material
(i.e., the total energy absorbed in breaking a specimen); they give concrete a
considerable amount of apparent ductility. The application of LEFM to a material with a highly non-linear stress-strain (σ-ϵ) curve, at least near the
ultimate stress, does not appear to be reasonable. On the other hand, PIC may
well behave in a more brittle fashion than plain concrete, though in practice
a monomer is added that gives a tough, flexible co-polymer. Again, the application of simple LEFM models to this very heterogeneous material would appear
to be inappropriate.

The purpose of the present work (based on a recent review (ref. 2)),is to
describe the available literature dealing with the applications of fracture
mechanics to both fibre reinforced and polymer impregnated cementitious systems.
The emphasis will be on whether the fracture mechanics models which have been
used to characterize these materials provide an adequate representation of
their fracture properties.

2 POLYMER IMPREGNATED CONCRETE

Although PIC has been studied for a number of years, relatively few researchers have applied fracture mechanics principles to these materials. Only LEFM
has been applied; no non-linear fracture mechanics models have been used.
Tazawa and Kobayashi (ref. 3) found that the improvements in strength obtained
with PIC could be predicted reasonably well using the classical Griffith theory;
they also suggested that the Weibull "weakest link" theory applied to concrete.
The subsequent fracture mechanics studies of PIC (refs. 4-10) showed that both
K_c and the fracture energy increased significantly for PIC, compared to plain
hcp or concrete. In all of these studies, polymerized methyl-methacrylate
(PMMA) was the impregnant, though a few tests were carried out by Cook and
Crookham (ref. 10) using styrene and butyl methacrylate as well. In general,
the investigators found that polymer impregnation increased the values of K_c by
a factor of about three (refs. 5-8,10); only with premix polymer additives was
K_c reduced (refs. 10,11). Cook and Crookham (ref. 10) showed that K_c depended
not only on the type of polymer used, but also on the type of polymerization;
thermally polymerized specimens had lower K_c values than did irradiation polymerized specimens. This reflected their observation that thermal treatment reduced the fracture toughness of plain concrete, while irradiation had no effect.
They made two other observations: (i) PIC was a notch sensitive material,
though cast notches gave different results from natural cracks; and (ii) as with
plain cement and concrete, K_c appeared to vary with the ratio of crack depth to
specimen depth, as shown in Fig. 14.1, though a limiting value was reached at a
crack length of about 35-40 mm. They maintained that LEFM could be used to
characterize PIC. Similarly, Evans et al. (ref. 6) found that K_c was independent of crack length, beyond a crack length of about 20 mm. On the other hand,
Auskern and Horn (ref. 4) found that the fracture energy was increased by a
factor of about eight by impregnation, whereas Griffith theory would have predicted an increase by a factor of only five.

The mechanisms by which polymer impregnation increases fracture toughness are
not completely understood. Evans et al. (ref. 6) suggested that impregnation
suppressed the onset of microcracking, leading to an enhanced fracture stress.
In addition, by bridging across the crack as shown in Fig. 14.2, the polymer
could increase the resistance to crack extension (similar to the effects of
fibre reinforcement). Aleszka and Beaumont (ref. 7) postulated that polymer
impregnation decreased the porosity and hence the tendency for crack initiation;
Cook and Crookham (ref. 10) similarly indicated that the polymers tended to
eliminate stress-inducing flaws. Munoz-Escalona and Ramos (ref. 9) argued that
polymers not only reduced the porosity, but also improved the bond between the
aggregate and the matrix. This view was supported by Clifton et al. (ref. 5),

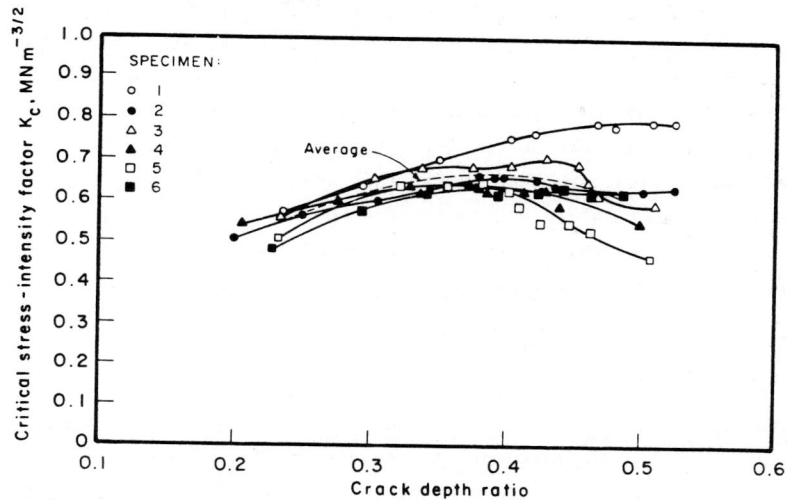

Fig. 1. K_c versus crack depth for polymer concrete (ref. 10).

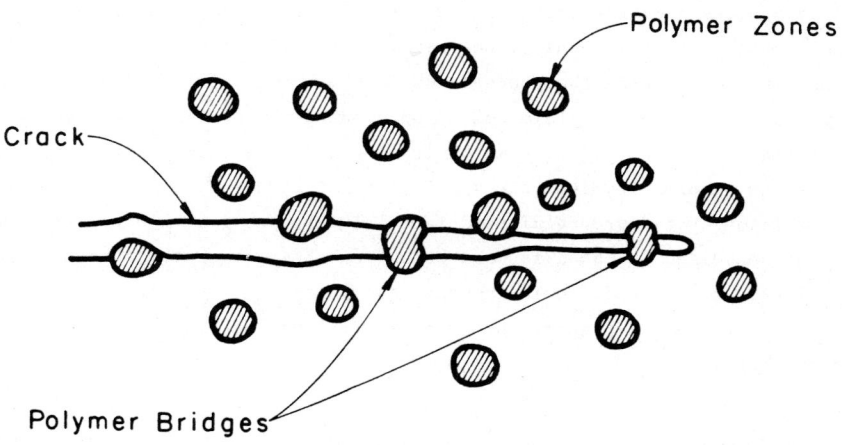

Fig. 2.   Schematic of a macrocrack "bridged" by polymer particles (ref. 6).

who found that plain mortar failed primarily through the hydrated cement phase, while with polymer impregnation the fracture surface passed through both the hydrated cement and the aggregate particles. Bhargava and Rehnstrom (ref. 12) confirmed this view. Using high speed photography on specimens loaded explosively, they found that for PIC there was more fracture through the aggregate particles, again suggesting better bond. They also found that the crack velocity in PIC was about 800 m/s, compared to about 180 m/s in plain concrete. This might indicate that crack tortuosity and microcracking are reduced in PIC.

From these few studies, it would appear that the determination of the linear elastic fracture parameters for PIC is subject to the same uncertainties and experimental difficulties which have been described for plain concrete (ref. 1). Since no non-linear fracture models have as yet been applied to PIC it is not known whether they would be an improvement.

3 FIBRE REINFORCED CONCRETE

Fibre reinforced cements and concretes have been studied extensively since the early 1960's. They were developed not to obtain high strength materials, but in order to overcome the problems brought about by the low tensile strength of plain concrete. These studies have resulted in the production of a range of composite materials, in which fibre reinforcement is used to inhibit the propagation of cracks through the brittle cementitious matrix. It was found that a variety of fibres could be used for this purpose: steel, glass, carbon, asbestos, and a wide range of organic fibres, both natural and manufactured. Since it was recognized from the beginning that the function of these fibres was to provide a crack arrest mechanism, the early investigators attempted to analyze these new materials using the concepts of fracture mechanics, which were just beginning to be applied to plain cement and concrete as well. Over the years, a substantial number of fracture mechanics studies have been carried out. Initially, these studies involved the application of LEFM. However, because of the recognition that fibre reinforced materials are not linearly elastic, and because of anomalies (to be discussed below) resulting from the application of linear elastic fracture criteria, some investigators have begun, in recent years, to apply elastic-plastic (non-linear) fracture mechanics to fibre reinforced concrete.

3.1 Applications of linear elastic fracture mechanics

The first applications of fracture mechanics to frc appear to have been made by Romualdi and Batson (refs. 13,14). Using continuous, aligned steel wires in plates with centrally located slots loaded in tension, and on the tension sides of beams in bending, they found that with appropriate wire spacings flaws could

be prevented from propagating. Both theoretical and experimental studies
showed that for a given G_c of concrete, the stress required to initiate cracking increased markedly as the wire spacing was decreased. This work was extended to concrete reinforced with randomly dispersed steel wires by Romualdi
and Mandel (ref. 15), who found that fibre reinforcement greatly increased the
post-cracking strength of concrete. Their postulate was that since the low
tensile strength of concrete is due to the propagation of cracks originating at
internal flaws, the function of the steel fibres is effectively to reduce K_I
for a given stress level, due to the restraint of the adjacent fibres. It was
later shown (ref. 16) that the effectiveness of the fibres is inversely proportional to the square of the fibre spacing. The increase in G_c is due to the
energy required to strip the fibres from the matrix ahead of the crack tip, as
modelled in Fig. 14.3 (ref. 17).

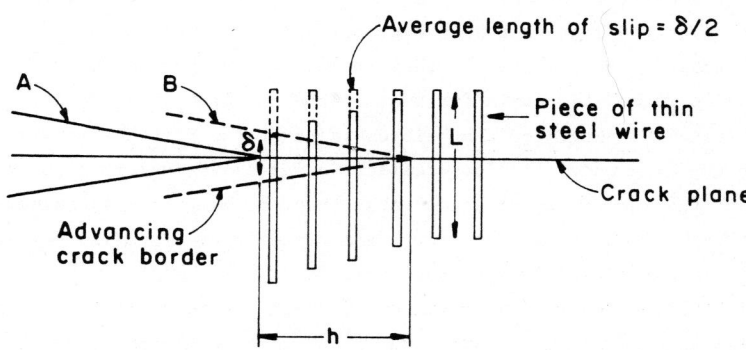

Fig. 3. Idealized version of wire slippage in zone of new crack surface
(ref. 17).

Romualdi (ref. 17) also derived a formula for G_c in terms of the reinforcing
parameters:

$$G_c = \frac{u\pi L \;\alpha p}{380d} \tag{1}$$

where u is the bond strength, L is the fibre length, d is the fibre diameter,
p is the volume percentage of fibres, and α is related to the angle between the
crack faces. However, based on a somewhat different model of the way fibres behave (Fig. 4), Parimi and Sridhar Rao (ref. 18) later derived a different
formula for G_c:

$$G_c = \frac{u^2 L^3 \; p}{314 \; E_f d^2} \tag{2}$$

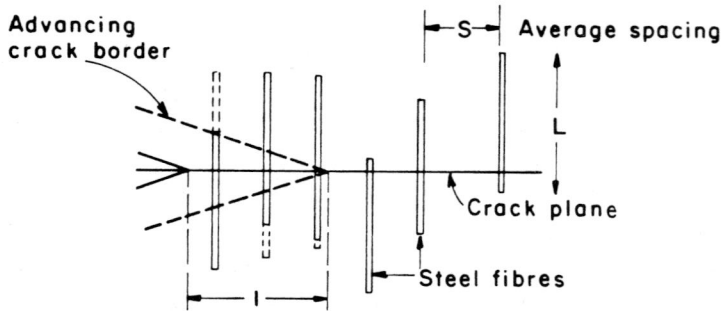

Fig. 4. Idealized version of advancing crack and random fibres (ref. 18).

where E_f is the elastic modulus of the fibres, and the other symbols are as defined above. Clearly, equations (1) and (2) give quite different predictions of G_c, in spite of the fact that both expressions were found by their authors to give reasonable approximations to their respective test results.

These studies formed the basis for most of the subsequent applications of LEFM. Shah (ref. 19) suggested that fibres increase the fracture toughness (measured as the area under the load-deflection curve) by improving the composite action of the material. He showed that fracture toughness increased with increasing fibre volume, fibre ductility, and aspect ratio; similar results were obtained by Parimi and Sridhar Rao (ref. 18).

Several durability studies were also carried out. Batson (ref. 20) examined cracked steel frc exposed to seawater. He found that salt water apparently reacted with both the steel fibres and the matrix, leading to a decrease in G_c with increasing exposure. Even after long-term storage in a dry environment, Harris et al. (ref. 21) found that K_c of steel frc decreased, perhaps due to the development of shrinkage cracks. Patterson and Chan (ref. 22) tested single-edge notched specimens of high alumina cement reinforced with E-glass fibres. With water curing, K_c increased with age for only about three weeks and then decreased, due to the effect of water on the cement-glass bond.

A number of fibres other than steel and glass have also been employed. By studying the total energy absorption during fracture, Swamy and Rao (ref. 23) showed that the order of effectiveness of different fibres was steel > glass > polypropylene. Andonian et al. (ref. 24) studied the strength and fracture characteristics of cellulose frc; the specific work of fracture and fracture toughness were found to depend mainly on the fibre pull-out characteristics. The effects of wood (primarily cellulose) fibres were also studied by Coutts and Campbell (ref. 25), who measured the energy to punch right through the samples. Again, the fibres were found to increase the "toughness" of the material. For asbestos cements as well, Mai, Foote and Cotterell (ref. 26) found that the fracture toughness of the composite came mainly from the work required to debond the fibres; Mai (ref. 27) estimated that this mechanism contributed about 95% to the total work of fracture. Akers et al. (ref. 28), however, showed that this mechanism depended on the diameter of the asbestos fibre bundles. When the strength of the bundles was greater than the bond strength, they did indeed act as crack arrestors, with subsequent microcrack failure (pull-out) at the ends of the bundles; when the failure strength of the bundles was about the same as the bond strength failure also occurred by fracture of the fibres. Briggs et al. (ref. 29) studied the work of fracture of cements reinforced with carbon fibres; for a given fibre content, the work of fracture increased as the elastic modulus of the fibres increased. Briggs (ref. 30) also reported that the impact fracture energy increased with increasing fibre content, though Ali et al. (ref. 31) found that a carbon fibre addition of 3% by volume made no improvement to impact resistance. Finally, in a closely related system, Walton and Majumdar (ref. 32) found that the fracture energy of gypsum plaster was also greatly improved by the addition of glass fibres.

Looking at even more complex systems, two investigations (refs. 7,8) showed that the work of fracture of PMMA impregnated steel fibre concrete was further increased by a factor of 3 or 4. This was explained in terms of the work done in overcoming friction during fibre pull-out; PMMA impregnation increased the interfacial frictional shear stress considerably. For ordinary fibre concrete, the energy of plastic deformation of the fibre is of the same order of magnitude as the fibre pull-out energy term; with PMMA impregnation, the latter term becomes dominant.

In addition to these studies, a number of other investigations, both experimental and theoretical, were carried out to explore the effect of fibres in modifying the fracture behaviour of plain cement and concrete. Weiss (ref. 33) assumed that fibres act as physical barriers to cracking, as well as greatly increasing the total energy for fracture by the energy required to debond or break the fibres. He developed a series of formulae to predict the effects of fibre reinforcement. Kelly (ref. 34) also studied the control of cracking in hardened cement by fibre additions; he showed that the work which must be done on the matrix as the crack spreads was approximately equal to $4J/m^2$. Wischers (ref. 35) suggested that much of the energy applied in testing concrete is dissipated by microcracking; steel fibres can bridge these microcracks, and thus contribute to the dissipation of a large proportion of the applied energy without a large loss in strength. Naaman et al. (ref. 36) suggested a stress distribution ahead of the crack as shown in Fig. 5(a). Using the model shown in Fig. 5(b) and assuming that the strength of concrete is controlled by the

488

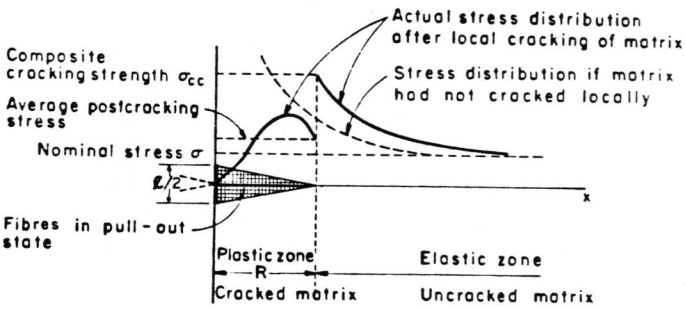

(a)

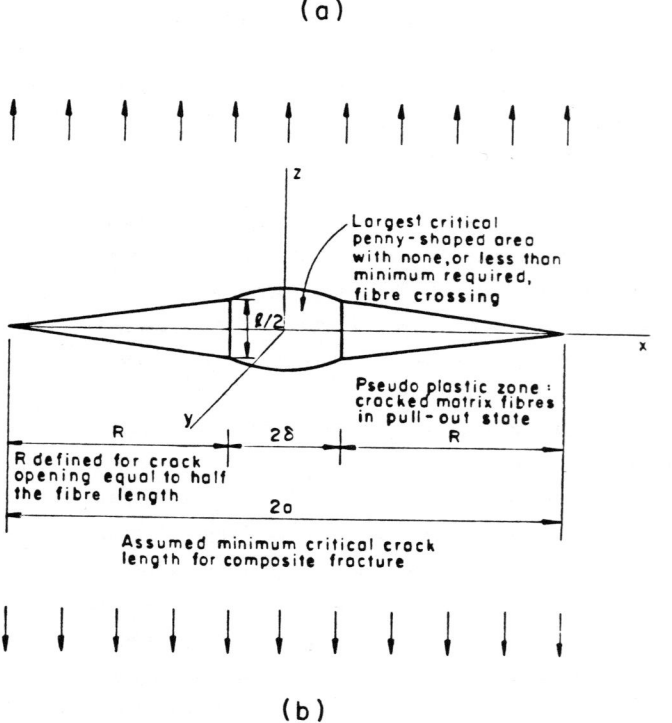

(b)

Fig. 5   (a) Distribution of longitudinal stress ahead of a crack in fibre-reinforced concrete;   (b) Assumed critical crack model controlling the fracture of fibre-reinforced concrete (ref. 36).

"worst" flaw, they derive a statistical fracture model, based on probabilistic considerations. Recently, Morton (ref. 37) estimated the work of fracture and post-cracking stress of concrete reinforced with short, high strength steel fibres, by consideration of the behaviour of individual inclined wires. He showed that for short fibres, or where the interfacial shear stress was low, much of the work of fracture was due to the plastic deformation of fibres not perpendicular to the crack face; random fibres were more effective in this regard than aligned fibres.

The development of cracking in frc was described by Swamy and Rao (ref. 23) as consisting of three stages: (i) at low loads, some of the applied energy goes in to crack initiation; (ii) at higher loads, up to the maximum load, energy is dissipated in sub-critical crack growth; and (iii) further energy is used up in unstable crack growth, corresponding to the descending branch of the σ-ε curve. They suggested a method of analysis to evaluate the energy absorption capacity of concrete in terms of the maximum load, specimen compliance, and deflection. The fracture mechanisms were examined in greater detail by Aleszka and Schnittgrund (ref. 38), who examined the fracture surfaces of frc using a scanning electron microscope. They found that, regardless of the type of loading, the presence of fibres increased the irregularity of the macroscopic fracture surface, though the microstructure remained unchanged. The fracture surface revealed cracking at both the cement-aggregate and cement-fibre interfaces. They suggested that in frc, fracture begins with cracks at the cement-aggregate interface, which then propagate into the matrix, where they are obstructed by fibres. These fibres force the cracks to go around them, increasing the energy requirement of the material.

Differences in behaviour between specimens with sawn or pre-formed notches and those with "natural" cracks were also explained. Djabarov et al. (ref. 39) noted that for steel frc, the increase in K_c for "natural" cracks was greater than for sawn notches, since in the former case fibres continue to bridge the gap. It appeared that matrix failure occurred at about the same load for both plain and fibre reinforced concretes; the function of the fibres was to inhibit the growth of the microcracks. Strange (ref. 40) found that fibres have no influence on the stress at which a pre-formed notch began to extend; Brown (ref. 41) and Patterson and Chan (ref. 22) also concluded that fibres did not inhibit crack initiation. Indeed, for asbestos cements, Mai (ref. 27) argued that since up to the first crack the material had a linear σ-ε curve, LEFM could be used to describe the failure of these materials.

However, as an increasing amount of experimental data became available, inconsistencies in the measured fracture parameters began to appear, in a similar fashion to the inconsistencies that were emerging for plain concrete. That is,

measurements of G_c or K_c appeared to be strongly dependent on the specimen geometry and the method of measurement. For example, Brown (ref. 41) tested small
glass fibre reinforced cement beams in 4-point bending, and calculated K_c
values based on the assumption that the stress concentration at the crack tip
was unaffected by the fibres. It was found that the "apparent" toughness (K_a)
was directly proportional to the crack growth over the range tested (4 to 12 mm)
as shown in Fig. 14.6. (It should be noted that a number of investigators have

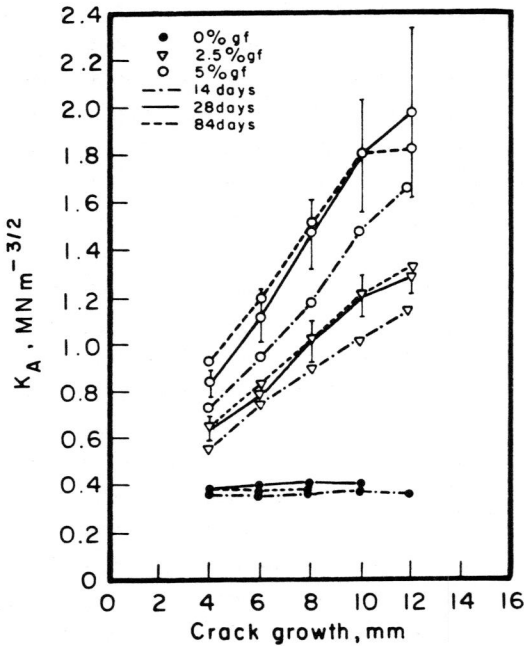

Fig. 6. Relationship between fracture toughness and crack growth. The limits
represent the standard error of the mean (ref. 41).

used the "apparent" toughness, K_a, rather than K_c, in order to make explicit
their view that these tests do not measure the true fracture toughness, but
only an apparent value, since the assumptions inherent in the calculation of
K_c are at best only approximately met in frc systems.) Javan and Dury (ref. 42)
using the CNRBB test (circumferentially notched round bar under bending), found
that both K_c and G_c increased with increasing notch depth. For autoclaved asbestos fibre reinforced mortars, Mai (ref. 27) showed that the specific work of
fracture increased with increasing crack length for double cantilever beam specimens. Conversely, Harris, Varlow and Ellis (ref. 21) found that for their
notched beams in bending, the work of fracture decreased with an increasing

ratio of notch depth to specimen depth. Finally, Brandt (ref. 43) showed that for his relatively small (50x50x300 mm) notched beams, K_c and G_c appeared to be independent of the relative crack depth. It was also shown (ref. 26) that, at least for asbestos cement notched beams in bending, a minimum beam depth of 50 mm was required to get valid K_c values; for smaller depths, the beams were notch insensitive.

Nishioka et al. (ref. 44) argued that LEFM could be applied to their steel frc notched beams in bending. However, their data (Fig. 7) suggest that K_c is affected by the specimen size, the span to depth ratio, and perhaps by whether the test is carried out in 3-point or 4-point bending. Ohigashi (ref. 45), on the other hand, found that fracture energies obtained from 3-point bending tests of notched specimens, Izod-type impact tests, and estimated from pull-out tests of fibres all were in reasonable agreement with each other. He found that the values obtained from impact tests decreased as the notch depth increased, presumably due to decreases in the extent of secondary cracking and in the irregularity of the fracture surfaces. In the bending tests, the fracture energy decreased slightly as the strain rate increased, but the notch depth had little effect, as long as the notch was deep enough compared to the specimen depth (i.e. notch depth/specimen depth $\approx$ 0.75). From the above, it is clear that different investigations may lead to different conclusions as to the effects of specimen geometry on the fracture parameters.

Other studies have revealed even greater complications. Visalvanich and Naaman (ref. 46) found that different techniques for calculating the fracture properties all gave about the same results for asbestos cement, but quite different results for both plain and steel fibre reinforced mortars. They concluded that LEFM could be applied to asbestos cement; the value of K_c reached a constant value after a crack extension of about 75 mm. However, for the steel fibre reinforced mortars, while K_c appeared to be independent of the initial crack length, it increased with increasing crack growth, and did not reach a constant value even for the 610 mm long specimens that were tested. Swamy (ref. 47) showed that slow crack growth could not be neglected in fracture toughness tests; K_c first increased linearly with crack growth, and then much more rapidly. The fracture toughness vs. crack growth curves gave a clear indication of the influence of slow crack growth on the fracture behaviour of fibre concretes.

In addition, Swamy (ref. 47) found that crack extension, as observed using optical microscopy, was complex; considerable branch cracking was observed during the tests. Similarly, Patterson and Chan (ref. 22), in their tests on specimens made of high alumina cement and E-glass fibres, noted a large zone of fine cracks around the crack tip (Fig. 8) which seemed to move with the

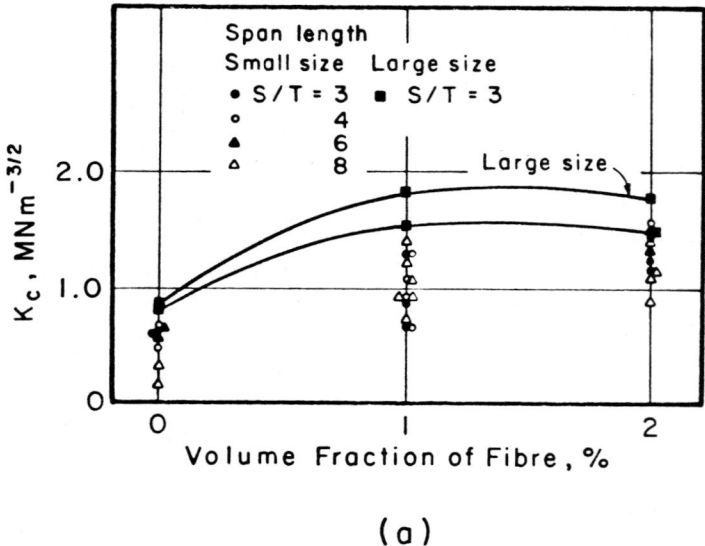

(a)

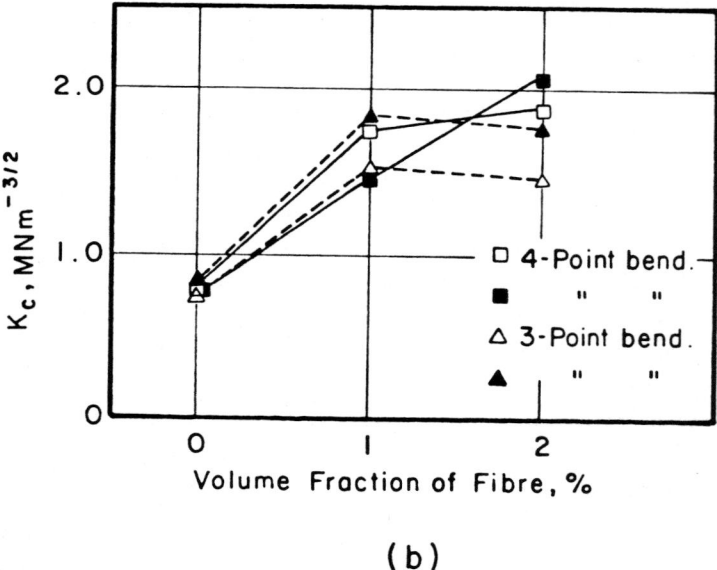

(b)

Fig. 7. (a) The effect of specimen size on K_c; (b) The effect of type of loading on K_c (ref. 44).

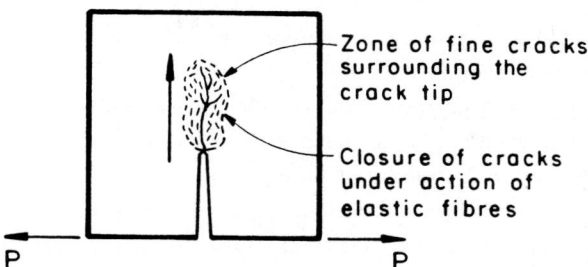

Fig. 8. Movement of the crack zone surrounding the crack tip (ref. 22).

crack tip as the crack extended, thus making it difficult even to define the crack tip.

Apart from the problems of determining the fracture parameters, there is also some evidence to indicate that linear elastic fracture parameters are not very sensitive to the addition of fibres to concrete, and therefore do not provide a useful means of characterizing the greatly increased energy required to fracture frc. For instance, Harris et al. (ref. 21) showed that while the work of fracture (calculated as the area under the load-deflection curve) was increased by about two orders of magnitude by the introduction of 2% by volume of fibres, K_c increased only by a factor of two. Mindess et al. (ref. 48) also found that K_c and G_c only increased marginally with fibre additions up to 2% by volume. In addition, Gustaffson (ref. 49) reported that for 2-dimensionally oriented steel fibre reinforced concrete, G_c was not proportional to the fibre volume.

Argon and Shack (ref. 50) discussed a number of other problems associated with the application of LEFM to frc, particularly the problems related to specimen size, and the large size of the pseudo-plastic (microcracked) zone ahead of the crack. They concluded that: "For the fracture mechanics approach....to be applicable, the width dimension of the notched specimens should be many times the combined length of the crack and the 'plastic process zone'. This condition has not been achieved by any investigator. It is therefore not very surprising that the reported discrepancy [between the improvements in K_c and in fracture energy due to fibre additions] occurs." They suggested that a non-linear crack model would be more appropriate for frc. More recently, Naaman and Shah (ref. 51) also pointed out that there are increasing needs to identify appropriate test methods, specimen geometries, and analytical techniques for evaluating the fracture properties of frc.

3.2 Applications of elastic-plastic fracture mechanics

As has been indicated above, there are considerable difficulties in characterizing the behaviour of frc by using LEFM. Therefore, a number of extensions of LEFM into the elastic-plastic region have been applied both to plain and to fibre reinforced cementitious materials.

(i) Crack Opening Displacement (COD). COD can be used to characterize fracture behaviour in the vicinity of a sharp notch by considering the opening of the notch faces (which can be measured using clip gauges). This can be shown to be analogous to G_c, and thus COD values can be related to K_c. COD measurements can be made even when there is plastic flow (or microcracking in the case of fibre cement composites) ahead of the crack. In the study already referred to, Nishioka et al. (ref. 44) used COD analysis on a variety of fibre concrete specimens, and ended by concluding that the ASTM test method for metals is applicable to frc. Studies carried out at the University of Illinois at Chicago Circle (refs. 52,53) compared a variety of fracture parameters: K_c, J_c, COD, and R-curve analysis. They found that, in general, COD values depended to some extent on specimen geometry, and concluded that COD did not appear to provide a useful critical value with which to characterize the fracture of fibre reinforced cementitious materials.

(ii) J-Integral. The J-integral is a path-independent energy line integral, which is a one-parameter average measure of the elastic-plastic field near a crack tip. It may be considered as the potential energy difference between two identically loaded bodies having differing crack lengths. For a non-linear elastic body, the J-integral may also be interpreted as the energy available for crack extension. For the linear elastic case, J_c is equivalent to G_c. The J-integral was applied to plain concrete and frc first by Mindess et al. (ref. 48), and subsequently by Halvorsen (refs. 54,55) and others. It was found that J_c could be correlated with fibre content, reflecting a change in material behaviour as shown in Fig. 9 (ref. 48), while G_c and K_c were insensitive to the fibre volume. This suggested that J_c might be an appropriate fracture criterion for frc. Halvorsen (refs. 54,55) extended this work to a study of 98 different frc mixes. Above a fibre content of about 0.75% by volume, J_c increased significantly; it also increased with increasing fibre length, aspect ratio, and end anchorage. For development of maximum toughness, a ductile pullout mode was required. However, the material appeared to be notch insensitive for the relatively small (76 x 76 x 381 mm) beams tested. In addition, he found J_c to be dependent on specimen dimensions, particuarly depth. Brandt (ref. 43) found that J_c could be used to characterize the first crack strength of steel frc, which he also found to be a notch-sensitive material. However, there was a great deal of scatter in the data, largely because it was not

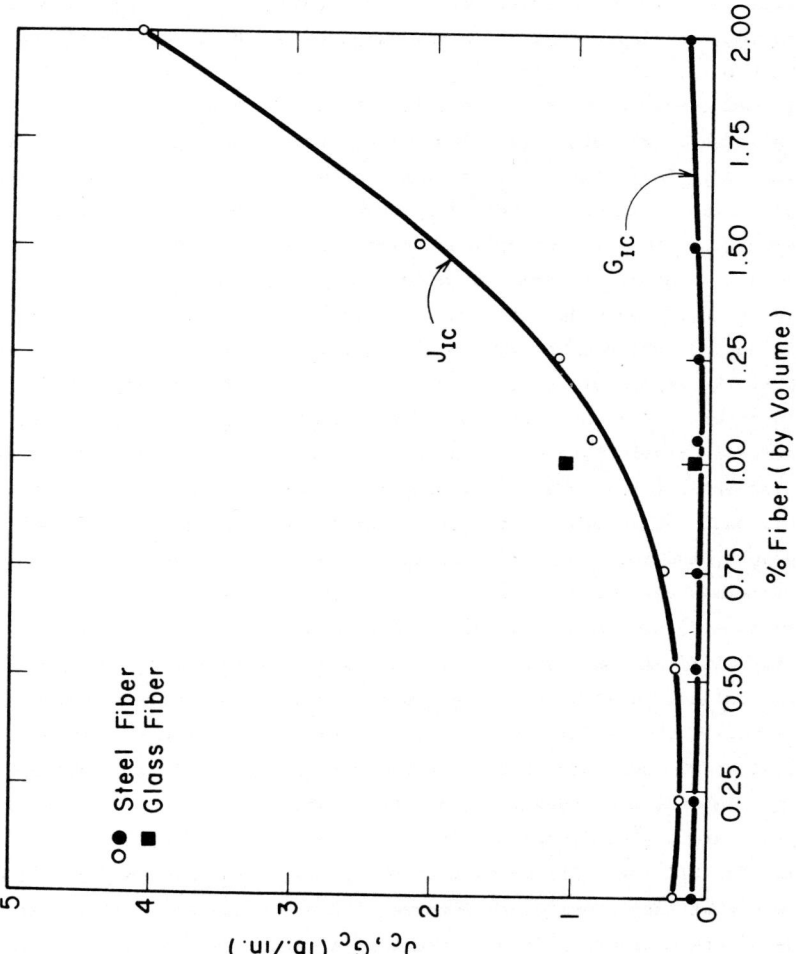

Fig. 9. Effect of fibre content on average values of J_c and G_c (ref. 48)

possible to obtain a truly brittle fracture in the steel frc. A series of studies carried out at the University of Illinois at Urbana-Champaign (Refs. 56-58) indicated that the J-integral could be used as a suitable fracture criterion for both plain and fibre reinforced concretes. Carrato (ref. 57) found that J_c was independent of notch geometry, and that notch depth became important only when the maximum aggregate size approached the size of the uncracked ligament. J_c also appeared to be independent of specimen size in the limited range of sizes studied, and did not vary between 3-point and 4-point bending. He suggested that the large scatter in the computed values of the J-integral could be drastically reduced if a displacement criterion was substituted for the maximum load criterion used in the calculations. Conversely, however, Velazco et al. (ref. 52) found that the J-integral was not independent of crack length. Moreover, Sih (ref. 59) has recently pointed out that, theoretically, the J-integral simply cannot be applied to composite systems such as frc.

(iii) <u>R-Curve Analysis</u>. R-curves are plots of the resistance to crack growth (K_R) as a function of crack extension. They therefore can be used to characterize the resistance to fracture of materials during stable crack growth. They provide an indication of toughness development as a crack is extended in a stable manner. K_R represents the driving force required to produce stable crack growth prior to the unstable crack growth at K_c. K_R is equivalent to K_c at a particular instability condition determined during an R-curve experiment. This technique appears to be very promising for characterizing frc. As shown in Fig. 10 (ref. 52), the observed R-curves appear to be independent of the initial notch depth, and to be sensitive to the fibre volume. Wecharatana and Shah (ref. 53), however, found that the calculated R-curves were influenced by the crack opening; larger openings led to higher strain energy release rates. Therefore, for different specimen geometries, it was necessary to compare strain energy release rates at the same crack openings. It was also found necessary to modify the definition of strain energy release rate to include both the elastic and inelastic strain energy components of the crack propagation process. In addition, Shah (ref. 60) has shown that R-curves too are influenced by different specimen geometries.

Lenain and Bunsell (ref. 61) found that the crack resistance of sheet asbestos cement could be characterized by R-curves. They were able to distinguish three stages of crack growth: (i) the creation of a zone of microcracks (of the order of 12-28 mm) in front of the main crack, (ii) the growth of this zone along with slow, stable crack growth, and (iii) the extension of the main crack, with the size of the microcracked region remaining constant. They proposed the model shown in Fig. 11 to explain their results; the fibres which bridged the microcracks induced a closing stress at the crack tip, and so increased the

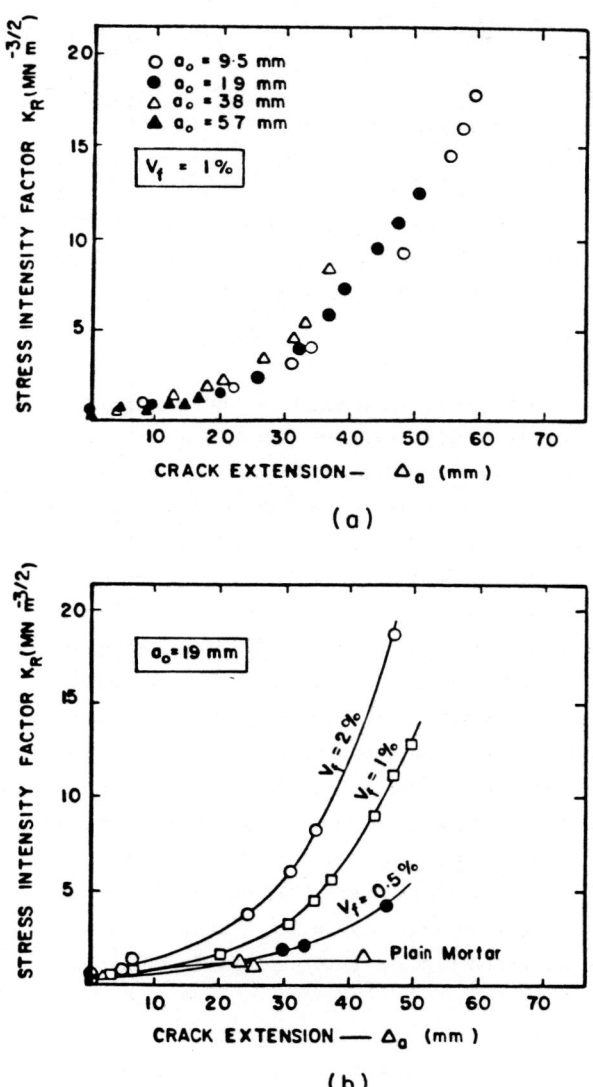

Fig. 10. (a) R-Curves for different initial notch depths; (b) R-Curves for different fibre volumes (ref. 52).

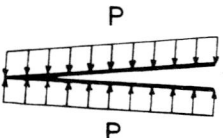

Fig. 11. Proposed model in which bridging fibres are replaced by uniformly distributed closing pressure (ref. 61).

work of fracture. Similarly, Mai et al. (ref. 26) found that K_R curves were the best means of characterizing the fracture behaviour of asbestos-cellulose cement composites. They later suggested (ref. 62) that the stress intensity factor for frc can be separated into two components,

$$K = K_R + K_r = K_i \tag{3}$$

where K_i corresponds to fracture initiation of the matrix, K_R is the stress intensity factor at the tip of the crack if there were no fibres bridging the crack, and K_r is the stress intensity factor due to the bridging fibres closing the crack faces. In comparing notch bend tests, compact tension tests, and double cantilever beam tests, they showed that K_R curves were reasonably independent of specimen geometry (contrary to Shah's (ref. 60) conclusions), and that K_R could therefore be used as a material property.

  (iv) Fictitious Crack Model. In the fictitious crack model, the entire fracture zone is assumed to act as a "fictitious" crack, which has the ability to transfer stresses across the crack, (similar to the tied crack model). Instead of conventional fracture parameters, the stress-displacement curve is introduced to describe processes in the fracture zone; for the rest of the material, conventional σ-ε relations are used. This model is suitable for analysis only by using computerized numerical methods. The fictitious crack model was first applied to frc by Hillerborg (ref. 63), who claimed that both crack stability and different types of fracture could be analyzed using this technique. This technique has also been used by Petersson (ref. 64).

4 CONCLUSIONS

    In spite of all the work that has been done to develop fracture mechanics techniques which might be used to characterize the fracture of hardened cement paste, plain concrete, and fibre reinforced concrete, there is as yet no agreement on the "best" way of characterizing the fracture processes in these materials. Indeed, there is still a controversy as to whether fracture mechanics

is even applicable to these materials. A great deal remains to be learned
about the application of fracture mechanics to frc. Only further work, both
theoretical and experimental, will reveal whether this avenue of research will
turn out to be a fruitful one.

REFERENCES

1   S. Mindess, in F.H. Wittmann (Ed.), Fracture Mechanics of Concrete,
    Elsevier, Amsterdam, 1982, pp.
2   S. Mindess, Int. J. Cem. Comp., 2 (1980) 3-11.
3   E. Tazawa and S. Kobayashi, in Polymers in Concrete, SP-40, American
    Concrete Institute, Detroit, 1973, pp. 57-92.
4   A. Auskern and W. Horn, Cem. Concr. Res., 4 (1974) 785-795.
5   J.F. Clifton, J.E. Fearn and E.D. Anderson, Polymer Impregnated Hardened
    Cement Pastes and Mortars, NBS-BSS 83, National Bureau of Standards,
    Washington, D.C., 1976.
6   A.G.Evans, J.R. Clifton and E. Anderson, Cem. Concr. Res., 6 (1978)
    535-548.
7   J.C. Aleszka and P.W.R. Beaumont, in Proc. 1st Int. Cong. Polymer Concretes,
    London, May 5-7, 1975, London, Construction Press, Lancaster, 1976,
    pp. 269-275.
8   P.W.R. Beaumont and J.C. Aleszka, J. Mat. Sci., 13 (1978) 1749-1760.
9   A. Muñoz-Escalona and C. Ramos, J. Mat. Sci., 13 (1978) 301-310.
10 D.J. Cook and G.D. Crookham, Mag. Concr. Res., 30 (1978) 205-214.
11 B. Hillemeier and H.K. Hilsdorf, Cem. Concr. Res., 7 (1977) 523-536.
12 J. Bhargava and A. Rehnstrom, Cem. Concr. Res., 5 (1975) 239-248.
13 J.P. Romualdi and G.B. Batson, ASCE Proc., J. Eng. Mech. Div., 89 (1963)
    147-168.
14 J.P. Romualdi and G.B. Batson, J. Amer. Concr. Inst., 60 (1963) 775-790.
15 J.P. Romualdi and J.A. Mandel, J. Amer. Concr. Inst., 61 (1964) 657-670.
16 J.P. Romualdi, M. Ramey and S.C. Sanday, in Causes, Mechanism and Control
    of Cracking in Concrete, SP-20, American Concrete Institute, Detroit, 1968,
    pp. 179-203.
17 J.P. Romualdi, in A.E. Brooks and K. Newman (Eds.), Proc. Int. Conf.,
    The Structure of Concrete, London, Sept., 1965. Cement and Concrete
    Association, London, 1968, pp. 190-201.
18 S.R. Parimi and J.J. Sridhar Rao, in Proc. Int. Conf. Mechanical Behavior
    of Materials, Kyoto, Aug. 15-20, 1971, The Society of Materials Science,
    Japan, 1972, Vol. V, pp. 176-186.
19 A.P. Shah, in M. Te'eni (Ed.), Proc. Southampton 1969 Civil Eng. Mat. Conf.
    Structure, Solid Mechanics and Engineering Design, Southampton, April, 1969,
    Wiley-Interscience, 1971, pp. 367-376.
20 G.B. Batson, "Strength of Steel Fibre Reinforced Concrete in Adverse
    Environments", Special Report M-218, Construction Engineering Research
    Laboratory, Champaign, Illinois, 1977.
21 B. Harris, J. Varlow and C.D. Ellis, Cem. Concr. Res., 2 (1972) 447-461.
22 W.A. Patterson and H.C. Chan, Composites, 6 (1975) 102-104.
23 R.N. Swamy and C.V.S.K. Rao, in Conf. Fibre Reinforced Materials, Institu-
    tion of Civil Engineers, London, 1977, pp. 77-86.
24 R. Andonian, Y.W. Mai and B. Cotterell, Int. J. Cem. Comp., 1 (1979)
    151-158.
25 R.S.P. Coutts and M.D. Campbell, Composites, 10 (1979) 228-232.
26 Y.W. Mai, R.M.L. Foote and B. Cotterell, Int. J. Cem. Comp., 2 (1980) 23-34.
27 Y.W. Mai, J. Mat. Sci., 14 (1979) 2091-2102.
28 S.S. Akers, G.G. Garrett and R.B. Tait, Proc. Electronmicroscopy Society of
    South Africa, 7 (1977) 57-58.

29 A. Briggs, D.Y. Bowen and J. Kollek, in Proc. 2nd Int. Conf. on Carbon
    Fibres, Their Place in Modern Technology, London, Feb. 1974, Unwin, Old
    Woking, 1974, Paper No. 17.
30 A. Briggs, J. Mat. Sci, 12 (1977) 384-404.
31 M.A. Ali, A.J. Majumdar and D.L. Rayment, Cem. Concr. Res., 2 (1972)
    201-212
32 P.L. Walton and A.J. Majumdar, J. Mat. Sci., 12 (1977) 831-836.
33 V. Weiss, Cem. Concr. Res., 3 (1973) 189-205.
34 A. Kelly, in G. Piatti (Ed.), Proc. Seminar on Advances in Composite
    Materials, Oct. 1976, Applied Science, London, 1978, pp. 113-129.
35 G. Wischers, in Betontechnische Berichte 1978, Beton-Verlag GmbH,
    Düsseldorf, 1979, pp. 31-56.
36 A.E. Naaman, A.S. Argon and F. Moavenzadeh, Cem. Concr. Res., 3 (1973)
    397-411.
37 J. Morton, Mater. Constr. (Paris), 12 (1979) 393-396.
38 J. Aleszka and G. Schnittgrund, Technical Report M-122, Construction
    Engineering Research Laboratory, Champaign, Illinois, 1975.
39 N. Djabarov, A. Wehrstedt, H.-J. Weiss and K. Yamboliev, in Proc. 2nd Nat.
    Conf. Mechanics and Technology of Composite Materials, Varna, 1979, Sofia,
    Bulgaria, 1979, pp. 593-596.
40 P.C. Strange, "The Fracture of Plain and Fibre Reinforced Concrete",
    Ph.D. Thesis, Dept. of Civil Engineering, University of Auckland, New
    Zealand, 1977.
41 J.H. Brown, Mag. Concr. Res., 25 (1973) 31-38.
42 L. Javan and B.L. Dury, Concrete, 13 (1979) 31-33.
43 A.M. Brandt, Int. J. Cem. Comp., 2 (1980) 35-42.
44 N. Nishioka, S. Yamakawa, K. Hirakawa and S. Akihama, in R.N. Swamy (Ed.),
    Testing and Test Methods of Fibre Cement Composites, RILEM Symposium,
    Sheffield, 1978, Construction Press, Lancaster, 1978, pp. 87-98.
45 T. Ohigashi, in R.N. Swamy (Ed.), Testing and Test Methods of Fibre Cement
    Composites, RILEM Symposium, Sheffield, 1978, Construction Press, Lancaster,
    1978, pp. 67-78.
46 K. Visalvanich and A.E. Naaman, in W.F. Chen and E.C. Ting (Eds.), Fracture
    in Concrete, Proc. ASCE session, Hollywood, Florida, Oct. 27-31, 1980.
    American Society of Civil Engineers, New York, 1980, pp. 65-81.
47 R.N. Swamy, Int. J. Cem. Comp., 2 (1980) 43-53.
48 S. Mindess, F.V. Lawrence and C.E. Kesler, Cem. Concr. Res., 7 (1977)
    731-742.
49 P.J. Gustafsson, "Brottmekaniska Studier; Lättbetong och Fiber-Armerad
    Betong", Rapport TVBM-5001, Dept. of Building Materials, The Lund Institute
    of Technology, 1977.
50 A.S. Argon and W.J. Shack, in Fibre Reinforced Cement and Concrete, RILEM
51 A.E. Naaman and S.P. Shah, in Fracture Mechanics Applied to Brittle Mate-
    rials, STP 678, American Society for Testing and Materials, Philadelphia,
    1979, pp. 183-201.
    rials, 1979, pp. 183-201.
52 G. Velazco, K. Visalvanich and S.P. Shah, Cem. Concr. Res., 10 (1980) 41-51.
53 M. Wecharatana and S.P. Shah, in W.F. Chen and E.C. Ting (Eds.), Fracture
    in Concrete, Proc. ASCE Session, Hollywood, Florida, Oct. 27-31 1980,
    American Society of Civil Engineers, New York, 1980, pp. 82-105.
54 G.T. Halvorsen, "Toughness of Portland Cement Concrete", Ph.D. Thesis,
    University of Illinois at Urbana-Champaign, 1979.
55 G.T. Halvorsen, Int. J. Cem. Comp., 2 (1980) 13-22.
56 K. Rokugo, "Experimental Evaluation of Fracture Toughness Parameters of
    Concrete", M.S. Thesis, Dept. of Civil Engineering, University of Illinois
    at Urbana-Champaign, 1979.
57 J.L. Carrato, "Experimental Evaluation of the J-Integral", M.S. Thesis,
    Dept. of Civil Engineering, University of Illinois at Urbana-Champaign,
    1980.
58 K. Rokugo, C.E. Kesler and F.V. Lawrence, in Proceedings, JCI (Japan Concr.
    Inst) Second Conference, 1980, pp. 125-128.

59 G.C. Sih, Personal communication.
60 S.P. Shah, in Advances in Cement-Matrix Composites, Proceedings, Materials
    Research Society, Symposium L, Boston, Nov. 1980, pp. 83-89.
61 J.C. Lenain and A.R. Bunsell, J. Mat. Sci., 14 (1979) 321-332.
62 R.M.L. Foote, B. Cotterell and Y.W. Mai, in Advances in Cement-Matrix
    Composites, Proceedings, Materials Research Society, Symposium L, Boston,
    Nov. 1980, pp. 135-144.
63 A. Hillerborg, Int. J. Cem. Comp., 2 (1980) 177-184.
64 P.E. Petersson, in Advances in Cement-Matrix Composites, Proceedings,
    Materials Research Society, Symposium L, Boston, Nov. 1980, pp. 95-106.

Fracture Mechanics of Concrete,
edited by F.H. Wittmann, 1983
Elsevier Science Publishers B.V., Amsterdam — Printed in The Netherlands

Chapter 7

NUMERICAL METHODS IN FRACTURE MECHANICS

by D.A. CHAMBERLAIN and L.F. BOSWELL

1. INTRODUCTION

Fracture mechanics considerations are an established aspect of the design
process for metallic components and structures. This is not, however, the case
for structural concrete. A brief reflection on the complex nature of concrete
must cast some doubt on the general applicability of a fracture mechanics
approach in structural concrete design. This complex material is a heterogeneous
composite for which non-linear behaviour is caused by such factors as crushing,
aggregate interlock, shrinkage, creep and moisture diffusion in addition to
cracking. The initiation and propagation of cracks in stages have been found to
have a major influence on the non-linear behaviour.

An area of uncertainty exists regarding the analytical representation of
the cracking process and controversy occurs over whether a crack should be
smeared or discretely modelled. It is probable that a combination of the two
approaches would provide the best representation.

The smeared approach leads to constitutive relationships, which reflect the
general extent and orientation of cracking. Although these may approximate
the overall state of fracture, no information is made available on individual
cracks. This approach, therefore, does not strictly align with the generally
accepted notion of fracture mechanics regarding the propagation of discrete
cracks.

The discrete crack approach places no limitations on the level at which the
material structure may be considered. The detailed particle matrix may be
modelled if adequate computing resources and relevant characteristic parameter
data are available. For use in structural design, however, fracture can only
be considered in terms of overall material properties, which in turn are
functions of the composition.

The main objective of this chapter is to review the application of the
finite element method to fracture problems. This method is the most widely
applicable numerical method of analysis. Since the results obtained from using
the method are inherently related to the principles of fracture mechanics,
it has been considered pertinent to review those aspects of fracture which
are later used in finite element modelling. Thus, the theory and application
of 'discrete crack' fracture mechanics for both the linear elastic and non-
linear cases are reviewed. In particular, reference is made to the fracture of
concrete. Whilst there is some uncertainty over the relevance of linear

elasticity, which of necessity refers to singular stress fields, it should be recognised that useful mixed mode propagation models have been developed.

In addition to reviewing the current state-of-the-art of the application of the finite element method to fracture, reference has been made to unpublished findings of a recently concluded research program described elsewhere (ref. 1,2).

2. THE DEVELOPMENT OF THEORETICAL FRACTURE MECHANICS

2.1 Elastic Fracture Mechanics

Early references to fracture phenomena were concerned with the fracture size effect observed in iron wires and rods (ref. 3,4). Perhaps the first most significant theoretical development, however, occurred in 1913 when Inglis (ref. 5), in response to comments made by Hopkinson (ref. 6), presented computations for the stress concentrations at elliptic holes and sharp corners. Muskhelishvili (ref. 7) also obtained similar effective solutions to these problems.

The theoretical foundation of continuum-based fracture mechanics, is usually attributed to Griffith (ref. 8). He postulated that a brittle solid contains numerous randomly distributed flaws of various sizes and that the largest flaw was the most significant. Furthermore, he proposed a crack propagation theory which states, that a crack will propagate if the release of elastic energy associated with the crack growth exceeds the energy necessary to create that growth. Unfortunately, Griffith presented an incorrect expression for the strain energy increment which he subsequently corrected (ref. 9) when he proceeded to consider complex loading cases. In this later work he used the Inglis stress distributions to determine rupture in accordance with a limiting tensile stress. In spite of the correction and subsequent clarification (ref. 10) there remains some confusion in published work regarding the original error.

Irwin (ref. 11) and then Orowan (ref. 12) found that the Griffith energy equation was inappropriate for situations where crack tip plasticity was significant. It was concluded, from the results of X-ray studies on low carbon steels (ref. 13) that the Griffith surface energy was less significant than the plastic work performed at the crack tip. As a consequence, the surface energy term in Griffith's equation was replaced by a plastic work term. This modification has been questioned (ref. 14), however, since Griffith considered only the global potential energy in the system. In fact, it is not until the more recent work of Rice (ref. 15,16) and others that some understanding of the elastic-plastic fracture process has been achieved.

Irwin (ref. 17,18) developed a basis for the application of fracture mechanics to practical design problems. He identified the three independent modes of internal free surface dislocation (see Fig. 1); these modes being associated with 'crack-extension-forces' and stress intensity factors.

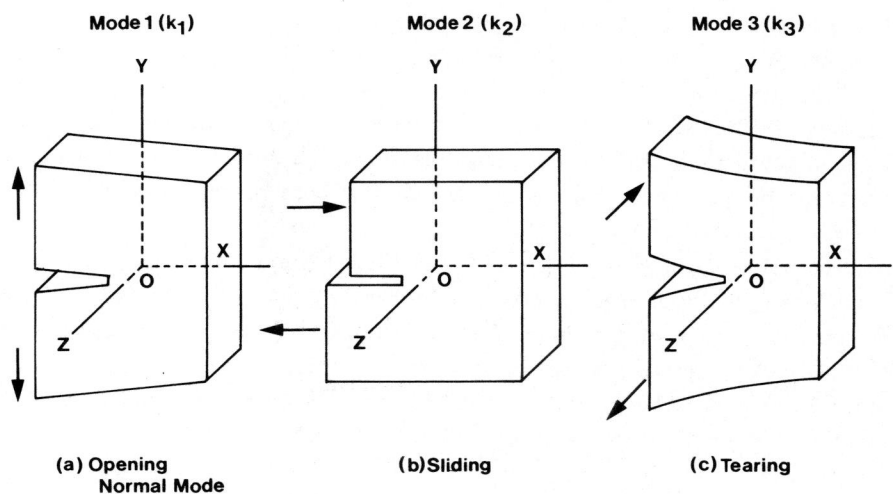

Mode 1 (k₁) Mode 2 (k₂) Mode 3 (k₃)

(a) Opening Normal Mode   (b) Sliding   (c) Tearing

FIGURE 1 CRACK PROPAGATION MODES

Irwin's work is based upon the stress function procedures suggested by Westergaard (ref. 19) and may be considered as a natural development of the work of Sneddon (ref. 20). Furthermore, he showed that a close relationship exists between the rate of strain energy release per crack extension and the intensity of the stress corresponding to the individual modes. The stress intensity factors characterise the level of stress relating to the individual modes and a critical combination is assumed to exist for crack extension to occur. The 'crack-extension-force' concept has been used to rank materials in order of fracture resistance (ref. 21). Such classification, however, is unreliable (ref. 14).

Until Sih (ref. 22) introduced the concept of the strain energy density factor (SEDF), few attempts (ref. 23,24) were made to consider modes other than pure mode I type of crack extension. The difficulties associated with the problem of a crack turning by a combination of modes were avoided by erroneously assuming coplanar crack extension. Sih's work represented a distinct departure from previous theoretical developments in two respects. These are, that the strain energy release rate method is of necessity concerned with energy increments, whereas the SEDF approach refers to the

critical state of the current energy level and, the direction of crack
propagation can be determined from the distribution of the strain energy
density around the crack tip. Sih states (ref. 14) that the two fundamental
hypotheses of crack extension are:

(i) 'The crack will spread in the direction of the maximum potential energy
density' (minimum strain energy density) and

(ii) 'the critical intensity, S_{CRIT}, of this potential field governs the onset
of crack propagation'.

A natural extension of this theory is to separate the volumetric and
deviatoric components of the strain energy density. In certain circumstances
the volumetric or deviatoric strain energy density factors can be adopted as
characteristic quantities known as S_V and S_D.

More recently, Hussain et al (ref. 25) have proposed a theory which is
the strain energy release rate counterpart of the SEDF method. According
to this theory, a crack will propagate in the direction of maximum elastic
energy release rate and that this occurs when a critical level G_{CRIT} is
attained. It is conceivable that the elastic energy release rate could also
be separated into critical volumetric and deviatoric characteristic
components.

2.2 Yielding Fracture Mechanics

It can be readily observed that crack growth is notionally related to the
level of applied load. In representing such cracks or line cuts, to which
they obviously approximate, the Inglis and Muskhelishvili approaches imply
infinite crack tip stress. By assuming that some limited tensile capacity
exists in a material, the existence of such infinite or near infinite stress
would result in cracks propagating at negligible loads. Such a phenomenon is
unrealistic and some localised yielding must occur. It may be argued in
defence of the linear elastic fracture mechanics (LEFM) approach, however,
that the correct order of magnitude of crack tip energy intensity can be
represented even though the associated stress level is unrealistic.

The established yield methods, ignoring the questionable use of the
plastic work factor in the Griffith surface energy term are, the 'crack
opening displacement' method and the J path independent integral method.
The former method was separately proposed by Dugdale (ref. 26) and
Barenblatt (ref. 27). The singularity from the crack tip strain field was
removed by Barenblatt by considering a cohesive zone ahead of the crack tip.
This cohesive zone is represented by restraining stresses, which are
independent of the applied load and have a maximum intensity, which is a
function of the crack opening for a given material under given conditions.

Dugdale's model differs in concept in that yielding ahead of the crack is represented by a fictitious crack having a length which is greater than the real length. 'The effective size' (ref. 28,29) of the crack has been defined as the sum of the actual crack size and the plastic zone correction factor.

The J path independent integral method for the approximate analysis of strain concentration at notches and cracks has been proposed by Rice (ref.16). The computed value of J is a measure of the crack tip strain intensity. Rice applied this value to non-linear elasticity with post-yield strain hardening. For 'small scale yielding' in which the plastic zone is small compared with the crack length and specimen thickness, etc., the J quantity is directly related to the elastic strain intensity factors, for the three modes of crack extension. Rice also showed that the Barenblatt cohesive model is equivalent to the Griffith surface energy theory.

More recently (ref. 30) the J integral method has been reviewed and some questionable practice has been highlighted regarding its application. The integral, which is essentially based on deformation plasticity, (non-linear elasticity), has been inappropriately used in conjunction with incremental plasticity based on the Prandtl-Reuss equations. The large scale yielding, which may be observed in experimental results does not provide a correct basis for the assessment of the small scale yielding occurring at crack tips. In order to overcome this difficulty other integrals may be used, including the F_e (ref. 31) and J_{EXT} (ref. 32) integrals. The F_e integral represents the generalised force on the singularity, whereas the J_{EXT} integral represents the potential energy release rate and thus has the same meaning as the previously mentioned J integral. During the evaluation of the J_{EXT} integral the distributed dislocations of the continuum theory of dislocation (ref.33) are taken as being equivalent to incompatibilities in the elastic strain field. Thus this integral, unlike the J integral, can be applied to incompatible strain fields. Non-homogeneous bodies may be considered (ref.34) by adopting two-path independent integrals, J_X and J_Y. For these bodies, J_X can be considered to be the force acting on the crack tip parallel to the crack surface and is, therefore, the Rice J integral. The J_Y integral is perpendicular to the crack surface.

3. THE APPLICATION OF FRACTURE MECHANICS TO CONCRETE

The adoption of LEFM implies that the fracture material has perfect elastic properties and furthermore is capable of sustaining stress concentrations. These unrealistic concentrations induce local yielding and it is the arguable extent of such yielding which continues to confuse the interpretation of results and the practical application of the method. In addition, there is the more fundamental issue of whether concrete and other similar materials can be reliably represented by an equivalent homogeneous mass. For 'small plasticity', the linear elastic approach can provide acceptable results. A growth in the plastic zone size, however, has the effect of increasing fracture toughness. The plastic zone increases with a decrease in the yield strength.

To examine the validity of LEFM the extent of plasticity at the crack tip must be assessed and can be compared with certain control dimensions, namely, crack length, specimen thickness and ligament (uncracked portion) length. Various estimates for the practical extent of the plastic zone have been made (ref. 26,28,35,36) by associating a simple yield criterion with the mode I crack extension. From these estimates there is some general agreement that the plastic zone size τ_p is given by

$$\tau_p = \alpha \left(\frac{k_{IC}}{\sigma_y} \right)^2 \tag{1}$$

in which α is a factor depending upon whether plane stress or plain strain conditions prevail. k_{IC} and σ_y are the critical stress intensity factor and the yield stress, respectively. Typical plastic zones for the plain strain and plane stress conditions (see Fig. 2) are given.

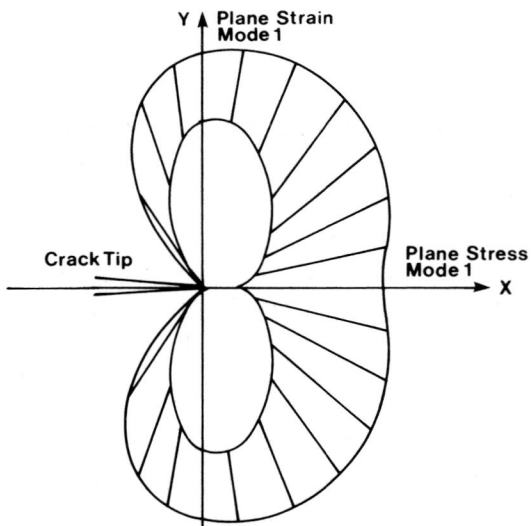

**FIGURE 2   THEORETICAL PLASTIC ZONE SHAPES
FOR PURE MODE 1**

Although reference is often made to the plane stress condition, it has been suggested (ref. 37) that this has no obvious relevance to fracture mechanics. Poisson's ratio for concrete, however, is sufficiently small for there to be no appreciable numerical difference in fracture calculations between the plane stress and plane strain states.

Considering concrete as the fracture material and including all the necessary assumptions, a typical plastic zone size may be calculated as follows for $\sigma_y = 3.5$ MN/m^2, $k_{IC} = .75$ MN/m$^{3/2}$ and $\alpha = 1/\Pi$

$$\tau_p = \frac{1}{\Pi} \left(\frac{.75}{3.5} \right)^2 \simeq 15 \text{ mm}$$

It should be noted that the value for k_{IC} is a high value taken from (ref. 38). For most brittle materials, $\tau_p = 0.1$ mm, but for concrete, however, the tensile yield stress is low.

The quantity defined as

$$\ell_{ch} = \left(\frac{k_{IC}}{\sigma_y} \right)^2 \tag{2}$$

is considered to be a characteristic parameter, which if small compared with the control dimensions, indicates the applicability of LEFM. From experimental observations it is generally accepted that the control dimensions should exceed 2.5 ℓ_{ch}. Thus from the above data, concrete specimens would have control dimensions in excess of 115 mm. Since there is a wide variation in quoted values of k_{IC}, no particular significance is placed on the value of 115 mm. For example, a value of $k_{IC} = 0.23$ MN/m$^{3/2}$ (ref. 38) would result in a corresponding control dimension of only 11 mm.

If the extent of the plastic zone size is reliably estimated using equation (1), it must be concluded that with few exceptions, the specimen sizes reported in fracture studies are too small for the results to be interpreted using LEFM. Furthermore, there is some evidence that an overall size effect exists for a considerable range of specimen sizes. The results of direct tension tests for concrete cylinders and prisms (ref. 39) indicate that the size effect still holds for control dimensions well in excess of the largest fracture specimens reported. Values of ℓ_{ch} between 100 mm and 200 mm have been suggested (refs.40-42).These values imply that corresponding control dimensions must be greater than the range 250 mm to 500 mm for a LEFM interpretation. It is interesting to note that some full scale structural components would not meet this requirement. In order to overcome this difficulty a method has been proposed (ref. 41) in which the results obtained from small specimens may be corrected for scaling up purposes. A direct method is used for the determination of G_{IC} in conjunction with a fictitious crack model.

In spite of the above discussion, there are reasons to doubt the large values of ℓ_{ch}. A value of ℓ_{ch} = 200 mm corresponds to a theoretical plastic zone size of the order of 60 mm, which could not reasonably develop in small specimens. An example of this (ref. 38) occurs when ℓ_{ch} = 275 mm thus implying an effective plastic zone size of about 90 mm to be associated with a concrete beam specimen having a depth of only 75 mm. Clearly ℓ_{ch} and the plastic zone size must exist within the constraint of the specimen dimensions.

Tests have been carried out (ref. 2) to assess the plastic zone size for notched concrete and mortar beams subject to pure bending. An experimental apparatus (ref. 2) similar to that used by other researchers, (ref. 43) but with considerably more instrumentation, was used. Surface demountable miniature LVDT displacement transducers (see Fig. 3) were used. The depth of crack was estimated from the average depth extrapolated from the transducer readings for both faces by determining the roots of the equation given by

$$d^2_c \left[a_0(d_1-d_2)+a_1d_2-a_2d_1 \right] -d_c \left[(d_1+d_2)(d_1-d_2)a_0+d^2_2a_1-d^2_1a_2 \right] +\left[a_0d_1d_2(d_1-d_2) \right] =0$$

$$(3)$$

An explanation of the notation (see Fig. 3) is given.

The results of tests for the variation of beam compliance with notch depth for a number of individual tests (see Fig. 4) are given. The depth determined from equation (3) is indicated for the corresponding case. The apparent notch depth was checked in some cases by allowing dye to penetrate the crack and instantly measuring the depth after breaking the specimen. From these results it would appear that the compliance method of crack depth assessment is reliable.

In order to assess the extent of the plastic zone size, the variation of crack depth with applied moment has been determined (see Fig. 5) Curves have been included in this figure to represent the Dugdale and the elastic non-singular stress distribution. The following relationships correspond to the stress distributions (see Fig. 6).

Dugdale

$$\bar{x} = \frac{(\ell_c - d - d_c)(d - d_c - \ell_c)}{2(d - d_c)}$$

$$(4)$$

$$\frac{M_{CRIT}}{M_0} = \frac{2}{hd^2} \left(\bar{x}^3 + h^3 + 3\ell_ch(h + \frac{\ell_c}{2}) \right)$$

where $h = d - \ell_c - d_c - \bar{x}$

and $M_0 = M_{CRIT}$ for $d_c = 0$

$$(5)$$

Elastic non-singular

$$\bar{x} = \frac{(d - d_c)}{2}$$

$$\frac{M_{CRIT}}{M_0} = \left(1 - \frac{d_c}{d}\right)^2 \tag{6}$$

where $M_0 = \sigma_T bd^2/6$

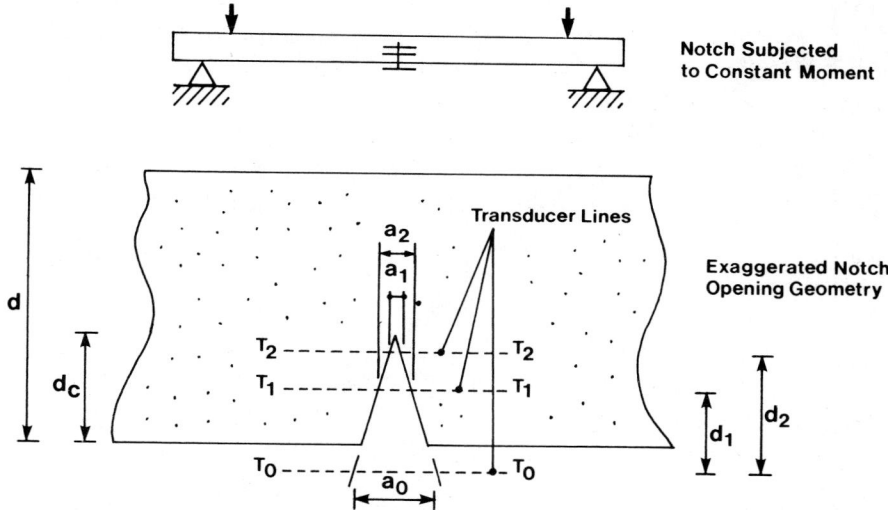

FIGURE 3 TRANSDUCER ARRANGEMENT FOR CRACK DEPTH ESTIMATION

Curves for various ℓ_c/d ratios (see Fig. 5) show the effect of the plastic zone size. Reference to the banded curves for notched beam tests shows that their behaviour is consistent with a small plastic zone size. Although severe crack tip concentrations cannot exist, it is interesting to note that the curve of singular linear elastic finite element solutions agree reasonably with the test specimen behaviour. Unfortunately, without further investigation, the matter of the plastic zone size is likely to remain controversial.

Whilst individual contributions show that fracture parameters, usually the critical stress intensity factor or strain energy release rate for mode I, can be identified, there is little general agreement (ref. 38). This is doubtless due, in part, to the comparative differences in such factors as specimen size and geometry, testing rigs and procedures, and the detailed material composition. In spite of these apparent complications and other more specific factors to be dealt with elsewhere in this book, the adoption

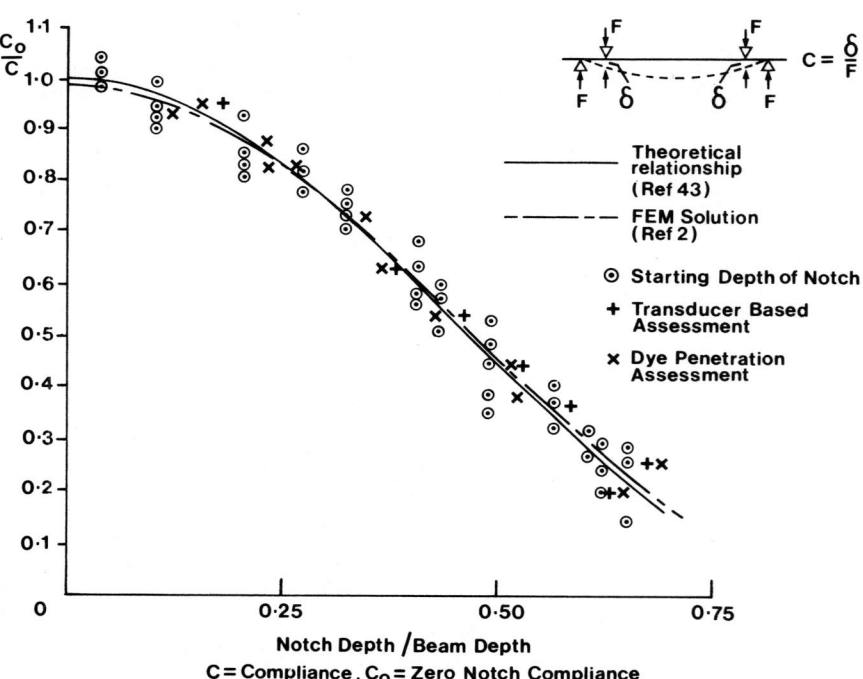

FIGURE 4 VARIATION OF COMPLIANCE WITH NOTCH DEPTH
FOR CONCRETE BEAMS IN PURE BENDING

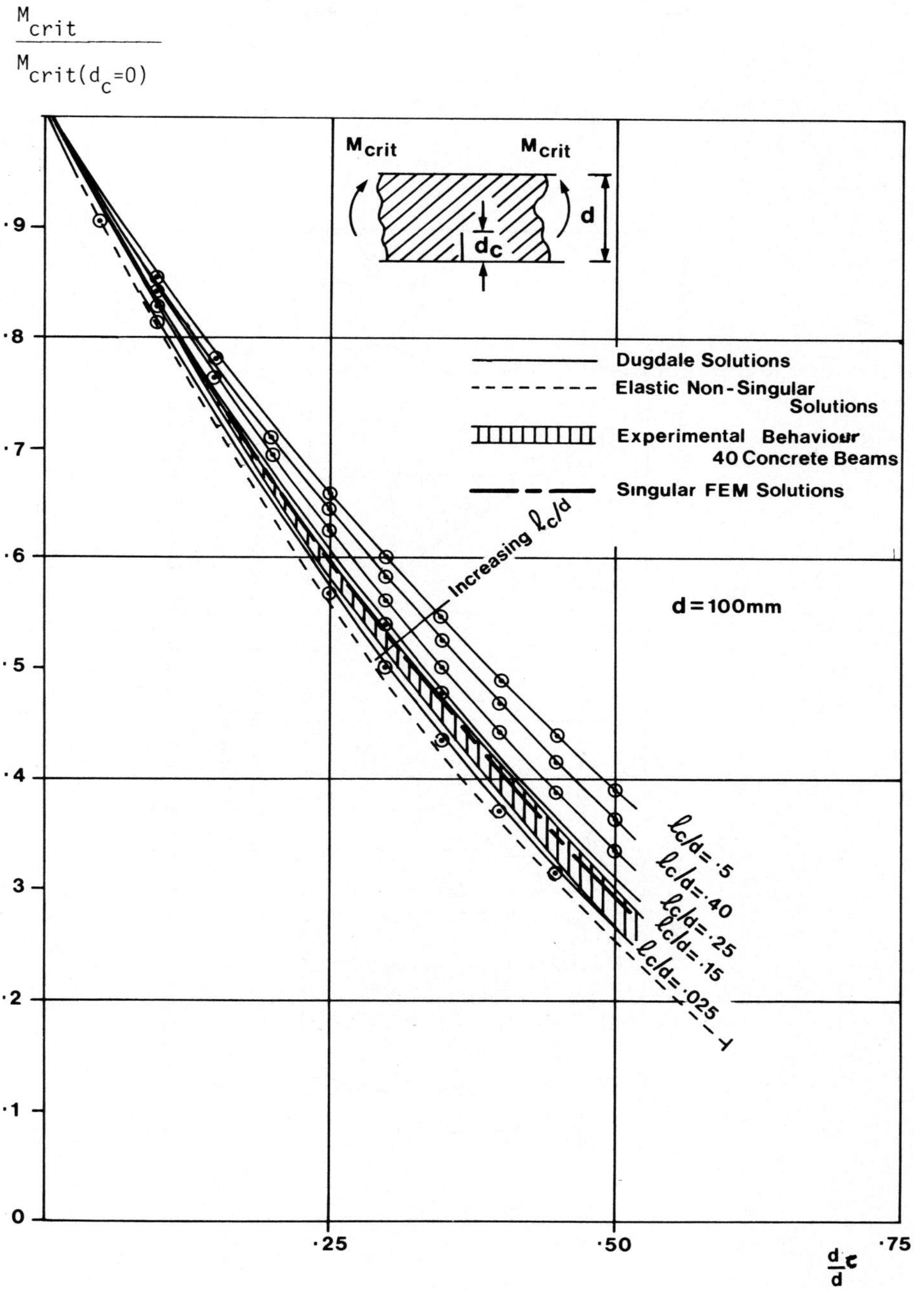

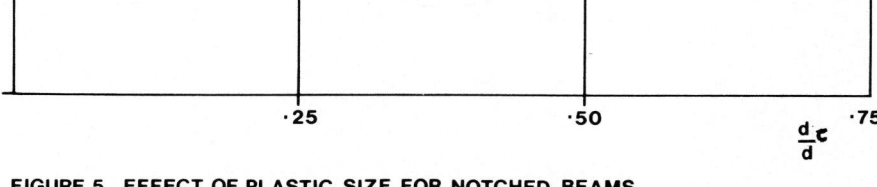

M_{crit} / $M_{crit(d_c=0)}$
·9
·8
·7
·6
·5
·4
3
·2
·1
0
·25
·50
·75
$\frac{d_c}{d}$
M_{crit}
M_{crit}
d
d_c
Dugdale Solutions
Elastic Non-Singular Solutions
Experimental Behaviour 40 Concrete Beams
Singular FEM Solutions
Increasing ℓ_c/d
d=100mm
$\ell_c/d=\cdot5$
$\ell_c/d=\cdot40$
$\ell_c/d=\cdot25$
$\ell_c/d=\cdot15$
$\ell_c/d=\cdot025$

FIGURE 5 EFFECT OF PLASTIC SIZE FOR NOTCHED BEAMS

514

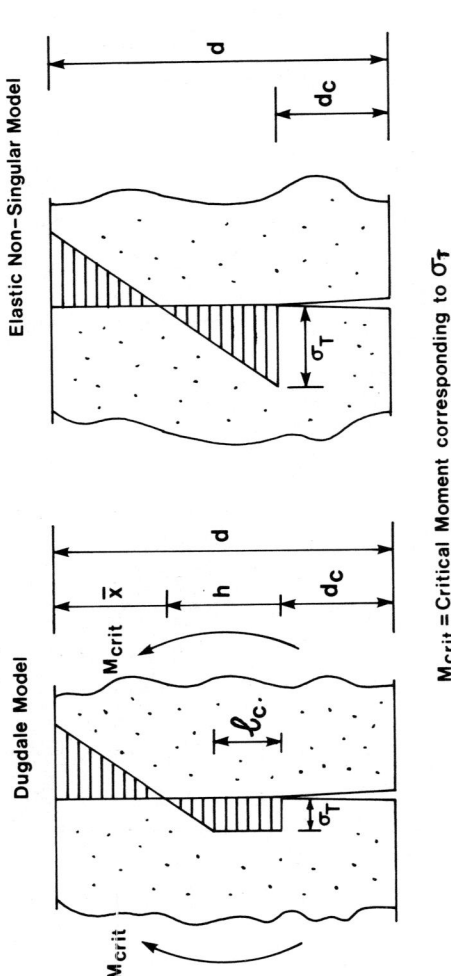

M_crit = Critical Moment corresponding to σ_T
b = Breadth

FIGURE 6   CRACK STRESS DISTRIBUTIONS FOR PURE BENDING

of the fracture mechanics approach anticipates the existence of quantifiable characteristic fracture parameters.

The stress intensity factor approach is clearly limited in respect of mixed mode crack propagation, on account of the difficulty in obtaining the modes II and III characteristics. For this reason, the single parameter SEDF approach has been applied to mixed mode crack propagation in concrete (ref. 1). Whilst further data will ultimately test the validity of such an approach, results so far indicate that the separation into volumetric and deviatoric component characteristics is relevant (ref. 2). However, the opinion has been expressed that for rock or concrete, cracks can only propagate in the mode I fashion, in spite of the curving crack path.

In respect of the elasto-plastic approach to fracture in concrete, it is difficult to assess the nature of the crack tip plasticity. For metals, microscopic examination is meaningful, whereas for concrete, the scale of its composition somewhat precludes similar consideration. The advisability of applying large scale observations in any way to the crack tip problem is questionable.

4. THE FINITE ELEMENT METHOD

4.1 Introduction

It is recognised that the most effective numerical method which may be applied to the fracture problem is the finite element technique (ref. 44). This technique may be combined with the established principles of fracture mechanics to provide a powerful analytical tool. There are four main areas for consideration during the application of the finite element technique in fracture and these are: (i) the nature of the crack tip elements, (ii) methods of interpreting and evaluating the results of the analysis, (iii) special aspects of the global analysis including crack propagation and (iv) implementation and interactive graphics.

Both singular and non-singular crack tip element schemes have been used in finite element fracture mechanics studies. Examples of the latter are the blunt (ref. 45) and smeared crack representations. Smeared cracking can be achieved by adjusting the element stiffness matrix in accordance with a simple crack mechanism and/or a constitutive law. The organisation of such computer programs is straightforward when using the well established non-linear solution schemes.

Since the actual separation of the finite element mesh is not necessary, it is not possible to model discrete crack propagation other than in a localised sense, therefore such propagation will not be considered any further.

The non-singular treatment, which is particularly linked to yielding fracture, will be considered after reviewing the singular stress intensity-based formulations. Within the context of elasto-plastic fracture mechanics it is important to take account of the crack tip stress singularity. It is evident in some published literature that this requirement is not always appreciated. Consideration will now be given to the form of the singular field.

4.2 The Classical Solution Function

Considering the general case of an infinite plate with a sharp crack inclined to a biaxial stress system (see Fig. 7) the displacement and stress fields can be obtained for example, using the Goursat functions proposed by Williams (ref. 46).

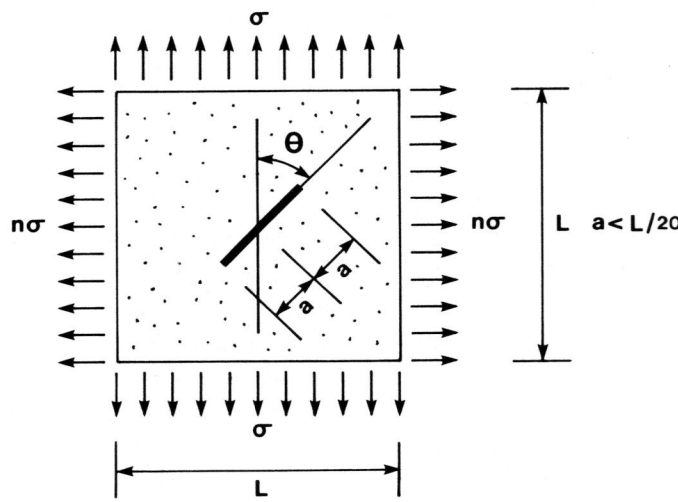

FIGURE 7   IDEALIZATION OF CRACKED INFINITE PLATE

In fact a variety of singular field functions are available, notably using the single algebraic function approach employed by Westergaard (ref. 19). In the above, two independent analytical functions are used in the solution of the biharmonic equation, $\nabla^4\phi$, as follows:

$$\psi(z) = \sum_{n=0}^{n=\infty} A_n z^{\lambda_n} \quad \text{and}$$

$$\chi(z) = \sum_{n=0}^{n=\infty} B_n z^{(\lambda_n+1)}$$

(7)

where eigenvalues $\lambda_n (n=0,1,2$ etc.) are assumed to be real, and the constants A_n and B_n are complex.

The real and imaginary parts of the functions are themselves conjugate harmonic functions, satisfying both the Cauchy-Riemann and Laplace equations thus

$$\phi = R_e \left[\bar{z}\psi(z) + \chi(z) \right] \qquad (8)$$

where R_e denotes the real part and z is the complex number

$$\begin{aligned} z &= x + iy \\ \bar{z} &= x - iy \end{aligned} \qquad (9)$$

Proceeding on a polar basis, and recognising the stress free boundary conditions on the free crack surface, the first few terms of the ϕ series are:

$$\phi = -r^{3/2} \left[a_1 (\sin (\tfrac{\beta}{2}) + \sin (\tfrac{3\beta}{2})) - b_1 (\cos (\tfrac{\beta}{2}) + \tfrac{1}{3} \cos (\tfrac{3\beta}{2})) \right]$$
$$+ r^2 \left[a_2 (1 - \cos(2\beta)) + b_2 (0) \right] + \dots \qquad (10)$$

where a_n and b_n are real constants and r, β the polar reference coordinates (see Fig. 8).

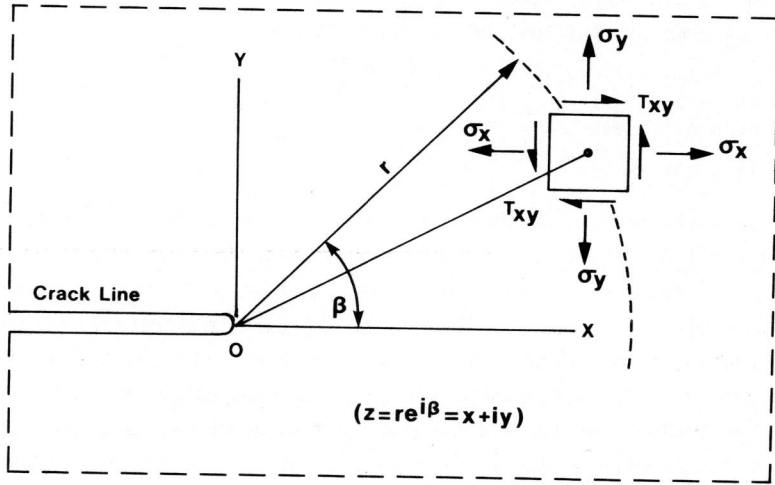

FIGURE 8 CRACK TIP GEOMETRY

Using the Airy stress functions and converting to rectangular coordinates, the stress field equations are established as follows

$$\sigma_x = \frac{K_1}{\sqrt{2r}} \cos(\tfrac{\beta}{2}) \left[1 - \sin(\tfrac{\beta}{2}) \sin(\tfrac{3\beta}{2}) \right] - \frac{K_2}{\sqrt{2r}} \sin(\tfrac{\beta}{2}) \left[2 + \cos(\tfrac{\beta}{2}) \cos\tfrac{3\beta}{2} \right] + 0(1)$$

$$\sigma_y = \frac{K_1}{\sqrt{2r}} \cos(\tfrac{\beta}{2}) \left[1 + \sin(\tfrac{\beta}{2}) \sin(\tfrac{3\beta}{2}) \right] - \frac{K_2}{\sqrt{2r}} \left[\sin(\tfrac{\beta}{2}) \cos(\tfrac{\beta}{2}) \cos(\tfrac{3\beta}{2}) \right] + 0(1)$$

$$\sigma_z = \frac{2v}{\sqrt{2r}} \left[K_1 \cos(\tfrac{\beta}{2}) - K_2 \sin(\tfrac{\beta}{2}) \right] + 0(1)$$

$$\tau_{xy} = \frac{K_1}{\sqrt{2r}} \left[\sin(\tfrac{\beta}{2}) \cos(\tfrac{\beta}{2}) \cos(\tfrac{3\beta}{2}) \right] + \frac{K_2}{\sqrt{2r}} \cos(\tfrac{\beta}{2}) \left[1 - \sin(\tfrac{\beta}{2}) \sin(\tfrac{3\beta}{2}) \right] + 0(1)$$

(11)

where K_1 and K_2 are the well-known stress intensity factors associated with the modes I and II components of crack extension (see Fig. 1). The factor K_1 used in the above has the same significance as that in the Griffith equation.

$$G_I = \frac{\pi(1 - v^2)K_1^2}{E}$$

(12)

which is the strain energy release rate.

In the case of biaxial loading, it can be shown that K_1 and K_2 relate to the applied stress state (see Fig. 7) as follows

$$K_1 = \sigma\sqrt{a} (\sin^2\theta + n \cos^2\theta)$$

$$K_2 = \sigma\sqrt{a} (1 - n)\sin\theta\cos\theta$$

(13)

where θ is the crack inclination and a its half length. It should be noted that equations (11) clearly embody the singular form of behaviour, possessing the predominant $1/\sqrt{r}$ term of the ϕ series. In finite element analysis the objective is to reproduce or to approximate as closely as possible the classical solution fields. This is most appropriately achieved by the use of special crack tip elements, though conventional elements may be adapted to advantage. Details of the various finite elements commonly used in fracture studies will be discussed after a brief outline of the overall formulation of the method.

4.3 Finite Element Formulations

4.3.1 The stiffness formulation

The displacement field within an element $\{d_{(x,y)}\}$ can be related to the nodal displacement vector $\{d_n\}$ as follows:

$$\{d_{(x,y)}\} = N \{d_n\} \tag{14}$$

where N denotes the displacement interpolation functions, called shape functions. These functions are often based on polynomial expressions providing for inter-element boundary continuity. By differentiating these functions as N^* in accordance with the strain-displacement relationships, the element stiffness matrix k_e is thus formed from the principal of virtual work as

$$[k_e] = \int_{vol} [N^*]^T [E] [N^*] \; dvol \tag{15}$$

where E is the elasticity matrix, and vol the volume of the element. The integration process is carried out numerically often using Gauss quadrature.

4.3.2 The hybrid formulation

The accuracy of the displacement formulation is dependent upon the allocation of sufficient degrees of freedom. In a static sense, equilibrium does not exist within an element, though of necessity it exists at the interconnecting nodal points. Whilst a force-based flexibility formulation could provide for internal equilibrium, such a single field would, by reciprocal consideration, be deficient in respect of continuity. In this case accuracy would depend on the number of independent stress function coefficients in the element system.

To improve the idealisation, a hybrid approach can be adopted (ref. 47), by means of which, two or more field functions coexist for a single element. For fracture mechanics studies, the assumed stress hybrid has been usefully employed. Two independent fields are used, one being a singular stress field which can be represented by equivalent boundary tractions, and the other a conforming boundary displacement field. Inter-element boundary equilibrium is enforced by the use of Lagrange multipliers. The stress field is character-ised by singular and non-singular components as

$$\{\sigma\} = M_0 \{\beta_0\} + M_s \{\beta_s\} \tag{16}$$

where M_0 are usually polynomial coefficients, $\{\beta_0\}$ the vector, M_s the singular field function coefficients and $\{\beta_s\}$ the stress intensity factors.

By equivalent surface tractions, we have on the element boundary:

$$\{T\} = N_0 \{\beta_0\} + N_s \{\beta_s\} \tag{17}$$

The compatible boundary displacements $\{\bar{U}\}$ are given by

$$\{\bar{U}\} = \{\bar{Y}\} \{d\} \tag{18}$$

where $\{d\}$ is the list of nodal point displacements, and $\bar{Y}$ the interpolation matrix presented for the boundary. The assumed stress complementary energy functional is given by

$$\Pi_c = \tfrac{1}{2} \int_{Vol} \{\sigma\}^T [E]^{-1} \{\sigma\} \, dVol - \int_S T.\bar{U} ds \tag{19}$$

where S is the entire boundary area for which displacements are prescribed.

Now substituting equations (16), (17) and (18) in (19) and taking variations on Π_c with respect to $\{\beta_0\}, \{d\}$ and $\{\beta_s\}$ give equations of the form:

$$\begin{bmatrix} K_{rr} & K_{rs} \\ K_{rs} & K_{ss} \end{bmatrix} \begin{Bmatrix} d \\ \beta_s \end{Bmatrix} = \begin{Bmatrix} F \\ 0 \end{Bmatrix} \tag{20}$$

which, on solution give both displacement and stress intensity factors. It is also interesting to note that by eliminating $\{\beta_s\}$, a conventional stiffness matrix can be obtained (ref. 48) in the form,

$$\begin{bmatrix} K \end{bmatrix} = \begin{bmatrix} K_{rr} \end{bmatrix} - \begin{bmatrix} K_{rs} \end{bmatrix} \begin{bmatrix} K_{ss} \end{bmatrix}^{-1} \begin{bmatrix} K_{rs} \end{bmatrix}^T \tag{21}$$

A simpler, more effective hybrid than the above has been proposed, (ref.49), leading to 5, 9 and 17, noded embedded-crack super elements. By the use of such super elements, the total degrees of freedom necessary to achieve a given accuracy can be considerably reduced. A more recent three dimensional assumed stress hybrid has also proved effective (ref. 48).

The alternative assumed displacement hybrid has been extensively developed for fracture studies (ref. 50), with interesting contributions for multilayer fracture in angle play laminate (ref. 51). A recently modified hybrid has also proved useful in the modelling of reinforced concrete (ref. 52).

There are two distinct classes of crack tip finite elements, those which have an embedded crack and those which border cracks. In the latter case groups of elements will have a common vertex defining the crack tip and the free surfaces corresponding to element sides. Whilst solutions are still occasionally presented using conventional, low order element schemes, it has been demonstrated (ref. 53) that the resulting output is of little value. The broad division of element types and schemes can be considered as those

which make use of (a) displacement or stress fields which embrace a complete
or part series of the form given by equation (10) for example,(b)
polynomial functions for displacement which possess forms of singularity,
(c) interpolated displacement functions which give the $1/\sqrt{r}$ stress
singularity, (d) displacement functions derived from the integration of
strain functions or (e) hybrids of stress and displacement functions giving
the $1/\sqrt{r}$ stress singularity.

Byskov (ref. 54) appears to have been the first to propose a finite element
based on the classical singular displacement function (ref. 55). The element
(see Fig. 9) uses many terms in the ϕ series, thus requiring a large
number of boundary nodes.

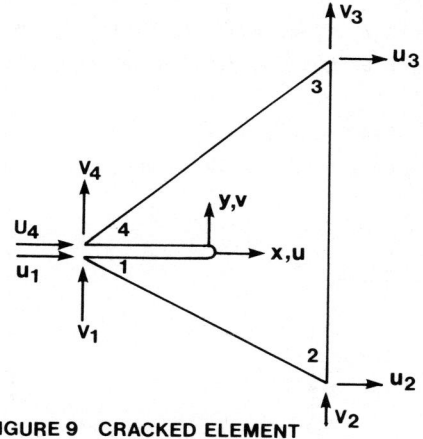

FIGURE 9   CRACKED ELEMENT

It can be used in conjunction with constant strain triangles, though
allowance must be made for the lack of inter-element continuity. Walsh has
developed a singular rectangular element (ref. 56), for which the fields are
expressed in terms of the stress intensity factors and nodal displacements.
This contribution also demonstrates the consideration of inter-element
discontinuity. Jones and Callman   (ref. 57) have used a least squares fit
technique to overcome boundary discontinuity for a circular element which has
been shown to perform well as a relatively large element. Wilson (ref. 58)
and Holston (ref. 59) have both developed similar, circular elements,
but have employed the Lagrangian multiplier method to enforce continuity.

A simpler approach to crack tip elements, lies in the use of polynomial-
based displacement functions, rather than the classical solution form. This
approach was adopted by Levy et al (ref. 60), who produced a $1/r$
singularity by collapsing one side of a bilinear rectangular element to form

a triangle. A more useful, $1/\sqrt{r}$ singular triangle element having boundary displacement continuity has been proposed by Tracey (ref. 61). This element, has a displacement field given by

$$d = (1 - \xi^m)d_1 + \xi^m(1 - n)d_2 + \xi^m n d_3 \tag{22}$$

where ξ, n are the natural coordinate set. A similar six-noded triangular element has been proposed by Blackburn (ref. 62), which unlike the previous types, can reproduce constant strain. For this element the displacement field is given by

$$d = b_1 + b_2\xi + b_3 n + (b_4\xi + b_5 n + b_6\xi n)/(\sqrt{\xi} + n) \tag{23}$$

Blackburn and Hellen (ref. 63) have extended the function for a three dimensional wedge element. Hughes and Akin (ref. 64) have recently described a method of developing crack tip elements to virtually any order, though it has been shown that a high element function order does not necessarily lead to greateraccuracy (ref. 65).

The well-established standard triangular and rectangular isoparametric elements (see Fig. 10) have been extensively employed in distorted forms (refs. 66,67,68,69,70).

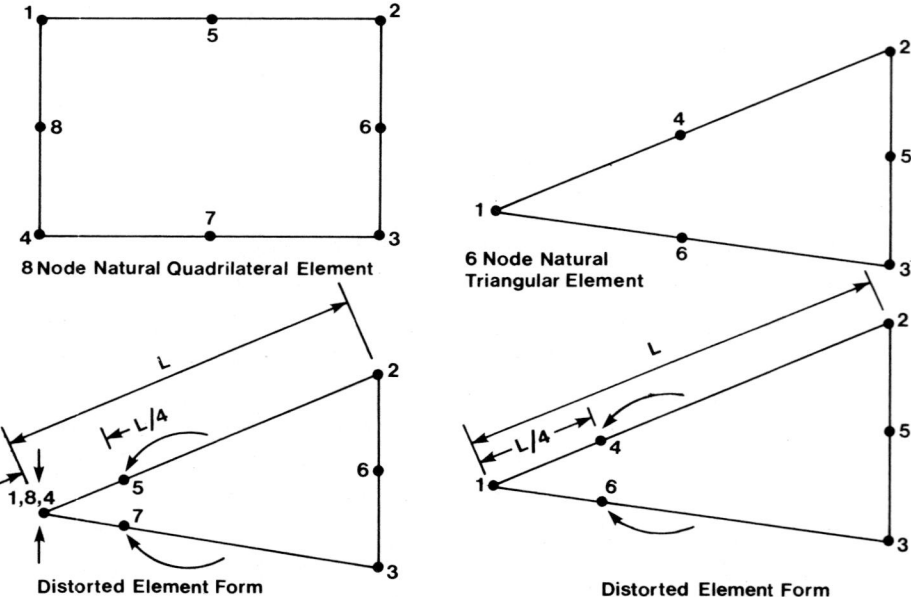

8 Node Natural Quadrilateral Element

6 Node Natural Triangular Element

Distorted Element Form

Distorted Element Form

FIGURE 10   ISOPARAMETRIC ELEMENTS

For these elements the opposite mid-side nodes are displaced to the quarter points. Although the single distortion results in a $1/\sqrt{r}$ singularity on the sides denoted by 1-2 and 4-3, the singularity, in the case of the rectangle, is not represented on intermediate radials. For this reason, a second distortion is applied to the rectangle by completely collapsing the side 1-4. Whilst the elements are comparable, the rectangular element gives inferior results where sides are not kept straight (ref. 69). The displacement function for these distorted elements is of the form

$$d = a_1 + b_1\sqrt{r} + c_1 r \tag{24}$$

Thus, rigid body motion and constant strain can be represented. Whilst these distorted elements can be surrounded by the standard elements, it has been shown (ref. 71) that transition elements, with side nodes moved to positions between the quarter and mid-side points, result in improved accuracy.

The use of the distorted isoparametric coordinate system permits a straight forward conformal mapping (ref. 55) of the difficult singular displacement field into a non-singular field. The ease with which potentially difficult fracture problems can be handled is readily demonstrated in published work.

It is interesting to note that a close examination of the shape functions for the commonly used doubly-distorted eight-noded quadrilateral reveals a deficiency (ref. 2). The shape functions for nodes 2, 6 and 3 (see Fig. 10) are as follows

$$
\begin{aligned}
N_2 &= \sqrt{r}\ C(T^2 + T - 2)/2 + r\ C^2(1 - T)\\
N_6 &= \sqrt{r}\ C(1 - T^2)\\
N_3 &= \sqrt{r}\ C(T^2 - T - 2)/2 + r\ C^2(1 + T)
\end{aligned}
\tag{25}
$$

where $C = \text{cosine}^{\frac{1}{2}}(\theta)$

and $T = \text{tangent}(\theta)/\text{tangent}(A)$ $\tag{26}$

These nodes would usually be in the vicinity of a common arc relative to the crack tip (common nodes 1, 4 and 8) and it would be expected that their form would be similar. Although N_6 contains the $\sqrt{r}$ term, at the appropriate distance from the crack tip, it would be expected that the more significant r term should be present. N_6 is enhanced by the introduction of a ninth central node. For the resulting nine-noded collapsed quadrilateral, the corresponding three shape functions are

$$
\begin{aligned}
N_2' &= T(1 - T)\left[\sqrt{r}(C) + r(2C^2)\right]/2\\
N_6' &= (1 - T^2)\left[\sqrt{r}(C) + r(2C^2)\right]/2\\
N_3' &= T(1 + T)\left[-\sqrt{r}(C) + r(2C^2)\right]/2
\end{aligned}
\tag{27}
$$

where C and T are defined in equations (26).

Whilst this modification necessitates an additional degree of freedom per tip element, improved radial distribution and transition is achieved for mixed mode cases.

Displacement shape functions for partially curved trapezoidal and triangular elements (ref. 72) have been obtained by the integration of assumed strain functions. Consideration of the trapezoidal element (see Fig. 11) shows that it can be divided into two subregions for which the polar strain functions f_r, f_θ and $f_{r\theta}$ are

$$f_r = \frac{\partial u}{\partial r} = \frac{1}{\sqrt{r}} \quad (b_1 + b_2\theta)$$

$$f_\theta = \frac{u}{r} + \frac{\partial v}{r\partial\theta} = \frac{1}{\sqrt{r}} (b_3 + b_4\theta) \qquad (28)$$

$$f_{r\theta} = \frac{\partial u}{r\partial\theta} + \frac{\partial v}{\partial r} - \frac{v}{r} = \frac{1}{r} (b_5 + b_6\theta)$$

in which b_1 - b_6 are constants.

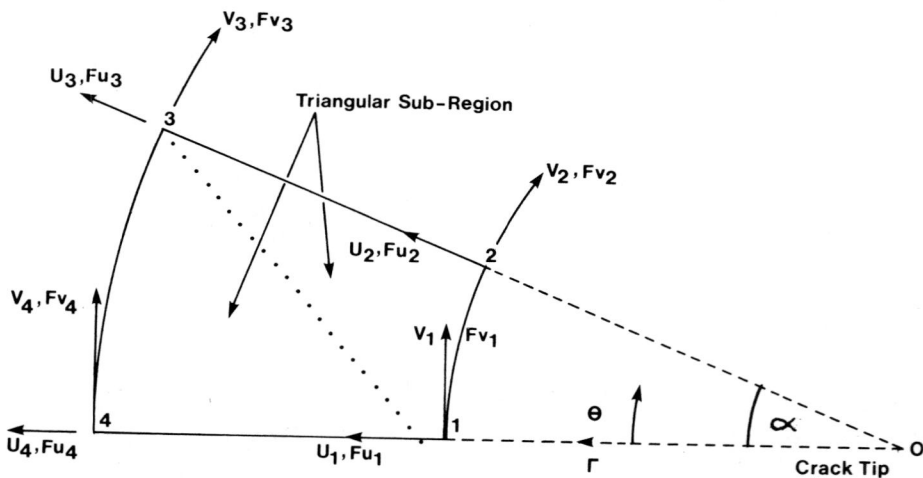

FIGURE 11 TRAPEZOIDAL ELEMENT

Integrating these functions gives

$$u = 2(b_1 + b_2\theta)\sqrt{r} \quad \text{and}$$
$$v = (b_3 - 2b_1)\theta\sqrt{r} \qquad (29)$$

After allowing for rigid body motion and simplifying the constants, the displacements become

$$u = a_1\sin\theta + a_2\cos\theta + a_4\sqrt{r} + a_5\sqrt{r}\theta$$
$$v = a_1\cos\theta - a_2\sin\theta + a_3r + a_6\sqrt{r}\theta \tag{30}$$

Using these displacement functions, the stiffness matrix is determined for each sub-triangle and these are subsequently combined to give the complete stiffness matrix for the trapezoidal element. The elements thus established have been shown to perform reasonably well.

5.   INTERPRETATION OF FINITE ELEMENT RESULTS

5.1 The Stress Intensity Factor Approach

The simplest interpretation of finite element results is by the determination of stress intensity factor values. For displacement field-based solutions, these can be obtained by extrapolating (ref. 73) values to the crack tip, i.e. at $r = 0$ or by what is essentially equation fitting (ref. 74). A least squares approach has been employed in mixed mode studies (ref. 2). In this approach the u and v displacement output for a radial is fitted to the equations

$$u = A_u + B_u\sqrt{r} + C_u r$$
$$v = A_v + B_v\sqrt{r} + C_v r \tag{31}$$

and the constants are determined during the fitting by extracting only the singular portion, the K_I and K_{II} values are determined from

$$\begin{Bmatrix} K_I \\ K_{II} \end{Bmatrix} = \begin{bmatrix} E & F \\ G & H \end{bmatrix} \begin{Bmatrix} B_u \\ B_v \end{Bmatrix} \tag{32}$$

where E, F, G and H are functions of θ, the radial reference angle to the radial under consideration. By examining various radials, the variation of the intensity factors with angle can be evaluated. In the context of the classical solution, these factors would be constant over the entire field. An examination of the K_I, K_{II} distribution, therefore, would appear more relevant than confining attention to a particular radial. Within the framework of a post-processing computer program (ref. 2), K_I and K_{II} values can be graphically presented in contour form.

The classical stress field solutions can be used to obtain the intensity factors as an alternative to the above. In the case of the displacement field-based formulations, the stresses are derived from the displacement field and can give less accurate solutions (ref. 75). This is not usually the case, however, for stress field and hybrid formulations. The efficiency of singular finite elements in the estimation of the stress intensity (ref. 65) has been examined for both modes I and II.

5.2 The Stress Energy Density Approach

The total volumetric and deviatoric strain energy densities in terms of stress may be written as follows

$$\frac{dW_T}{dV} = \frac{(1+\nu)}{2E}\left[\sigma_x^2+\sigma_y^2+\sigma_z^2 - \frac{\nu}{(1+\nu)}(\sigma_x+\sigma_y+\sigma_z)^2+2(\tau_{xy}^2+\tau_{yz}^2+\tau_{xz}^2)\right]$$

$$\frac{dW_V}{dV} = \frac{(1-2\nu)}{6E}\left[\sigma_x+\sigma_y+\sigma_z\right]^2 \tag{33}$$

$$\frac{dW_D}{dV} = \frac{(1+\nu)}{6E}\left[(\sigma_x-\sigma_y)^2+(\sigma_y-\sigma_z)^2+(\sigma_z-\sigma_x)^2\right] + \frac{(1+\nu)}{E}\left[\tau_{xy}^2+\tau_{xz}^2+\tau_{yz}^2\right]$$

Sih (ref. 76) has proposed that the strain energy density factor

$$S = \frac{dW_T}{dV} \cdot r \tag{34}$$

where r is the radial distance from the track tip to the area under considera-tion and can be used as a critical quantity. The direction of minimum S i.e. S_{min}, determines the direction of possible crack propagation, which occurs only when S_{min} reaches a critical level S_{crit}. The S theory has been used in both brittle and ductile fracture studies (ref. 77). Considering the stress field equations(11), which possess the $1/\sqrt{r}$ common factor, it can be seen that the S parameter is, in the context of linear elasticity, independent of the radial distance. This is clearly restricted to the small region adjacent to the crack where the $1/\sqrt{r}$ terms dominate. Thus

$$S = f(\text{stress state}, \nu, E) \tag{35}$$

A predominance of dilation may be associated with a brittle fracture during failure whereas a predominance of distortion tends to be associated with yield fracture. For this reason, it may be expedient to separate the volumetric strain energy density S_V and the distortional strain energy density S_D using equation (33) in the form of the product equation (34).

The three density factors may be related to the stress intensity factors as follows:

$$\begin{Bmatrix} S \\ S_V \\ S_D \end{Bmatrix} = \begin{bmatrix} a_{11} & 2a_{12} & a_{22} & a_{33} \\ b_{11} & 2b_{12} & b_{22} & b_{33} \\ c_{11} & 2c_{12} & c_{22} & c_{33} \end{bmatrix} \begin{Bmatrix} K_I^2 \\ K_I K_{II} \\ K_{II}^2 \\ K_{III}^2 \end{Bmatrix} \tag{36}$$

where a_{ij}, b_{ij} and c_{ij} are functions of θ and ν. Whilst the stress intensity factors can obviously be determined, it is not necessary to do so where the densities are treated as the crucial quantities. Again, computer graphics can usefully be employed in the assessment of the density fields. Examples of the elastic strain energy density distribution have been recently published (ref. 78).

5.3 Strain Energy Release Rate Approach

The methods discussed so far are based on the interpretation of a single solution i.e. the static crack problem. Utilizing the Griffith's strain energy release rate concept, crack propagation can be handled by obtaining two separate solutions, the first corresponding to the postulated crack configuration, and the second to the increment of crack development. The strain energy release rate G can thus be determined from the strain energy increment δU which is associated with an increase in crack surface area δA as

$$G = \frac{\delta U}{\delta A} \tag{37}$$

It is still commonly thought that the finite element mesh representation is relatively unimportant because only the difference in strain energy is examined. The authors consider this to be an unsound philosophy however, which may encourage the misrepresentation of the crack problem. Stress intensity factors can be related to the corresponding strain energy release rates (ref. 79) although virtually no data exists for other than the mode I value G_I. G_{II} and G_{III} values are usually calculated from the G_I values when required, but this is unsound practice.

To overcome the double solution requirement, a differential stiffness (ref. 80) method may be employed as follows

$$[K] + [\Delta K] \ \{\{d\} + \{\Delta d\}\} = \{P\} \tag{38}$$

where $[K]$ is the global stiffness matrix which is subject to a change $[\Delta K]$ on account of the crack growth. The applied force system $\{P\}$ being conservative thus

$$[K]\{d\} = \{P\} \tag{39}$$

Combining equations (38) and (39) and eliminating secondary products gives

$$\{\Delta d\} = -\left[K\right]^{-1}\left[\Delta K\right]\{d\} \tag{40}$$

Thus, having computed $\{d\}$, the sparsely populated matrix $\left[\Delta K\right]$ can be used to estimate the displacement increments $\{\Delta d\}$. The evaluation of G thus commences. A fundamental difficulty remains, however, in that the incremental model operates on the assumption that the direction of crack propagation is known. In some early studies crack advancement was determined on the basis of the pure mode I development irrespective of the nature of the applied load system. It is commonly observed that cracks turn and bifurcate, thus the above theory is of limited use, other than for contrived solutions such as for test specimens.

A substantial improvement has recently been proposed in the form of the maximum energy release rate theory (ref. 25) which like the strain energy density method has the radial aspect. It is suitable for mixed mode studies where the differential stiffness approach is employed.

5.4 The Line Integral Method

It can be shown that a path-independent integral exists for contours which include the crack tip (see Fig. 12)

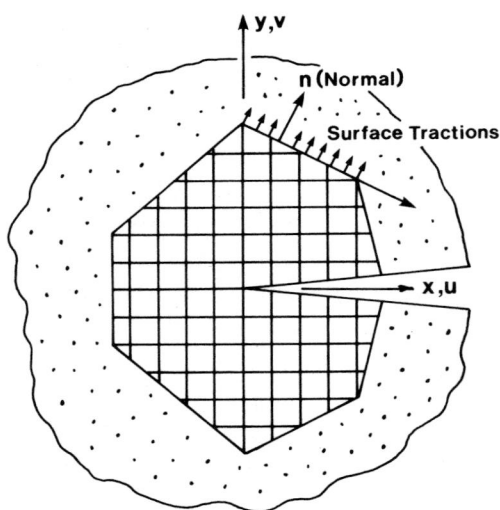

FIGURE 12 J-INTEGRAL PATH

Several such integrals have already been discussed. A consideration of the Rice J integral shows that

$$J = \int_S (W_{dy} - T \frac{\partial u}{\partial x} ds) \tag{41}$$

where W is the strain energy density, T the boundary traction vector and u the displacement vector. S is the complete boundary loop of which ds is an elemental part. For the pure mode I case of crack propagation, the stress intensity factor K_I can be evaluated as follows

$$K_I = \left[\frac{JE}{(1-\mu^2)} \right]^{\frac{1}{2}} \tag{42}$$

The integral is evaluated from the finite element results and experience has proved this to be an efficient means of determining the stress intensity factors. In practice, the nature of the finite element approximation results in some apparent path dependency and considerable judgement must therefore be exercised.

6. NON-SINGULAR DISCRETE CRACK MODELS

Non-singular discrete-crack finite element studies have been carried out for concrete at various levels of complexity. In the simplest form the cracking is reproduced by mesh separation, which is governed by a no-tension criterion. More detailed models allow for normal and shear stresses in the crack layer due to aggregate interlock with the material possessing a finite tensile capacity. The crack surfaces are usually assumed to be stress free in singular studies and the results correspond to the classical solutions. A recently reported equilibrium based model (ref. 52), however, offers the possibility of singular strain fields and does incorporate crack surface considerations.

The 'Fictitious Crack Model' is perhaps the most commonly used non-singular based model and is appropriate for both non-yielding and yielding materials. The micro-cracked zone ahead of the apparent crack is represented by relation-ships between transfer stresses, strain and opening displacement. Since the work done at the crack tip may be calculated, an apparent stress intensity factor (ref. 81), given by

$$K_c' = \sqrt{GE} \tag{43}$$

may be determined.

Special elements are not necessary for the finite element model. General plate elements can be used for the continuum and the stress transfer is achieved by negative stiffness bar elements. To facilitate the requirement

for the additional nodes which are necessary to propagate the crack, schemes
such as the 'nodal grafting' technique (ref. 82) could be used.

7. THE CONSIDERATION OF ACCURACY IN THE USE OF THE FINITE ELEMENT METHOD

Numerous studies have been made of the factors affecting the accuracy of the
finite element type solutions. These factors include the actual element
properties, the degrees of freedom assigned, the numerical integration and the
element mesh idealisation of the problem being considered. Further considerations arise in non-linear applications where such factors as step size, time
increment and the constitutive law implementation are relevant. For crack
studies in which singular fields are adopted the representation of the crack
tip region is of prime importance.

A detailed investigation (ref. 83) has been made into accuracy of the
distorted isoparametric elements which are commonly used in crack studies. It is
usual for the K_I values for the pure mode I case to be used to assess accuracy
and convergence. The size and number of elements about the crack tip prove to
have a major effect on accuracy, whereas the density of integration points is
relatively unimportant. A group of twelve equal triangles about the crack tip
show a notable improvement on a similar group of eight. If many more than
twelve are used in the group, however, the radial solution field begins to
deteriorate due to the excessive aspect ratio of the individual elements
(ref. 2). Whilst the group of eight are satisfactory for the pure mode I case,
the more accurate field of the twelve-element group is necessary for mixed
mode cases. Accuracy does improve as the crack tip group are reduced in size,
although a compromise must be made as more distributed elements will then be
necessary. For dynamic applications the K_I values are shown to increase with
a decrease in time step.

Finite element programs may provide output data for node and integration
point locations and this information is used to determine the stress intensity
factors, etc. Unfortunately, this can result in a potentially valid solution
being incorrectly interpreted. Whilst the pattern and density of the
integration points may not significantly affect the solution, the output
sample at different patterns of output points can yield different K_I values.
Where field functions are incorporated in some interpolated form, output can
readily be obtained for a network of points. Such output can then be
examined using computer graphics or curves fitted to individual radials
(see equation 31).

In recent years there has been a trend towards more sophisticated computer
programming incorporating convergence features such as the 'self-adaptive'
scheme (ref. 84). In this scheme, automatic control is exercised over the

discretization error and thus the necessity for repeated complete solutions is eliminated. With the crack tip being the region of interest, the two distinct approaches to the problem are the 'p' and 'h' convergence concepts. The assessment of the 'p' convergence is by the introduction of additional degrees of freedom into a fixed element mesh configuration (ref. 85). This can be achieved by the use of graded polynomial functions. The transition to higher orders is achieved by releasing variables which have been previously constrained to artificially achieved lower order elements. The program logic identifies elements where the approximation is inadequate and new degrees of freedom are introduced as necessary.

In the 'h' convergence scheme, the diameter of the crack tip group of elements is reduced towards zero to achieve convergence with the order of field functions being preserved. Since convergence could generally be obtained by the 'h' process, it was thought that the use of higher order elements was inefficient (ref. 53). Subsequent investigation into the 'p' convergence, however, generally proves this to be a more efficient basis for convergence in static and dynamic problems.

8. FINITE ELEMENT DATA PROCESSORS

In the early days of finite element work, program software was generally confined to the main formulation and solution aspects. The user was required to prepare the input and interpret the resulting output. This was inevitably a time-consuming process in which errors were easily made. Subsequently, input data generation and error detection software was incorporated in the widely available finite element packages. More recently, improvements in computer time sharing facilities together with the falling costs of micro-computers and visual display peripherals, have given the impetus for developments in user-orientated graphics. In the context of finite element programming, such facilities exist as both pre-and post-processors relative to the main solution.

The function of the pre-processor is to provide an efficient means of establishing a valid input data base. A comprehensive two dimensional, graphical finite element pre-processor has recently been reported (ref. 86) (see Fig. 13). This is a fully interactive program, which generates meshes by means of the transfinite mapping technique (ref. 87). Elsewhere, the Laplacian (ref. 88) and Isoparametric (ref. 89) methods have been employed for mesh generation. Such systems usually employ direct digitization hardware, commonly a digitizing tablet and pen, in conjunction with a high-speed refresh vector display terminal for verification. This arrangement provides for immediate detection of defective data with a provision for interactive editing. Therefore, final data base can be rapidly established.

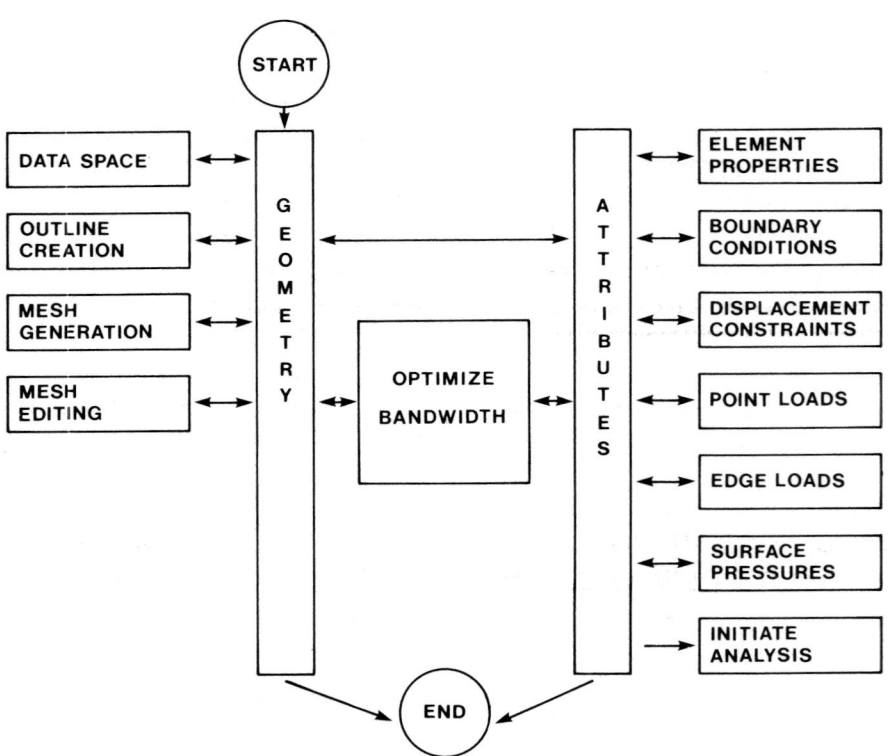

FIGURE 13 SYSTEM FLOW CHART FOR GRAPHICAL FINITE ELEMENT PRE-PROCESSOR

The output of finite element programs can typically consist of many thousands of parameter values and it is the processing of this material which is conducted by the post-processors (ref. 90). Where a general purpose graphics package is available, such programs may be easily devised. A simple program has been developed (see Fig. 14) for use in crack propagation studies. In this, the co-ordinates and solution parameters are written sequentially to disc file, which is then accessed by the post-processor acting in an inter-active mode. The user may select individual or combinations of parameters, that can be displayed as contour or isometric block projections over the zone of interest. By this means, the distribution of the strain energy density about a crack tip, for example, is readily assessed. In mixed-mode crack propagation studies, where the radial field distributions are of interest, such a facility is of obvious value.

9. CONCLUSIONS

The overall behaviour of concrete is determined by many factors and some of these are discussed in other chapters. Whilst constitutive laws have been used to represent non-linear behaviour, they provide little information on the formation and propagation of cracks. For this reason considerable interest has arisen in applying the theory of fracture mechanics to concrete.

The finite element technique has been the most successful numerical method for the investigation of fracture in concrete. A range of elements are available to represent a crack tip, however, some of these do not perform well in mixed-mode cases.

Further work is required to identify the factors which affect the fracture parameters of concrete such as the critical strain energy density. This improved experimental information could then be used with the already established numerical methods to provide a basis for design.

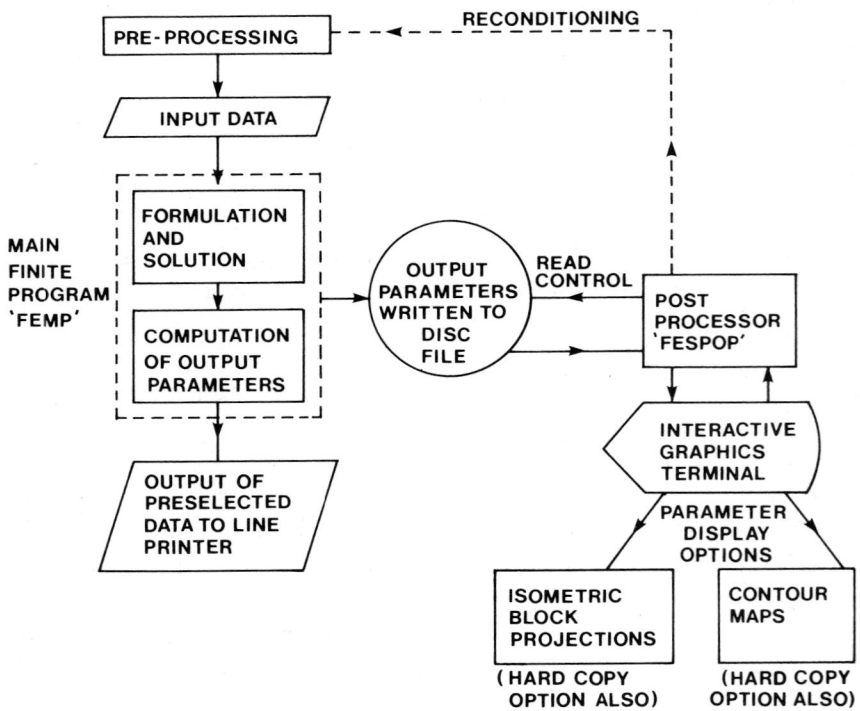

FIGURE 14   FLOW CHART FOR INTERACTIVE
GRAPHICS POST-PROCESSOR

REFERENCES

1   D.A. Chamberlain, An Investigation into Mixed Mode Crack Propagation in Cement Paste, Mortar and Concrete Specimens, Cement and Concrete Association, Research Seminar, July 1981.

2   D.A. Chamberlain, Crack Propagation in Concrete, Ph.D thesis, The City University, London, to be submitted.

3   S.P. Timoshenko, History of the Strength of Materials, McGraw-Hill, New York, 1953.

4   I. Todhunter and K. Pearson, History of the Theory of Elasticity and of the Strength of Materials, from Galilei to the Present Time, Cambridge University Press, 1886.

5   C.E. Inglis, Stresses in a Plate due to the Presence of Cracks and Sharp Corners, 54th Session of the Institution of Naval Architects, March 14, 1913.

6   B. Hopkinson, Observations on Stresses around a Crack, Sheffield Society of Engineers and Metallurgists, January 1910.

7   N. Muskhelishvili, Sur l'Integration de l'Equation Biharmonique, Izvestiya Ross. Akad, Nauk 13,6, 1919, pp. 663-686.

8   A.A. Griffith, The Phenomena of Rupture and Flow in Solids, Philosophical Transcript of the Royal Society London, Series A, Volume 221, 1920, p 163.

9   A.A. Griffith, Proceedings of the First International Congress on Applied Mechanics, Delft, 1924, p 55.

10   G.C. Sih and H. Liebowitz, Mathematical Fundamentals of Fracture, Academic Press, New York, 1968, p 67.

11   G.R. Irwin, Fracture of Metals (ASMS 1947), ASM, Cleveland, 1948.

12   E. Orowan, Proceedings of the Symposium on Fracture of Metals, Wiley, New York, 1952, p. 139.

13   E. Orowan, Energy Criterion of Fracture, Welding Research Supplement, 1955, p. 157.

14   G.C. Sih, A Special Theory of Crack Propagation. Methods of Analysis and Solution of Crack Problems. Ed. G.C. Sih, Noordhoff International Pub., Leyden, 1973, pp. XX-XLV.

15   J.R. Rice, Elastic-Plastic Fracture Mechanics. The Mechanics of Fracture. Ed. F. Erdogan, Applied Mechanics Division, Vol. 19, ASME, New York, 1976, pp. 23-53.

16   J.R. Rice, Mathematical Analysis in the Mechanics of Fracture, Fracture, Vol. II, Ed. Leibowitz, H. Academic Press, New York, 1968, pp. 191-311.

17   G.R. Irwin, Relation of Stresses near a Crack to the Crack Extension Force, Proceedings of the 9th International Congress of Applied Mechanics, 101 (11), Brussels, 1956.

18   G.R. Irwin, Fracture Mechanics, Proceedings of the 1st Symposium on Naval Structural Mechanics, Ed. H.J. Goodier and H.J. Hoff, 1958, pp. 557-594.

19   H.M. Westergaard, Journal of Applied Mechanics, Vol. 24, No.3, 1937, p 361.

20   I.N. Sneddon, The Distribution of Stress in the Neighbourhood of a Crack in an Elastic Solid, Proceedings of the Royal Society, London, A187, 1946, pp. 229-260.

21   G.R. Irwin et al, Proceedings of the American Society for Testing Materials, Vol. 58, 1958, pp. 640-657.

22   G.C. Sih, Strain Energy Density Factor Applied to Mixed Mode Crack Problems, Institute of Fracture and Solid Mechanics Technical Report, Lehigh University, 1972.

23   F. Erodogan and G.C. Sih, Journal of Basic Engineering, Vol. 85, 1963, p 519.

24   W.G. Knauss, International Journal of Fracture Mechanics, Vol. 6, 1970, p 183.

25 M.A. Hussain, et al, Strain Energy Release Rate for a Crack Under Combined Mode I and Mode II, Fracture Analysis, ASTM, STP 560, 1974, pp. 2-28.

26 D.S. Dugdale, Yielding of Steel Sheets Containing Slits, Journal of the Mechanics and Physics of Solids, Vol. 8, 100, 1960.

27 G.I. Barenblatt, Advances in Applied Mechanics Vol. 7, 1962, p 55.

28 G.R. Irwin, Proceedings of the 7th Sagamore Research Conference. Ordance Materials, US Office of Technical Services, Washington D.C., Vol. IV, 1961.

29 F.A. McClintock and J.A.H. Hult, Proceedings of the 9th International Congress on Applied Mechanics, 8(51), 1956.

30 J.R. Rice, Some Computational Problems in Elastic-Plastic Crack Mechanics. Proceedings of the 1st International Conference on Numerical Methods in Fracture Mechanics, Swansea, 1978, pp. 434-449.

31 J.D. Eshelby, The Continuum Theory of Lattic Defects, Philosophical Transcript of the Royal Society, London, Vol. 244, 1957, pp. 87-144.

32 H. Miyamoto and K. Kageyama, Extension of J. Integral to the General Elasto-Plastic Problem and Suggestion of a New Method of its Evaluation. Proceedings of the 1st International Conference on Numerical Methods in Fracture Mechanics, Swansea, 1978, pp. 479-495.

33 E. Kroner, Kontinums Theorie der Versetzungen und Eipenspanningen, Springer, 1958.

34 H. Miyomoto and M. Kikuchi, Evaluation of J_K Integrals for a Crack in Two Phase Materials, Proceedings of the 2nd International Conference on Numerical Methods in Fracture Mechanics, Swansea, 1980, pp. 359-370.

35 G.R. Irwin, Encyclopaedia of Physics, 6, 1958, p 551.

36 J.E. Srawley and W.F. Brown, Fracture Toughness Testing ASTM-STP 381, American Society for Testing Materials, Philadelphia Pa, USA, 1965, p 133.

37 D.A. Chamberlain, Discussion Paper, Royal Society Symposium on Fracture Mechanics in Design and Service - Living with Defects, December 1979

38 R.N. Swamy, Developments in Concrete Technology, Ed. F.D. Lydon, Chapter 6, 1978, pp 221-280

39 C.V.S. Kameswara, and R.N. Swamy, A Statistical Theory for the Strength of Concrete, Cement and Concrete Research, Vol. 4, Pergamon Press Inc. USA, 1974, pp 669-681.

40 A. Hillerborg, et al, Analysis of Crack Formation and Crack Growth in Concrete by means of Fracture Mechanics and Finite Elements, Cement and Concrete Research Vol. 6, 1976, pp. 773-782.

41 P.E. Peterson, Fracture Energy of Concrete, Method of Determination, Cement and Concrete Research, Vol. 10, No. 1, 1980, pp. 79-89.

42 P.E. Peterson, Fracture Energy of Concrete, Practical Performance and Experimental Results, Cement and Concrete Research, Vol. 10, No. 1, 1980, pp. 91-101.

43 J.H. Brown, Measuring the Fracture Toughness of Cement Paste and Mortar, Magazine of Concrete Research, Vol. 24, December 1972, pp. 185-196.

44 O.C. Zienkiewicz, the Finite Element Method in Engineering Science, McGraw-Hill Pub. 1971.

45 Z.P. Bazant, Blunt Crack Bond Propagation in Finite Element Analysis, Journal of Engineering Mechanics Division, ASCE, Vol. 105, No.EM2,April, 1979, pp. 297-315.

46 M.L. Williams, On the Stress Distribution at the Base of a Stationary Crack, Journal of Applied Mechanics, Vol. 24, 1957, pp. 109-114.

47 T.H.H. Pian, Crack Elements, Proceedings of the World Congress on Finite Element Methods in Structural Mechanics, Robinson and Associates, Verwood, Dorset, England, Vol. 1, 1975 pp. 1-39.

48 T.H.H. Pian and K. Mariya, Three-Dimensional Analysis by Assumed Stress Hybrid Elements, Proceedings of the 1st International Conference on Numerical Methods in Fracture Mechanics, Swansea, 1978, pp.363-373.

49 P. Tong, et al, A Hybrid-Element Approach to Crack Problems in Plane Elasticity, International Journal for Numerical Methods in Engineering, Vol. 7, 1973, pp. 297-308.

50 S.N. Atluri, Hybrid Finite Element Models for Linear and Non-Linear Fracture Analysis, Proceedings of the 1st International Conference on Numerical Methods in Fracture Mechanics, Swansea 1978 pp. 52-56.

51 T. Nishioka and N. Atluri, Multilayer Stress Hybrid Finite Element Method for Fracture Analysis of Angle Ply Laminate , Proceedings of the 2nd International Conference on Numerical Methods in Fracture Mechanics, Swansea, 1980, pp. 195-206.

52 J. Blaauwendraad and H.J. Grootenboer, Essentionals for Discrete Crack Analysis, IABSE Colloquium, Advanced Mechanics of Reinforced Concrete, Delft 1981, pp. 442a-442j.

53 P. Tong and T.H.H. Pian, On the Converence of the Finite Element Method for Problems with Singularity, International Journal of Solids and Structures, Vol. 9, 1973, pp. 313-321.

54 E. Byskov, The Calculation of Stress Intensity Factors Using the Finite Element Method with Cracked Elements, International Journal of Fracture Mechanics, Vol. 6, 1970, pp. 159-167.

55 N. Muskhelishvili, Some Basic Problems of Mathematical Theory of Elasticity Pub. Nooroff Groningen, Netherlands 1963.

56 P.F. Walsh, The Computation of Stress Intensity Factors by a Special Finite Element Technique, International Journal of Solids and Structures, Vol. 7, 1971, pp.1333-1342.

57 R. Jones and R.J. Callinan, On the use of Special Crack Tip Elements in Cracked Elastic Sheets, International Journal of Fracture, Vol 13, No.1 1977, pp. 51-64.

58 W.K. Wilson, Some Crack Tip Finite Elements for Plate Elasticity in Fracture Toughness, Proceedings of the 5th Material Symposium on Fracture Mechancs, ASTM, STP514, 1971.

59 A. Holston, A Mixed Mode Crack Tip Finite Element, International Journal of Fracture Mechanics, Vol.12, No. 6, 1976, pp. 887-899.

60 N. Levy, et al, Small Scale Yielding near a Crack in Plane Strain: A Finite Element Analysis, International Journal of Fracture Mechanics, Vol. 7, No. 2, 1971, pp. 143-157.

61 D.M. Tracey, Finite Elements for Determination of Crack Tip Elastic Stress Intensity Factors, Engineering Fracture Mechanics, Vol. 3, 1971, pp. 255-266.

62 W. Blackburn, Calculation of Stress Intensity Factors at Crack Tips using special Finite Elements in 'The Mathematics of Finite Elements, J.R. Williams Ed., Academic Press, New York 1973, pp. 327-336.

63 W. Blackburn and T.K. Hellen, Calculation of Stress Intensity Factors in Three Dimensions by Finite Element Methods, International Journal of Numerical Methods in Engineering, Vol. 11, No. 2, 1977, pp. 211-230.

64 T.J. Hughes and J.E. Akin, Techniques for Developing Special Finite Element Shape Functions with Particular Reference to Singularity, International Journal for Numerical Methods in Engineering, Vol. 15, 1980, pp. 733-751.

65 G.H. Staab and C.T. Sun, Singular Finite Element Efficiency in Estimating Stress Intensity Factors, International Journal for Numerical Methods in Engineering, Vol. 17, 1981, pp. 557-572.

66 R.D. Henshall and K.G. Shaw, Crack Tip Elements are Unnecessary, International Journal for Numerical Methods in Engineering, Vol. 9, 1975, pp. 475-507.

67 J.M. Bloom, An Evaluation of a New Crack Tip Element - The Distorted 8 Node Isparametric Element, International Journal of Fracture, Vol. 1 No. 4, 1975

68 R.S. Barsoum, On the use of Isoparametric Elements in Linear Fracture Mechanics, International Journal for Numerical Methods in Engineering, Vol. 10, 1976, pp. 25-37.

69 C.E. Freeze and D.M. Tracey, The Natural Isoparametric Triangle Versus Collapsed Quadrilateral for Elastic Crack Analysis, International Journal of Fracture Vol. 12, 1976, pp. 767-770.

70 A.N. Nayfeh and E.A.M. Nassar, Mathematical Simulation for Isoparametric Elements of Arbitrary Orders, International Journal for Numerical Methods in Engineering, Vol. 17, 1981, pp. 465-470.

71 P. Lynn and A.R. Ingraffea, Transitional Elements to be used with Quarter Point Crack Tip Elements, International Journal for Numerical Methods in Engineering, Vol. 11, 1977.

72 B.B. Sabir and B. Barr, New Finite Elements for Fracture Analysis, Proceedings of the 2nd International Conference on Numerical Methods in Fracture Mechanics, Swansea, 1980, pp. 25-40

73 S.K. Chan, et al, On the Finite Element Method in Linear Fracture Mechanics, Engineering Fracture Mechanics, Vol. 2, 1970, pp. 1-17.

74 J.R. Rice and D.M. Tracey, Computational Fracture Mechanics, Numerical and Computer Methods in Structural Mechanics, Ed. S.J. Ferwes et al, Academic Press 1978

75 M. Broekhoven, Computation of Stress Intensity Factors for Nozzle Corner Cracks by Various Finite Element Procedures, 3rd S.M.I.R.T. Conference, London G4/6 1975.

76 G.C. Sih, A Three Dimensional Strain Energy Density Theory of Crack Propagation, Three Dimensional Crack Problems, Vol II, Noordhoff International Publishers, Leydon, 1975, p. XV.

77 G.C. Sih, Mechanics of Ductile Fracture, Proceedings Fracture Mechanics and Technology, Vol. II, Noordhoff International Publishing, Leydon, 1977, pp. 767-784.

78 P.S. Theocaris and G. Papadopoulos, The Distribution of the Elastic Strain Energy Density at the Crack Tip for Modes I and II, International Journal of Fracture Vol. 18, 1982 pp. 81-112.

79 M. Ichikawa and S. Tomaka, A Critical Analysis of the Relationship between Energy Release Rate and the Stress Intensity Factors for Non-Coplanar Crack Extension under Combined Loading, International Journal of Fracture, Vol. 18, 1982, pp. 19-28.

80 T.K. Hellen, On the Method of Virtual Crack Extension, International Journal for Numerical Methods in Engineering, Vol. 9, No. 1, 1975 pp. 187-208.

81 P.E. Petersson and P.J. Gustavsson, A Model for Calculation of Crack Growth in Concrete-like Materials, Proceedings of the Second International Conference on Numerical Methods in Fracture Mechanics, 1980 pp. 707-719.

82 A. Ingraffea, Nodal Grafting for Crack Propagation Studies, International Journal for Numerical Methods in Engineering, Vol. II, No. 2, 1922.

83 L.P. Harrop, Linear Elastic Stress Intensity Factors Using a Distorted Isoparametric Finite Element, Proceedings of the First International Conference on Numerical Methods in Fracture Mechanics, 1978, pp. 302-314.

84 A. Peano, et al, Self Adaptive Convergence at the Crack Tip of a Dam Buttress, Proceedings of the First International Conference on Numerical Methods in Fracture Mechanics, 1978, pp. 268-280.

85 B.M. Irons, Engineering Applications of Numerical Integration in Stiffness, AIAA Journal Vol. 4, No. II, 1966, p 2035.

86 R. Haber et al, A General Two Dimensional Graphical Finite Element Preprocessor Utilizing Discrete Transfinite Mappings, International Journal for Numerical Methods in Engineering, Vol. 17, 1981, pp. 1015-1044.

87 C.A. Hall, Transfinite Interpolation and Applications to Engineering Problems, in 'Theory of Approximation', Ed. Law and Sahney, Academic Press, 1976, pp. 308-331.

88 W.R. Buell and B.A. Bush, Mesh Generation - a Survey, Journal Eng. Indust. ASME, Ser. B, 95, 1973.

89 O.C. Zienkiewicz and D.V. Phillips, An Automatic Mesh Generation Scheme for Plane and Curved Surfaces by Isoparametric Coordinates, International Journal for Numerical Methods in Engineering, Vol. 3, 1971, pp. 519-528.

90 F.T. Tracey, Graphical pre and post processors for Two Dimensional Finite Element Method Programs, SIGGRAPH 77 II(2) pp. 8-12.

Fracture Mechanics of Concrete,
edited by F.H. Wittmann, 1983
Elsevier Science Publishers B.V., Amsterdam — Printed in The Netherlands

Chapter 8

THE CRACKING AND FRACTURE OF CONCRETE:
AN ANNOTATED BIBLIOGRAPHY 1928-1981

by Sidney MINDESS

1   INTRODUCTION

The cracking and fracture of concrete have been studied seriously at least
since the pioneering work of Richart, Brandtzaeg and Brown in 1928, who
investigated the development of cracks in concrete under load. However, the
first applications of fracture mechanics to concrete were by Neville in 1959,
who attempted to relate the size effect on concrete strength to the distribution
of Griffith flaws, and then by Kaplan in 1961, who carried out the first
experimental investigation of the application of fracture mechanics to concrete.
Also, with the advent of the scanning electron microscope, a number of
investigators have considered the morphology of cracks and fracture surfaces,
and the phases through which cracks preferentially propagate.

The focus of this annotated bibliography is on the application of fracture
mechanics to cementitious systems: hardened cement paste, mortar, concrete,
fibre reinforced concrete, and polymer impregnated concrete. However, to put
these studies in context, two other groups of papers are also included. One
group deals with the investigations that laid the groundwork for the fracture
mechanics studies, but examining in detail the crack patterns that developed
in cement and concrete under load, and by studying such phenomena as rate-of-
loading effects and sustained loading. The second group deals with the
morphology of cracks and fracture surfaces, in an attempt to understand the
microstructural phenomena involved in crack propagation. In this bibliography,
the references to fracture mechanics are as complete as it has been possible
to make them. The other two groups of references are, of necessity, much more
selective: each is worthy of an annotated bibliography of its own.

Several qualifications to the above must be made at the outset. First, this
bibliography deals primarily with papers available in the English language
journals and literature. However, there has certainly been a significant
body of work reported in other languages (but unfortunately mostly inaccessible
to the author); its inclusion would have made this bibliography much more
complete. Second, because of the very large number of journals and conference
proceedings in which cement and concrete research have been reported, even the
English language references are probably far from complete. The author would

like to thank the many people who submitted additional references to him after a preliminary version of this bibliography was circulated; these have certainly helped make the bibliography more comprehensive.

The papers included here are arranged chronologically by year of publication; within each year, the papers are listed alphabetically, by author. An author index is also provided. Since all of the papers deal with essentially the same topic, it was not possible to provide a useful subject index which could properly discriminate between papers. A uniform system of units has not been adopted; instead, units are given in the form in which the data were presented in the original papers.

There is still considerable controversy as to which fracture mechanics criteria best described the fracture of cementitious systems, and indeed over the very applicability of fracture mechanics to these systems. The aim of this compilation of abstracts is to make access to the literature easier for current researchers in this field, and to provide the background for the future research which will, eventually, settle this controversy.

2   CONVERSION FACTORS

Quantity	To convert from SI	To English (or cgs)	Multiply by
G (strain energy release rate)	J/m^2	$in.-lb/in^2$	0.0057
K (stress intensity factor)	$MN\ m^{-3/2}$	$psi\sqrt{in}$	9.100×10^{-4}
Length	mm	in	0.039
Stress	MPa	psi	145.0
Surface energy	J/m^2	$in.-lb/in^2$	0.0057
	J/m^2	$ergs/cm^2$	1000
Surface tension	N/m	$dynes/cm^2$	0.0010
Temperature	$^\circ C$	$^\circ F$	$9/5\ ^\circ C + 32$

3 ABBREVIATIONS

The following abbreviations and symbols are used in the text:

a/c	aggregate/cement ratio	K_R	stress intensity factor from R-curve analysis
AE	acoustic emission		
CH	calcium hydroxide	LEFM	linear elastic fracture mechanics
CNRBB	circumferentially notched round bar in bending		
		ℓ_{ch}	characteristic length
COD	crack opening displacement	MMA	methyl methacrylate
C-S-H	calcium silicate hydrate (or "cement gel")	n	slope of $V-K_I$ plot
		PIC	polymer impregnated concrete
DCB	double cantilever beam		
DEN	double edge notch	PMMA	polymethyl methacrylate
DT	double torsion	r.h.	relative humidity
E	modulus of elasticity	SEM	scanning electron microscope
f'_c	cylinder compressive strength	SEN	single edge notch
		sfrc	steel fibre reinforced concrete
FEM	finite element method		
frc	fibre reinforced concrete (or cement)	sfrm	steel fibre reinforced mortar
F-δ	force-displacement (curve)	$V-K_I$ plot	plot of crack velocity vs. Mode I stress intensity factor
G_c	critical strain energy release rate		
G_I	Mode I strain energy release rate	w/c	water/cement ratio
		γ, γ_F	surface free energy
G_{II}	Mode II strain energy release rate	ε	strain
		$\dot{\varepsilon}$	strain rate
hcp	hardened cement paste	σ	stress
J_c	critical J-integral	$\dot{\sigma}$	stress rate
K	stress intensity factor		
K_c	critical stress intensity factor	σ_c	compressive stress
		σ_t	tensile stress
K_{IC}	Mode I critical stress intensity factor	σ_{ult}	ultimate stress

542

1   F.E. Richart, A. Brandtzaeg and R.L. Brown, A Study of the Failure of Concrete Under Combined Compressive Stresses, Bulletin No. 185, Engineering Experiment Station, University of Illinois, 1928.

     This work appeared to be the first recognition that substantial microcracking occurred before the stresses and strains became large enough to cause (tensile) failure under compressive loading. Compressive load leads to tensile stresses. Failure starts with bond failure; then lateral splitting begins.

2   F.E. Richart, A. Brandtzaeg and R.L. Brown, The Failure of Plain and Spirally Reinforced Concrete in Compression, Bulletin No. 190, Engineering Experiment Station, University of Illinois, 1929.

     By studying spirally reinforced concrete, three regions in the σ-ε curve were noted: linear, non-linear, and highly non-linear. The first region corresponds to elastic behaviour; in the second, bond failure begins; in the third region ($\sim 75\%$ σ_{ult}) there is a large volume increase.

3   F.A. Blakey and F.D. Beresford, A Note on Strain Distribution in Concrete Beams, Civil Engineering and Public Works Review, 50 (1955) 415-416.

     Concrete beams, 30 in. long, 6 in. deep and 2 in. wide, were tested with strain gauges attached to measure strains on the underside of the beams. Results showed that in the constant moment region, strains varied by up to a factor of two for higher loads. From the σ-ε curves plotted for different points along the beam, it was found that E was about the same everywhere for low loads. However, at higher loads, the values of E varied considerably. It was therefore suggested that this behaviour was due to microcracking on the bottom of the beam. These cracks, however, were stable, and did not lead immediately to failure.

4   F.A. Blakey, Some Considerations of the Cracking or Fracture of Concrete, Civil Engineering and Public Works Review, 52 (1957) 1000-1003.

     The author provided a summary of the then current state of knowledge on fracture. The conclusions reached were:

i) Failure is related to the volumetric strain, and to the shear strain energy absorbed in volumetric strain.

ii) Cracking may start by: (a) fracture of the paste; (b) fracture of the aggregate; (c) bond failure.

5   P. Dantu, Etude des Contraintes dans les Milieux Hétérogènes. Application au Béton, Annales de L'Institut Technique de Batiment et des Travaux Publics, 11 (1958) 55-77.

Concrete was considered as a multi-phase material, and the upper and lower bound solutions for E were established. Using photoelastic coatings on the surface of a concrete specimen in compression, it was found that strains in the mortar matrix were higher than those in the coarse aggregate, as was the case for stresses as well. Maximum stresses and strains tended to occur near the mortar-coarse aggregate interface.

6   M. Hori, Statistical Aspects of Fracture in Concrete, I. An Analysis of Flexural Failure of Portland Cement Mortar from the Standpoint of Stochastic Theory, Journal of the Physical Society of Japan, 14 (1959) 1444-1452.

It is suggested that most of the scatter observed in the results of mechanical measurements on concrete must be considered to be due to the inherent characteristics of the material. Two such characteristics are the static strength and the time to failure under sustained load. It is assumed that the phenomenon of rupture may be interpreted as a kind of Markov process. About 100 mortar specimens, 40x40x160 mm, were tested in flexure under sustained or increasing load. It is concluded that the theory is at least approximately applicable to the static failure of such inhomogeneous materials as concrete.

7   A.M. Neville, Some Aspects of the Strength of Concrete, Civil Engineering (London), Part I: 54 (1959) 1153-1156; Part II: 54 (1959) 1308-1310; Part III: 54 (1959) 1435-1439.

An attempt was made to show that size effects on concrete strength are related to the random distribution of Griffith flaws, such as voids. The Griffith theory, which postulates microscopic failure at the location of a flaw, was combined with Weibull's "weakest link" theory. The Griffith theory was extended to biaxial and triaxial cases. It was argued that the Griffith theory is only approximate in the case of concrete, since concrete strength may be governed by a limiting strain rather than a limiting stress criterion.

1961

8   M.F. Kaplan, Crack Propagation and the Fracture of Concrete, Journal of
the American Concrete Institute, 58 (1961) 591-610.

This was the first experimental study of the applicability of fracture
mechanics to concrete. Three different mixes were tested in both 3-point
and 4-point bending, using SEN beams of two sizes: 3x3x16 inches, and
6x6x20 inches. The notch depth was varied. In the analysis, it was
assumed that there was no crack growth before fracture. Kaplan found
that:

i) G_c was approximately constant for different notch depths.

ii)   4-point loading gave G_c values about 15% lower than those
obtained in 3-point loading.

iii) G_c for the smaller beams was about 38% less than for the larger
beams.

iv) G_c was about 12 times as large as that estimated from the
surface energy of concrete.

Nonetheless, Kaplan concluded that LEFM provided an appropriate
fracture criterion for concrete.

1962

9   F.A. Blakey and F.D. Beresford, Discussion of the paper Crack
Propagation and the Fracture of Concrete, by M.F. Kaplan, (JACI, 58
(1961) 591-610), Journal of the American Concrete Institute,
59 (1962) 919-923.

In a discussion of Kaplan's work (ref. 8), the writers reject Kaplan's
interpretation (that LEFM can be applied to concrete). They suggest that
Kaplan's data can be better explained by "simpler" ideas, such as the
size effect, variations in the type of loading, and the energy dissipated
as heat.

10   S. Brunauer, Tobermorite Gel - The Heart of Concrete, American Scientist,
50 (1962) 210-229.

The structure of "tobermorite gel" (C-S-H) is reviewed. Based on
heat of solution data it is concluded that the specific surface energy
of C-S-H is 386 ± 20 ergs/cm^2. This is compared with the specific surface
energies of calcium hydroxide (1180 ergs/cm^2) and hydrous amorphous
silica (129 ergs/cm^2).

11   J. Glucklich, Discussion of the Paper Crack Propagation and the Fracture
      of Concrete, by M.F. Kaplan (JACI, 58 (1961) 591-610), Journal of the
      American Concrete Institute, 59 (1962) 919-923.

In a discussion of Kaplan's work (ref. 8), the writer supports
Kaplan's interpretation. However, he points out three limitations to
this approach:

i) Slow (subcritical) crack growth.

ii) Uncertainty with regard to the most severe crack in a large
concrete member.

iii) The effect of the degree of indeterminacy of the system.

12   G.R. Irwin, Discussion of the Paper Crack Propagation and the Fracture
      of Concrete, by M.F. Kaplan (JACI, 58 (1961) 591-610), Journal of the
      American Concrete Institute, 59 (1962) 929.

In a discussion of Kaplan's work (ref. 8), the writer supports
Kaplan's approach. He suggests that a procedure for estimating the
effective crack depth at the critical load is necessary for proper
evaluation of G_c. He also suggests that "a smaller degree of sophis-
tication in measurement procedures than seems necessary for metals
would be appropriate for crack toughness measurements in concrete".

13   T. Yu. Lyubimova and E.R. Pinus, Crystallization Structure in the
      Contact Zone Between Aggregate and Cement in Concrete, Kolloidnyi
      Zhurnal, 24 (1962) 578-587.

The contact zone between cement and different aggregates was
studied, using microhardness tests. It was found that for carbonate
aggregates, there was no apparent alteration within the aggregate itself;
the thickness of the interfacial zone extended about 50 μm into the hcp.
For siliceous aggregates, the aggregates themselves appeared to undergo
some reaction, with a disturbed surface layer about 50 μm thick; the zone
of disturbance extended about 100 μm into the hcp.

1963

14   J. Glucklich, Fracture of Plain Concrete, Journal of the Engineering
      Mechanics Division, ASCE, 89 (1963) 127-138.

It is argued that in concrete the strain energy released by crack
growth is transformed mainly to surface energy; however, the fracture of
concrete is not limited to a single crack, but covers the entire highly
stressed zone in the form of a multitude of microcracks. Thus, the
newly formed surface may be much larger than the apparent fracture area.

This is complicated by the fact that in concrete, bond strength
< matrix strength < aggregate strength. Some bond separation occurs
before any load is applied. Up to about 0.70 σ_{ult}, cracks develop
mainly in the interfacial region; above this load, cracks develop in the
matrix as well. Cracks rarely penetrate the aggregate, but tend to
bridge between the larger aggregate particles. Areas of "higher
strength" act as crack arrestors, because they increase the energy
demand.

It is further argued that G_c is a material constant, and is perhaps
the best fracture criterion. In tension, G_c increases with crack length;
in compression, it remains constant. Compressive strength > tensile
strength because of the much greater microcracking that occurs in
compression.

15   T.T.C. Hsu, F.O. Slate, G.M. Sturman and G. Winter, Microcracking of
Plain Concrete and the Shape of the Stress-Strain Curve, Journal of the
American Concrete Institute, 60 (1963) 209-224.

The internal cracking of concrete was observed using microscopic and
x-ray techniques as the concrete was loaded axially to strains of up to
0.003. It was observed that very little cracking occurs through the
aggregate. Bond cracks exist before any load is applied, probably due
to shrinkage, settlement, and hydration. The bond strength is less than
the mortar strength. Bond cracks increase in size and number with
increasing strain, but the increase is negligible below about 0.3 σ_{ult}
(ε=0.0006). Bond cracks occur first around the larger aggregate
particles. Cracks in the mortar begin to increase and form continuous
crack patterns at strains of about 0.0012 to 0.0018 (70-90% of σ_{ult}).
These mortar cracks appear always to bridge between nearby bond cracks.
On the descending branch of the σ-ε curve, the concrete is very cracked.
It was concluded that bond cracking does not in itself cause failure;
concrete has a very redundant structure. Considerable interconnection
between bond cracks is needed before failure occurs.

16   M.F. Kaplan, Strains and Stresses of Concrete at Initiation of Cracking
and Near Failure, Journal of the American Concrete Institute, 60 (1963)
854-879.

Microcracking in concrete was investigated using an electrical
resistance strain gauge technique. Cracking was found to occur at loads
considerably less than those required to cause failure. The results

suggested that the initiation of cracking may be more dependent on strain than on stress. Tensile stresses and strains at or near failure depended on the method of test.

17   J.P. Romualdi and G.B. Batson, Mechanics of Crack Arrest in Concrete, Journal of the Engineering Mechanics Division, ASCE, 89 (1963) 147-168.

Concrete plates, 132-in. long, 24-in. wide and 2.5-in. thick, with centrally located slots ranging from 2 to 12 inches in length, were tested in tension, and G_c computed. It was found that G_c increased with increasing crack length (with no correction made for slow crack growth). Up to 7% by volume of continuous, aligned fibres were then added, in order to examine the crack arrest mechanism. It was found that with appropriate wire spacings, flaws could be prevented from propagating.

18   J.P. Romualdi and G.B. Batson, Behaviour of Reinforced Concrete Beams with Closely Spaced Reinforcement, Journal of the American Concrete Institute, 60 (1963) 775-790.

A series of beams (3 x 5 x 72 in.) were reinforced with varying amounts of aligned steel fibres and were tested in 4-point bending. The results were analyzed in terms of a fracture arrest concept; at some critical reinforcement spacing, cracks could be contained between adjacent reinforcing elements.

19   J.P. Romualdi and J.A. Mandel, Tensile Strength of Concrete Affected by Uniformly Distributed and Closely Spaced Short Lengths of Wire Rein- forcement, Journal of the American Concrete Institute, 61 (1964) 657-670.

Fracture arrest concepts were applied to wire reinforced concrete. Splitting tension tests were carried out on 3.88-in. diameter cylinders, 8 inches long; bending tests were performed on 38 x 3 x 1.75 in. beams. The speciments were reinforced with varying amounts of 1.5-in. steel wires. It was found that the closely spaced fibres increased the tensile strength, and also contributed to considerable post-cracking strength. The function of the steel wires was to effectively reduce K for a given stress, due to the restraining effects of adjacent fibres.

20   W. Wright and J.G. Byrne, Stress Concentration in Concrete, Nature, 203 (1964) 1374-1375.

Tensile tests were carried out on concrete specimens, in order to ascertain the effects of holes of various shapes. It was found that neither round, square, nor diamond-shaped holes led to any effective difference in failure stress. That is, there was no stress concentrating effect. Elliptical holes also had no effect. It was therefore concluded

that inherent stress concentrations in concrete are much higher than
those artifically introduced.

21   H. Yokomichi, K. Matsuoka, and N. Takada, Some Tests on Cracking of
      Concrete, in Review of the Nineteenth General Meeting, Cement
      Association of Japan, Tokyo, 1965, pp. 246-249.

      Compression tests were carried out on polished concrete prisms,
      50 x 50 x 150 mm. It was observed that:

      i)   cracks occur at the same time as the load is applied.

      ii)   for repeated loading, hardly any new cracks are noted until
      the previous maximum load is reached.

      iii)   the first cracks to form are bond cracks, particularly around
      large aggregate particles.

      iv)   bond cracks occur parallel to the load for low w/c ratios, or
      at the underside of aggregate particles for high w/c ratios.

      v)   mortar cracks bridge between the bond cracks.

      vi)   only at high stresses do isolated mortar cracks appear.

                              1966

22   J. Glucklich, Static and Fatigue Fractures of Portland Cement Mortar in
      Flexure, in Proceedings of the First International Conference on
      Fracture, Japan, 1965, The Japanese Society for Strength and Fracture
      of Materials, 1966, Vol. 3, pp. 1343-1382.

      Both static and fatigue tests were carried out on 42 x 4 x 2 in.
      mortar beams in flexure. It was found that G_c was about 7.5% lower in
      fatigue than in static tests, perhaps due to creep effects. It was
      concluded that the strain-energy release rate was the appropriate
      fracture criterion for both static and fatigue loadings. It was also
      stated that: (i) the non-linearity of the σ-ε curve in flexure on un-
      notched beams was due to slow crack growth; and (ii) G_c is approximately
      the same for both notched and un-notched specimens.

23   S. Kamiyama, An Effect of Notch for Uni-Axial Tensile Strength of Mortar
      and Concrete, in Review of the Twentieth General Meeting, The Cement
      Association of Japan, Tokyo, 1966, pp. 153-156.

      Uniaxial tensile tests were carried out on mortar and concrete using
      notched specimens. Notches decreased the tensile strength by about 24%.
      Strain concentrations occurred near the notches, but the decrease in
      strength was not proportional to the degree of strain concentration.
      The strength did not appear to be affected by the details of the notch
      geometry.

549

24   J. Lott and C.E. Kesler, Crack Propagation in Plain Concrete, in
      Symposium on Structure of Portland Cement Paste and Concrete, Special
      Report 90, Highway Research Board, Washington, D.C., 1966, pp. 204-218.

The fracture of concrete was described using LEFM. The authors define
K_c' as the pseudo-fracture toughness, since they include in this term the
modifications of the elastic field near a crack tip due to aggregates.
SEN beams (4 x 4 x 12 inches) were tested in 4-point bending, using
different notch depths. K_c' ($\sim$0.3 ksi$\sqrt{in}$) was approximately independent
of w/c ratio, but increased with coarse aggregate content. The fracture
process was modelled as the growth of a zone of micro-cracking till some
critical size at which fracture occurs.

25   S. Ya. Yarema and G.S. Krestin, Determination of the Modulus of Cohesion
      of Brittle Materials by Compressive Tests on Disc Specimens Containing
      Cracks, Fiziko-Khimicheskaya Mekhanika Materialov, 2 (1966) 10-14;
      English translation in Soviet Materials Science, 2 (1966) 7-10.

The critical equilibrium of a disc in diametral compression with a
symmetrical centre crack was studied. Mortar specimens were used, with
a diameter of about 58 mm, and two different thicknesses: $\sim$ 190 mm and
$\sim$ 34 mm; the average sand size was 0.81 mm. Three different crack
lengths were used: 10, 15 and 30 mm. A value of $K = 0.732$ kg/mm$^{-3/2}$
was obtained.

26   J.W. Dougill, A Mathematical Model for the Failure of Cement Paste and
      Mortars, Magazine of Concrete Research, 19 (1967) 135-142.

A mathematical model for hcp and mortar is proposed, in which
progressive fracture, or microcracking, occurs during all stages of
loading. The mode of failure depends on the method of test. It is
shown that, if the failure mode is ductile rather than brittle, the
ultimate strength may be almost unaffected by micro-inhomogeneity.

27   J. Glucklich and L.J. Cohen, Size as a Factor in the Brittle-Ductile
      Transition and the Strength of Some Materials, International Journal of
      Fracture Mechanics, 3 (1967) 278-279.

A theoretical discussion was presented which suggested that highly
unstable equilibrium exists between the respective rates of strain-
energy release and energy demand. When the system has a lot of stored
strain energy, any sudden drop in energy demand (as cracks grow into
locally weak regions) creates an excess of released energy which takes
the form of kinetic energy capable of doing work against the remaining
resistance, thus leading to a lower fracture load and reduced ductility.

Specimen size is important only in so far as it governs the amount of stored energy. It was asserted that these considerations are applicable to concrete.

1968

28   R.H. Evans and M.S. Marathe, Stress Distribution Around Holes in Concrete, Matériaux et Constructions, 1 (1968) 57-60.

Tensile tests on concrete were carried out, using three types of openings in the specimens; square, circular, and circular with a steel ring. Stresses and strains around the openings were obtained. For the circular holes, stress concentration factors of ~1.6 (rather than the theoretical 3.0) were obtained. However, these stresses were redistributed, and it was found that the holes did not particularly affect the structural behaviour of the concrete. The authors argued that inherent flaws in the concrete introduced stress concentrations that could not be offset by artificially induced stress concentrators.

29   R.H. Evans and M.S. Marathe, Microcracking and Stress-Strain Curves for Concrete in Tension, Matériaux et Constructions, 1 (1968) 61-64.

Complete σ-ε curves for concrete in tension were obtained. It was found that large strains were due to the initiation of microcracks. Specimens failed along the falling branch of the σ-ε curve, at about 25 - 40% of the maximum stress.

30   J. Glucklich, The Effect of Microcracking on Time-Dependent Deformations and the Long-Term Strength of Concrete, in A.E. Brooks and K. Newman (eds.), The Structure of Concrete, Proceedings of an International Conference, London, 1965, Cement and Concrete Association, London, 1968, pp. 176-189.

The microcracking of concrete was examined with regard to its influence on both time-dependent and time-independent properties of concrete. Pre-existing cracks tend to propagate under load. Due to microcracking, and the influence of the aggregate, energy demand is increased as the crack grows, leading to progressive microcracking. There is also a stress corrosion effect: water adsorption on cement surfaces reduces surface tension and the energy demand, so that stable cracks can again propagate.

31   T.C. Hansen, Cracking and Fracture of Concrete and Cement Paste, in
      Causes, Mechanism, and Control of Cracking in Concrete, SP-20, American
      Concrete Institute, Detroit, 1968 pp. 43-66.

      The value of the surface energy of C-S-H gel ($\sim$386 ergs/cm^2) obtained
      by Brunauer (ref. 10) was inserted into the Griffith equation. This led
      to a calculated critical flaw size of $\sim$0.02 mm to 0.2 mm, the same order
      of magnitude as the size of some unhydrated cement grains. It was
      suggested that this might be the "weak link" causing failure.

32   M.F. Kaplan, The Application of Fracture Mechanics to Concrete, in
      A.E. Brooks and K. Newman (eds.) The Structure of Concrete, Proceedings
      of an International Conference, London, 1965, Cement and Concrete
      Association, London, 1968, pp. 169-175.

      Griffith's theory of fracture was discussed. It was suggested that
      capillary pores might be the strength-controlling flaws in concrete,
      while gel pores ($\sim$20A) were too small for this. Cracks existing in
      concrete might also grow and cause failure.

33   K. Kato, Cracking Patterns in Plain Concrete, in Review of the Twenty-
      Second General Meeting, The Cement Association of Japan, Tokyo, 1968,
      pp. 177-178.

      In addition to bond cracks, mortar cracks, and aggregate cracks, "void
      cracks" were also distinguished as a fourth type of crack. It was argued
      that these are important in affecting the visco-plasticity of concrete.

34   K. Newman, Criteria for the Behaviour of Plain Concrete Under Complex
      States of Stress, in A.E. Brooks and K. Newman (eds.), The Structure of
      Concrete, Proceedings of an International Conference, London, 1965,
      Cement and Concrete Association, London, 1968, pp. 255-274.

      A detailed examination of failure criteria for concrete is presented.
      An analysis of the Griffith criterion is also presented. It is suggested
      that the Griffith theory, extended to a generalized three-dimensional
      case, might provide criteria from which the onset of fracture could be
      predicted.

35   G.S. Robinson, Method of Detecting the Formation and Propagation of
      Microcracks in Concrete, in A.E. Brooks and K. Newman (eds.), The
      Structure of Concrete, Proceedings of an International Conference,
      London, 1965, Cement and Concrete Association, London, 1968, pp. 131-145.

      Acoustic emission, pulse velocity and X-ray techniques were used to
      study the formation and propagation of cracks in concrete under com-
      pression. The failure process seemed to be :

i) microcracking due to shrinkage;

ii) additional microcracking under low loads, mostly at the cement-aggregate interface;

iii) at some load, these cracks spread through the mortar phase and interconnect. These cracks tend to be parallel to the applied load, and tend to separate the specimen into loosely connected "columns";

iv) eventually, more cracks develop and failure occurs.

36 J.P. Romualdi, The Static Cracking Stress and Fatigue Strength of Concrete Reinforced with Short Pieces of Thin Steel Wire, in A.E. Brooks and K. Newman (eds.) The Structure of Concrete, Proceedings of an International Conference, London, 1965, Cement and Concrete Association, London, 1968, pp. 190-201.

After a review of the implications of using LEFM to explain concrete behaviour, the role of steel fibres in arresting crack growth through the energy required to strip them from the matrix was discussed. An expression for G_c for frc was developed:

$$G_c = \frac{u \pi L}{380} \, \alpha \, \frac{p}{d}$$

where u = bond strength, L = fibre length, α is related to the crack opening, p = volume percent of fibres, and d = fibre diameter. (See Parimi and Rao, ref. 92)

37 J.P. Romualdi, M. Ramey and S.C. Sanday, Prevention and Control of Cracking by Use of Short Random Fibres, in Causes, Mechanism and Control of Cracking in Concrete, SP-20, American Concrete Institute, Detroit, 1968, pp. 179-203.

It was found that steel fibres oppose crack growth, and that the resistance to crack growth is proportional to the inverse square of the fibre spacing. G_c increased with fibre content, because of the energy required to cause the short fibres to slip, ahead of the crack tip. It was also found that fibres increase both the impact and fatigue resistance of concrete.

38 W. Ruetz, The Two Different Physical Mechanisms of Creep in Concrete, in A.E. Brooks and K. Newman (eds.), The Structure of Concrete, Proceedings of an International Conference, London, 1965, Cement and Concrete Association, London, 1968, pp. 146-153.

Acoustic emission was used to monitor cracking in specimens of concrete loaded in compression. It was found that:

i) microcracking occurs at a stress level of about 0.5 σ_{ult};

ii) the rate of crack formation is related to the inelastic strain rate;

iii) in fully dried specimens, the rate of crack development diminishes to zero under high sustained stress.

39 S.P. Shah and F.O. Slate, Internal Microcracking, Mortar-Aggregate Bond and the Stress-Strain Curve of Concrete, in A.E. Brooks and K. Newman (eds.), The Structure of Concrete, Proceedings of an International Conference, London, 1965, Cement and Concrete Association, London, 1968, pp. 82-92.

A stereo-microscope at 40x and X-ray techniques were used to examine microcracks in strained concrete. It was noted that bond cracks exist before any load is applied, and that the interfacial region is the "weak link" in concrete. Bond cracking appeared to be caused by heterogeneity induced by the aggregate.

40 S.P. Shah and G. Winter, Inelastic Behaviour and Fracture of Concrete, in Causes, Mechanism and Control of Cracking in Concrete, SP-20, American Concrete Institute, Detroit, 1968, pp. 5-28.

It was shown that for concrete, the heterogeneity introduced by the aggregate leads to inelastic behaviour. Evidence was presented that the bond between the aggregate and the mortar is the weak link in concrete.

41 G.W.D. Vile, The Strength of Concrete Under Short-Term Static Biaxial Stress, in A.E. Brooks and K. Newman (eds.), The Structure of Concrete, Proceedings of an International Conference, London, 1965, Cement and Concrete Association, London, 1968, pp. 275-288.

Biaxial tests were carried out on concrete, and a strength criterion based on the Coulomb-Mohr theory and a limiting tensile strain was developed. The order of failure in normal concrete was claimed to be: (i) tensile bond failure, (ii) shear bond failure; (iii) shear and tensile matrix failure, and (iv) tensile matrix (and occasional aggregate) failure.

42 F.H. Wittmann, Surface Tension, Shrinkage and Strength of Hardened Cement Paste, Matériaux et Constructions, 1 (1968) 547-552.

It was shown that the surface tension and the surface energy of solids were reduced by the presence of an adsorbed water film, decreasing the strength of materials with large surface areas. Using the Griffith theory, and from tests on hcp, it was found that the surface energy was 1750 erg/cm^2 at w/c=0.3; 1370 erg/cm^2 at w/c=0.45; and 657 erg/cm^2 at w/c=0.6.

554

43   F.H. Wittmann, Zum Einfluss der Oberflachenenergie auf die Festigkeit
      eines porosen Stoffes, Z. Agnew, Phys., 25 (1950), 160.

The surface tension and the surface energy of a solid are reduced by
the presence of an adsorbed water film. When the surface tension of a
porous material with a large interior surface is reduced, the length
increases, whereas the strength decreases. With the help of Griffith's
theory of crack propagation, it is possible to calculate the surface
energy. For hardened cement paste with a water/cement ratio of 0.45 and
0.6, the surface energy is found to be 1370 erg/cm^2 and 657 erg/cm^2
respectively. The cement paste was allowed to hydrate for 28 days without
loss of moisure at a temperature of 20°C. The results are in agreement
with the surface energy of porous glass with a similar internal surface.

44   A. de Sousa Coutinho, Note sur la Rupture de Beton Maintenu a une
      Contrainte Constante, Materiaux et Constructions, 2 (1969) 49-57.

Creep tests were carried out on concrete. The principal findings
were:
      i)   A sustained load a little below that required to cause static
fatigue caused a small increase in the ultimate stress.
      ii)   The increase in strength on sustained loading was greater when
the cement was loaded at an earlier age. It was hypothesized that this
increase in strength was due to additional and forced hydration due to
the external pressure.

45   A.D. Husak, Static Fatigue of Portland Cement Concrete, Ph.D. Thesis,
      Carnegie-Mellon University, Pittsburgh, 1969.

The static fatigue of hcp was studied as a function of r.h. Sand-
cement beams, 5.5 x 0.5 x 0.5 inches, with a small circular notch
(r = 0.03 in.) were loaded in flexure. The beams were first moist-cured
for 7 days, and then were equilibrated at various relative humidities.
The mean fracture load under short-term loading was 10.6 lbs. The
specimens were subjected to sustained loads of 9.5 lbs and 10.0 lbs.
It was found that an increase in r.h. decreased the time to failure.
A stress corrosion mechanism, with water as the corrosive agent, was
postulated, in which the Si-O-Si bonds are attacked by OH$^-$ ions. It
was also noted that stress altered the surface structure of hcp: the
needle-like particles disappeared.

555

46   D.R. McCreath, J.B. Newman and K. Newman, The Influence of Aggregate
      Particles on the Local Strain Distribution and Fracture Mechanism of
      Cement Paste During Drying Shrinkage and Loading to Failure, Materiaux et
      Constructions, 2 (1969) 73-84.

      Using 2-dimensional concrete plates loaded in compression, the
      progressive nature of crack propagation and failure were studied. Up to
      about 60% of σ_{ult}, there is local fracture and crack initiation, but the
      cracks are stable. Beyond this point, cracks begin to interconnect
      through the matrix, but remain stable till about 80% of σ_{ult}. Finally,
      the cracks become continuous and failure occurs. It was found that the
      strain necessary for crack initiation decreased as the volume fraction
      of coarse aggregate increased.

47   F. Moavenzadeh and R. Kuguel, Fracture of Concrete, Journal of Materials,
      4 (1969) 497-519.

      The Griffith theory was applied to a variety of hardened cement pastes,
      mortars and concretes. Specimens were made with different w/c ratios;
      they consisted of 25x25x305 mm SEN beams tested in 3-point bending. The
      ratio of notch length to specimen depth was 0.4 A microscopic technique
      was used to estimate the "true" fracture area. It was found that G_c and
      K_c did not change very much with age. The fracture energy obtained was
      about 10 times that calculated from theoretical considerations. This
      implied a zone of microcracks, or plastic flow, at the crack tip.
      Aggregates were found to make the cracking more stable.

48   D.J. Naus and J.L. Lott, Fracture Toughness of Portland Cement Concretes,
      Journal of the American Concrete Institute, 66 (1969) 481-489.

      The apparent fracture toughness of concrete was measured on SEN beams
      (2x2x14 inches and 4x4x12 inches) loaded in 4-point bending, with cast
      flaws. It was found that:
      i) K_c decreased with increasing w/c ratio for hcp and mortar,
      but there was no effect in concrete.
      ii) K_c decreased with increasing air content.
      iii) K_c increased with age.
      iv) K_c increased with increasing maximum aggregate size.
      v) K_c increased with increasing gravel/cement ratio.
      vi) K_c increased with increasing sand/cement ratios for mortars;
      it decreased with increasing sand/cement ratios for concrete.

49   Y. Niwa, S. Kobayashi, W. Koyanagi and K. Nakagawa, Microcracks of
        Concrete Under Triaxial Compression, in Review of the Twenty-Third
        General Meeting, The Cement Association of Japan, Tokyo, 1969,
        pp. 168-172.

        Uniaxial and triaxial tests were carried out on 105 mm mortar cubes.
        For the triaxial tests, it was found that most bond cracks appear to
        start at the bottom of aggregate particles, initiating from voids or
        areas of bleeding. At 60-90% of σ_{ult}, mortar cracks develop from the
        tips of bond cracks, and bridge between them. In uniaxial compression,
        mortar cracks become extensive only at about 0.9 σ_{ult}, and are approxi-
        mately parallel to the load direction.

50   Y. Niwa, S. Kobayashi and A. Miyaji, Some Considerations on Fracture
        Criteria of Cement Mortar Subjected to Multiaxial Compression, in Review
        of the Twenty-Third General Meeting, The Cement Association of Japan,
        Tokyo, 1969, pp. 163-168.

        Triaxial tests were carried out on 105 mm mortar cubes. It was found
        that fracture initiation began at stresses of about 0.6-0.7 σ_{ult}. In
        addition, the experimental data closely fit the Griffith fracture
        criterion.

51   S. Popovics, Fracture Mechanism in Concrete: How Much Do We Know?
        Journal of the Engineering Mechanics Division, ASCE, 95 (1969) 531-544.

        A review (with 45 refs.) was prepared on the experimental evidence
        which shows that concrete failure is due to progressive internal cracking,
        which starts at the coarse aggregate-mortar interface at less than σ_{max}.
        The strain energy is transformed to surface energy by the creation and
        propagation of internal cracks. However, the numerical application of
        the Griffith criterion is not reliable, since a crack propagates not as
        a single crack, but as a zone of microcracks. Also because of inhomo-
        geneities, we do not really know either E or γ_s at the crack tip.

52   G.B. Welch and B. Haisman, The Application of Fracture Mechanics to
        Concrete and the Measurement of Fracture Toughness, Materiaux et
        Constructions, 2 (1969) 171-177.

        The failure of concrete in uniaxial tension and compression was
        explained in terms of the energy concepts of fracture mechanics. Based
        on a review of the literature, the difficulties associated with the
        measurement and evaluation of K_c were discussed, and basic assumptions
        relating to slow crack growth and stress concentrations due to notches
        were proposed. It was concluded that LEFM can be applied to concrete,
        even though K_c and G_c appeared to depend on the test methods.

53   H. Yokomichi, Y. Kakuta and K. Ayuta, On Critical Points in Deformation of Concrete, in Review of the Twenty-Third General Meeting, The Cement Association of Japan, Tokyo, 1969, pp. 156-160.

Compression tests and short-time (5 minute) creep tests were carried out on concrete cylinders 200 mm by 100 mm in diameter. The stress under sustained load at which microcracking began was in the range of 0.5-0.7 σ_{ult}; the stress at which unstable crack growth began was in the range of 0.8-0.95 σ_{ult}.

1970

54   D. Darwin and F.O. Slate, Effect of Paste-Aggregate Bond Strength on Behaviour of Concrete, Journal of Materials, 5 (1970) 86-98.

Coarse aggregate was coated with a thin layer of polystyrene to reduce paste-aggregate bond strength, and the resulting concrete was compared to concrete made with untreated aggregate for strength, stiffness, and type and amount of microcracking. It was concluded that large reductions in bond strength caused only small changes in E and σ_{ult} for compressive loading. Variations in bond strength had little effect upon the amount of interfacial microcracking before loading; upon loading, there was no effect on interfacial microcracking, but the coated specimens exhibited a small increase in mortar cracking.

55   R.F. Feldman and P.J. Sereda, A New Model for Hydrated Portland Cement and its Practical Implications, Engineering Journal (Canada), 53 (1970) 53-59.

A model for the structure of hcp is presented. Amongst many other things, it is suggested that failure will occur at Griffith cracks where there are large stress concentrations. The Si-0-Si bonds will be strained, and this strain energy will contribute to a greater ease of formation of hydroxyl groups (-Si-OH HO-Si-) in the presence of water vapor. When the rate of diffusion is such that it will deliver some minimum amount of water into a spreading crack, no further decrease in strength will occur. It is also suggested that tobermorite crystals containing interlayer water may decompose under stress, leading to creep.

56   K.P. George, Theory of Brittle Fracture Applied to Soil Cement, Journal of the Soil Mechanics and Foundations Division, ASCE, 96 (1970) 991-1010.

The applicability of the Griffith theory to soil-cement was examined, using Type I cement and various Mississippi soils. SEN beams (76x76x286 mm) were tested. G_c was found to be independent of notch depth for crack/depth ratios ranging from 0.08 to 0.33. It was also

558

found that: (i) G_c increases as the clay content of the soil increases; (ii) G_c increases rapidly as the temperature falls below $0^{\circ}C$; (iii) G_c increases slightly as the loading rate increases. It was concluded that G_c is a true materials parameter for this system.

57 G.C. Hoff, Crack Extension Force Concept Applied to the Compressive Failure of Portland-Cement-Based Mortars, Miscellaneous Paper C-70-19, U.S. Army Engineer Waterways Experiment Station, Vicksburg, Mississippi, 1970.

An attempt was made to characterize the compressive failure of mortar using G_c. Tests were carried out on 3- by 6-in cylinders. After loading, some of the cylinders were sawn in half and examined for the number and length of cracks. Analysis of the data showed only that G_c was approximately proportional to σ_{ult}. It was concluded that G_c was not helpful in characterizing mortars under compressive loading.

58 I.D.C. Imbert, The Effect of Holes on Tensile Deformations in Plain Concrete, Highway Research Record, No. 324, Highway Research Board, Washington, D.C., 1970, pp. 54-65.

The effect of holes on the tensile deformation of thin concrete plates (1200x200x35 mm) was studied. Extensive inelastic deformation occurred in the vicinity of hole edges, while strains further away remained elastic for most of the loading range. This behaviour was explained in terms of the energy-release concept of fracture mechanics: inelastic deformation is due to progressive microcracking. Such microcracking, in which the cracks remain relatively short, occurs when G_c is retarded, in this case by the strain gradients due to the holes. These gradients localize cracking in the vicinity of the hole edges, and inhibit it elsewhere.

59 W.R.A. Knox, Fracture Mechanisms in Plain Concrete Under Compression, Fracture Toughness of High Strength Materials, Iron and Steel Institute, London, 1970, pp. 158-162.

Based on studies of both perspex and concrete, three possible compressive fracture mechanisms in concrete were postulated: (i) a shear (Mode II) displacement; (ii) a first crack (Mode I) mechanism; and (iii) a progressive cracking mechanism. Which mechanism predominates depends on the boundary conditions imposed on the specimens. Since G_I is an order of magnitude smaller than G_{II}, the shear mechanism is unlikely. The progressive cracking mechanism appeared to be most likely to be encountered. Cracks stabilize as they encounter aggregate particles; the application of increasing stress causes the formation of numerous stable cracks which eventually link up to cause failure.

60   E.M. Krokosky, Strength vs. Structure. A Study for Hydraulic Cements, Materiaux et Constructions, 3 (1970) 313-323.

A state-of-the-art report on strength vs. structure for hcp was presented. The effects of porosity were discussed. The weak bonds resulting from the inter-growth of hydration products were also discussed. It was argued that the discrepancy between tensile and compressive strengths does not depend on bond strength within the C-S-H; instead, it can be explained entirely on the basis of stress concentrations.

61   I. Rosenthal and J. Glucklich, Strength of Plain Concrete Under Biaxial Stress, Journal of the American Concrete Institute, 67 (1970) 903-914.

Hollow concrete cylinders, with an outer diameter of 305 mm, a wall thickness of 27.5 mm, and a height of 350 mm, were tested under biaxial and uniaxial stresses, and stress combinations. Under biaxial and uniaxial tension, and under combined biaxial tension and compression, failure was by splitting; under biaxial compression, sliding or shear failure occurred. For the first case, a critical tensile strain criterion was proposed; for the second case, a critical strain energy release rate criterion was postulated. A comparison of the Griffith theory and the test results showed general agreement, except in the biaxial compression zone. It was suggested that closer agreement would have been obtained if the Griffith criterion were modified to take into account the brittle, multiphase character of concrete:

i)   The specific surface tension energy should be a function of the isotropic stress, rather than a constant.

ii) Slow crack growth should be taken into consideration.

62   S.P. Shah and S. Chandra, Fracture of Concrete Subjected to Cyclic and Sustained Loading, Journal of the American Concrete Institute, 67 (1970) 816-825.

Both cyclic and sustained compressive loading of concrete was carried out. It was found that the static fatigue limit was about 70% of σ_{ult}. Volumetric, ultrasonic and optical microscopic measurements showed that progressive internal microcrack propagation occurred under both sustained and cyclic loading. Crack growth under sustained loading occurred because of the phenomenon of stress corrosion, and was significantly influenced by the presence of water. Sustained loading also appeared to have a strengthening effect, probably because it led to increased van der Waals bonding in the hcp.

1971

63   S. Diaz, Fracture Mechanisms of Concrete Under Static, Sustained, and
      Repeated Compressive Loads, Ph.D. Thesis, University of Illinois,
      Urbana, 1971.

      Crack propagation in concrete under static, sustained, or repeated
      compressive loading was determined, using fluorescent ink and photography
      to monitor crack growth. It was found that in compression, initial
      crack extension could be in the radial direction, or in the direction of
      maximum radial shear. The interaction of collinear or parallel cracks
      increased the stress intensity at the crack tip. In general, however,
      initial crack extension was approximately in the direction of the
      maximum tensile stress. Below about 0.85 σ_{ult}, cracking was mainly
      around voids and at the cement-aggregate interface. Beyond this point,
      cracks became unstable. It was hypothesized that failure occurs when
      some critical configuration of microcracks is reached.

64   J.W. Dougill, Further Consideration of a Mathematical Model for
      Progressive Fracture of a Heterogeneous Material, Magazine of Concrete
      Research, 23 (1971) 5-10.

      Uniaxial σ-ε curves are derived for a mathematical model designed to
      represent a heterogeneous material (simulating mortar and concrete) that
      fails in tension by progressive fracture. The results suggest that the
      behaviour is dependent on the energy required to fracture the material.
      For high fracture energies, stable fracture propagation occurs; for low
      energies, failure occurs by instability under reducing stress.

65   A.D. Husak and E.M. Krokosky, Static Fatigue of Hydrated Cement Concrete,
      Journal of the American Concrete Institute, 68 (1971) 263-271.

      Mortar beams with the dimensions 5.5x0.5x0.5 in., with a circular
      notch in the tensile face, were used for static fatigue tests in 3-point
      bending. Loads were 90% and 95% of the static strength, and the
      relative humidity was controlled at various levels between 0% and 90%.
      It was found that times to failure increased with decreasing relative
      humidity; at 0% r.h., the beams did not fail during the tests.
      Application of stress altered the surface structure of hcp regardless of
      humidity level, but structural evidence of stress corrosion could not be
      observed using the SEM at resolutions down to 1000 A. A failure
      mechanism was proposed which assumed that preferentially oriented
      surfaces in the hcp are highly stressed, with the stress contributing
      energy to break bonds at the sites of defects and flaws. Bringing water

in contact with the stressed bonds brings about a reaction between the Si-O bonds and OH⁻; the OH⁻ ions become bonded to the Si atoms by splitting of the Si-O-Si bonds.

66   W. Koyanagi and K. Sakai, Observations on the Crack Propagation Process of Mortar and Concrete, in Review of the Twenty-Fifth General Meeting, The Cement Association of Japan, Tokyo, 1971, pp. 153-157.

The crack arrest action of the cement-aggregate interface was investigated by measuring K_c and G_c, using SEN beams 47 x 100 x 388 mm, with a 10 mm notch, loaded in 3-point bending. Both strength and E increased with increasing aggregate volumes, suggesting that they play a crack arresting role in the fracture process. It was assumed that almost all of the strain energy was transformed into surface energy. It was found that G_c increased with increasing aggregate volume in mortars; in concrete, the effect was slight. For lightweight aggregates, G_c decreased with increasing aggregate volume. K_c also increased with aggregate content; the effect was more pronounced in concrete than in mortar.

67   K.T. Krishnaswamy, Mechanism of Failure and Microcracking of Plain Concrete Under Uniaxial Tensile Loading, Indian Concrete Journal, 45 (1971) 204-208,222.

Direct tensile tests were carried out on concrete. It was found that at ultimate stress $\sigma_t/\sigma_c \simeq 0.10$, though this ratio changed somewhat for different concrete strengths. It was observed that bond cracks existed even before loading, mostly near large aggregate particles. The width of these bond cracks was greater under tensile loading than under compressive loading, due to higher stresses at the interface in the former case. Failure was due to the interconnection of the bond cracks.

68   F. Moavenzadeh and T.W. Bremner, Fracture of Portland Cement Concrete, in M. Te'eni (ed.), Structure, Solid Mechanics and Engineering Design, Proceedings of the Southamption 1969 Civil Engineering Materials Conference, Wiley-Interscience, 1971, pp. 997-1007.

Cracking of the hcp and concrete in both tension and compression was studied using an optical microscope and small (1x1x0.04 in.) polished specimens. It was found that cracks tended to form either at the cement-aggregate interface (bond failure) or as internal cracks in the aggregate which then propagated into the matrix. It was argued that elimination of CH from the interfacial region would improve the bond strength.

69   K. Newman and J.B. Newman, Failure Theories and Design Criteria for
      Plain Concrete, in M. Te'eni (ed.), Structure, Solid Mechanics and
      Engineering Design; Proceedings of the Southamption 1969 Civil Engineering
      Materials Conference, Wiley - Interscience, 1971, pp. 963-995.

      The failure criteria for concrete are discussed. Crack initiation is
      due to the elastic incompatibility of the hcp and aggregate on drying;
      these cracks will then propagate under tensile stress. Their formation
      relieves stress concentrations, and equilibrium is restored. As the
      load increases, the initially stable cracks begin to grow, till the
      system becomes unstable. The Griffith criterion, which predicts that
      at the ultimate strength $\sigma_c = 8\sigma_t$, is about right for concrete.

70   H.R. Sasse, On the Problem of Fracture Behaviour of Concrete-Like Two-
      Phase Systems (in German). Forschungsberichte des Landes Nordrheim-
      Westfalen, Nr. 2192, Westedeutscher Verlag-Opladen, 1971, 93 pp.

      The fracture mechanics of concrete is analyzed by taking into account
      its internal structure. Concrete is described as an agglomeration of
      round aggregate particles cast in a homogeneous matrix. Six structural
      models are proposed, and they are applied to results of tests on 65
      different mortar mixes, in which steel, glass, polystyrol, etc., were
      used as aggregates. The strength is described as a function of the
      strength and elastic parameters of the matrix and the aggregate, mix
      composition, and some regression coefficients. There is good agreement
      between experimental and calculated values.

71   S.P. Shah, Micromechanics of Concrete and Fibre Reinforced Concrete, in
      M. Te'eni (ed.) Structure, Solid Mechanics and Engineering Design,
      Proceedings of the Southampton 1969 Civil Engineering Materials Conference,
      Wiley-Interscience, 1971, pp. 367-376.

      The cracking of plain and fibre reinforced concrete was studied. In
      plain concrete, the non-linear behaviour was assumed to be due to
      composite action between the hcp and aggregate; volume changes were
      related to internal microcrack propagation. For steel frc, the "fracture
      toughness" was taken to be the area under the load-deflection curve. It
      was found that this toughness increased with increasing fibre volume,
      fibre aspect ratio, and fibre ductility.

72   S.P. Shah and F.J. McGarry, Griffith Fracture Criterion and Concrete,
      Journal of the Engineering Mechanics Division, ASCE, 97 (1971) 1663-1676.

      Since the first cracks on loading of concrete do not lead to failure,
      an attempt was made to determine the notch sensitivity and critical crack
      length for concrete, made with different types, volumes and sizes of

aggregate. In flexure, 22x2x2 in. beams with a triangular notch were tested in 3- point loading at 7 days, using different notch depths. Toughness was measured as the area under the σ-ε curve. It was found that mortar and concrete are notch-insensitive, since toughness was independent of crack length, while hcp was notch sensitive. Toughness increased with increasing aggregate volume and maximum aggregate size.

For tension tests, DEN specimens with elliptical notches were used, and the observations made during the flexure tests were confirmed. Concrete was found to be notch insensitive at least to a notch depth of 1 inch (for specimens 12x3x0.5 inches). The fracture of concrete is gradual, due to progressive deterioration rather than the unstable growth of a single crack. Microcracks can be arrested at aggregate particles by debonding, or by running into a different phase.

73   R.N. Swamy, Aggregate-Matrix Interaction in Concrete Systems, in M. Te'eni (ed.) Structure, Solid Mechanics and Engineering Design, Proceedings of the Southampton 1969 Civil Engineering Materials Conference, Wiley-Interscience, 1971, pp. 301-315.

The forces binding the cement and aggregate phases, and the effect of aggregate on the failure and strain distribution in hcp, mortar and concrete were discussed. It was stated that cracks at the interface exist even before loading; discrete microcracking occurs as cracks grow and are arrested by different phases. It was concluded that aggregate particles behave as stress-raising discontinuities. The steep stress gradients and strain concentrations within the material are due to the differential stiffness of the matrix and the aggregate, and inherent imperfections in the interfacial region.

74   F.H. Wittmann and J. Zaitsev, Versuche zur Bestimmung der Dauerfestigkeit des Zementsteins, Zement-Kalk-Gips, 24 (1971) 160.

The behaviour of specimens of hardened cement paste (water/cement ratio: 0.4) under high constant load was investigated. The deformation and the length of time until the occurence of failure were measured. To enable the shrinkage stresses to be neglected, all the measurements were performed in air with 100% relative humidity and at $20^{\circ}C$. The age of the specimens at the start of loading was varied between 1 day and 28 days.

564

75   Ju. V. Zaitsev, Deformation and Failure of Hardened Cement Paste and Concrete Subjected to Short Term Load, Cement and Concrete Research, 1 (1971) 123-137.

A simplified model of a brittle, porous material was used to discuss the application of fracture mechanics to hcp and concrete in compression. The solution of the problem of crack propagation in a material with a random distribution of pores was analyzed using Monte Carlo methods. The theoretical predictions compared reasonably well with experimental results.

76   Ju. V. Zaitsev, Deformation and Failure of Hardened Cement Pastes and Concrete Under Sustained Load, Cement and Concrete Research, 1 (1971) 329-344.

Fracture mechanics was applied to concrete and hcp in compression and tension. The problem of crack propagation near pores was also analyzed. It was shown that the time to failure under sustained load could be determined from a knowledge of creep, E, and the short-term strength.

77   Ju. V. Zaitsev, Experimental Investigation to Determine the Behaviour of Hardened Cement Paste and Plaster of Paris Under High Load, Cement and Concrete Research, 1 (1971) 437-447.

Both hcp and plaster of paris were studied under high loads, and the strain and time to failure determined. Experimental results agreed well with theoretical considerations based on fracture mechanics.

78   J. Zaitsev and F.H. Wittmann, Zur Dauerfestigkeit des Betons unter konstater Belastung, Bauingenieur, 46 (1971) 84.

A model to describe a porous structure is presented. Stress analysis, applied to the model, allows calculation of the crack propagation from a single cylindrical pore in an infinite homogeneous plate. The time dependance of the crack length is discussed.

1972

79   M.A. Ali, A.J. Majumdar and D.L. Rayment, Carbon Fibre Reinforcement of Cement, Cement and Concrete Research, 2 (1972) 201-212.

High modulus, high strength carbon fibres were used as reinforcement for cement. Izod impact tests were carried out on 50 x 25 x 6 mm specimens. At a fibre content of about 3% by volume, it was found that the impact strength was scarcely improved over that of the unreinforced matrix.

80   J.E. Barrick, II, The Effects of Temperature and Relative Humidity on
      Static Fatigue of Hydrated Portland Cement, Ph.D. Thesis, Carnegie-
      Mellon University, Pittsburgh, 1972.

The effect of temperature and relative humidity on the static fatigue
of hcp was studied, using 0.5 x 0.5 x 5.5 in. mortar SEN beams. A stress-
corrosion mechanism was postulated dependent on the presence of CH. It
was concluded that direct water attack on the silicate network is not
what causes stress corrosion, but rather OH^- attack. The results showed
that the time to failure increased with decreasing r.h. It also
increased with increasing temperature, the static fatigue effect almost
disappearing at 60°C. The hypothesis was therefore that there is
hydroxyl attack on the silicate network, with the OH^- coming from the CH.

81   R.L. Berger, Calcium Hydroxide: Its Role in the Fracture of Tricalcium
      Silicate Paste, Science, 175 (1972) 626-629.

The influence of CH on the fracture of C_3S pastes was studied by
casting specimens between glass slides, sealing the edges to prevent
drying, and then breaking the specimens in 3-point bending. Optical
microscopy was then used to examine the fracture path. The CH crystals
represented low porosity areas. The fracture path changed as a function
of curing time:
1 day: The fracture path goes around CH crystals and preferentially
through the "outer product" C-S-H. The path is very tortuous.
3 days: The crack still goes preferentially around CH, but in some
cases it terminates at these crystals. In other cases, several new
branch cracks are created. That is, there is some crack arrest due to
CH crystals.
8 days: The fracture path is still preferentially through the C-S-H.
However, some CH crystals are fractured, generally near the extremities
of the crystals.
>22 days: The differences between the CH and C-S-H phases decrease, and
the fracture path becomes more direct. At higher w/c ratios, the time
during which cracks go around CH crystals increases.

82   E.T. Brown and J.A. Hudson, Discussion of S.P. Shah and F.J. McGarry,
      Griffith Fracture Criterion and Concrete, J. Eng. Mech. Div., 97 (1971)
      1663-1676; Journal of the Engineering Mechanics Division, ASCE, 98 (1972)
      1310-1312.

The authors argue that the load-deflection curves presented in the
original paper (ref. 72) are incorrect, because they were obtained in a
soft testing machine, and therefore the work of fracture calculated
included a considerable component due to energy released by the machine
on unloading.

83   J.H. Brown, Measuring the Fracture Toughness of Cement Paste and Mortar, Magazine of Concrete Research, 24 (1972) 185-196.

      K_c of hcp and mortar was measured using both SEN beams (38 x 38 x 250 mm) and DCB specimens (50 x 100 x 350 mm). A compliance technique was used to determine crack growth. For hcp, K_c ($\sim$0.48 MNm$^{-3/2}$) appeared to be independent of crack growth; for mortars, K_c increased with increasing crack growth. It was found that the DCB and SEN beam specimens gave somewhat different results. For both hcp and mortar, crack growth initiated at a lower value of K than that needed to keep the crack running.

84   A.G. Cooper and J. Figg, 1. Fracture Studies of Set Cement Paste, Journal of the British Ceramic Society, 71 (1972) 1-4.

      Slow bend tests were carried out on hcp beams of two sizes: 10 x 10 x 40 mm, and 25 x 50 x 200 mm prisms, with sawn triangular notches. The work of fracture and surface energy were estimated from the area under the load-deflection curves. It was found that for 40-day old specimens, the fracture surface energy of dry specimens was 14.9 J/m^2; for wet specimens it was 12.4 J/m^2. From the Griffith equation, the critical flaw size was then calculated to be about 7 mm for dry specimens and 6 mm for wet specimens.

85   R.K. Chir and C.M. Sangha, A Study of the Relationships Between Time, Strength, Deformation and Fracture of Plain Concrete, Magazine of Concrete Research, 24 (1972) 197-208.

      Concrete cylinders (50 mm dia. by 150 mm long) were tested in compression at strain rates ranging from 2.5×10^{-3}/s to 5.0×10^{-8}/s. It was found that at strain rates lower than 2.5×10^{-4}/s, strength was independent of strain rate. The critical stress for unstable fracture propagation did not correspond to the onset of dilation. At the highest loading rates, there was a tendency for inclined rather than vertical cracks to form.

86   J. Glucklich, The Strength of Concrete as a Composite Material, in Mechanical Behaviour of Materials, Proceedings of the International Conference on Mechanical Behaviour of Materials, Kyoto, 1971, The Society of Materials Science, Japan, 1972, Vol. IV, pp. 104-112.

      An attempt was made to formulate a "law" to assess the contribution of the aggregate to concrete strength, with particular reference to the bond strength itself. The problem was treated by satisfying both the stress and energy conditions for crack propagation. G increases with crack length because:

i)   In more heterogeneous materials, there are more microcracks, and thus a greater energy demand.

ii)   Cracks go around aggregate particles, therefore creating more surface area.

iii)   As the cracks grow, the stress-free zone begins to contain more aggregate, and contributes less energy to crack extension.

Therefore, the addition of aggregate increases both K_c and G_c.

87   B. Harris, J. Varlow and C.D. Ellis, The Fracture Behaviour of Fibre Reinforced Concrete, Cement and Concrete Research, 2 (1972) 447-461.

The work of fracture and K_c for frc samples made with glass, mild steel and high carbon steel fibres were determined using SEN beams in 3-point bending. The work of fracture, γ_F, was determined from the area under the load-deflection curve. γ_F could be increased by 2 orders of magnitude with 2% by volume of fibres, while K_c increased by only a factor of two. It was found that:

i) γ_F decreased with increasing notch/depth ratio; it was approximately equal to 10 J/m^2.

ii) K_c and γ_F were not significantly different in wet or dry specimens.

iii) K_c decreased with long term storage in a dry environment, perhaps due to shrinkage cracks. Mild steel gave the highes K_c values after long term storage.

88   C.E. Kesler, D.J. Naus and J.L. Lott, Fracture Mechanics - Its Applicability to Concrete, in Mechanical Behaviour of Materials, Proceedings of the International Conference on Mechanical Behaviour of Materials, Kyoto, 1971, The Society of Materials Science, Japan, 1972, Vol.IV, pp. 113-124.

The applicability of LEFM to hcp, mortar and concrete was studied using plate specimens 2 in. deep, 12 in. high, and varying in length from 18 to 36 inches, with a center crack. It was found that K_c varied with flaw size; it was concluded that K_c was not a valid fracture criterion for any of the materials studied. It was also found that large localized strains could occur at or near flaws.

89   M.S. Lamkin and V.I. Paschenko, Determination of the Critical Stress Intensity Factor of Concrete (in Russian). Izvestiya VNIIG im. B.E. Vedeneeva, Leningrad, 99 (1972) 234-239.

The critical stress intensity coefficient N_c (using the Barenblatt notation) was measured by splitting cylindrical concrete specimens with a preformed crack in the plane of the splitting forces. The specimen dimensions (ℓ x d) were 100x150 mm and 300x700 mm. The change of N_c with time was observed, and was similar to the results of Naus and Lott (ref. 48). It was concluded that the loading scheme could substantially influence the process of microcracking at the crack tip.

90   T.C.Y. Lim, A.H. Nilson and F.O. Slate, Stress-Strain Response and Fracture of Concrete in Uniaxial and Biaxial Compression, Journal of the American Concrete Institute, 69 (1972) 291-295.

Square plates of an idealized model and of real concrete were studied under uniaxial and biaxial compression. Crack development was monitored by X-ray techniques. It was found that for the model concrete in uniaxial compression:

   i)   shrinkage cracks and bond cracks existed before loading; their growth under load led to failure.

   ii)   bond cracks occur first around the larger aggregate particles; they do not grow appreciably below about 65% of σ_{ult}.

   iii)   mortar cracks were initiated at about 85% of σ_{ult}, tending to bridge bond cracks between the larger aggregate particles.

   iv)   a splitting mode of failure occurred.

The increase in strength and stiffness under biaxial loading was due to the confinement of potential microcracking.

91   K. Okada and W. Koyanagi, Effect of Aggregate on the Fracture Process of Concrete, in Mechanical Behaviour of Materials, Proceedings of the International Conference on Mechanical Behaviour of Materials, Kyoto, 1971, The Society of Materials Science, Japan, 1972, Vol. IV, pp. 72-83.

The effect of aggregate content on the fracture process in mortar and concrete was considered, using the Griffith theory. Compressive strength, modulus of rupture, and splitting tensile strength were determined, and SEN beams (4.7 x 10 x 38.8 mm) were tested in 4-point bending. In general G_c and K_c increased with aggregate content for normal weight concrete; for lightweight concrete, G_c and K_c decreased with increasing aggregate content. It was concluded that aggregates have a crack arresting function in normal weight concrete.

92   S.R. Parimi and J.J.S. Rao, Effectiveness of Random Fibres in Fibre-Reinforced Concrete, in Mechanical Behaviour of Materials, Proceedings of the International Conference on Mechanical Behaviour of Materials, Kyoto, 1971, The Society of Materials Science, Japan, 1972, Vol. V, pp. 176-186.

Assuming a random, uniform distribution throughout the specimen, the increase in fracture toughness of steel frc was investigated. A formula for strain energy release rate was derived:

$$G_c = \frac{u^2 L^3 p}{314\ E_f d^2}$$

where E_f = E of the fibres, d = fibre diameter, L = fibre length,
p = fibre volume %, u = bond strength. It was shown that G_c increased
with fibre L/d ratio and with fibre volume. Values ranged from ∿20 N/m
to ∿1470 N/m. (See Romualdi, ref. 36)

93   C.V.S.K. Rao and J.K.S. Rao, Statistical Aspects of Strength and Fracture
     Behaviour of Concrete, in Mechanical Behaviour of Materials, Proceedings
     of the International Conference on Mechanical Behaviour of Materials,
     Kyoto, 1971, The Society of Materials Science, Japan, 1972, Vol. IV,
     pp. 53-62.

     The effects of microcracks on the mechanical behaviour of concrete
     are described. Using a theory which describes concrete as a two-phase
     material, with one phase being cracks, the size effect in concrete is
     explained. However, a full quantitative evaluation of the theory is not
     possible, since this would require an accurate estimate of the amount of
     microcracking.

94   S. Stockl, Strength of Concrete Under Uniaxial Sustained Loading, in
     Concrete for Nuclear Reactors, SP-34, American Concrete Institute,
     Detroit, 1972, Vol. I, pp. 313-326.

     The strength of concrete under sustained load was investigated. It
     was found that strength decreased under sustained load, and then increased
     due to further hydration; however, the results showed a great deal of
     scatter. It was postulated that the decrease in strength was due to
     stress concentrations at the edges of pores, which induced slow crack
     growth until some critical crack length was reached.

95   R.N. Swamy, Fracture Phenomena of Hardened Paste, Mortar and Concrete,
     in Mechanical Behaviour of Materials, Proceedings of the International
     Conference on Mechanical Behaviour of Materials, Kyoto, 1971, The Society
     of Materials Science, Japan, 1972, Vol. IV, pp. 132-142.

     The fracture mechanisms in hcp, mortar and concrete were examined.
     The failure mode depends on the volume and geometry of aggregates, which
     tend to prevent fast crack propagation because of the discontinuities in
     the contact zone. Interfacial cracking may occur on drying. External
     loading produces random, non-uniform strains internally. Material
     failure begins at the cement-aggregate interface. In the ascending part
     of the σ-ε curve, the cohesive strength of the hcp is the main load

bearing mechanism. In the descending branch, changes in surface energy plus inelastic deformation at the interface bring about stable fracture.

96 P.F. Walsh, Fracture of Plain Concrete, Indian Concrete Journal, 46 (1972) pp. 469-470, 476.

LEFM was applied to large concrete SEN beams, 3 to 15 inches deep, 10 to 50 inches long, tested in 3-point bending. K_c was calculated. It was concluded that the minimum beam depth for LEFM to be valid is 9 inches. Smaller members may be notch insensitive, and failure loads can simply be related to the modulus of rupture of the net section.

97 F.H. Wittmann and Ju. Zaitsev, Behaviour of Hardened Cement Paste and Concrete Under High Sustained Load, in Mechanical Behaviour of Materials, Proceedings of the International Conference on Mechanical Behaviour of Materials, Kyoto, 1971, The Society of Materials Science, Japan, 1972, Vol. IV, pp. 84-89.

A theory of crack propagation in porous, viscoelastic materials was developed, and used to describe the loss of strength with duration of load for hcp and concrete. However, it was found that the decrease in strength under tensile load was less than expected. This was ascribed to creep near the crack tip, which brings about stress relaxation. Therefore, the critical stress needed to cause crack growth is increased. This stress redistribution occurs in both tension and compression.

98 A. Yoshimoto and M. Kawakami, Microcracking in Cement Paste Under Flexure-Tension, in Review of the Twenty-Sixth General Meeting, The Cement Association of Japan, Tokyo, 1972, pp. 165-166.

Small mortar beams were loaded in 4-point bending, and red ink was used after loading to impregnate the cracks. The cracks were found to be either hair-shaped or void-shaped. As the strain increased, only the void-shaped cracks propagated.

99 A. Yoshimoto, S. Ogino and M. Kawakami, Microcracking Effect on Flexural Strength of Concrete After Repeated Loading, Journal of the American Concrete Institute, 69 (1972) 233-240.

Flexural tests under repeated loading were carried out on 100 x 100 x 500 mm beams. It was found that the strength increased on repeated loading, due to microcrack propagation. These microcracks were believed to relieve the stress concentrations at the tips of bond cracks, thereby increasing strength. The observed microcracks were smaller than the critical Griffith crack, and hence did not lead to failure, which requires the growth of bond cracks.

1973

100   R.L. Berger, F.V. Lawrence, Jr. and J.F. Young, Studies on the Hydration
       of Tricalcium Silicate Pastes II. Strength Development and Fracture
       Characteristics, Cement and Concrete Research, 3 (1973) 497-508.

       The fracture of C_3S pastes was studied using both optical microscopy
and SEM. In young pastes, the fracture was observed to pass primarily
through the high porosity C-S-H, avoiding low porosity areas where CH
had crystallized in the pores. That is, the C.H. areas acted as rigid
inclusions which caused the fracture path to deviate. In more mature
pastes, this discrimination was lost as the matrix became more homogeneous. However, the different fracture paths did not appear to affect
the tensile strength, which was proportional to the cube of the gel/
space ratio (i.e., it depended primarily on porosity).

101   J.H. Brown, The Failure of Glass-Fibre-Reinforced Notched Beams in
       Flexure, Magazine of Concrete Research, 25 (1973) 31-38.

       Glass frc SEN beams (250 x 38 x 38 mm) were tested in 4-point loading.
The notch depth was about 14 mm, and a compliance technique was used to
determine crack length. It was found that for the unreinforced matrix,
K_c was approximately constant, at about 0.4 $MNm^{-3/2}$. However, the
apparent K_c for the glass frc increased with increasing crack growth.
K_c also increased with increasing fibre content. It was approximately
independent of age.

102   J.H. Brown and C.D. Pomeroy, Fracture Toughness of Cement Paste and
       Mortars, Cement and Concrete Research, 3 (1973) 475-480.

       Both SEN beams and DCB specimens were used to measure the fracture
toughness (K_c) of concretes made with small aggregate particles. It was
found that the addition of aggregate increased K_c, and also resulted in
a progressive increase in toughness with crack growth. Toughness
increased with aggregate content. For hcp, toughness increased with
decreasing w/c ratio, almost irrespective of age. In hcp, the unhydrated
clinker acts in a similar manner to aggregates in mortar.

103   G.D.T. Carmichael and K. Jerram, The Application of Fracture Mechanics
       to Prestressed Concrete Pressure Vessels, Cement and Concrete Research,
       3 (1973) 459-467.

       A fracture mechanics approach was used to predict the approximate
extent of cracking in prestressed concrete pressure vessels subjected to
overpressure. The stress intensity factor, K_c, was calculated by means
of the crack closure work concept, using finite element analysis for

different crack lengths. It was concluded that the pseudo-fracture toughness approach is suitable for predicting fracture of concrete.

104   D.H. Clyde, On Crack Direction in Relation to Griffith's Bi-Axial Failure Criterion, Cement and Concrete Research, 3 (1973) 537-547.

A theoretical treatment of the Griffith biaxial failure criterion was carried out. It was concluded that the trend to axial cleavage depends on the crack trajectory away from the defect rather than the departure direction from the defect surface. In fact, the system is much more complicated, since cracking in compression may involve arrest of many cracks before final failure.

105   E.M. Krokosky, Static Fatigue in Hydrated Portland Cement, Materiaux et Constructions, 6 (1973) 447-452.

Theories concerning the static fatigue of hcp, mortar and concrete were reviewed. Based on new experimental evidence, it was shown that the time to failure decreases with increased relative humidity and lower temperatures. It was concluded that a stress-corrosion mechanism was responsible for static fatigue in cementitious materials. Though the details are not clear, it appears that there is chemical attack on the Si-O bonds, dependent on the hydroxyl ion concentration derived from the dissolution of CH.

106   M.S. Lamkin, V.I. Paschenko and L.P. Trapesnikov, Application of the Brittle Fracture Theory to Determination of the Size of the Thermal Cracks in Elements of Concrete Structures (in Russian). Trudy Koordinatsyionnykh Sovyeshchanii po Gidrotekhnikie, Energya, Leningrad, 82 (1973) 68-73.

The applicability of LEFM is supported by experimental observations on the effect of the geometrical similarity in the critical loads of matched specimens of different sizes loaded in splitting. A linear relation was observed between the splitting strength of concrete and K_{IC}. It was assumed that the length of the thermal crack followed from the size of the region in which K_I (obtained from the temperature distribution) was bigger than its critical value. The value of K_{IC} used in the numerical example was 0.627 MNm$^{-1.5}$. Three different crack configurations were studied in the case of a wall element in which the temperature field was described using orthonormal Legendre polynomials.

107   K.A. Mal'tsov and L.A. Shiryaeva, On the Character of Concrete Failure in Compression and in Tension (in Russian). Trudy Koordinatsyonnykh Sovyeshchanii po Gidrotekhnikie, Energya, Leningrad, 82 (1973) 29-33.

The process of the fracture of concrete is discussed, without using fracture mechanics notions. To obtain complete σ-ε curves, specimens with cross-sections of 50 x 50 mm and 100 x 100 mm, at different humidities, were tested in axial tension and compression at the strain rate of 300 x 10^{-6}/h. In "dry" concrete the observed region of dispersive microcracking is smaller, although the maximum stress is higher than in concrete of higher humidity. Three fracture stages were distinguished, and the problem of the ultimate strain was discussed. An accelerated text method for the long term strength of concrete was suggested, based on the observation of the increase of concrete deformation at the same value of stress for increasing humidity values.

108 S.R. Naaman, A.S. Argon and F. Moavenzadeh, A Fracture Model for Fibre Reinforced Cementitious Materials, Cement and Concrete Research, 3 (1973) 397-411.

A statistical model is developed to predict the tensile properties of frc. The first part of the model, simulating ductile failure, is based on the mechanics and statistics of composite materials. The second part of the model, covering brittle failure, incorporates a fracture mechanics criterion. The "weakest link" theory is then applied to bound the overall model and to introduce the problem of size effects. Upper and lower bounds to the model are also provided.

109 D. Naus, Fracture Mechanics Applicability to Portland Cement Concretes, Technical Manuscript M-42, Construction Engineering Research Laboratory, Champaign, Illinois, 1973.

The applicability of LEFM to hcp, mortar and concrete was investigated by testing plate specimens with a precast flaw. The dimensions of the specimens were 2 x 12 x W in., where W ranged from 18 to 36 in. The results indicated that at least for these tests LEFM did not apply to the materials tested. It was found that large localized strains occurred near flaws.

110 F. Radjy and T.C. Hansen, Fracture of Hardened Cement Paste and Concrete, Cement and Concrete Research, 3 (1973) 343-361.

This is a review of the literature in the field of fracture mechanics of cement and concrete, with 44 refs. The application of the Griffith theory is discussed, and various fracture parameters from the literature are tabulated. As well, cracking and fracture are discussed from a phenomenological point of view.

111 K. Rajagopalan, Discussion of the Paper Fracture of Plain Concrete, by J.F. Walsh, (Ind. Conc. J., 46 (1972) 469-470,476) Indian Concrete Journal, 47 (1973) 211, 224.

It was suggested that Walsh's (ref. 96) analysis, which was confined to crack/depth ratios of 1/3, should be extended to other crack/depth ratios. It was also suggested that other test methods should be used, and that a better stress analysis around cracks is also needed.

112   S.D. Santiago and H.K. Hilsdorf, Fracture Mechanisms of Concrete Under Compressive Loads, Cement and Concrete Research, 3 (1973) 363-388.

A conceptual model was formulated to describe progressive crack growth in plain concrete under compressive loading. At the macroscopic level, cracking was observed in prisms under compressive loads. In addition, fracture mechanics concepts were used to analyze the behaviour of isolated cracks under compressive loading. It was concluded that below about 0.85 σ_{ult}, the cracking is concentrated at the cement-aggregate interface and at voids. At higher loads, cracks extend into the mortar.

113   D.C. Spooner and C.D. Pomeroy, Energy Dissipating Processes in the Compression of Cement Paste and Concrete, Cement and Concrete Research, 3 (1973) 481-486.

The behaviour of hcp and concrete under cyclic compressive loading was examined. The energy lost is due in part to creep, in part to micro-cracking. Measurements of strain softening were used to distinguish between creep and cracking. It was postulated that progressive damage occurs prior to failure.

114   P. Stroeven, Some Aspects of the Micromechanics of Concrete, Ph.D. Thesis, Stevin Laboratory, Technological University of Delft, 1973.

Stereological techniques were used to monitor cracking in concrete. Amongst other things, a discussion of crack propagation was presented for concrete, and it was concluded that the experimental techniques described could be used to add to our understanding of the cracking process in concrete.

115   R.N. Swamy and C.V.S.K. Rao, Fracture Mechanism in Concrete Systems Under Uniaxial Loading, Cement and Concrete Research, 3 (1973) 413-427.

The principal features of the cracking, fracture and deformation of concrete are discussed, and a theoretical model is developed which correlates size effects, stiffness, and ductility with the behaviour of concrete under load. The model is based on concrete as a 2-phase material, one phase being microcracks, and predicts the non-linear σ-ϵ behaviour, size-dependent strength, and size effects on stiffness and ductility.

116   E. Tazawa and S. Kobayashi, Properties and Applications of Polymer
       Impregnated Cementitious Materials, in Polymers in Concrete, SP-40,
       American Concrete Institute, Detroit, 1973, pp. 57-92.

Polymer impregnated concrete specimens (impregnated with various
polymers) were cast in the form of 40 x 40 x 160 mm beams. It was found
that strength could be predicted reasonably well using the Griffith
theory. It was also suggested that the Weibull theory could be applied
to concrete.

117   K. Togawa, T. Satoh and K. Araki, Parameters on the Fracture Toughness of
       Mortar and Concrete, in Review of the Twenty-Seventh General Meeting, The
       Cement Association of Japan, Tokyo, 1973, pp. 117-120.

SEN beams, 100 x 100 x 400 mm, with notch depths ranging from 5 to
60 mm were tested to measure K_c and G_c of mortar and concrete. It was
found that for concrete K_c increased with increasing aggregate volume;
apparently, the strength of the coarse aggregate was more important than
the strength of the cement-aggregate interface. With lightweight
aggregates, K_c decreased with increasing aggregate volume.

For mortars, K_c depended more on the cement-aggregate bond than on the
aggregate strength. K_c increased with increasing w/c ratio, and with
increasing maximum size of aggregate. K_c did not appear to depend
particularly on notch depth.

118   P.F. Walsh, Discussion of the paper Measuring the Fracture Toughness of
       Cement Paste and Mortar, by J.H. Brown (Mag. Concr. Res., 24 (1972)
       185-196). Magazine of Concrete Research, 25 (1973) 220-221.

In discussing Brown's paper (ref. 83), it was pointed out that for
reliable fracture tests, it is necessary to use a specimen depth of at
least 150 mm. More generally,

$$\left(\frac{K_c}{\sigma_t}\right)^2 \; > \; 50 \text{ mm}$$

for LEFM to apply, where σ_t = nominal stress at failure. For smaller
specimens, fracture is likely to be notch insensitive.

119   V. Weiss, Crack Development in Concrete with Closely-Spaced Reinforcement
       and in Similar Materials, Cement and Concrete Research, 3 (1973) 184-205.

Cracks are defined in terms of stability. The effectiveness of
reinforcement in preventing cracking depends on the degree of bond, the
volume of reinforcement, and G_c of the matrix. Structural cracks grow
slowly under increasing load, until the energy release is large enough

to cause them to propagate. Reinforcement acts as a physical barrier to crack growth. It also greatly increases the energy demand, by the energy required to break or debond the reinforcement.

120 H. Yokomichi, Y. Fujita and N. Saeki, Experimental Researches on Crack Propagation of Plain Concrete, in Review of the Twenty-Seventh General Meeting, The Cement Association of Japan, Tokyo, 1973, pp. 144-147.

SEN beams (130 x 150 x 450 mm) of mortar and concrete with varying notch depths, were used to study crack propagation and G_c. Mortar plates with a crack-like slit were tested in uniaxial compression. Acoustic emission was used to monitor crack initiation. It was found that G_c decreased with increasing w/c ratio. However, notch depth appeared to have little effect. G_c was lower for concrete than for mortar. G_c measured on the plates in compression was about fifteen times as high as from the SEN beams. This was considered to be due to slow crack growth in the compressive tests. It was ascertained using FEM that the early crack propagation was arrested and a new stress concentration occurred at the tip of a slit.

121 Ju. V. Zaitsev and F.H. Wittmann, Fracture of Porous Viscoelastic Materials Under Multiaxial State of Stress, Cement and Concrete Research, 3 (1973) 389-395.

Crack growth in porous and viscoelastic materials was studied and this approach was used to describe the decrease in strength of hcp under high sustained load. Stress relaxation near the crack tips was also considered. There was good agreement between theoretical and experimental results.

1974

122 A. Auskern and W. Horn, Fracture Energy and Strength of Polymer Impregnated Cement, Cement and Concrete Research, 4 (1974) 785-795.

Polymer impregnated cement SEN beams (76x13x13 mm) were tested in 3-point bending, and the results were analyzed to provide a measure of the effective surface energy (or fracture energy). The fracture energy of the impregnated material was considerably higher than that of the unimpregnated hcp (7.5×10^4 vs. 0.91×10^4 ergs/cm^2). The increase appeared to be entirely due to the polymer contribution. The Griffith equation would predict an increase in fracture strength of about 5 times.

123   D.E. Beskos, Fracture of Plain Concrete Under Biaxial Stress, Cement and
       Concrete Research, 4 (1974) 979-985.

       An attempt was made to explain the behaviour of concrete under biaxial
stresses, using modifications of both the Coulomb-Mohr and Griffith
theories. A modified Griffith-type criterion was proposed in the
compression range:

$$(\sigma_1-\sigma_2)^2 \ (1-\alpha) + 8\sigma_t \ (\sigma_1+\sigma_2) = 0 \quad 0 < \alpha < 1$$

where α is the factor of reduction of the maximum tensile stress at the
most critical crack. In the tension zone, $\sigma_1 = \sigma_t$.

       This makes it possible to adjust α in order to get the measured
ratios of the ultimate σ_t/σ_c, which are about 10 in practice, and not
8 as predicted by Griffith.

124   A.G. Bishara and P. Tantayonondkul, Use of Latex in Concrete Bridge Decks,
       Final Report EES 435, Engineering Experiment Station, Ohio State
       University, Columbus, Ohio, 1974.

       It was found that the fracture toughness of latex modified concrete
was about 17% higher than that of ordinary air-entrained concrete.

125   G.E. Blight, Possible Applications of an Energy Approach to Design in
       Non-Metallic Construction Materials, in First Australian Conference on
       Engineering Materials, University of New South Wales, 1974, pp. 757-789.

       From a review of the literature, it was concluded that the concepts
of fracture mechanics were applicable to concrete. Two questions were
raised:

       i)   Is strength controlled by introduced notches, or by inherent
(intrinsic) cracks in the concrete?

       ii)   To what extent is the fracture energy affected by a transition
from a plane stress to a plane strain condition?

126   A. Briggs, D.H. Bowen and J.J. Kollek, Mechanical Properties and Dura-
       bility of Carbon-Reinforced Cement Composites, in Proceedings, Inter-
       national Conference on Carbon Fibres, Their Place in Modern Technology.
       The Plastics Institute, London, 1974, Paper No. 17.

       Four types of carbon fibres, with E-values ranging from 77 to
385 GNm^{-2}, were used to prepare carbon fibre reinforced cements. It was
found that the work of fracture increased with fibre content; for a
given fibre content, the work of fracture increased as E of the fibre
increased. With increased consolidation pressure (improving the fibre-
matrix bond), the work of fracture decreased, probably due to the fact
that with better bond, fibres break rather than pulling out of the
matrix. Impact tests were also carried out (Izod).

127   D.J. Cook and M.N. Haque, The Effect of Sorption on the Tensile Creep
       and Strength Reduction of Dessicated Concrete, Cement and Concrete
       Research, 4 (1974) 367-379.

       It was argued that strength increases on dessication due to con-
solidation of the C-S-H system, and an increase in the physical forces
of attraction in the C-S-H. On resaturation, strength reductions are
due to fluid re-entry into the C-S-H structure. Sorption dilates the
gel structure, reduces the internal bonding energy, and thus assists the
propagation of pre-existing microcracks.

128   D.J. Cook and M.N. Haque, Strength Reduction and Length Changes in
       Concrete and Mortar on Water and Methanol Sorption, Cement and Concrete
       Research, 4 (1974) 735-744.

       From tests carried out on specimens which had been dried and then
resaturated, it was found that strength decreased on sorption by both
water and methanol. They argued that this strength reduction was due to
the fact that surface energies were reduced by sorption.

129   D.J. Cook and M.N. Haque, The Tensile Creep and Fracture of Dessicated
       Concrete and Mortar on Water Sorption, Matériaux et Constructions,
       7 (1974) 191-196.

       The dynamic effects of water sorption on the strength reduction and
tensile creep of mortar and concrete were investigated. It was concluded
that moisture assisted crack growth is the major mechanism of tensile
creep and/or fracture of a material that had been dried and was then
resaturated.

130   R.K. Dhir and C.M. Şangha, Development and Propagation of Microcracks in
       Plain Concrete, Matériaux et Constructions, 7 (1974) 17-23.

       Cored and sawn cylindrical specimens of concrete were loaded to pre-
determined stress levels at different strain rates; after unloading,
polished and strained slices were observed with an OM at 30x, and crack
patterns were studied. It was found that:
       i)   Crack density was greatest in the horizontal direction, due to
the relative weakness of the cement-aggregate bond at the lower portions
of aggregate particles.
       ii)   Microcracking was primarily due to bond failure. Beyond about
50% of σ_{ult}, the upper parts of the aggregate particles also began to
contribute to the crack density.
       iii) The degree of matrix cracking was found to increase with
increasing strain rates.

131   P.L. Domone, Uniaxial Tensile Creep and Failure of Concrete, Magazine of
      Concrete Research, 26 (1974) 144-152.

      Uniaxial tensile tests were carried out on cylindrical concrete
      specimens with expanded ends. The specimen gauge length was 165 mm, with
      a 65 mm diameter. Short-term stress and strain to failure were measured,
      as well as tensile creep, for 28-day old specimens. The evidence
      suggested that a fracture limit envelope exists. The strength under
      sustained load of sealed concrete was about 85% of the short-term
      ultimate strength; for concrete immersed in water, the long-term strength
      was about 75% of the short-term strength. The immersed concrete also
      showed a larger increase in strain than the sealed concrete, indicating
      that the creep and fracture processes may be moisture sensitive.

132   J. Glucklich, U. Korin and F. Shauer, The Reinforcement of Concrete by
      Polymers, Final Report, TDM 74-14, Technion, Israel Institute of
      Technology, Department of Mechanics, Haifa, Israel, 1974.

      G_c of polymer impregnated hcp and mortar was determined. Tests were
      carried out on small (approximately 2.5 x 2.5 x 25 cm) SEN beams loaded
      in 4-point bending. Polymer impregnation was found to bring about an
      increase in G_c of about a factor of 3-4. The increases were due both to
      improvements in the hcp by filling of voids and in the cement-aggregate
      bond. However, impregnation reduced the extent of microcracking, and
      this reduced the amount of energy dissipated during fracture. This
      accounted for the fact that while bonding was improved by a factor of
      about 5, and the hcp by a factor of about 4.7, the overall effect in
      mortar was only an improvement of about 3.2.

133   S. Mindess, J.S. Nadeau and J.M. Hay, Effects of Different Curing
      Conditions on Slow Crack Growth in Cement Paste, Cement and Concrete
      Research, 4 (1974) 953-965.

      Double torsion tests were carried out on plates (228 x 76 x 13 mm)
      made of cement pastes, but cured in different environments: fresh water,
      sea water, and low and high pressure steam. Plots were made of crack
      velocity versus stress intensity. It was found that the steam cured
      specimens had a lower fracture toughness and were somewhat more sensitive
      to static fatigue than the room-temperature cured specimens. There
      was some evidence that crack growth is aided by the presence of water in
      the environment or by increased w/c ratio.

134   J.S. Nadeau, S. Mindess and J.M. Hay, Slow Crack Growth in Cement
       Paste, Journal of the American Ceramic Society, 57 (1974) 51-54.

       Double torsion tests were used to determine the dependence of crack
       velocity on stress intensity for hcp. The specimens had the dimensions
       228 x 76 x 13 mm. Fracture toughness was measured both by the double
       torsion method and using SEN beams; the agreement between the two methods
       was very good. The slope of the log V-log K_I curve was about 35. The
       critical flaw size for hcp was found to be about 0.1 mm.

135   D.J. Naus, G.B. Batson and J.L. Lott, Fracture Mechanics of Concrete,
       in Fracture Mechanics of Ceramics, Vol. 2, Plenum Press, New York, 1974,
       pp. 469-482.

       This is a selective review (with 15 references) of fracture mechanics
       applications to cement and concrete up to 1972. It is concluded that
       the Griffith criterion is applicable to concrete if microcracking is
       considered. They argue that most studies yield relative results, rather
       than valid measures of K_c; a unique K_c for cement, mortar or concrete
       does not exist. The apparent K_c is greatest for concrete, smaller for
       mortar, and least for hardened cement paste.

136   F. Radjy, Fracture of Hardened Cement Paste in Relation to Surface Forces
       and Porosity, Journal of the American Ceramic Society, 57 (1974) 88-89.

       The fracture surface energy (γ_F) was interpreted in terms of two
       limiting fracture mechanisms:
          i)   all C-S-H particles encountered by the propagating crack break; or
          ii)   C-S-H particles are assumed to be whisker-like and do not break;
       the fracture surface proceeds solely along the internal surface of the
       hcp.
       It was argued that the "whisker" model was more reasonable, and that
       therefore a direct comparison of γ_F with the surface energy of tober-
       morite was not relevant.

137   M.J. Setzer and F.H. Wittmann, Surface Energy and Mechanical Behaviour of
       Hardened Cement Paste, Applied Physics, 3 (1974) 403-409.

       The surface energy of hcp was modified by adsorbing or desorbing
       water. It was possible, using the Griffith criterion, to explain
       quantitatively the decrease in strength of hcp as the water content
       increased. It was also possible to determine the contribution of surface
       energy changes to shrinkage and swelling.

138   D. Walsh, M.A. Otooni, M.E. Taylor, Jun., and M.J. Marcinkowski, Study
      of Portland Cement Fracture Surfaces by Scanning Electron Microscopy
      Techniques, Journal of Materials Science, 9 (1974) 423-429.

      The fracture surfaces of hcp hydrated for various times were studied
      using the SEM. They found that cleavage through the hcp occurred
      mostly through the weakly bonded basal planes of the CH, and secondarily
      through the CH - C-S-H interface. The strongest bonding appeared to
      occur between spherulites of C-S-H. At early ages, fracture was inter-
      granular, going around the C-S-H grains. At later ages, the amount of
      transgranular fracture increased.

139   F.H. Wittmann, Bestimmung physikalischer Eigenschaften des Zementsteins,
      Schriftenreihe Deutscher Ausschuss fur Stahlbeton, 232 (1974).

      Creep and shrinkage of hardened cement pastes are studied experi-
      mentally and theoretically. Rate theory is applied to explain creep.
      Strength is related to surface free energy. The influence of moisture
      content on strength can be satisfactorily described.

140   F.H. Wittmann and J. Zaitsev, Verfomung und Bruchvorgang poroser
      Baustoffe bei kurzzeitiger Belastung und Dauerlast, Schriftenreihe
      Deutscher Ausschuss fur Stahlbeton, 232 (1974).

      A model to describe a porous structure is presented. Stress analysis
      is applied first to one cylindrical pore in an homogeneous and isotropic
      plate, then to two interacting neighbour pores. The resulting cracks are
      calculated in the two cases. The crack propagation in a statistical
      porous structure is developed. Numerous experiments have been carried
      out with hardened cement paste.

141   J. Zaitsev and F.H. Wittmann, A Statistical Approach to the Study of the
      Mechanical Behaviour of Porous Materials Under Multi-axial State of
      Stress, in Proceedings of the 1973 Symposium on Mechanical Behaviour of
      Materials, The Society of Materials Science, Japan, 1974, pp. 105-109.

      The stress distribution around pores with different shapes in a
      homogeneous and isotropic material is discussed. If the probability of
      occurrence of different pore shapes and pore sizes is taken into con-
      sideration, a random pore can be mathematically developed. The prob-
      ability of occurrence of stresses in the material around a random pore
      is calculated. The resulting stress field corresponds to the prob-
      ability of occurrence of stresses in the matrix of the porous material
      to be studied. The stress field can be analyzed.

142   J. Zaitsev and F.H. Wittmann, Eine theoretische Studie des Verhaltens
      von Beton unter kurzzeitiger ein-und zweiachsiger Belastung, Material-
      spruf., 16 (1974) 170.

      The stress distribution around pores with different shapes and different orientations in a homogeneous and isotropic material is discussed. If the relative frequency of different pore shapes is taken into consideration a "random pore" can be mathematically developed. A microscopic analysis of pore shape frequency in hardened cement paste is described. The frequency of stress peaks in the materials around a random pore is calculated. The probability of crossings of a given stress level can be studied.

1975

143   J. Aleszka and G. Schnittgrund, An Evaluation of the Fracture of Plain Concrete, Fibrous Concrete, and Mortar Using the Scanning Electron Microscope, Technical Report M-122, Construction Engineering Research Laboratory, Champaign, Illinois, 1975.

The fracture surfaces of both plain and steel fibre-reinforced mortar and concrete, broken in splitting tension, flexure, and compression showed that there were no microstructural differences between samples which were fractured wet or dry. On a microscale, the fracture surface appearance appeared to be independent of the type of loading.

144   M.A. Al-Kubaisy and A.G. Young, Failure of Concrete Under Sustained Tension, Magazine of Concrete Research, 27 (1975) 171-178.

Cylindrical concrete specimens with a gauge length of 200 mm, and 100 mm in diameter, were tested under sustained axial tension. Ultrasonic pulse velocity and strain distribution measurements were made. It was found that interfacial cracks form in the concrete at an early stage of loading, but that failure was governed by the growth of cracks in the cement matrix. Under sustained load, the strength decreased to about 0.7 σ_{ult}. A log-log plot of sustained stress vs. time to failure was linear, with the fracture parameter n ≈ 33.6.

145   A.S. Argon and W.J. Shack, Theories of Fibre Cement and Fibre Concrete, in Fibre Reinforced Cement and Concrete, Rilem Symposium 1975, The Construction Press, Ltd., Lancaster, England, 1975, pp. 39-53.

The problems of applying fracture mechanics to frc were discussed. These included size effects, and the large size of the pseudo-plastic zone ahead of the crack. A non-linear crack model was recommended.

146   J. Bhargava and A. Rehnstrom, High Speed Photography for Fracture Studies of Concrete, Cement and Concrete Research, 5 (1975) 239-248.

High speed photography was used to study the fracture of both plain concrete and PIC, using 20-40 mm thin concrete prisms fractured by detonation of an explosive in contact with the top surface of the specimen. The principle conclusions were:

i)   Initial cracks formed along the vertical mortar-aggregate interface, as in static loading.

ii)   Second, some horizontal cracks as well as cracks along the grain boundaries formed.

iii) The velocity of crack propagation was ∿180 m/s, though there were large local variations, perhaps due to crack branching.

iv)   Failure was eventually by cleavage.

For PIC, there was more fracture through the aggregate, indicating better bond. The crack velocity in PIC was also higher, ∿800 m/s.

The authors suggested that in concrete, cracking started during the passage of the compression wave, due to secondary tensile stresses. However, there was no great increase in strength over the static strength for PIC, since there was little inelastic deformation.

147   A.S. Salah El-Din and M.M. El-Adawy Nassef, A Modified Approach for Estimating the Cracking Moment of Reinforced Concrete Beams, Journal of the American Concrete Institute, 72 (1975) 356-360.

An expression to predict the cracking moment of reinforced concrete beams was derived, using an energy balance cracking criterion and the finite element method. It was found that there was good agreement between theoretical and experimental results.

148   V.M. Entov and V.I. Yagust, Experimental Investigation of the Regularities of Quasistatic Development of Macrocracks in Concrete, Izvestiya Akademii Nauk SSR, Mekhanika Tverdogo Tela., 4 (1975) 93-103.

An experimental investigation was carried out on the development of cracks in low- and medium-strength concrete subjected to short-term loading. The boundaries of the applicability of linear fracture mechanics to concrete were determined. The failure energy was also evaluated approximately. Using specimens up to 100 x 2000 x 2500 mm, a micro-fracture process region of about 80-100 mm in length was observed.

149   T. Esaki and Y. Tokumitsu, Time-Dependent Deformation Fracture of Concrete Under High Sustained Stress, Review of the Twenty-Ninth General Meeting, The Cement Association of Japan, Tokyo, 1975, pp. 207-209.

A study was carried out of the effects of sustained compressive load on 400 mm long x 100 mm diameter concrete cylinders. The specimens were loaded at an age of 28 days to between 34% and 90% of the short term ultimate strength. An empirical relationship between deflection, stress and time was developed. Time-to-failure envelopes were also determined.

150   J. Glucklich and U. Korin, Effect of Moisture Content on Strength and
        Strain Energy Release Rate of Cement Mortar, Journal of the American
        Ceramic Society, 58 (1975) 517-521.

Small mortar beams, SEN beams, and compression specimens of mortar
were tested to determine the influence of the moisture content on the
strength and G_c as the specimens were dried from the saturated state.
It was found that the bending strength first increased with drying, and
then decreased at very low r.h., probably due to drying cracking. The
compressive strength increased continuously on drying, particularly
below 50% r.h. G_c decreased slightly on drying to 80% r.h., then
increased gradually on further drying to 20% r.h., and increased rapidly
on drying from 20% r.h. These changes were attributed primarily to
increased surface energy as water is removed.

151   W.A. Patterson and H.C. Chan, Fracture Toughness of Glass Fibre-Rein-
        forced Cement, Composites, 6 (1975) 102-104.

Compact tension single edge cracked specimens were made using high-
alumina cement reinforced with E-glass fibres, with a glass content of
about 15% by weight. Using a quasi-static approach, the work to extend
the crack was calculated. It was found that for air cured specimens,
K_c increased with age up to about 5 weeks; for water cured specimens,
K_c increased with age for about 3 weeks and then decreased, due to the
effect of water on the fibres and on the bond. A large zone of fine
cracks was noted around the principal crack when the specimens were
loaded; this zone moved with the crack tip, making it difficult to define
the crack tip. K_c appeared to be independent of specimen size. Fracture
toughness only developed after the initial cracking of the brittle matrix.

152   D.C. Spooner, C.D. Pomeroy and J.W. Dougill, The Deformation and
        Progressive Fracture of Concrete, Proceedings of the British Ceramic
        Society, 25 (1975) 101-107.

Concrete specimens were tested in compression to various strains (one
or more loading cycles) until specimens either failed or were well into
the descending branch of the $\sigma-\varepsilon$ curve. The higher the volume of
aggregate, the larger the changes in E that occurred with each loading
cycle. Cement paste specimens exhibited brittle fracture, but concretes
with a high aggregate volume yielded $\sigma-\varepsilon$ curves with descending branches.
It was concluded that fracture was progressive; for a given strain,
higher aggregate contents led to more damage. It was found that for
loads up to 80-90% of σ_{ult}, damage was detected during the first load
application. On reloading, no further damage occurred until the
previous load had been exceeded.

153 Y. Tanigawa and Y. Kosaka, Mechanism of Fracture and Failure of Concrete as a Composite Material, Memoirs of the Faculty of Engineering, Nagoya University (Japan), 27 (1975) 208-263.

A general review is presented of the fracture of concrete, based mainly on the Japanese literature. The fracture mechanics approach is not discussed per se.

154 P.F. Walsh, Cracks, Notches and Finite Elements, in J.M. Corum and S.E. Benzley (eds.), Computational Fracture Mechanics, Symposium of the Computational Technical Committee, 2nd National Congress on Pressure Vessels and Piping, ASME, New York, 1975, pp. 49-61.

A numerical finite element procedure for the computation of stress intensity factors was presented; the method is applicable to orthotropic materials. As an example, the method was applied to the initiation of cracking at a re-entrant notch in a concrete member. The effect of specimen size was also noted.

155 A. Yoshimoto, K. Kawasaki and M. Kawakami, Studies of Formation of Continuous Crack in Concrete by Using a Slow-Motion Picture, in Review of the Twenty-Ninth General Meeting, The Cement Association of Japan, Tokyo, 1975, pp. 243-244.

Concrete compression specimens, 70 x 70 x 140 mm, were tested to study cracking in 60-day old specimens. At some "critical" stress, longitudinal mortar cracks suddenly occurred, due to tensile splitting. It was concluded that continuous mortar cracks were not caused by the propagation of bond cracks, but were due to spontaneous tensile splitting.

1976

156 J.C. Aleszka and P.W.R. Beaumont, The Work of Fracture of Concrete and Polymer Impregnated Concrete Composites, in Proceedings, First International Congress on Polymers in Concrete, 1975, The Construction Press Ltd., Lancaster, England, 1976, pp. 269-275.

SEN beams (125x25x25 mm) were used to assess K_c and the work of fracture of concrete impregnated with MMA, and of polymer-impregnated frc and plain concrete. Impregnation decreased the porosity and the tendency for crack initiation. For polymer impregnated frc, the work of fracture was increased considerably, due to the work done at the crack tip to overcome the frictional drag existing at the wire-matrix interface. For ordinary frc, the energy of plastic deformation of the fibre was of the same order of magnitude as the fibre pull-out energy term. In impregnated frc, the improvement in interfacial shear strength by a factor of three made the fibre pull-out term the dominant one.

157   B. D. Barnes, Morphology of the Paste-Aggregate Interface, Ph.D. Thesis, School of Civil Engineering, Purdue University, West Lafayette, Indiana, 1976.

An extensive study was carried out on the nature and morphology of the interfacial zone between cement and aggregate and on crack development in concrete. It was found that a thin "duplex film" of CH backed with a layer of C-S-H particles deposits directly on the aggregate surface; a few large CH crystals then develop and extend into the hcp. A secondary deposit of thick CH crystals deposits later and effectively ties the duplex film to the hcp. In mortars, fine cracks ($\sim$ 1 μm) were found within a zone of 40 μm from the aggregate prior to loading, and likely existed before drying. On loading in compression, these cracks extended near the interface until high stress levels initiated unstable crack propagation.

158   B.D. Barnes and S. Diamond, Initiation and Propagation of Cracks near Portland Cement Paste-Aggregate Interfaces, in Proceedings of the 2nd International Conference on Mechanical Behaviour of Materials, ASM, Metals Park, Ohio, 1976, pp. 1414-1417

Both Portland Cement and mortar were cast against glass microscope slides. After curing, the hcp was allowed to dry, causing separation due to shrinkage between the hcp and the glass; the mortar specimens were fractured. Three boundary zones were observed using the SEM:

    i)   An aggregate - "duplex film" boundary zone occuring at the original aggregate interface and a $\sim$1.5 μm thick layer of CH overlain or intergrown with C-S-H.

    ii)   A duplex film-secondary lime boundary zone; the secondary lime layer served to join the duplex film to the paste.

    iii)   A secondary lime-paste boundary zone.

Cracks were generally observed within 30-40 μm of the aggregate surface, typically within one of the first two boundary zones.

159   B. Barr and T. Bear, A Simple Test of Fracture Toughness, Concrete, 10 (1976) 25-27.

A new test for fracture toughness was described, which used a circumferentially notched round bar in bending (CNRBB). Mortar specimens were 150 mm long, and either 41 or 54 mm in diameter. It was concluded that K_c increased as the notch depth increased up to 8 mm. K_c increased as the displacement rate increased. For these notches, K_c was essentially independent of notch width. It was claimed that for deep notches, K_c would be independent of notch depth.

160   J.E. Barrick, II, and E.M. Krokosky, The Effects of Temperature and
      Relative Humidity on Static Strength of Hydrated Portland Cement, Journal
      of Testing and Evaluation, 4 (1976) 61-73.

      Small, notched, mortar beams (0.5x0.5x5.5 in.) were tested in static
      fatigue at various temperatures and relative humidities and the time to
      fracture measured. It was found that the time to failure increased
      with decreasing r.h. However, the time to failure increased with
      increasing temperature; static fatigue was almost non-existant at 60°C.
      A stress-corrosion mechanism dependent on the presence of CH was postu-
      lated, i.e., static fatigue in hcp is a result of chemical attack on
      Si-O bonds. A very high stress is required to activate this stress
      corrosion mechanism. It was further postulated that the stress corrosion
      was due to a hydroxyl ion attack mechanism (from dissolution of CH
      rather than direct water attack.

161   Z.P. Bazant, Instability, Ductility, and Size Effect in Strain-Softening
      Concrete, Journal of the Engineering Mechanics Division, ASCE, 102 (1976)
      331-344.

      An analysis of the descending branch of the σ-ε curve ("strain-
      softening") is presented. It is concluded that this phenomenon cannot
      occur in a continuum; rather, a heterogeneous material is required (like
      concrete). Failure occurs by an unstable localization of strain, in
      which stored strain energy is transferred into a small (several times
      the aggregate size) strain-softening region. A stability analysis may
      be carried out using FEM, but the size of the elements to be used is
      critical

162   J.R. Clifton, J.E. Fearn and E.D. Anderson, Polymer Impregnated Hardened
      Cement Pastes and Mortars, NBS-BSS 83, National Bureau of Standards,
      Washington, D.C., 1976.

      Polymer impregnated (PMMA) hcp and mortar specimens were prepared,
      and compared with regard to their strength and fracture parameters.
      Fracture tests were carried out on double torsion specimens,
      105 x 57 x 6.3 mm. It was found that impregnation greatly increased the
      measured K_c values, and the strength under sustained load appeared also
      to be increased. In terms of fracture characteristics, the unimpreg-
      nated mortars failed by fracture through the hcp, with the sand grains
      still covered by a layer of hcp. The fracture surface of the impregnated
      mortars passed through both the hcp and the sand grains.

163   A.G. Evans, J.R. Clifton and E. Anderson, The Fracture Mechanics of
       Mortars, Cement and Concrete Research, 6 (1976) 535-548.

       Double torsion specimens (105x57x6 mm) and acoustic emission were
used to study the fracture of plain and polymer impregnated mortars. It
was found that:
       i)   Fracture mechanics parameters such as K_c were independent of
crack length beyond a crack length of 20 mm.
       ii) K_c was not particularly affected by w/c ratio and curing time.
       iii)   Polymer impregnation increased K_c.
       iv)   From the acoustic emission results, it was found that extensive
microcracking occurred in both materials. However, impregnation
suppressed the onset of microcracking, leading to an enhanced fracture
stress.
       v)   Microcracking appeared to occur within a zone around the
macrocrack tip, and contributed to crack propagation resistance.

164   D.D. Higgins and J.E. Bailey, A Microstructural Investigation of the
       Failure Behaviour of Cement Paste, in Hydraulic Cement Pastes: Their
       Structure and Properties, Proceedings of a Conference at University of
       Sheffield, 8-9 April 1976, Cement and Concrete Association,
       Wexham Springs, 1976, pp. 283-296.

       A special optical microscopy technique was used to study structural
changes in stressed hcp. Stable microcracks were observed, and these
were used to provide an explanation for fracture. It was found that
stable cracks do occur in hcp, and seem to be composed of smaller microcracks. Cracks either go around anhydrous cores, or are arrested by
them. Cracks are due in part to localized regions of high w/c ratio,
which occur during hydration, and alternate with denser regions. Growth
and coalescence of microcracks occurs. The authors argued that the
evidence supports a tied-crack fracture model.

165   D.D. Higgins and J.E. Bailey, Fracture Measurements on Cement Paste,
       Journal of Materials Science, 11 (1976) 1995-2003.

       Both flexural tests (small SEN beams) and direct tension tests (small
DEN specimens) were carried out on hcp. It was found that in flexure:
(i) the material is notch sensitive; (ii) K_c appeared to be independent
of notch depth; (iii) K_c increased with age to about 20 days, then
remained constant; (iv) K_c increased with decreasing w/c ratio; (v) K_c
increased somewhat with increased rate of loading; (vi) K_c increased
with increasing notch width, but remained constant with notch widths
below 0.5 mm; (vii) K_c increased with increasing beam depth, but appeared
to be independent of changes in beam width or length.

In tension, K_c increased with specimen size, suggesting that LEFM was
not applicable, at least up to a specimen size of 600 mm in length and
110 mm in depth. The addition of sand made the size effect even more
severe. It was suggested that a tied crack model (similar to the Dugdale
model) should be used to represent the fracture of hcp.

166   B. Hillemeier, Bruchmechanische Utersuchungen des Rissfortschritts in
Zementgebundenen Werkstoffen, Dr-Ing. Thesis, Karlsruhe University, 1976.

Wedge-loaded compact tension specimens were used to study the fracture
of hcp. The results were analysed using finite element methods. K_c was
found to decrease with increasing crack length, at least up to a limiting
crack/depth ratio of about 0.6, where it appeared to stabilize. K_c of
the cement-quartz interface was found to be considerably less than that
of hcp.

167   A. Hillerborg, M. Modeér and P.E. Petersson, Analysis of Crack Formation
and Crack Growth in Concrete by Means of Fracture Mechanics and Finite
Elements, Cement and Concrete Research, 6 (1976) 773-782.

Fracture mechanics concepts were introduced into finite element
analysis using a tied crack model (similar to the Dugdale model) for
fracture. This is a way of expressing the energy absorption, G_c, in the
energy balance approach. The method has been used to explain the
difference between bending and tensile strengths, and also the variation
in bending strength with beam depth. In the model, it is assumed that as
the crack opens the stress does not fall to zero at once, but decreases
with increasing crack width. This region corresponds to a microcracked
zone with some remaining ligaments for stress transfer, thus absorbing
energy as the crack grows.

168   K. Kishitani and K. Maeda, Compressive Fracture of Cracked Mortar, in
Review of the Thirtieth General Meeting, The Cement Association of Japan,
Tokyo, 1976, pp. 216-217.

Values of G_c versus crack length were obtained from splitting tension
tests on mortar. For crack lengths greater than 15 mm,
$G_c \simeq 8000$ erg/cm; this value decreased for shorter cracks. It was
concluded that the mechanism of compressive fracture was due to tensile
failure of the crack tip.

169   H. Kitagawa, S. Kim and M. Suyama, Determination of Fracture Toughness
of Concrete Materials by Diametral Compression Tests, in Proceedings,
Nineteenth Japan Congress on Materials Research, The Society of Materials
Science, Tokyo, 1976, pp. 160-163.

K_c for concrete was measured using diametral compression (splitting
tension) specimens with a center slit. K_c appeared to increase with

increasing specimen diameter, and was affected by the slit width.

170 H. Kitagawa and M. Suyama, Fracture Mechanics Study on the Size Effect for the Strength of Cracked Concrete Materials, in Proceedings, Nineteenth Japan Congress on Materials Research, The Society of Materials Science, Tokyo, 1976, pp. 156-159.

SEN beam tests were carried out on mortar and concrete, with a specimen depth of 80 mm. COD was measured. It was found that K_c decreased with increasing crack/depth ratio, but was more-or-less independent of specimen depth.

171 A. Maher and D. Darwin, A Finite Element Model to Study the Microscopic Behaviour of Plain Concrete, CRINC Report-SL-76-02, The University of Kansas Center for Research, Inc., Lawrence, Kansas, 1976.

A finite element model of concrete was developed to study the strength and deformation of plain concrete under uniaxial and biaxial compressive stresses. It was concluded that the strength of the mortar-aggregate interface on the behaviour of the model of concrete used was small compared with the effect of nonlinearity of the mortar itself. It was assumed that mortar cracks are initiated only in the presence of bond cracks; the pattern of crack propagation was similar whether cracks were present at the interface before loading or were caused by loading.

172 S. Mindess and J.S. Nadeau, Effect on Notch Width on K_{IC} for Mortar and Concrete, Cement and Concrete Research 6 (1976) 529-534.

Tests were carried out on mortar and concrete to determine whether K_c depends on the length of the crack front. Notched beams were prepared with a constant length of 203.2 mm and depth of 50.8 mm, but with the width (length of crack front) varying from 45 to 254 mm. The beams were tested in 3-point bending. It was found that there was no dependence of fracture toughness upon the length of the crack front. The tests implied that the crack front experiences a plain strain stress condition.

173 E.N. Scerbakov and Ju. V. Zaitsev, Voraussage der Verformungs - und Festigkeitseigenschaften von Beton unter Dauerlast, Cement and Concrete Research, 6 (1976) 515-528.

It was shown that the decrease in strength of concrete under high sustained load can be predicted using fracture mechanics methods and a statistical estimation of linear creep strain.

174 M.J. Setzer, A Method for Description of Mechanical Behaviour of Hardened Cement Paste by Evaluating Adsorption Data, Cement and Concrete Research, 6 (1976) 37-48.

A thermodynamic approach was used to interpret sorption data, which permitted calculation of changes in the surface free energy, using the

Griffith equation. It was proposed that

$$\left(\frac{\sigma}{\sigma_o}\right)^2 = 1 - \frac{\Delta\gamma}{\gamma_o}$$

where σ_o and γ_o are the strength and surface free energy respectively, at saturation, $\Delta\gamma$ is the change in surface free energy, and σ is the strength at a given relative humidity.

175   D.C. Spooner, C.D. Pomeroy and J.W. Dougill, Damage and Energy
       Dissipation in Cement Pastes in Compression, Magazine of Concrete
       Research, 28 (1976) 21-29.

       Cement paste specimens (200 x 75 x 75 mm) were tested in compression.
It was concluded that hcp is not an elastic, brittle material. Damage,
characterized by changes in E and by energy dissipation during sequential
loading and unloading cycles was detected at strains of about
500×10^{-6}. Damage was progressive. For a given strain, the quantity
of "solid material" fractured appeared to be independent of the w/c ratio.
The fracture mechanism appeared to be the same for hcp as for concrete.

176   J. Spurrier and A.R. Luxmoore, The Critical Constrained Crack Length in
       Fibre-Reinforced Cementitious Matrices, in Fibre Science and Technology,
       Vol. 9, Applied Science Publishers Ltd., England, 1976, pp. 225-236.

       Some of the theories of the effects of fibres in cementitious materials
are brought together (Romualdi & Batson, ASCE, 1963; Griffith).
Romualdi's work is modified to make it compatible with the Griffith
criterion. A set of equations are presented to give an indication of
fibre size and concentration required for improvement in the "first
crack strength" of the reinforced brittle matrix.

177   M.S. Stucke and A.J. Majumdar, Microstructure of Glass Fibre-Reinforced
       Cement Composites, Journal of Materials Science, 11 (1976) 1019-1030.

       The microstructure and fracture surfaces of glass fibre-reinforced
cement composites were examined by SEM. The toughness of grc in dry
environments was attributed to the formation of a complex structure of
subsidiary cracks, resulting in a large failure surface area; in wet
conditions, the amount of subsidiary cracking was reduced, leading to a
reduction in toughness. It was suggested that glass fibre failure occurs
when the bending stresses in the fibres at the surface of a transverse
crack exceed the Griffith theoritical strength σ_f. Assuming for glass
fibres, $\sigma_f \simeq 1.8$ GNm^{-2}, $\gamma \simeq 5$ Jm^{-2} and $E \simeq 70$ GNm^{-2}, a critical flaw
size of 0.06 μm was calculated. However, SEM observations of the surface
of corroded fibres revealed considerably larger flaws. The tensile

strength of corroded fibres, ≈ 1.1GPa, implied a critical flaw size of ~ 0.15 μm, and flaws of about this size were observed.

178   L.P. Trapesnikov, On the Question of Creep Effect in the Crack Development in Concrete (in Russian). Trudy Koordinatsyionnykh Sovyeschanii po Gidrotekhnike, Energya, Leningrad, 112 (1976) 64-67

The crack development problem is treated within the framework of the quasi-static linear theory of viscoelastic bodies, taking into account the general thermodynamic relations. The crack propaqates if sufficient energy is available. The energy demand can be determined through calculations of stresses and displacements using the theory of viscoelasticity. It is concluded that both K_I and K_{IC} are time-dependent.

179   P.F. Walsh, Crack Initiation in Plain Concrete, Magazine of Concrete Research, 28 (1976) 37-41.

LEFM was applied to a variety of notch configurations in concrete. The effect of beam size was particularly considered, using SEN beams tested in 3-point loading. A failure criterion was formulated in terms of K_c. It was concluded that specimens must be large enough so that the zone of disturbance near a crack tip is small compared to the specimen dimensions. It was suggested that for proper fracture toughness testing, the beam must be of sufficient depth so that the crack length is at lease 78 f_c'/σ_c^2, where f_c' is the cylinder compressive strength, and σ_c is the modulus of rupture. For the concrete tested, this yielded a minimum beam depth of about 230 mm.

180   F.H. Wittmann, The Structure of Hardened Cement Paste - A Basis for a Better Understanding of the Materials Properties, in Hydraulic Cement Pastes: Their Structure and Properties, Proceedings of a Conference held at University of Sheffield, 1976; Cement and Concrete Association, Wexham Springs, 1976, pp. 96-117.

The structure of hcp was modelled as a xerogel (a 3-dimensional network of colloidal particles), in which both primary and secondary bonds contribute to strength. Above about 50% r.h., the disjoining pressure of the water separates surfaces held together only by van der Waals bonding.

Using the Griffith criterion, the change in strength with r.h. can be related to the change in interfacial energy. The strength should decrease as the moisture content (and hence the disjoining pressure) increases, at least in the range where changes in r.h. affect surface energy.

181 A. Yoshimoto, K. Kawasaki and M. Kawakami, Microscopic Cracks in Cement Matrix and Deformation Behaviour of Concrete, in Proceedings, Nineteenth Japan Congress on Materials Research, The Society of Materials Science, Japan, 1976, pp. 126-131.

From plotting log-log σ-ε relationships for cement and concrete, two "kinks" were found in the linear plots. The first one was related to the propagation of cracks in the paste, while the second one was associated with longitudinal cracks bridging between the aggregate particles. This occurred at a crack size of about 0.1 mm.

1977

182 S.A.S. Akers, G.G. Garrett and R.B. Tait, "In Situ Scanning Electron Microscope Observations of Flexural Failure of Asbestos Cement, Electronmicroscopy Society of South Africa, 7 (1977) 57-58.

A four-point loading method was used to study the flexural failure of asbestos cement, made with an asbestos/cement ratio of 0.1. It was found that the fibre bundles acted as crack arrestors, with subsequent microcrack failure at the ends of the bundles. When the failure strength of the bundle was about the same as the bond strength, failure also occurred by fracture of the bundle; this was dependent on the diameter of the fibre bundles. The strength of asbestos cement is thus a function of the bundle size and the asbestos/cement ratio.

183 B. Barr and T. Bear, Fracture Toughness, Concrete, 11 (1977) 30-32.

Further work on the circumferentially notched round bar under bending (CNRBB) was reported. For tests on mortar, it was found that eccentric loading gave essentially the same results as ordinary bending tests. However, a smaller specimen could be used in the eccentric loading tests. With eccentric loading, K_c decreased with increasing notch depth.

184 G.B. Batson, Strength of Steel Fibre Reinforced Concrete in Adverse Environments, Special Report M-218, Construction Engineering Research Laboratory, Champaign, Illinois, 1977.

An investigation of crack steel frc subjected to a saltwater environment was examined, using wedge opening loaded specimens 10 in. long, 4.6 in. wide and 4 in. deep, with 1.0, 1.5 and 2.0 percent steel fibres by volume. The strain energy release rate of specimens exposed to saltwater for various times was compared with that of control specimens. It was found that G_c decreased on exposure to saltwater; higher fibre contents reduced the amount of degradation. It appeared that the saltwater reacted with both the steel fibres and the concrete matrix.

594

185 T.J. Bear and B. Barr, Fracture Toughness Tests for Concrete, International
     Journal of Fracture, 13 (1977) 42-46.

     Two fracture mechanics specimens were described: the circumferentially
     notched round bar in four-point bending, and the same bar subjected to an
     eccentric longitudinal load. It was concluded that both tests gave similar
     K_c values for concrete, and that the variability was lower than for other
     fracture mechanics tests.

186 A. Briggs, Review. Carbon Fibre-Reinforced Cement, Journal of Materials
     Science, 12 (1977) 384-404.

     Impact resistance tests (Izod) were reported for various carbon fibre
     reinforced cements. The impact fracture energy was several times greater
     than the fracture energy calculated from slow bend tests. This was
     accounted for by the much more complicated fracture pattern of the impact
     specimens.

187 P. Desayi, Fracture of Concrete in Compression, Matériaux et Constructions,
     10 (1977) 139-144.

     Concrete prisms, 6 x 6 x 12 in., were cast with preformed centrally
     located notches of different lengths and at different inclinations, and
     loaded in compression. Generally, a single crack propagated from each end
     of the notch, but sometimes several cracks formed, probably because of the
     chance location of large aggregate particles. It was found that cracks
     initiated at the notches, but propagated in the direction of loading.
     Notches less than 0.5 in. long did not reduce the strength. G_c varied with
     notch length and notch inclination; it seemed to increase with increasing
     length.

188 T. Esaki and Y. Tokumitsu, Study on Mechanism of Time-Dependent Deformation
     and Failure of Concrete, in Review of the Thirty-First General Meeting, The
     Cement Association of Japan, Tokyo, 1977, pp. 145-146.

     Cylindrical specimens were subjected to constant sustained loads at
     various stress levels. The relationship between stress level (σ_S) and
     time-to-failure (t_R) was found to be

     $$\log \sigma_S = -0.014 \log t_R + 1.98$$

     where σ_S is given as a percentage of the ultimate short-term stress, and t_R
     is in minutes.

189 Y. Fujita, N. Saeki, N. Takada, and H. Nara, On Properties of Cracking of
     Plain Concrete, in Review of the Thirty-First General Meeting, The Cement

Association of Japan, Tokyo, 1977, pp. 147-149.

Prisms 15 x 15 x 30 mm were broken in tension, and the fracture surfaces were then examined using an electron-probe microanalyser. It was found that:

i) Microcrack initiation in hcp due to drying shrinkage occurs primarily in the CH phase.

ii) Well-crystallized CH is also found at the interface between hcp and aggregate, where it is the phase in which cracks initiate.

190 K.P. George, Fracture Mechanics Approach to Cracking in Pavements, presented at the 2nd ASCE Engineering Mechanics Division Specialty Conference, Raleigh, North Carolina, May 23-25, 1977.

Cracking in pavements was modelled using a fracture mechanics approach. Elastic strain energy, surface energy, and energy expended in the "plastic" zone were considered, and this theory was then applied to a soil-cement slab. Experimental crack patterns agreed reasonably well with those predicted.

191 O.E. Gjørv, S.I. Sørensen and A. Arnesen, Notch Sensitivity and Fracture Toughness of Concrete, Cement and Concrete Research, 7 (1977) 333-344.

The notch sensitivity and fracture toughness of concrete made with different types and sizes of aggregate were investigated, using 50 x 50 x 550 mm SEN beams in 3-point bending, with various notch depths up to 25 mm. It was found that hcp was more notch sensitive than either mortar or concrete. Neither concrete nor mortar showed a constant K_c with varying notch depth, though hcp did. It was concluded that:

i) While the fracture of hcp may be governed by LEFM, as may the fracture of lightweight aggregate concrete, LEFM does not apply to small mortar or concrete specimens.

ii) Even small mortar and concrete specimens are notch sensitive.

192 P.J. Gustafsson, Brottmekaniska Studier; Lattbetong och Fiber-Armerad Betong. Rapport TVBM-5001, Dept. of Building Materials, The Lund Institute of Technology, 1977, 131 pp.

A condition for stable crack growth of brittle materials in 3-point bending was studied theoretically. Crack growth was shown to be stable when the total energy in the system divided by the fracture surface area was less than G_c. K_c and G_c were determined for nine different lightweight concretes, and were found to be about linearly related to density.

However, different methods of calculating G_c and K_c gave different results. One useful approach was to determine G_c as the area under the load-deflection curve divided by the fracture surface area. G_c was also determined for 2-dimensionally oriented steel fibre reinforced concrete: G_c was not proportional to the fibre volume. Work on glass fibres was also carried out.

193 N.M. Hawkins, A.N. Wyss and A.H. Mattock, Fracture Analysis of Cracking in Concrete Beams, Journal of the Structural Division, ASCE, 103 (1977) 1015-1030.

An energy balance criterion based on LEFM was used to predict loads at which rapid crack propagation would begin in concrete beams. An unstable crack would develop when the rate of energy released by continued crack growth exceeded the energy input to the system. G_c was determined directly, and was used to predict diagonal tension cracking. It was suggested that crack growth could be blunted by aggregate, and affected by adjacent cracks.

194 B. Hillemeier and H.K. Hilsdorf, Fracture Mechanics Studies on Concrete Compounds, Cement and Concrete Research, 7 (1977) 523-536.

The fracture parameters of hcp, aggregate, and the cement-aggregate interface were studied using wedge-loaded compact tension specimens. It was found that:

i)   hcp is notch sensitive.

ii) in hcp, K_c decreases with increasing crack length; a characteristic K_c occurs only after the crack has propagated a certain distance. The decrease of K_c with crack growth may be due to the formation of microcracks around the initial notch.

iii) K_c of the cement-quartz interface decreases very steeply with crack length, and levels off at a value of K_c about 1/3 that for hcp.

iv) behaviour of concrete in bending depends largely on the K_c of the interface.

195 A.A. Khrapkov, L.P. Trapesnikov, G.S. Geinats, V.I. Paschenko and A.P. Pak, The Application of Fracture Mechanics to the Investigation of Cracking in Massive Concrete Construction Elements of Dams, in Fracture 1977, Vol. 3, ICF4, Waterloo, Ontario, 1977, pp. 1211-1217.

LEFM was used to study cracking in large concrete dams. The object was to analyze the problems of thermal cracking. K_c data was obtained from the literature. It was found that thermal stresses achieved their maximum values just at the moment of crack closing. From their own experimental data, K_c was found to vary from 0.15 to 0.61 $MNm^{-3/2}$ as the maximum aggregate size increased from 20 to 40 mm.

196 S.C. Kim and H. Kitagawa, A Method of Determination of Mixed Mode Fracture Toughness of Brittle Materials Under Compression, in Fracture Mechanics and Technology, Vol. II, Proceedings of an International Conference, Hong Kong. Sijthoff and Noordhoff International Publishers, 1977, pp. 1011-1019.

The diametral compression test was applied to radially cracked discs. Stress intensity factors were obtained by approximate and photoelastic methods. It was concluded that the ratio K_I/K_{II} could be obtained by changing the angle between the line of action of the compressive loads and the crack line. The values of K_{II} at fracture were larger than the values relative to the first mode fracture toughness (for $K_{II} = 0$) predicted from the existing criteria.

197 M.D. Kotsovos and J.B. Newman, Behaviour of Concrete Under Multiaxial Stress, Journal of the American Concrete Institute, 74 (1977) 443-446.

It was postulated that, under complex stress states, the fracture mechanism of concrete is that of crack initiation, followed by stable crack growth which eventually becomes unstable. These different stages can be identified. Both upper and lower bound failure criteria were proposed.

198 A. Maher and D. Darwin, Microscopic Finite Element Model of Concrete, in Proceedings, First International Conference on Mathematical Modelling, Vol. III, St. Louis, Missouri, 1977, pp. 1705-1714.

A finite element model of concrete was developed to study the strength and deformation of plain concrete under uniaxial and biaxial compressive stress. Separate representations were employed for mortar, aggregate, and the mortar-aggregate interface. The interfacial bond strength was represented by a Mohr-Coulomb envelope. It was assumed that mortar cracks

initiated only in the presence of cracks at the mortar-aggregate interface. It was concluded that the pattern of crack propagation is similar, whether cracks are present at the interface before loading or are caused by loading.

199 B. Marchese, SEM Topography of Twin Fracture Surfaces of Alite Pastes 3 Years Old, Cement and Concrete Research, 7 (1977) 9-18.

The SEM was used to study the fracture surfaces of 3 year old C_3S pastes. It was found that the fracture path was transparticle, even in the wide gel areas. In lamellar CH areas, the fracture went along the cleavage planes. There was some evidence of water attack of the CH masses, but not of the C-S-H masses. There apeared to be strong adhesion between the C-S-H and CH, including perhaps some chemical bonding.

200 J. Mazars, Existence of Critical Strain Energy Release Rate for Concrete, in Fracture 1977, Vol. 3, ICF4, Waterloo, 1977, pp. 1205-1209.

The Griffith criterion was applied to concrete, in an attempt to show that G_c is a specific concrete parameter. Tests were carried out on 600 x 340 x 80 mm plates with different crack widths, using both central cracks and single-edge cracks. Also, SEN beam tests were carried out on specimens 150 x 100 x 1000 mm. It was found that $G_c \cong 10.2$ J/m^2; it appeared to be independent of test method and crack geometry.

201 H. Mihashi and M. Izumi, A Stochastic Theory for Concrete Fracture, Cement and Concrete Research, 7 (1977) 411-422.

A stochastic theory for concrete fracture was proposed, in which the concrete fracture is considered as a series of crack propagation processes within the hcp. The theory also takes into consideration the statistical nature of the stress distribution as a result of material defects. The theory can be used to explain the dependence of strength upon temperature, strain rate, and specimen size.

202 S. Mindess, F.V. Lawrence, Jr., and C.E. Kesler, The J-Integral as a Fracture Criterion for Fibre Reinforced Concrete, Cement and Concrete Research, 7 (1977) 731-742.

SEN beams (3x3x14 inches) were tested in 4-point bending, using hcp, plain concrete, and eight different steel and glass fibre reinforced concretes. K_c, G_c and J_c were evaluated. It was found that fibre additions of less than 0.75% by volume did little to improve the fracture behaviour. J_c could be correlated with the fibre volume at higher fibre contents, reflecting a change in material behaviour; K_c and G_c were insensitive to the fibre content. It was suggested that the J-integral might be an appropriate fracture criterion for concrete.

203 S. Mindess and J.S. Nadeau, Effect of Loading Rate on the Flexural Strength of Cement and Mortar, American Ceramic Society Bulletin, 56 (1977) 429-430.

Both hcp and mortar were tested as a function of the rate of loading (over about 6 orders of magnitude in $\dot{\varepsilon}$) to see if the same crack growth

parameter, n, (the slope of the V-K_I plot) could be obtained from rate of
loading tests as it could be from double torsion tests. It was found that
the n-values obtained were much lower than those obtained for concrete by
other authors. There was good agreement between the rate-of-loading and
double torsion tests for the mortar, but not for the hcp. It was concluded
that hcp and mortar were more susceptible to subcritical crack growth than
concrete. In addition, to achieve very high strain rates, some specimens
were tested using an instrumented impact machine.

204 S. Nagamatsu and Y. Sato, Experimental Studies of Creep Rupture and
Deformation of Cement Mortar Under High Structural Compressive Load, in
Review of the Thirty-First General Meeting, The Cement Association of
Japan, Tokyo, 1977, pp. 196-198.

Small prismatic specimens with an inclined center slit were loaded in
compression, and cracking was observed using OM; creep strain was also
measured. At 50x magnification, microcracks were first noted at from 30 -
70% of σ_{ult}. Under sustained loads, it was found that at low loads,
microcracks began to grow, but then stabilized; at higher loads, they
continued to propagate until failure occurred.

205 K. Okada, W. Koyanagi and K. Rokugo, Energy Transformation in the Fracture
Process of Concrete, in Memoirs of the Faculty of Engineering, Kyoto
University, Kyoto, Japan, 34, Part 3 (1977) 389-402.

Acoustic emission was used to study cracking of concrete in
compression, and in the 3-point loading of SEN beams. Both reversible and
irreversible energies in the fracture process were calculated. The
irreversible energy was assumed to consist of that consumed by crack
formation, and that consumed by viscous friction. Crack growth in the SEN
beams was proportional to energy dissipation in the beam. It was found
that energy dissipation in dry beams was greater than that in wet beams.

206 A.P. Pak, L.P. Trapesnikov, T.P. Sherstobitova and E.N. Yakovleva,
Experimental and Analytical Determination of the Critical Length of Crack
in Concrete (in Russian). Inzvestiya VNIIG im. B.E. Vedeneeva, 116 (1977)
50-54.

It is assumed that under certain conditions the critical crack is a
material (or specimen) constant. In some cases, K_c was found to be
proportional to the nominal strength. Tests for determining the critical
crack size in concrete were carried out using a split cyclinder test, using
cylinders 0.4×0.4 m, with a vertical central notch varying in length from
20 to 150 mm. Aggregate sizes ranged from 20 to 60 mm. The correlation

600

between the critical crack size and the specimen size was obtained from the theory of elasticity. A pseudo-plastic zone at the crack tip was assumed, but slow crack growth was not considered. It was found that if the crack size is more than twice the aggregate size, LEFM can be applied; the critical crack and maximum aggregate size are linearly related.

207 P.C. Strange, The Fracture of Plain and Fibre Concrete, Ph.D. Thesis, Department of Civil Engineering, University of Auckland, New Zealand, 1977.

The initiation and propagation of cracks in plain and fibre reinforced concrete were studied, with regard to specimen dimension and the role of the aggregate. SEN beams of sizes ranging from 6 x 12 x 60 mm to 100 x 200 x 1000 mm were tested in 3- point bending. For plain concrete, it was observed that K_c increased with increasing specimen size, and with increasing maximum aggregate size. Values ranged from ~0.5-1.1 MNm$^{-3/2}$. Slow crack growth was also noted, beginning at about 70% of the ultimate load. It was concluded that aggregate particles can cause the cracks to form, and can also impede their extension. For fibre concrete, it was found that fibres had no influence on the stress at which a pre-formed notch began to extend.

208 P. Stroeven, Structural Loosening of Plain Concrete in Uniaxial Compression, in Proceedings of the First National Conference on Mechanics and Technology of Composite Materials, Varna, 1976; Bulgarian Academy of Sciences, Sofia, 1977, pp. 627-632.

Stereomicroscopic observations at 5-10x were used to describe the cracking mechanisms of concrete in uniaxial compression. A dense structure of microcracks was found, particularly around aggregate particles. Cone-shapes "poles" were also observed around the aggregate particles. The cracks eventually joined and became unstable under increasing load. A model for crack formation in concrete under uniaxial compression was developed, similar to that proposed by Vile [Ref. 41].

209 R.N. Swamy and C.V.S.K. Rao, Toughness and Ductility of Fibre Reinforced Concrete Composites in Flexure, in Fibre Reinforced Materials, Institution of Civil Engineers, London, 1977, pp. 77-86.

The post-cracking behaviour of frc was studied in flexure on SEN beams (100 x 100 x 500 mm) in 4-point loading. Steel, polypropylene and glass fibres were used. Three stages in the σ-ε behaviour were defined: (i) a linear region, in which the energy goes in part into initiating cracking; (ii) in the second region, up to the maximum load, energy is used up in sub-critical crack growth; (iii) the descending branch, in which there is energy absorbed in unstable crack growth. A method of analysis was suggested to evaluate the energy absorption capacity of concrete in terms of the maximum load, specimen compliance, and deflection. The total energy absorption capacity increased with increasing fibre content. The order of effectiveness of the different fibres was steel > glass > polypropylene.

210 R.B. Testa and N. Stubbs, Bond Failure and Inelastic Response of Concrete, Journal of the Engineering Mechanics Divsion, ASCE, 103 (1977) 296-310.

A model was proposed to study the effect of matrix-aggregate bond failure on the σ-ε response of concrete. A tensile crack extension criterion based on LEFM was used to predict the stress at the interface for progressive bond separation. It was assumed that cracks would be arrested by branching into the mortar. The model predicts that:

i) Decreasing the size of the inclusions is the same as increasing the bond strength.

ii) Increasing only the bond strength increases the stress at which crack initiation begins.

iii) Increasing only the mortar strength leads to a larger area of debonding.

211 P.L. Walton and A.J. Majumdar, Fracture Energy of Plain and Glass-Reinforced Gypsum Plaster, Journal of Materials Science, 12 (1977) 831-836.

Glass fibre-reinforced gypsum boards were cast using a spray-suction technique. The specimens tested had about 5% by volume of 32 mm long fibres in a random 2-dimensional distribution; the specimen dimensions were 10 mm x 17 mm x 130 mm and 18 mm x 17 mm x 130 mm SEN beams, tested in center-point loading. Plain mortar specimens had notch/depth ratios ranging from 0.1 to 0.8; the fibre reinforced specimens all had a notch/depth ratio of 0.43. The work of fracture was calculated from the area under the σ-ε curve. It was found that both G_c and the work of fracture were approximately constant for the plain gypsum in the range of notch/depth ratios from about 0.25 to 0.60; at shallower notch depths, the values were higher, and for deeper notches the values were lower. G_c decreased with increasing water/plaster ratios. The addition of the glass fibres increased G_c by a factor of about 4, and the work of fracture by several orders of magnitude. The apparent "critical crack length" increased from about 0.33 to 0.93 mm. Charpy impact tests gave somewhat higher work of fracture values.

602

212 F.H. Wittmann, Grundlagen eines Modells zue Beschreibung charakteristischer
     Eigenschaften des Betons, Schriftenreihe Deutscher Ausschuss für
     Stahlbeton, 290 (1977) 43-101.

     A new model for the microstructure of hardened cement paste is
     developed. In doing so simplifying assumptions have to be introduced.
     Conventional models usually provide qualitative indications, such as creep
     increases, as the temperature rises. On the basis of the new model
     suggested in this report, however, it is possible to relate the macroscopic
     behaviour of concrete to its physical origin, i.e. the corresponding
     processes in the microstructure of hardened cement paste, in a quantitative
     way.

213 Ju. V. Zaitsev and E.N. Scerbakov, Creep Fracture of Concrete in
     Prestressed Concrete Members During Manufacture, in Fracture 1977, Vol. 3,
     ICF4, Waterloo (Canada), 1977, pp. 1214-1222.

     Fracture mechanics principles (based on probabilistic concepts) were
     used to try to analyze cracking in concrete subjected to variable sustained
     load, as in prestresed concrete members. A function for the total crack
     length was developed, and a function for the total time to failure was also
     presented, based on the concept of a measure of the amount of material
     destruction that had taken place.

214 Ju. V. Zaitsev and F.H. Wittmann, Crack Propagation in a Two-Phase Material
     Such as Concrete, in Fracture 1977, Vol. 3, ICF4, Waterloo, (Canada), pp.
     1197-1203.

     An attempt is made to simulate crack growth in concrete using the
     Monte Carlo method. This is done by randomly placing inclusions in a
     matrix, and modelling the system so that each "aggregate" particle has one
     interfacial bond crack. As the load is increased, these cracks grow until
     a composite crack runs through the material, with a slight overall inclina-
     tion. The same technique can also be used to analyze concrete under high
     sustained load or impact load.

                                    1978

215 P.W.R. Beaumont and J.C. Aleszka, Cracking and Toughening of Concrete and
     Polymer-Concrete Dispersed with Short Steel Wires, Journal of Materials
     Science, 13 (1978) 1749-1760.

     The cracking stress and the fracture toughness of steel frc and PMMA
     impregnated frc were measured. For the fracture tests, SEN beams (25 x 25
     x 125 mm, notched to a depth of 5 mm) were used. The fracture surface

energy (2γ) and work of fracture were also determined.

It was found that for plain concrete, $2\gamma=1.75$ J/m^2; for PMMA impregnated concrete, $2\gamma=20$ J/m^2. K_c increased from 0.5 to 1.5 MNm$^{-3/2}$ as the PMMA content increased from 0 to 18% by volume. The work of fracture was explained in terms of the work done in overcoming friction during the fibre pullout. The greater toughness of the PMMA concrete was due to an increase in interfacial frictional shear stress.

216 G.Yu. Berdichevskii and V.L. Chernyavskii, On Modelling of Cracking in Concrete Constructions (in Russian). Materialy Konf. i Soveschanii po Gidrotkhnike. Energya, Leningrad, 119 (1978) 83-97.

About 15 gypsum paste and gypsum-lime-sand mortar prismatic dog-bone shaped specimens (about 40x40x240 mm) were tested in eccentric compression. The preformed crack in the central part of the specimen was from 0.1 to 0.75 of the cross-sectional dimension. The K_{IC} values obtained ranged from 14 to 280 kNm$^{-1.5}$, with 50% differences observed between K_{IC} at the initiation of crack growth and at further growth. The authors are unaware of the reasons for the differences in formulae for K_I calculations cited in various sources.

217 A. Carpinteri, A. Di Tommaso and E. Viola, Stato Limite Di Frattura Nei Materiali Fragili, Giornale del Genio Civile (Italy), (1978) 201-224.

An attempt was made to provide a theoretical justification for the experimental results obtained for the fracture of rock and concrete. Fracture mechanics was used to provide a failure model for these materials, and the analysis was used to explain some apparent discrepancies in experimental results.

218 D.J. Cook and G.D. Crookham, Fracture Toughness Measurements of Polymer Concretes, Magazine of Concrete Research, 30 (1978) 205-214.

The fracture toughness of polymer concretes was measured using 100 x 100 x 500 mm SEN beams in 4-point loading. It was found that premix polymer additives reduced K_c, while polymer impregnation increased K_c by a factor of about three. PIC was a notch sensitive material. Compliance techniques for monitoring crack length were found to be useful. Cast notches gave somewhat different results from natural cracks. K_c increased with crack growth to a limiting value at a crack length of about 35-40 mm.

219 D.J. Cook and G. Crookham, A Discussion of the Paper Notch Sensitivity and Fracture Toughness of Concrete, by O.E. Gjørv, S.I. Sørensen and A. Arnesen (CCR 7, 333, (1977), Cement and Concrete Research, 8 (1978) 387-388.

In their discussion, Cook and Crookham confirm that concrete is notch sensitive, based on their tests of 100 x 100 x 500 mm SEN specimens in 4-point loading; notch sensitivity was greater for larger specimens. In addition, slow crack growth occurs before failure. They argued that LEFM may be used with small specimens as long as slow crack growth is accounted for, and if the notch depth is not very large.

220 D. Darwin, Discussion of Bond Failure and Inelastic Response of Concrete, by R.B. Testa and N. Stubbs (J. Eng. Mech. Div., ASCE, 103 (1977) 246-310), Journal of the Engineering Mechanics Division, ASCE, 104 (1978) 507-509.

In a discussion of the work of Testa and Stubbs (ref. 210), it is argued that the emphasis on bond failure between coarse aggregate and mortar to explain the inelastic behaviour of concrete may be incorrect. Instead, it is suggested that hcp is itself a highly nonlinear material, and that it is the nonlinearity of the hcp that is the most important factor in controlling the axial σ-ϵ behaviour of concrete.

221 I.A. Dikovskii, On Fracture Toughness of Concrete (in Russian). Rabotosposobnost' Stroit. Materialov v Usloviakh Vozd. Razlichnykh Expluat. Fakturov. Mezhvuzovsky Sbornik, Kazan, K.Kh.T.I. in S.M. Kirova, 1 (1978) 17-18.

SEN beams, 100x100x310 mm, made of keramzite and of normal aggregate were loaded in 3-point bending. K_{IC} was found to be about 0.3 $MNm^{-1.5}$.

222 A. Hillerborg, A Model for Fracture Analysis, Report TVBM-3005, Division of Building Materials, Lund Institute of Technology, Lund, Sweden, 1978. An account of the fictitious crack model for fracture is presented.

In this model, it is assumed that concrete has a linear σ-ϵ curve, which is valid only till some limiting strain (or stress) is reached. Beyond this, a fracture zone begins to develop, and additional deformation is then confined to the fracture zone. This gives rise to a σ- crack width (σ-w) relationship. The area under the σ-w curve is considered to be equal to G_c. It is necessary to assume the exact shape of the σ-w curve, and this can affect the results. This model was then applied to the bending of uncracked beams, and crack growth.

223 M. Kawamura, Internal Stress and Microcrack Formation Caused by Drying in Hardened Cement Pastes, Journal of the American Ceramic Society, 61 (1978) 281-283.

It was shown (by theoretical calculations) that $Ca(OH)_2$ particles in hcp may be under high triaxial compressive stresses at the time of drying at a relatively high r.h. These stresses can be relaxed by plastic or viscoelastic strains, or by microcracking. The calculated relatively high tensile and shear stresses around these dense particles are likely to produce microcracks.

224 A. Kelly, Fibre Reinforcement of Brittle Matrixes, in G. Piatti (ed.), Proceedings of a Seminar on Advanced Composite Materials, 1976, Applied Science, Barking, England, 1978, pp. 113-129.

Fracture mechanics and crack motion between obstacles were studied both theoretically and experimentally. Crack control in cement using fibre reinforcement was also studied.

225 G.V. Marchukaitis, The Calculation of Strength Increase of the Polymer Impregnated Concrete (in Russian). Rabotosposobnost' Stroit. Materialov v Usloviakh Vozd. Razlichnykh Expluataatsionnykh Faktorov. Mezhuvuzovsky Sbornik, Kazan, 1 (1978) 82-85.

The degree of strength increase due to polymer impregnation was calculated by dividing two Griffith formulae specified respectively for impregnated and non-impregnated concrete. The formulae for E of the impregnated material included the cement, water and air contents, the degree of hydration, and the strength and elastic modulus of plain concrete. Some formulae were also taken from the theory of porous bodies. The agreement between analytical and experimental data was in the range ±15%. It was concluded that impregnation of concrete to less than 30% of the pore volume is uneconomical.

226 K. Mizutani, An Indirect Tension Test with Applications to Fracture of Brittle Materials, Ph.D. Thesis, Purdue University, West Lafayette, Indiana, 1978.

An indirect tensile test was developed for fracture studies of, amongst other materials, hcp. The test specimen consisted of a tension ring, with a central ligament 0.25 x 1.25 x 11 inches (i.e. a modified theta specimen). It was shown that:
    i)   hcp is notch sensitive;
    ii) fracture toughness increased with decreasing w/c ratio, and with increasing age; and
    iii) γ_F was greater for wet than for dry samples.

227 K. Morita and K. Kato, Fundamental Study on Fracture Toughness and Evaluation by Acoustic Emission Technique of Concrete, in Review of the Thirty-Second General Meeting, The Cement Association of Japan, Tokyo,

606

1978, pp. 138-139.

An attempt was made to determine the relationship between the fracture
strength and fracture toughness of concrete, using acoustic emission
techniques. SEN beams (100 x 100 x 420 mm) with notches of varying depths
and widths were tested in 3-point bending. It was found that:

    i) K_c was independent of notch width.

    ii) K_c was essentially independent of notch depth.

    iii) The material was notch insensitive for notch/depth ratios less than
about 0.04.

    iv) The acoustic emissions showed a rapid increase at about 94% of the
ultimate load.

228 A. Muñoz-Escalona and C. Ramos, Fracture Morphology and Mechanical
Properties of Thermocatalytically Polymerized MMA-Impregnated Mortar,
Journal of Materials Science, 13 (1978) 301-310.

Various mortar mixes were impregnated with MMA. The fracture surfaces
were studied using the SEM. It was found that the polymer acted in two
ways:

    i) it filled the pores and microcracks; and ii) it formed an anisotropic
irregular network, improving the bond between the matrix and the aggregate.
Fracture occured both around and through C-S-H.

It was claimed that the MMA inhibited fracture initiation and
propagation.

229 K. Nishioka, S. Yamakawa, K. Hirakawa, and S. Akihama, Test Method for the
Evaluation of the Fracture Toughness of Steel Fibre Reinforced Concrete, in
Testing and Test Methods of Fibre Cement Composites, RILEM Symposium 1978,
The Construction Press, Lancaster, England, 1978, pp. 87-98.

The fracture toughness of steel frc was evaluated using SEN plate bend
specimens in 3- and 4-point bending. Several sizes were used: 60 x 60 x
530 or 260 mm and 150 x 150 x 530 or 260 mm. Fibre volumes ranged from 0-
2%. COD measurements were made. K_c was determined on the basis of maximum
load, ignoring slow crack growth. It was concluded that the ASTM method
for fracture testing of metals is also suitable for concrete. K_c of the
frc was 2-3 times greater than K_c for plain concrete.

230 T. Ohigashi, Measurement of Effective Fracture Energy of Glass Fibre
Reinforced Cement, in Testing and Test Methods of Fibre Cement Composites,
RILEM Symposium 1978, The Construction Press, Lancaster, England, 1978, pp.
67-78.

The effective fracture energy (γ) of glass fibre reinforced cement

(GRC) was studied, using 3-point bending on beam specimens with notches of different depths on both the top and the bottom faces. For plain mortar, the fracture energy decreased with increasing notch depth. For different methods of measuring γ, the following results were obtained:

$$3\text{-point bending} \qquad : \gamma \cong 5.0 \text{ KJ}/m^2$$
$$\text{Izod (impact) tests} \qquad : \gamma \cong 6.5 \text{ KJ}/m^2$$
$$\text{pull-out tests} \qquad : \gamma \cong 4.4 \text{ KJ}/m^2$$

231 A.P. Pak, L.P. Trapesnikov, T.P. Sherstobitova and E.N. Yakovleva, The Verification of the Hypothesis of Generalized Normal Fracture for the Sand-Cement Specimens (in Russian). Materialy Konf. i Soveschanii po Gidrotekhnike, Energya, Leningrad, 119 (1978) 66-70.

Prismatic specimens 50×50×100 mm, with through thickness central notches of lengths 2c = 4-30 mm, inclined to the horizontal by an angle φ varying from 0° to 90° were tested in direct tension, the load being applied through steel end plates glued with epoxy. A formula was proposed for the notched specimen strength as a function of K_{IC}, $\lambda = K_{II}/K_I$, and some geometrical parameters. It was found that this formula applied for $2c \geqslant 15$ mm. The limiting values for φ were also determined.

232 A.P. Pak, L.P. Trapesnikov and E.N. Yakovleva, On the Determination of the Critical Crack Length for Concrete (in Russian). Izvestyia VNIIG im. B.E. Vedeneeva, 120 (1978) 26-29.

Continuing the work reported in ref. 206, this investigation was an attempt to relate the critical crack size and the maximum aggregate size and their dependency on specimen shape. The new specimens were paddle-shaped, with a length of 1.7 m and a central cross- section of 0.2 x 0.2 m. The specimens, with a maximum aggregate size of 20 mm, had SEN type notches 20 or 40 mm long. The conclusions of ref. 206 were verified.

233 E.N. Peresypkin, Stress Intensity Factors and Crack Opening in Reinforced Concrete Elements (in Russian). Beton i Zhelezobeton, 12(1978) 27-34.

A rectangular reinforced concrete beam was loaded in combined tension and bending. In the tensile zone, there were through thickness vertical cracks of regular spacing. Besides the force and moment equilibrium equations, there was an additional relation between stresses in the beam and deflections, calculated using the influence functions for a rectangular elastic area loaded on its edge. The stresses at the crack front were assumed to be in the form of Jacobi polynomials, with a factor added to take into account the singularity at the crack tip. A relation was obtained for K_I due to the bending moment and the normal force. A

608

numerical example was presented.

234   E.N. Peresypkin and L.P. Trapesnikov, Stress Intensity Factors in Cracked
      Reinforced Concrete Elements (in Russian). Izvestiya VNIIG im. B.E.
      Vedeneeva, 121 (1978) 13-18.

      The problem of K_I in a rectangular reinforced concrete beam having
      periodically distributed cracks in the tensile zone was considered using
      the analysis presented in ref. 233. However, in the present work, the
      influence functions for a rectangular elastic area loaded on its edge were
      obtained experimentally using models of organic glass. The bond stress
      distribution was approximated by a linear function. Calculated values of
      K_I were given for various values of the degree of reinforcement, crack
      spacing, and crack length; an empirical coefficient related to the type of
      concrete and type of reinforcement was also introduced.

235   C. Sok, Etude de la Propagation d'une Fissure dans un Béton non Armé, Bull.
      Liaison Lab. Ponts Chaussees, 98 (1978) 73-84.

      Crack growth was studied in large double cantilever beams, constructed
      so that the rate of crack growth was constant. The DCB specimens permitted
      1.8 m of crack growth; longitudinal prestressing was used to keep the
      crack from deviating. Acoustic emission was used to help monitor cracking
      in the specimen. R-curve analysis was used to describe the results. It
      was concluded that for concrete, large specimens are needed, as values of
      K_c increased with crack length till about 500 mm of crack growth was
      observed, after which K_c stabilized. Using acoustic emission, it was
      concluded that a damage zone existed 200-500 mm ahead of the apparent end
      of the crack. Therefore, very small specimens cannot be used for fracture
      mechanics studies on concrete.

236   N.H. Stein, De Sterkte van Verhard Cement, (in Dutch) Kleien Keramiek,
      28 (1978) 22-33.

      This work represents a review of the fundamental processes in the
      fracture of hcp, dealing with evidence concerning whether the hydrate
      crystals break or are merely separated during fracture.

237   S.E. Swartz, K-K. Hu and G.L. Jones, Compliance Monitoring of Crack Growth
      in Concrete, Journal of the Engineering Mechanics Division, ASCE, 104
      (1978) 789-800.

      It was found possible to monitor crack growth in plain concrete beams
      subjected to repeated loads by compliance measurements. COD measurements

were also made. The failure mechanism was found to be a combination of aggregate fracture and bond failure. It was concluded that fracture toughness is not a fundamental material property. However, an "effective" K_C may be significant if related to aggregate size, gradation, and content.

238 K.L. Watson, The Estimation of Fracture Surface Energy as a Measure of the "Toughness" of Hardened Cement Paste, Cement and Concrete Research, 8 (1978) 651-656.

The fracture surface energy of hcp was obtained from the work of fracture method on beams (20 x 20 x 250 mm) with different notch geometries. It was suggested that this could be best achieved by using deep rectangular notches and a large beam span/depth ratio. For these experiments, a notch/depth ratio of 0.62-0.75 and a span/depth ratio of 10.5 were found to be suitable. γ_F was found to be 3.5 J/m^2 at 2 days, rising to 3.8 J/m^2 at 10 days.

239 B. Zech and F.H. Wittmann, A Complex Study on the Reliability Assessment of the Containment of a PWR, Part II: Probabilistic Approach to Describe the Behaviour of Materials, Transactions of the 4th International Conference on Structural Mechanics in Reactor Technology, San Francisco, California, 15-19 August 1977, Vol. J(a), Paper J 1/11 (published also in: Nuclear Engineering and Design, 48 (1978) 575-584).

The intention of this paper is to contribute to the development of methods to be used for the quantification of the risk of nuclear power plants. For this purpose a reliability analysis of a structural component, i.e. a reactor containment structure is carried out. Detailed information in various fields had to be developed and compiled.

1979

240 N. McN. Alford and A.B. Poole, The Effect of Shape and Surface Texture on the Fracture Toughness of Mortars, Cement and Concrete Research, 9 (1979) 583-589.

Aggregate shape and texture were expressed in terms of an "angularity factor" for 17 different aggregates. SEN beams (1 x 1 x 6 inches) were tested in flexure. It was found that G_c did not depend on the porosity, density, or water absorption of the aggregates. However, in general, G_c increased with increasing angularity factor, for several reasons: (i) angular aggregates are better crack arrestors; (ii) angular aggregates cause more microcracking; (iii) angular aggregates have better bond to the cement.

241 R. Andonian, Y.W. Mai and B. Cotterell, Strength and Fracture Properties of
     Cellulose Fibre Reinforced Cement Composites, International Journal of
     Cement Composites, 1 (1979) 151-158.

     The strength and fracture characteristics of random cellulose fibre
     reinforced cement composites, with fibre mass fractions in the range 2-10%,
     were studied. E, tensile and flexural strengths, and the specific work of
     fracture (R) were calculated. The fracture tests were carried out on
     specimens 150x20x8 mm. The origin of fracture toughness was found to come
     mainly from fibre pull-out; the predicted R-values seemed to compare well
     with the experimental results.

242 Anonymous, Study on the Cracking Strength and Maximum Crack Width in
     Reinforced Concrete Members, Journal of the Dalian Institute of Technology
     [China], (1979) 67-80.

     The existing methods for calculating the cracking strength and crack
     width in reinforced concrete members are presented. Equations for
     calculating the cracking strength are proposed.

243 A.N. Bakhtibaev, V.I. Betekhtin, A.D. Kadyrbekov and V.R. Regel, Time and
     Temperature Dependences of the Strength of Cement With Contacts of
     Crystallization or Coagulation Type, Fiz. Prochn. Kompoz. Mater., [Mater.
     Vses. Semin.] 3rd, (1979) 112-117. Chemical Abstracts, 94: 126389, 1981.

     The durability of hcp with w/c = 0.33 was determined by sustained
     tensile loading, for cements with crystallization-type and coagulation-type
     bonds. The fracture activation energies for these two types of cements
     were 95 kcal/mol and 24 kcal/mol, respectively. The lower fracture
     activation energy for the coagulation-type bonds was attributed to the
     hydrolytic rupture of the Si-O bonds.

244 B.D. Barnes, S. Diamond and W.L. Dolch, Micromorphology of the Interfacial
     Zone Around Aggregates in Portland Cement Mortar, Journal of the American
     Ceramic Society, 62 (1979) 21-24.

     The interface between cement and Ottawa (silica) sand was examined
     using the SEM. It was observed that:
     i)   The fracture surface produced in dried mortar specimens followed the
     outline of the sand grains, which were never cleaved; the fracture path
     was at or near the interface.
     ii) A duplex film about 1 μm thick formed at the interface. The bottom
     of the film was a layer of CH, overlain by a layer of C-S-H.

iii) Large CH crystals formed in the interfacial zone.

245   Z.P. Bazant and L. Cedolin, Blunt Crack Band Propagation in Finite Element
      Analysis, Journal of the Engineering Mechanics Division, ASCE, 105 (1979)
      297-315.

      Crack propagation in concrete is modelled using FEM, combined with a
      distributed ("smeared") crack model within the finite elements. It is
      argued that a blunt (or smeared) crack front is a more realistic assumption
      than is a sharp crack for concrete.

246   Z.P. Bazant and A.B. Wahab, Instability and Spacing of Cooling or Shrinkage
      Cracks, Journal of the Engineering Mechanics Division, ASCE, 105 (1979)
      873-889.

      The propagation of cooling cracks and shrinkage cracks was considered.
      FEM techniques were used to determine the critical states at which first
      every other crack is arrested, and then is closed. It was found that the
      crack depth-to-spacing ratio at which the critical state is reached is very
      sensitive to the temperature profile.

247   S. Benouniche, A Model for the Compressive Microcracking Damage of
      Concrete, These 3ème Cycle, University Paris VI, December, 1979 (in
      French).

      When the damage theory is used to describe the cracking and failure of
      concrete under tensile stress, the damage parameter is a scalar. Under
      compressive stress, because of the direction of the main cracks, an aniso-
      tropic three-dimensional formulation must be used, with the aid of a damage
      tensor. Some results in support of this formulation are presented.

248   R.C. Bradt, Fracture of Refractory Concretes, in Proceedings of the 4th
      Annual Conference on Materials for Coal Conversion and Utilization, 1979,
      pp. IV/39-IV/41.

      The fracture of castable refractory concrete being considered for the
      lining of coal gasifiers was studied, in the temperature range of 25-
      1350°C. Work on determining the slow crack growth parameters was
      discussed.

249   A.M. Brandt and J. Kasperkiewicz, Crack Propagation Energy in SFRC (Steel
      Fibre Reinforced Concrete), in Seminar on Fracture Mechanics of Fibre Rein-
      forced Cement-Based Composites, Delft, 1979, pp. D1-3].

      A review is given of the basic notions of linear and non-linear
      fracture mechanics, particularly as applied to the fracture testing of

concrete-like composites. Experimental data (as reported in ref. 296) are presented, and recommendations for future research are given. For sfrc elements the analysis should cover the whole fracture process, and should not be limited to various limit state parameters.

250 A.R. Bunsell, The Nature of Crack Growth in Composite Materials, in G.C. Sih and V.P. Tamuzs (eds.) Fracture of Composite Materials, Proceedings of the First USA-USSR Symposium, Riga, 1978, Sijthoff and Noordhoff, the Netherlands, 1979, pp. 349-359.

Slow crack growth in asbestos-cement was studied by examining the load-deflection curve obtained from the testing of compact tension specimens. Acoustic emission was also used, to identify four stages of crack propagation: (1) creation of a zone of microcracks ahead of the notch; (2) stable crack propagation after this zone has reached a certain size; (3) increase in resistance to cracking due to straining and pollutant fibres; (4) failure after a maximum length of the crack is bridged.

The stress intensity factor required to cause propagation, $K = K_o + K_r$, where K_o is the stress intensity factor of the matrix in the presence of fibres, and K_r is the inhibiting factor due to the fibres. It is concluded that LEFM can be used to characterize fracture in asbestos cement.

251 R.S.P. Coutts, Wood Fibre Reinforced Cement Composites, in Institute of Industrial Technology Research Review, C.S.I.R.O., Division of Chemical Technology, Australia, 1979, pp. 1-16.

Up to 12% by weight of wood fibres were incorporated into concrete. It was found that:

i) The fracture energy of the composites decreased with time of curing.

ii) The impact toughness of the composites could be considerably increased by vacuum dewatering, reaching a maximum at a fibre weight percent of about 8%.

iii) SEM examination indicated that fibre pullout was the main fracture mechanism. Surface treatment of the fibres using various coupling agents, however, was able to bring about a failure mode in which many of the fibres fractured.

252 R.S.P. Coutts and M.D. Campbell, Coupling Agents in Wood Fibre-Reinforced Cement Composites, Composites, 10 (1979) 228-232.

The effects of surface treatment of discrete wood fibres on the fracture energy and flexural strength of wood-cement composites were measured. A variety of coupling agents were used: titanium alkoxides, titanates, silanes and dihalides. The specimens for impact toughness were in the form of thin plates, 150mm x 150mm x 7mm thick. Energy was measured using a G.E. Puncture tester, in terms of the energy required to go right through the sample. It was found that some coupling agents could increase the energy for fracture, though perhaps at the cost of strength. In general, the fracture energy decreased with time of curing, both for the controls and for the treated samples. However, the values obtained were higher than those for plain portland cement.

253 N. Djabarov, A. Wehrstedt, H.-J. Weiss and K. Yamboliev, Investigations on the Fracture Behaviour of Steel Fibre Reinforced Concrete, in Proceedings of the Second National Conference on Mechanics and Technology of Composite Materials, Varna 1979, Sofia, Bulgaria, 1979, pp. 593-596.

Steel frc flexural specimens, 40 x 40 x 160 mm, were tested in 3-point bending; acoustic emission was used to monitor cracking. It was found that with plain concrete, there was no acoustic warning before failure; with frc, there was considerable noise. The steel fibres greatly improved the fracture toughness. It was noted that the increase in K_c with "natural" cracks was greater than with sawn notches, since in the former case fibres bridge the gap. It appeared that matrix failure occurred at about the same load for both plain and fibre reinforced concrete; the function of the fibres was to inhibit the growth of the microcracks.

254 G.G. Garrett, H.M. Jennings and R.B. Tait, The Fatigue Hardening Behaviour of Cement-Based Materials, Journal of Materials Science, 14 (1979) 296-306.

The fracture morphology of cement-based materials under fatigue loading was studied using re-hydration techniques and the SEM. It was found that static fracture results in a largely intergranular fracture path, through the C-S-H. Fatigue produces extensive regions of exposed, unhydrated cement particles. These particles are apparently exposed by an attrition process subsequent to the passage of the main microcrack fronts.

Strength increases following fatigue loading seem to result from damage introduced in the form of microcracks in the early part of the fatigue loading. At high stress levels, this effect is seen only if the test is stopped within the first 20% of the fatigue life. Fatigue hardening requires the presence of free H_2O within the matrix; dried samples do not show this effect. The mechanism appears to involve the hydration of freshly exposed unhydrated surfaces. The observed strength increases result from a decrease in total porosity, and reductions in localized stress concentrations. Compressive fatigue produces microcracks which remain open even at the minimum of the load cycle.

255 K.P. George, Application of Fracture Mechanics to Crack Growth Damage in Pavements, in G.C. Sih and S.R. Valluri (eds.), International Conference on Fracture Mechanics in Engineering Applications, Bangalore, 1979, Sitjthoff, Aalphen aan den Rijn, 1979, pp. 849-860.

Assuming a plane state of stress, and using the energy balance during crack growth, a formula is derived for the crack growth in a pavement slab. The parameters are: E, surface energy, width of the plastic region, yield strength of the material, and the increment of stress due to thermal or shrinkage processes. Computer simulation results for cracking in a soil-cement slab show satisfactory agreement with the laboratory studies. A mechanistic fatigue design model is proposed on the basis of the Griffith criterion. The life of the slab is defined as the number of load cycles associated with crack growth to the surface. To apply the Paris law for fatigue, it is assumed that the size and distribution of starter flaws can be estimated by statistical analysis on specimens prepared in the laboratory or cut from the pavement. The pavement crack increment is presented as a function of K_I and K_{II}, and a formula for the fatigue life is proposed. The mechanics of crack growth in the surface overlay, caused primarily by the stresses and deformations in the base layer and/or by stresses in the overlay itself are investigated.

256 A. Grudemo, Microcracks, Fracture Mechanism and Strength of the Cement Paste Matrix, Cement and Concrete Research, 9 (1979) 19-34.

A review of the data on fracture strength, porosity, etc., for hcp is presented. There is then an attempt to relate strength, microstructure and fracture mechanics. The author argues that it is necessary to consider the arrangement of the different phases, the size and shape of the different particles and their grain boundaries, and the strengths of different bonds. Four possible crack paths are suggested: (i) lamellar cracks between CH particles; (ii) radial cracks through "outer" and "inner" gel envelopes; (iii) spherical cracks around anhydrous nuclei; and (iv) ruptured contact zones between adjoining outer gel coats. It is also possible that these zones may act as crack arrestors.

257 Y.M. de Haan, Basic Elements of Fracture Mechanics, in Seminar on Fracture Mechanics of Fibre Reinforced Cement-Based Composites, Delft, 1979, pp. A1-12.

Basic LEFM notions are presented, and an estimation is given of the orders of magnitudes of surface energy and K_{IC} for some components of concrete-like composites.

258 G.T. Halvorsen, Toughness of Portland Cement Concrete, Ph.D. Thesis,
University of Illinois, Urbana, 1979.

The J-integral was used to provide a quantitative measure of the
toughness of 98 different hcp, concrete, and steel frc mixes. Tests were
carried out on SEN beams of two sizes: 3 x 3 x 15 in. and 6 x 6 x 21 in.
It was found that the J-integral is dependent on specimen dimensions,
especially depth. In addition, steel frc is notch-insensitive for the
small 3 x 3 x 15 in. specimens tested. Aggregate increases toughness.
For frc, only above about 0.75 - 1.0% by volume of fibres did the J-
integral increase significantly. The J-integral increased with fibre
content, fibre length, aspect ratio, and end anchorage. It was concluded
that the J-integral might be useful in characterizing fracture in
cementitious materials.

259 A. Hillerborg, The Fictitious Crack Model and its Use in Numerical
Analyses, presented at the International Conference on Fracture Mechanics
in Engineering Applications, Bangalore, India, 1979.

The fictitious crack model was used to characterize the σ-ϵ
relationships in the fracture zone of concrete. This model is particularly
suitable for FEM analysis. The whole fracture zone is assumed to act as a
"fictitious" crack, which has the ability to transfer stresses according to
the σ-crack width (σ-w) curve for the real fracture zone. It is necessary
to estimate (or assume) a particular shape of the σ-w curve.
An example of the use of the method is given. It is claimed that the
model can give the complete behaviour of a non-precracked structure: crack
formation, crack growth, and rupture.

260 L. Javan and B.L. Dury, Fracture Toughness of Fibre Reinforced Concrete,
Concrete, 13 (1979) 31-33.

The CNRBB test (circumferentially notched round bar under bending) was
used to determine the fracture toughness of cylinders 200 mm long, 100 mm
in diameter, with notches of different depths. Both steel and polypropylene fibres were used. It was found that cracks at the cement-aggregate
interface existed even before the specimens were loaded. Both K_c and
G_c were found to increase with increasing notch depth. Both parameters
increased with increasing fibre content; steel fibres were about twice as
effective as polypropylene fibres.

261 A. de S. Jayatilaka, Concrete, in Fracture of Engineering Brittle

616

Materials, Applied Science Publishers Ltd., London, 1979, pp. 281-328.

A review of the structure and mechanical properties of cement and concrete was presented. As a part of this, a brief review of fracture mechanics applications to cement and concrete was included.

262 K. Jujii, K. Haga and S. Fujii, A Study on Progressive Failure of Brittle Materials by Fracture Mechanics, Memoirs of the Faculty of Engineering, Tokushima University, 24 (1979) 1-11.

The fracture mechanism of aggregative materials was studied by the application of the technique of LEFM. Mortar specimens (w:c:s = 1:2:6) were tested to obtain K_{IC} and K_{IIC} values. Although K_{IC} values were influenced only a little by the width under three point loading, K_{IIC} values were about 30% decreased by the change of the width from 50 to 75 mm with a specimen size of 200 x 150 mm. $K_{IC} \simeq 56$ kg cm$^{-3/2}$ was obtained with the thicker specimens. Rectangular plates and discs made of mortar with a slant notch were also tested and the crack propagation was simulated by the finite element method.

263 O.A. Kayyali, C.L. Page and A.G.B. Ritchie, Frost Action on Immature Cement Paste - Effects on Mechanical Behaviour, Journal of the American Concrete Institute, 76 (1979) 1217-1225.

A limited number of freezing and thawing cycles were applied to hcp at early (<1 day) stages of hydration. After further normal curing, it was found that both the strength and G_C for non-air entrained pastes (measured on 100 x 100 x 200 mm SEN beams in 4-point bending) showed irreversible losses. The failure mode was explosive. For air-entrained pastes, both strength and G_C were slightly increased.

264 J.C. Lenain and A.R. Bunsell, The Resistance to Crack Growth of Asbestos Cement, Journal of Materials Science, 14 (1979) 321-332.

The crack resistance of asbestos cement was studied, using compact tension specimens (120 x 50 x 6.3 mm), with 5.4% by volume of asbestos. Acoustic emission was recorded. Three stages of crack growth were distinguished:

i) creation of a zone of microcracks in front of the main crack within the elastic region of the σ-ε curve; the size of this zone was 12-28 mm.

ii) growth of this microcracked zone, with slow, stable crack growth, which became unstable when the microcracked zone reached a size of 28 mm.

iii) propagation of the crack with the size of the microcracked region remaining constant. The fibres which bridge the microcracks induce a closing stress at crack tips, and so increase the work of fracture. This phenomenon could be best explained using R-curve analysis.

K_C was found to be $\cong 1.7$ MNm$^{-3/2}$; acoustic emission was found to be proportional to K^6.

265 Y.W. Mai, Discussion of the Paper The Estimation of Fracture Surface Energy as a Measure of the "Toughness" of Hardened Cement Paste, by K.L.Watson (CCR, 8(1978) 651); Cement and Concrete Research, 9 (1979) 537-539.

It was suggested that Watson's data (ref. 238) on fracture surface energy be re-analyzed using a different procedure for calculating G_C. Some data of Shah and McGarry (ref. 72) was re-analyzed in this way. It was suggested that LEFM might be used to get useful G_C values with the proposed technique. A quasi-static crack propagation technique was also suggested.

266 Y.W. Mai, Strength and Fracture Properties of Asbestos-Cement Mortar Composites, Journal of Materials Science, 14 (1979) 2091-2102.

The fracture of asbestos frc was studied, with fibre contents ranging from 5 - 20% by weight. Both grooved DCB specimens (180 x 76 x 6 mm) and SEN beams in 3-point bending (190 x 6 mm strips) were tested. The specimens were autoclaved. The results obtained on the two types of specimens were in general agreement. It was concluded that a modified law of mixtures was adequate to predict bending and tensile strengths. The specific work of fracture (R) increased with increasing crack length for the DCB specimens, due to fibres bridging the crack tips. K_C was also measured. The fibre pull-out mechanism contributed about 95% to the total work of fracture, and was the major source of fracture toughness.

Since up to the first crack the material had a linear σ-ε curve, LEFM could be used. For asbestos cements, J_c agreed quite well with $R_{initiation}$. Both K_c and R_i were approximately independent of the crack/depth ratios. The fracture parameters increased with increasing fibre content up to about 15%.

267 H. Mihashi, T. Sasaki and M. Izumi, Failure Process of Concrete: Crack Initiation and Propagation, in Proceedings of the Third International Conference on Mechanical Behaviour of Materials, Cambridge, Vol. 3, Pergamon Press, London, 1979, pp. 97-107.

A photoelastic coating technique was used to observe local strains caused by microcracking in model concretes to simulate either single pores or single aggregate particles. Fracture in uniaxial compression was considered as a series of crack processes created in the local tensile stress fields caused by material defects. Four stages in the failure process were identified: (i) bond cracking at the cement-aggregate interface; (ii) extension of these cracks into the mortar matrix, with cracks developing in the loading direction, or linking the weakest path along the interfaces; (iii) linking of bond cracks by oblique shear cracks; and (iv) rupture after the development of an unstable, dominant oblique crack.

268 S. Mindess, Application of Fracture Mechanics to Cement and Concrete, in Mechanics of Concrete-Like Composites, Proceedings of a Study Session at Jablonna, Poland, 1979, in press.

This work is a review (with 98 refs.) of the applications of fracture mechanics to cement and concrete. The phenomenological aspects of the fracture process are described in terms of the structure of hcp and of concrete. The difficulties in applying LEFM to cement and concrete are described, in particular the as yet unsolved problem of defining a valid specimen geometry. It is concluded that, at least for small specimens, LEFM does not provide an adequate fracture criterion for cement or concrete; however, non-linear fracture models may eventually be more useful.

269 S. Mindess, The Fracture of Cement and Concrete, in Proceedings of the Engineering Foundation Conference on Cement Production and Use, Rindge, N.H., 1979, pp. 175-185.

In a brief review (with 47 refs.) of the literature on the applications of fracture mechanics to cement and concrete, it was concluded that LEFM probably does not apply to these materials, because there is as yet no agreement on a valid test specimen, and because the data in the literature are highly contradictory.

270 M. Modéer, A Fracture Mechanics Approach to Failure Analysis of Concrete Materials, Report TVBM-1001, Division of Building Materials, University of Lund, Sweden, 1979.

The cracking of concrete was described by a "fictitious crack model", by which it is possible to transmit some load across the microcracked region. This was related to the Dugdale and Barenblatt "tied crack" models. A method for measuring the fracture energy (G_c) of concrete was presented, which was suitable for analysis using FEM. It was shown that K_c varied with specimen depth. It was concluded that for LEFM to be applicable, the specimens have to exceed a characteristic length (ℓ_{ch}) defined as

$$\ell_{ch} = G_c \, E/\sigma_t^2$$

This yields the following values:

> concrete: 200-300 mm
>
> mortar: 100-200 mm
>
> hcp: 5-10 mm

LEFM specimens should have sizes of at least 10 ℓ_{ch}!

271 K. Morita and K. Kato, Fundamental Study on Evaluation of Fracture Toughness of Artificial Lightweight Aggregate Concrete, in Review of the Thirty-Third General Meeting, The Cement Association of Japan, Tokyo, 1979, pp. 175-177.

The relationship between the fracture strength and the fracture toughness of artificial lightweight aggregate concrete was studied, using acoustic emission techniques. SEN beam 100 x 100 x 420 mm, with notches of different depths, were tested in 3-point bending. It was found that:

   i)   For this lightweight concrete, K_c was about 30% lower than for ordinary concrete.

   ii) K_c was essentially independent of the notch/depth ratio.

   iii) Values of K_c estimated by acoustic emission techniques were about 10% lower than those measured.

272 J. Morton, The Work of Fracture of Random Fibre Reinforced Cement, Matériaux et Constructions, 12 (1979) 393-396.

The work of fracture and post-cracking stress of frc were estimated, using considerations of the pull-out behaviour of individual inclined wires. For short fibres, or where the interfacial shear stress was low, it was found that much of the work of fracture was due to plastic deformation of fibres not perpendicular to the matrix crack face. Random fibres were more effective in this regard than aligned fibres.

273 A.E. Naaman and S.P. Shah, Fracture and Multiple Cracking of Cementitious Composites, in S.W. Frieman, (ed.) Fracture Mechanics Applied to Brittle Materials, ASTM STP 768, A.S.T.M., Philadelphia, PA, 1979, pp. 183-201.

After a review of the attempted applications of LEFM to fibre reinforced cementitious materials, three series of tests on steel frc were reported; tension tests on notched prisms and plates, DCB tests, and SEN beams in 4-point bending. It was found that depending on the specimen dimensions and type of test, brittle failure might not occur. A pseudo-plastic zone occurred in front of the crack tip, due to the action of the fibres. It was concluded that experimental techniques to measure fracture properties must still be developed for frc.

274 S. Nagamatsu and Y. Sato, Study on Creep Fracture Strength. Probability
     Distribution of Creep Fracture Strength and the Effects of Water Content,
     in Review of the Thirty-Third General Meeting, The Cement Association of
     Japan, Tokyo, 1979, pp. 230-232.

     The probability distributions of failure times under sustained load
     were obtained for 25 x 25 x 250 mm mortar beams tested in 4-point bending;
     both dry and saturated specimens were tested. It was concluded that:
     i) The statistical distribution of strengths under sustained loads can be
     derived by the same theoretical methods as are used for static strengths,
     based on fracture mechanics.
     ii) As the water content of the specimens decreases the resistance to
     static fatigue increases.
     iii) For completely dried specimens, the statistical distribution curve
     of strength under sustained load is approximately the same as that for
     static strength.

275 K. Rokugo, Experimental Evaluation of Fracture Toughness Parameters of
     Concrete, M.S. Thesis, Dept. of Civil Engineering, University of Illinois
     at Urbana - Champaign, 1979.

     The J-integral was investigated for mortar, concrete, and fibre-
     reinforced concrete, using notched beams in bending. Different methods of
     evaluating the J-integral were examined, as well as other fracture
     criteria, such as the unit dissipated energy during a test, K_c, and G_c.
     Different specimen sizes, aggregate sizes, and notch depths were studied.
     It was concluded that the J-integral could be used as a suitable fracture
     criterion for concrete and fibre-reinforced concrete. However, the unit
     dissipated energy was more sensitive to variations in fibre content.

276 N. Saeki, N. Takada and S. Hataya, On Studies for Cracking and Failure of
     Concrete by Acoustic Emission Techniques, in Review of the Thirty-Third
     General Meeting, The Cement Association of Japan, Tokyo, 1979, pp. 234-
     235.

     The cracking of concrete was studied using acoustic emission techniques
     on concrete specimens made with varying w/c ratios and sand contents.
     Three different energy levels of acoustic emission were defined: (i) bond
     cracking, (ii) crack arrest; and (iii) mortar crack formation, with (i) and
     (iii) having the highest energy levels. It was also concluded that the
     failure process in concrete could be analyzed using a Weibull distribu-
     tion.

277   S.P. Shah, Whither Fracture Mechanics in Concrete Design? in Proceedings of
      the Engineering Foundation Conference on Cement Production and Use, Rindge,
      N.H., 1979, pp. 187-199. Also in Proceedings of a Seminar on "Fracture
      Mechanics of Fibre-Reinforced Cement-Based Composites," Delft, July 5,

      The applications of fracture mechanics to cement and concrete were
      reviewed, with a focus on the problem of determining a valid fracture
      mechanics parameter for cementitious composites. It was suggested that R-
      curve analysis might provide a unique fracture mechanics parameter for
      these systems.

278   C. Sok, J. Baron and D. Francois, Mecanique de la Rupture Appliquee au
      Beton Hydraulique, Cement and Concrete Research, 9 (1979) 641-648.

      Large DCB specimens of concrete (2.8 x 1.1 x 0.3 m) were tested. It
      was found that an extensive damage zone existed ahead of the crack tip, as
      shown by acoustic emission methods. The damage zone was greater than 0.2 m
      in size initially, and grew to a size greater than 0.5 m for a visible
      crack extension of only 0.03 m. R-curve analysis was used to characterize
      the energy necessary for crack propagation; this energy increased as the
      crack propagated, and reached much greater values than the crack initiation
      value at the beginning of the test. It was concluded that specimens must
      be large enough so that more than 0.5 m of crack propagation can take place
      in order to get valid fracture parameters.

279   P.C. Strange and A.H. Bryant, Experimental Tests on Concrete Fracture,
      Journal of the Engineering Mechanics Division, ASCE, 105 (1979) 337-343.

      Notched and unnotched concrete beams of various sizes (up to 1000 x
      200 x 100 mm) were tested in 3-point bending, and notched specimens were
      also tested in tension. It was found that K_c increased with increasing
      specimen size and with increasing maximum aggregate size. The authors
      argued that the dependence of K_c on size must be due to the
      nonapplicability of linear stress analyses in the crack tip region. Slow
      crack growth was evident in tension, but occurred to a lesser degree in
      bending. Due to inelastic behaviour, the relevance of flexural testing to
      Mode I fracture toughness was questioned.

280   P.C. Strange and A.H. Bryant, The Role of Aggregate in the Fracture of
      Concrete, Journal of Materials Science, 14 (1979) 1863-1868.

      After a brief review of the fracture of concrete, 3-point bending tests
      on both unnotched and notched SEN beams, with notches of different depths,
      were described. Four different sizes of beams were tested, ranging from 12

x 12 x 48 mm to 100 x 200 x 800 mm. A cast notch was used. It was found
that:

   i)   Fracture toughness increased with increasing aggregate size.

   ii) Matrix cracks were initiated by aggregate particles.

   iii) Different aggregates gave different K_c values.

   iv) Fracture toughness increased with increasing specimen size.

281 P. Stroeven, Mechanics of Microcracking in Concrete Subjected to Fatigue
Loading, in Proceedings of the Third International Conference on Mechanical
Behaviour of Materials, Cambridge, Vol. 3, Pergamon Press, London, 1979,
pp. 141-150.

    Stereological techniques were used to examine crack morphologies in
fatigue. A model for crack development was proposed for compressive
loading: (i) at low stresses, interfacial cracks grow due to high shear at
the interface; (ii) as the stress increases, this cracking is inhibited by
surface roughness of the aggregate; (iii) eventually, a stable, cone-like
zone of triaxially compressed material develops at the "poles" of the
aggregate particles, which prevents further crack growth at the interface.
These zones are surrounded by axially oriented bond cracks of the smaller
particles. Subsequently, axial cracks are initiated due to cracking in the
shear areas, which eventually bridge between the isolated microcracks and
lead to failure.

282 P. Stroeven, Geometric Probability Approach to the Examination of
Microcracking in Plain Concrete, Journal of Materials Science, 14 (1979)
1141-1151.

    A geometric probability approach was used to describe the
microstructural characteristics of the cracking of plain concrete.
Concrete was considered to be a 2-phase material. Image analysis
procedures, using an automic scanner, permitted quantification of the
microcracking (to a minimum crack length of 1 mm). Fluorescent dye was
used to penetrate the cracks. Crack lengths and crack specific surface
areas were obtained as a function of load level.

283 P. Stroeven, Crack Development in Concrete as Influenced by the Addition of
Short Steel Fibres, in Seminar on Fracture Mechanics of Fibre Reinforced
Cement-Based Composites, Delft, 1979, pp. C1-45.

    Formulae are derived for the direct tensile strength of fibre
reinforced materials, taking into account the strength of the matrix, the
fibre-matrix bond strength, the size and orientation of the fibres, and the
concrete mix parameters. The fracture mechanics approach of Romualdi et

al. (refs. 36, 37) is used. G_c is expressed in terms of the crack opening, but no method of determing the crack opening is presented. There is an attempt to estimate theoretically the reinforcing effect of fibres, simultaneously taking into account the effect of the coarse aggregate particles.

284 R.N. Swamy, Fracture Mechanics Applied to Concrete, in F.D. Lydon, (ed.), Developments in Concrete Technology-1, Applied Science Publishers Ltd., London, 1979, pp. 221-281.

This is a review, with 81 references, of current applications of fracture mechanics to concrete. An extensive review of the principles of fracture mechanics is also provided. It is concluded that a non-linear fracture parameter would be more suitable than LEFM to describe the fracture behaviour of concrete.

285 J. Takeda and H. Hinayama, Crack Propagation in Concrete Subjected to High Rate of Loading, in Review of the Thirty-Third General Meeting, The Cement Association of Japan, Tokyo, 1979, pp. 232-233.

The characteristics of the dynamic extension of cracks in concrete were studied, using rectangular prisms with lengths ranging from 130 to 500 mm, with a central crack 75 mm long, and loaded perpendicular to the crack at loading rates ranging from 3 kg/sec to 2×10^6 kg/sec. It was found that K_c changed as a function of the loading rate.

286 L.P. Trapesnikov, Two-Parameter Model of Fracture of Concrete Under Tension, Taking into Account the Structure and the Creep. Description of the Model (in Russian). Izvestyia VNIIG im. B.E. Vedeneeva, 128 (1979) 93-103.

At the tip of a propagating crack, a zone of intensive microcracking (the pre-fractured zone) is formed. The amount of energy involved in this process is expected to be as much as 90% of the total fracture energy. A review of the analytical and experimental investigations of crack propagation is given. Since the previous models do not properly describe the cracking of concrete, their modification is proposed. The formation of the pre-fractured zone is due to the work of the tensile stresses on the corresponding strains. A necessary condition for crack propagation is the increase of the size of the pre-fractured zone to a limiting value, which is of the order of the maximum aggregate size. A theory is presented based on the linear theory of viscoelasticity of aging materials; the crack formation criterion corresponds to a criterion of ultimate normal stress at the crack front.

624

287 L.P. Trapesnikov, Two-Parameter Model of Fracture of Concrete. Application to the Long Term Strength Under Axial Tension (in Russian). Izvestyia VNIIG im. B.E. Vedeneeva, 129 (1979) 101-108.

An application of the theory proposed in ref. 286 is given for the case of a strip with an edge notch loaded in direct tension. Short term and long term strengths are analyzed. Curves for the size of the pre-fractured zone are presented in terms of the crack length and the stress level. Curves are also presented for the maximum values of K_{IC}. In addition, an estimate is made of the effects of time on the development of the pre-cracked zone. It is concluded that for the same initial crack length, the pre-cracked zone size is smaller in the case of the long-term process than in the short-term case.

288 K.L. Watson, Reply to a Discussion of The Estimation of Fracture Surface Energy as a Measure of the 'Toughness' of Hardened Paste, by Y.W. Mai (CCR, 9(1979) 537); Cement and Concrete Research, 9 (1979) 541-544.

In reply to Mai's discussion (ref. 265), Watson recalculated his values using Mai's suggestion. For span/ depth ratios of 10.5 and 4, he obtained G_C values of 10 and 20 J/m^2, respectively, indicating that the span/ depth ratio is very important. Watson also found that the G_C values obtained using his method gave quite different results from those obtained using Mai's method.

289 G. Wischers, Aufnahme und Auswirkungen von Druckbeans-pruchungen auf Beton, (in German), in Beton Technische Berichte 1978, Beton-Verlag GmbH, Düsseldorf, 1979, pp. 31-56.

A review of the strength and cracking of concrete is presented. It is argued that concrete is not brittle because localized stress concentrations lead to the formation of microcracks. A large proportion of the applied energy is dissipated by these microcracks and the breaking of small particles. Steel fibres in concrete bridge these microcracks, and lead to a mechanism for dissipating a large proportion of the applied energy without a large loss in strength.

1980

290 ACI Committee 224, Control of Cracking in Concrete Structures, Concrete International, 2 (1980) 35-76.

The principal causes of cracking in concrete and recommended crack control procedures are presented. The current state of knowledge of microcracking and fracture mechanics is discussed, with 44 references.

291 B. Barr, W.T. Evans, R.C. Dowers and B.B. Sabir, Fracture Toughness of
     Concrete, in Numerical Methods in Fracture Mechanics, Proceedings of the
     2nd International Conference, Swansea, Wales, 1980; Pineridge Press,
     Swansea, 1980, pp. 737-749.

A simple test is described to measure K_c of concrete. A concrete
cube is modified by the introduction of two slits running the full width of
the cube on opposite sides; the slits are introduced at mid-height of the
cube. The cube is then eccentrically loaded, along one of the top edges
(parallel to the slits). A finite element solution for K_c was developed.
It was found that K_c appeared to be independent of the notch depth.

292 Z.P. Bazant, Material Behaviour Under Various Types of Loading, in S.P.
     Shah (ed.), High Strength Concrete, Proceedings of a Workshop held at the
     University of Illinois at Chicago Circle Dec. 2-4, 1979, Published 1980,
     pp. 79-92.

Amongst other topics, the application of fracture mechanics to concrete
is discussed. Because of the large microcracked zone, LEFM can only be
applied to structures whose cross-sectional dimensions are at least 100
times the size of the maximum aggregate particles. Since most specimens
are much smaller, K_I and G_I are not unique, but depend upon the
experimental conditions. Crack propagation is treated in terms of crack
bands. However, for high strength concretes, because the mortar is
stronger and the material behaves in a more brittle fashion, the specimen
size for which LEFM applies is decreased.

293 Z.P. Bazant and L. Cedolin, Fracture Mechanics of Reinforced Concrete, in
     W.F. Chen and E.C. Ting (eds.), Fracture in Concrete. Proceedings of an
     ASCE Session, Florida, 1980. American Society of Civil Engineers, New
     York, 1980, pp. 28-35.

A finite element analysis of cracking in reinforced concrete is
presented. It is shown that the extension of cracks in concrete cannot be
predicted properly using a constant strength criterion for crack extension.
Instead, the authors use the energy release rate as their criterion for
crack extension in FEM analysis of reinforced concrete. Expressions for
calculating the energy release rate in FEM programs are also presented.
For reinforced concrete, the bond slip between the steel reinforcement and
the concrete is also taken into account. Instead of a single sharp crack,
a broad or "smeared" crack representation is used, in which the crack front
is assumed to consist of a diffuse zone of microcracks, the size of which
is related to the maximum aggregate size.

626

294 Z.P. Bazant and L. Cedolin, Fracture Mechanics of Reinforced Concrete, Journal of the Engineering Mechanics Division, ASCE, 106 (1980) 1287-1306.

An energy criterion for the propagation of a blunt "smeared" crack in reinforced concrete was formulated. This energy criterion was generalized for non-linear material behaviour. However, the method requires that bond-slip near the crack band also be taken into account.

295 Z.P. Bazant and A.B. Wahab, Stability of Parallel Cracks in Solids Reinforced by Bars, International Journal of Solids and Structures, 16 (1980) 97-105.

The effect of one layer of steel reinforcement on the instability of a system of parallel equidistant shrinkage or cooling cracks in a concrete halfspace, or on parallel equidistant cracks due to beam bending, was investigated using FEM techniques. It was found that instabilities of cracks in reinforced concrete are strongly affected by the presence of reinforcement. The presence of reinforcement greatly increases the penetration depth of cooling or drying at which the instability occurs, but does not prevent the instability from occurring deeper beneath the reinforcement.

296 A.M. Brandt, Crack Propagation Energy in Steel Fibre Reinforced Concrete, International Journal of Cement Composites, 2 (1980) 35-42.

SEN steel fibre reinforced concrete beams, 50x50x300 mm, were tested in 4-point bending. Deflection, concrete strain, and crack opening were measured during the tests, and K_c, G_c, J-integral, and the surface energy were calculated. It was found that fracture mechanics appeared to be useful in characterizing this material, particularly the J-integral approach. A great deal of scatter was found in the data, largely because it was not possible to obtain a truly brittle fracture with the sfrc. Within the size range tested, G_c and K_c did not appear to be sensitive to the relative crack depth.

297 C.B. Brown, Micromechanics of Achieving High Strength and Other Superior Properties, in S.P. Shah (ed.), High Strength Concrete, Proceedings of a Workshop Held at the University of Illinois at Chicago Circle, Dec. 2-4, 1979. Published 1980, pp. 31-35.

The use of fracture mechanics applied to concrete was considered. It was suggested that the matrix cracking as load increases and the difficulties in measuring fracture parameters makes modelling the process difficult. An alternative approach involving the use of entropy in the statistical description of crack density was presented. However, the relationship between the probability distribution of the crack density and the phenomenological parameters has yet to be worked out.

298 A. Carpinteri, Notch Sensitivity in Fracture Testing of Aggregative Materials, Nota Tecnica n.45, I.S.C.B.-Gennaio 1980, Universita Di Bologna, Facolta Di Ingegneria. Engineering Fracture Mechanics, in press.

Using dimensional analysis, notch sensitivity effects on the fracture toughness of aggregative materials were studied, including the effects of specimen size and crack size. The application of Buckingham's theorem allows the definition of a non-dimensional parameter (the test brittleness number, s) which governs the notch sensitivity phenomenon. $s = K_{IC}/\sigma_u b^{1/2}$, where b is the specimen depth; for $s \geqslant 0.50$, test results lose their significance. This theory is then used to explain variations in K_c with crack length, specimen size, and specimen type. It is concluded that notch sensitivity is not an intrinsic material property for mortar and concrete, but depends on the specimen size.

299 J.L. Carrato, Experimental Evaluation of the J-Integral, M.S. Thesis, Dept. of Civil Engineering, University of Illinois at Urbana-Champaign, 1980.

The J-integral was studied using notched beams in bending on mortar, concrete and fibre-reinforced concrete. It was found that J_C was independent of the type of notch, notch tip sharpness, or notch thickness. Notch depth became important only when the maximum aggregate size approached the size of the uncracked ligament. J_C was also independent of specimen size within the limited range of sizes studied (3x3x15 in., 3x6x15 in., and 4x6x15 in.), and did not vary between 3-point and 4-point loading. It was therefore concluded that the J-integral was a useful fracture criterion for both plain and fibre-reinforced concrete. However, it was suggested that scatter could be drastically reduced if the J-integral was calculated using a displacement criterion rather than a maximum load criterion.

300 L. Cedolin and Z.P. Bazant, Effect of Finite Element Choice in Blunt Crack Band Analysis, in Computer Methods in Applied Mechanics and Engineering 24, North-Holland Publishing Company, 1980, pp. 305-316.

The method of element-wide blunt smeared crack bands, which allows an effective FEM analysis of cracks which may propagate in any direction, was developed. It was shown that meshes of different size gave substantially the same results, indicating satisfactory convergence. However, the computed stress intensity factors by this method only approximated the exact solutions.

301 L. Cedolin and Z.P. Bazant, Fracture Mechanics of Crack Bands in Concrete, in ASTM STP 745, American Society for Testing and Materials, Philadelphia, 1980 (in press).

The criterion for propagation of an element-wide crack band into the finite element in front of the crack band was expressed in terms of the energy consumed to create the cracks. It was shown that the currently used tensile strength criterion is not objective, as it gives incorrect convergence as the finite element mesh is refined. However, it was concluded that it is possible to use an equivalent strength which can be made to be consistent with the energy criterion. It was further concluded that for reinforced concrete the bond slip of bars near the crack band must be taken into account.

302 A.F. Chtchourov, Microtexture et Résistance Mécanique du Ciment Durci, in Proceedings of the Seventh International Congress on the Chemistry of Cement, Vol. IV, Paris, 1980, Editions Septima, 1981, pp. 404-410.

Studies were made of crack formation in hcp and concrete, using acoustic emission and electronic fractography, in order to investigate the role of the microtexture and the statistical nature of the structure in determining strength. K_c and γ_F were determined, and the effects of admixtures were also studied. It was concluded that in compression, instability is controlled by the coalescence of pores. In tension, the crack length is the controlling factor.

303 B.J. Dalgleish, P.L. Pratt and R.I. Moss, Preparation Techniques and the Microscopical Examination of Portland Cement Paste and C_3S, Cement and Concrete Research, 10 (1980) 665-676.

SEM was used in conjunction with energy dispersive analysis to study the structure of hardened pastes. It was shown that in young pastes, interparticle fracture takes place through the porous regions of C-S-H, and around CH that has crystallized in the remaining pore space. As the paste matures, the extent of transparticle fracture increases, with the crack propagating through both C-S-H and CH regions along a much straighter path.

304 S. Diamond and S. Mindess, Scanning Electron Microscopic Observations of Cracking in Portland Cement Paste, in Proceedings of the Seventh International Congress on the Chemistry of Cement, Vol. III, Paris, 1980, Editions Septima, 1980, pp. VI. 114-119.

A small, wedge-loaded compact tension specimen of hardened cement paste was loaded in the specimen chamber of a scanning electron microscope, using a specially designed loading frame. It was observed that cracks are approximately parallel-sided, mostly unbranched except near the tip, and only approximately straight. Cracks appear to occur as assemblages of relatively short straight segments set at small angles to each other. The crack tip itself cannot be definitely located.

305 R.M.L. Foote, B. Cotterell and Y.W. Mai, Crack Growth Resistance Curve For a Cement Composite, in Advances in Cement-Matrix Composites, Proceedings, Materials Research Society, Symposium L, Boston, Nov. 1980, pp. 135-144.

It was suggested that the stress intensity factor for frc can be separated into two components,
$$K = K_R + K_r = K_i$$
where K_i corresponds to fracture initiation of the matrix, K_R is the stress intensity factor at the tip of the crack if there were no fibres bridging the crack, and K_r is the stress intensity factor due to the bridging fibres closing the crack faces. It was shown that K_R curves are reasonably independent of geometry in comparing notch bend, compact tension, and double cantilever beam specimens, and that K_R can therefore be used as a material property.

306 Z. Guofan, G. Junsheng, L. Wanqing and W. Qingxiang, Experiments and Calculating Method for the Cracking Strength and Maximum Crack Width in Reinforced Concrete Members, Journal of Building Structures, Beijing, China, 1 (1980) 1-17.

The cracking strengths of reinforced concrete members subjected to bending, eccentric compression and eccentric tension were determined. Equations for calculating these cracking strengths are presented, which involve a parameter which takes into account the plastic behaviour of the tension zone.

307 K. Gylltoft, Bond Failure in Reinforced Concrete Under Cyclic Loading. A Fracture Mechanics Approach, Research Report Tulea 1980:29, Division of

630

Structural Engineering, University of Lulea, Sweden, 1980.

A non-linear finite element analysis of bond failure of smooth steel bars in concrete was carried out. Fracture mechanics was used to model the progressive fracture of the contact zone between the steel and the concrete. Two directions of failure were considered: parallel to and perpendicular to the steel bar. The "fictitious" crack model was used to represent the crack tip. It was suggested that biaxial fracture mechanics be used to model this problem.

308   G.T. Halvorsen, J-Integral Study of Steel Fibre Reinforced Concrete, International Journal of Cement Composites, 2 (1980) 13-22.

The J-integral was used to measure the toughness of steel fibre reinforced concretes. Notched beams 75 x 75 x 380 mm were loaded in 4-point bending, and the J-integral calculated from the maximum load. This technique was able to show variations in toughness with fibre type and content. For most fibres, 0.75-1.0% by volume is needed to produce significant increases in toughness above that of the plain concrete. However, variability in the results was very large.

309   G.T. Halvorsen, J_m Toughness Comparison for Some Plain Concretes, International Journal of Cement Composites, 2 (1980) 143-148.

The J-integral was used to characterize cement pastes, mortar, and concrete. It was found that the mere presence of aggregate is more important than the aggregate geometry or type in determining resistance to crack growth. It was concluded that the J-integral may supplement strength measurements. However, it is subject to considerable experimental variability.

310   A. Hillerborg, Analysis of Fracture by Means of the Fictitious Crack Model, Particularly for Fibre Reinforced Concrete, International Journal of Cement Composites, 2 (1980) 177-184.

An approach to analyzing the fracture of plain concrete and frc using the "fictitious crack model" is described. Instead of conventional fracture parameters, the stress-displacement curve is introduced to describe the crack process in the fracture zone. For the rest of the material, conventional σ-ε relations are used. Computations for this model can only be carried out using computerized numerical methods. It is claimed that both crack stability and different types of fracture can be analyzed using this technique.

311 O.A. Kayyali, C.L. Page and A.G.B. Ritchie, Frost Action on Immature Cement Paste - Microstructural Features, Journal of the American Concrete Institute, 77 (1980) 264-268.

The effect of freeze-thaw cycling on immature cement pastes was studied. It was concluded that this modified the hydration sequence and caused a redistribution of hydration products to occur. For non-air-entrained pastes, the fracture surface was characterized by large CH crystals surrounded by pores. It was suggested that these CH crystals constituted flaws in the material, and that the surrounding pores were large enough to create intensified stress fields which interacted with the CH and weakened the concrete.

For air-entrained pastes, small uncleaved CH crystals were found in the air voids. It was suggested that entrained air voids also provide sites for the segregation of CH crystals.

312 M.M. Kim, H.-Y. Ko and K.H Gerstle, Determination of Fracture Toughness of Concrete, in W.F. Chen and E.C. Ting (eds.), Fracture in Concrete, Proceedings of an ASCE Session, Florida, 1980, American Society of Civil Engineers, New York, 1980, pp. 1-14.

Fracture studies were carried out on SEN mortar beams of two different sizes: 3 in. wide x 3 in. deep x 11.5 in. long, and 3 in. wide by 1.5 in. deep x 11.5 in. long. Ratios of notch depth to specimens depth (a/w) varied from 0.04 to 0.69. Mix proportions were w/c = 1.2, and a/c = 4.28. The test age was > 28 days. The results were:

i) For both beam sizes, the nominal flexural strengths of the notched beams were about 75% of those of the unnotched beams.

ii) K_c showed a high variability, but appeared to be independent of a/w.

iii) The initial K_c was higher for the 3 in. deep specimens (476 psi$\sqrt{in}$. vs. 353 psi$\sqrt{in}$.).

iv) After slow crack growth during loading was estimated by compliance techniques, the recalculated K_c values were 651 psi$\sqrt{in}$. for the 3 in. deep beams, and 450 psi$\sqrt{in}$. for the 1.5 in. deep beams, with a larger scatter in results than for the K_c values based on the initial crack depth.

The authors therefore concluded that:

i) The compliance method for estimating slow crack growth is inaccurate.

632

ii) Notch effects can be measured even in shallow beams.

iii) The existence of a true K_c for concrete is uncertain.

313   J. Lemaitre and J. Mazars, A Model for the Behaviour and Fracture of
Concrete, in 4eme Symposium Franco-Polonais de Mécanique, Marseille, June
1980 (in French).

The authors argue that the mechanical properties of concrete can be
studied using irreversible process thermodynamics. A model is developed
based on elastic theory and the damage theory. Some experimental results
are presented to verify the proposed model.

314   K.E. Løland, Continuous Damage Model for Load-Response Estimation of
Concrete, Cement and Concrete Research, 10 (1980) 395-402.

A semi-empirical damage model was used to estimate the σ-ε relationship
for concrete in uniaxial tension. Damage was considered to be a measure of
the relative proportions of pores and cracks in the concrete. Energy
considerations regarding cracking were used to establish the descending
branch of the σ-ε curve. The model was related to the σ-ε curves
published by Evans and Marathe (ref. 29).

315   K.E. Løland and O.E. Gjørv, Ductility of Concrete and Tensile Behaviour,
Research Report BML 80.613, University of Trondheim, NTH, Divison of
Building Materials, Trondheim, Norway, 1980.

A method was presented which combined fracture mechanics and damage
mechanics to describe the ductility and tensile behaviour of concrete.
Ductility was given as the maximum strain capacity in uniaxial tension, and
was calculated from G_c data obtained by testing notched beams in 3-point
bending. The ascending branch of the σ-ε curve was obtained by tensile
tests. The results indicated a good relationship between ductility and
tensile strength.

316   K.E. Løland and T. Hustad, Load Response of C-25 Concrete With and Without
Addition of Silica Fume, Research Report STF65 A80048, Cement and Concrete
Research Institute, University of Trondheim, NTH, Trondheim, Norway, 1980
(in Norwegian).

The mechanical properties of concrete (strength, stiffness and
ductility) were determined for concretes made with and without the addition
of silica fume. Ductility was defined as the ultimate strain capacity in
uniaxial tension. It was calculated by a method which combined LEFM and
damage mechanics, and included the determination of the ascending branch of

the σ-ε curve in direct tension, along with a measure of G_c. The results indicated that concrete with silica fume additions had a slightly lower ductility than normal concrete. The silica fume concrete contained 10% of silica fume by weight of cement, and about 30% less portland cement than the normal concrete.

317   A. Maher and D. Darwin, Mortar Constituent of Concrete Under Cyclic Compression, Structural Engineering and Engineering Materials SM Report No. 5, The University of Kansas Center for Research, Inc., Lawrence, Kansas, 1980.

The behaviuor of the mortar constituent of concrete under cyclic compression was studied, using mixes with w/c = 0.5 and w/c = 0.6, at ages ranging from 5 to 70 days. Complete monotonic and cyclic σ-ε envelopes were obtained. The degradation was found to be a continuous process, and a function of both total strain and load history. It was suggested that the nature of the damage in mortar may dominate the behaviour of concrete.

318   Y.W. Mai, R.M.L. Foote and B. Cotterell, Size Effects and Scaling Laws of Fracture in Asbestos Cement, International Journal of Cement Composites, 2 (1980) 23-34.

SEN beams were tested in 3-point bending in order to determine size effects on the fracture load, fracture stress, and strain energy density in asbestos-cellulose cement composites. They found that to get valid K_C values, a minimum beam depth of 50 mm had to be used; when the beam depth was only 25 mm, the beams were notch insensitive. K_R curves were found to be best to characterize the fracture behaviour. It was found that fracture parameters followed the laws of scaling predicted from fracture mechanics theories. K_C and R (specific work of fracture) were reasonably independent of size for beams of depth greater than 50 mm. The toughness of the composite came mainly from the work required to debond the fibres. K_C was found to be about 1.3 $MNm^{-3/2}$.

319. H. Mihashi and F.H. Wittmann, Stochastic Approach to Study the Influence of Rate of Loading on Strength of Concrete, HERON (The Netherlands), 25 (1980).

A stochastic theory for the fracture of concrete was presented, based on probability models. This theory can be used to describe fracture under both monotonically increasing load and time-dependent loading conditions, such as sustained load and repeated load. This theory was compared to experimental results on rate-of-loading tests carried out on different

concretes, with the loading rate varied over three orders of magnitude. It
was concluded that low strength materials experience a more pronounced
strength increase as the rate of loading increases, and that the
distribution of strength (at least in compression and in 3-point bending)
can be described by Weibull's distribution function.

320 S. Mindess, The Fracture of Fibre-Reinforced and Polymer Impregnated
     Concretes, International Journal of Cementitious Composites, 2 (1980) 3-11.

     The literature on the application of fracture mechanics to fibre
reinforced concrete, polymer concrete and polymer impregnated concrete was
reviewed (with 37 refs.). The difficulties of applying fracture mechanics
principles to these systems was discussed, as were some of the models that
have been proposed for the way in which polymers or fibres improve the
fracture toughness of cementitious systems. It was concluded that a non-
linear fracture mechanics criterion would be more suitable for describing
the fracture of these materials.

321 S. Mindess and S. Diamond, The Cracking and Fracture of Mortar, in W.F.
     Chen and E.C. Ting (eds.), Fracture in Concrete, Proceedings of an ASCE
     Session, Florida, 1980, American Society of Civil Engineers, New York,
     1980, pp. 15-27.

     The propagation of cracks in mortar (w/c=0.4, a/c=0.9, age=49 days) was
studied using a wedge-loaded compact tension specimen, of dimensions 31.8 x
24.0 x 12.7 mm. The specimen was tested within the sample chamber of an
SEM, so that direct observations of crack growth could be made at
magnifications up to about 2000x. The study revealed that:
     i) The process of crack propagation in mortars is very complicated; the
simple fracture mechanics models which have been applied to mortars greatly
oversimplify the geometric features of the crack extension process.
     ii) A considerable amount of energy is dissipated in the large amount of
multiple cracking that is seen to occur, as well as in the creation of very
tortuous crack surfaces. These energy dissipative processes should be
considered in developing fracture models for mortar.
     iii) Cracks develop and propagate preferentially in the interfacial
regions between sand and cement, but are not limited to these regions.

322 S. Mindess and S. Diamond, A Preliminary SEM Study of Crack Propagation in
     Mortar, Cement and Concrete Research, 10 (1908) 509-519.

     The process of cracking was observed in mortar specimens, using a device which permitted the testing of wedge-loaded compact tension specimens within the sample chamber of an SEM. It was found that the process of crack extension in mortars is very complicated: the crack is tortuous, there is some branch cracking, discontinuities in the cracks are observed, and there is some tearing away of small bits of material in some areas of cracking. The results suggest that the simple fracture mechanics models oversimplify the geometric features of the crack extension process.

323 K. Morita, Evaluation of Fracture Toughness by Means of the Work of Fracture, The 35th Annual Meeting of the Civil Engineering Institute of Japan, Vol. V, 1980, pp. 269-270.

The work of fracture, γ_f, was measured from the diagram of the load-COD curve and used as an equivalent energy release rate. γ_f was almost independent of water-cement ratio, crack-depth ratio and the maximum load.

324 K. Morita, Influence of Concrete Parameters on Fracture Toughness, Proceedings of the 34th General Meeting, The Cement Association of Japan, Tokyo, 1980, pp. 246-249.

Three point bending tests were carried out to study the influence of concrete parameters on K_{IC}. The studied parameters were the contents of cement, fine and coarse aggregate, and water-cement ratio.

325 K. Okada, W. Koyanagi and K. Rokugo, Energy Approach to Flexural Fracture Process of Concrete, Transactions of the Japan Society of Civil Engineers, 11 (1980) 301-304.

SEN beams, 4.7 x 10 x 39 cm, were tested in 3-point bending, and load-deflection curves obtained. The fracture energy was obtained from the area under the load-deflection curves, and was compared to that in compression. Crack growth was examined using dye impregnation, and acoustic emission was also recorded. It was found that:

i)   Crack growth was proportional to the dissipated energy calculated from the load-deflection diagrams.

ii)   Cracks grow extensively after the peak load is reached.

iii) The locations of crack sources obtained by acoustic emission coincided well with the observed fracture surface.

326 P.E. Petersson, Fracture Energy of Concrete: Method of Determination, Cement and Concrete Research, 10 (1980) 78-89.

A number of methods for determining G_c of concrete are discussed. It

636

is concluded that:

i) The determination of G_c indirectly from 3-point bend tests on notched beams is too dependent on specimen size to be useful; for valid results, the specimen depth must be greater than 10 times a "critical length", defined such that $d > 10 \ G_c \ E/\sigma_t^2$.

ii) A stable 3-point bend test on a notched beam can be used for a direct calculation of G_c based on the area under the force-displacement curve. This method permits much smaller specimens to be used.

iii) LEFM cannot be used unless $d > 10 \ \ell_{ch}$ because the influence of the microcracked zone in front of the crack tip is too great.

iv) The fictitious crack model (cf. Hillerborg; Modeer) is useful for fracture mechanics calculations for concrete.

327 P.E. Petersson, Fracture Energy of Concrete: Practical Performance and Experimental Results, Cement and Concrete Research, 10 (1980) 91-101.

The results of G_c tests are analyzed using the fictitious crack model. Experiments were carried out on 50x50x600 mm specimens in 3-point bending, with a 25 mm sawn notch at mid-span. Conditions for the stability of a 3-point bend test are also calculated. It is concluded that:

i) G_c is strongly dependent on aggregate quality, with strong aggregates giving higher G_c values. G_c increases with age and with decreasing w/c ratio.

ii) For normal concrete, G_c is in the range 60-100 N/m.

iii) For normal concrete, $\ell_{ch} (=G_c E/\sigma_t^2)$ is in the range 200-300 mm. ℓ_{ch} decreases with age and with decreasing w/c ratio.

iv) G_c is not affected by maximum aggregate size, because the tensile strength decreases.

v) G_c increases somewhat with increasing volume fraction of aggregates.

328 P.E. Petersson, Fracture Mechanical Calculations and Tests for Fibre-Reinforced Cementitious Materials, in Advances in Cement-Matrix Composites, Proceedings, Materials Research Society, Symposium L, Boston, Nov. 1980, pp. 95-106.

The fictitious crack model was applied to frc. That is, the fracture zone near a crack tip was modelled by a "tied" crack. This is characterized by a σ-w (stress vs. crack widening) curve. A very stiff tensile test machine was also described, with which stable tensile tests could be carried out on frc, and σ-w curves obtained.

329   P.E. Petersson and P.J. Gustavsson, Model for Calculation of Crack Growth in Concrete-Like Materials, in Numerical Methods in Fracture Mechanics, Proceedings of the 2nd International Conference, Swansea, Wales, 1980; Pineridge Press, Swansea, 1980, pp. 707-719.

The fictitious crack model, based on fracture mechanics and finite element techinques, was developed to analyze the fracture of concrete. It was shown that the method of determining K_C from a 3-point bend test on a SEN beam is strongly affected by specimen dimensions and notch depth.

330   A. Piva and E. Viola, Two Arc Cracks Around a Circular Rigid Inclusion, Meccanica, (1980) 166-176.

The problem of two symmetric arc cracks around a rigid circular inclusion embedded in an infinite matrix under biaxial loading was considered. The aim was to generalize the results obtained by Testa and Stubbs (ref. 210), who had considered such a system under compressive loading as a simplified model for concrete. A stress criterion was developed, which could accommodate crack extension at the interface, or crack deviation into the matrix; this was used to study the fracture response of the system.

331   A. Piva and E. Viola, Stress-Strain Response of a Concrete Mathematical Model, in Proceedings, 5th Congresso Nazionale Di Meccanica Teorica ed Applicata, Palermo, Italy, 1980, pp. 237-248.

A mathematical model of concrete, consisting of a partially bonded elliptic cylindrical piece of aggregate embedded in a mortar matrix, was studied. Since debonding was considered to be the principal cause of the inelastic response of concrete, particular attention was paid to the effect of bond separation on the σ-ε curve under biaxial loading.

332   K. Rokugo, C.E. Kesler and F.V. Lawrence, Evaluation of Fracture Toughness of Concrete by J-Integral Method, in Proceedings, JCI (Japance Concrete Institute) Second Conference, (1980) pp. 125-128.

The fracture toughness parameters J_C and G_C of concrete were calculated using notched beams in bending. J_C was found to increase with increases in fibre content, maximum aggregate size, and concrete strength. It appeared that J_C was independent of specimen dimensions.

333   N. Saeki and N. Takada, Crack Propagation and COD of Concrete, The 35th

Annual Meeting of the Civil Engineering Institute of Japan, Vol. V, 1980, pp. 267-268.

Rectangular plates (300 x 150 x 50 mm) with a 30 mm crack, of hardened cement paste, mortar and concrete, were tested under a direct tensile load in order to study the cracking behaviour and the crack tip opening displacement (COD). The critical values of COD were 304 μm for hardened cement paste (w/c = 0.50), 10-12 μm for mortar (c/s = 0.5) and 13-16 μm for concrete.

334 V.E. Saouma, A.R. Ingraffea and D.M. Catalano, Fracture Toughness of Concrete - K_{IC} Revisited, Report 80-9, Department of Structural Engineering, Cornell University, Ithaca, New York, 1980, accepted for publication in the Journal of the Engineering Mechanics Division, ASCE.

The data of Kesler, Naus and Lott (ref. 88) are re-analyzed, using more modern computational techniques and stress intensity analysis. It is shown that the conclusion of Kesler, Naus and Lott (that LEFM does not apply to hcp, mortar, or concrete) is incorrect, and that proper interpretation of their data suggests that LEFM can indeed be applied to cement, mortar and concrete.

335 S.P. Shah, Fracture in Fibre Reinforced Concrete, in Advances in Cement-Matrix Composites, Proceedings, Materials Research Society, Symposium L, Boston, Nov. 1980, pp. 83-89.

It was suggested that R-curve analysis might be useful in quantifying the resistance to slow crack growth in frc. However, both the elastic and inelastic strain energy components of the crack propagation process must be included in the analysis. To this end, a modified definition of the strain energy release rate was proposed. It was shown that R-curves were influenced by different specimen geometries.

336 S. Somayaji, Influence of Notch Dimensions on the Effective Surface Energy and Notch-Sensitivity of Cement Compounds, in W.F. Chen and E.C. Ting (eds.) Fracture in Concrete, Proceedings of an ASCE Session, Florida, 1980, American Society of Civil Engineers, New York, 1980, pp. 36-49.

The results of many of the experimental studies which have been carried out to measure K_c or G_c are compared. It is shown that the results obtained are incompatible with each other, as the values obtained appear to depend on (i) beam dimensions, (ii) notch dimensions, (iii) concrete properties, and (iv) the method of analysis used. A new method of analysis based on energy considerations is proposed and is then applied to the SEN beam in 3-point bending. This new approach, when applied to data in the literature, appears to reduce the discrepancies in K_c and G_c values from these different investigations. In some cases, notch dimensions appear to be significant in determinations of K_c.

337 S. Somayaji, A Discussion of the Papers: Fracture Energy of Concrete: Practical Performance and Experimental Results; Method of Determination, by P.E. Petersson, (CCR, 10(1980) 101), Cement and Concrete Research, 10 (1980) 471-474.

The author disagrees with Petersson's conclusion that G_c can be determined from a measurement of the area under the force-displacement curve for a notched beam loaded in 3-point bending:

i) All of the external energy is not spent in creating a new crack, since elastic strain energy continues to be stored in the beam.

ii) The F-δ relationship depends on the type of loading. This could give rise to different F-δ curves.

338 T. Sugama, L.E. Kukacka and W. Horn, The Effects of β-C_2S/Class H Cement Mixed Fillers on the Kinetics and Mechanical Properties of Polymer Concrete, Cement and Concrete Research, 10 (1980) 413-424.

The effective surface energy of different polymer concretes was determined using 3-point bending tests of 1.25 x 1.25 x 7.50 cm bars notched to a depth of 0.63 cm at mid-span. The total energy was calculated analytically. Measured fracture energies ranged from 0.34×10^3 to 4.06×10^3 ergs/cm^2. It was concluded that a copolymer polymer concrete structure, consisting of a Ca-polymer complex and the hydration products of cement formed during exposure to hydrothermal conditions had a higher energy absorption consumed during crack propagation when compared to a system which does not have a Ca-copolymer complexed structure and hydration products.

339 R.N. Swamy, Influence of Slow Crack Growth on the Fracture Resistance of Fibre Cement Composites, International Journal of Cement Composites, 2 (1980) 43-53.

SEN fibre concrete beams, 100x100x500 mm (using steel, glass or polypropylene fibres) were tested in 4-point bending. Crack extension as observed using optical microscopy was shown to be complex. It was shown that the fracture toughness vs. crack growth curves gave a clear characterization of the influence of slow crack growth on the fracture behaviour of fibre concretes, and that slow crack growth cannot be neglected in fracture toughness tests.

Considerable branch cracking was observed during the tests, and even low fibre volumes imparted some stable crack growth to the system. The apparent K_c at first increased linearly with crack growth, and then much more rapidly.

340 R.B. Tait and H. Bohm, In-Situ Scanning Electron Microscope Observations of Double Torsion Fracture of Concrete, Proceedings, Electron Microscopy Society of Southern Africa, 10 (1980) 17-18.

Double torsion tests were carried out on small (33 x 11 x 1.2 mm) mortar specimens, 6 months old, in the sample chamber of an SEM. The major crack, 2-3 μm wide, could be followed as it propagated, and could be arrested at any stage. It was found that the crack propagated both around and through sand grains; fracture was characterised by multiple microcracking and crack branching. It was concluded that the stress intensity in the vicinity of the crack tip strongly determines the final fracture path, and that significant amounts of energy are needed to develop both the side and the main crack.

341 K. Togawa and J. Nakamoto, Study on the Surface Fracture and the Impact Wear of Mortar and Concrete, Transactions of the Japan Society of Civil Engineers, 11 (1980) 289-290.

The application of the energy balance criterion to the surface fracture mechanism of hcp, mortar and concrete was discussed, as was the validity of G_c as an index of wear resistance of the impact wear of mortar and concrete. It was concluded that:
i) Surface crack propagation (under static or impact loading through a steel sphere) obeys an energy balance concept.
ii) G_c can be used as index of wear resistance applicable for studded tires and tire chains; it is not useful for wear due to abrasive powders.
iii) G_c is more sensitive to variations in concrete quality than is wear resistance.

342 F. Velazco, K. Visalvanich and S.P. Shah, Fracture Behavior and Analysis of Fibre Reinforced Concrete Beams, Cement and Concrete Research, 10 (1980) 41-51.

Steel fibre reinforced concrete beams with varying notch depths and different fibre volumes were tested in 4-point bending. The specimens were 38mm wide, 76mm deep, and 457mm long. The results were analyzed to examine the applicability of various fracture mechanics approaches, including: K_c, J-integral, COD, R-curve analysis, and the compliance technique for determining slow crack growth. The R-curve analysis seemed to be the most promising fracture parameter, independent of specimen geometry, and able to predict the effects of fibres. Neither K_c, J-integral, or COD were both independent of crack length and sensitive to the fibre volume fraction.

343 K. Visalvanich and A.E. Naaman, Evaluation of Fracture Techniques in
Cementitious Composites, in W.F. Chen and E.C. Ting (eds.), Fracture in
Concrete, Proceedings of an ASCE Session, Florida, 1980, American Society
of Civil Engineers, New York, 1980, pp. 65-81.

Contoured DCB specimens, 610 mm long, made of asbestos cement, mortar
and sfrm were tested. Crack extension, COD, and the crack opening of the
cantilever's arms were measured as a function of load. Several different
methods were used to determine fracture properties:
1) three different analytic procedures for K_c
2) compliance measured strain energy technique
3) quasi-static energy measuring technique
4) an energy measuring technique that takes into account the permanent
    deformation on unloading.

All of these techniques gave "about" the same results for asbestos cement,
but not for mortar or sfrm. It was found that for these last two
materials, methods (2) and (3) underestimated the true fracture energy,
while (4) gave more realistic results. The main conclusions of the study
were:
   i) LEFM can be applied to asbestos cement; it can be applied to mortar
only if the specimen is large enough to accommodate crack growths greater
than 200 mm.
   ii) K_c for sfrm is independent of the initial crack length, but does not
appear to reach a plateau for the 610 mm specimens tested.
   iii) K_c for asbestos cement is about 4.5 ksi$\sqrt{in}$ (4.9 MNm$^{-3/2}$); K_c for
plain mortar is about 1.2 ksi$\sqrt{in}$ (1.3 MNm$^{-3/2}$). These values were reached
after a crack extension of ~75 mm in asbestos cement, and ~200 mm in
mortar.

344 M. Wecharatana and S.P. Shah, Resistance to Crack Growth in Portland Cement
Composites, in W.F. Chen and E.C. Ting (eds.), Fracture in Concrete,
Proceedings of an ASCE Session, Florida, 1980, American Society of Civil
Engineers, New York, 1980, pp. 82-105.

The authors used 813 mm long DT specimens and 610 mm long DCB specimens
to study crack growth in mortar, concrete and sfrc. The definition of
strain energy release rate was modified to include both the elastic and
inelastic strain energy absorbed during crack growth. Using this
definition, R-curves were calculated for the various materials. It was
found that:
   i) The calculated R-curves seemed to be independent of the test

configuration.

ii) For sfrc, R-curves were influenced by the crack openings; larger openings led to higher strain energy release rates.

iii)For the same crack opening, strain energy release rates were reasonably comparable for the DT and DCB specimens.

iv) The values of the R-curve were increased by increasing the volume or maximum size of aggregate, or by adding fibres.

345 M. Wecharatana and S.P. Shah, Double Torsion Tests for Studying Slow Crack Growth of Portland Cement Mortar, Cement and Concrete Research, 10 (1980) 833-844.

Slow crack growth in mortar was studied using 813 mm long double torsion specimens. It was found that the strain energy release rate (G_c) calculated conventionally did not accurately represent the fracture behaviour of the material. A modified definition of G_c was therefore developed, which included both the elastic and the inelastic strain energy absorbed during crack extension. For the double torsion specimens studied, unstable crack propagation occurred at a crack length of about 305-380 mm, at a value of G_c=9.77x10^{-2} N/mm (0.56 lb/in.). A plot of the V-K_I curve indicated rapid crack growth at a value of K_{IC}=1.32MNm$^{-3/2}$ (1200 psi√in.)

346 F.H. Wittmann, Micromechanics of Achieving High Strength and Other Superior Properties, in S.P. Shah (ed.), High Strength Concrete, Proceedings of a Workshop Held at the University of Illinois at Chicago Circle, Dec. 2-4, 1979, published 1980, pp. 8-30.

The structure of concrete is divided into three levels: (i) micro-level, consisting of the porous matrix of hcp; (ii) mezzo-level, consisting of mortar, including sand inclusions, microcracks, and capillary pores; (iii) macro-level, consisting of the concrete structure, including coarse aggregate, air voids, and interfacial cracks. Crack propagation in a homogeneous, infinite plate with one cylindrical hole is described, and this approach is then extended to cracked and porous materials. Finally, the cracking of composite materials is described. The failure of normal and high strength concrete is studied using computer simulations. In normal concrete, cracks tend to go around aggregate particles; in high strength concretes, cracks may go through aggregate particles as well.

347 B. Youdovitch, B.D. Klichanis, Ou. I. Papiachvili and V.G. Abramova, Portland Cement Strength Depending on Technological Factors. Micromechanism of Destruction, in Proceedings of the Seventh International Congress on the Chemistry of Cement, Vol. IV, Paris, 1980, Editions Septima, 1980, pp. 216-219.

The microstructure of hcp with w/c ratios of 0.25 and 0.35 was studied by both TEM and SEM techniques. It was found that, up to an age of three days, microcracks due to loading concentrate at pores. After seven days, cracks form preferentially around "submicrostructural zones", particularly regions of low symmetry. It was concluded that hcp should be modelled as a composite material, containing statistically distributed fibres.

348 Yu. Zaitsev, Influence of Structure on Fracture Mechanism of Hardened Cement Paste, in Proceedings of the Seventh International Congress on the Chemistry of Cement, Vol. III, Paris, 1980, Editions Septima, 1980, pp. VI. 176-180.

The influence of the pore structure of hcp on its fracture mechanism and strength was investigated, and a mathematical model of the fracture process was developed. The model takes into account different pore size distributions in hcp. Crack propagation is analyzed using the Monte Carlo method.

349 B. Zech and F.H. Wittmann, Variability and Mean Value of Strength of Concrete as Function of Load, Journal of the American Concrete Institute, 77 (1980) 358-362.

An experimental study was carried out on the effect of rate of loading on the flexural strength of mortar. The results were explained in terms if the stochastic theory for concrete fracture proposed by Mihashi and Izum , in which the dependence of strength on loading rate is governed by the equation

$$\frac{f_d}{f_s} = \left[\frac{\dot{\sigma}}{\dot{\sigma}_o} \right]^{1/(1+\beta)}$$

where f_d = dynamic strength
       f_s = static strength
       $\dot{\sigma}$ = load rate corresponding to f_d
       $\dot{\sigma}_o$ = load rate corresponding to f_s
       β = material constant

The maximum loading rate used was $5 \times 10^4 N/mm^2/sec$ (7252 x 10^3 psi/sec). In these tests, β was found to be in the range 20-25. It was also found that as the rate of loading increased, the coefficient of variation remained essentially unchanged, as predicted by the theory. Stronger

644

specimens were less affected by loading rate than were weaker specimens.

350 S. Ziegeldorf, H.S. Müller and H.K. Hilsdorf, A Model Law for the Notch
Sensitivity of Brittle Materials, Cement and Concrete Research, 10 (1980)
589-599.

A model law for the notch sensitivity of cementitious materials is presented, assuming that LEFM is applicable to these materials. It indicates that the notch sensitivity of the specimen is a necessary but not sufficient condition for the applicability of LEFM. The model law predicts:

i) Maximum notch sensitivity will occur at a relative crack length of 0.246.

ii) Notch sensitivity will increase with increasing size.

iii)Notch sensitivity will decrease as the ratio of fracture toughness to tensile strength (K_c/σ_t) increases.

The model law is then used to analyze data from the literature, as well as some data obtained by the authors.

1981

351 N. McN. Alford, A Theoretical Argument for the Existence of High Strength
Cement Pastes, Cement and Concrete Research, 11 (1981) 605-610.

This is a discussion of whether the volume of porosity or whether the size of individual pores is responsible for controlling strength in cement paste. It is suggested (see also Birchall, Howard and Kendall, ref. 359) that the conventional strength vs. total porosity relation is largely fortuitous. Emphasis is placed upon the maximum pore size as being responsible for the strength characteristics of hcp. The influence of pore linkage is also discussed.

352 J.H. Argyris, G. Faust and K.H. William, Finite Element Analysis of
Concrete Cracking, Inst. Fuer Statik und Dynamik, Stuttgart University,
Report ISD-256, 1979, 28 pp. Chemical Abstracts, 94: 89134 (1981).

Two methods of using finite element analysis to predict the tensile cracking of concrete were studied: (i) smeared crack approach (strength model) and (ii) discrete crack approach (fracture mechanics model). Using the example of a thick walled concrete ring, the limitations of these two models were discussed.

353 J.E. Bailey and D.D. Higgins, Discussion of Flexural Strength and Porosity
of Cements, by J.D. Birchall, A.J. Howard and K. Kendall, (Nature, 289

(1981) 388-390); Nature, 292 (1981) 89.

In a discussion of ref. 359, the authors argue that the increase in flexural strength that was reported was not due solely to the elimination of large flaws; rather, it was due to a reduction in w/c ratio, good packing, compaction and drying out. To improve strength, it is necessary to increase the total number of effective chemical bonds per unit area of the material.

354 B.I.G. Barr, W.T. Evans and R.C. Dowers, Fracture Toughness of
     Polypropylene Fibre Concrete, International Journal of Cement Composites
     and Lighweight Concrete, 3 (1981) 115-122.

The fracture toughness of polypropylene fibre concrete was measured using both cubes with two slits in eccentric compression, and also compact tension specimens. It was found that the introduction of polypropylene fibres did not particularly change the fracture toughness. It was concluded that LEFM could be applied to these materials.

355 Z.P. Bazant, Anelasticity and Fracture of Concrete, in A.P.S. Selvadurai
     (ed.), Mechanics of Structured Media, Part B, Proceedings of the Interna-
     tional Symposium on the Mechanical Behaviour of Structured Media, Ottawa,
     1981; Elsevier Scientific Publishing Company, 1981, pp. 3-35.

A review of recent progress in mathematical models for nonlinear triaxial behaviour and fracture of concrete is presented. Particular attention is given to the plastic-fracturing theory, modelling of friction and dilatancy, propagation of crack bands, interlock of crack surfaces, and the strain-rate effect. Numerical applications are discussed.

356 Z.P. Bazant and L. Cedolin, Propagation of Crack Bands in Heterogeneous
     Materials, in Advances in Fracture Research, Vol. 4, Proceedings of the 5th
     International Conference on Fracture, Cannes, 1981. Pergamon Press, 1981,
     pp. 1523-1529.

The crack front in concrete is treated as a large diffuse zone of microcracks, the size of which is related to the aggregate size. This also permits effective FEM modelling of crack propagation. It is concluded that a constant strength limit cannot be used as a crack extension criterion; rather an energy release rate criterion must be used. It is also necessary, in reinforced concrete, to take into account the bond-slip between steel reinforcement and concrete.

357 Z.P. Bazant and B.H. Oh, Concrete Fracture Via Stress-Strain Relations,

Report No. 81-10/665C, Centre for Concrete and Geomaterials, The Technological Institute, Northwestern University, Evanston, Illinois, 1981.

A fracture theory for a heterogeneous material for which the fracture zone is <u>not</u> small compared to structural dimensions, and which exhibits gradual strain-softening due to microcracking, is developed. The material fracture properties are characterized by only three parameters: fracture energy, uniaxial strength, and the width of the crack band (i.e. the fracture process zone). A method of determining the fracture energy from measured complete σ-ε relations is also given. It is concluded that the fracture of concrete must be analyzed in terms of both strength criteria and fracture energy criteria. A simple formula is derived to predict from the tensile strength and aggregate size the fracture energy for a fully developed fracture process zone; the width of this process zone in plain concrete is about 3 times the maximum aggregate size.

358 J. Bergues and M. Terrien, Study of Concrete's Cracking Under Multiaxial Stresse, in Advances in Fracture Research, Vol. 5, Proceedings of the 5th International Conference on Fracture, Cannes, 1981, Pergamon Press, 1981, pp. 2253-2260.

Samples of microconcrete (cylinders 32 mm in diameter, 72 mm long) were tested in triaxial loading. Cracking was monitored using acoustic emission. It was possible to define a domain of "non-microcracking" of the concrete. In the brittle case, this coincided closely with the elastic limit; as the confining pressure increased, giving more apparent ductility, these two limits diverged. Tests showed also that the domain defined by the microcracking limit was stable.

359 J.D. Birchall, A.J. Howard and K. Kendall, Flexural Strength and Porosity of Cements, Nature, 289 (1981) 388-390.

From tests on unnotched and on SEN beams of hcp in 3-point bending, it was concluded that the flexural strength of ordinary hcp is controlled by the presence of large voids. By removing these flaws, the flexural strength could be increased greatly. From tests on beams with different notch depths, it was shown that the Griffith theory applies adequately to ordinary hcp. For ordinary hcp, a fracture energy of 19 J/m^2 was determined; when large voids were removed by better sample preparation, the fracture energy increased to 30 J/m^2.

360 D.R. Bressi and G. Ferrara, Fracture Toughness Parameters for Concrete,

Theoretical and Experimental Analysis (in Italian), L'Energia Elettrica, (1981) 478-484.

Analysis was carried out, using numerical methods, in order to estimate the influence of the thickness, shape and size of a concrete specimen on K_{IC} and G_{IC} evaluation. It was found that the use of the test brittleness number (cf. Carpinteri) was a promising approach to the problem.

361 A. Carpinteri, A Fracture Mechanics Model for Reinforced Concrete Collapse, in Colloquium on Advanced Mechanics of Reinforced Concrete, Delft, June, 1981.

Fracture mechanics concepts are used to consider the effects of cracks in reinforced concrete beams. It is shown that the stability of the fracture process depends on the mechanical and geometrical (including scale) properties of the beam cross-section. It is concluded that either by increasing the steel percentage or the beam depth, the concrete fracturing process becomes stable, and the fracture sensitivity of the system increases.

362 A. Carpinteri, Experimental Determination of Fracture Toughness Parameters K_{IC} and J_{IC} for Aggregative Materials, in Advances in Fracture Research, Vol. 4, Proceedings of the 5th International Conference on Fracture, Cannes, 1981, Pergamon Press, 1981, pp. 1491-1498.

K_{IC}, G_{IC} and J_{IC} were determined for marble, mortar, and two concretes with different aggregate sizes, using SEN beams in 3-point bending. The mortar and concrete specimens were 15×15×60 cm, with varying notch depths. Buckingham's Theorem for physical similitude and scale modelling was applied to fracture mechanics. It was found that K_{IC} increased with increasing notch/depth ratio up to about 0.25-0.30, and then became about constant. A non-dimensional parameter (the test brittleness number) was defined, which governs the fracture-sensitivity phenomenon:

$$s = K_{IC}/\sigma_u b^{1/2}$$

where b is the specimen depth. For s > 0.50, the tests lose their meaning.

363 A. Carpinteri, Static and Energetic Fracture Parameters for Rocks and Concretes, Matériaux et Constructions, 14 (1981) 151-162.

The parameters K_{IC}, G_{IC} and J_{IC} were evaluated for several aggregative materials, using concrete with maximum aggregate sizes of 9.5 and 19.1 mm, and mortar. SEN beams (15×15×60 cm) were tested in 3-point bending.

Buckingham's theorem for physical similitude and scale modelling was used to analyze the results. A non-dimensional parameter (s, the test brittleness number) was defined, which governs the fracture sensitivity phenomenon. $s = K_{IC}/\sigma_u b^{1/2}$, where b is the specimen depth; for $s \geqslant 0.50$, tests lose their meaning.

364   R.L. Carrasquillo, F.O. Slate and A.H. Nilson, Micro-cracking and Behaviour of High Strength Concrete Subject to Short-Term Loading, Journal of the American Concrete Institute, 78 (1981) 174-186.

The progressive microcracking of concretes with σ_c ranging from 4500 to 11000 psi (31 to 76 MPa) was studied by loading 4 x 8-in. cylinders in compression in a very stiff machine, and then examining the crack patterns formed using X-ray techniques. It was found that the development of combined cracks (combinations of bond and mortar cracks) was an essential step in the failure process. Normal strength concretes began to develop combined cracks at about 70% of the strain at maximum load; high strength concretes did not develop significant combined cracks till about 90% of the strain at maximum load. High strength concrete allowed less microcracking than normal strength concrete; it failed more suddenly, with fewer planes of failure.

365   J.F. Chappell and A.R. Ingraffea, A Fracture Mechanics Investigation of the Cracking of Fontana Dam, Department of Structural Engineering, Cornell University, Ithaca, New York, Report No. 81-7, 1981.

The concepts of LEFM were applied to analyze a large crack which was discovered in the Fontana Dam. A 3-dimensional finite element analysis confirmed that the crack was due to a combination of thermal expansion and thermally induced crack growth. Theories of mixed-mode fracture propagation were incorporated into a 2-dimensional finite element analysis to perform a fracture analysis on a cross-section of the cracked portion of the dam. A lower bound prediction of the current crack front location was obtained.

366   S. Chatterji, Discussion of Flexural Strength and Porosity of Cements by J.D. Birchall, A.J. Howard and K. Kendall, (Nature, 289 (1981) 388-390); Nature, 292 (1981) 89.

In a discussion of ref. 359, the author argued that the reported increase in flexural strength was probably due to an alteration in the nature of the paste through the use of rheological aids (plasticizers) and polymers.

367 S. Chhuy, M.E. Benkirane, J. Baron and D. Francois, Crack Propagation in
     Prestressed Concrete. Interaction with Reinforcement, in Advances in Frac-
     ture Research, Vol. 4, Proceedings of the 5th International Conference on
     Fracture, Cannes, 1981, Pergamon Press, 1981, pp. 1507-1514.

The extent of the microcracked zone ahead of the main crack and the
fracture energy of concrete were measured as a function of crack
propagation in large (1.1 x 0.3 x 3.5 m) DCB specimens prestressed to
various levels. The fracture toughness was found to change as the crack
propagated, as a function of the prestressing force. Using strain gauges,
a micro-cracked region was defined as the region where $\epsilon > 10^{-4}$; this
region remained approximately the same size as the crack propagated, having
a length typically of 15-20 cm. K_C reached a constant value only after the
crack propagated about 1 m. When steel bars were introduced perpendicular
to the crack path, they appeared not to affect the microcracked zone. A
slight increase in fracture energy was noted, presumably due to the closing
force opposing the crack opening.

368 D.J. Cook and P. Chindaprasirt, A Mathematical Model for the Prediction of
     Damage in Concrete, Cement and Concrete Research, 11 (1981) 581-590.

A mathematical model was developed to describe the compressive $\sigma-\epsilon$
curves for cement, mortar, and concrete, and this model was then used to
determine the energy dissipated in damage on loading. Good correlation was
shown between predicted and experimental behaviour.

369 D.J. Cook and G.D. Crookham, Evaluation of Polymer Concretes Using the
     Method of Energy Dissipated in Damage, The International Journal of Cement
     Composites and Lightweight Concrete, 3 (1981) 247-254.

The influence of polymer addition to concrete is investigated using a
technique which measures the energy dissipated in damage from a cyclic
loading sequence. It is concluded that as the strength and stiffness of
the hcp increase, the strain decreases in the paste and increases in the
aggregate. Thus, the likelihood of fracture of the aggregate increases and
the magnitude of the strain energy release rate increases due to the
increased strain energy of the aggregate. This leads to an increase in the
magnitude of damage.

For low strength concretes, a greater distribution of short cracks
(bond cracking) develops, with a more uniform distribution of fracture
through the specimen and a low rate of dissipation of energy in damage with
increasing strain.

For the PIC series, improvement in strength is primarily due to the
increased load carried by the aggregate. Thus, the rate of energy

dissipated in damage will increase, and the amount of damage per unit of strain will also increase.

370   M.N. Haque, Influences on Flexural Strengths, Concrete, 15 (1981) 26.

        The dependence of flexural strength on r.h. is described. It is
    concluded that, if drying shrinkage cracks can be avoided, the flexural
    strength increases with the degree of drying because of the increase in
    surface free energy of the concrete. This increased surface free energy is
    lost on wetting, resulting in a considerable decrease in strength for
    saturated specimens.

371   M. Harada, S. Niizeki and M. Satake, Analyses of Crack Propagation in
    Brittle Materials, The 36th Annual Meeting of the Civil Engineering
    Institute of Japan, Vol. III, (1981).

        Twelve mortar plates (160 x 135 x 30 mm) with a hole were tested and
    G_c values were estimated by means of an acoustic emission technique. The
    crack propagation, which initiated from the top of the hole, was simulated
    with a finite element method using an isoparametric model. The simulated
    cracking process was in good agreement with the experimental results.

372   A. Hillerborg and P.E. Petersson, Fracture Mechanics Calculations, Test
    Methods and Results for Concrete and Similar Materials, in Advances in
    Fracture Research, Vol. 4, Proceedings of the 5th International Conference
    on Fracture, Cannes, 1981, Pergamon Press, 1981, pp. 1515-1522.

        A calculation model, based on a combination of fracture mechanics and
    FEM is presented. The fracture zone in front of a crack is represented by
    a "fictitious" crack that is able to transfer stress. The effects of
    specimen dimensions are also accounted for. The complete tensile σ-ϵ curve
    is introduced as a fracture mechanics parameter, and a new, very stiff
    testing machine for tensile testing is described.

373   H.K. Hilsdorf and S. Ziegeldorf, Fracture Energy of Concrete, in P.C.
    Kreijger (ed.), Proceedings of the NATO Advanced Research Institute,
    Adhesion Problems in the Recycling of Concrete, Saint-Rémy-Les-Chevreuse,
    1980, Plenum Publishing Corp. New York, 1981, pp. 101-124.

        This is a review (with 22 refs.) of the fracture of concrete, and its
    relationship to the demolition of concrete structures. The authors
    describe concrete as a three-phase material, consisting of hcp, aggregate,
    and the interfacial region betwen the cement and the aggregate. There

follows a discussion of the fracture process in concrete. The fracture energies of the three phases are described, as are the variations in their values due to: test methods, moisture state, temperature, and strain rate. The fracture energies of the phases are in the range

hcp	$\sim$ 3-15 J/m^2	
aggregate	$\sim$ 50-150 J/m^2	tension
interfaces	$\sim$ 0.5-35 J/m^2	
concrete	$\sim$ 30-60 J/m^2	
concrete	$\sim$ 5x10^3 - 200x10^3 J/m^2 compression	

Based on a comparison with the energy required to demolish concrete, it is concluded that most of the energy used to fracture concrete is transformed to other forms of energy, such as heat; only a relatively small amount goes into actual fracture of the concrete.

374   Chen-Ming J. Huang, Finite Element and Experimental Studies of Stress-Intensity Factors for Concrete Beams, Ph.D. Thesis, Kansas State University, Manhattan, Kansas, 1981.

SEN beams, 4-in. high, 3-in. wide and 15-in. long were tested in 3-point and 4-point loading, with a variable notch depth. Two different concrete mix designs were used. Stress intensity values were calculated using a finite element program. It was concluded that:

i) Notched beams do not yield stress intensity values comparable to those obtained from pre-cracked beams.

ii) Therefore, some previously published test results using notched beams are not valid, and questions concerning size effects and the influence of mix proportions should be re-evaluated.

375   S. Hungspreug, Local Bond Between a Reinforcing Bar and Concrete Under High Intensity Cyclic Load, Ph.D. Thesis, Department of Structural Engineering, Cornell University, Ithaca, New York, 1981. [Department of Structural Engineering Report 81-6].

The local bond behaviour of ribbed reinforcing bars in concrete under high intensity cyclic load was studied, using simple pullout tests. Amongst other things, internal cracking was found to stabilize about one bar diameter from the center of the test bar. This phenomenon could be explained using LEFM.

376   K. Kato, Microcracks, Deformation and Physical Properties of Plain Concrete, in Advances in Fracture Research, Vol. 5, Proceedings of the 5th International Conference on Fracture, Cannes, 1981, Pergamon Press, 1981, pp. 2275-2280.

The relationship between microcracking and deformation of concrete in compression was studied, using acoustic emission techniques. New equations relating stress to strain, Poisson's ratio, and volumetric strain were developed.

377 A.A. Khrapkov and V.A. Seiliger, On Crack Propagation in Rock Foundations of Massive Concrete Dams, in Advances in Fracture Research, Vol. 2, Proceedings of the 5th International Conference on Fracture, Cannes, 1981, Pergamon Press, 1981, pp. 661-666.

The transition of concrete dams and their foundations to the limit equilibrium state on the basis of brittle fracture theory is investigated. Stress intensity factors are obtained using finite element methods. An example calculation is presented evaluating the stress intensity factors along an experimental trajectory of a crack formed in a model structure on a rock foundation. It is concluded that both experimental and calculation data are still needed in order to establish more reliable safety factors for dams.

378 M.D. Kotsovos and J.B. Newman, Fracture Mechanics and Concrete Behaviour, Magazine of Concrete Research, 33 (1981) 103-112.

It was postulated that the fracture mechanism of concrete was that of crack extension due to initiation of crack branches, followed by propagation of these branches which is at first stable but eventually becomes unstable and leads to compelte disruption. The extension and propagation processes create voids within the material and relieve high predominantly tensile stress and strain concentrations which exist near the crack tips. The fracture mechanism is valid under any state of stress, and leads to a description of the fracture processes under generalized loading conditions.

379 J. Mazars, Mechanical Damage and Fracture of Concrete Structures, in Advances in Fracture Research, Vol. 4, Proceedings of the 5th International Conference on Fracture, Cannes, 1981, Pergamon Press, 1981, pp. 1499-1506.

The progressive deterioration of concrete under stress is expressed by a combination of elastic theory and a tensile strain damage criterion. These concepts are incorporated into FEM analysis. Numerical and experimental results for the case of plain and reinforced concrete beams in bending are compared. It is concluded that the proposed model can explain microcracking, crack formation and crack propagation, without making use of fracture mechanics.

380 M.D. Meiser and R.E. Tressler, Mechanical Properties of a Low Density
     Aluminous Cement/Perlite Composite, American Ceramic Society Bulletin,
     60 (1981) 901-905.

The fracture surface energy of low density (176-465 kg/m^3) aluminous cement composites, made with expanded perlite, was measured using two different techniques. The work of fracture and notched-beam tests were carried out on specimens 55.9 x 2.5 x 15.9 mm. Machine displacement rates for the two tests were 4.23 x 10^{-8} m/s and 2.117 x 10^{-6} m/s, respectively. It was found that the work of fracture increased with increasing bulk density, in the range 2-8 J/m^2. Work of fracture values were about an order of magnitude higher than the notched beam values. Using the Griffith equation, critical flaw sizes in the range of about 1.8 - 7.4 x 10^{-4} m were calculated; these values could not be correlated with the controlled microstructural variables.

381 H. Mihashi, and F.H. Wittmann, Probabilistic Concept to Describe the
     Influence of Rate of Loading on Strength of Concrete, Transactions, 6th
     International Conference on Structural Mechanics in Reactor Technology
     (SMIRT), Paris, France, Vol. J(b), Paper J 6/4, 1981.

The structure of concrete is composed of aggregates and a binding agent i.e. porous hardened cement paste. Under normal conditions there are already big pores and cracks in the matrix as well as in the interface before a specimen is loaded. All structural defects can be assumed to be statistically distributed all over the specimen. If a load is applied cracks start to grow from the most critical structural defects. For a realistic estimation of the reliability of a secondary containment under impact loading conditions the influence of rate of loading on mean value and variability of strength is of major interest. The presented theory predicts that the mean strength increases with a power law as the rate of loading increases, while the coefficient of variation remains constant.

382 S. Mindess and S. Diamond, Observed Energy-Dissipative Features of
     Crack Propagation in Mortar, in P.C. Kreijger (ed.), Proceedings of the
     NATO Advanced Research Institute, Adhesion Problems in the Recycling of
     Concrete, Saint-Rémy-Les-Chevreuse, 1980, Plenum Publishing Corp.,
     New York, 1981, pp. 217-224.

Cracking in mortar was studied using small wedge-loaded compact

654

tension specimens, tested in the sample chamber of an SEM. It was
noted that cracks developed preferentially along the sand-cement
interfaces. It was concluded that a considerable amount of energy must
be dissipated in the large amount of branch cracking that was seen to
occur, as well as in creating the tortuous crack surfaces; these energy
dissipative processes should be considered when developing fracture
mechanics models.

383 J.S. Nadeau, R. Bennett and S. Mindess, Acoustic Emission in the Drying
     of Hardened Cement Paste and Mortar, Journal of the American Ceramic
     Society, 64 (1981) 410-415.

     Acoustic emission (AE) was monitored during the drying of hcp and
mortar, as well as during the fracture of notched hcp beams (100 mm
long, 37 mm high, 12.7 mm wide). For hcp, K_c increased from zero to
about 0.35 $MNm^{-3/2}$ at 40 days; there was a parallel increase in AE. It
was assumed that AE resulted from cracking. Since hcp gave more AE
than did mortar during drying, it was suggested that cracking at the
interface between hcp and aggregate was not the principal source for
AE. Rather, AE appeared to be associated with cracking in the hcp
matrix.

384 A. Neerhoff, Correlation between Fracture Toughness and Zeta Potential of
     Cement Stone, in P.C. Kreijger (ed.), Proceedings of the NATO Advanced
     Research Institute, Adhesion Problems in the Recycling of Concrete, Saint-
     Remy-Les-Chevreuse, 1980, Plenum Publishing Corp., 1981, pp. 267-284.

     The zeta-potentials of cement, portland blast furnace slag, and quartz
were measured in saturated $Ca(OH)_2$ solutions. Potassium ferrocyanide and
potassium ferrocyanate were used to change the charge in the liquid from
negative to positive. DCB fracture tests were then carried out in the lime
solution with increasing concentrations of potassium ferrocyanide. The
specimens had the dimensions 40x40x160 mm, and had a chevron shaped cast
notch. They were cast with w/c=0.3, and tested at an age of 28 days. Very
small changes in K_c were observed; the maximum appeared to occur at the
iso-electric point for both types of cement. It was suggested that this
might be due to a combination of two effects:
     i) attack of siloxane bonds by OH^- as the pH increased, along with
dissolution of Ca^{2+} from the lattice.
     ii) Rebinder effect, probably connected with the movement of bonding
charges from their minimum-energy positions in the C-S-H when the zeta
potential $\neq 0$.

385   A.S. Ngab, F.O. Slate and A.H. Nilson, Microcracking and Time-Dependent
       Strains in High Strength Concrete, Journal of the American Concrete
       Institute, 78 (1981) 262-268.

       Prisms of both ordinary and high strength concrete (89 x 89 x 267 mm)
       were loaded to stress-strength ratios of 0.30 to 0.85. Slices were then
       cut, and the optically observed cracks were mapped. Results indicated
       that:
       i)   The amount of microcracking in high strength concrete is
             significantly less than in normal strength concrete, for both short-
             term and sustained loading.
       ii)   The amount of microcracking appears to be approximately linearly
             related to strain.
       iii) The σ-ε relationship for high strength concrete is linear to a higher
             percentage of the compressive strength than that of normal strength
             concrete.

386   A.P. Pak and L.P. Trapeznikov, Experimental Investigations Based on the
       Griffith-Irwin Theory Processes of the Crack Development in Concrete, in
       Advances in Fracture Research, Vol. 4, Proceedings of the 5th International
       Conference on Fracture, Cannes, 1981, Pergamon Press, 1981, pp. 1531-1539.

       The fracture toughness of concrete was investigated by testing about
       3000 specimens, with varying types of cement, types of aggregate, maximum
       aggregate size, specimen age, specimen temperature, specimen size, and type
       of test (SEN beam in tension; prism with a central slit; compressive
       cylinder with a central notch). It was found that K_c increased with
       increasing crack length, depended on the maximum aggregate size, and
       depended on the temperature. The critical crack length was also found to
       increase linearly with increasing maximum aggregate size.

387   V.V. Panasyuk, L.T. Berezhnitsky and V.M. Chubrikov, Estimation of Cement
       Concrete Crack Resistance According to the Failure Viscosity (in Russian),
       Beton i Zhelezobeton (Moscow), (1981) 19-20.

       K_{IC} for hardened cement and mortar was studied. It was found that K_{IC}
       increased from about 0.12 to 0.75 $MNm^{-3/2}$ as the cement:sand ratio changed
       from 1:5 to 1:0. K_{IC} also increased from about 0.25 to about 1.30 $MNm^{-3/2}$
       as the strength of the cement increased from about 200 to 500 kg/cm^2.

388   P.E. Petersson, Crack Growth and Development of Fracture Zones in Plain
       Concrete and Similar Materials, Ph.D. Thesis, Lund Institute of
       Technology, Sweden, 1981; Division of Building Materials, Report TVBM-
       1006, Lund, Sweden, 1981.

656

The applicability of LEFM to concrete is analyzed using the Fictitious Crack Model. It is shown that LEFM cannot be applied directly to concrete unless the specimen dimensions are of the order of several meters. The complete tensile σ-ε curve is introduced as a fracture mechancis parameter. The fictitious crack model is then used to analyze crack propagation in concrete, using the finite element method for computation.

389 L.N. Popov and E.N. Ippolitov, Fracture of Fine-Corned Concrete Under Short-Time Compressive Load, in Advances in Fracture Research, Vol. 5, Proceedings of the 5th International Conference on Fracture, Cannes, 1981, Pergamon Press, 1981, pp. 2287-2291.

Four different cementitious systems were tested in compression: ordinary concrete; mortar; fine-grained concrete, containing aggregate particles in the 2-10 mm size range combined with fine quartz sand, and neat cement paste. It was observed that the cracks were mainly mortar cracks and bond cracks. It was concluded that fine-grained concrete, made with waste products from the mining industry, can be used to replace ordinary concrete.

390 C.M. Sangha, M.K. Isles, F.H. Hubbard and R.K. Dhir, Fracture Micromechanics of Plain Concrete, in Second Australian Conference on Engineering Materials, Sydney, 1981, pp. 73-83.

Fracture analysis was carried out using crack maps prepared using reflected light microscopy for concrete sections cut from test cylinders loaded to various points on the σ-ε curve. It was found that the aggregate-mortar bond controls concrete failure. There is not much subsequent development of bond cracks, but mortar cracks increase logarithmically, accompanied to a lesser degree by aggregate fracturing, towards ultimate disruptive failure.

391 V. Saouma, Interactive Finite Element Analysis of Reinforced Concrete: A Fracture Mechanics Approach, Ph.D. Thesis, Department of Structural Engineering, Cornell University, Ithaca, New York, 1981. [Department of Structural Engineering Report 81-5].

The concepts of LEFM were used to predict the trajectory and stability of cracks in reinforced concrete. Analysis was performed using interactive computer graphics.

392 V.E. Saouma, FEFAP. Finite Element Fracture Analysis Program Version 2-0, SETEC-CE-81-055, Department of Civil Engineering, University of Pittsburgh, Pittsburgh, PA, 1981.

This finite element fracture analysis program analyses 2D quasi-static, stable, and unstable crack growth in arbitrary structures, based on the assumption that LEFM prevails. The program is capable of automatically nucleating and extending a discrete crack at an arbitrary location, angle, and length within the finite element mesh. The program can also be used to model shear stiffness along the crack and to model crack closure under stress redistribution. The program was developed particularly for the analysis of reinforced concrete structures, but is applicable to other areas as well.

393 V.E. Saouma and A.R. Ingraffea, Fracture Mechanics Analysis of Discrete Cracking, in Advanced Mechanics of Reinforced Concrete, Proceedings of the IABSE Colloquium, Delft, 1981.

A comprehensive solution to the problem of discrete crack modelling using the concepts of LEFM is described. The solution predicts the trajectory and stability of any number of cracks propagating in mixed-mode. Analysis is performed in an interactive computer graphics environment. The example reinforced concrete problem analyzed shows that:
  i) K_c does not play a dominant role in the load-displacement response;
 ii)   Aggregate interlock must be modelled;
iii)   Most of the structural non-linearity stems from concrete cracking.

394 T.K. Sia, Y.W. Mai and B. Cotterell, Strength and Fracture Properties of Epoxy-Cement Composites, in Second Australian Conference on Engineering Materials, Sydney, 1981, pp. 515-529.

The strength and fracture properties of epoxy-resin modified mortars and cellulose-cement composites were studied, using tensile tests and 3-point bending tests. For stable cracking, the specific work of fracture was calculated; for unstable cracking, G_c was also calculated. For the mortars, it was found that G_c increased by about 85% compared with control samples, at about 10% epoxy content. From SEM observations, it was concluded that the mechanism of strengthening is not due to void filling, but to the formation of a continuous epoxy phase within the cement matrix, and/or the improvement of bond strength. In the cellulose-cement composites, however, only marginal improvements in fracture resistance were found.

395 S. Somayaji, A Discussion of the Paper, A Model Law for the Notch Sensitivity of Brittle Materials, by S. Ziegeldorf, H.S. Muller and H.K. Hilsdorf, (CCR 10, (1980) 589-599), Cement and Concrete Research, 11 (1981) 479-482.

In a discussion of ref. 350, it was argued that the proposed model law could not be used to predict quantitatively the failure strength of concrete in the presence of cracks. This was due to the fact that K_c was not a material constant, but was dependent on size, amongst other things.

396 R.A. Tait and W. Keenliside, Toughness of Cellulose Cement Composites, in Advances in Fracture Research, Vol. 2, Proceedings of the 5th International Conference on Fracture, Cannes, 1981, Pergamon Press, 1981, pp. 1099-1108.

Three systems: ordinary concrete, coir reinforced cement (20% by volume of coir) and a pressure molded wood particle - cement composite (20% by volume of wood) were tested for E and for K_{IC}, using SEN beams in 3-point bending, with varying notch depths. It was found that K_{IC} of the pressure molded product was significantly higher than that of the other two systems. It was suggested that the loss of stiffness resulting from a low density filler was offset by an increase in toughness, due to a high energy fracture path during failure. It was found that for the relatively small specimens used, K_c was independent of the notch/depth ratio.

397 Y. Tanigawa, K. Yamada and S.-I Kiriyama, Power Spectra Analysis of Acoustic Emission Wave of Concrete, Proceedings of the Second Australian Conference on Engineering Materials, Sydney, 1981, pp. 97-108.

An experimental study was carried out in order to obtain some fundamental data on the frequency characteristics of the AE wave caused by the microfracture of concrete. Mortar specimens, two-phase model specimens consisting of a mortar matrix and a mortar aggregate, and concrete specimens were tested under uniaxial compression, shear-compression and splitting tension. The location search of local fracture was also carried out by using the time lags of the onset of the AE wave monitored by five sensors. Acoustic power spectra of the AE waves were determined by the fast Fourier transformation method, and the effects of the frequency components and so on were examined. It was found that:

i) The number of AE events of the shear-slip type increases gradually with the increase of relative stress, independently of the loading pattern, and that of the tensile crack type tends to increase slightly under splitting tension.

ii) The fracture process of concrete may be accurately traced by the location search of local fracture and the detailed examination of the onset directions of AE waves.

iii) Under uniaxial compression, the number of high frequency components gradually increases with the increase of stress level, and the number of low freqency components also increases at the maximum stress.

iv) The frequency characteristics of AE waves of concrete seem to be affected by the stress level rather than the fracture type.

398 K. Visalvanich and A.E. Naaman, Frcture Methods in Cement Composites, Journal of the Engineering Mechanics Division, ASCE, 107 (1981) 1155-1171.

   K_{IC} and the fracture energy of concrete, frc and asbestos cement were compared, using different analytic techiques and experimental tests. DCB specimens allowing crack lengths of up to 406 mm were used. It was concluded that:

1. LEFM applies to asbestos cements, and to plain mortar under certain conditions.
2. K_{IC} is meaningful for mortar as long as crack growths greater than 200 mm are permitted.
3. Fracture energies from compliance measurements underestimate the true fracture energy for mortar and frc.
4. K_{IC} for asbestos cement was ~4.9 $MNm^{-3/2}$, compared to about 1.3 $MNm^{-3/2}$ for mortar.
5. There was evidence that a crack in mortar starts to propagate at a critical opening angle of 0.13°.
6. A method for measuring the fracture energy which accounts for the existence of the permanent deformation at unloading was developed.

399 F.H. Wittmann, Mechanisms and Mechanics of Fracture of Concrete, in Advances in Fracture Research, Vol. 4, Proceedings of the 5th International Conference on Fracture, Cannes, 1981, Pergamon Press, 1981, pp. 1467-1487.

   Based on a discussion of the structure of hcp and concrete, crack initiation and propagation in porous composite materials are explained. The different failure processes in hcp, ordinary concrete and high strength concrete are explained using a complex computer simulation. A stochastic approach to describe the failure of hcp and concrete is introduced, and this probabilistic concept is used to describe the relationship between strength and loading rate. It is concluded that classical fracture mechanics cannot be applied to concrete, and that fracture toughness cannot be characterized sufficiently by a single parameter.

400   F.H. Wittmann and Yu. V. Zaitsev, Crack Propagation and Fracture of Composite Materials Such as Concrete, in Advances in Fracture Research, Vol. 5, Proceedings of the 5th International Conference on Fracture, Cannes, 1981, Pergamon Press, 1981, pp. 2261-2274.

The structure of concrete is subdivided into three levels: hcp, mortar, and coarse aggregate in a mortar matrix. Crack propagation is studied at all three levels; first by examining crack propagation in a porous, homogeneous material (hcp); then, by looking at crack growth in a pre-cracked matrix (mortar); finally, by examining cracking in a composite material (concrete). Using the models, crack growth in both normal and high-strength concretes can be simulated using computer techniques. Finally, the behaviour of both hcp and concrete under high sustained load is predicted.

401   A. Yam, Effect of Fibre Reinforcement on the Crack Propagation in Concrete, M.A.Sc. Thesis, Department of Civil Engineering, University of British Columbia, Vancouver, British Columbia, 1981.

Double torsion specimens (1219 x 406 x 51 mm) made of hcp, and concretes reinforced with glass and steel fibres, were tested at an age of about 3 years to study the effect of fibre reinforcement on crack propagation in concrete. It was concluded that resistance to rapid crack growth and K_{IC} increased with fibre content up to about 1.25 - 1.5% fibres by volume, and then decreased, probably due to the incomplete compaction of the concrete at higher fibre contents. It was found that in this test geometry, fibres did not significantly restrain crack growth.

402   Yu.V. Zaitsev, Fracture Mechanism and Strength of Concrete Under Triaxial Compression, in Advances in Fracture Research, Vol. 5, Proceedings of the 5th International Conference on Fracture, Cannes, 1981, Pergamon Press, 1981, pp. 2281-2286.

A model of concrete, consisting of polygonal aggregate inclusions randomly dispersed in a homogeneous matrix, is introduced. Pre-existing, random microcracks are also considered. It is shown that interfacial cracks start to propagate under triaxial load in Mode II (shear), regardless of the value of the confining pressure. The simulation is done using Monte Carlo methods.

403 J. Zaitsev and F.H. Wittmann, Simulation and Crack Propagation and Failure of Concrete, Matériaux et Constructions, 14 (1981) 357-365.

The heterogeneous structure of concrete is described in terms of a multi-level system. To take different effects of crack propagation and crack arresting into consideration, four different levels have been introduced. Conditions for crack propagation in the porous structure of hardened cement paste and in a matrix with inclusions are studied analytically. By using the derived formulae crack formation and failure of lightweight, normal, and high strength concrete are discussed. Computer experiments are described.

404 S. Ziegeldorf, H.S. Müller and H.K. Hilsdorf, A Reply to S. Somayaji's Discussion of A Model Law for the Notch Sensitivity of Brittle Materials, (CCR 11(1981) 479-482), Cement and Concrete Research, 11 (1981) 483-484.

In their reply to ref. 395, the authors state that their proposed model law requires that LEFM be applicable to the specimens in question.

405 S. Ziegeldorf, H.S. Müller and H.K. Hilsdorf, Effect of Aggregate Particle Size on Mechanical Properties of Concrete, in Advances in Fracture Research, Vol. 5, Proceedings of the 5th International Conference on Fracture, Cannes, 1981, Pergamon Press, 1981, pp. 2243-2251.

Two models are presented: the first describes the dependence of notch sensitivity of brittle materials on specimen size, fracture toughness and tensile strength; the second describes grain size dependent crack formation in concrete. This second model indicates that cracks due to internal dessication exist in concrete, and are formed at aggregate particles with a diameter larger than some given diameter.

NAME INDEX

Numbers following a name indicate pages where an author is mentioned in this volume. If underlined, they refer to a contribution to this volume. Numbers given in brackets refer to the reference number of publications listed in the annotated bilbiography on pp. 542-661.

DHIR, R.K. (85, 130, 390)

DIAMOND, S. 3, 76, (158, 244, 304, 321, 322, 382)

DIAZ, S.J. 37, (63)

DIEDERICHS, U. 157, 207

DIKOVSKII, I.A. (221)

DI TOMMASO, A. (217)

DJABAROV, N. 489, (253)

DOLCH, W.L. (244)

DOMONE, P.L. (131)

DOUGILL, J.W. 341, (26, 64, 152, 175)

DOWERS, R.S. 437, 449, (291, 354)

DUDDERAR, T.D. 142

DUFFY, J. 100

DUGDALE, D.S. 242, 506, 510

DURY, B.L. 447, 490, (260)

EITEL, W. 76

EL-DIN, A.S. Salah (147)

ELLIS, C.D. 424, 432, 447, 490, (87)

ENTOV, V.M. 23, (148)

ESAKI, T. (149, 188)

EVANS, A.G. 16, 18, 482, (163)

EVANS, R.H. (28, 29)

EVANS, W.T. 437, 449, (291, 354)

FANINGER, K.H. 197

FAUST, G. (352)

FEARN, J.E. (162)

FELDMAN, R.F. 50, (55)

FERRARA, G. (360)

FIGG, J. 399, 422, 423, (84)

FLOOD, E.A. 53

FOOTE, R.M.L. 448, 487, (305, 318)

FRANCOIS, D. (278, 367)

FREUDENTHAL, A.M. 301, 305

FUJII, S. (262)

FUJITA, Y. (120, 189)

GARRETT, G.G. (182, 254)

GEINATS, G.S. 435, (195)

GEORGE, K.P. 421, 435, (56, 190, 255)

GERMAIN, P. 344

GERSTLE, K.H. (312)

GJØRV, O.E. 10, 18, 19, 22, 343, (191, 315)

GLUCKLICH, J. 8, 11, 14, 22, 400, 416, 418, 420, 422, (11, 14, 22, 27, 30, 61, 86, 132, 150)

GREEN, A.T. 168

GREGORY, D.A. 132

GRIFFITH, A.A. 1, 8, 9, 11, 14, 54, 258, 301, 303, 304, 384, 411, 413, 504, 506, 507, 518, 527

GROSS, G.E. 376

GRUDEMO, Å. 48, 65, 66, 76, (256)

GRÜNBERG, J. 102

GUOFAN, Z. (306)

GURNEY, C. 452

GUSTAFFSON, P.J. 492, (192, 329)

GYLLTOFT, K. (307)

HAGA, K. (262)

HAHN, G.J. 421, 435

HAISMAN, B. 18, 398, 419, 431, (52)

HALVORSEN, G.T. 24, 443, 444, 451, 494, (258, 308, 309)

HANSEN, T.C. 16, (31, 110)

HAQUE, M.N. 5, 8, (127, 128, 129, 370)

HARADA, M. (371)

HARDER, D. 400

HARRIS, B. 424, 432, 447, 486, 490, (87)

HATAYA, S. (276)

HAWKINS, N.M. (193)

HAY, J.M. 382, 395, 398, 399, 406, 433, (133, 134)

HELLEN, T.K. 522

HERMITE, R.G. 168

HIGGINS, D.D. 10, 18, 20, 21, 35, 59, 61, 385, 392, 393, 394, 395, 398, 406, 434, (164, 165, 353)

HILLEMEIER, B. 10, 19, 21, 260, 394, 395, 399, 403, 434, (166, 194)

666

SUBJECT INDEX

Numbers following a key-word indicate pages where this subject is mentioned in this volume. Numbers in brackets refer to the reference number of publications listed in the annotated bibliography and the corresponding page number. Example : (269 p618) means reference number 269 in the annotated bibliography and it can be found on page 618.